THE RHS ENCYCLOPEDIA OF HOUSE PLANTS

INCLUDING GREENHOUSE PLANTS

THE RHS ENCYCLOPEDIA OF HOUSE PLANTS

INCLUDING GREENHOUSE PLANTS

Kenneth A. Beckett

In association with the Royal Horticultural Society

American consultant Victoria Jahn

Salem House

Key to Symbols

Cultural requirements and overall plant shape/growth habit are summarized in the form of at-a-glance symbols throughout the A-Z section of this book. These provide quick reference and supplement the main text.

Environment

Suitable for the home

Suitable for the greenhouse

For home or greenhouse

Temperature requirements

Tropical/warm (minimum 15–18°C [60–65°F])

Temperate (minimum 10–15°C [50–60°F])

Cool (minimum 5–10°C [40–50°F]; 0–15°C [32–60°F] if noted as hardy down to −3°C [27°F])

Light requirements

Full sun

Partial shade

Full shade

Watering requirements

Light – Allow rooting medium to completely dry out between waterings.

Medium – Allow surface of rooting medium to dry out between waterings.

Heavy – Keep entire rooting medium moist at all times.

Plant Habit/Shape

Erect

Spreading/Prostrate

Mat-forming

Bushy

Weeping

Climbing/Scrambling

Rosette-forming

Pendent/Trailing

Tufted/Fan-like/ Clump-forming

Globular (or cylindrical)

Author's acknowledgments

The great majority of pictures in this book are of plants growing in botanical and private gardens. My wife Gillian, who took most of the colour transparencies, and I are particularly grateful to the following people for their cooperation in allowing us access to their plant collections behind the scenes: Mr C. D. Brickell, formerly Director of the RHS Gardens, Wisley, Dr R. D. Shaw, Curator of the Royal Botanic Garden, Edinburgh and Mr J. B. E. Simmons, Curator of the Royal Botanic Gardens, Kew. We are especially indebted to Mr L. Maurice Mason for allowing us unlimited time among his treasures at Talbot Manor, Fincham, Norfolk over the years. As a result, more of his plants feature in this book than anyone else's.

First published in the United States by
Salem House Publishers 1987,
462 Boston Street, Topsfield,
Massachusetts 01983

Filmset in Goudy by Wyvern Typesetting Limited,
Central Trading Estate, 277 Bath Road, Bristol BS4 3EH
Printed and bound in Italy by Arnoldo Mondadori Editore,
Via G. V. Zeviani 2, 37100 Verona

Library of Congress Catalog Card Number 87–4546

Beckett, Kenneth A.
The RHS encyclopaedia of house plants.
1. House plants — Dictionaries
I. Title II. Royal Horticultural Society
635.9'65'0321

ISBN 0 88162 285 0

Editor: Brian Carter
Design: Mick Keates
Production: Hugh Allan

This book was designed and produced by
Swallow Editions Limited, Swallow House,
11–21 Northdown Street, London N1 9BN

Contents

Introduction

During the past thirty years or so there has been a spate of books on house plants, some at the most basic and popular level, others styled as dictionaries and encyclopedias. Of those published in the English language, not one has been most truly comprehensive and with its information presented in a consistent, technical or even semi-technical fashion. This encyclopedia is intended for the serious collector of plants, not just for the home but also for the greenhouse and sunroom. It is fair to say that all the plants in this book will thrive just that bit more successfully in a greenhouse, even those long recognized as reliable house plants. Nevertheless, in the past a sizeable list of plants suitable for the home has been compiled, mainly by trial and error, and continues to grow larger year by year. From a commercial point of view, a successful house plant is one that can be easily propagated in quantities. Many potentially excellent house plants do not come into this category and are seldom seen in other than a greenhouse setting.

During my more than forty years as a professional horticulturist, I have been much involved in growing tender plants under glass, both in nurseries and in botanic gardens. Frequently I have tried such plants in the home with varying degrees of success. As a result I am willing to risk my reputation and say that most of the plants in this encyclopedia can be attempted in the home. You must of course, be willing to experiment and expect a degree of failure in the long term. So much depends on one's skill as a cultivator. The ideal situation is to have a sunroom or greenhouse. From these, plants not usually recommended for the home can be tried out indoors and returned if they start to look unhappy.

The introductory section which follows includes all the essential general cultural details, plus separate items or specialized groups, i.e. trees, shrubs, climbing plants, perennials, water plants, etc. For convenience, all the plants are classified into four temperature categories: tropical, temperate, cool and hardy. What these mean in absolute terms of day and night temperatures and daily and seasonal care is described at the end of the cultivation section.

Each generic entry also contains a set of symbols. These provide a ready means of displaying the vital information needed for any plants you have or are keen to acquire.

In writing the species, variety and cultivar entries, the endeavor has been to provide description with enough technical detail to confirm doubtful names and even to identify unnamed plants. While all the plants mentioned are in cultivation somewhere, many do not have a readily available commercial source. However, the section *Species Which Play Hard to Find* should help the reader to overcome this problem.

Well tried house plants occupy a very wide range of climates and soils, but nowhere in the wild are conditions so sterile or basically inimical to plant growth as those in the home: no direct sunlight, change of air, mist or rain. It says much for the plant kingdom that it is adaptable enough to produce members which will not only survive, but thrive there.

Just as some people are adaptable to climate and living conditions, so are certain plant species. Those suitable for the house must be able to stand fluctuating temperatures, extremes of light and shade, drought and flood and an atmosphere of desert dryness (though limited humidification is now easily provided). Of course, no one plant should have to tolerate all these extremes. It is up to us to choose the most suitable species or cultivars for the conditions we have.

As might be expected, some of the most successful long-term house plants are those from extreme climatic conditions. Desert plants, especially cacti and other succulents, can be highly satisfactory, as are epiphytes (many bromeliads and orchids) which perch on exposed tree branches, and the forest-floor dwellers adapted to poor illumination.

A BRIEF HISTORY OF HOUSE PLANTS

From the available historical evidence it seems fairly conclusive that the ancient Egyptians can be credited with the beginnings of house plant culture. The effect and motivation were, of course, in no way comparable to those of today. The containerized plants were mostly formally pruned small trees and the areas they graced were more like enclosed courtyards than rooms. The Greeks and Romans cultivated plants in a similar way, but the ancient Chinese developed it further. They grew a larger selection of plants in containers and originated the tree-dwarfing technique perfected as *bonsai* by the Japanese. The Arabs (Moors) followed Roman traditions and grew a variety of plants in containers, bringing the practice to southern Europe.

It was not until the advent of glass windows and some form of heating that the culture of house plants as we now know it could begin in colder climates. This could have happened several hundred years ago, but the time was not ripe. What did begin about 600 years ago was the development of the greenhouse, beginning in Europe as little more than a pavilion with a glazed side and heated by an open stove. The incentive was the growing of tender evergreens, oranges and other citrus fruits. Gradually, the plant's needs for light and heat without fumes became understood, and by the middle years of the 18th century greenhouses, much as we know them today, were in operation. Tropical plants then became the rage and collectors were sent to Africa, Asia and South America in search of new exotic plants. As a result, material was at hand for experimental growing in the home. But something further was needed to trigger a desire for house plants. Growing urbanization cut people off from the

The Victorians were real house plant enthusiasts. TOP A showcase of ferns takes pride of place. ABOVE Careful watering of a hanging basket. LEFT Ornamental stands for plants, jardinières, were most popular additions to the drawing room.

countryside and a desire for growing things around the home was quickly born. Gas and oil stoves and open coal fires give off fumes lethal to many plants. To create suitable conditions for growth, the glass cases made by Maconochie and Ward for shipping plants from abroad were brought into the home. These provided the quite humid conditions ideal for many plants, but for ferns in particular, and before long the Victorian fern craze was in full swing.

Since World War II a whole new nursery industry has arisen to meet the tremendous demand for pot plants. Such has been the competition that nursery empires have been made and lost in the process. Nurserymen, their representatives and friends, continue to look out for likely plants. The seeking goes on in various ways and, quite often, visits to botanic gardens and other public and private collections will yield a new or relatively unknown plant worth trying out. Botanic gardens are continually increasing their genera and species, either by exchange with other institutions or via the collections made in the wild by their staff members. The more enterprising nurserymen employ collectors or go themselves to likely countries, often mounting small expeditions to remote jungle or desert areas. The plants so collected have to be established in the nursery, then tried out in a few homes to see how they react. Many do not come up to expectations, but every now and then a new one catches on and becomes a firm favorite.

ABOVE LEFT A traditional courtyard in Lima, Peru is mimicked to some extent by the modern interior (CENTER) with palms giving an indoor/outdoor garden effect. ABOVE Wonder plants of the 1980s, hybrid hippeastrums have become extremely popular since Dutch growers mass-produced the bulbs.

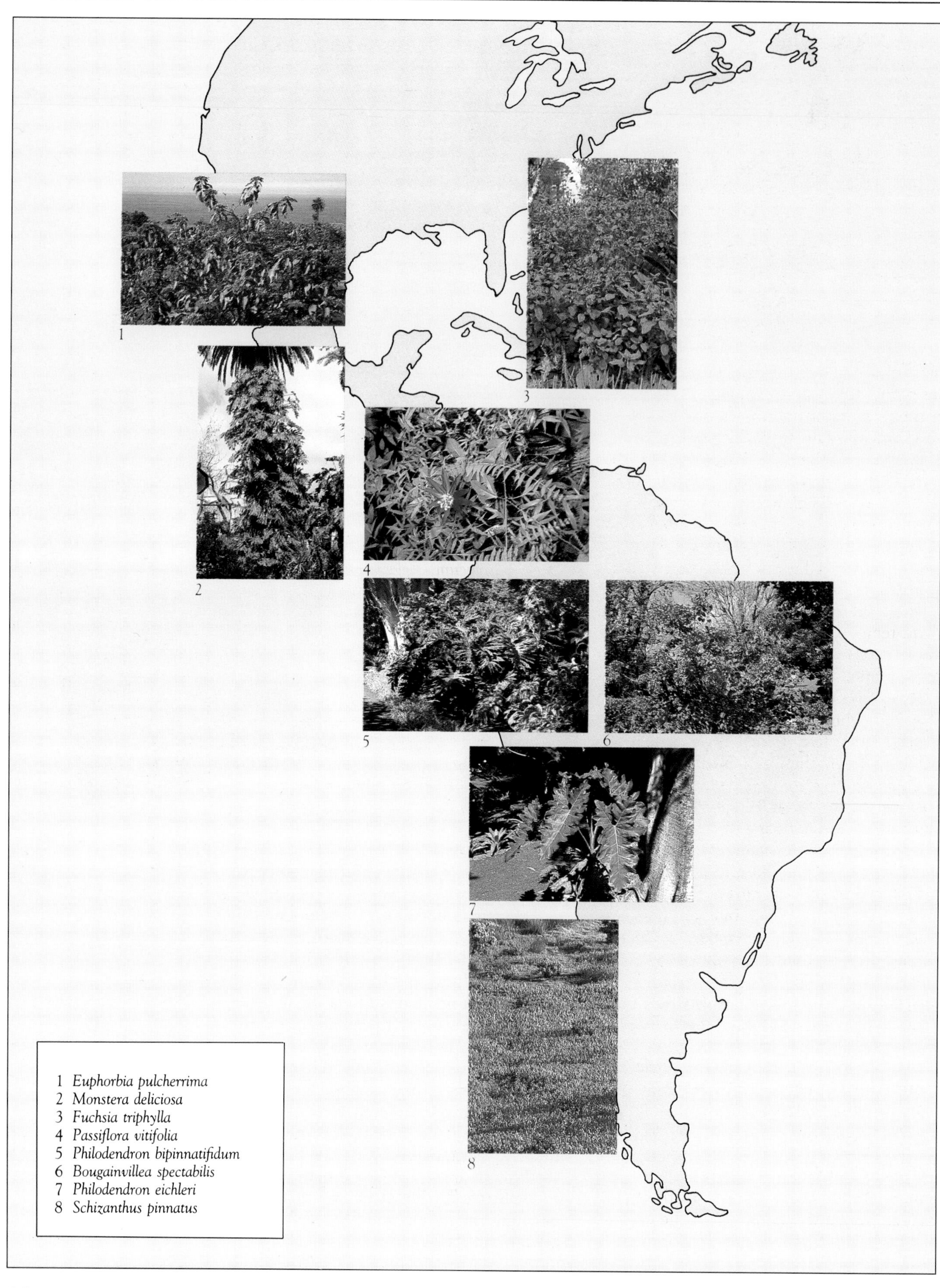

1
2
3
4
5
6
7
8
1 *Euphorbia pulcherrima*
2 *Monstera deliciosa*
3 *Fuchsia triphylla*
4 *Passiflora vitifolia*
5 *Philodendron bipinnatifidum*
6 *Bougainvillea spectabilis*
7 *Philodendron eichleri*
8 *Schizanthus pinnatus*

WHERE HOUSE PLANTS COME FROM

Most of the plants we grow in our homes are native to the forests, jungles and deserts of the tropics and subtropics, but this is by no means the full story. Such is the adaptability of the plant kingdom that it is perfectly possible to see representatives of all but the arctic climate zones growing happily together on the same window sill. Let us do a brief world tour to see where some of our best known house plants come from (maps facing and on p. 12–13).

The Mediterranean region ranges from warm temperate to subtropical, but on the whole it has yielded few good long-term house plants. Among exceptions are the sweet bay (*Laurus nobilis*), giant reed (*Arundo*), European fan palm (*Chamaerops humilis*), and olive (*Olea*). If we include the Canary Islands, then a number of good succulent genera are included, e.g. *Aeonium*, *Ceropegia* and *Euphorbia*. Moving south into the drier parts of Africa we add more indispensable succulents, notably *Aloe*, *Haworthia*, *Gasteria*, *Sansevieria*, *Kalanchoe*, a great variety of *Euphorbia* and many *Mesembryanthemum* allies (among them, the living stones *Lithops*, *Dinteranthus* and *Pleiospilos*). The humid African tropics yield *Dracaena* (especially *D. godseffiana* and *D. sanderiana*), *Mussaenda*, *Clerodendrum*, *Begonia*, *Peperomia* and *Thunbergia*. From drier sites come *Streptocarpus*, *Cissus*, *Clivia*, *Cyanotis*, *Pelargonium* and *Strelitzia*. South Africa in particular is the home of many beautiful bulbous plants, notably *Gladiolus*, *Lachenalia*, *Sparaxis*, *Nerine*, *Freesia*, *Haemanthus* and *Zantedeschia*.

Central and South America are the homes of many house plants. All the true cacti are found there, mostly in the desert regions, but with a few epiphytes in the forests and jungles, notably *Epiphyllum*, *Schlumbergera*, *Rhipsalis* and *Rhipsalidopsis*. The bromeliads are also restricted to the Americas, from *Billbergia* to *Tillandsia* and *Vriesea*. All the best known aroids come from these countries too, in particular *Dieffenbachia*, *Anthurium* and *Philodendron*. Other specialities are *Columnea*, *Kohleria*, *Calathea*, *Fuchsia* and *Aphelandra*. Mexico has a rich tropical flora and has given our homes a surprising number of good plants. Poinsettia (*Euphorbia pulcherrima*) is perhaps the most familiar of all, but there are also *Zebrina*, *Rhoeo*, *Bouvardia*, *Maranta*, *Achimenes*, *Episcia*, *Monstera* and many more. There are also such well known succulents as *Echeveria* and *Pachyphytum*.

The warmer parts of Asia have also yielded many house plants. Various species of *Eranthemum*, *Hoya*, *Colocasia*, *Zingiber*, *Serissa*, *Hibiscus* and *Ficus* come from India, while Burma, Assam and Thailand have provided the unique *Nepenthes*, *Platycerium*, *Musa* and *Nelumbo*.

Scindapsus, *Spathiphyllum*, *Pandanus*, *Ixora*, *Codiaeum*, *Aglaonema* and many orchids are some of the specialities of the East Indies (Malaysia, Java, Sumatra, Borneo, Philippines).

China and Japan have provided a wealth of plants for temperate gardens, but fewer for the house. In the house category are *Camellia*, *Cycas*, *Ardisia*, *Aucuba*, *Chrysanthemum*, *Ophiopogon*, *Liriope*, *Acorus* and *Aspidistra*.

Last, but by no means least, we come to Australasia with its many species of *Eucalyptus*, *Melaleuca*, *Pittosporum*, *Grevillea*, *Callistemon*, *Leptospermum* and *Acacia*. One of the most familiar of large indoor plants, *Brassaia actinophylla*, the Queensland umbrella tree is also native to this area, as is *Livistona australis*, aptly known as the Australian fan palm.

By hybridising and careful selection, man has added many plants to our homes and conservatories which are totally unknown in the wild. A familiar example is the florists' chrysanthemum in its vast array of flower forms and sizes. Many orchids are the result of the plant breeders' skill some of which have resulted in the creation of special hybrid generic names, e.g. × *Wilsonara* and × *Vuylstekeara*. Most of the showy orchid cacti (*Epiphyllum* hybrids) are of complex parentage involving the crossing of several genera and species.

9
10
11
12
13
14
15
16
17
18
19
20
21

9 *Narcissus poeticus*
10 *Rosmarinus officinalis*
11 *Euphorbia canariensis*
12 *Chamaerops humilis*
13 *Nerium oleander*
14 *Mussaenda erythrophylla*
15 *Dracaena surculosa*
16 *Begonia sutherlandii*
17 *Aristea major*
18 *Gladiolus orchidiflorus*
19 *Kalanchoe fedtschenkoana*
20 *Haemanthus* sp.
21 *Citrus medica*
22 *Dendrobium* sp.
23 *Ficus benghalensis*
24 *Hibiscus rosa-sinensis*
25 *Rhaphiolepis umbellata*
26 *Trachelospermum asiaticum*
27 *Camellia japonica*
28 *Aeschynanthus speciosus*
29 *Melaleuca fulgens*
30 *Rhododendron lochae*
31 *Eucalyptus* sp.
32 *Howeia belmoreana*

HOW TO CHOOSE AND BUY

House plants offered for sale may be young plants recently raised from seeds or cuttings, or fairly mature specimens in full growth. Two points should be considered before buying; suitability for the purpose intended, and whether the plant is a healthy specimen that will last if properly cared for.

It is not always easy for the beginner to decide what is a healthy, well-grown plant, particularly if it is unfamiliar and has coloured or variegated leaves. Nevertheless, there are a few points to look for. Unless the plant is a climber, it should be compact and of sturdy growth, not thin and weak with long stretches of bare stem between the leaves. The leaves should stand firmly on their stalks, not hang limply which could mean either lack of water or a root rot disease. Apart from cacti and such plants as mother-in-law's tongue (*Sansevieria*) with a narrow, erect habit, a well-grown pot plant should at least marginally overlap the rim of its container. Tall and slim plants like the *Sansevieria* should be twice as tall as their pot. If the plant is bought in flower, make sure there are buds to come. Give it a gentle shake and avoid the specimen that sheds blooms in abundance. Having chosen and bought the plant, take it to its new home as soon as possible, particularly in winter. Greenhouse plants subjected to long periods of cold, especially in frosty weather, may drop their leaves or have their growth checked in other ways. They may even die. It is advisable not to purchase plants which are outdoors during cold winter weather.

The scorching effect of direct sun through glass can be attractively counteracted by the installation of adjustable blinds, as in this functional conservatory. Choose gauzy fabrics for an ideal soft shade.

LIGHT

The quality of light in the home is extremely deceptive to the human eye. Often it seems very bright when the sun shines, but this is only in contrast to the gloom within the room. In addition, glass filters out the ultraviolet rays and reduces light intensity which also drops drastically further into the room. In the northern and southern temperate zones, the light varies greatly with the seasons, being brightest in summer and dullest in winter. In winter the sun rises and sets to the south of east and west (to the north in the southern hemisphere), and the low-angle light strikes almost horizontally into windows on that side of the house. In summer, the sun shines directly into east-facing windows until late morning, then into west-facing ones from early afternoon. In winter and summer, directly north-facing windows (south in the southern hemisphere) never get direct sunlight. However, this only applies to houses aligned exactly north south and even a few degrees to the east or west makes a great difference in the amount of seasonal light. Trees and buildings nearby also obscure the light.

It is important to bear in mind these aspects of light when positioning plants around the house. For example, succulents must have plenty of light and will only thrive close to a window. Plants that grow mainly or entirely in winter, e.g. *Lachenalia* and *Freesia*, need a south-facing window (north in the southern hemisphere). East and west windows from late spring to early autumn can be hot and bright, and with the exception of most succulents, plants will need screening from the direct rays of the sun with a net curtain or blind.

SITING PLANTS IN THE HOME

When deciding where best to put a particular plant, a number of considerations must be borne in mind as well as the more obvious question of where it looks best. Perhaps most important is whether the plant demands full sun, or whether it is a shade lover and can happily be put at some distance from a window. Temperature too is important: many plants will not thrive if kept permanently in the hot, dry atmosphere produced by central heating, and moving them to a cooler place at night may be necessary. The majority of plants grown in the home are tropical and can stand only brief spells with the temperature below 10°C(50°F). Near-freezing temperatures are lethal to many house plants and it is important not to leave them between the window and the room curtains at night unless the window is double-glazed, which will insulate them against the cold.

If the window ledge does not provide enough space for the plants needing good light, a high table in front of it can be used. Alternatively, one or more glass shelves can be fitted across the window resting on brackets on either side. Such light-demanding plants as cacti will thrive particularly well here. Containers hanging from hooks securely fitted above the window serve a similar purpose and will be ideal for growing hanging plants. Climbers grown in pots on the windowsill can be trained up the sides of the window and even induced to meet across the top.

Many custom-built plant display units are now available, the open, backless bookcase kind being useful both as a room divider and against a wall. Plants can be encouraged to climb from this type of unit, either on trelliswork or on strings fixed to the wall. Another attractive way of covering a wall with plants is to fix shelves high up – especially if there is a source of light nearby so that trailing plants can be encouraged to hang down from them.

Groups of plants standing together in the corners of the room, either in separate pots or planted in a single container, can also be most effective. For these, some form of artificial lighting may be necessary if they are far from the window. Fluorescent strip lights are excellent as they give light without heat, and can make a dark corner or an unused fireplace into the focal point of the room by illuminating an attractive group of plants. Special units containing two or three strip lights can be bought, as can plant stands into which they are fitted. If ordinary bulbs are used for lighting, be careful to keep the plants at a safe distance from them, otherwise they will get scorched.

Enjoying the benefit of full light by day, with supplementary artificial light in the evening, a selection of foliage house plants thrive in this modern setting, linking the interior decor with the patio beyond.

CULTIVATION

SOIL

The health and vigor of a plant depends largely upon the rooting medium. It is possible to use garden soil, but this is seldom rich enough for the concentrated root system of pot plants and frequently gets compacted and sour. The best medium for pot plants is one which holds moisture yet is well aerated, free draining and contains plenty of plant food. Modern potting soils fulfil all these requirements and are readily available from garden centers and hardware stores prepacked in small or large amounts.

These potting soils fall into two categories, those with and those without loam (a fertile soil blending sand and organic matter, such as is found under old pastures). Several loam-based mixtures can be obtained but they vary widely in percentage of soil in the mix. There are even several potting mixes on the market which have soil in their name, yet there isn't a speck of soil in the mix! Quality also varies widely, which is why the clean, light, and easy to handle all-peat or peat and sand mixes are very popular nowadays. Many house plants grow in them exceedingly well. The only real disadvantage of the all-peat potting mixes is that they are very light even when moist. This means that large plants can easily become top heavy. The addition of sand or gravel to these mixes gives extra weight, and also helps water to percolate. All-peat potting mixes also shrink when they get very dry and tend to part from the sides of the pots. They are then difficult to water properly, as water passes rapidly down the crack without being absorbed. If this does happen, submerge the pots in water for at least half-an-hour, then let the potting mix drain, firming it when the operation is finished. Peat mixes normally need little firming during potting. Light pressure with the finger tips is sufficient. Compaction makes watering difficult and cuts down aeration.

* A bushel is a bulk measure now technically obsolete but still useful. A box 55x25x25cm(22x10x10in) or one of equal capacity holds one bushel when loosely filled.

MAKING UP YOUR OWN POTTING MIX

Perfectly acceptable potting can be made up at home using peat, sand and fertilizer with or without the addition of garden soil. Suitable formulas are: 3 parts garden soil (4 or 5 if it is sandy), 2 parts moss peat and 1 part coarse sand or grit. If soil is not available use 3 parts peat to 1 part sand or grit. To each bushel* of either formula add 8 tablespoons 5–10–5 fertilizer and 5 tablespoons of ground limestone. Omit the lime if the soil is chalky or if lime-hating plants such as azaleas are being grown.

If garden soil is used, expect weed seedlings to appear. Most commercially available potting soils are partially sterilized by heat to kill pests, diseases and weed seeds.

For long-term pot plants, that is those staying in the same pot for one year or more, it is best to use the loam and soil-based mixes.

PLANTS IN WATER

Several well-known plants can be grown by hydroculture (in water) for quite long periods.

Hydroculture is based upon the established practice of hydroponics, used

Nowadays, containers are readily available in every shape and size, suitable for the daintiest fern to the sturdiest shrub or tree. Terracotta tubs and troughs (ABOVE) are very stable with a natural look; plastic pots (LEFT) offer the widest choice of color, are lightweight and highly durable.

for growing vegetable crops in arid regions. Hydroponic culture proper is carried out in beds of gravel or vermiculite flooded at intervals with water containing essential minerals. Hydroculture usually does away with the rooting medium and sometimes even the mineral solution.

To grow plants this way, take a container that will hold water, preferably no more than 10cm(4in) deep, as plant roots must have oxygen and this is diffused more rapidly in shallow water. Use crumpled chicken wire or washed pebbles or gravel to support the plant material, and make an arrangement of suitable plants such as *Aglaonema*, *Dracaena*, *Philodendron*, × *Fatshedera*, *Tolmiea*, *Coleus* or *Dieffenbachia*. Fill the container with water, which must be changed weekly and topped up as necessary. Once a month add a few drops of any proprietary liquid fertilizer to the water, then a week later pour it away and replace with clean water.

CONTAINERS

A wide variety of containers in which to grow plants is available, from straightforward plastic or clay pots to ornamental, sometimes glazed, bowls and jardinières. For the general run of house plants, plastic pots are to be preferred. Clay pots too are good but, being porous, soil moisture is easily lost and the plants need watering more frequently. With the exception of those for plants which need wet soil conditions, all containers should have a bottom hole to allow water to drain through freely. Stagnant soil conditions can be death to some house plants. This is particularly true of the bromeliads, orchids and other epiphytes. For these there are clay and plastic containers with side holes or latticework walls. Wood rafts, baskets or pieces of cork bark are even better and make pleasing and effective homes for their occupants.

POTTING

It will be necessary to pot initially only if plants have been raised from seeds or cuttings in the home, or are rooted cuttings or seedlings donated by a friend.

Pots should be new or should be cleaned before use by scrubbing in hot water to remove any pest or disease organisms. Plastic pots which have several small holes in the bottom and are up to 12.5cm(5in) in diameter do not need drainage material in them. Clay pots of the same size, or plastic ones with a single large hole in the bottom should have this covered with a small square of wire screen or a crock (a piece of broken flower pot). All pots above this size should have 2.5cm(1in) or so of crocks, gravel or small stones to provide drainage. Before putting it in the pot, make sure that the root system of the plant is moist and the new soil is neither too wet nor too dry. As a test, take a handful and squeeze tightly. If it falls apart as soon as the hand is opened it is too dry and if it remains tight and compacted then it is too wet. Ideally it should cohere, but crumble easily when agitated. If it is too wet, spread the potting mix out for a few hours in the sun or a warm place. If it is too dry, add water a little at a time and mix well, until it responds to the squeeze test.

Choose a pot that will accommodate the root system of the plant to be potted, leaving 2.5cm(1in) all round for later growth. Place the plant in the pot, making sure it sits at the same level as it did in the pot or box from which it was removed. Fill all round with potting mix. If an all-peat mix is being used, firm lightly. Top up the pot leaving space between the soil surface and the pot rim for drainage. A space of about 1cm($\frac{1}{2}$in) is enough for pots up to 12.5cm(5in) size; increase the gap proportionally up to 2.5cm(1in) for a 20cm(8in) pot. If a loam-based potting mix is used, press firmly after potting.

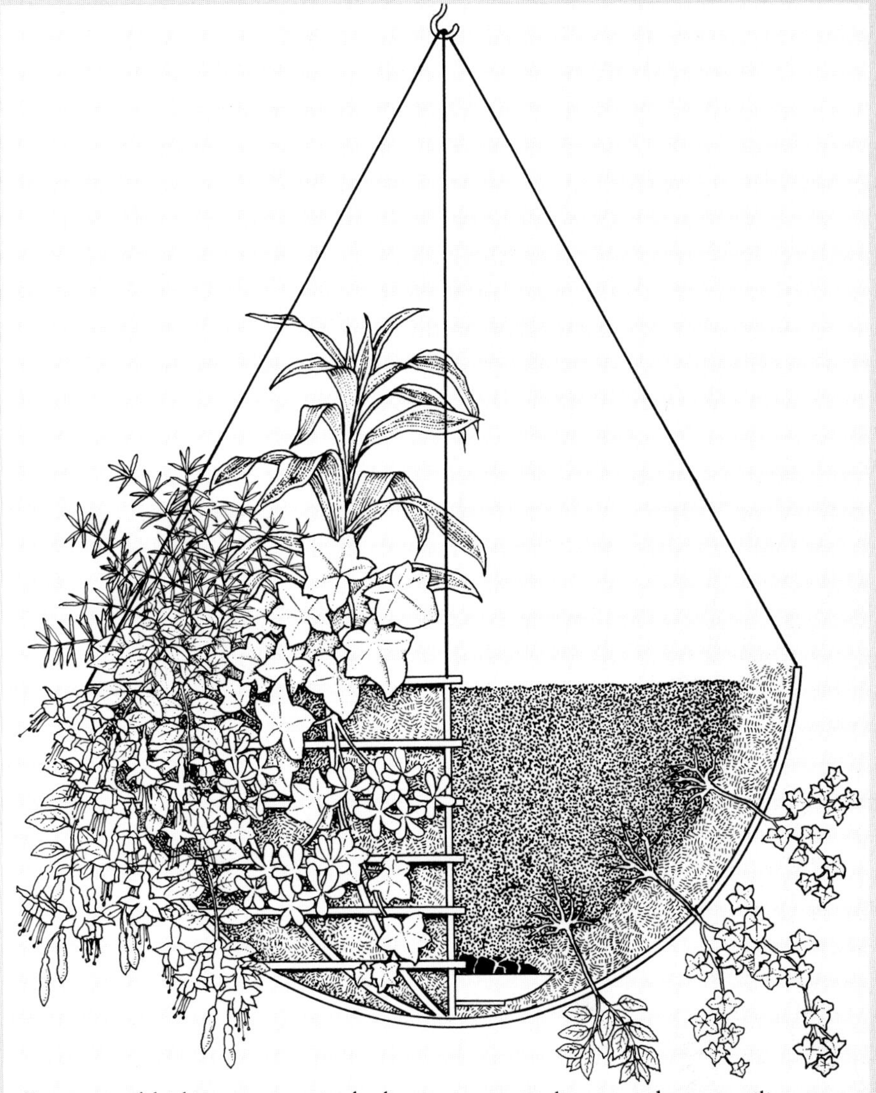

For a natural-looking hanging basket, first line the frame with sphagnum moss, then place a saucer in the bottom to retain the draining water; fill with compost. Position upright plants in the centre and pendent or trailing ones around the edge – small trailers can be inserted into the basket walls.

POTTING ON

This is a similar procedure to initial potting. Make sure that the plant to be potted on (transferred) is moist, then remove it from the pot by inverting the pot across the upturned hand with the plant hanging between the fingers. Now hold the pot and tap the rim against a hard surface. It should then be possible to lift off the pot. If it does not come free the compost may be too dry, so water it and try again.

The size of the new pot should be related to the plant, allowing at least 2.5cm(1in) of space all round. Sit the plant on a layer of potting mix sufficiently deep for the root ball to be covered with about 1cm(½in) of new mix when the operation is completed. Then fill in, firming with the finger or a blunt stick until this level is reached.

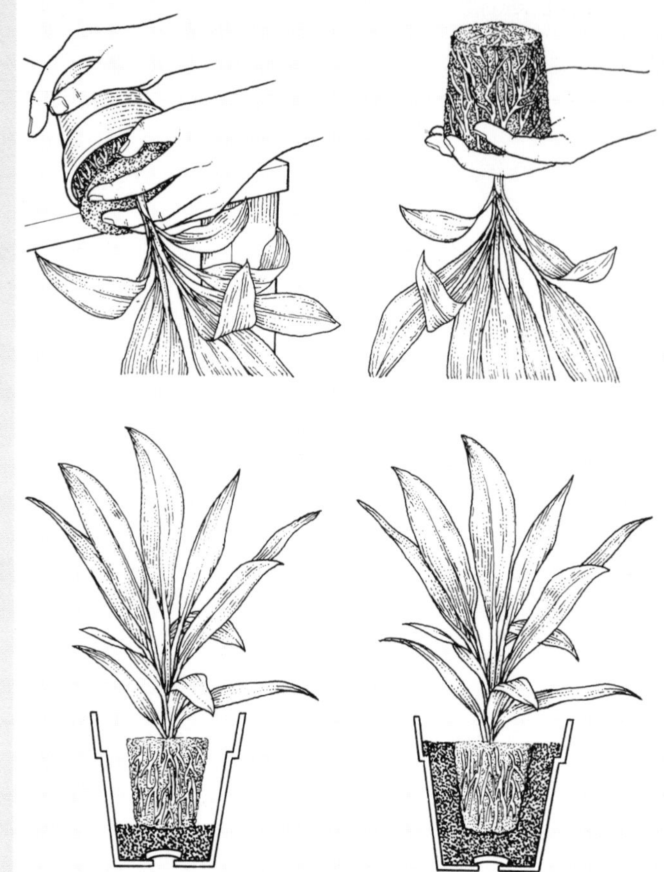

Healthy plants will soon outgrow their pots and should be potted-on regularly if they are to maintain a good shape and increase in size, but always progress in modest increments to the final container dimension. Over-potting will simply encourage a weak root system and the plant will look out of proportion.

REPOTTING

Although the terms repotting and potting on are often interchanged, each has a separate meaning. Potting on, as explained above, involves moving a plant into a larger pot. Repotting is a means of giving a plant fresh soil while retaining it in the same sized container. It is used for long-term pot plants after they have been in the same sized pot for several years and are not required to grow much bigger.

First remove the plant from its pot then strip off soil and roots so that the actual root ball is reduced by a quarter or a little more. A small handfork is useful for this job and where roots are dense and tough an old bread knife is useful. Replace the plant in the same pot (cleaned first) or a clean container of identical size, using the same technique outlined under potting on.

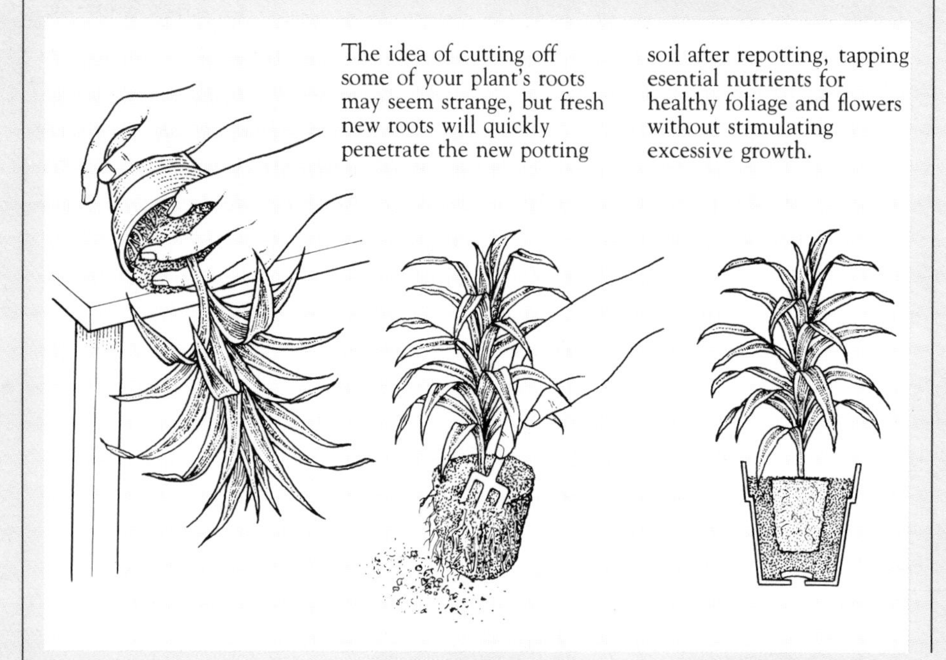

The idea of cutting off some of your plant's roots may seem strange, but fresh new roots will quickly penetrate the new potting soil after repotting, tapping esential nutrients for healthy foliage and flowers without stimulating excessive growth.

TOP-DRESSING

Less drastic than repotting is the practice of top-dressing. Here the plant is left in its container, but the top layer of roots and soil is removed to a thickness of one quarter of the root ball depth. A kitchen fork with the top half of the tines bent over at right angles is an ideal tool for this job. The old soil is replaced with fresh potting soil and gently firmed. Plants growing in beds and borders in the greenhouse also benefit from top-dressing, but here it is not necessary to remove so much of the old top soil beforehand or even any at all. Late winter is the best time for both kinds of top-dressing.

WATERING

The frequency of watering depends upon the time of the year, the temperature and whether the plant is root bound (the pot being filled with roots as is the case just before repotting). For example, an actively growing plant in a sunny window may need watering once, and sometimes twice, a day during warm spells in summer, while in winter once or twice a week will be plenty.

Probably the easiest method of deciding when the plant needs water is to scratch into the soil surface with the forefinger. If the soil feels barely moist about 1cm(½in) down, then water should be given. During the summer or in a warm room, if in doubt, give water. At other times and in a cooler room, do not. It will be safer for the plant to wilt slightly before watering than to be kept soggily wet.

HUMIDIFYING

As we have seen, many of the plants used in the home are native to tropical forests and jungles where the air is humid for much or all of the yearly cycle. Although when cultivated in containers they may grow well enough without it, most house and sun room plants appreciate some form of regular humidity around their leaves. A few will not really thrive without it.

The easiest way to provide humidity is to use shallow, waterproof trays filled with flooded gravel or small pebbles upon which the plants are stood. Deeper trays or containers can be filled with peat moss into which the pots can be plunged up to the rim. The peat is kept moist and not only evaporates to create humidity, but keeps the root systems cooler and helps to cut down on watering.

Alternatively, use one of the small hand-operated mist sprayers, damping the foliage daily. This is particularly useful for epiphytic plants grown on sections of tree branches.

FEEDING

However good a potting mix may be, sooner or later the plant roots exhaust the main nutrients and need some supplementary feeding. Various kinds of plants reach this stage at different times. A slow-growing plant such as *Clivia* may not need feeding for a year or so, whereas a quick-growing plant such as *Cineraria* will need it within six weeks of the final potting.

Fertilizers for pot plants are best applied in a liquid state. As a rule, quick-maturing annuals will need a first feed when the flower buds appear. Foliage plants will need it when the newest leaves fail to reach the size of those below. Once feeding has started, it should continue at intervals of 10 to 14 days while the plant is actively growing or flowering. During the cooler months or when growth slows or ceases, feeding can be stopped.

For a quick tonic as well as for regular feeding, the modern foliar feeds are worth trying. These are water soluble fertilizers which, when dissolved in water, can be sprayed onto the foliage where they are rapidly absorbed and used by the plant. They are particularly valuable if the more traditional methods of feeding have been forgotten and the plant needs to be brought to health and vigour speedily.

The indoor environment is all too often dry, dusty, polluted with smoke or draughty – undesirable conditions for the majority of plants, especially those native to tropical zones. Bottle gardens are a popular and attractive means of maintaining a humid micro-climate for moisture-loving species, especially ferns. Water vapor released by the leaves and rising from the damp soil condenses on the sides of the bottle and trickles back to the roots – very little escapes. Only species with waxy leaves will tolerate being stoppered, however. The initial planting procedure demands some patience, but a mature bottle garden is most rewarding.

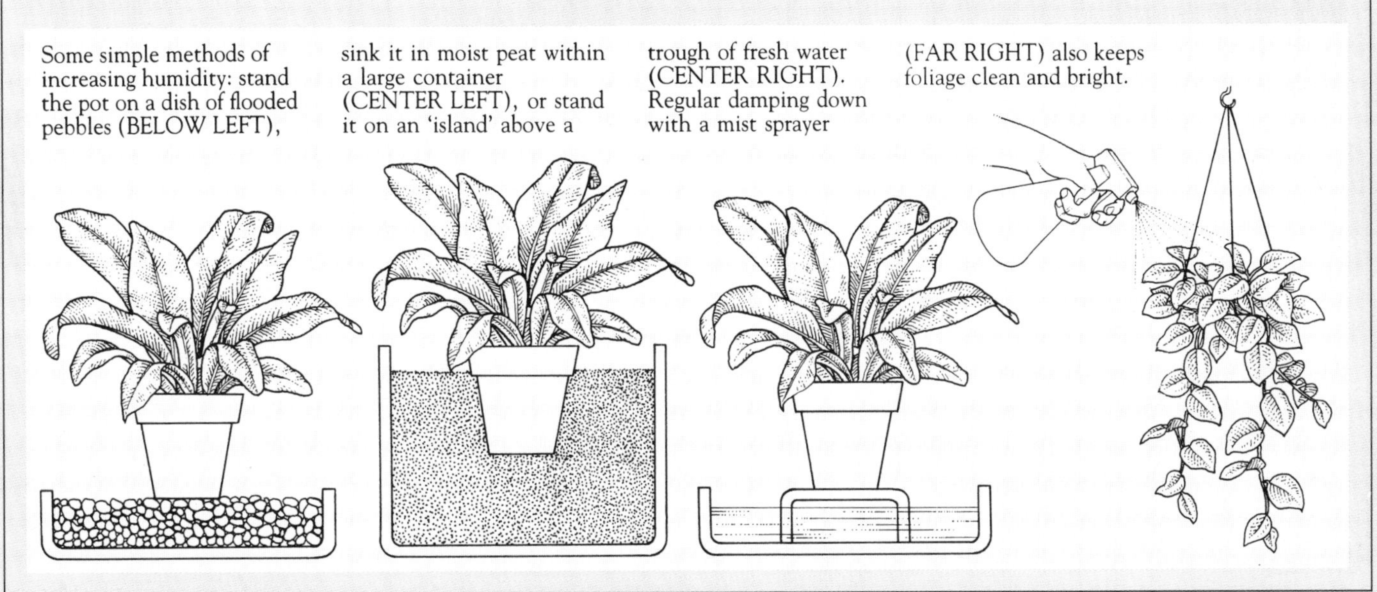

Some simple methods of increasing humidity: stand the pot on a dish of flooded pebbles (BELOW LEFT), sink it in moist peat within a large container (CENTER LEFT), or stand it on an 'island' above a trough of fresh water (CENTER RIGHT). Regular damping down with a mist sprayer (FAR RIGHT) also keeps foliage clean and bright.

TEMPERATURES

TROPICAL

In the tropics and subtropics temperatures never, or only very briefly, fall below 15–18°C(60–65°F), and may rise to 30°C(86°F) or above by day. With the exception of the orchids, bromeliads and succulents, most plants from this region need regular watering and humidity throughout the year, and bright but not direct sunshine. Any exceptions are quoted in the A–Z section.

Propagation is carried out in late spring and summer in a heated propagating case set at a minimum of 21°C(70°F). Potting and repotting are done in late spring or early summer, and all proprietary potting soils are acceptable. For techniques, *see* p. 17–18.

TEMPERATE

Plants in this category come from the cooler parts of the subtropics and the warm temperate zone, where night temperatures may fall to 10–13°C (50–55°F) during the winter, rising to 18–21°C(65–70°F) by day, but are higher in summer. With the exception of the succulents and the bulbous plants, the majority of house plants from this region need regular watering from late spring to early autumn, but much less in cooler

A characteristic tropical scene (ABOVE RIGHT) and temperate landscape (BELOW) – the natural habitats of many popular house plants, foliage and flowering.

months. They also require plenty of winter light, but screening from the hottest summer sun. Feeding is carried out during the main spring–summer growing season and humidity is also appreciated at this time.

Potting and repotting are done mainly during spring and all proprietary potting soils are acceptable. Propagation is carried out in spring and summer (*see* p. 34–39).

COOL

This category covers the plants from the edge of the warm temperate and the warmer parts of the cool temperate zones where winter night temperatures range around 5–10°C(40–50°F) rising by day to 13°C(55°F) or above. In summer, temperatures approximate those for the temperate category described above. Light, watering, humidity, feeding and repotting are also as for the temperate category. The propagation season extends from late spring to early autumn (*see* p. 34–39).

HARDY

All the comments made under the *Cool* heading apply here, with the addition that all plants so classified are reasonably frost–hardy and will tolerate winter temperatures down to −3°C(27°F) or just below in an unheated room, sun room or greenhouse.

Cool climates (BELOW CENTER) and hardy climates (BELOW RIGHT) are the origin of numerous tolerant house plants as well as those for outdoor cultivation.

Special groups

Trees

Podocarpus montanus

Dracaena 'Warneckii'

Schefflera arboricola var.

Eriobotrya japonica

Albizzia julibrissin

ABOVE RIGHT
Ficus elastica 'Zulu Shield'

A surprising number of trees from tropical to warm temperate regions are used as decorative foliage pot plants when young. Familiar examples are to be found in such genera as *Brassaia*, *Coccoloba*, *Dracaena*, *Cordyline*, *Ficus*, *Jacaranda*, *Grevillea*, *Laurus*, *Pittosporum* and *Radermachera*. Most palms also come into this category. These and several others are especially valuable for sizeable rooms, hallways and greenhouses where tall plants with largely handsome or otherwise decorative foliage can be used so effectively. Even so, it is only in the largest conservatories and in the lofty foyers of big office blocks where they can begin to approach their full potential.

Provided a recommended temperature range is maintained, there are no special cultural problems. All the proprietary potting mixes stocked by reputable garden centers are suitable. However, a loam or soil-based type is to be preferred as it not only has a longer lasting plant food reserve, but provides weight and stability. It must be realized that a tall plant heavy with foliage can become top-heavy in a light all-peat or peat and perlite mix. If such a potting mix is used it should be blended with a quarter part by volume of washed coarse sand, grit or gravel.

With the exception of such deciduous trees as *Jacaranda*, watering must be carried out regularly throughout the year. The aim is to maintain a moist but not wet potting mix. Over-dry conditions at the roots are likely to result in yellowing or falling of the leaves, thus spoiling the plant's appearance.

Plants kept in the same sized containers for a year or more will benefit from applications of liquid feed at two- to three-week intervals during the late spring to early autumn period.

Healthy young trees will not need supporting. They do, however, lean towards the light and if not turned regularly may become lopsided. For this reason it is sometimes worthwhile tying the main stem to a slim cane at least until it reaches the desired height and has become woody.

If large specimens are required, young ones will need potting on annually, ideally in early spring. Once the plant is big enough, the *status quo* can be maintained by pruning and annual top-dressing in early spring. Additionally, every three to five years, repotting will be necessary to maintain a plant which is healthy but not over-vigorous.

It is usual to allow young trees to grow naturally until the desired height is reached. When this point is attained, remove the top quarter to one third of the main stem and allow one or more side branches to grow up. These in turn can be cut back by up to half annually in spring. (This does not apply to palms.) Very vigorous species can be cut back more drastically. In time, the plant may get too large and the hard pruning destroy its attractive juvenile habit. When this happens a new plant should be propagated and the old one discarded. Propagation is usually by seed or air layering, but some species, e.g. *Jacaranda*, *Ficus* and *Laurus*, can be increased by summer cuttings.

SHRUBS

Many shrubs from warm temperate to tropical regions are suitable for culture in containers in the home and the conservatory. Familiar examples are to be found in such genera as *Acalypha*, *Callistemon*, *Camellia*, *Hibiscus*, *Leptospermum*, *Pentas*, *Medinilla*, *Pisonia* (*Heimerliodendron*), *Pseuderanthemum*, etc. Most shrubs are grown for their flowers, but some, for instance the first and last examples, provide handsome foliage with year round appeal. Some shrubs are naturally small, others become large if grown in tubs or similar containers. All are attractive when young. Big specimens, whether they are grown for their leaves or flowers, have their place in a large room or conservatory, providing focal points of great charm.

A number of shrubs can also be used as climbers, tying them out flat, especially to cover a wall or screen in the conservatory. Camellias respond well to this treatment and look very effective, as do *Callistemon*, *Hibiscus* and *Prostanthera*.

Given a suitable temperature range, there are no special cultural problems. All the proprietary potting soils are suitable unless one is attempting the tender rhododendrons, heaths (*Erica*) and their allies. These require an acid rooting medium and this is catered for under the proprietary named rhododendron or ericaceous potting mixes. When all-peat mixes are used it is advisable to blend them with about a quarter part by bulk of washed coarse sand or grit. This not only improves the texture for such fine rooted plants as *Erica* and *Rhododendron* but lends weight and stability to the root-ball. It must be realized that a tall plant heavy with leaf and/or blossom, can easily become top-heavy in an all-peat or peat and perlite mix.

Evergreen shrubs, and these are greatly in the majority, must be regularly watered throughout the year. Aim to maintain a moist but not wet rooting medium. Deciduous species, e.g. fuchsias, overwintered under very cool conditions, and those which are cut back hard after blooming, e.g. *Bouvardia*, poinsettia (*Euphorbia pulcherrima*) and *Streptosolen*, must be watered sparingly during the coldest months. Allow the potting soil to just about dry out between applications and only resume normal watering when new growth is underway. Plants kept in the same sized containers for a year or more should be given liquid feed at two- to three-week intervals in the late spring to early autumn period when most growth is made.

If large specimens are wanted, young ones will need potting on annually, ideally in early spring. Once the plant is big enough the *status quo* can be maintained by pruning and annual top-dressing in spring. Additionally, every three to five years, re-potting will be necessary to maintain a plant which is healthy but not over-vigorous.

Shrubs grown for their blossoms are better after an annual pruning, removing half to two-thirds of the flowered stems. This will promote a compact habit and a free-blooming state. Prune summer and autumn flowering species in winter or early spring. Those which bloom in winter and spring should be cut back as soon as the last flower fades. Shrubs grown entirely or mainly for their foliage can be cut back as required, in spring.

Camellia japonica var.

Fuchsia 'Checkerboard'

Euphorbia pulcherrima var.

Hibiscus rosa-sinensis var.

Pittosporum tobira var.

ABOVE LEFT
Datura × *candida* 'Plena'

Syngonium podophyllum var.

Thunbergia grandiflora

Allamanda cathartica

Mandevilla splendens

Eccremocarpus scaber

CLIMBING PLANTS

ABOVE RIGHT
Clematis cirrhosa balearica

Climbers form an important plant group for house and sun room decoration. They may be annuals, herbaceous perennials or woody vines. All are typified by their flexible, usually fast-growing stems. All need a means of support to which they cling by twining, aerial roots or by specially modified stems, leaves or stipules known as tendrils. No one plant family has a monopoly on climbing plants, but *Convolvulaceae*, *Leguminosae* and *Vitaceae* contain many important genera and species. Climbers may be hardy, cool, temperate or tropical and their cultivation needs are basic in all but supporting and pruning.

Small climbers in pots and tubs are supported by canes or the sort of twiggy stems called pea or bean sticks. Alternatively, sections of trellis or plastic netting can be used, or one of the specially designed plant supports. The large, permanent woody-stemmed species must have a stronger support system. Such plants are primarily for plant rooms, sun rooms and greenhouses where walls can be drilled for eye bolts (vine eyes). Fix these in rows 45–60cm(1½–2ft) apart each way and join them up with galvanized wire. A natural and very effective way of growing some of the smaller climbers is to let them ramble through a shrub. For example, a tub-grown *Pittosporum tenuifolium*, 1.2–1.5m (4–5ft) tall makes a splendid support for *Tropaeolum tricolorum*, *Billardiera longiflora* or *Ipomoea quamoclit*. Support shrub and climber must be grown in their own containers stood side by side.

All the woody-stemmed climbers need pruning every now and then, ideally annually. Where the height of the support is less than the growth potential of the plant, some early pruning is recommended. Start at planting time or the spring following, by cutting back the young plant by a half to two-thirds. This will induce several young stems to form low down which in turn can have their tips pinched out when 30–45cm(1–1½ft) tall. Tie the subsequent stems to the support, spacing them evenly. This will provide a basic framework covering the allotted area. Subsequently, annually after flowering or when dormant, excess growth is thinned out and the rest tied to its support. This is very much a counsel of perfection because some climbing plants, notably the twiners and self-clingers, are difficult to deal with. Sometimes it is best to do very little for two or three years, then to prune drastically and expect to see very little blossom in the following season.

Watering, feeding and propagating are as for ordinary pot plants. As most climbers are vigorous, fast-growing plants however, watering and feeding must never be neglected, especially in the growing season.

PERENNIALS

Some of the best known and most useful pot plants for the home and the greenhouse are evergreen perennials, for example, *Asparagus densiflorus* 'Sprengeri', *Aspidistra elatior*, most begonias and ferns, *Nertera granadensis*, *Rhoeo spathacea*, *Zebrina pendula*, etc. All bromeliads and bulbous plants and most succulents are also perennials, but as they have special cultural considerations they are dealt with separately.

In the horticultural sense, and as used here, perennials are non-woody plants which grow from year to year. They can be grouped into two broad categories: crown (clump-forming and tufted) and radiant (prostrate or trailing). Crown plants produce all stems and/or leaves in a sheaf-like formation direct from ground level, e.g. *Asparagus densiflorus* and *Aspidistra*. Radiant plants start off as a single stem but this soon branches and forms a mat, each stem rooting as it runs, e.g. *Soleirolia* (*Helxine*), *Tradescantia fluminensis* and *Callisia elegans*.

As with shrubs and annuals, some perennials are grown primarily or entirely for their flowers, others for their foliage. A lesser number are attractive on both counts. A few, e.g. *Nertera* and *Rivina*, are grown mainly for their colorful fruits. On the whole they are easily grown, providing a suitable temperature range is maintained. All the proprietary potting soils stocked by reputable garden centers and nurseries are suitable. Soil- (loam-) based mixes are to be preferred as they have a longer-lasting reserve of plant food. However, if regularly fed, perennials thrive in all-peat mixes.

Perennials, as defined above, must be watered regularly throughout the year. Aim to maintain a moist but not wet rooting medium. If temperatures are cooler than optimum, as is usually the case of a greenhouse in winter, let the plants start to dry out between watering. Over-wet soil conditions and low temperatures provide ideal conditions for root rot fungi. Plants actively growing and which have been in their containers for six to eight weeks will greatly benefit from liquid feeding. Apply this at two-week intervals until growth slows down in autumn or early winter.

If big specimen plants are required, young ones must be potted on annually in late winter or early spring. Once the plant is big enough, a *status quo* can be achieved by annual top dressing in early spring. Additionally, about every three to four years repotting will be necessary to maintain vitality. Plants that look congested at this time are best divided, retaining only the most vigorous outer parts of the plant. If the plant is not easily divided, then take cuttings and grow on a young specimen to eventually replace the old one.

Some perennials, for example the taller begonias and chrysanthemums, may need supporting. For this purpose use split, or naturally slim canes, preferably dyed green. Insert these so that they are as unobtrusive as possible and use green wire or plastic ties. Remember that clumsy or over-obvious support systems greatly detract from the beauty of the plant.

Most perennials can be allowed to develop naturally. If necessary, tall or spindly stems may be pruned to encourage bushiness. Pinch out the tips of young stems, or cut off the top one-third or more, of more mature stems during the growing season. Exceptions to this generalization are mentioned under the appropriate A–Z entry.

Perennials which produce new flowering stems annually, as typified by *Chrysanthemum* and *Rehmannia*, must be cut back to soil level once all flowers have fallen, or at least before new growth starts from the base. Other species, for example *Rivina*, *Ruellia* and *Strobilanthes*, get untidy if allowed to grow naturally and the flowering or fruiting stems are best cut back by at least half once faded.

Ruellia mackoyana

Spathiphyllum hybrid

Codiaeum pictum

Asparagus densiflorus var.

Clivia nobilis 'Citrina'

ABOVE LEFT
Streptocarpus 'Helen'

Azolla caroliniana

Eichhornia crassipes

Thalia dealbata

Cyperus papyrus

WATER PLANTS

ABOVE RIGHT
Nyphaea caerulea var.

Many plant families contain members which either live in water or need wet soil to thrive. Some of these are decorative and desirable plants for the home and sun room. They are clearly divisible into species which are submerged, those which float, and those which are marginal. The submerged and floating groups are usually best grown in aquariums, though some floating kinds such as water lilies (*Nymphaea*) perhaps look best in ceramic bowls or half tubs. All the submerged plants described in this book have highly decorative foliage. They also thrive in low light intensities and can be grown under artificial illumination. A nicely arranged, well-lit tank of plants will brighten up a shady room or corner in a most effective way.

A few fish can be an added attraction but make sure they are not bottom foragers which can disturb newly set plants and make the water cloudy.

Wherever positioned, tank arrangements of aquatics must be screened from direct sunlight, except in winter. Floating and marginal plants need some direct light each day. Temperature requirements depend on the provenance of the species and this is given in the appropriate A–Z entry.

Aquatic plants can be grown in no more than 5–7.5cm(2–3in) of clean sand. However, for really healthy growth, place a 2.5cm(1in) layer of loam or good garden soil in the bottom of the tank and cover with a 4–5cm(1½–2in) layer of sand or grit before planting. Plant water lilies and other aquatics with floating leaves in pots or specially designed baskets of loam, a loam-based potting mix or good garden soil. After potting, cover the soil with a layer of gravel to prevent it from mixing with the water and making it cloudy. Feeding is only necessary when the plants cease to grow satisfactorily, and is effective in containerized plants with one of the proprietary pellets made especially for water lilies. It is best not to fertilize tanks as this encourages the unicellular green algae and blanket weed. Complete cleaning out is the only way to get rid of these.

Propagation is mainly by division, in spring. Submerged aquatics with leafy stems are best propagated by cuttings taken in small bunches and pushed immediately into the rooting medium. If seed is available this is sown in the usual way but with the pots stood just awash in containers of water. Sow in spring or as soon as the seed is ripe and maintain a temperature just above that required for growth of adult plants.

Marginal aquatics are grown as for ordinary pot plants, but are kept permanently wet in trays or deep containers of water. Some of the larger ones, e.g. *Cyperus papyrus* and *Thalia dealbata*, make splendid accent plants for the large house or sun room.

A half tub makes an attractive container for a conservatory water garder Shown here in section, th planting medium can be banked up to suit shallow-water plants, leaving the center free for deeper species such as water lilies (*Nymphaea*).

FERNS

Ferns are primitive plants which inhabited the earth long before flowering plants evolved. They differ from the higher plants most markedly in their mode of reproduction, starting life as microscopic, dust-like specks called spores. When a frond is mature, it produces, on the backs or edges structures called sori, the so-called 'fern seeds'. Within these are tiny bodies called sporangia which contain the spores. Each spore grows into a tiny, leaf-like body known as a prothallus. This is the reproductive stage in the life of all ferns. On the underside of each prothallus, male and female cells are borne, and on their fusion, a tiny plant begins to grow.

Most of the ferns that grow well in the home, e.g. *Pteris*, *Platycerium*, *Pellaea* and *Adiantum*, are comparatively tough and thrive under the same conditions as the usual range of flowering and foliage plants. Looking at ferns in general however, they need more shade and humidity than flowering plants. For this reason, some of the smaller, more delicate kinds are best in bottle gardens, terrariums or plant cases. Ferns are renowned for the elegance of their dissected, bright green leaves, but, as a glance at the bird's nest fern (*Asplenium nidus*) and the staghorn fern (*Platycerium bifurcatum*) will show, they vary widely. Most of the 10,000 known species are tropical, but those suitable for the house occur in all the climatic zones. Temperature requirements for each genus and species will be found in the A–Z entries.

Ordinary potting soil can be used for most ferns, but the all-peat potting mixes are to be preferred. Water regularly throughout the year, though less when the temperature is below average, but do not keep the soil permanently wet. Epiphytic ferns, e.g. *Davallia* and *Platycerium*, must be allowed almost to dry out between watering. Humidity is recommended for all ferns, though the epiphytic sorts will grow without. Ideally, stand pots on trays of flooded pebbles, and/or mist daily. All ferns, once they are a year established in their containers, should be fed at monthly intervals from spring to autumn. Most ferns are ideal for shady window where direct sunlight does not penetrate. Epiphytic ferns are best with direct light from late autumn to mid-spring and all ferns can take, and even appreciate, winter sunlight.

Most ferns are easily propagated by division in late winter or spring. Choose well-clumped specimens and separate carefully using a knife if necessary. Pot immediately and keep well shaded for a week or so until new roots form.

A very rewarding and intriguing exercise is to raise plants from spores. These can be purchased or collected from mature fronds showing the brown sporangia. Fill a pot or pan with seed-sowing soil or sphagnum moss and soak with boiling water to kill pests and diseases. When cool, dust the surface with the spores and either place in a plastic bag or stand in a shallow tray of water and cover with a sheet of glass. Keep out of direct sunlight and maintain a temperature somewhat above that required for the adult plant. In due course, the tiny green prothalli will develop and these must be syringed every few days to facilitate fertilization. Once the tiny fern plants have a few recognizable fronds they can be pricked off and potted on, as for seed-raised pot plants. Make sure they are kept humid and shaded until growing well.

Pteris quadriaurita var.

Athyrium goeringianum var.

Platycerium hillii

Adiantum raddianum var.

Drynaria quercifolia

ABOVE LEFT
Blechnum gibbum

Aechmea hybrid

Vriesia hieroglyphica

Billbergia vittata

Cryptanthus bivattatus var.

Guzmannia hybrid

Bromeliads

ABOVE RIGHT
Neoregelia carolinae var.

Also known as air plants, these are all members of the *Bromeliaceae*, a family almost entirely restricted to the Americas. Most species live as epiphytes in forests and jungles where they cling tightly to the branches and trunks of trees, often fully exposed to the elements. The species used as house plants have leathery, strap-shaped leaves arranged in architectural tubular or bowl-shaped rosettes, some of which act as reservoirs. Most sorts are tropical but stand a wider range of temperatures than this category suggests. They are also generally much more tolerant of fluctuating cultural conditions. For this reason they make excellent house plants.

The best way to display bromeliads is to secure them with moss and potting soil to branches using green PVC coated wire. Alternatively, grow them in small containers of all-peat potting soil mixed with equal parts of sphagnum moss or the bark or plastic chips sold for orchid growing. Water throughout the year unless the temperatures are lower than normal, but always allow the soil to dry out between applications. Although humidity is not essential, growth is finer if a daily misting is given from spring to autumn. Most bromeliads are best grown lightly screened from direct sunlight, at least in summer; though some, e.g. *Tillandsia*, will take direct sun at all times. Feeding is best done during the summer months only. Those species which have a water-tight cup or tube in the center should have this topped up regularly with rain or soft water; but if temperatures are low in winter it is best to empty the plants until warmer conditions return. Propagation is by well-grown offsets removed in early summer and potted singly in little containers of potting mix. A propagating case is not needed, but may aid a more rapid establishment.

ORCHIDS

The orchid family (*Orchidaceae*) is a huge one, even the most conservative estimate allowing 17,000 species. These have a worldwide distribution, but the majority are tropical. From a cultural and ecological point of view they can be divided into two main groups: those that grow on trees (epiphytic) and those that live rooted in the soil (terrestrial).

The epiphytic group is by far the largest and the one usually meant when orchids for the greenhouse or home are discussed. They are mostly distinguished by possessing bulb-like swollen stems known as pseudo bulbs. These act as stores for water and food, but also bear roots, leaves and flowers. The roots are thick and fleshy, clinging tightly to bark crevices or mossy rocks. In some species a percentage of the roots hang in the air, absorbing atmospheric moisture. Some even carry out photosynthesis *in lieu* of leaves. For these reasons they cannot be grown like ordinary pot plants. However, if the right rooting conditions are provided, a surprising number of orchids will grow in the home.

Until comparatively recently the root fibers of the royal fern (*Osmunda regalis*) and varying amounts of sphagnum moss were the basic ingredients of all orchid-growing mixes. The osmunda fiber is now scarce, but bark and plastic chips have proved to be suitable substitutes. Some orchids will grow in bark chips alone, e.g. *Cattleya*, others need something richer. A good all-purpose potting mix can be made up of 2 parts sphagnum moss, 2 parts plastic or bark chips and 1 part rough peat moss. Slatted baskets or perforated pots are advisable. Ordinary clay pots may also be used but should be one-third filled with crocks or similar drainage material.

When potting, the plant is positioned in the container, if necessary winding the roots round to get them in. Pour the mix around the roots and tap the container to settle the material, but do not firm in any other way. If the plant is too heavy, wind heavy-gauge galvanized wire around the rim of the pot, twist to secure, and leave an end sticking up like an aerial to act as a stake. Potting or re-potting is best done in spring or autumn but not while the plant is actually blooming. A very effective way to grow epiphytic orchids is to place them on attractively-shaped tree branches as described under Bromeliads.

Orchids need regular watering but this must never be overdone. After watering, always allow the compost to just dry out before repeating. The best way is to dunk each pot in a bucket of water, holding it there for about 10 seconds. Some species need a dry resting period of up to two months at the end of the growing period and this is mentioned in the A–Z entries. Although some orchids will grow and even flower satisfactorily in the dry atmosphere of a room, e.g. *Dendrobium nobile*, some sort of humidification is necessary for the majority.

Most orchids are best screened from direct sunlight from late spring to mid-autumn. Those species which need a dry rest also respond to full sun, which ripens the growth and initiates flowering.

Feeding orchids is still somewhat controversial. They will certainly grow and flower without feeding of any kind, but most sorts, especially the hybrids, respond to doses of standard liquid feed at half strength with each watering during the short season of active growth.

Propagation is not difficult by division. Only well-established plants must be used, ideally those that have been growing undisturbed for three years. Make sure each division has several pseudobulbs, leaves and roots. It will be necessary to use scissors or a sharp knife to divide most orchids efficiently. Late spring or immediately after flowering is the best time. Pot each division at once and water sparingly until new growth appears.

Cymbidium Dwarf hybrid

Dendrobium nobile

Paphiopedilum hybrid

Odontoglossum hybrid

Pleione bulbocodioides

ABOVE LEFT
Vanda × rothschildiana

Echeveria 'Blue Curls'

Lithops karasmontana

Euphorbia pugniformis

Astrophytum ornatum

Mammillaria bocasana

SUCCULENTS (INCLUDING CACTI)

ABOVE RIGHT
Rhipsolidopsis gaerntneri

The name cactus, or its plural cacti, is often used for all the fleshy-leaved and fleshy-stemmed plants, but strictly speaking only members of the American family *Cactaceae* are eligible. Cacti are stem succulents and are comparable to the fleshy-stemmed spurges (*Euphorbia*) from Africa. In both cases the leaves are absent or very tiny. Stonecrops (*Sedum*), living stones (*Lithops*), *Haworthia*, *Aloe*, etc., are leaf succulents. Here the stem is normal, but the leaves are thick with water storage tissue. Both stem and leaf succulents are largely inhabitants of low rainfall areas with plenty of sunshine, mainly semi-deserts and mountain slopes. There are some exceptions however. These are often known as forest cacti and cling to tree trunks and branches in tropical forests. Best known are *Epiphyllum*, *Rhipsalis*, *Rhipsalidopsis* and *Schlumbergera*. The culture of all these is the same as outlined under Bromeliads.

Contrary to views held by many non-gardening people, the desert succulents do not thrive on neglect. Plenty of sun is needed especially during the active growth period in summer. High night temperatures are not required and even best avoided. Most species will survive at a minimum winter temperature of 5°C(41°F) and many can take slight frost. Some genera, e.g. *Rebutia*, *Mammillaria*, and *Chamaecereus*, will not flower well or at all if kept too warm in winter. Summer temperatures can soar to 32°C(90°F) and above by day, but there should be a good air flow around the plants. Fancy potting mixes are not needed, but they must be sharply drained. Use any proprietary potting soil and mix it with a third by bulk of grit or coarse, washed sand. Pot or repot in spring, firming the soil only lightly. Allow the compost to just dry out, then water thoroughly. This should be the pattern of watering during the growing season. All succulents must have plenty of water while they are actively growing, generally in the summer, but the soil must never stay wet for more than a few days. Water sparingly in autumn and give none at all in winter and early spring unless shrivelling is observed. From late spring onwards water regularly. Established plants respond greatly to liquid feed applied at two- to three-week intervals during the late spring to late summer period.

Extra humidity is not necessary, but a light misting over in the summer is beneficial. Shading also is not essential, though some succulents with thinner leaves are best screened from the hottest summer sun.

Propagation is easily carried out by division, stem or leaf cuttings in summer. No propagating case is necessary. Leaf cuttings of *Sedum*, *Graptopetalum*, some *Crassula* species and similar leaf succulents are simply laid on the surface of damp, coarse sand or potting soil. Stem cuttings are laid on the window ledge for a few days so that the base can heal over. They are then inserted in sand or equal parts sand and potting soil.

Many cacti carry a fearsome armory of spines, demanding considerable care when handling. A collar of rolled up newspaper or thin cardboard provides an ideal solution.

ANNUALS

Primula obconica

Calceolaria × *herbeohybrida*

Matthiola Brompton var.

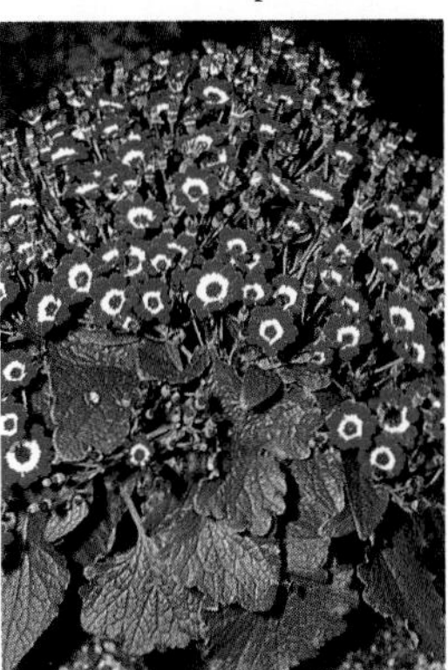

Senecio bicolor cineraria var.

Lobelia erinus var.

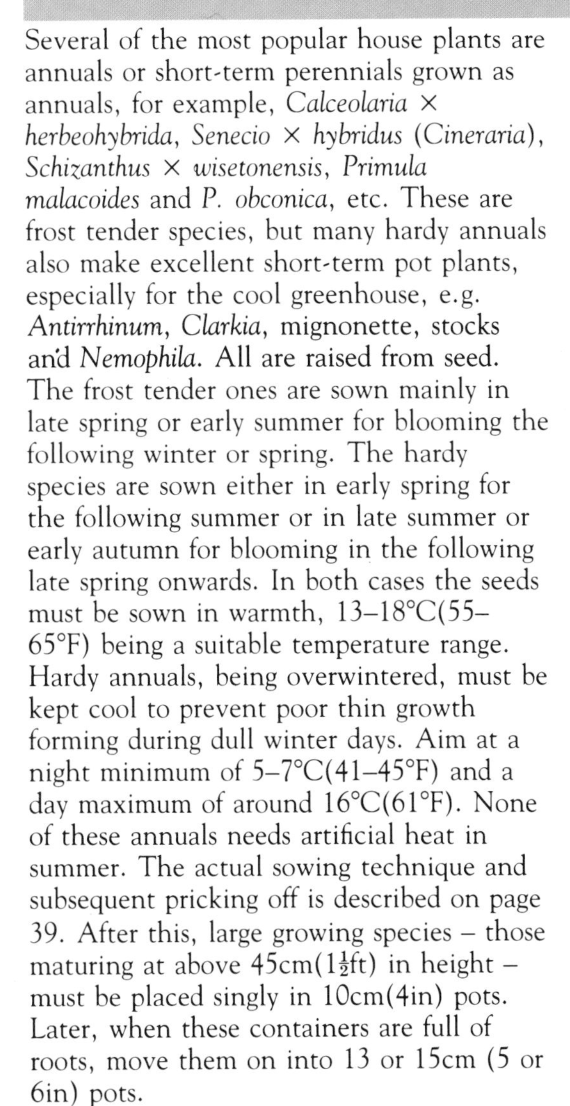

Several of the most popular house plants are annuals or short-term perennials grown as annuals, for example, *Calceolaria* × *herbeohybrida*, *Senecio* × *hybridus* (*Cineraria*), *Schizanthus* × *wisetonensis*, *Primula malacoides* and *P. obconica*, etc. These are frost tender species, but many hardy annuals also make excellent short-term pot plants, especially for the cool greenhouse, e.g. *Antirrhinum*, *Clarkia*, mignonette, stocks and *Nemophila*. All are raised from seed. The frost tender ones are sown mainly in late spring or early summer for blooming the following winter or spring. The hardy species are sown either in early spring for the following summer or in late summer or early autumn for blooming in the following late spring onwards. In both cases the seeds must be sown in warmth, 13–18°C(55–65°F) being a suitable temperature range. Hardy annuals, being overwintered, must be kept cool to prevent poor thin growth forming during dull winter days. Aim at a night minimum of 5–7°C(41–45°F) and a day maximum of around 16°C(61°F). None of these annuals needs artificial heat in summer. The actual sowing technique and subsequent pricking off is described on page 39. After this, large growing species – those maturing at above 45cm(1½ft) in height – must be placed singly in 10cm(4in) pots. Later, when these containers are full of roots, move them on into 13 or 15cm (5 or 6in) pots.

Smaller growing annuals, e.g. *Viscaria*, pansy, toadflax (*Linaria*), *Nemesia*, etc., are more effective if three plants are spaced out directly into final 13 or 15cm (5 or 6in) pots. Watering must be carried out regularly. Aim to keep the potting soil just moist. Dryness at the root can result in buds withering and foliage spoiling. During the sunniest summer weather light shade from direct sunlight is necessary plus all the ventilation possible. All these annuals are best grown in a frame or greenhouse where light and humidity are good, and only brought into the home when they start to flower. Nevertheless it is possible to grow them from seed in the home providing one has a window which is sunny in winter, i.e. south-facing in the northern hemisphere, north in the southern hemisphere.

Taller annuals, especially those grown through a northern temperate winter may be a bit weak in the stem and in need of support. This should be done while the plants are still half grown. Twiggy sticks are good. Alternatively, split bamboo dyed green can be used.

Whatever method is adopted it must be unobtrusive when the plant is in bloom. About one month after the annuals have been placed in their final sized containers, liquid feeding with a general fertilizer should commence. Apply this at 7–10 day intervals until flowering eases off and few if any buds are left.

To help further prolong the flowering season, regularly remove spent blooms and any seed pods that form – the formation of seeds drains the plant's energy.

ABOVE LEFT
Petunia 'Prelude Velvet'

Tulipa 'Jockey Cap'

Tigridia pavonia

Hyacinthus 'Ostara'

Gloriosa rothschildiana

ABOVE RIGHT
Hymenocallis narcissiflora

Place Dutch hyacinths in bulb fiber with their neck just clear of the surface. For early flowers, plant cold treated bulbs and keep them covered in a warm, dark place until the flowers buds are just visible.

BULBOUS PLANTS

This includes all the soil-borne storage organs—true bulbs, corms and tubers. True bulbs are composed of fleshy scales derived from modified leaves, attached to a very short, thickened stem called a basal plate. Corms are greatly enlarged stem bases generally globular or disk-shaped. Tubers may be greatly enlarged stems or roots.

Bulbs, corms and tubers occur in a variety of plant families but the best-known ones are mostly in the lily family (*Liliaceae*), amaryllis family (*Amaryllidaceae*) and iris family (*Iridaceae*). From a cultural point of view they can be divided into hardy and tender groups. The hardy ones are very short-term house plants, being brought in as they bloom and taken out as soon as flowering ceases. Tender bulbs are potentially long-term house plants, though some may be completely dormant and leafless for several months of each year. Cultivation of each group is as follows.

HARDY BULBS

Daffodils, narcissi, tulips, hyacinths, crocuses, scillas and other hardy bulbs all make cheerful pot plants to brighten up winter days. Purchase plump, healthy bulbs and set them close together in pots or pans as early in autumn as possible. Use a commercial potting soil for the best results. Bury small bulbs 2.5cm(1in) deep; bigger ones can have the neck of the bulb protruding above the soil. Water thoroughly, then put in a cool dark place. A shady site outside is best, ideally with the containers buried in moist peat. Alternatively, keep them under the stairs or in an unheated north facing room, ideally in plastic bags to conserve moisture. This initial cool period is important as it encourages a good root system to form which later will keep leaves and flower buds developing healthily. About six to eight weeks after potting, or when shoots are 2.5–5cm(1–2in) high, bring the containers into a window, but screen from direct sunlight for the first week or so. Avoid high temperatures, which above 21°C(70°F) may cause tiny flower buds to wither. Water regularly, but do not overdo it. Over-wet or dry soil conditions are the primary causes of failure to bloom properly. Mist daily to provide extra humidity. If in a warm room by day, move to a cool one at night if at all possible. This will prolong the floral display. If liquid feed is applied at two-week intervals once flower buds show and the plants are kept going after blooming ceases, the bulbs can be used year after year, though the quality will be less good after the first time.

TENDER BULBS

The culture of tender bulbs is much like that of the hardy ones but differs in the following ways. Potting and repotting may take place also in winter or spring, occasionally also in late summer, e.g. *Lachenalia* and *Freesia*. The time to pot is mentioned under the A–Z generic descriptions. The bulbs are usually completely buried (*Nerine* and *Hippeastrum* [amaryllis] are exceptions) and are spaced

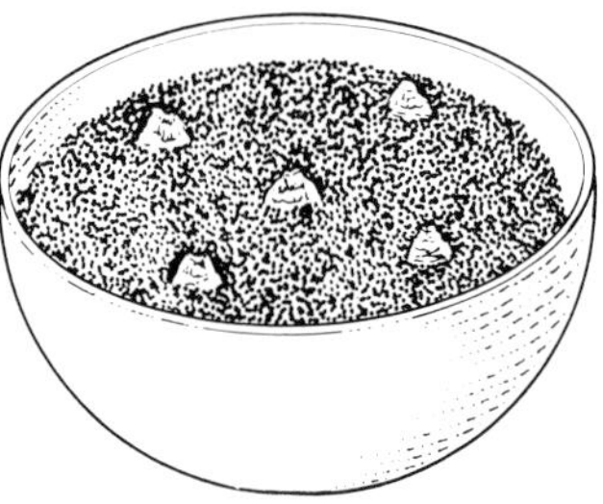

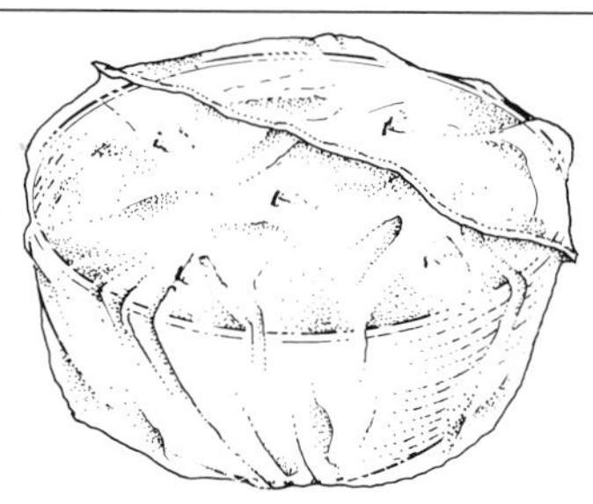

more widely to allow room for growth of offsets. Feeding and humidifying should be regularly carried out. Some species, e.g. *Hippeastrum*, *Nerine* and *Hymenocallis* are best left in their containers for two or three years before repotting. These and others must be dried off in autumn and kept dry until the following spring. All the cultivation variants are noted in the A–Z section.

Propagation is easily effected by separating offsets at repotting time, and growing these on to maturity in smaller pots (*see* p. 39). Seed may also be sown in spring and provides a fascinating means of increase for those with patience, most bulbous plants taking 3–6 years to reach flowering size.

BELOW Dutch hyacinths can be grown entirely in water in specially designed bulb glasses. Keep the water level topped up to just below the bulb, and to keep the water sweet add a few chips of charcoal.

BOTTOM *Narcissus* 'Cragford' in flower in a trough of pebbles. This is a satisfactory alternative to growing in bulb fiber, especially for the larger species, keeping the roots well aerated.

THE ORIENTAL ART OF BONSAI

The word *bonsai* simply means *plant in a container*. The plant can be a tree – the most familiar type – or a shrub. The basic principle of this unusual form of cultivation is to severely restrict root extension and in so doing stunt the growth of the stems and leaves without impairing their health and beauty. The elegance of a full-size forest tree or shrub is transformed into miniature, though the ageing process is not accelerated and bonsais still take many many years to gain their adult appeal.

It is very important to remember that if the species selected is a hardy outdoor plant in its natural size, the bonsai equivalent will need exactly the same environment for most of the year. Only bring it into the home for short stays. Since this book deals exclusively with plants suited to permanent indoor cultivation, most bonsais are outside our scope. But, why not try an exotic tree or shrub such as mentioned on p. 22–23.

You will need a shallow container with drainage holes – glazed or unglazed ceramic troughs are readily available nowadays. Select a good compost which is very well drained. Initial propagation follows the same principles as for any other tree or shrub, but subsequent training and pruning (both of top growth and roots) needs careful attention. Young plants must be pinched back hard. When new shoots develop, only the weakest one should be retained. From now on growth in a small container should be slow. Using copper wire, the shoots can be twisted and contoured into shapes which mimic an aged and gnarled adult. Branches and roots have to be pruned regularly and with deciduous species the first flush of leaves is sometimes thinned or removed entirely to encourage regrowth of smaller, neater ones. Pinch the growth shoots to two or three buds as necessary. Root pruning should be carried out at repotting – every couple of years. Cut back old or dead roots and reduce the rest by one- or two-thirds according to the age of the plant, then repot with fresh potting soil in the same (or slightly larger) container.

For fuller instructions, consult a specialist publication on bonsai.

A bonsai *Malus baccata* 'Manchuria'.

MAINTENANCE AND PROPAGATION

PRESERVING PLANTS WHILE ON VACATION

What to do with house plants while one is on vacation is a frequent problem especially if they are to be left untended for two to three weeks. Mature or root-bound specimens that need daily watering in summer are a particular headache. Fortunately there are several things that can be done to tide plants over a period of absence.

The first essential is to devise a means whereby water lost by transpiration (the equivalent in plants to perspiration in man) is conserved. One of the easiest methods is to water the plant thoroughly and then enclose it in a plastic bag large enough to take the whole plant. Put the pot in the bag and tie it around the base of the plant so that the root ball is entirely enclosed. Then loosely secure the mouth of the bag above the plant so that an opening large enough to take two or three fingers is left for aeration. Place the plant where it is screened from all direct sunlight, preferably in the coolest room in the house. A plant treated in this way should survive in good condition for about three weeks in summer, and longer in winter. Some fast-growing plants such as *Impatiens* may make soft, pale shoots which are unhealthy during this period, but these can be carefully cut away on return home and the plant will be none the worse.

Really root-bound pots are best set in a bowl of water allowing 2.5–4cm(1–1½in) of water for each week away. It is not, however, advisable to keep plants going like this for more than three weeks as there is always the danger of the soil becoming sour and the roots rotting.

There are now several devices on the market for watering plants while one is away. One of these uses absorbent wicks to convey a steady flow from a jar or bucket directly to the soil. The best wicks are made of glass fiber and do not rot. It is possible to improvise on this idea by placing a container of water (a pint jug is ideal), just above the pots with thick wool wicks leading to it from each pot. The wicks can be made from four or five strands of thick wool twisted together. Anchor the end of each wick to the soil surface with a small stone or a hairpin.

On wide, strong window sills, one or more sub-irrigation trays can be installed. These are filled with sand kept wet with a feeder bottle (such as is used to dispense water automatically for poultry and pets).

Plants may be grown permanently by such means and never watered by hand. If this method is employed it is important that the pots are not cracked as this may prevent the water from entering them and you must make sure the pots are pushed into the surface of the sand with a screwing motion to ensure that the soil and the wet sand are in contact.

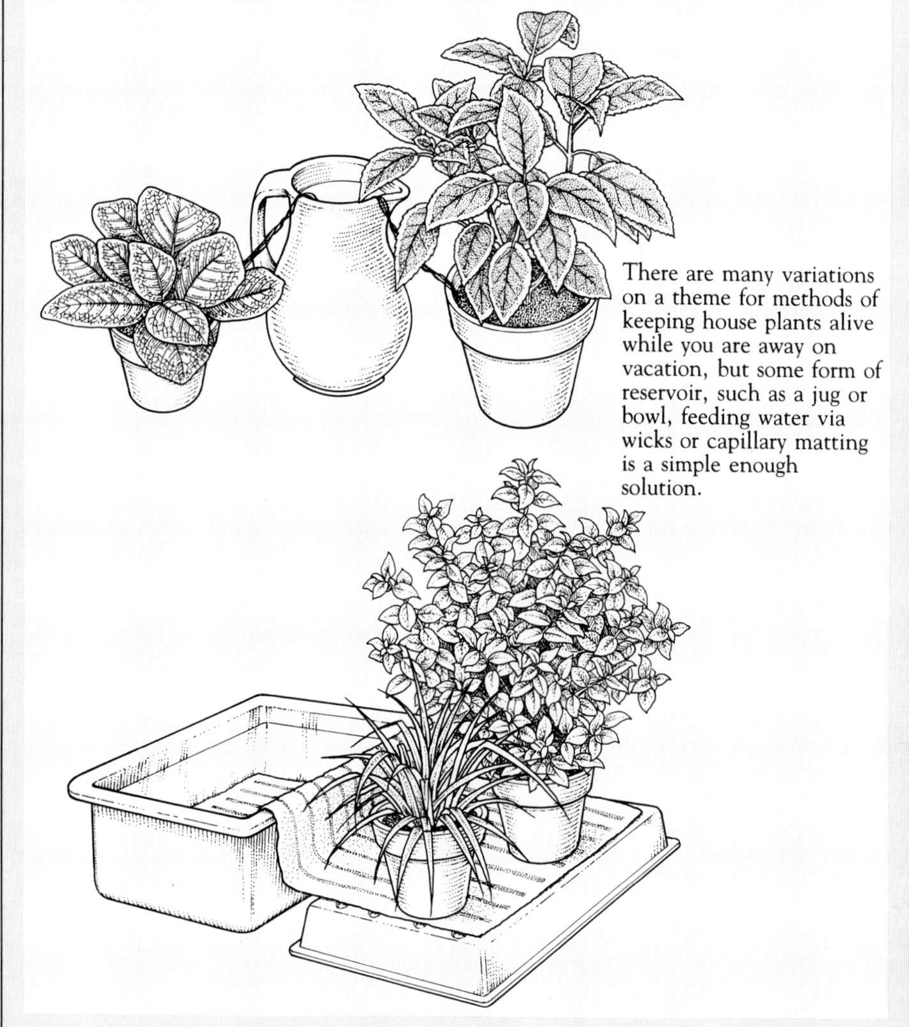

There are many variations on a theme for methods of keeping house plants alive while you are away on vacation, but some form of reservoir, such as a jug or bowl, feeding water via wicks or capillary matting is a simple enough solution.

PROPAGATION

Although a wide variety of house plants is readily available there is much satisfaction to be gained from raising them in the home from seeds or cuttings, especially if they are a gift from a friend or relative. No special equipment is needed, though the plastic seed or propagating trays or pots that have rigid transparent covers are very useful. Plastic bags or sheeting may be used instead, supported on inverted U-shaped ribs of galvanized wire.

Rooting is hastened or made more certain if the pots of cuttings are heated from below. For this reason there is much to be said for using one of the custom-built propagating cases with a built-in thermostatically controlled heating element. The easiest rooting medium for cuttings is coarse sand, though when rooted they must be potted without delay to prevent starvation. The addition of up to half by volume of peat moss will conserve moisture and prevent rapid starvation. If potting is delayed use a liquid or foliar feed.

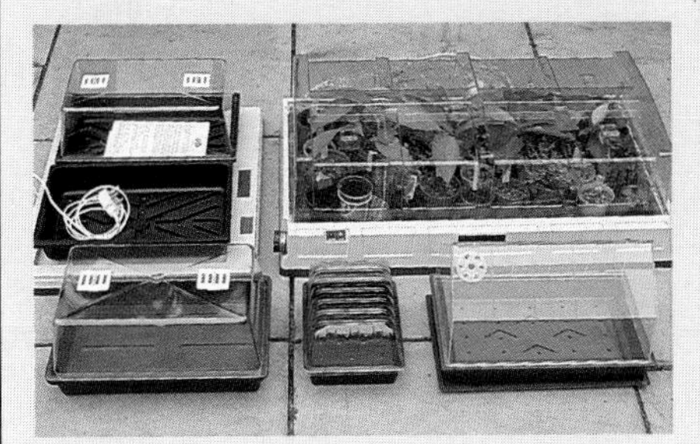

Selection of simple and electric propagators

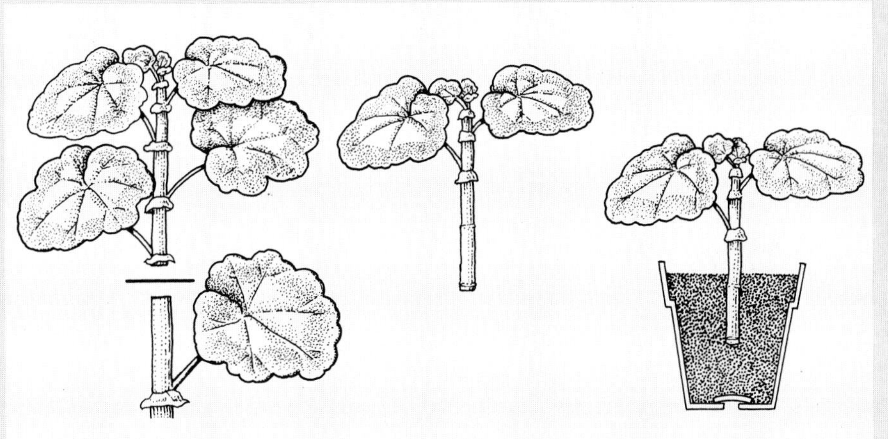

Geraniums (zonal *Pelargonium*) are easily propagated from stem cuttings following the steps shown here. Be sure to carefully peel away the small leaf-like stipules from the basal nodes, which would soon rot and infect the cutting. Rooting should be under way in ten to twenty days.

STEM CUTTINGS

Also known as slips, these are detached pieces of stem, with or without leaves. They can be divided for house plants into two categories, the softwood cuttings (cuttings taken from the soft growing tips of shoots) and semi-hardwood cuttings (sections of older, firmer stems).

Take tip or softwood cuttings in summer, selecting sturdy shoots with several leaves and severing the end with a sharp knife just above a bud or leaf. The exact length of the cutting will depend on the species. For example, a robust-stemmed plant such as a geranium (zonal *Pelargonium*) will be about 10cm(4in) long, with two or three leaves left at the top. The smaller, wiry shoots of *Erica* × *hyemalis* (the popular winter-blooming winter heath or heather) need be no more than 4cm(1½in) long, with maybe a dozen needle-like leaves. Whatever the species, carefully break or cut off the leaves from the lower half of the cutting and sever cleanly at the point just below where the lowest leaf grew out. Insert one-third to one-half of the stem's length in the rooting medium and firm by pressing down with the point of the dibber. To make sure of success, one of the rooting hormone powders or liquids available on the market can be used. Dip the bases of the cuttings into the powder or liquid before insertion. The spacing of the cuttings also varies with the species, but a distance equal to half the length of the cuttings before insertion is adequate.

Large cuttings such as those of geranium (*Pelargonium*) and × *Fatshedera* may be set singly in 6 or 8cm (2½ or 3in) pots, but better use is made of the space if the cuttings are put in rows or circles in the boxes or larger pots before being placed in the propagator. Water the cuttings with a fine-rosed watering can as soon as they are inserted, and place the propagator in a temperature of 18°–21°C (65–70°F) out of direct sunlight. Rooting may take anything from 10 to 30 days, sometimes longer. As soon as growth is seen to start again and new leaves begin to unfold, it is fairly certain that the cutting has rooted and

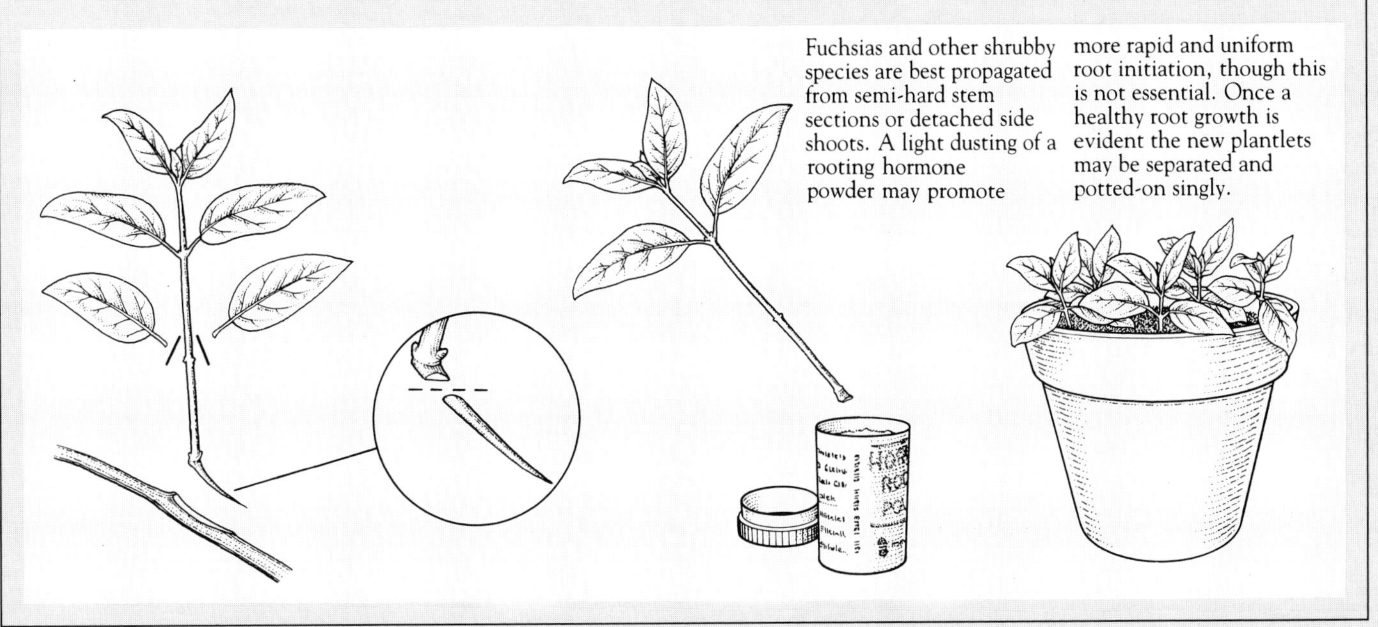

Fuchsias and other shrubby species are best propagated from semi-hard stem sections or detached side shoots. A light dusting of a rooting hormone powder may promote more rapid and uniform root initiation, though this is not essential. Once a healthy root growth is evident the new plantlets may be separated and potted-on singly.

will be ready to be potted. Set each small plant singly in a pot of 1 part coarse sand and one part peat moss or a proprietary equivalent. Water thoroughly and keep in a shady place for a few days until new roots form.

Semi-hard cuttings are made from the older, firmer growth of shrubby plants. They may be made from sections of long stems, or from shorter lateral (side) shoots that are just becoming woody at the base. They are trimmed and treated in the same way as soft cuttings, though they usually take longer to root.

Some shrubs, such as the *Erica* mentioned above, root best from small lateral stems pulled off so that a sliver of the stem they are borne on remains at their base. To do this, grasp the shoot at the base with the thumb and forefinger and firmly but carefully pull back and down until it comes free. If a long tail of the older tissue remains, trim it with a knife or scissors. Then divest the cutting of lower leaves and insert it as described.

LEAF CUTTINGS

Some plants, notably African violets (*Saintpaulia*), *Streptocarpus*, most begonias and *Dieffenbachia* can be reproduced by treating single leaves or even parts of leaves as cuttings. African violets, small-leaved begonias and peperomias are grown from single leaves detached with 2½–5cm(1–2in) of stem. Insert about half the stem into the rooting medium and treat as for stem cuttings.

The long narrow leaves of *Sansevieria* and *Streptocarpus* can be cut transversely into 8cm(3in) sections, each one being inserted for a third of its length. Larger-leaved peperomias such as *P. argyreia* (*P. sandersii*) and *P. magnoliifolia* can either to be inserted as whole leaves or cut into sections. To do this, take a sharp knife or razor blade and cut off the leaf stalk. Now lay the leaf on a flat surface and cut wedges from it like slices from a cake, the inner point of each wedge tipped by part of the leaf stalk. Three to four wedges can be cut from each leaf. Insert the lower third of each wedge at an oblique angle (about 45°) into the rooting medium.

The large leaves of begonias, particularly *B. rex* and its hybrids, can be cut into 2.5cm(1in) squares and these laid upon the surface of the rooting medium, almost touching each other. Each section will produce one to three tiny plantlets if a temperature of 23°C(74°F) and high humidity can be maintained. Another method with large begonia leaves is to sever a leaf without a stalk and nick the main veins underneath at 2.5cm(1in) intervals with the point of a knife. Then place the leaf right way up on the surface of the potting soil. If it does not sit flat, a few pebbles or hairpins will hold it in place. Plantlets arise from the nicks. A temperature of 21–23°C (70–74°F) is needed for the best results.

Some plants make propagating simple for the gardener by producing easily detached plantlets on the leaves or stems. Spider plant (*Chlorophytum*) develops small plants on the flowering stem, usually as the flowers fade. Pick-a-back plant (*Tolmiea*) produces a plantlet at the junction of the leaf stalk and blade, and the hen and chickens fern (*Asplenium bulbiferum*) bears many fernlets on its lacy fronds. In each case let the plantlets produce at least three or four of their own before detaching them. Thereafter, treat them as cuttings. If only a few young plants are needed, peg the leaves, fronds or stems down into pots or boxes of potting soil. The plantlets will root and can then be detached and potted-on when well grown.

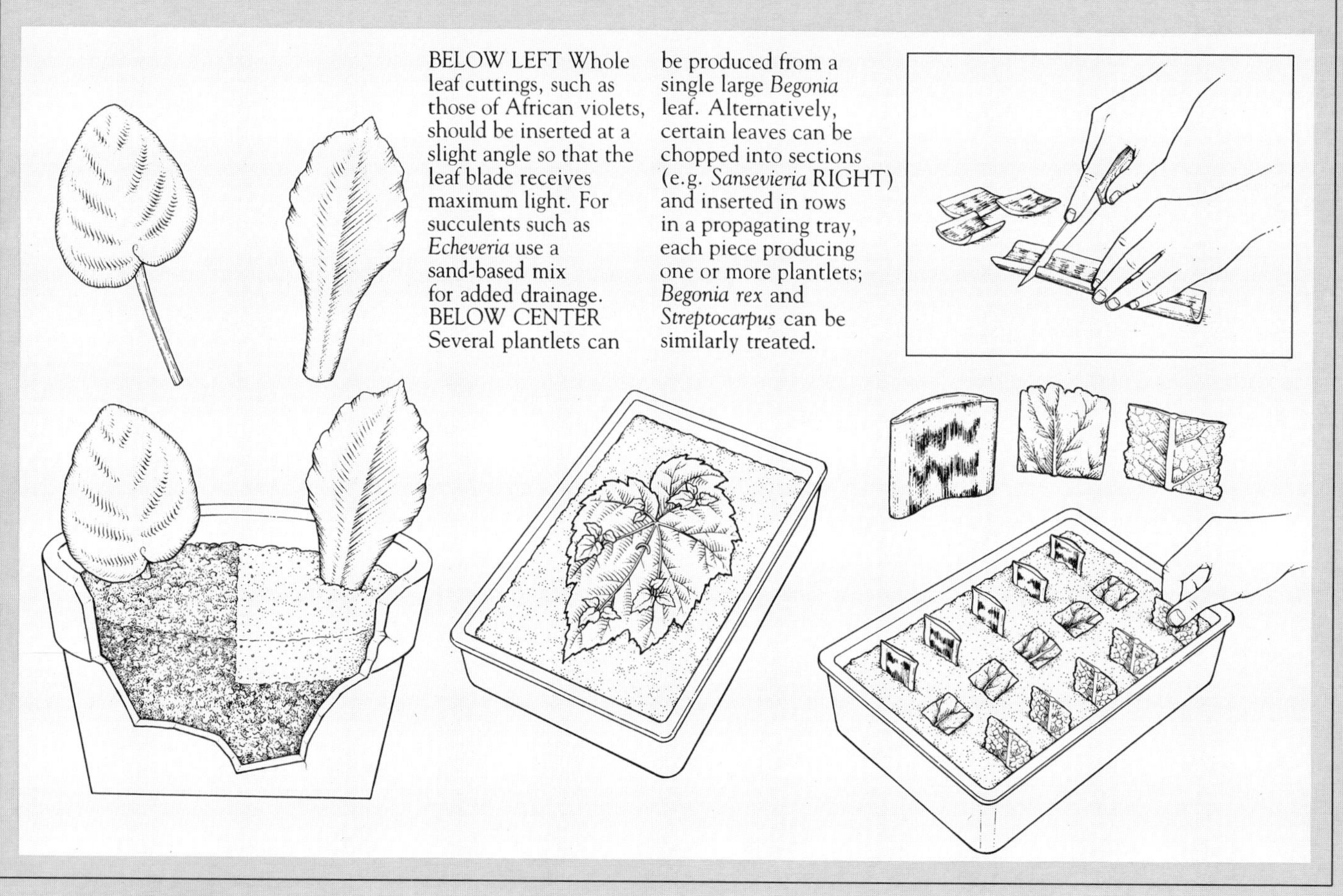

BELOW LEFT Whole leaf cuttings, such as those of African violets, should be inserted at a slight angle so that the leaf blade receives maximum light. For succulents such as *Echeveria* use a sand-based mix for added drainage. BELOW CENTER Several plantlets can be produced from a single large *Begonia* leaf. Alternatively, certain leaves can be chopped into sections (e.g. *Sansevieria* RIGHT) and inserted in rows in a propagating tray, each piece producing one or more plantlets; *Begonia rex* and *Streptocarpus* can be similarly treated.

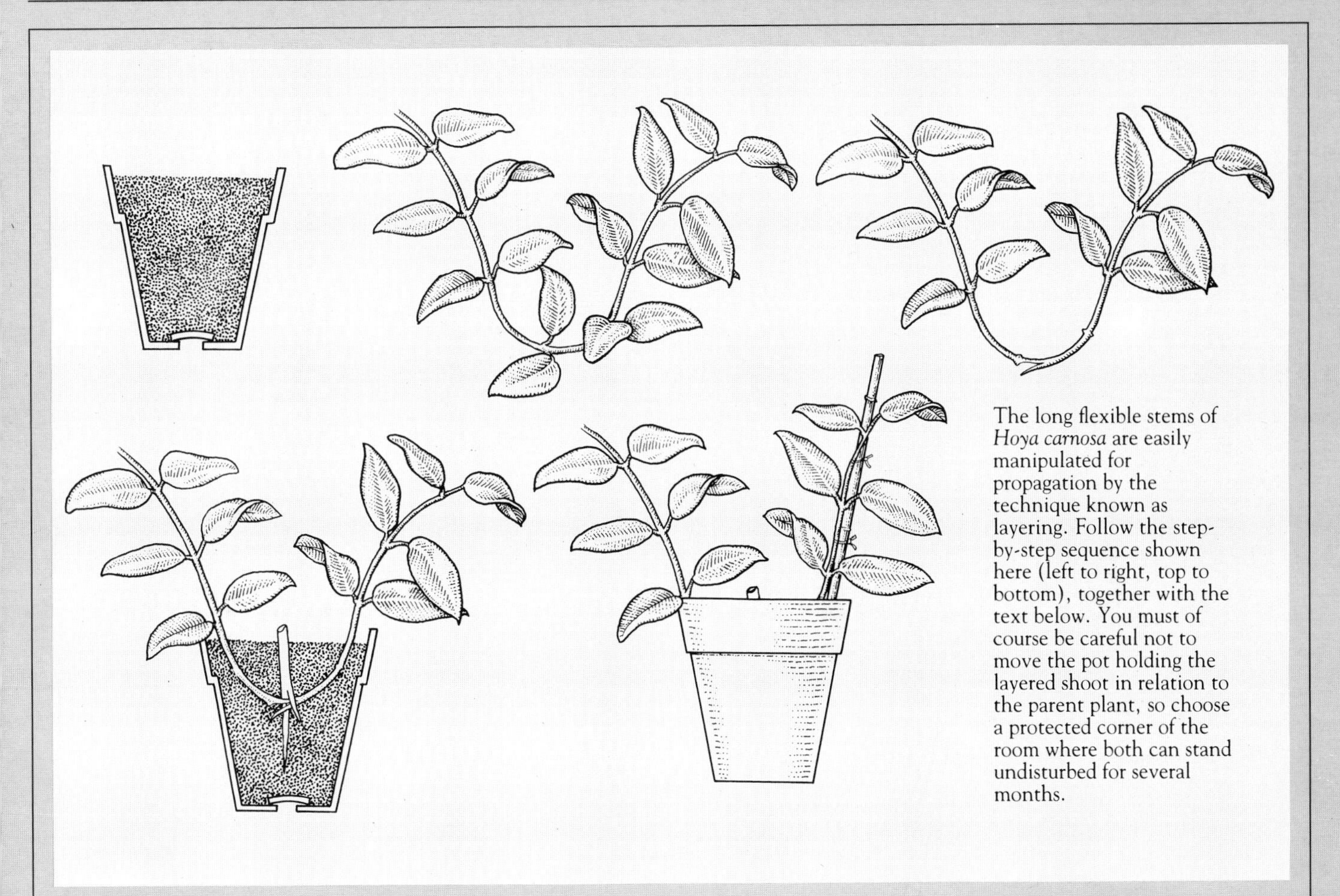

The long flexible stems of *Hoya carnosa* are easily manipulated for propagation by the technique known as layering. Follow the step-by-step sequence shown here (left to right, top to bottom), together with the text below. You must of course be careful not to move the pot holding the layered shoot in relation to the parent plant, so choose a protected corner of the room where both can stand undisturbed for several months.

LAYERING

Covering part of a young stem with soil often stimulates the formation of roots and this is the basis of the almost fool-proof method of propagation known as layering. Plants with flexible stems are required. (For rigid-stemmed sorts see *Air-layering*, below.

Start by preparing some boxes or small pots of potting or seed-sowing soil, and place them near the plant to be layered. Now select a healthy one-year-old stem and bend it downwards into a U-shape so that the base of the U can be buried in the soil of one of the pots or boxes. Now remove any leaves from the vicinity of the U, and nick or slice half through it at its base. Make a depression in the soil and bury the base of the U, securing it in place with a forked stick or hairpin. It is advisable to secure the free end of the layer to a slim cane. To hasten or make sure of rooting, dust the wound with a rooting powder. Keep the soil just moist and rooting will take place in from 6 weeks to 6 months or more, depending on the species. When well rooted, sever the layer from the parent plant and move it into a larger container.

Several plants are capable of self-propagation. Some form small plantlets complete with root initials. These can be severed and potted or rooted like layers. Others, particularly the self-clinging climbers, produce root initials along their stems. Once in contact with compost these quickly grow out. New shoots emerge from the rooted section, which can be separated.

AIR LAYERING

Where it is not possible to bend a stem down for ordinary layering, the following alternative method can be used. Choose a healthy one-year-old stem and remove the leaves from a zone about 20–30cm(8–12in) below the tip. Make a notch or slice in this leafless region and treat with a rooting powder. Form a loose tube around the cut with a piece of plastic wrap 20×10cm(8×4in). Secure the tube at the bottom end with water-proof tape then fill with moist sphagnum moss or a mixture of moss and potting soil. Pack firmly but not tightly, and

Although the process of air layering is a familiar one, people tend to be reluctant to use it, probably because they are afraid of mutilating their favorite house plant. Actually, the technique is simple provided one starts off with a sharp knife and carefully follows the sequence illustrated (left to right, top to bottom). Perhaps the main difficulty is the patience required – at least one growing season – before the air-layered stem produces adequate roots to be cut away from the main stem for potting-up. With *Ficus elastica* air layering provides a useful means of rejuvenating an old, lanky plant.

secure the top end with tape. A slender stem may need a cane support. When roots show against the plastic, sever the stem and remove the tube. Pot immediately without removing moss or soil.

Place in a propagating case or large plastic bag, and keep shaded from direct sunlight for a few weeks until new root growth is resumed.

DIVISION

Plants which have a crown (several stems or shoots coming together from or below ground level) can be divided or split into two or more sections. Most ferns come into this category, as do several of the cacti and succulents. Do this in spring just as the new growth starts. Take the plant from its pot and if possible remove some of the soil by shaking or by using a small fork. A knife may be needed to sever some of the stems which lace the crown together. Large crowns can be parted by inserting two handforks back to back in the center and then pulling the handles away from each other. The separate parts can be repotted at once, and treated as for mature specimens.

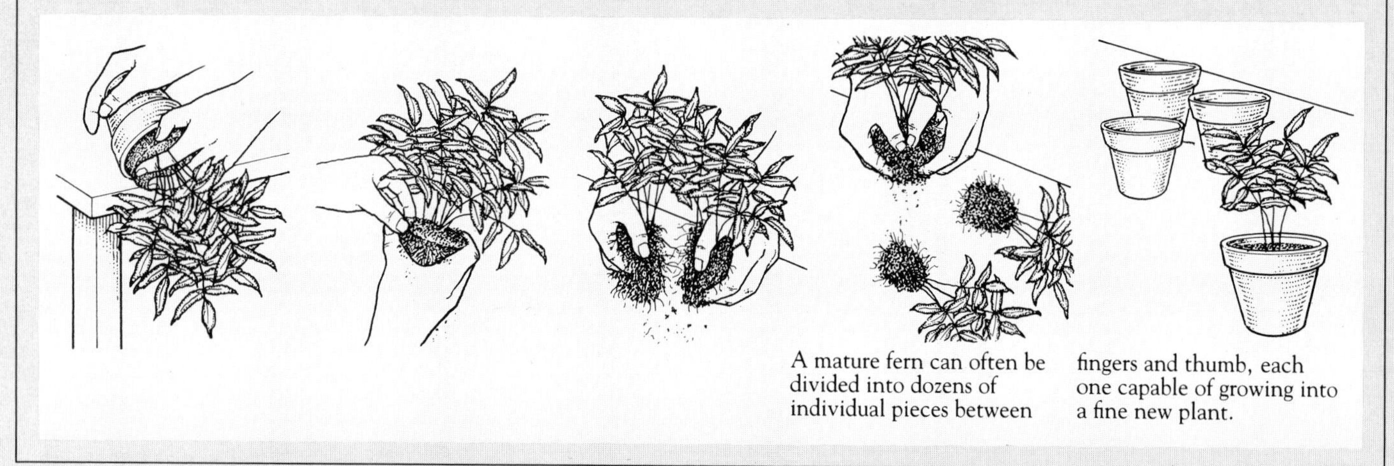

A mature fern can often be divided into dozens of individual pieces between fingers and thumb, each one capable of growing into a fine new plant.

OFFSETS

Bulbous plants and some others produce offsets (small versions of themselves) at the side of the main bulbs or plant. At potting time these can be separated and potted individually.

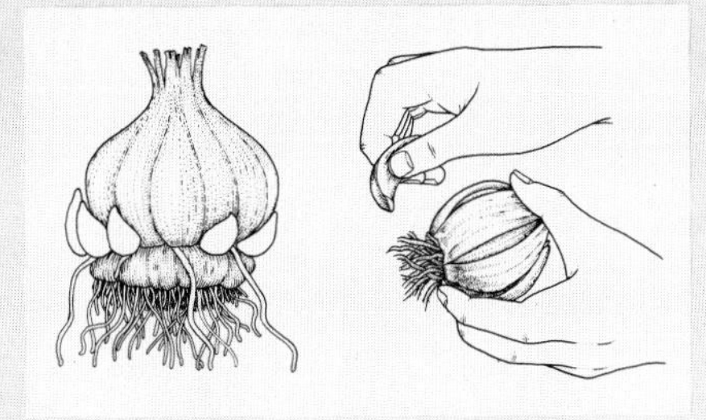

Propagate lilies by detaching scales.

SEEDS

Although comparatively few house plants are regularly raised from seeds, this can be a most rewarding way of obtaining a favorite plant. In order to germinate and grow into healthy plants, seeds require moisture, air, an adequate temperature and minerals in the rooting medium to insure rapid growth. Several seed-starting mixes such as Jiffy-mix are available commercially and all provide a suitable germinating and growing medium. Fill the pan or box with soil mix to within 1–2cm($\frac{1}{2}$–$\frac{3}{4}$in) of the rim, firming it lightly but evenly. Seeds must be sown thinly so that the resulting seedlings are not over-crowded, starved, and therefore prone to disease. The larger seeds are best spaced out 2.5–4cm(1–1$\frac{1}{2}$in) apart with tweezers. Deep sowing is a common cause of failure; seeds should be covered with a layer of soil no thicker than their own diameter. Very fine seeds such as those of begonia are best sown on the surface and, at the most, dusted over with sand.

Water very fine seeds by immersing the container to just below the rim in a bowl of water. Allow to drain for an hour or so and then place in a plastic bag or cover with a sheet of glass. Keep at a temperature between 16°C and 21°C (60°F and 70°F); the more tropical species should have a temperature of between 21°C and 23°C (70°F and 74°F). If the seed containers are placed on a very sunny window sill, cover with a sheet of newspaper. As soon as the seedlings appear, remove all covers. If water is needed use a fine-rosed can or immerse as described above.

The first leaves to appear when the seed germinates are usually the seed leaves and are different from the adult or true leaves. When the first true leaf is well developed, the seedlings must be separated and potted singly if big or spaced out about 2.5–5cm(1–2in) apart in a seed tray of potting soil. Seedlings are usually transplanted with a dibber (a pencil-shaped peg). This is used to make a narrow hole deep enough to accommodate the seedling's roots at full length. Insert the seedlings, firming with the tip of the dibber and water with a fine-rosed can. Put them in a lightly shaded site for a day or two until new roots form.

When the young plants are several inches tall, or the leaves of neighboring plants overlap each other in the pots or boxes, they are ready for separating and potting individually.

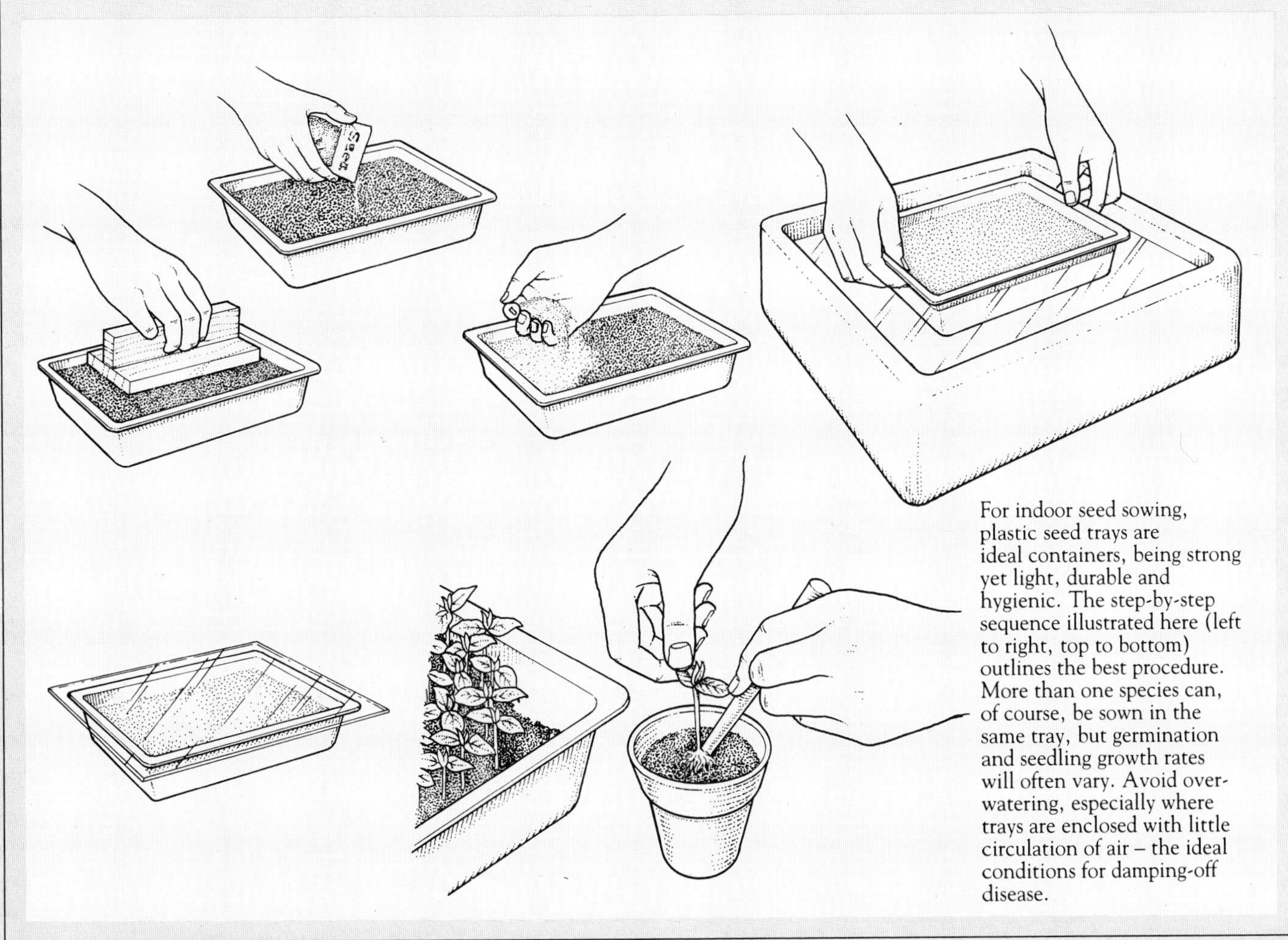

For indoor seed sowing, plastic seed trays are ideal containers, being strong yet light, durable and hygienic. The step-by-step sequence illustrated here (left to right, top to bottom) outlines the best procedure. More than one species can, of course, be sown in the same tray, but germination and seedling growth rates will often vary. Avoid over-watering, especially where trays are enclosed with little circulation of air – the ideal conditions for damping-off disease.

PESTS AND DISEASES

Fortunately plants in the home suffer very little from the attacks of pests and diseases. The few that do occur are usually brought in on new plants, so it is advisable to examine these carefully before placing them with existing plants. The troubles listed below are those which are most likely to occur. Where chemical cures are given, even innocuous ones, it is best to do the spraying outside on a window ledge or in the garden if possible, or else in a sink which can be washed down afterwards. Since drought-stressed plants can be damaged by pesticides, make sure that they are well watered at least a day in advance and always follow the makers' instructions to the letter. An overdose may damage the plant, and an underdose will almost certainly not wipe out the pest or disease. For specific advice on pesticide use in your state, contact your county cooperative extension agent listed in the telephone directory.

SYMPTOMS, CAUSES AND TREATMENTS

Leaves deformed Mildly to badly deformed leaves, particularly the young ones, are usually the result of aphids (greenfly) feeding. Aphids are tiny, oval, soft-bodied creatures, usually green but sometimes black or other colors. They feed and live communally, often in great numbers, and multiply fast. There are winged and wingless forms. Sponging infected leaves and shoots with sudsy detergent water will often remove small infestations. If they persist, spray with malathion or use one of the systemic insecticides which are watered onto the soil. See also *Leaves with reddish streaks* entry.

Leaves eaten Pieces removed from leaves usually mean that a stray caterpillar or earwig is at work. The caterpillar should be searched for and destroyed. Earwigs are nocturnal and hide by day. They can be trapped in a small roll of corrugated cardboard, or a flower pot loosely filled with crumpled paper. Inspect daily.

Leaves falling prematurely There are several reasons why leaves fall before their time. A common cause is over-watering which causes root rot; leaves tend to yellow, wither and fall. Root mealy bugs may be responsible. Tap the plant from its container and check for small whitish insects. If the falling leaves are also mottled and show a faint webbing (not always very apparent), glasshouse red spider mite will be the cause. These are microscopic creatures, generally red-spotted, but a hand lens is needed to see them properly. Spray with a miticide and try to spray daily with clean water as this discourages re-infection. See also the next entry.

Leaves with scales Undersides of leaves may become infested with small mussel-shaped or oval scale insects, which suck the sap sometimes causing leaf fall or weakening of the plant. They may be brown, yellowish-brown or whitish, but if the scales are very small and are accompanied by tiny flies like miniature snow white moths, they are of the glasshouse whitefly. This is difficult to control but spraying with malathion will kill the adults. Reapplication will probably be necessary to kill newly hatched scales. For larger scales, see *Stems with scales*—page 42.

Leaves with sinuous pale tracks Cineraria and chrysanthemum leaves bearing this damage are infested by the tiny grubs of the chrysanthemum leaf-mining fly. This feeds on the soft tissue between the upper and lower leaf skins. Hand picking is the most effective cure.

Leaves with reddish streaks Rusty or reddish streaks on *Hippeastrum* (amaryllis) leaves, especially if accompanied by distortion, are caused by the bulb mite. This is a microscopic colorless creature that lives within the bulb. Control is difficult and the bulbs are best discarded.

Leaves wilting Persistent wilting of leaves and shoot tips denotes trouble at the root or stem base: if these are rotten the plant should be destroyed. Severe attack by root mealy bug may also induce wilting. See *Leaves falling prematurely*; see also *Stems with brown or blackened areas.*

Peach-potato aphids (*Myzus persicae*) infesting and deforming cineraria

Red spider mites (*Tetranychus urticae*) spin very fine webs

Soft scale (*Coccus hesperidum*) seen on the underside of a bay leaf

Whitefly (*Trialeurodes vaporariorum*) and nymphs infest leaf undersides

Damage by chrysanthemum leaf miner grubs (*Phytomyza syngenesiae*)

Bulb scale mites (*Steneotarsonemus laticeps*) scar young narcissus leaves

Legless grubs of the vine weevil (*Otiorhynchus sulcatus*) destroy roots

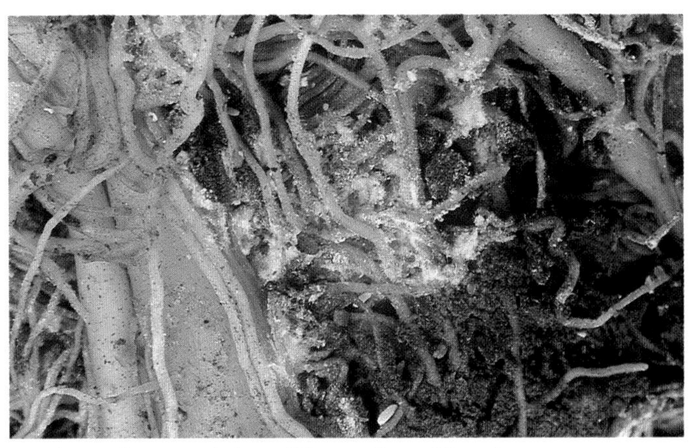

A colony of root mealy bugs (*Rhizoecus*) on *Chlorophytum*

Yellow mosaics and mottling on leaves are often symptoms of a virus infection – here pelargonium virus – transmittable through generations.

White coatings of powdery mildew, here on a begonia leaf, may be discrete at first, but spread to cover much of the plant if not treated.

Oleander scale insects (*Aspidiotus nerii*) on stems of asparagus fern.

A mealy bug (*Pseudococcus*) infestation on *Schlumbergera*.

Leaves yellowing Premature leaf yellowing often accompanies the attacks of sucking insects, and root damage by insects or fungi. See the *Leaves falling prematurely* and *Leaves with scales* entries. Yellowing, especially between the veins, is also caused by lack of iron and is known as chlorosis. It often occurs on azaleas (*Rhododendron*) when grown in a soil with high PH. A temporary cure can be effected by watering with sequestrene (iron chelate).

Stems with brown or blackened areas Cuttings, and sometimes older plants of geranium (*Pelargonium*), may die from the base upwards, the stem blackening and rotting. This is blackleg disease, thriving under over-wet, cold conditions. There is no cure, though it may be possible to take fresh cuttings from the green tips. Make sure these are not kept too wet and provide sufficient warmth. Over-cool, moist or humid conditions also favour the growth of grey mold. This starts as dead, brown areas which later bear a whitish or grey fluffy mold. Cut out or remove damaged tissue, then dust with captan. Maintain warmer, drier conditions.

Stems with scales Tiny brown scale insects, which may be flattened or humped, sometimes appear on leaves and particularly stems of woody plants. They are unsightly and in time weaken the plant. They also secrete honeydew which allows the growth of a sooty mold. Swab woody stems with strongly soapy water, or plain rubbing alcohol. Softer stems and foliage can be sprayed with malathion.

Stems with pink-tinged or whitish fluffy masses Shrubby or woody climbing plant stems may be infested by small soft insects covered with waxy wool. These are mealy bugs which suck sap, weaken plants, and secrete honeydew which enables the unsightly sooty mould to form. Control as for *Stems with scales*.

Stems of seedlings collapsing Overcrowded seedlings are prone to collapse at the base followed by death. This is caused by damping-off disease. Water with captan. In future, avoid over-watering, and sow seeds thinly and use sterile pots and potting mixes.

SPECIES WHICH PLAY HARD TO FIND

Reading about gardening is a favorite occupation of the true gardener and enthusiastic descriptions of plants often spark off the desire to grow them. Asking at garden centers and nurseries nearby may soon reveal the fact that one or more of the desirable plants is 'unobtainable'. The search does not have to stop there, however. If one has a tenacity of purpose, there is a good possibility that a source of supply will be found.

Having systematically searched all the nearby garden centers and nurseries, a countrywide search is the next step. Happily this is a much less daunting task than it might seem since the publication of such source guides as *Gardening by Mail: A Source Book* (a guide to nurseries in the United States) by Barbara J. Benton, published by Tusker Press in 1986. As the plant you are searching for is now known to be uncommon or rare, it is best to concentrate on firms which specialize in the relevant plant group, e.g. trees and shrubs, perennials, cacti. If there are many to contact, draft a letter and get it duplicated. Telephoning is another possibility but not always fully satisfactory.

Another approach is to join a specialist horticultural society or a major national one, for example The American Horticultural Society. Most societies hold plant sales and produce seed lists annually. All societies produce a newsletter, bulletin or yearbook in which a request could be published for the plant you seek. Attending meetings puts you in touch with fellow enthusiasts who might be able to help directly or suggest further contacts. Societies also organize visits to private gardens and collections. If the plant you seek is there, a friendly word with the owner may well result in cuttings or seed or even a young plant, either for purchase or exchange. This same approach can be used when visiting gardens which open to the public from time to time. Botanic gardens should not be neglected when the search seems hopeless. Some of them will provide propagating material if approached diplomatically.

If the book or article which sparked off the desire is not too old, do not hesitate to contact the author via the publisher.

If the search in your own country is fruitless it is usually possible to carry it on abroad, especially if the specialist society you belong to has an international membership.

For the really dedicated searcher, there is the possibility of journeying to the country of origin and collecting seed or plant material in the wild. For this approach one will need the aid of floras and perhaps the help from the staff of a national or major botanic garden to locate the exact area in which to search. It is also important to find out whether the country concerned allows its native plants to be taken, and that the country in which you live allows them in. For complete information regarding the importation of plants into the US, contact the United States 'Department of Agriculture'.

A plant import permit will be required at the point of entry and should be obtained well in advance by writing to: Permit Unit, USDA, APHIS, PPQ, Federal Building, Room 368, Hyattsville, Maryland 20782. No permit is necessary to import seeds and bulbs.

Regular shows at the RHS in London (BELOW LEFT), and collections at public gardens such as the Edinburgh Royal Botanic Gardens (BELOW – Desert House) and nurseries or garden centers (BOTTOM LEFT) provide a valuable means of viewing house plants, including rarer species and varieties, before making a purchase or propagating your own.

HOUSE PLANTS A–Z

A

ABROMEITIELLA

Bromeliaceae

Origin: Temperate regions of South America. A genus of five species of terrestrial bromeliads. They are evergreen perennials with narrowly triangular, grey-tinted leaves in rosettes which in turn are densely packed into cushions or mats. The 3-petalled greenish flowers are carried close to the leaves. These plants are essentially grown for their interesting hummocky growth habit and for attractive foliage. They are best grown in big pans in a conservatory or planted out with a collection of succulents (*Abromeitiella* species are best cultivated as for succulents), but are also suitable for the home in a really sunny window. Propagate by division or cuttings of single rosettes in spring. *Abromeitiella* was named for the German botanist J. Abromeit.

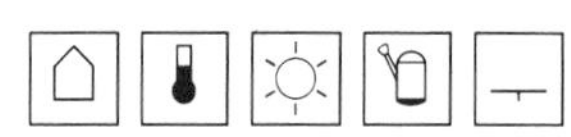

Species cultivated

A. brevifolia S. Bolivia and N.W. Argentina (Andean slopes)
Leaves 2–3cm($\frac{3}{4}$–$1\frac{1}{4}$in) long, ovate-triangular, tapering to a hard point, grey scaly. Slender, intense green flowers about 3cm($1\frac{1}{4}$in) long are borne in clusters of one to three, but only on mature plants. Although slow growing, this species can eventually fill a 30cm(12in) pan and is then an impressive sight.

A. chlorantha Argentina (Andean slopes)
Much like *A. brevifolia* in general appearance, but rosettes a little smaller with leaves to 2cm($\frac{3}{4}$in) long. Leaf margins usually with minute teeth which are not present in *A. brevifolia*. However, these small characters can be variable and the bromeliad expert W. Rauh includes plants named *A. chlorantha* in *A. brevifolia*.

Abromeitiella chlorantha

ABUTILON

Malvaceae
Flowering maples

Origin: Tropical to warm temperate areas of the world, especially South America where they grow in light woodland and scrub. A genus of 100 evergreen shrubs, annuals and perennials which are grown for their pendulous, bell-shaped flowers and long-stalked, maple-like leaves. All flower well when small, either grown on year after year or propagated annually from stem cuttings in spring or late summer. *Abutilon* derives from the Arabic name for mallow.

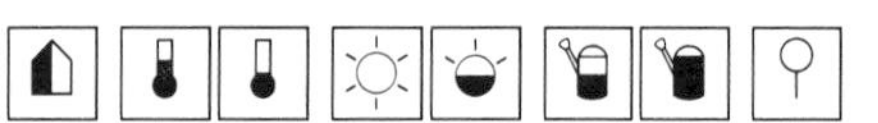

Species cultivated

A. darwinii Brazil
Tender shrub to 2m($6\frac{1}{2}$ft) or more with dark-veined, orange-red flowers opening from spring to summer. For temperate conditions.

A. globosom See the following.

A. × hybridum (*A. globosom* of gardens)
Hybrids between *A. darwinii* and *A. striatum* and other species. Vigorous plants similar to *A. darwinii* but with flowers in shades of red, orange, yellow and white. For temperate conditions. Among those available are: 'Ashford Red', crimson; 'Boule de Neige', white; 'Canary Bird', soft glossy yellow ageing reddish; 'Fireball', orange-red; 'Golden Fleece', golden yellow; 'Nabob', deep waxy crimson; 'Orange Glow', orange-yellow; 'Savitzii', small slender growth, leaves heavily splashed with white; 'Souvenir de Bon', pale orange-yellow, veined light red-purple, leaves broadly margined creamy yellow.

A. megapotamicum Brazil
Slender-stemmed pendulous shrub to 2m($6\frac{1}{2}$ft), the flowers having a crimson calyx and yellow petals. For temperate conditions. *A.m.* 'Variegatum' has yellow blotched leaves.

A. × milleri
A hybrid of *A. megapotamicum* and very similar to that species, but more robust with red-veined, yellow flowers. 'Kentish Belle' (*A. megapotamicum* × 'Golden Fleece') has orange-red petals and a grey-red calyx. For temperate conditions.

A. ochsenii Chile
An almost hardy shrub which will flower in an unheated sun room, growing to 2m($6\frac{1}{2}$ft) or more in height with 3-lobed leaves to 9cm($3\frac{1}{2}$in) long. Saucer-shaped, blue-purple flowers 5–6cm(2–$2\frac{1}{2}$in) wide open in spring. For cool conditions.

A. striatum Brazil
Tender shrub to 3m(10ft) or more. Leaves 3- to 5-lobed, 8–15cm(3–6in) long. Flowers 7–10cm($2\frac{3}{4}$–

4in) long, red, veined orange, from spring to autumn. For temperate conditions. *A.s.* 'Thompsonii' has somewhat smaller leaves mottled yellow.

A.* × *suntense (*A. ochsenii* × *A. vitifolium*)
A hybrid with blended characters of each parent, having saucer-shaped mauve-blue flowers to 6–8cm(2½–3in) wide. *A.* × *s.* 'Jermyns' has violet-purple flowers. For cool conditions.

ACACIA

Leguminosae

Origin: Tropics and sub-tropics; those listed below from Australia. A genus of about 800 evergreen or deciduous trees and shrubs. They are cultivated mostly for their very small pale yellow flowers which are aggregated into fluffy pompons or bottle brushes. In some species the flowers are fragrant. The leaves of many acacias are decorative, especially those which are bipinnate and almost ferny. Sometimes they are of this form in the seedling stage but are soon lost, being displaced by flattened leaf-like stalks known as phyllodes. All will flower in containers and may be propagated by seed in spring. *Acacia* is derived from the Greek *akis*, a point, referring to the fact that some species are spiny.

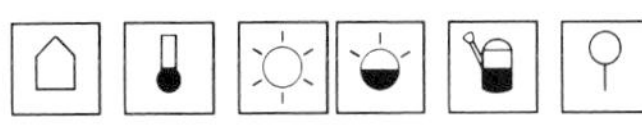

Species cultivated

A. armata Kangaroo thorn
Prickly shrub to 3m(10ft) with narrowly oblong evergreen phyllodes which can reach 1m(3ft). Rich yellow flowers are borne in globular heads to 1cm(⅜in) wide, opening in spring.

A. baileyana Cootamunda wattle
Small pendulous tree in the wild with 5cm(2in) long, blue-grey leaves and, in spring, an abundance of 6mm(¼in) wide, rich yellow, globose heads of flowers.

A. cultriformis Knife acacia
Small shrub reaching only 2m(6½ft). It has grey-green phyllodes which are obliquely obovate to triangular and up to 2cm(¾in) in length. Spring-borne yellow flowers are carried in globose heads up to 4mm($\frac{3}{16}$in) across.

ABOVE LEFT *Abutilon megapotamicum*
ABOVE *Abutilon* × *milleri*

A. cyanophylla Blue-leaved wattle
Eventually a tall shrub with bluish-green stems and phyllodes. The latter vary greatly in size and shape, being linear to lanceolate, sometimes oblanceolate and curved sideways; they are usually to 15cm(6in) long, but can be double this. Globular yellow flower heads are carried in conspicuous axillary and terminal panicles 15–30cm(6–12in) long in spring.

A. dealbata Silver wattle
The familiar 'mimosa' of florists, this species can make a tree to 30m(100ft) in mild climates, but grows well in a tub and can be flowered at shrub size. The leaves are silvery blue-green, finely pinnate and up to 13cm(5in) long when well grown. Lemon-yellow flowers 6mm(¼in) across are carried in globose clusters from late winter to spring.

A. decora
A shrub reaching 2m(6½ft) in height, this species develops narrow lanceolate phyllodes to 5cm(2in) in length. Deep yellow, 5mm($\frac{3}{16}$in) globose flower heads open in spring.

A. discolor
In the wild, either a shrub or small tree depending upon the conditions, but in a tub, only of shrub size. This species retains typical leaves which are pinnate and up to 6cm(2½in) in length. Flowers deep yellow, carried in globose heads which are larger than in most species, reaching 12mm(½in) across.

A. longifolia Sydney golden wattle
Slender willow-like tree or shrub to 4.5m(15ft) or more in the wild. Phyllodes lanceolate, bright green, to 15cm(6in) long. Bright yellow, globular heads of flowers about 6mm(¼in) across open in spring.

ABOVE *Acacia verticillata*
RIGHT *Acacia armata*

A. lophantha See *Albizia lophantha.*

A. melanoxylon Blackwood acacia
Large tree in the wild reaching 40m(130ft), but when flowering in a container, much smaller. It is unusual in that it bears both pinnate leaves and scimitar-shaped phyllodes which can reach 14cm($5\frac{1}{2}$in) in length but usually only half this. Flowers cream, carried in 7mm($\frac{1}{4}$in) wide globose heads in spring.

A. neriifolia
Large shrub or small tree with narrowly lanceolate phyllodes 8–13cm(3–5in) long. Flowers yellow in globose heads 8mm($\frac{1}{3}$in) wide, borne in spring.

A. podalyriifolia Queensland silver wattle, Mt. Morgan wattle
Decorative shrub which can reach 3m(10ft). Its foliage attraction lies in the silvery blue-green colour of the ovate phyllodes which are 4cm($1\frac{1}{2}$in) long. Flowers rich yellow, in globose heads 8–13mm($\frac{1}{3}$–$\frac{1}{2}$in) across, opening in spring.

A. pravissima Oven's wattle
Small tree to 6m(20ft) in the wild, but remaining smaller in a tub. Phyllodes small, grey-green and triangular, only 12mm($\frac{1}{2}$in) long and borne so closely along the stems that their bases overlap. Tiny 4mm($\frac{3}{16}$in) globose flower heads open in spring.

A. prominens Golden-rain/Gosford wattle
Tree to 6m(20ft) or more in the wild, less in a tub. Long narrow phyllodes to 2.5–5cm(1–2in). Flowers light golden, carried in globose heads 5mm($\frac{3}{16}$in) across.

A. retinodes Wirilda
10mm(33ft) tree in the wild. Narrowly lanceolate, curved, dark green phyllodes 8–15cm(3–6in) long. Globose heads of pale yellow flowers 7mm($\frac{1}{4}$in) across, fragrant, opening in late winter.

A. riceana
Small weeping tree to 10mm(33ft), its branches bearing pointed, awl-shaped phyllodes 3–5cm($1\frac{1}{4}$–2in) long. Globose heads of pale yellow flowers 5mm($\frac{3}{16}$in) across, opening in spring.

A. verticillata Prickly Moses
Slender shrub or tree to 10m(33ft) with whorls of six or more phyllodes, each 1–2cm($\frac{3}{8}$–$\frac{3}{4}$in) in length and tipped with a slender awl-shaped prickle. Although not unpleasantly sharp, the plant is best in a position where it will not be brushed against too often. Flowers pale yellow, clustered into 2.5cm(1in) long bottle brushes, freely borne in spring.

ACALYPHA

Euphorbiaceae

Origin: Tropics and sub-tropics. A genus of 450 species most of which eventually become shrubby. Some species are foliage plants cultivated for their ornamental leaves, others are grown for their long, rather catkin-like flower spikes. Most tend to get straggly as they age and are usually replaced from cuttings in summer annually or every second year. Best for the warm conservatory, but also suitable as short term house plants. *Acalypha* derives from *akalephes*, Greek for nettle.

Species cultivated

A. godseffiana Lance copperleaf New Guinea
Smaller and more bushy than the better known *A. wilkesiana*. Ovate-lanceolate, lustrous green leaves, each one with a broad serrated margin of creamy-

white. Inconspicuous greenish-yellow, tufted catkin-like flower spikes. *A.g. heterophylla* has most of the leaves narrowly lanceolate.

A. hispida Red-hot cat's tail, Chenille plant New Guinea
Shrub reaching 3m(10ft) in height, but usually smaller than this. Ovate leaves to 15cm(6in) long and tiny deep crimson flowers borne in dense pendulous tassels which can reach 50cm(20in) in length. A white-flowered form is occasionally seen.

A. wilkesiana Copperleaf Pacific Islands
Shrub to 2m($6\frac{1}{2}$ft) grown for its decorative ovate leaves which are normally a coppery green and can be mottled or splashed with shades of red. Flower spikes are small and insignificant. Several named forms are available including 'Macafeeana', 'Macrophylla' and 'Musaica', but vary only in the amount of coloration in the leaves. *A.w.* 'Obovata' has pink-margined, obovate leaves.

ACANTHOCALYCIUM See LOBIVIA
ACANTHORHIPSALIS See RHIPSALIS

ACANTHOSTACHYS

Bromeliaceae

Origin: South America. A genus of one bromeliad species which is closely related to the pineapple (*Ananas*) but usually grows epiphytically and looks effective in a hanging basket. Propagate by cuttings of mature, but non-flowering rosettes or by division in early summer. *Acanthostachys* derives from the Greek *akantha*, a thorn and *stachys*, a spike, referring to the prickle-tipped bracts of the flower spike.

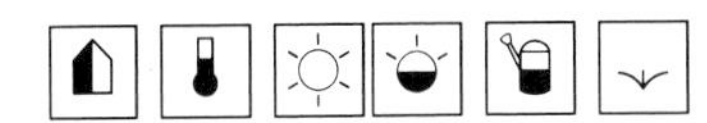

Species cultivated

A. strobilacea Brazil, Paraguay and N. Argentina
Leaves 30–90cm(1–3ft) long, very slender, with marginal, spine-tipped teeth and a tapered point; they are strongly channelled and arch outwards, forming loose, open rosettes. Flowering stems are usually somewhat shorter than the leaves, topped by one or two long, leaf-like bracts and a dense, cone-like flower spike to 5cm(2in) long. Flowers 3-petalled, yellow, to 2.5cm(1in) long, each one in the axil of a red or orange bract. After flowering, sweet, edible, yellow berries may develop in a closely pressed spike resembling a tiny pineapple.

ACCA See FEIJOA

ACHIMENES

Gesneriaceae
Hot water plants

Origin: Tropical areas of America (Mexico unless stated). A genus of 50 species of herbaceous perennials growing from small cone-like rhizomes made up of a mass of closely packed, fleshy scale leaves. They are sometimes incorrectly referred to as tubers. These rhizomes are usually bought dry, or stored in this way in winter and started into growth in spring. The spreading species and cultivars make good hanging basket plants. In pots, some form of support is necessary and twiggy sticks are satisfactory. The name 'hot water plant' probably came from instructions not to water them with cold water as they come from tropical climates, and in the days of irregularly heated houses were difficult to grow. With modern heating they are easy and rewarding. The vivid, tubular or bell-shaped flowers open freely in summer. Cultivate as for hardy bulbs and propagate by natural increase of rhizomes. *Achimenes* is probably derived from the Greek word *achaemenis*, a magic plant.

Species cultivated

A. andrieuxii
A dwarf tufted plant, its height equivalent to the length of its leaves – 7–10cm($2\frac{3}{4}$–4in). Flowers violet-blue, spurred and somewhat bell-shaped, up to 12mm($\frac{1}{2}$in) wide.

A. antirrhina
Erect, reaching 38cm(15in) with 6cm($2\frac{1}{2}$in) long leaves. Flowers 4cm($1\frac{1}{2}$in) long, funnel-shaped, reddish-orange with a yellower throat. 'Red Cap' is a redder-flowered selection.

A. bella
Erect-growing but small, reaching only 8cm(3in). Stems and leaves are woolly and the bell-shaped flowers are 2.5cm(1in) long, pale bluish-violet with yellow markings in the throat.

A. candida Panama northwards to Mexico
Spreading habit with stems to 30–45cm(1–$1\frac{1}{2}$ft). Leaves 5–7cm(2–$2\frac{3}{4}$in) long. Flowers creamy-white marked with red and yellow, tubular and expanding at the mouth to 12mm($\frac{1}{2}$in).

BELOW *Acanthostachys strobilacea*
BOTTOM *Achimenes cettoana*

A. cettoana
Bushy, erect species to 30cm(1ft) tall with whorls of narrowly lanceolate 7cm(2¾in) long leaves. Flowers tubular with spreading lobes, in shades from light to dark violet with a small white eye; they are 2.5cm(1in) long and 12mm(½in) across.
A. coccinea See A. *erecta*.
A. dulcis
Erect to spreading with stems that can reach 60cm(2ft). Leaves 10cm(4in) long. Flowers funnel-shaped and milky-white, 3cm(1¼in) long.
A. ehrenbergii (A. *lanata*)
Dwarf, almost prostrate species with 10–15cm(4–6in) long leaves which are glossy above and woolly

RIGHT *Achimenes* 'Schneewitschen'
BELOW *Achimenes* 'Elke Michelssen'

beneath. Flowers bell-shaped, large, to 4cm(1½in) across the mouth and pale purplish-pink with orange markings on the white throat.
A. erecta Central America and Jamaica
The first achimenes to be introduced, reaching England in 1778, still frequently referred to by its older incorrect name of A. *coccinea*. It has trailing stems to 45cm(1½ft) in length, long narrow leaves and vivid red tubular flowers spreading at the mouth to 1cm(⅜in) across.
A. fimbriata
Small and spreading with stems only 10cm(4in) long. Leaves 4cm(1½in) long, broadly ovate. Flowers bell-shaped and white with purple margins, the petal edges being fringed.
A. flava
Rather similar to A. *erecta* with stems to 45cm(1½ft), narrow leaves to 8cm(3in) long and ochre-yellow tubular flowers with red spots within the throat.
A. grandiflora Central America to Mexico
Erect, with hairy stems 30–60cm(1–2ft) in height. Leaves ovate, rough-hairy. Flowers deep rose-purple, tubular to 5cm(2in) long with a pure white eye. A.g. 'Maduna' has paler flowers; 'Robusta' has larger, more glossy leaves, but smaller flowers.
A. gymnostoma See *Gloxinia gymnostoma*.
A. heterophylla Guatemala
Erect, to 30cm(1ft) with 10–13cm(4–5in) leaves and narrow, bright red-orange flowers opening to a deeply lobed, wider mouth.
A. lanata See A. *ehrenbergii*.
A. longiflora Mexico southwards to Panama
Trailing with stems reaching 30cm(1ft) or more and leaves 8cm(3in) long. Large tubular flowers to 7cm(2¾in) across varying in colour from lavender to dark purple-blue, with a white throat. A.*l. alba* is white with a few purple markings. 'Guerrero' is a pale violet-blue with wavy petal lobes; 'Major' has 8cm(3in) wide violet-blue flowers – the largest achimenes.
A. mexicana (A. *scheeri*)
Erect, to 35cm(14in) with 8cm(3in) long, velvety leaves. Flowers large, bell-shaped 4–5cm(1½–2in)

long, in shades of blue-purple with a white throat.
A. misera Guatemala
A small-flowered trailing species with stems up to 25cm(10in). Tubular white, 1cm($\frac{3}{8}$in) long flowers open later in the season than most achimenes.
A. patens
Erect to spreading with stems to 30cm(1ft) long. Leaves 11cm($4\frac{1}{2}$in), hairy and toothed. Flowers long-spurred, tubular and purple with a white throat, to 3cm($1\frac{1}{4}$in) across.
A. pedunculata Mexico south to Honduras
Tall, erect plant to 1–1.5m(3–5ft) in height. Flowers orange-scarlet, bell-shaped, marked with yellow and red.
A. scheeri See *A. mexicana*.
A. skinneri Guatemala
Erect plant growing to 75cm($2\frac{1}{2}$ft) with broadly ovate leaves 4–7cm($1\frac{1}{2}$–$2\frac{3}{4}$in) long. Flowers tubular, rose-purple with an orange-yellow throat, opening with spreading lobes to 4cm($1\frac{1}{2}$in) across.
A. tubiflora See *Sinningia tubiflora*.
A. warscewicziana El Salvador to Mexico
Stems 30cm(1ft) long, trailing and hairy. White, purple spotted, 2.5cm(1in) long, tubular flowers borne singly.

Cultivars

Many named cultivars have been raised by hybridizing the best of the species, their parentage often being uncertain or complex. A selection is listed alphabetically:
'Ambroise Verschaffelt', of trailing habit, the flowers white with strong purple veining.
'Andersonii', of trailing habit, mauve-purple flowers.
'Burnt Orange', erect to spreading, to 45cm($1\frac{1}{2}$ft) tall, the flowers orange-red with a yellow throat.
'Camille Brozzoni', erect with white-eyed, pale purple flowers.
'Elke Michelssen', a compact plant with purplish-pink flowers.
'Little Beauty', erect with crimson-pink flowers having a yellow eye.
'Master Ingram', spreading habit with very large, velvety red flowers.
'Paul Arnold', deep purple flowers with a white throat which is marked with yellow and purple.
'Peach Blossom', a trailing plant with magenta-pink flowers.
'Pink Beauty', spreading with pale crimson-pink flowers.
'Purple King' ('Pulcherrima', 'Patens Major'), a compact plant to 38cm(15in) with rich, reddish-purple flowers.
'Queen of Sheba', a trailer with rich pink flowers.
'Schneewitschen', a spreading plant with long-tubed, pure white flowers.
'Yellow Beauty', erect with primrose-yellow flowers; the first really yellow-flowered achimenes hybrid.

ACIDANTHERA See GLADIOLUS
ACMENA See EUGENIA

ACOKANTHERA

Apocynaceae

Origin: Southern to tropical East Africa and Arabia. A genus of evergreen trees and shrubs characterized by opposite pairs of lanceolate to ovate leaves with clusters of 5-lobed, tubular, fragrant flowers borne in their axils. Flowers set to form ovoid to globose, poisonous berries. The species described are best grown in large containers and, once bloomed, may have their flowered stems cut back by half. They are best in the conservatory, being more effective in bloom when of a good size. Propagate by seed in spring or by cuttings in summer. *Acokanthera* derives from the Greek words *akis* and *anthera* meaning spike and anther, referring to the pointed nature of the anthers.

  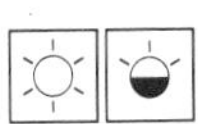  

Species cultivated

A. spectabilis Wintersweet South Africa
Tree to 5m(16ft) in the wild, but smaller and shrubby in cultivation. Oblong to lanceolate, firm textured, glossy deep green leaves provide a foil for the 2cm($\frac{3}{4}$in) wide, white flowers which open from winter to spring. They are followed by poisonous, purple-black berries, produced freely even when plants are small.
A. venenata South Africa
This species is similar to *A. spectabilis* but smaller in all its parts and having ovate leaves and pink-tinted flowers. Its berries are also poisonous.

TOP *Acokanthera spectabilis*
ABOVE *Acorus gramineus*

ACORUS

Araceae

Origin: North temperate and sub-tropical zones. A genus of two species of hardy evergreen perennials grown for their foliage – narrow, iris-like leaves which arise from creeping rhizomes. The flowers are minute and greenish, carried on a spadix low down among the leaves. These are durable pot plants ideal for the unheated room or porch. Propagate by division in spring. *Acorus* comes from the Greek *akoron*, their name for the yellow flag iris (through later confusion it was used for the sweet flags).

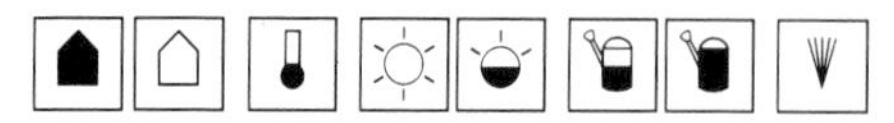

Species cultivated

A. gramineus Sweet flag India to Japan
Neat species with grassy leaves 30–50cm(12–20in) long, and a slender 5–10cm(2–4in) spadix in spring. *A.g.* 'Variegatus' is a cultivar with leaves striped longitudinally with cream.

A.g. pusillus
A smaller, more tufted plant and the one most usually available as a pot plant. Smaller forms are sometimes seen listed as *A.g. pusillus minimus*.

ACROCLINIUM See HELIPTERUM

ACROSTICHUM

Pteridaceae

Origin: Widespread in tropical swamps around the world. A genus of two or three ferns. Although eventually of large size, small specimens make attractive pot plants. Propagate by division in spring. *Acrostichum* derives from the Greek *akros*, terminal and *stichos*, a row, referring to the sporangia being produced on the upper leaflets only.

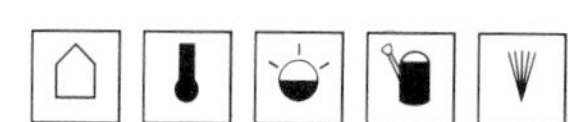

Species cultivated

A. aureum Leather fern Pantropical
Leaves (fronds) simple-pinnate, 60–180cm(2–6ft) in height, 30–60cm(1–2ft) across, the narrowly oblong pinnae green and leathery. Sporangia reddish-brown, completely covering the lower surfaces of the topmost pinnae on fertile fronds. For successful cultivation, the soil must be kept very moist; potted specimens are best kept in saucers or trays of water. Large specimens in tubs form very handsome, dense clumps of erect rhizomes and fronds.

A. crinitum See *Elaphoglossum crinitum*.

ACTINIOPTERIS

Actiniopteridaceae

Origin: Tropical Africa and Asia. A genus of five species of xerophytic ferns, only one of which is cultivated. Although it has a reputation for being difficult to grow successfully, this is mainly due to keeping the soil too wet. Sharply drained soil is required, for example equal parts grit and an all-peat compost. Water regularly, but only when the soil has almost dried out; being xerophytes, dry atmospheric conditions within the home are tolerated. Propagate by spores (larger plants can also be divided) in spring. *Actiniopteris* (sometimes spelled *Actinopteris*) derives from the Greek *aktis*, a ray and *pteris*, a fern, from the shape of the leaves.

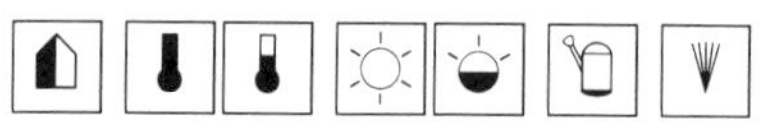

Species cultivated

A. semiflabellata (*A. australis*)
A dainty tufted species, not unlike a very tiny fan palm. Leaves 8–20cm(3–8in) tall, the actual frond to 4cm($1\frac{1}{2}$in) long and wide, fan-shaped and cut into many ray-like segments. Sporangia in two rows on the undersides of partially rolled leaf segments.

ADA

Orchidaceae

Origin: Northern South America. A genus of about eight species of epiphytic orchids, only one of which is at all popular. The genus as a whole is clump-forming with very distinctive, almost bell-shaped flowers in arching racemes. The species described below grows best in a conservatory but can be brought indoors when in bloom. Propagate

BELOW L-R *Acrostichum aureum, Ada aurantiaca, Adenium obesum*

by division in spring after flowering. *Ada* presumably derives from the feminine Christian name.

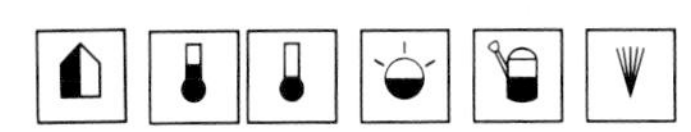

Species cultivated

A. aurantiaca Colombia
Pseudobulbs clustered, narrowly ovoid and somewhat flattened laterally, 6–10cm(2½–4in) long. Leaves one or two per bulb, narrowly strap-shaped, up to 30cm(1ft) long, slightly curved. Racemes arching, up to 50cm(20in) long, the terminal part bearing seven to 12 or more closely set, vivid cinnabar-red flowers in winter and spring; each bloom about 2.5cm(1in) long, composed of five similar, slender-pointed sepals and enclosing a much smaller labellum.

ADENIUM

Apocynaceae

Origin: Tropical and sub-tropical Africa and arid regions of Arabia. A genus of 15 species of semi-succulent shrubs, most of which have thickened main stems sometimes with the appearance of a tuber. The somewhat thickened leaves are generally short-lived on plants in the wild, falling during the dry season. In cultivation, however, they remain for a year or more and, as they are often a lustrous deep green, provide a good setting for the colourful, tubular and 5-petalled flowers. Propagate by cuttings in summer or by seed in spring. The species described below flower when young. *Adenium* derives from the Arabian country Aden (now part of the Yemeni Arab Republic), the country from which the first species was described.

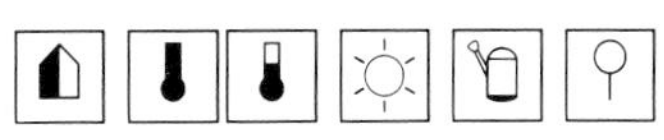

Species cultivated

A. obesum Desert rose East Africa
A shrub to 1.8m(6ft), but slow growing and much smaller in containers. Main stem usually grossly swollen and tuber-like (then known as a caudex), but the branches no more than slightly thickened. Leaves at the ends of the stems only, ovate and glossy deep green, 4–10cm(1½–4in) long. Flowers bright pink, 5cm(2in) long, carried in terminal, umbel-like cymes in summer. Perhaps the most attractive member of its genus and usually free-flowering.

A.o. multiflorum
A taller plant with white flowers edged pink.

A. somalense East Africa
This species resembles A. *obesum* in form, but has wavy-margined, glaucous leaves and smaller pink or white blooms.

ADIANTUM

Adiantaceae

Maidenhair ferns

Origin: Cosmopolitan, especially tropical America. A genus of 200 species of evergreen and deciduous ferns. They are generally typified by blackish glossy leaf stalks (stipes) and deeply dissected leaves which can be pinnate, bipinnate or multipinnate. The individual leaflets (pinnae or pinnules) vary from oblong to triangular to fan-shaped, occasionally rounded and usually lobed. The sori which bear the spores are borne on the margins of the pinnae. Plants require shade from direct sun at all times of the year. Propagate either by division or by sowing spores in spring. Some species are tufted, others rhizomatous; all are among the best ferns for the home. *Adiantum* comes from the Greek *adiantos*, dry or unwetted, alluding to the water repellent surface of the fronds, most species being very resistant to dry air.

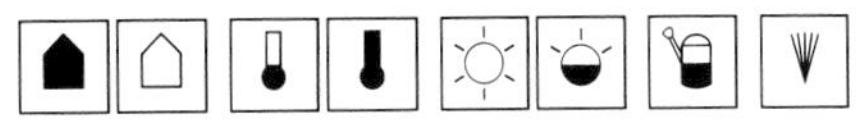

Species cultivated

A. aemulum See A. *raddianum*.

A. capillus-veneris Common maidenhair
Cosmopolitan in distribution
Rhizomatous, the blades of the fronds narrowly triangular, 15–30cm(6–12in) long, bi- or tripinnate with the pinnules rounded to fan-shaped, 1–2.5cm(⅜–1in) long. Cool.

A. caudatum Tropical Asia and Africa
Tufted, blades of fronds sparsely hairy, linear, simply pinnate, 15–30cm(6–12in) long, strongly arching, often rooting at the tip. Pinnae rhomboid to fan-shaped, 1–2cm(⅜–¾in) long, the upper margin prominently lobed. Makes an attractively unusual specimen for a hanging basket. Cool.

A. concinnum Mexico to northern South America and West Indies
A shortly rhizomatous species, forming dense clumps. Blades of fronds narrowly triangular to lanceolate-ovate, 30–45cm(1–1½ft) long, bi- or tripinnate, with the pinnules rhomboid to fan-shaped, about 1cm(⅜in) long. A distinctive feature of this fern is the way some of the pinnules overlap all the connecting stalks. Cool.

ABOVE *Adiantum diaphanum*
RIGHT *Adiantum* 'Double Leaflet'

A. cuneatum See A. *raddianum*.
A. decorum See A. *raddianum* 'Decorum'.
A. diaphanum New Zealand, Australia, southern China to Melanesia
Clump-forming, the roots with small ovoid tubers. Blades of fronds linear to lanceolate, 13–18cm(5–

7in) long, simply pinnate (or with two or more basal branches and then appearing bipinnate). Pinnae oblong, with crenate teeth, to 1.5cm($\frac{5}{8}$in) long with scattered whitish and/or blackish hairs. Cool.
A. edgeworthii Himalaya, China
Much like A. *caudatum* but hairless and pinnae without or having only shallow lobes. Cool.
A. formosum Australian maidenhair
Australia and New Zealand
Rhizome creeping eventually forming colonies. Blades of fronds broadly triangular, to 60cm(2ft) long, three to four times pinnate. Pinnules 6–20mm($\frac{1}{4}$–$\frac{3}{4}$in) long, narrowly triangular at the base, rounded and toothed or lobed above. Cool.
A. fragrantissimum See A. *raddianum* 'Fragrantissimum'.
A. gracillimum See A. *raddianum* 'Gracillimum' (A.*g. micropinnulum* see A.*r.* '*Micropinnulum*').
A. henslovianum Venezuela, Galapagos
Clump-forming, blades of fronds ovate, tripinnate, 15–30cm(6–12in) long, the pinnules fan-shaped, 1.2–2cm($\frac{1}{2}$–$\frac{3}{4}$in) wide, 6–10mm($\frac{1}{4}$–$\frac{3}{8}$in) long. The plant under this name in cultivation is distinctive in having green instead of black leaf stalks. Temperate.

A. hispidulum (A. *pubescens*) Australian/rose maidenhair Asia, Australasia, Pacific Islands
Tufted, blades of the fronds broadly ovate, pedate and hairy, 20–60cm(8–24in) long, the pinnae themselves pinnate, the pinnules rounded to oblong, to 1.5cm($\frac{5}{8}$in) long; young fronds tinted copper-pink. Tropical.
A. macrophyllum Mexico to South America
Rhizomatous, in time forming wide, loose clumps. The blades of the fronds are ovate-oblong, 15–30cm(6–12in) long and simply pinnate. Each pinna is 5–10cm(2–4in) long, ovate, with two rounded lobes at the base, smooth, rich green and of good substance, quite unlike those of any other maidenhair fern. Tropical.
A. pubescens See A. *hispidulum*.
A. raddianum (A. *cuneatum*, A. *aemulum*)
Tropical America
Tufted with the blade of the frond roughly triangular, 15–30cm(6–12in) long, 2- to 4-pinnate. The pinnules are rhomboid to oblong, about 7mm($\frac{1}{4}$in) long; an enormously variable species giving rise to most of the commonly grown tender maidenhairs. These include (all tropical):
'Bridal Veil', almost identical to 'Micropinnulum'.
'Brilliant Else', like 'Elegantissimum' but the young fronds are light red.
'Decorum', much like true A. *raddianum* but more robust with thicker frond stalks and broader, blunter pinnules.
'Deflexum', pinnae and pinnules pointing outwards and downwards giving the fronds a distinctive elegance.
'Double Leaflet', pinnules cleft into two to four tiny pinnule-like narrow segments, creating highly ornamental, mist-like foliage.
'Elegantissimum', drooping pale green fronds, the pinnules to 4mm($\frac{3}{16}$in), fan-shaped.
'Fragrantissimum' ('Fragrans' of some catalogues), larger than the true species and more vigorous, pinnules to 1.5cm($\frac{5}{8}$in) long.
'Fritz Luthi', neat, lacy fronds, the 1cm($\frac{3}{8}$in) pinnules carried at various angles.
'Gracillimum', pinnules variable but only 2–4mm($\frac{1}{16}$–$\frac{3}{16}$in) long, rarely larger. A very elegant cultivar.
'Grandiceps', tip of frond and lateral pinnae semi-cristate, the whole frond arching strongly; good for hanging baskets.
'Kensington Gem', a very robust cultivar with a total height of 1m(3ft). In all respects like a super *raddianum*, but sterile and possibly of hybrid origin.
'Lawsoniana', the frond blade more sharply triangular, the pinnules 8–20mm($\frac{1}{3}$–$\frac{3}{4}$in) long, narrowly wedge-shaped and usually deeply cleft. It is close to 'Elegantissimum' and probably synonymous with the A. *decorum* 'Wagneri' of catalogues and other books.
'Legrandii', similar to 'Gracillimum' but pinnules more numerous and congested, especially towards the middle.
'Micropinnulum' (A. *gracillimum micropinnulum*), pinnae again divided into tiny segments up to

Adiantum raddianum 'Fritz Luthi'

3mm(⅛in) long; whole frond has an airy grace which is heightened by being pinkish-bronze when young.
'Mist', almost identical to 'Micropinnulum'.
'Morgan', similar to 'Double Leaflet'.
'Pacific Maid', pinnae large, up to 2cm(¾in) long, overlapping and forming a dense frond.
'Pacottii' (Double maidenhair), pinnae, pinnules and smaller segments very dense and overlapping.
'Tuffy Tips', akin to 'Pacific Maid', but blades of fronds irregular in outline and strongly crested.
'Weigandii', blade of fronds narrowly triangular, generally longer than the species, the pinnae well spaced out, terminal pinnules somewhat bunched and overlapping.
A. rubellum Bolivia
Clump-forming, blades of fronds triangular, to 15cm(6in) or more long, bipinnate. Pinnules more or less fan-shaped but variable, 12mm(½in) long, light green with a hint of pink. Whole of fronds crimson-pink when unfolding. Some authorities make this fern yet another form of *A. raddianum*. Temperate.
A. scutum See *A. tenerum*.
A. tenerum (*A. scutum*) Florida, West Indies, Mexico to Peru
Tufted, with the blade of the fronds triangular to ovate, 3- 4-pinnate and 40–75cm(16–30in) long. Pinnules rounded to rhomboid, 1–2cm(⅜–¾in) long and asymmetrical. Cultivars include (all tropical):
'Farleyense', mostly fan-shaped pinnules which are lobed and ruffled; they are up to 4cm(1½in) in length and overlap.
'Scutum', pinnules both larger and broader than the type.
'Scutum Roseum', similar to 'Scutum' but the young fronds are reddish.
A. trapeziforme Tropical America
Tufted with triangular 3- 4-pinnate fronds to 45cm(1½ft) long. Pinnules mostly trapeziforme (irregularly diamond-shaped) and 2–5cm(¾–2in) long. Tropical.

ADROMISCHUS

Crassulaceae

Origin: South Africa. A genus of 50 species of succulents closely related to *Cotyledon*, *q.v.* They are small plants making good subjects for the home, having attractive foliage all the year round – clusters of fleshy, lanceolate to ovate leaves which are

Adromischus festivus

often patterned or marbled in darker shades. The flowers are small and 5-petalled, reddish or white in colour and borne in summer. Propagate by seed in spring or by leaf cuttings in summer. *Adromischus* derives from the Greek *hadros*, thick and *mischos*, a stalk, referring to the plant's short, thick stems.

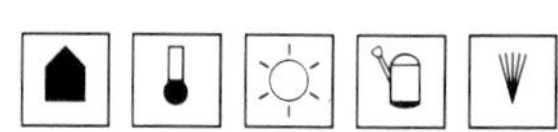

Species cultivated

A. alveolatus Cape, Namaqualand
Tuberous rooted, forming small clumps of erect stems. Leaves globular to obovoid or narrower, up to 4cm(1½in) long, whitish green, rough-textured to tuberculate. Flowers green and maroon, 1.5cm(⅝in) long, several on a stem 15cm(6in) or more tall.

A. cooperi (of gardens) See *A. festivus*.

A. cristatus
Leaves 2.5–4cm(1–1½in) long, short-stalked with wavy margins, paddle-shaped and green. Flowers whitish-red on erect stems 15–20cm(6–8in) long.

A. festivus
Leaves fleshy, rounded to spathulate, to 2.5cm(1in) long held more or less erect, wavy at the edges especially near the tips, grey-green with dark blotches. Flowers pinkish, 6mm(¼in) long.

A. maculatus (*Cotyledon maculata, Crassula maculata*) Calico-hearts
A shrublet to 10cm(4in) tall. Leaves rounded to spathulate with a horny margin, green with brown blotches. Flowers small, pinkish-white.

A. marianiae Cape
Clump forming shrublet (the semi-erect, bare stems developing with age). Leaves about 5cm(2in) long, lanceolate, slightly convex to channelled above, sharply edged, green with dark spots or mottling, flushed pinkish or reddish-brown in bright light. Flowers small, red and green or white in clusters above the leaves.

A. tricolor Cape (widespread)
Clump-forming. Leaves to 6cm(2½in) long, cylindrical to narrowly oblong-ovoid, grey-green with purplish markings. Flowers white or purple tinted, 1.6cm(⅝in) long, in panicles to 25cm(10in) high.

A. trigynus Cape (widespread)
Clump-forming. Leaves fairly crowded, almost orbicular, to 4cm(1½in) long, bright green with brownish-purple spotting. Flowers reddish to purple-brown, about 1.2cm(½in) long in clusters on stems to 25cm(10in) tall.

A. umbraticola Cape, Transvaal
Clump-forming shrublet. Leaves up to 5cm(2in) long, obovate-cuneate to elongate-oblong, rounded or waved at the tip, glaucous, with or without a light marbling. Flowers small and purplish on simple or branched stems well above the leaves.

AECHMEA

Bromeliaceae

Origin: West Indies and South America. A genus of 150 mainly epiphytic species many of which make extremely durable pot plants. They are rosette forming plants with sword-shaped, toothed or spiny, arching to erect leaves, the bases of which fit together forming a water-holding cup. The 3-petalled flowers are carried in spike-like panicles, usually on long stems, often bearing coloured bract leaves. The berry-like fruits can be brightly coloured. After blooming the rosette slowly dies as young offsets form around the base. Propagate by offsets in late spring. *Aechmea* derives from the Greek *aichme*, a point, the sepals having rigid points.

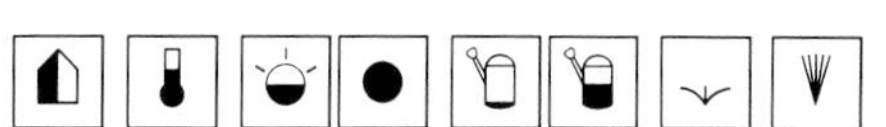

Species cultivated

A. bracteata Mexico to Central America
Flowering plant 1.5m(5ft) or more in height. Leaves pale green, grey scaly, to 90cm(3ft) long, in dense rosettes. Flowering stem up to 60cm(2ft) tall. Flowers in autumn, pale yellow in a pyramidal panicle set with red bracts. Fruits blue.

A. bromeliifolia Wax torch Honduras to Argentina
Flowering plant 60–90cm(2–3ft) in height. Leaves 50–90cm(20–36in), strap-shaped, tapering to a spine tip, dark grey-green and waxy white-scaly. Flowers in dense clusters, yellowish green within greeny-white bracts.

A. calyculata Brazil and Argentina
Flowering plant to 60cm(2ft). Leaves 30–50cm(12–20in), linear with rounded ends terminating in a prickle, green with grey scales. Flowers yellow in a dense spike with red bracts.

A. caudata Brazil
Flowering plant to 1m(3ft). Leaves to 60cm(2ft) or more, linear, dark grey with greenish scales, round-ended with a short spine. Flowers yellow in a short spike to 5cm(2in) long with brownish-pink bracts becoming yellow. *A.c.* 'Variegata' (*Billbergia forgetii*) has leaves boldly banded with pale yellow.

A. chantinii Brazil to Peru
Flowering plant 40–100cm(16–39in) in height. Leaves to 40cm(16in), green with pinkish to grey-white cross-banding, terminating in a short spine. Flowers yellow in a loose spike to 5cm(2½in) long with red bracts.

A. coelestis Brazil
Flowering plant to 70cm(28in) in height. Leaves grey-scaly with a rounded tip terminated by a short spine. Flowers in winter, sky-blue in a dense pyramidal cluster to 20cm(8in) long terminating an erect stem 30–40cm(12–16in) tall. Fruits green, tinted pink.

A. comata (*A. lindenii*) Brazil
Flowering plant to 1m(3ft) in height. Leaves more or less erect, up to 60cm(2ft) long in a slim rosette. Flowers with yellowish-red sepals and yellow petals, borne in a short, dense spike terminating a stem 60cm(2ft) or more in height; usually winter blooming.

A. cylindrata Brazil
Flowering plant to 60cm(2ft) high. Leaves 30–40cm(12–16in) long, prickly margined with a rounded tip terminated by a short spine, dark green above, paler beneath. Flowers with pale blue petals and pink sepals in a dense spike 15–20cm(6–8in) long, topping a stem to 40cm(16in) tall; usually winter blooming.

A. dealbata Brazil
Flowering plant to 80cm(32in) in height. Leaves 50–70cm(20–28in) long, lanceolate, pointed tipped, banded silvery-grey, forming an erect cylindrical rosette. Flowers violet to red with grey/white-felted, red bracts. Flowers and bracts form a dense cone-like spike on a stem 60cm(2ft) tall. Closely related and similar to the more familiar *A. fasciata*.

A. discolor See *A. fulgens discolor*.

A distichantha Brazil to Argentina
Flowering plant to 70cm(28in) in height. Leaves arching, rigid, 90–150cm(3–5ft) long, pointed, more or less channelled, grey-scaly with brown prickly margins. Flowers purple or blue in the axils of white-felted, pink bracts forming a rigid, spiky panicle topping a 30–60cm(1–2ft) stem.

A.d. albiflora
Flowers white.

A.d. glaziovii (*A. glaziovii*)
Inflorescence shorter and thicker with fewer flowers; plant usually smaller with round-tipped leaves.

A. fasciata (*Billbergia rhodocyanea*) Silver vase, Urn plant Brazil
Flowering plant 50cm(20in) in height. Leaves broad, densely grey-scaled and usually banded with silvery-white. Plants with brightly marked leaves have been selected by nurserymen. Purple-blue flowers and pink bracts form a dense pyramid. *A.f.* 'Variegata' has creamy yellow, longitudinal stripes.

A. fulgens Coralberry Brazil
Flowering plant 40–50cm(16–20in) in height. Leaves to 45cm(1½ft), grey-green with some cross-banding. The long inflorescence to 20cm(8in) has violet-purple flowers becoming red. Fruits red.

A.f. discolor (*A. discolor*)
Undersides of leaves dark red.

A. gamosepala Brazil, Argentina
Flowering plant to 50cm(20in) in height. Leaves strap-shaped to 35cm(14in), green with a grey-scaly back and a shortly pointed or spined tip. Flower spikes loose, pale blue with brownish-red bracts.

A. glaziovii See *A. distichantha glaziovii*.

A. lindeni See *A. comata*.

A. luddemanniana Guatemala and Honduras
Flowering plant 30–70cm(12–28in) in height. Leaves are strap-shaped to 50cm(20in), rounded at the ends, green to reddish-brown with grey scales. Flowers blue, later red forming a short spike to 4cm(1½in). Fruits white and purple.

A. mariae-reginae Costa Rica
Leaves up to 80cm(32in) long with densely toothed margins and a rounded, shortly spined tip; green to grey-green on both surfaces. Flowers white, purple-tipped, arranged in a very dense, 15–20cm(6–8in) long, cone-like spike on a stem to 60cm(2ft) or more tall. The most striking aspect of a plant in bloom is the large, lance-shaped, bright red bracts on the flowering stem. These are particularly dense towards the top of the stem and are down-swept, forming an inverted shuttlecock beneath the flower spike. Flowering is in summer.

A. mexicana Mexico to Ecuador
Leaves 70–90cm(28–36in) long, pointed or round-tipped, spiny-toothed, pale green, blotched dark green. Flowers in winter, usually red, sometimes lilac, in a narrowly pyramidal, stiff panicle 30–

ABOVE LEFT *Aechmea fulgens discolor*
ABOVE *Aechmea cylindrata*
LEFT *Aechmea fasciata* 'Variegata'

60cm(1–2ft) long on a short stem among the leaves. White fruits follow.

A. miniata Brazil
Leaves to 45cm(1½ft) long with an abrupt, stiff point and small prickly teeth. Flowers red, borne in a broad panicle to 10cm(4in) long, followed by red fruits.

A. miniata var. discolor
Leaves flushed red-purple beneath.

A. nudicaulis Mexico, West Indies to Brazil
Flowering plant 30–70cm(12–28in) tall. Leaves olive green, cross-banded with grey beneath, to 90cm(3ft) long. Flowers carried in loose spikes, yellow with small, bright red bracts. Free-flowering and often recommended as a good plant for beginners.

A.n. cuspidata
Flowers and bracts yellow.

A. orlandiana Brazil
Flowering plant to 50cm(20in) – the height of the flowering spike. Leaves linear, to 35cm(14in), light green with strong dark purplish cross-banding and mottling, ending in a spine. Flowers yellow to white with long red bracts carried in a dense spike to 10cm(4in) long.

A. paniculata See *A. penduliflora.*

A. penduliflora (*A. paniculata, Billbergia paniculata*) Costa Rica to Peru and Brazil
Leaves 40–60cm(16–24in), widening towards the tip, densely spiny-toothed though the upper part may be smooth, grey-scaly. Flowers orange; fruits yellow to red, in a stiff, open spiral panicle to 6cm(2½in) long. Each panicle tops a 30–45cm(1–1½ft) tall stem partly sheathed by red bracts.

A. pineliana Brazil to Peru
Leaves 30–60cm(1–2ft) long, semi-erect with a rounded end and terminal spine. The well-spaced marginal teeth are reddish-brown and the dark green surfaces silvery grey-scaly, often in cross-bands. Flowers yellow, in a dense cone-like spike to 7cm(2¾in) long. The stem which bears it can reach 45cm(1½ft) and the spike is enclosed by long, overlapping, waxy-red bracts.

Aechmea × 'Foster's Favorite'

A. racinae Brazil
Flowering plant up to 50cm(20in) in height. Leaves 30cm(1ft) long, prickly toothed or smooth-margined, bright green. Flowers with red sepals 12cm(4¾in) long. Each raceme terminates an arching, pendent-tipped stem up to 45cm(1½ft) long, emerging among or beneath the leaves. A good subject for a hanging basket.

A.r. erecta
Flowering stem shorter and upright.

A. ramosa Brazil
Flowering plant to 1m(3ft) or more. Leaves strap-shaped to 60cm(2ft) or more long, green with grey scaliness, sometimes with a pinkish coloration. Yellow flowers and later fruits are carried on distinctive red or orange stalks.

A. recurvata Brazil
Leaves very narrow, 1cm(⅜in) wide, tapering to a slender point, spiny toothed, recurved, turning red when exposed to the sun. The leaf bases are greatly expanded and closely overlapping, creating a rosette with a 'bulbous' base. Flowers with red sepals and red to lilac-red petals, carried in a dense red-bracted spike to 6cm(2½in) long just above the leaves.

A.r. ortgiesii
Flowering stem very short or missing.

A.r. benrathii
Like *A.r. ortgiesii* but leaf bases purple-spotted and leaf blades with very few or no teeth.

A. tillandsioides Mexico to Colombia
Leaves narrow, 2–5cm(¾–2in) wide and 50–90cm(20–36in) long, margined with brown spiny teeth. Flowers yellow, in a stiff 10–30cm(4–12in) long open panicle, topping a red-bracted stem about 30cm(1ft) tall.

A. veitchii Costa Rica, Panama and Colombia
Flowering plant 1m(3ft) in height. Leaves tongue-shaped, 30–90cm(1–3ft), pale green above and grey-scaly beneath. Flowers white with pink-tipped sepals and red bracts.

A. weilbachii Brazil
Flowering plant 60–70cm(24–28in) in height. Leaves to 45cm(1½ft) or more, dark green and shiny, having a spined tip. Flowers lilac-purple with red bracts, in loose spikes to 15cm(6in) long.

Hybrids

A. × 'Foster's Favorite' (*A. victoriana* × *racinae*)
Lacquered wine-cup
This popular hybrid favours the small *A. racinae* parent though it can be larger, to 60cm(2ft) in height. Its pendent flower spike followed by red fruits is also somewhat longer. The main point of difference is the lustrous wine-red leaves which make the plant a strikingly decorative subject the year round.

A. 'Foster's Favorite Favorite'
A sport of 'Foster's Favorite', similar, but with the leaves bordered with a broad cream band and an overall red flush.

AEONIUM

Crassulaceae

Origin: Arabia, Ethiopia, the Mediterranean region and the North Atlantic Islands. A genus of 40 species of succulents, many of which thrive in the home, being very decorative when in bloom. They have fleshy, obovate leaves in rosettes – some species form a solitary rosette, others produce a colony. Many species are shrubby, their rosettes standing on a woody stem. The starry flowers have six to 12 petals and are borne in terminal, rounded or pyramidal panicles. Stems die after flowering and are best cut out (solitary species die completely after flowering). Propagate by seed or by leaf or stem cuttings in spring or summer. *Aeonium* is the Latin name for what is now *A. arboreum*.

Species cultivated

A. arboreum Portugal, Spain, Morocco, Sicily and Sardinia
Sparsely branched shrub to 1m(3ft) in height. Leaves glossy, bright green, fleshy and 5–9cm(2–3½in) long. Panicles of bright yellow flowers open at the ends of the branches from winter to spring. *A. a.* 'Atropurpureum' is similar to the type, but with very striking dark, glossy-purple leaves.

A. caespitosum See *A. simsii*.

A. canariense Tenerife (Canary Is.)
Short, thick, sparingly branched stems and dense cup-shaped rosettes of leaves. Each leaf is ovate to spoon-shaped, light green and downy, with finely hairy margins. Flowers bright yellow carried in panicles on stalks about 60cm(2ft) tall opening in spring.

A. × domesticum See *Aichryson × domesticum*.

A. gomerense Gomera (Canary Is.)
Sparingly branched to 90cm(3ft) with slender, rather lax branches. Leaves 6–10cm(2½–4in), fleshy and glaucous with a red margin. Flowers white, opening in spring.

A. haworthii Tenerife (Canary Is.)
Small, freely branched shrub to 60cm(2ft). Leaves glaucous, red-edged and borne in loose rosettes. Yellow to rose-pink flowers appear in spring.

A. holochrysum Tenerife, Gomera, La Palma, Hierro (Canary Is.)
Shrub to 1m(3ft) with stout, fleshy stems. Leaves 10cm(4in) long, shining green and narrow, carried in dense rosettes to 20cm(8in) across. Flowers yellow, making a conical inflorescence.

A. lindleyi Tenerife (Canary Is.)
Rounded shrublet to 30cm(1ft) with many thin branches. Leaves thickly succulent, stickily hairy and glossy, bright green but often with a reddish tinge. Golden-yellow flowers open from summer to autumn.

A. manriqueorum Gran Canaria
Branched shrub to 1m(3ft). Leaves long, spathulate, to 9cm(3½in), shining green and streaked with red, forming rosettes, conical at first, later flat, to 20cm(8in) across. Flowers yellow in pyramidal inflorescences.

ABOVE LEFT *Aeonium simsii*
ABOVE *Aeonium urbicum*

A. simsii (*A. caespitosum*) Gran Canaria
Tufted perennial forming cushions of more or less stalkless rosettes. Leaves linear-lanceolate, 3–6cm(1¼–2½in) long, green, and prominently margined with soft white hairs. Golden-yellow flowers are freely borne in spring making this a most attractive species.

A. tabulaeforme Tenerife (Canary Is.)
Stalkless, solitary rosettes up to 50cm(20in) across, flat and plate-like made up of grass-green spathulate leaves which are finely hairy along the margins. When 2–3 years old the centre of the rosette elongates to form a flowering stem with yellow flowers; the plant then dies. Propagate by seed or leaf cuttings.

A. undulatum Gran Canaria
Sparsely branched shrub to 1m(3ft) with silvery grey stems having red-brown leaf scars. The rosettes are up to 30cm(1ft) across. Leaves 9–15cm(3½–6in) long, wide and frilled at the ends. Flowers bright yellow in pyramidal inflorescences.

A. urbicum Tenerife and Gomera (Canary Is.)
Shrub, 1–2m(3¼–6½)ft tall when in bloom. Stem robust, erect, unbranched, terminated by a solitary rosette to 25cm(10in) or more wide. Leaves spathulate, ciliate, pale green to glaucous with red edges. Flowers white or pink-tinted in a huge terminal panicle up to 60cm(2ft) or more in height. Plants take five to seven or more years to reach maturity, then bloom, set seed and die (monocarpic).

AERANGIS See ANGRAECUM

AERIDES

Orchidaceae
Fox-tail orchids

Origin: India, S.E. Asia, Japan. A genus of 40 species of epiphytic orchids. They have erect stems which bear two parallel rows of usually strap-shaped, leathery leaves with bilobed tips. The fragrant flowers have a waxy texture and are borne in axillary, pendent racemes. The petals and sepals are similar in form, the three-lobed labellum having a forward curving, horn-like spur. Propagate by cuttings of basal shoots or the tops of old plants with aerial roots in spring. Best in the warm conservatory, but suitable as a short term pot plant in the home. *Aerides* derives from the Greek *aer*, air – the plants growing on trees.

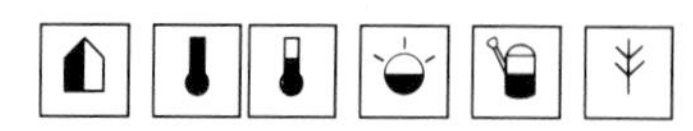

Species cultivated

A. ampullaceum See *Ascocentrum ampullaceum*.

A. crassifolium Burma
Stem to 60cm(2ft); leaves to 17cm(6¾in). Flowers bright rose-purple 4cm(1½in) wide, borne in summer.

A. falcatum Burma, Thailand, Laos
Stem to 2m(6½ft); leaves 15–40cm(6–16in) wide. Flowers white, tipped with amethyst-purple, 2.5–3cm(1–1¼in) long.

A. fieldingii Fox-brush Himalaya
Stem to 1m(3ft); leaves 20–25cm(8–10in). Flowers white, mottled with bright rose, 2.5–4cm(1–1½in) wide, carried in dense, tail-like racemes 60cm(2ft) or more long, borne in early summer.

A. lawrenceae Philippines
Stem to 1.5m(5ft); leaves 22–30cm(9–12in). Flowers white or cream, tipped with crimson-purple, borne in autumn to winter; spur bright green, 4cm(1½in) long.

A. longicornu India, Burma
Stem to 35cm(14in); leaves spindle-shaped. Flowers often solitary, white, flushed with purple, 2.5cm(1in) wide; spur 2cm(¾in) long.

A. quinquevulnerum Philippines
Stem to 1m(3ft) or more, often well branched; leaves to 30cm(1ft), glossy light green. Flowers white, purple-tipped, fragrant, to 4cm(1½in), borne in summer to autumn.

A. vandarum China, Himalaya, N. India
Stem slender, often branched, to 1m(3ft); leaves spindle-shaped, grooved, 10–22cm(4–9in) long. Flowers borne in twos or threes, white, lightly flushed pink, 5cm(2in) wide, borne in spring. Needs support.

Aerides falcatum

AESCHYNANTHUS

Gesneriaceae

Origin: India, China and Malaysia. A genus of 80 species of evergreen trailing and climbing subshrubs many of which grow epiphytically; they make splendid, colourful indoor plants. The lanceolate to ovate leaves are often somewhat fleshy and are carried in pairs. The showy, tubular, sometimes hooded flowers are red or orange and form terminal clusters. They are followed by berry-like fruits which are often white in colour. The trailing species are most effective grown in hanging baskets, but those which climb are best grown in a pot or pan supported by a moss stick. Propagate by cuttings in spring or summer. *Aeschynanthus* derives from the Greek *aischune*, shame and *anthos*, a flower, suggesting rather oddly that the flowers blush for shame!

Species cultivated

A. boschianus Java
Epiphyte of weakly climbing or trailing habit, to 30cm(1ft) tall. Leaves ovate, flushed with purple. Flowers scarlet, to 8cm(3in), from a tubular, purple-brown calyx in spring. Included by some botanists in *A. radicans*.

A. bracteatus Assam, Sikkim
Trailing, sparsely branching to 60cm(2ft) or so. Leaves to 10cm(4in), elliptic, pointed, somewhat fleshy. Flowers to 3.5cm(1⅓in) long, orange-red, in clusters of up to seven, with red bracts, in spring.

A. ellipticus New Guinea
Trailing to pendent, eventually to 60cm(2ft) long. Leaves ovate, 6–11cm(2½–4½in) long, thick-textured and lustrous. Flowers 8cm(3in) long, glowing reddish-orange, marked with dark red stripes, emerging from tubular yellow-green calyces; borne in clusters at the shoot tips.

A. evrardii Vietnam
Semi-erect to erect, to 30cm(1ft) or more in height. Leaves lanceolate, 12cm(4¾in) long, thick and smooth. Flowers in terminal clusters from the upper leaf axils, 9cm(3½in) long, reddish-orange with dark red striping on the lobes.

A. grandiflorus See *A. parasiticus*.

A. javanicus Java
Similar to *A. lobbianus* but with scarlet and yellow flowers within green calyces with reddish mouths. Botanists now consider *A. boschianus*, *A. javanicus* and *A. lobbianus* to be forms of a variable *A. radicans*.

A. lobbianus Lipstick vine
Epiphyte, trailing or climbing weakly to 60cm(2ft) in height. Leaves elliptic. Flowers to 4cm(1½in) long, crimson with a yellow throat and a silky, dusky-red calyx. A good hanging basket plant flowering in spring and summer. Considered by some to be a form of *A. radicans*.

A. longiflorus Java
Trailing, fairly robust to 90cm(3ft) or so. Leaves ovate, slender pointed, 5–10cm(2–4in) or more long, lustrous rich green. Flowers 6–10cm(2½–4in) long, orange-red to crimson from dark red calyces and with pink stamens; they are carried in erect clusters of two to seven at the tips of the shoots in summer.

A. marmoratus (*A. zebrinus*) Burma, Thailand
Epiphytic, trailing to 60cm(2ft). Leaves elliptic, 7–10cm(2¾–4in) long, netted yellow-green above, purple beneath. Flowers pale green with chocolate-brown mottling, 3cm(1¼in) long. A good subject for a hanging basket, flowering from summer to autumn.

A. micranthus Himalaya (low altitudes only)
Trailing, to about 45cm(1½ft) long. Leaves lanceolate, 4–5cm(1½–2in) or more long, light green and thick-textured. Flowers narrowly tubular, deep red-

purple, 2.5cm(1in) or a little more in length, usually in twos and threes in the upper leaf axils. Not showy, but one of the most easy and dependable bloomers.

A. nummularius New Guinea
Trailing, to about 20cm(8in) long; whole plant slightly velvety-hairy. Leaves 1–2cm($\frac{3}{8}$–$\frac{3}{4}$in) long, heart-shaped and densely borne, often just overlapping. Flowers red to purple-red, about 2cm($\frac{3}{4}$in) long, emerging from an urn-shaped calyx. Needs shade and warmth to thrive.

A. obconicus Malaya
Stiffly trailing to about 60cm(2ft) long. Leaves 5–10cm(2–4in) long, broadly ovate, blunt-tipped, thick-textured and somewhat leathery. Flowers about 5cm(2in) long, red, the lower half hidden in a bowl-shaped, deep purple, hairy calyx.

A. parasiticus (*A. grandiflorus*, *A. parviflorus*) India
Trailing, sometimes with semi-erect stems to 1.5m(5ft), but usually less. Leaves oblong-lanceolate, slender-pointed, toothed, somewhat fleshy, to 10cm(4in) long. Flowers 4–5cm($1\frac{1}{2}$–2in) long, dark crimson, tipped with orange, carried in terminal clusters in late summer.

A. parviflorus See *A. parasiticus*.

A. pulcher Royal red bugler Java
Epiphytic climber or trailer. Leaves ovate, about 4cm($1\frac{1}{2}$in) long, thick-textured. Flowers hooded, bright orange-red with a yellow throat, 4–6cm($1\frac{1}{4}$–$2\frac{1}{2}$in) long, opening from summer through to winter.

A. radicans Java, Malaysia
Epiphytic with trailing or weakly climbing stems, compact in habit. Leaves elliptic to ovate, about 4.5cm($1\frac{3}{4}$in) long. Flowers to 5cm(2in) long, red with a green calyx streaked with purple-red. Botanists now consider that *A. boschianus*, *A. lobbianus* and *A. javanicus* are all forms of this very variable species.

A. speciosus (*A. splendens*, *Trichosporum splendens*) Java, Borneo, Malaysia
Trailer to 60cm(2ft). Leaves often borne in whorls, pale green, waxy in texture and 7–10cm($2\frac{3}{4}$–4in) long. Flowers 5–8cm(2–3in) long, bright orange, marked in the throat with red-brown, opening from summer to autumn.

A. splendens See *A. speciosus*.

A. × splendidus (*A. parasiticus* × *A. speciosus*)
Similar to *A. speciosus* but more robust with larger leaves. Flowers bright orange, blotched maroon.

A. tricolor Malaysia
Trailing, to about 30cm(1ft) long. Leaves 2.5cm(1in) long, ovate-cordate, dark green above, paler beneath. Flowers to 4cm($1\frac{1}{2}$in) long, red with black and yellow stripes on the upper lobes and emerging from a dark bowl-shaped calyx, usually in pairs from the upper leaf axils.

A. zebrinus See *A. marmoratus*.

TOP LEFT *Aeschynanthus speciosus*
ABOVE *(top to bottom) Aeschynanthus longiflorus*, *A. radicans*, *A. tricolor* (in bud)

AGAPANTHUS

Alliaceae
African lilies

Origin: South Africa. A genus of five to 12 species (depending upon the classification followed). They are clump-forming, evergreen or deciduous perennials, having strap-shaped leaves and rounded umbels of showy, funnel-shaped, 6-tepalled flowers. They need large pots or, ideally, tubs to flower well and can be stood outside during the summer, or grown in the conservatory all the year. Propagate by division in spring. Deciduous species are generally hardy. *Agapanthus* derives from the Greek *agape*, love and *anthos*, a flower.

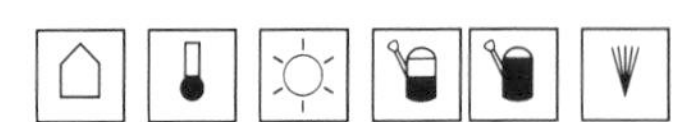

Species cultivated

A. africanus (*A. umbellatus*)
Evergreen. Leaves to 35cm(14in), channelled. Flowering stems to 60cm(2ft). Flowers blue-purple, opening in autumn, 3–5cm(1¼–2in) long, carried in loose umbels of five to 25 blooms, rarely more. Cool. Rare in cultivation. *A. campanulatus* and its hybrids are often grown under this name.

A. campanulatus
Deciduous. Leaves to 45cm(1½ft). Flowering stems to 1m(3ft). Flowers mid-blue or white, 3–4cm(1¼–1½in) long. Hardy.

A.c. patens
Smaller and more slender, having dense umbels of flowers with widely spreading tepals. Probably identical with *A. umbellatus globosus.*

A. caulescens
Evergreen. Similar to *A. africanus*, but leaves arising from a stem-like base. Flowering stem to 1m(3ft). Flowers bright to deep blue, 3–4cm(1¼–1½in) long, opening widely in early autumn. Cool.

A. comptonii
Evergreen. Narrow, channelled leaves rarely more than 60cm(2ft) long. Flowering stem 1m(3ft) or more, carrying a loose umbel of narrowly bell-shaped, mid to deep-blue flowers. A dwarf form sometimes listed is possibly derived from *A. praecox minimus*. Cool.

A. inapertus
Deciduous. Leaves to 60cm(2ft) long and 2.5cm(1in) wide from a short stem. Flowering stems can reach 1.7m(5½ft), but are often less. Flowers pendent and markedly tubular with the tube longer than the flared mouths, violet to dark blue. A white form is also known. Hardy.

A.i. pendulus ('Pendulinus' of gardens)
Leaves to 5cm(2in) wide; deep blue, narrowly tubular flowers.

Agapanthus one of the Headbourne Hybrids

A. praecox (*A. umbellatus*)
Evergreen. Once included in *A. africanus*. Leaves to 90cm(3ft) long and 6cm(2½in) wide. Arching flower stems to 1m(3ft) carry dense umbels of light to mid lilac-blue or white flowers, each to 7cm(2¾in) long in summer and early autumn. Cool.

A.p. minimus
Smaller flower stems to 60cm(2ft) with shorter clumps of narrower foliage.

A.p. orientalis
Smaller than the species with stems to 75cm(2½ft), but wider leaves than *A.p. minimus*.

A.p. praecox
Under this name are included many variants, some of which are of hybrid origin. They cover a range of shades of blue, and some with variegated foliage which are very attractive for pot culture; these include 'Aureovittatus' with yellow-striped leaves, 'Argenteovittatus' with white stripes and 'Golden Rule' with gold lines.

Hybrids

A. Headbourne Hybrids
A group of hybrid cultivars of garden origin, derived mainly from *A. campanulatus* and *A. inapertus*. Deciduous and hardy, but equally good as conservatory tub plants. In height they can be from 60–120cm(2–4ft) and are available in shades of blue and white. Hardy.

AGAPETES

Ericaceae

Origin: Himalaya to China, S.E. Asia. A genus of 80 shrubby, evergreen species some of which are epiphytic. They have mainly lanceolate leaves and urn-shaped flowers with marked angles between the five joined petals. The fruits are berry-like. Support is necessary for the lax stems. Propagate by seed in spring or by cuttings (with a heel) in late summer. Plants are best in the conservatory, but will survive for a limited period in the home. *Agapetes* derives its name from the Greek *agapetos*, beloved or desirable, referring to the attractive patterning on the flowers.

Species cultivated

A. macrantha Burma
Stems lax, to 2m(6½ft) or more. Leaves dark green, lanceolate. Flowers white, closely patterned with yellow and red, 4cm(1½in) long, opening in winter. Tropical.

A. rugosa (*Pentapterygium rugosum*)
Stems lax, reaching 3m(10ft) in length. Leaves lanceolate, 7–10cm(2¾–4in) long, bright green with a wrinkled texture. Flowers white with purplish or reddish patterning, 2.5(1m) long, opening in spring. Temperate.

A. serpens (*Pentapterygium serpens*) China
Pendulous stems to 3m(10ft) or more in length.

Leaves 1.2–2.5cm($\frac{1}{2}$–1in) long, shiny green. Flowers rosy-red covered with regular, dark red, V-shaped markings, opening in spring. Temperate.

AGATHAEA See FELICIA

AGATHOSMA

Rutaceae

Origin: South Africa. A genus of 140–180 species of mainly small, wiry, evergreen shrubs. In general they have a somewhat heath-like appearance with small simple leaves and tiny 5-petalled flowers in profuse terminal umbels. They bloom when young and make pleasing pot or tub plants. Although unsatisfactory as house plants, they can be brought indoors from the conservatory when in bloom. During the summer they can be stood outside in a sheltered site. An acid, free-draining soil is essential. Propagate by cuttings in late summer or early autumn. *Agathosma* derives from the Greek *agathos*, pleasant or good and *osme*, fragrance, alluding to the sweetly aromatic smell of some species.

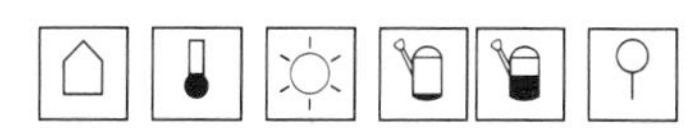

Species cultivated

A. ciliata
Erect-stemmed shrub 30–60cm(1–2ft) in height. Leaves 0.7–2cm($\frac{1}{4}$–$\frac{3}{4}$in) long, linear to narrowly ovate, pointed tipped and with a marginal fringe of long white hairs. Flowers white, in umbels 2cm($\frac{3}{4}$in) wide, borne in spring.

A. corymbosa (*Diosma purpurea*)
Bushy rounded shrub to 60cm(2ft) high and wide. Leaves linear-lanceolate, up to 9mm($\frac{1}{3}$in) long. Flowers white, lilac or purple, in umbels 1.2–2cm($\frac{1}{2}$–$\frac{3}{4}$in) wide, borne in spring.

A. pulchella See *Barosma pulchella*.

A. punctata
Downy-stemmed shrub, 60–90cm(2–3ft) tall. Leaves oblong-lanceolate, 4–9mm($\frac{3}{16}$–$\frac{1}{3}$in) long, closely set. Flowers white or purple-tinted, in umbels up to 2cm($\frac{3}{4}$in) wide, borne in spring.

AGAVE

Agavaceae

Century plants

Origin: Southern North America to northern South America. A genus of 300 species of succulent evergreen perennials, grown indoors chiefly as foliage plants – young specimens make good house plants whilst larger ones make imposing subjects for the conservatory. They form stemless rosettes of thick, fleshy, usually spine-tipped leaves (beware of their sharpness). Flowers do not develop on containerized specimens unless they are grown in very large tubs; nor do leaves reach full size where the roots are restricted. When they do appear, the flowers are numerous, greenish or brownish and individually small, but carried in towering racemes or candelabra-like clusters at the tops of long stems. Once flowered the rosette always dies, but is replaced by offsets or stem bulbils from which new plants can be propagated. Seed, when available, should be sown in spring. *Agave* derives from the Greek *agavos*, admirable, from the appearance of a plant in full bloom, which is worthy of admiration.

LEFT *Agapetes macrantha*
BELOW LEFT *Agapetes serpens*
BELOW *Agathosma ciliata*

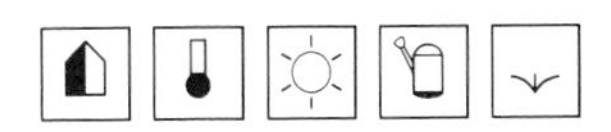

Species cultivated

A. americana Century plant, Maguey, American aloe Mexico, naturalized elsewhere
Leaves grey-green, to 1.5m(5ft) and up to 20cm(8in) in width. In warm climates flowering stems can reach 8m(26ft). Fermented sap from the cut flowering stem gives the Mexican national drink, pulque. There are several variegated forms.

A.a. marginata
Leaves with two pale yellow marginal bands.

A.a. marginata-alba
Leaves with two white to cream marginal bands and a pink suffusion when young.

A.a. medio-picta
Leaves with a central yellow stripe.

A.a. variegata
Dark green, somewhat twisted leaves with broad yellow marginal bands.

A. angustifolia Habitat unknown
Rosettes dense, eventually forming clumps, each

ABOVE *Agave angustifolia* 'Marginata'
ABOVE RIGHT *Agave americana medio-picta*

rosette borne on a short trunk to 40cm(16in) high. Leaves sword-shaped, spine-tipped and leathery, 50–90cm(20–36in) long, varying from light green to grey-green. Flowers greenish, in a narrowly pyramidal panicle, the top of which can reach 2.7m(9ft) in height. *A.a.* 'Marginata' has leaves broadly edged cream.

A. atrovirens South Mexico
Rather like *A. americana* in shape and habit, but a little smaller and the leaves very dark green.

A. attenuata (*A. glaucescens*) Mexico
Stemless when young, the rosettes grow from short trunks as the plant matures. Leaves ovate, to 70cm(28in) long, tapering to a point, sometimes waxy white beneath. Flowers on larger, older specimens in a dense, cylindrical raceme to 2m(6½ft) or more in length, arching strongly. Often viviparous.

A. bracteosa N.E. Mexico
Rosettes eventually forming clumps. Leaves slender, arching, not spine-tipped but finely and sharply toothed, 30–35cm(12–14in) long, grey to pale green. Flowers creamy-yellow in a 2m(6½ft) long, dense, cylindrical inflorescence of imposing appearance.

A. deserti California (Colorado Desert)
Rosettes crowded, in the wild forming dense, wide clumps. Leaves spine-tipped and toothed, triangular-lanceolate, to 30cm(1ft) long, channelled and grey-green. Flowers chrome-yellow, in panicles 2–3m(6½–10ft) tall.

A. falcata Mexico
Rosettes dense. Leaves 40–50cm(16–20in) long, very slender, stiff and sharp pointed, usually grey-green, variously flushed purple or reddish-brown especially in sunny sites. Flowers yellowish, in 2m(6½ft), open, bracted panicles.

A. ferdinandi-regis (*A. victoriae-regina laxior*, *A.v. nickelsii*) New Mexico
Close to *A. victoriae-reginae*, but each rosette with fewer leaves which taper gradually to a dark, terminal spine.

A. filifera Thread agave Mexico
Numerous linear to narrowly lanceolate leaves, edged with whitish, horny threads. Flowers form a dense cylindrical raceme to 2.5m(8ft) tall. A good pot plant because of its lack of spines.

A. franzosinii Mexico
Similar to *A. americana*, but leaves longer and slimmer, elegantly arching, blue-white to blue-grey.

A. glaucescens See *A. attenuata*.

A. horrida Mexico
Leaves to 40cm(16in), stiff, glossy dark green with strong, thick marginal spines. Flowers yellow, carried on 3–4m(10–13ft) stems in the wild. As it does not produce offsets, propagation is by seed only.

A. × leopoldii (*A. filifera* × *A. schidigera*)
Similar to *A. filifera*, but leaves longer and narrower, liberally hung with curling, white horny threads.

A. palmeri S.W. USA
Dark green to grey-green leaves to 1.5m(5ft) long in dense rosettes. Flowers green to yellow-green and, in the wild, carried at the top of 2.5–6.5m(8–21ft) tall stems. Cultivated for its dark green foliage.

A. parrasana N. Central Mexico
Rosette compact, dense, almost spherical when mature. Leaves obovate, to 35cm(14in) long, margined in the upper part with spiny teeth and terminated by a hard spine, deep grey-green. Flowers yellow, in a stiff, tufted panicle to 3m(10ft) tall.

A. parryi Arizona, New Mexico
Rosettes of grey-green leaves to 30cm(1ft) in length, margined with prickles. Flowers pale yellow on a 2.5m(8ft) stem.

A. parviflora Arizona, Mexico
Dense rosettes of dark green, 10cm(4in) long leaves, marked with white above and with short, white horny threads along the margins. Flowers tubular, borne in a simple spike on a 1.5m(5ft) stem.

A. polyacantha Probably from Mexico
Dense rosettes, sometimes clump-forming but usually only after flowering. Leaves 50–70cm(20–28in) long, broadly lanceolate, tapering to a slender, sharp, terminal spine and bearing small, dark, spiny teeth; dark green or slightly grey-green. Flowers greenish, in a dense raceme 2m(6½ft) or more high.

A. striata Mexico
Rosettes of long, narrow leaves to 45cm(1½ft), grey-green, striped with darker green – so dense that they form an almost rounded head. Green flowers are carried on 3.5m(12ft) stems. A.s. 'Nana' is a dwarf form with leaves eventually to 30cm(1ft), but very slow growing.

A. stricta Mexico
Rosettes at first solitary, but when mature on a short, branching trunk. Leaves to 40cm(16in) long, dense, very slender, rigid and tapering to a spiny tip, radiating out from the base to create a spherical outline. Flowers greenish, in dense spikes 1.5–3m(5–10ft) tall.

A. utahensis Utah, California, Colorado
Stemless rosettes of grey-green leaves to 30cm(1ft) long, tapering to a point and with hooked, marginal prickles. Flowers yellowish, carried in a cylindrical raceme on a stem to 2.5m(8ft) high.

A. victoriae-reginae Mexico
Shapely rosettes rather resembling a globe artichoke, with 15cm(6in) long, broad, dark green, blunt leaves with white lines and horny margins, ending abruptly in a short spine. The flowering spike can reach 4m(13ft). See also *A. ferdinandi-regis*.

AGLAONEMA

Araceae

Chinese evergreens

Origin: Tropical and sub-tropical S.E. Asia. A genus of 21 species of evergreen perennials. They are tufted plants, having short stems which give some species an almost shrubby appearance. The ovate to oblong-lanceolate leaves are often patterned and the small petalless flowers are carried on a short spadix with an arum-like spathe. The bright red fruits are not often formed. Propagate by division or by summer cuttings. A genus of well-tried houseplants which thrive in poor light. *Aglaonema* derives from the Greek *aglaos*, bright and *nema*, a thread, referring to the stamens.

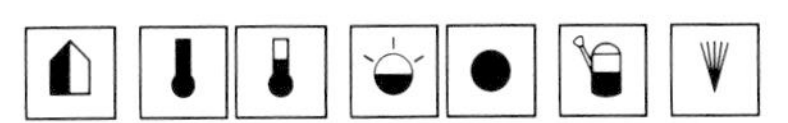

Species cultivated

A. acutispathum See *A. modestum*.

A. brevispathum hospitum Thailand
Leaves ovate, up to 20cm(8in) in length, deep glossy green, irregularly freckled with white, borne on stalks as long as or longer. Spathe pale green, 4cm(1½in) long, usually with a pointed tip. *A. brevispathum* itself has plain green leaves and a form

BELOW *Agave victoriae-reginae*
BOTTOM LEFT *Agave utahensis*
BOTTOM RIGHT *Agave filifera*

is known with a white midrib. This species has a creeping stem like that of *A. costatum* to which it is very closely related.

A. commutatum Molucca islands
Very variable. Thick-textured oblong-lanceolate leaves to 25cm(10in) long, dark green with grey feathering. Greeny-white spathes are carried on 8cm(3in) stems.

A.c. pseudobracteatum
Stems white with green marbling. Longer, narrower leaves with yellow and green mottling and distinct white veins. *A.c.* 'Malay Beauty' is a more robust form. *A.c.* 'Treubii' is smaller and neater, with bluish-green leaves marbled with silvery grey. *A.c.* 'Tricolor' has broadly elliptical leaves, variegated on both surfaces and carried on pinky-white leaf stalks.

BELOW (top to bottom) *Aglaonema commutatum* 'Treubii', *A.* 'Silver Queen', *A. commutatum*

A. costatum Malaya
Leaves broadly ovate, to 13cm(5in) long, dark green spotted with white and with a conspicuously white central vein. Pale green spathes carried on 5cm(2in) stalks.

A. modestum (*A. acutispathum*) Chinese evergreen South China
Leaves pointed, broadly ovate, shining dark green, 15–20cm(6–8in) long. Spathes pale green, 7.5cm(3in) high. The hardiest species, withstanding temperatures to 7°C(44°F). *A.m.* 'Variegatum' has irregularly creamy-white variegated leaves.

A. nitidum (*A. oblongifolium*) Malaya
Stiffly erect, dark glossy green, elliptic leaves to 20cm(8in) long. Small green spathe with white margins. *A.n.* 'Curtisii' has leaves neatly marked with a silvery herringbone pattern.

A. pictum Sumatra
30–60cm(1–2ft) tall. Leaves broadly ovate, to 20cm(8in) long, bluish-green with irregular silver-grey blotching. Spathes cream.

A.p. tricolor
Leaves two shades of green, mottled with silver-grey; pinkish stalks.

A. rotundum Sumatra
Leaves ovate to 15cm(6in) or more long, short stalked, deep green with a coppery flush and pink veins above, purple-red beneath, the whole with a metallic lustre. Spathes not recorded. A short compact plant and one of the most decorative.

A. simplex Java
Similar to *A. modestum*, but with somewhat larger, thinner-textured leaves. *A.s.* 'Angustifolium' has very narrow, slender-pointed, somewhat corrugated leaves of distinctive appearance.

Hybrids

'Fransher', a slender plant with lanceolate, pale green leaves, variegated with paler green and cream.
'Silver Queen', ovate leaves, overlaid with a paler green and silvery variegation which almost totally covers the leaf surfaces.

AICHRYSON

Crassulaceae

Origin: Islands of Macronesia. A genus of ten species of small, succulent annuals or perennials resembling *Aeonium*. They have spathulate to rounded leaves in rosettes which terminate solitary or branched stems. The starry, yellow, 6- to 12-petalled flowers are carried in terminal panicles. Propagate by seed in spring or by cuttings of single rosettes in summer. *Aichryson* is the classical Greek name for *Aeonium arboreum* transferred by the botanists Webb and Berthelot to this related genus.

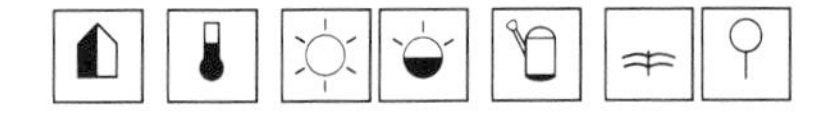

Species cultivated

A. × domesticum (*A. tortuosum, Aeonium × domesticum*) Origin unknown, possibly a hybrid from the true *A. tortuosum, q.v.*
A softly hairy shrublet forming a loose mound 10cm(4in) or more high and much wider as it ages. Leaves rounded, forming rosettes 2–5cm(¾–2in) wide. Flowers yellow, 7- or 8-petalled, freely produced in summer. *A. × d.* 'Variegatum' has white-bordered leaves.

A. laxum Canary Isles
Erect annual or biennial to 30cm(1ft) tall, though usually less, the stems repeatedly forking. Leaves rounded to spathulate, densely hairy, up to 1.5cm(⅝in) long, in loose, elongate rosettes. Flowers 9- to 12-petalled, pale yellow, in airy panicles in summer.

A.l. foliis-purpureis
Leaves suffused purple; breeds true from seed.

A. tortuosum Lanzarote, Fuerteventura (Canary Isles)
A dense hummock-forming shrublet to 10cm(4in) tall. Leaves obovate, 1.2–1.5cm(½–⅝in) long, sticky hairy. Flowers 8-petalled, golden-yellow, expanding in summer.

A. villosum Madeira, Azores
A spreading, bushy, white sticky-hairy annual or biennial up to 20cm(8in) tall. Leaves broadly

rhomboidal, to 2.5cm(1in) or so long. Flowers 6- to 9-petalled (usually 8), rich yellow, in late spring.

ALBIZIA (ALBIZZIA)

Leguminosae

Origin: Tropics and sub-tropics of Asia, Africa, Australia; one species in Mexico. A genus of 100 or more species of deciduous trees and shrubs closely related to *Acacia*. They have very decorative, bipinnate leaves and tiny flowers which are aggregated together in fluffy heads or spikes. In a small pot in the home they will not flower, but are well worth growing for their elegant, ferny foliage alone. If seeds are sown in early spring they can be treated as annuals making good short term house plants for a sunny window ledge. *Albizia* was named for F. degli Albizzi, an Italian nobleman who introduced *A. julibrissin* into cultivation in about 1759.

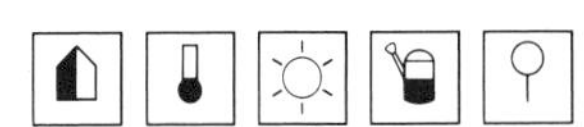

Species cultivated

A. distachya See *A. lophantha.*

A. julibrissin Silk tree Iran to China

In the wild, a large shrub or small tree with globular heads of pink flowers. Grown indoors for its ferny, bipinnate leaves with oblong-oblique leaflets 6–12mm($\frac{1}{4}$–$\frac{1}{2}$in) long.

A. lophantha (*A. distachya, Acacia lophantha*) Plume albizia S.W. Australia

A large shrub or small tree. Flowers white, pink or pale yellow in a bottlebrush-like spike. Leaves bipinnate, made up of several hundred 8mm($\frac{1}{3}$in) linear leaflets.

ALBUCA

Liliaceae

Origin: Africa; those cultivated, from warm temperate regions. A genus of 50 species of bulbous perennials with tufts of rather fleshy, linear leaves. The flowers are white or yellow, the inner three tepals always shorter than the outer; they are carried in racemes. Propagate by removing offsets, or by seed in spring. They are best grown in the conservatory, but can be brought into the home when in bloom. *Albuca* derives from *Albucus*, Latin for asphodel, presumably because the racemes of white flowers resemble those of asphodel.

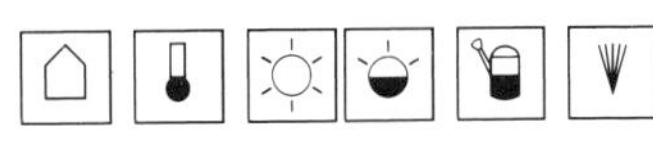

Species cultivated

A. canadensis (*A. major, A. minor*) South Africa

Leaves narrowly lanceolate, grey-green, in a spreading tuft. Flowering stems 30–90cm(1–3ft) tall, the upper part a raceme of two to 12 nodding blooms. Flowers about 2.5cm(1in) long, petals yellow with a central green stripe, opening usually in late spring or early summer. The name is confusing, the botanist Linnaeus who first described the plant being under the mistaken impression that it came from Canada.

A. cooperi South Africa

Much like *A. canadensis* though sometimes smaller with fewer flowers per stem. The easiest distinction is in the bulb, this species having the bulb scales with fibrous tips.

A. fastigiata Swaziland

Leaves narrowly sword-shaped, arching, 30–90cm(1–3ft) long. Flowering stems above the leaves, bearing a short raceme of 5cm(2in) wide, white, green-striped flowers which face upwards, usually in early summer.

A. major (*A. minor*) See *A. canadensis.*

A. nelsonii Natal

Leaves narrowly lanceolate, to 90cm(3ft) long and 3–5cm($1\frac{1}{4}$–2in) wide. Flowering stem 60–150cm(2–5ft) in height. Flowers white, 3cm($1\frac{1}{4}$in) long, their outer petals having a dull red central stripe and borne in a many-flowered raceme in early summer.

TOP LEFT *Aichryson × domesticum* 'Variegatum'
TOP RIGHT *Albizzia julibrissin*
ABOVE CENTRE *Albuca canadensis*
ABOVE *Albuca nelsonii*

ALLAMANDA

Apocynaceae

Origin: Tropical areas of northern South America and the West Indies. A genus of 15 species of evergreen climbers and sprawling shrubs. The leaves are ovate to lanceolate, usually in whorls or occasionally in opposite pairs. Large, trumpet-shaped flowers are borne in terminal clusters. To do well they need a large pot or tub, and the stems will need some form of support. Propagate by cuttings in spring. Allamandas are spectacular as large plants in a conservatory, but when small can be grown as short term house plants. *Allamanda* was named for Dr Frederick Allamand who lived from about 1735 to after 1776 – a Swiss botanist who collected in Surinam.

  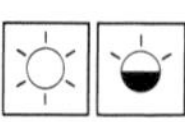  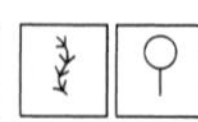

Species cultivated

A. cathartica Golden trumpet Brazil, Guyana
A vigorous, evergreen climber which is very variable in the wild where it scrambles through larger shrubs, reaching a height of 5m(16ft). In cultivation it is amenable to pruning. Leaves lanceolate, 10–15cm(4–6in) long, usually in whorls of four. Flowers 10cm(4in) wide, yellow, opening from summer to autumn. *A.c.* 'Grandiflora' has paler yellow flowers and the plant is a little less vigorous. *A.c.* 'Hendersonii' has rich golden-yellow flowers, 10–13cm(4–5in) across.

A. neriifolia Brazil
Shrub, 1–2m(3–6½ft) in height. Leaves oblong-lanceolate 7–13cm(2¾–5in) long, in whorls of three to five. Flowers golden-yellow, streaked with deeper reddish-orange in the throat, 4cm(1½in) across, opening from spring to autumn.

ABOVE *Allium schoenoprasum*
ABOVE RIGHT *Allamanda neriifolia*

ALLIUM

Alliaceae

Origin: Northern hemisphere. A genus of 450 mainly bulbous species, many of which are hardy. Most have linear, keeled, flattened or tubular leaves; a few are broader. The starry or bell-shaped flowers have six tepals and are borne in terminal umbels or, rarely, tiered whorls. Most species smell of onion or garlic when bruised. A few make ornamental or useful pot plants. Pot the bulbs and keep in a cool but not cold place and they will flower well in advance of those outside. Propagate by division, offsets or seeds. They are best in the conservatory but can be brought into the home when in bloom. *Allium* is the classical Latin name for garlic.

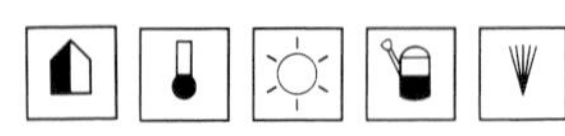

Species cultivated

A. neapolitanum Southern Europe
Leaves almost without onion scent, linear and keeled. Flowers glistening white, starry, to 12mm(½in) across, carried on stems to 25cm(10in) high. Forced, they will flower in late winter. Hardy.

A. schoenoprasum Chives Northern Europe, N. Asia, North America
Leaves cylindrical from stem-like bulbs (scallions). Flowers small, rosy-purple in rounded heads on 15–25cm(6–10in) stems in summer. For use as a herb in winter, pot several clumps in October and bring indoors successionally every 4–6 weeks for a continuous supply. Hardy.

A. triquetrum Western Europe
Leaves linear, channelled and keeled. Flowers white, bell-shaped, 1.5cm(½in) long, pendent and fragrant; they are carried on 20cm(8in) high stems which are triangular in section, in spring. Hardy.

ALLOPHYTUM See TETRANEMA

ALOCASIA

Araceae

Origin: Tropical Asia. A genus of 70 evergreen perennials with thick, fleshy rhizomes which can be erect and stem-like, or occasionally tuberous. The handsome leaves are peltate, sagittate to cordate, sometimes deeply lobed. Petalless flowers are borne on a spadix within an arum-like spathe, but are not decorative. Propagate by suckers, or cuttings of the rhizomes in spring. *Alocasia* derives from the Greek *a*, without and *Colocasia*, a closely related genus from which species were removed to form *Alocasia*.

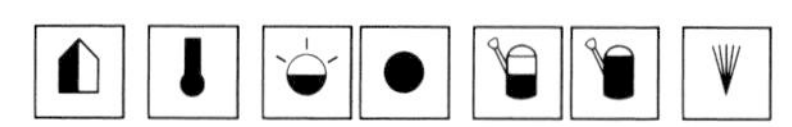

Species cultivated

A.* × *amazonica (*A. sanderiana* × *A. lowii*)
Leaves 30–40cm(12–16in) long, saggitate-peltate, with a wavy margin, very dark, glossy green strikingly contrasting with wide white veins and a narrower white margin.

A.* × *argyraea (*A. longiloba* × *A. pucciana*)
Leaves ovate, peltate with a heart-shaped base, 30–50cm(12–20in) long, firm and leathery with a silvery patina.

A. cucullata Bengal, Burma
Leaves to 30cm(1ft) long, heart-shaped, pointed, glossy rich green with prominent veins and more or less upturned margins.

A. cuprea Borneo.
Leaves ovate, peltate cordate, to 30cm(1ft) long with depressed veins, dark metallic green above and purplish beneath.

A. indica For plants grown under this name see *A. macrorrhiza*.

A. indica metallica For plants grown under this name see *A. plumbea*.

A. korthalsii Malaya
Leaves ovate-cordate, up to 60cm(2ft) long, the upper surface deep olive-green with a hint of grey, midribs and veins greyish to grey-white, lower surface purple. Akin to *A. putzeyi* and *A. watsoniana*.

A. longiloba Malaysia, Borneo, Java
Leaves sagittate, narrowly triangular in outline, 30–65cm(12–26in) long, the two basal lobes more than half as long as the main lobe; upper surface deep lustrous green with the margins and main veins grey; lower surface purple. *A.l.* 'Magnifica' has larger leaves.

A. lowii Borneo
Leaves broadly sagittate, to about 50cm(20in) long, deep green with a metallic sheen and silvery veins above, purple beneath. The two basal lobes are joined for about one quarter of their length.

A. macrorrhiza Giant elephant's ear Sri Lanka, Malaya
Rhizome trunk-like, in the wild to 4.5m(15ft) tall. Leaves broadly sagittate, to 60cm(2ft) or more. Spathe yellow-green, about 20cm(8in) long. *A.m.* 'Variegata' has leaves blotched with cream and grey-green.

A.m. rubra See *A. plumbea*.

A. micholitziana Philippines
Leaves narrowly sagittate, to 25cm(10in) long, wavy, main veins and midrib white, contrasting boldly with the rich velvety-green surface. A neat species, making a good pot plant.

A. odora Tropical Asia
Much like and confused with *A. macrorrhiza*, but leaves peltate and the main lobe has only six to ten pairs of veins (*A. macrorrhiza* has nine to 12).

A. plumbea (*A. macrorrhiza rubra*, *A. indica metallica* [but misapplied]) Java
Much like a smaller version of *A. macrorrhiza*, but variably purple flushed, sometimes deeply so; leaves to 75cm(2½ft) long.

A. portei Philippines
Leaves ovate-sagittate, to 2m(6½ft) long; those of mature plants deeply lobed (pinnatisect); those of juveniles, deeply sinuous only; deep green with a metallic lustre. Leaf stalk about the same length as the blade, very robust, green, marbled red-purple. Suitable only for the largest conservatory.

A. putzeyi For the plant grown under this name see *A. watsoniana*.

A. sanderiana Kris plant Philippines
Leaves 30–40cm(12–16in), narrow, sagittate, lobed, metallic silver-green with a grey vein pattern and white margins.

LEFT *Alocasia* × *amazonica*
TOP *Alocasia cuprea*
ABOVE *Alocasia korthalsii*

A. × sedenii (*A. lowii* × *A. cuprea*)
Tends to resemble the first parent in general appearance, but the basal lobes are joined for about two thirds of their length.

A. veitchii (*A. lowii veitchii*) Malaya
Leaves sagittate, narrowly triangular in outline up to about 45cm(1½ft) long; upper surface deep green with light grey margins and main veins; lower surface purple.

A. violacea See *Xanthosma violaceum*.

A. watsoniana Sumatra
Leaves ovate-sagittate, the basal lobes joined for about half their length, up to 90cm(3ft) long, upper surface glaucous tinted, all the veins silvery. The plant grown as *A. putzeyi* is like a smaller version but it seems likely that it is incorrectly named.

A. wentii (*A. whinckii*) New Guinea
Leaves ovate, peltate or sagittate, pointed-tipped, the basal lobes rounded and joined for half to two thirds of their length, deep glossy green and slightly wavy margined.

A. whinckii See *A. wentii*.

A. zebrina Philippines
Leaves to about 35cm(14in) long, sagittate, leathery textured, green above and beneath. This species is known by the brownish-black zebra stripes on the prominent, 50–90cm(20–36in) leaf stalks.

ALOE

Liliaceae

Origin: Africa, particularly south of the equator, Malagasy and Arabia where they grow in arid and semi-desert areas often among or beneath sparse scrub. A genus of 275 species of evergreen leaf succulents grown both for their ornamental leaves and red, yellow or orange flowers. The largest are erect and tree-like, the smallest are clump- or mound-forming perennials, e.g. the familiar *A. aristata*. In between are shrubby species of varying size and vigour and a group of semi-climbers or sprawlers with greatly elongated stems and well spaced leaves. Apart from the latter, the generally lance-shaped, sometimes mottled or patterned fleshy leaves are carried in tufts or rosettes. The tubular flowers, composed of six narrow tepals, are arranged in simple or branched racemes well above the leaves. They are often very showy and open mainly in late spring and summer. *Aloe* derives from the Arabic name *Alloeh*.

Species cultivated

A. abyssinica See *A. camperi*.

A. acutissima Malagasy
Stems to 1m(3ft), often reclining. Leaves about 30cm(1ft), tapering to a long slender point, grey-green tinged red. Flowers 3cm($1\frac{1}{4}$in) long, scarlet, in branched racemes.

A. aethiopica Abyssinia
Stemless, forming a dense rosette at ground level. Leaves sword-shaped, 60–70cm(24–28in) long, broadly grooved, glaucous, the margins with a red, horny edge and bearing sharp, triangular teeth. Flowers 2.5cm(1in) long, dark red, in branched racemes 1.2–1.35m(4–$4\frac{1}{2}$ft) tall.

A. africana Transvaal
Erect stem usually unbranched, 2–4m(6–13ft) high in the wild, remaining covered with old dry foliage. Leaves 45–60cm($1\frac{1}{2}$–2ft) long, dark to bluish-green, tapered and grooved, with a row of reddish spines down the back and along the margins. Flowers yellow to orange, up to 3.5cm($1\frac{1}{3}$in) long, in usually branched racemes.

A. arborescens South Africa
Erect, 1–4m(3–13ft) tall. Leaves to 60cm(2ft) long, usually grey-green. Flowers red, 4–4.5cm($1\frac{1}{2}$–$1\frac{3}{4}$in) long, in simple racemes borne in winter. *A.a.* 'Variegata' has attractive foliage.

A. aristata South Africa
Stemless, clump-forming. Leaves 8–10cm(3–4in) long, dark green, bearing white wart-like spines and an awned tip. Flowers orange-red, 4cm($1\frac{1}{2}$in) long.

A. bainesii South Africa
Tree-like; stem to 20m(65ft) in the wild. Leaves 60–100cm(2–3ft) long, sword-shaped, recurved. Flowers 3–4cm($1\frac{1}{4}$–$1\frac{1}{2}$in), salmon-pink, red or yellow. One of the largest species, slow-growing and a handsome pot specimen.

A. barbadensis Medicine plant or Burn plant
(*A. vera*) Original country of origin now uncertain, perhaps N.E. Africa and/or Arabia
It has long been cultivated by man, both for its magical and medicinal uses. As a result it is now widely naturalized (appearing as if wild) in many countries from Asia to the Mediterranean and westwards to the Canary Islands and West Indies. Plant forming clumps, stems short. Leaves in compact rosettes, narrowly lanceolate, 40–60cm(16–24in) long, grey-green, sometimes red-tinted, spiny-toothed. Flowers yellow, about 2.5cm(1in) long, in compact, simple or branched racemes 60–90cm(1–2ft) high.

A. bellatula Malagasy
Stemless, tufted. Leaves linear, 10–13cm(4–5in) long, minutely warted and spotted pale green. Flowers coral-red, 12mm($\frac{1}{2}$in) long in usually simple racemes.

A. × bortiana
Parents unknown, one probably *A. striata*. Short-stemmed when mature. Leaves lanceolate, 25cm (10in) long, blue-green. Flowers not recorded.

A. branddraaiensis Transvaal
Stemless. Leaves lanceolate, to 35cm(14in) in length, longitudinally white-striped. Flowers deep scarlet, 2.5cm(1in) long, in branched racemes.

A. brevifolia South Africa
Shortly stemmed, clump-forming. Leaves triangular-oblong, 7–18cm($2\frac{3}{4}$–7in) long, grey-green. Flowers red, 1.5cm($\frac{5}{8}$in) long. Sometimes confused with *A. humilis*..

A.b. depressa
Leaves broader, blue-green. Flowers orange-scarlet. Produces many offsets and is the more popular in cultivation.

A. bulbilifera Malagasy
Usually stemless. Leaves lanceolate, tapered, to 60cm(2ft) long. Flowers scarlet, 2.5cm(1in) long, in well branched panicle-like racemes. Bulbils form on the main flowering stem.

A. cameronii Malawi, Zimbabwe
Shrubby, erect to spreading, up to 1m(3ft) or more in height. Leaves 40–50cm(16–20in) long, the margins prominently toothed, usually green, but often flushed coppery-red in winter when dry and cool. Flowers rich bright scarlet, slightly incurved,

BELOW LEFT *Aloe arborescens* 'Variegata'
BELOW RIGHT *Aloe cameronii*

about 4.5cm(1¾in) long, in branched racemes. Variable in form and length of inflorescence.

A. camperi (*A. eru*) Ethiopia, Kenya
Shrubby, more or less erect, each stem with a dense rosette of foliage. Leaves sword-shaped, 40–60cm(16–24in) long, very fleshy, grooved above, margins with large, reddish teeth, the tip keeled and spinous, glossy dark green with white oblong spots. Flowers orange-yellow, about 1.5cm(⅝in) long, in branched racemes.

A. candelabrum Natal
Erect, stem usually unbranched, 2–4m(6–13ft) tall. Leaves 60–90cm(2–3ft) long, green and glaucous, with marginal spines and a row down the middle of the back, long persistent even when dead. Flowers scarlet, pink or orange, about 3cm(1¼in) long, in branched racemes.

A. castanea South Africa
Stem 1–5m(3–16ft), branched. Leaves to 40cm(16in). Flowers 2cm(¾in), red-brown.

A. chabaudii Rhodesia to Transvaal
Stemless, clump-forming. Leaves lanceolate, grey to blue-green, to 50cm(20in). Flowers dull red, 3cm(1¼in) long, in panicle-like racemes.

A. ciliaris Climbing aloe South Africa
Stem flexible, branched, scrambling or prostrate, to 5m(16ft) or more. Leaves linear-lanceolate, to 15cm(6in). Flowers 3cm(1¼in) long, scarlet, yellow-green tipped, in usually simple racemes.

A. comptonii Cape
Stemless, or rarely shortly stemmed, forming clumps or colonies of rosettes. Leaves up to 30cm(1ft) long, 9cm(3½in) broad at the base, with triangular, pale brown teeth at the margins and a few down the back, glaucous, sometimes red-flushed. Flowers dull scarlet, 3.5–4cm(1⅓–1½in) long, in branched racemes.

A. cooperi Cape
Stemless, erect, the rosettes distichous, either solitary or in clumps. Leaves 30–50cm(12–20in) long, slender and tapered, channelled, bright green with white spots at the base and white-margined teeth. Flowers pale orange to pinkish-green in simple, conical racemes.

A. cryptopoda Mozambique, Zimbabwe, Transvaal
Stemless, in small clumps or singly. Leaves to 60cm(2ft) long, somewhat glossy deep green. Flowers 3–4cm(1¼–1¾in) long, reddish-orange.

A. distans W. Cape
Stems prostrate, creeping, branching from the base. Leaves to 9cm(3½in) long, broadly ovate, glaucous, with large whitish teeth and a few off-white tuberculate spots. Flowers dull scarlet, 2.5–4.5cm(1–1¾in) long, in branched racemes. An easy going species deserving to be grown more often.

A. eru See *A. camperi.*

A. esculenta Zambia, Botswana
Stemless. Leaves to 40cm(16in) or more, lanceolate, grey-green, sometimes reddish flushed, thickly white-spotted. Flowers reddish-orange.

A. ferox South Africa
Stem 2–3m(6½–10ft) or more, unbranched. Leaves

Aloa ciliaris

bronzy-green, lanceolate, to 1m(3ft) long, spiny on back and margins. Flowers rich orange, about 3cm(1¼in), densely borne in branched racemes.

A. fosteri Transvaal
Usually stemless. Leaves linear-lanceolate, 40–50cm(16–20in), dark grey-green with longitudinal pale striping. Flowers 3–4cm(1¼–1¾in) long, orange-red and yellow.

A. humilis South Africa (Cape)
Stemless, forming clumps by offsets. Leaves 10–15cm(4–6in) long, linear-lanceolate, glaucous, with prominent whitish, almost transparent spine-like teeth to 3mm(⅛in) long. Flowers coral red with green markings, borne in a simple raceme 25–40cm(10–16in) tall. *A.h.* 'Globosa' is a smaller plant with blue-green incurved leaves.

A.h. echinata
Leaves warted below, softly spiny above.

A. jucunda Somalia
Stemless, or sometimes very shortly stemmed, eventually forming dense clumps. Leaves broadly ovate, 4cm(1½in) long, spreading to recurved, dark green above with pale green to white transparent spots, margins bearing red-brown, sharp teeth. Flowers 2cm(¾in) long, pink, in a simple raceme about 30cm(1ft) tall.

A. krapohliana South Africa
Stemless. Leaves incurved, grey-blue, 6–10cm(2½–4in) long with reddish margins. Flowers scarlet, tipped green.

A. latifolia (*A. saponaria latifolia*) South Africa (E. Cape)
Much like *A. saponaria*, but more robust, the leaves larger and greener with less numerous and more elongated spots.

A. marlothii Botswana to Natal
2–4m(6½–13ft) tall in the wild. In a container it has a handsome rosette of broadly lanceolate, pale or bluish-green leaves 50–60cm(20–24in) long, studded with small, reddish spines. Flowers orange, carried in branched racemes.

A. mitriformis South Africa
Stem sprawling or ascending at the tips. Leaves somewhat incurved, ovate-lanceolate, green to

Aloe succotrina

blue-green, 15–20cm(6–8in) long. Flowers 4–5cm($1\frac{1}{2}$–2in) long, scarlet, carried in loose panicles on stems 38–60cm(15–24in) in height.

A. ortholopha S. Zimbabwe
Stemless; rosettes densely leafy. Individual leaves lanceolate to narrowly ovate, more or less erect, up to 55cm(22in) long, with a horny, toothed margin, glaucous beneath. Flowers about 4cm($1\frac{1}{2}$in) long, yellowish-red, in a panicle to 90cm(3ft) tall.

A. parvula Central Malagasy
Stemless, leaves spreading to almost prostrate, 7–12cm($2\frac{3}{4}$–$4\frac{3}{4}$in) long, lanceolate, gradually tapering to a slender point, grey-green, with soft, whitish teeth and tubercles, the latter particularly on the upper surface. Flowers 2cm($\frac{3}{4}$in) long, red, in a simple loose raceme 20–30cm(8–12in) tall.

A. plicatilis South Africa, Cape
Shrubby (to small tree-forming in the wild), 1–5m(3–16ft) tall; stems branching by repeated forking (dichotomous), each one terminated by a tuft of leaves arranged in two parallel ranks (distichous). Each leaf is smooth and strap-shaped with a rounded tip, green to grey-green, about 30cm(1ft) long. Flowers about 2.5cm(1in) long, red, in dense, simple racemes.

A. polyphylla Lesotho
Stem short, usually less than 10cm(4in) tall; rosettes solitary or clustered. Leaves broadly lanceolate, tapering to a tough, red-brown tip; upper surface grey-green with a cartilaginous edge set with a few 5–8mm($\frac{3}{16}$–$\frac{1}{3}$in) long, pale, triangular teeth. Flowers 4cm($1\frac{1}{2}$in) long, green with purple tips.

A. rauhii Madagascar
Very short stem bears rosettes of triangular lanceolate leaves, 7–10cm($2\frac{3}{4}$–4in) long, grey-green, with pale, broken stripes giving a mottled effect. Flowers 2.5cm(1in) long, rose-scarlet, carried in simple racemes.

A. saponaria South Africa
Leaves 15–20cm(6–8in) from a short, suckering stem, blue-green, sometimes tinged with red, having wavy cross-bands of oblong, matt-white spots. Flowers drooping, orange-red, carried in a dense head.

A. striata Coral aloe South Africa
Almost stemless. Leaves grey-green, sometimes red-tinged, 40–50cm(16–20in) long, with completely smooth surfaces and margins; they are longitudinally striped with paler green. Flowers coral-red, borne in branched racemes.

A. succotrina South Africa
Almost stemless. Leaves to 50cm(20in) long, blue to grey-green with faint paler striping and spotting. Flowers red, tipped green, 3.5cm($1\frac{1}{3}$in) long, carried in simple racemes.

A. suzannae Malagasy
Eventually forming a thick stem. Leaves leathery to 1m(3ft). Flowers 3cm($1\frac{1}{4}$in) long, white, tinged with pink, in a long, cylindrical raceme.

A. transvaalensis Transvaal
Stemless, forming small clumps. Leaves narrowly lanceolate, 20–25cm(8–10in) long, dull pale green with oval white spots in transverse bands. Flowers pink to coral, 1.5cm($\frac{5}{8}$in) long, borne in branched racemes.

A. variegata South Africa
Clump-forming, eventually forming a 15cm(6in) stem. Leaves 11cm($4\frac{1}{2}$in) long, triangular-ovate, keeled and closely overlapping, dark green, strongly cross-banded with white spots. Flowers tubular, salmon-red, in loose racemes.

A. vera See *A. barbadensis.*

A. zebrina S.W. Africa
Eventually forming dense clumps. Leaves 15–30cm(6–12in) long, linear-lanceolate, deep green overlaid with a powdery, glaucous patina and marked with whitish dots arranged in irregular transverse bands; margins with brown-tipped, sharp horny teeth. Flowers red, inflated at the base, well spaced in branched racemes, borne in spring and summer.

ALOINOPSIS

Aizoaceae

Origin: South Africa. A genus of 16 species of succulents similar to *Titanopsis*. They are small tufted plants with a tuberous rootstock and thick fleshy, keel-tipped leaves shaped or marked to blend with their rocky habitats. Attractive daisy-like flowers, which only expand either in the afternoon or at dusk, are borne just above the leaf tips. Propagate by seed or division in spring. They can be grown in the home if kept cool in winter. *Aloinopsis* derives from *Aloe*, a genus of succulents in the Liliaceae family, and *opsis*, like – a curious name as there is really no resemblance at all!

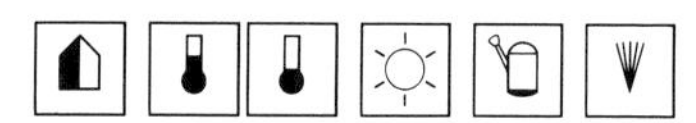

Species cultivated

A. malherbei (*Nananthus malherbei*) Cape
Leaves very broadly spathulate, erect, to 2.5cm(1in) long and almost as broad, grey-green with bold white tubercles on the margins and obscure ones on the upper and lower surfaces. Flowers 2.5cm(1in) wide, flesh-coloured to pinkish-brown, opening in autumn. Likes a limy soil.

A. peersii (*Cheiridopsis noctiflora, Nananthus soehlemannii*) Cape (Karroo region)
Leaves glaucous, narrowly ovate-spathulate, to 2cm($\frac{3}{4}$in) long, recurved. Flowers 2.5cm(1in) wide, yellow, opening in autumn after sunset.

A. rosulata Cape (Willowmore Division)
Leaves spathulate, up to 3cm($1\frac{1}{4}$in) long, deep lustrous green, the margins of the tips with whitish tubercles. Flowers yellow, 3–3.5cm($1\frac{1}{4}$–$1\frac{1}{3}$in) wide, borne in autumn.

A. setifera (*Titanopsis setifera*) Cape (Namaqualand)
Leaves narrowly spathulate, about 2cm($\frac{3}{4}$in) long, bristle-toothed at the tips and uniformly covered with closely set tubercles, glaucous to red-tinted. Flowers 2.5cm(1in) wide, yellow, in autumn.

A. spathulata (*A. crassipes, Titanopsis crassipes, T. spathulata*) Cape (Sutherland Division)
Leaves 2cm($\frac{3}{4}$in) or more long, very broadly spathulate, grey-green, covered with minute tubercles, particularly at the terminal, often red-tinted margin. Flowers about 3cm($1\frac{1}{4}$in) wide, deep pink, paler in bud, opening in autumn.

ALONSOA

Scrophulariaceae

Origin: South America; chiefly from the Andean region. A genus of six species of sub-shrubs and perennials having 4-angled stems, leaves held in opposite pairs or in whorls of three, and flowers in terminal racemes. The flowers are curious in that their stems are twisted so that each bloom is held upside down. Plants are best grown in a conservatory, but also tolerate home conditions. Propagate by seed in spring. *Alonsoa* is named for Alonzo Zanoni, Secretary of State of Colombia in the 18th century when it was a Spanish colony.

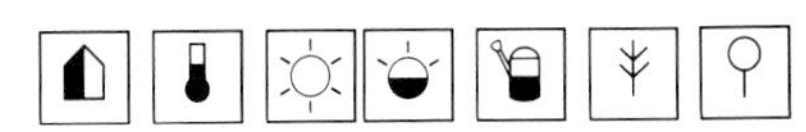

Species cultivated

A. warscewiczii (*A. compacta*) Mask flower
Erect, bushy perennial, 45–60cm($1\frac{1}{2}$–2ft) in height. Leaves cordate lanceolate, doubly toothed. Flowers in racemes, scarlet, 2–2.5cm($\frac{3}{4}$–1in) wide with five broad corolla lobes. Several cultivars are grown including some which are dwarf in stature, not exceeding 22cm(9in) in. Can be grown as an annual from a late winter or early spring sowing.

ALOPHIA See HERBERTIA
ALOYSIA See LIPPIA

ALPINIA

Zingiberaceae

Origin: Tropical and sub-tropical Asia and Polynesia. A genus of 250 handsome species of mainly evergreen, rhizomatous perennials. They form clumps of erect stems clad with lanceolate to ovate, simple leaves. The inflorescence, which can be a raceme or panicle, usually terminates the stem, but in some cases it is borne on a much shorter, leafless stem direct from the rhizome. Individual flowers are somewhat orchid-like in appearance and may or may not be sheltered by a colourful bract. Propagate by division in early summer. *Alpinia* commemorates the Italian Prospero Alpino (1553–1616), Professor of Botany at Padua.

Species cultivated

A. calcarata Indian ginger India
Stems clustered, slender, 1–1.5m(3–5ft) tall.

ABOVE LEFT *Aloinopsis setifera*
ABOVE *Alonsoa warscewiczii*

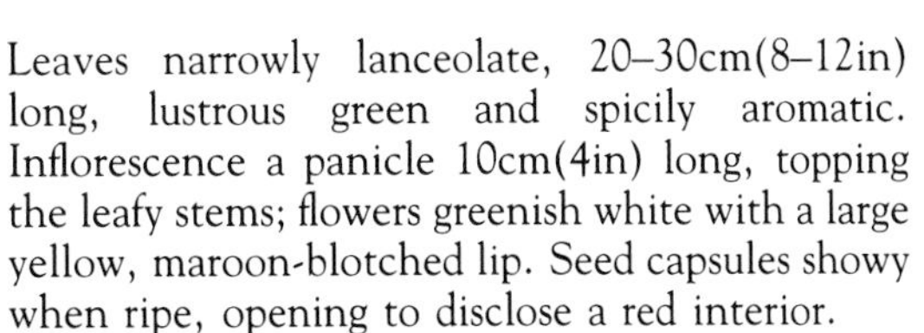

Leaves narrowly lanceolate, 20–30cm(8–12in) long, lustrous green and spicily aromatic. Inflorescence a panicle 10cm(4in) long, topping the leafy stems; flowers greenish white with a large yellow, maroon-blotched lip. Seed capsules showy when ripe, opening to disclose a red interior.
A. magnifica See *Nicolaia elatior.*
A. nutans See *A. zerumbet.*
A. purpurata Red ginger Pacific Islands
Stems colony forming, 1.5–2m(6–$6\frac{1}{2}$ft) tall. Leaves lanceolate, to 60cm(2ft) or so long, lustrous rich green. Inflorescence a raceme 30–90cm(1–3ft) long, erect at first, elongating and arching over with age. Flowers white, small, completely hidden by very much larger, glossy, bright red bracts.
A. rafflesiana Malaysia
Stems clustered, 1.2–1.8m(4–6ft) tall. Leaves narrowly lanceolate, to 60cm(2ft) long, shortly hairy. Inflorescence compact, about 9cm($3\frac{1}{2}$in) long terminating the leafy stems. Flowers yellow and red, about 2.5cm(1in) long.
A. sanderae Variegated ginger New Guinea
Stems clustered, slender, 45–60cm($1\frac{1}{2}$–2ft) tall. Leaves to 20cm(8in) long, lustrous pale green, boldly striped and edged with pure white. Rarely, if ever, blooming as a pot plant. The status of this plant is uncertain but it may be a dwarf, variegated

Alpinia sanderae

ABOVE *Alpinia zerumbet*
RIGHT *Alstroemeria pelegrina*

form of *A. rafflesiana*.

A. speciosa See *A. zerumbet*.

A. zerumbet (*A. speciosa*, *A. nutans*) Shell ginger, Pink porcelain lily E. Asia
Stems forming clumps, 1.5–3cm(5–10ft) tall. Leaves lanceolate, to 60cm(2ft) long. Inflorescence a pendulous raceme 20–30cm(8–12in) long terminating the leafy stems. Flowers with a bell-shaped, white, purple-red-tipped calyx and a 2.5–4cm(1–1½in) long, yellow lip petal spotted and striped red; fragrant.

ALSOPHILA See CYATHEA

ALSTROEMERIA

Alstroemeriaceae

Origin: Warm to temperate areas of South America. A genus of 50 species of tuberous-rooted perennials. They are erect, with lanceolate leaves which have a twisted stalk and are therefore held upside down. The flowers have six tepals, the three inner narrower than the three outer; they are often streaked or patterned with contrasting colours and are carried in terminal clusters. Best in the conservatory. Propagate by seed or by division in spring. *Alstroemeria* was named for Baron Claus Alstroemer (1736–94), a friend of Linnaeus.

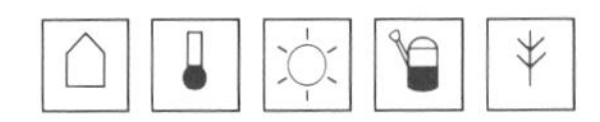

Species cultivated

A. brasiliensis Chile
To 1.2m(4ft) in height. Leaves on non-flowering shoots lanceolate; on flowering ones, linear, to 10cm(4in). Flowers yellow, red-flushed, to 4cm(1½in) long, the inner tepals spotted with brown.

A. pelegrina Chile
30–100cm(1–4ft) in height. Leaves lanceolate 5cm(2in) long. Flowers either solitary or in small clusters, crimson to lilac-purple, the upper tepals paler to almost white with purple-red streaking. *A.p.* 'Alba' (Lily of the Incas) has pure white flowers in clusters.

A. psittacina See *A. pulchella*.

A. pulchella Brazil
To 1m(3ft), with oblong-spathulate leaves. Flowers crimson, 4cm(1½in) long, tipped with green, the upper tepals brownish, spotted beneath with mahogany red.

A. violacea Chile
30–60cm(1–2ft) tall, with 2–5cm(¾–2in) long leaves. Flowers mauve-purple about 5cm(2in) across, the upper tepals spotted with deep purple.

ALTERNANTHERA

Amaranthaceae

Origin: World-wide tropics and sub-tropics. A genus of 200 species, mostly perennials. They can be tufted and mat-forming, or bushy and erect, having opposite pairs of narrow leaves which are often colourfully patterned. Flowers are tiny and insignificant. Propagate by cuttings or division in spring. Plants are best in a conservatory, but can also be grown in the home. *Alternanthera* derives from the Latin *alternans* and *anthera*, anthers, alluding to the alternate barren anthers.

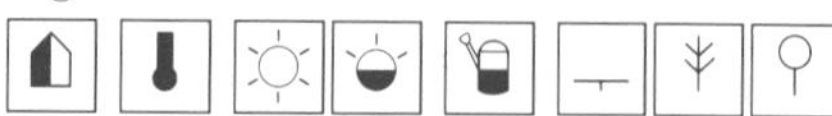

Species cultivated

A. amoena Brazil
Sometimes included in *A. ficoidea*. Mat or hummock-forming, with lanceolate to elliptic leaves 2.5–7.5cm(1–3in) long, veined and blotched red, orange or yellow.

A. bettzickiana Calico plant Brazil
Also sometimes included in *A. ficoidea*. Somewhat sub-shrubby to 30cm(1ft) or more in height. Leaves narrowly spathulate, olive-green and red. *A.b.* 'Aurea' ('Aurea Nana') has bright yellow leaves.

A. dentata (*A. ramosissima*) Northern South America, West Indies
Erect and bushy, 30–60cm(1–2ft) tall. Leaves ovate, pointed-tipped, up to 9cm(3½in) long. Flower heads white to greenish-white, about 2.5cm(1in) long, not showy, but more conspicuous than in the other species described here. *A.d.* 'Rubiginosa' has red-purple leaves.

A. ficoidea Parrot leaf Brazil
Mat-forming with broadly lanceolate, wavy margined leaves which are green, variously marked with red and purple.

A. ramosissima See *A. dentata*.

A. versicolor Brazil
Erect and bushy to 30cm(1ft), sometimes more. Leaves obovate, bronze-green, shaded and margined with red, pink and sometimes yellow.

AMARANTHUS

Amaranthaceae

Origin: Tropical, sub-tropical and temperate regions of the world. A genus of 60 species of annuals. Many are weedy and coarse-growing, but a few make handsome subjects for the summer conservatory, these having boldly coloured foliage or unusual tail-like clusters of densely packed, tiny flowers. Those described are erect plants with lanceolate to ovate, alternate leaves. Propagate by seed sown in spring. *Amaranthus* derives from the Greek *amaranthos*, unfading, the flower clusters of some species retaining their colour like everlastings.

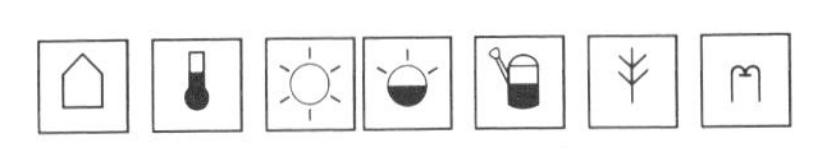

Species cultivated

A. caudatus Love-lies-bleeding Found throughout the tropics but probably originating from South America
Leaves ovate, green in the true species. Flowers red in clustered, pendent, catkin-like tails, to 40cm(16in) or more in length, opening through summer and autumn. Hybrids include:
'Viridis', pale green flowers.
'Green Balls', ball-like clusters of green flowers.

A. gangeticus See *A. tricolor*.

A. hypochondriacus (*A. hybridus*) Prince's feather Tropical America
Leaves oblong-lanceolate, green, up to 15cm(6in) long. Flowers in an erect panicle, red, the individual clusters arching at the tips. *A.h. erythrostachys* with reddish-purple leaves is the form normally grown.

A. tricolor Tropics
Leaves to 20cm(8in), ovate and flushed with red, yellow or bronze. Flowers inconspicuous, hidden amongst the foliage. Hybrids include:
'Joseph's Coat', leaves patterned in red, golden-yellow, orange and dark green.
'Molten Fire', leaves scarlet, turning to copper.
'Salicifolius', leaves lanceolate, drooping, orange-red and bronze.

Alternanthera ficoidea

× AMARCRINUM

(× CRINDONNA, × CRINODONNA)

Amaryllidaceae

Origin: A bigeneric garden hybrid between *Amaryllis belladonna* and *Crinum moorei*, raised independently by Mr Howard of Los Angeles, USA and Dr Ragioneri of Florence, Italy. A bulbous-rooted perennial, it blends the characters of both parents, making a very decorative pot plant. When in bloom it can be brought from the conservatory in to the home. Propagate by offsets in spring.

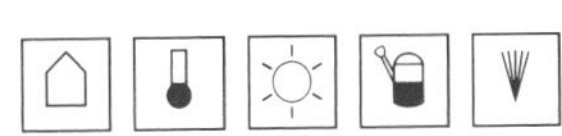

Species cultivated

× ***A. memoria-corsii*** (× *A. howardii*, × *Crinodonna corsii*)
Bulb long-necked, to 10cm(4in) in diameter. Leaves to 60cm(2ft) long, evergreen, strap-shaped, tapering to a point. Flowers funnel-shaped, to 10cm(4in) wide, pink with a white base, carried in umbels of four to 12 on stout stems 60–90cm(2–3ft) tall in autumn.

AMARYLLIS

Amaryllidaceae

Origin: South Africa. A genus of one single bulb species. (The popular pot amaryllis are all correctly *Hippeastrum* species, *q.v.*). It is best grown in the conservatory, but can be brought in to the home when in bloom. Propagate by offsets at planting

Amaryllis belladonna

time. *Amaryllis* was the name of a beautiful Greek shepherdess.

Species cultivated

A. belladonna Belladonna lily
Strap-shaped, deciduous leaves 45–75cm(1½–2½ft) long, medium to deep green, developing after the flowers have faded. Stout, solid stems to 75cm(2½ft) bear one to four pink, trumpet-shaped flowers 10–15cm(4–6in) wide, in late summer or autumn.
A. equestre See *Hippeastrum equestre*.
A. formosissima See *Sprekelia formosissima*.
A. reticulata See *Hippeastrum reticulatum*.

AMBYLOPETALUM See OXYPETALUM

AMICIA

Leguminosae

Origin: Warm temperate areas of Mexico to South America. A genus of eight species of shrubs with pliant stems and alternate, pinnate leaves. Pea flowers are borne in short racemes from leaf axils. The species described is grown chiefly as an unusual foliage plant. Propagate by heel cuttings from young shoots in late spring to summer. *Amicia* was named for Giovanni Battista Amici (1786–1863), Professor of Astronomy and Microscopy at Florence.

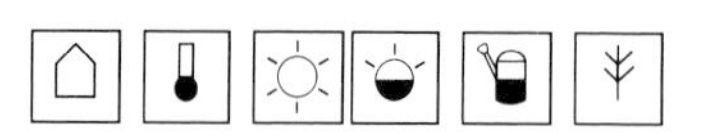

ABOVE RIGHT *Amicia zygomeris*
RIGHT *Amomum cardamomum*

Species cultivated

A. zygomeris Mexico
Erect, deciduous shrub to about 2m(6½ft). Leaves pinnate, of four or six obcordate, 5cm(2in) long leaflets, with two large, inflated, leaf-like stipules at the base. Yellow and purple, 2.5cm(1in) long flowers appear in autumn.

AMOMUM

Zingiberaceae

Origin: Old world tropics. A genus of 150 species of evergreen, rhizomatous perennials, one of which is cultivated for its decorative foliage – a surprisingly durable pot plant for the home. Propagate by division. *Amomum* is Greek for an Indian spice.

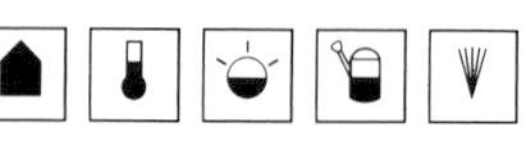

Species cultivated

A. cardamomum (*A. compactum*) Java
Clump-forming with leafy, cane-like stems to 90cm(3ft) or more, bearing stalkless, linear-lanceolate leaves to 25cm(10in) long. The flowers, which arise from the rhizome, are tubular, pale yellow, 1.5cm(⅝in) long and open in summer; they are rather hidden among the leaves.

AMORPHOPHALLUS

Araceae

Origin: Tropical regions of Africa and Asia. A genus of 100 species of unusual deciduous perennials characterized by large corm-like rhizomes, long-stalked, normally solitary, palmate leaves and small petalless flowers which are carried on a fleshy spadix which is surrounded by a large, spreading arum-like spathe. Propagate by removing any offsets when dormant, or by seed in spring. *Amorphophallus* derives from the Greek *amorphos*, shapeless or deformed, and *phallus*, penis, referring to the tip (appendix) of the flowering spadix.

Species cultivated

A. bulbifer India
Leaves trifoliate to 30cm(1ft), pinnately lobed, appearing after the flowers on a 1m(3ft) olive-green and whitish marbled stalk. Spathe 30cm(1ft) high, reflexed, flesh-pink inside, pinky-white marbled with olive-green outside. The pinkish spadix is truncheon-shaped. Spring flowering.
A. rivieri Devil's tongue S.E. Asia
Leaves trifoliate with an oblong, slender pointed blade, on a 75cm(2½ft) white leaf stalk, spotted with bronze-green. Spathe 20–30cm(8–12in), purple inside, pale green outside with a purple spadix. When mature the flower smells of rotten meat and cannot be kept indoors. It is cultivated, apart from its curiosity interest, for its handsome foliage.

AMPELOPSIS

Vitidaceae

Origin: Central and East Asia, North America. A genus of about 20 species of deciduous, climbing shrubs, with pinnate, palmate leaves and twining tendrils. The flowers are small and greenish and carried in clusters, later developing into grape-like berries. Propagate by cuttings in late summer, or by seed when ripe. *Ampelopsis* derives from *ampelos*, a vine and *opsis*, like; it is closely allied to the grape vine.

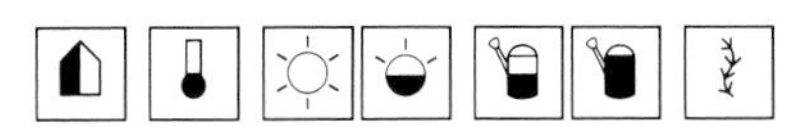

Species cultivated

Ampelopsis brevipedunculata (*A. heterophylla*)
Porcelain berry N.E. Asia
To 5m(16ft) or more. Leaves broadly ovate-cordate, 3- to 5-lobed, 6–13cm($2\frac{1}{2}$–5in) long. Flowers small, greenish followed by bright blue fruits. Hardy. For the home the following form is normally seen:
'Elegans' ('Tricolor', 'Variegata'), very much smaller, leaves densely variegated with white, sometimes with a pink tinge; can be restricted successfully in a small pot making a surprisingly good house plant.

Amorphophallus bulbifer

ANACAMPSEROS

Portulaceae

Origin: Dry areas of South Africa (one species from Australia). A genus of 71 species of xerophytic and succulent plants which are small, slow-growing and often with tuberous roots. The leaves are fleshy, often with long, hair-like stipules from their axils – in some species the smaller leaves are hidden by large, membranous stipules. Often showy, but short-lived, 5-petalled flowers open only in sunshine, sometimes for only a few hours. Propagate by seed or cuttings in spring or summer. *Anacampseros* derives from the Greek *anakampseros*, probably for a kind of stonecrop which, reputedly, brought back love when touched.

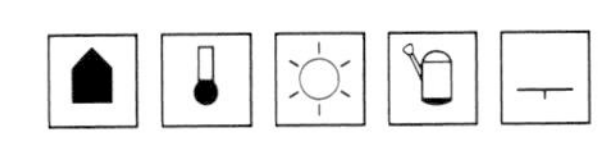

Species cultivated

A. albissima Cape to Namibia
Stem decumbent to prostrate forming small mats to 10cm(4in) wide. Leaves closely packed, minute, covered with white membraneous stipules. Flowers cream to white, about 6mm($\frac{1}{4}$in) or more wide.
A. karasmontana South Africa
About 2cm($\frac{3}{4}$in) tall. Leaves wedge-shaped, sometimes reddish, 1.2cm($\frac{1}{2}$in) long, stipular hairs curled. Flowers pink, 2cm($\frac{3}{4}$in) wide.
A. namaquensis South Africa
To 13cm(5in) tall. Leaves, pear- or wedge-shaped, white-felted, 4–5mm($\frac{3}{16}$–$\frac{1}{4}$in) long; stipular hairs yellowish. Flowers white, 8–10mm($\frac{1}{3}$–$\frac{3}{8}$in) wide.
A. papyracea Cape to Namibia
Stems prostrate or decumbent, about 5cm(2in) long, in clusters or small clumps. Leaves crowded, tiny, covered by shining white membranous stipules. Flowers very small, greenish-white, insignificant.
A. rufescens South Africa
Stems 5–8cm(2–3in), erect or prostrate. Leaves obovate to ovate-lanceolate, to 2cm($\frac{3}{4}$in) long, red or bronzed-flushed in good light; stipular hairs white, often wavy, almost 2cm($\frac{3}{4}$in) long. Flowers purple or pink, 3–4cm($\frac{1}{8}$–$\frac{3}{16}$in) wide, on stalks to 10cm(4in) long.
A. telephiastrum Cape (widespread)
Mat-forming with age, then composed of a number

Anacampseros papyracea

Ananas bracteatus 'Striatus'

of rosettes. Leaves thickly fleshy, ovate, about 2cm(¾in) long, smooth, green, with bristly hairs. Flowers 3–4cm(1¼–1½in) wide, rose-carmine, solitary or two to four on a stem, well above the leaves.

ANANAS

Bromeliaceae
Pineapples

Origin: Tropics of Central and South America. A genus of five perennial evergreen species. In the wild, unlike most bromeliads, these plants grow on the ground rather than being epiphytic. They are clump-forming with erect rosettes of sword-shaped, spiny-edged leaves. 3-petalled, red or blue flowers are borne in spikes and are followed by fleshy fruits which fuse to form a syncarp – the familiar pineapple. The species described are well-tried and durable pot plants. Propagate by suckers or by the leafy shoots which cap the fruits (slice off and plant in a propagating frame at 21–26°C[70–80°F]). *Ananas* is the South American (Tupi) name for these plants.

Anchusa capensis

Species cultivated

A. bracteatus Wild/Red pineapple Brazil, Paraguay
Leaves dark green, to 1.2m(4ft) in the wild, less in pots, with well spaced marginal spines. Flowers lavender with pink-red bracts on the flowering stems and beneath the florets. Fruits brownish-red. *A.b.* 'Striatus' ('Tricolor') has leaves margined with yellow and a bronze-green centre with red spines.
A. comosus (*A. sativus*) Pineapple Brazil
Shorter, channelled leaves to 90cm(3ft), less in pots, grey-green with small, closely set marginal spines. Flowers purple-blue with green bracts. Improved forms are commercially grown for their fruits, and variegated forms for ornament. *A.c.* 'Porteanus' is olive-green with a central yellow stripe. *A.c.* 'Variegatus' (Ivory pineapple) has broadly creamy-margined leaves.
A. sativus See *A. comosus*.

ANASTATICA

Cruciferae

Origin: Morocco to S. Iran. A genus of only one species. It is an annual which, in the wild, breaks free from the ground when it dies and blows around like a tumble weed, shedding its seeds as it goes. It has the curious property of opening flat again when wet, being hydroscopic. Propagate by seed sown at a temperature of about 18°C(64°F). This is essentially a fun plant of curiosity value only. *Anastatica* derives from the Greek *anastasis*, resurrection, because of its apparent return to life when wet.

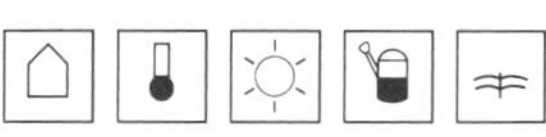

Species cultivated

A. hierochuntica Rose of Jericho, Resurrection plant
Prostrate to semi-erect; branches to 15cm(6in) long. Leaves obovate, stellate hairy. Flowers small, white, 4-petalled in short spikes in summer.

ANCHUSA

Boraginaceae

Origin: Europe, W. Asia, Africa. A genus of 50 species of annuals and perennials, those in cultivation being frost hardy plants for the open garden. One annual species, however, makes a colourful pot plant to enliven the winter, spring and summer conservatory. For winter blooming a minimum night temperature of 7–10°C(45–50°F) must be maintained. Propagate by seed sown in warmth in September and January or February. *Anchusa* derives from the Greek *Ankousa*, alkanet, some species yielding the red dye rouge.

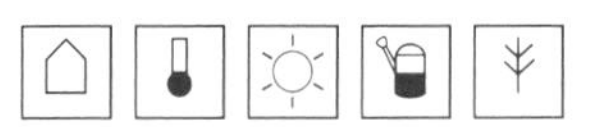

Species cultivated

A. capensis South Africa
Erect branching habit, 30–60cm(1–2ft) tall. Leaves narrowly lanceolate, 10cm(4in) or more long. Flowers tubular with five petal lobes, 6mm(¼in) wide, profusely borne in terminal cymes. Blue is the basic colour but the throat of each flower

is white and the petals are red-edged. *A.c. alba* is white and 'Blue Bird' a clear indigo blue.

ANEMIA (ANEIMIA)

Schizaeaceae

Origin: Widespread in tropics and sub-tropics. A genus of 90 species of small to medium-sized evergreen ferns. Tufted or creeping in habit, they have simply bi- or tripinnate leaves (fronds). Fertile leaves have the lowest pair of leaflets (pinnae) transformed into an elongated stalk bearing a panicle-like structure of sporangia which resemble tiny flower buds, hence one of their vernacular names – the flowering fern. Propagate by spores or division in spring. *Anemia* derives from the Greek *aneimon*, naked and *heima*, clothing, alluding to the unprotected sporangia.

 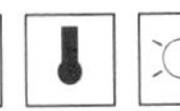  

Species cultivated

A. adiantifolia Pine fern Tropical South America to Florida, USA
Rootstock creeping. Leaves 45–90cm($1\frac{1}{2}$–3ft) tall, bi- or tripinnate, the segments (pinnules) oblong-obovate to narrowly wedge-shaped.
A. dregeana South Africa
Leaves 20–30cm(8–12in) long; individual pinnae 2.5–4cm(1–$1\frac{1}{2}$in) long, ovate-deltoid.
A. phyllitidis Mexico to Brazil
Tufted to clump-forming. Leaves 30–60cm(1–2ft) tall, simply pinnate, the individual leaflets ovate-oblong, 5–15cm(2–6in) long, firm-textured and somewhat lustrous. A semi-xerophyte in the wild, this is the best for the dry atmosphere of the home.
A. rotundifolia Brazil
Basically clump-forming, but when grown in a conservatory border the leaf tips root and produce plantlets, creating a colony. Leaves simply pinnate, 25–45cm(10–18in), arching; leaflets somewhat rounded, generally bluntly rhombic to fan-shaped.

ANEMONE

Ranunculaceae

Origin: Cosmopolitan, but most frequent in the northern hemisphere. A genus of 150 rhizomatous-rooted perennial species, their rootstocks being either slender and erect or irregularly tuber-like. The leaves are lobed or compound. The brightly coloured flowers are actually petalless, having instead five to 20 showy, petal-like sepals. Flower stems are erect and bear a characteristic whorl or involucre made up of three to four, usually dissected, green bracts that may be sepal- or leaf-like. Plants are best in a conservatory, but may be brought in to the home when in bloom. Propagate by division or by seed sown, preferably, as soon as ripe. *Anemone* has been long believed to derive from the Greek *anemos*, wind (hence common name wind flower), but it is now thought to be a Greek version of the Semitic *Naamen* (Adonis) whose blood produced *A. coronaria* (or *Adonis*).

Species cultivated

A. coronaria Poppy anemone S. Europe, Turkey
Tuberous-rooted. Leaves trifoliate, deeply lobed and toothed. Stems to 25cm(10in) or more, bearing 5- to 8-tepalled, red, blue, purple or pale yellow flowers 4–7cm($1\frac{1}{2}$–$2\frac{3}{4}$in) across. Often replaced in cultivation by its hybrids with *A.* × *fulgens*.
A.* × *fulgens (*A. pavonina* × *A. hortensis*)
Flowers with 15 narrow sepals, bright red.

Hybrids

Available in every colour except yellow:
De Caen strain, semi-double flowers with a single row of broadly obovate tepals.
St. Brigid, semi-double to double flowers made up of several rows of narrow tepals.

ANEMOPAEGMA

Bignoniaceae

Origin: Tropical America, mainly Brazil. A genus of 40 species of woody stemmed tendril climbers, only one of which is much cultivated. Propagate by cuttings in summer. *Anemopaegma* derives from the Greek *anemos*, the wind and *paigma*, sport, though the intended meaning is somewhat obscure.

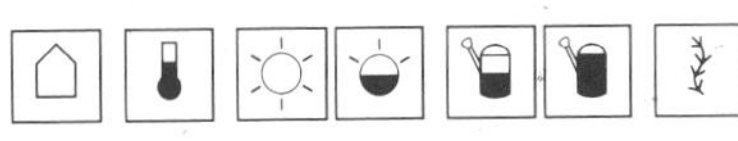

Species cultivated

A. carrerense Guyana, Trinidad, Venezuela
Stems slender, to about 5m(16ft), usually less in containers. The leaves are composed of two lustrous, ovate, pointed-tipped leaflets, each one 5–7.5cm(2–3in) long and a trifid tendril arising between them. Flowers about 5cm(2in) long, pale yellow, paler on the five broad lobes, carried in

TOP *Anemia dregeana*
ABOVE CENTRE *Anemone coronaria* cv.
ABOVE *Anemopaegma carrerense*

short axillary racemes, mainly in summer. Seldom cultivated but well worth seeking out.

A. chamberlaynei (*Bignonia chamberlaynei*) Brazil

Stems to 6m(20ft) or so, less in containers, fast growing. Leaves composed of two ovate, slender-pointed, somewhat wavy leaflets, each up to 10cm(4in) long. The tendril is the elongation of the leaf midrib and grows between and beyond the leaflets, terminating in three hooks. Flowers primrose-yellow, foxglove-shaped, 6–7cm(2½–2¾in) long, carried in pairs in the upper leaf axils, mainly in summer. An attractive plant, occasionally confused with *Doxantha unguis-cati* (*q.v.*).

ANGELONIA

Scrophulariaceae

Origin: Mexico and West Indies to Brazil. A genus of 30 species of perennials, sub-shrubs and annuals. Those described below are perennials, but are best grown as annuals. They have pairs of simple leaves and terminal racemes of flowers which, though smaller, are like those of monkey musk (*Mimulus*) when viewed from the front. These plants are excellent for the conservatory and can be brought in to the home when in bloom. Propagate by seed or by cuttings (from cut back plants) in spring. *Angelonia* derives from the South American vernacular name *Angelon*.

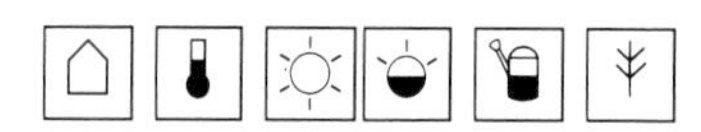

Species cultivated

A. angustifolia Mexico

Stems erect, 30–45cm(1–1½ft) in height. Leaves narrowly lanceolate, to 8cm(3in) long, finely toothed and slender-pointed. Flowers late summer, deep violet, in long racemes. *A.a.* 'Alba' is white.

A. grandiflora See next species.

A. salicariifolia (*A. salicariaefolia*) West Indies and northern South America

Whole plant sticky downy. Stems erect, 60–90cm(2–3ft) tall. Leaves lanceolate to narrowly ovate, variably toothed, pointed, to 8cm(3in) long. Flowers 2cm(¾in) wide, blue-purple or white in long, simple or branched racemes, autumn to early winter. *A.s.* 'Grandiflora' (*A. grandiflora* has larger flowers; 'Grandiflora Alba' is white.

Angraecum sesquipedale

ANGRAECUM

Orchidaceae

Origin: Tropical and sub-tropical areas of Africa, Malagasy, Indian Ocean islands and the Philippines. A genus of 220 species of epiphytic orchids which do not produce pseudobulbs. They generally have erect stems bearing two rows of fleshy, strap-shaped to narrowly ovate leaves. Flowers can be quite large, are usually carried in racemes but are occasionally solitary, and are mostly white and green, or sometimes yellow; they are starry and have a lip with a slender spur which is often elongated to great length. Plants can be grown in pans or baskets and are best in a conservatory since they need more humidity than can normally be provided in the home. Propagate by cuttings of well-grown side shoots in late spring. *Angraecum* derives from the Malaysian name *angurek*.

Species cultivated

A. eburneum (*A. superbum*) Mascarene Is.

Single stemmed, to 2m(6½ft) high. Leaves 90cm(3ft) long, lanceolate-ovate, bi-lobed at the tip. Flowers fragrant, 7cm(2¾in) long, green to greenish-white with a white labellum having a spur 10cm(4in) long; they are borne in two ranks on a stem to 90cm(3ft) or more in height and open in autumn and winter.

A. eichlerianum Tropical West Africa

Slender stems, pendulous or climbing, to 1.2m(4ft) or more in length. Leaves leathery, oblong-elliptic with notched tips. Flowers 8–9cm(3–3½in), yellow-green with a green-marked white labellum having a spur to 4.5cm(1¾in) long; they are usually borne singly, sometimes in twos or threes, are strongly fragrant and appear in autumn to winter.

A. infundibulare Tropical West Africa

Stems climbing or pendulous, to 1.2m(4ft), much more in the wild. Leaves leathery, oblong-elliptic to 10cm(4in). Flowers fragrant, yellow-green to 9cm(3½in) across with a white labellum; they are usually solitary, borne in autumn and winter.

A. rhodostictum (*Aerangis rhodostictum*) Cameroons, Ethiopia, Kenya, Tanganyika

Stems short, usually pendent if grown on a slab of bark or tree fern stem. Leaves to 15cm(6in) long, strap-shaped, bright green, bi-lobed at the tip. Flowering stems arching down, 25–35cm(10–14in) long, bearing up to 20 flowers arranged in two parallel rows; each bloom 2.5–3cm(1–1¼in) wide, white to cream or palest yellow with a red column and green-tipped spurs, borne in winter and spring.

A. sesquipedale Malagasy

Stem erect, robust, to 60cm(2ft) or more. Leaves dark, almost blue-green, to 30cm(1ft) in length, tip bi-lobed, the two lobes of unequal size. Flowers

fragrant, fleshy textured, ivory white to 18cm(7in) across with a heart-shaped lip and 25cm(10in) spur, borne in winter.
A. superbum See *A. eburneum.*

ANGULOA

Orchidaceae

Origin: Tropical South America, chiefly in the Andean forests. A genus of ten species or terrestrial or epiphytic orchids. They are clump-forming with ovoid to oblong pseudobulbs and boldly veined lanceolate leaves. The solitary flowers are erect, rounded to cup-shaped, with broadly overlapping tepals fancifully resembling a tulip. Propagate by division after flowering. Plants are best grown in a conservatory but can be brought in to the home for flowering. *Anguloa* was named for Don Francisco de Angulo, a Spanish botanist of the 18th century.

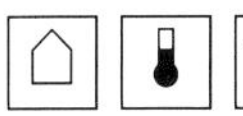    

Species cultivated

A. clowesii Tulip orchid Colombia, Venezuela
Pseudobulbs 11–15cm(4½–6in) tall; leaves broadly oblanceolate, 45–60cm(1½–2ft) long. Flowers 8cm(3in) long, appearing in spring; the tepals, which are pale to bright golden-yellow, having a waxy texture; labellum white to orange-yellow, hinged and boat-like, hairy and intricately lobed.
A. ruckeri Colombia
Similar to *A. clowesii* but somewhat smaller. Flowers fragrant to 9cm(3½in) long, greenish-brown on the outside, usually yellow closely spotted with red within, though sometimes white or blood-red, appearing in spring to summer.
A. uniflora Tulip orchid Colombia to Peru
Somewhat angular pseudobulbs 10–18cm(4–7in) tall; leaves broadly lanceolate, pleated, 45–60cm(1½–2ft) long. Flowers cup-shaped with a waxy texture and a somewhat sickly scent, white to cream, sometimes brown-spotted on the outside, pink-spotted within, appearing in early spring.

ANIGOZANTHOS

Haemodoraceae

Origin: Western Australia. A genus of ten strikingly unusual species of evergreen, clump-forming to tufted perennials with narrowly sword-shaped leaves. The woolly, long-tubular flowers are split at the mouth beneath and open out into six claw-like segments, fancifully like a kangaroo's paw; they are borne in simple or, more usually, branched racemes well above the foliage. Propagate by seed in spring, or by division after flowering or in spring. *Anigozanthos* derives from the Greek *anoigo*, to open and *anthos*, a flower, alluding to the way the tubular blooms are split on one side.

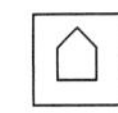  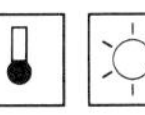  

Anigozanthos manglesii

Species cultivated

A. flavidus Tall kangaroo paw
Clump-forming; stems 90–150(3–5ft) tall or sometimes more. Flowers about 4cm(1½in) long, yellow-green, in well branched racemes, late summer to spring. Red forms are known. A moisture loving species and easily grown.
A. humilis Catspaw
Tufted; stems usually unbranched, 30–45cm(1–1½ft) tall, hairy. Flowers red and yellow, 4cm(1½in) or more long, clustered towards the stem tips in spring. Requires sharply drained sandy soil and careful watering. Probably short-lived.
A. manglesii Common green/Mangle's kangaroo paw
Clump-forming; stems up to 90cm(3ft) tall. Flowers 8cm(3in) long, bright red at the base, the rest vivid green, borne on red stems in branched racemes in spring and summer. The best known species and the floral emblem of Western Australia.
A. preissii Albany cat's paw
Tufted to small clump-forming; stems 45–60cm(1½–2ft), forked at the tips. Flowers to 6cm(2½in) long, clustered towards the stem tips, mainly yellow with an orange-red base, widely expanded at the mouth and the most claw-like in the genus, borne mainly in spring. Needs a sandy, sharply drained soil.
A. rufus Red kangaroo paw
Clump-forming; stems 60–75cm(2–2½ft) tall. Flowers about 6cm(2½in) long, dark red, in branched racemes in spring. Needs sandy, freely draining soil.

A. viridis Green kangaroo paw
Clump-forming; stems 45–60cm(1½–2ft) tall. Flowers about 6cm(2½in) long, bright green, in simple or branched racemes in spring.

ANISODONTEA

Malvaceae

Origin: South Africa. A genus of 19 species of shrubs and perennials, some of which make good, pot plants for the conservatory. The species described here are evergreen, soft-stemmed shrubs with an erect but bushy growth habit and freely produced, small, mallow-like flowers. Propagate by cuttings in summer or early autumn and seed in spring. *Anisodontea* derives from the Greek *anisos*, unequal and *odontion*, a small tooth, possibly a reference to the small projections on the mature carpels of certain species.

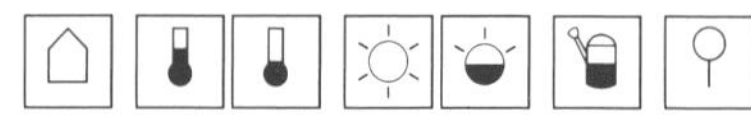

Species cultivated

A. capensis (*Malvastrum capense*)
About 90cm(3ft) high and wide, but easily kept to half this size in containers. Leaves 2–3cm(¾–1¼in) long, broadly ovate, deeply 3- to 5-lobed, the lobes again cut or toothed. Flowers one to three together in the upper leaf axils, about 2.5cm(1in) wide, pale rosy-magenta with a pattern of darker, radiating lines at the base of each petal, borne in spring to autumn.

A. hypomandarum (*A. hypomadarum* of gardens, *Malvastrum hypomandarum*)
Reputedly of hybrid origin. To 1.8m(6ft) or more in the conservatory border. Leaves ovate, to 4cm(1½in) long, usually 3-lobed but sometimes 5-lobed. Flowers 2.5–4cm(1–1½in) wide, white or pink with a pattern of magenta lines at the base of each petal, spring to autumn. Plants under this name in British gardens are usually *A. capensis*.

A. scabrosa (*Malvastrum scabrosum*)
To 1.8m(6ft) tall or more. Leaves linear to broadly ovate, 3- to 5-lobed and up to 8cm(3in) long. Flowers about 4cm(1½in) wide, white to magenta with darker spotting in the centre, spring to autumn. Whole plant usually slightly resinous and aromatic.

ANNONA

Annonaceae

Origin: Tropics, mostly from South America but some species from Africa and Asia; widely cultivated. A genus of 120 species of deciduous and evergreen shrubs and trees. They have alternate, simple, usually lanceolate to ovate leaves, somewhat cup-shaped flowers of three to six fleshy petals and large, many-celled (carpels), heart- or cone-shaped fruits. They are sometimes grown for interest as pot plants from seeds taken from imported fruits, but need large pots or tubs to reach fruiting size. Propagate by seed in spring or by cuttings in summer, at a minimum temperature of 21°C(70°F). *Annona* derives from a S. American Indian name for the custard apple or sow sop.

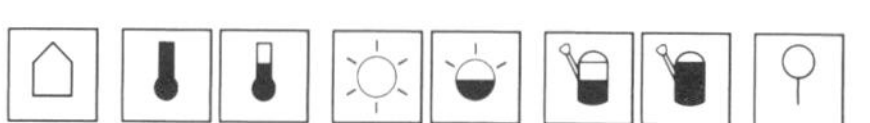

Species cultivated

A. cherimolia Cherimoya Ecuador, Peru
Deciduous tree to 6m(20ft) or so. Leaves broadly lanceolate, brown velvety hairy beneath. Flowers to 3cm(1¼in) long, greenish-yellow to reddish-brown and fragrant. Fruits green, rounded to heart-shaped, 8–15cm(3–6in) in diameter with a pattern of depressions rather like finger prints. Can be grown cooler than the other species (temperate).

A. reticulata Custard apple, Bullock's heart Tropical America
Deciduous tree to 8m(26ft), leaves oblong-lanceolate. Flowers olive-green or yellowish, often purple flushed. Fruits rounded, 7–12cm(2¾–4¾in) in diameter, yellowish-red; pulp sweet but rather insipid. Tropical.

A. squamosa Sweet sop, Sugar apple Tropical America
Deciduous tree to 6m(20ft). Leaves lanceolate. Flowers like those of *A. reticulata*. Fruits yellowish-green, heart-shaped, to 7cm(2¾in) in diameter, covered with rounded, fleshy tubercles. Tropical.

ANOECTOCHILUS

Orchidaceae

Origin: Tropical Asia, Australia, Polynesia. A genus of 25 species of semi-epiphytic or terrestrial orchids, not always easy to grow but when well suited, growing quite rapidly. They are small, having creeping rhizomes and fleshy, yet velvety, oval to elliptic leaves which are often brightly variegated, giving their name of jewel orchids. The flowers are of less significance, being small and borne in an erect terminal spike. Some growers

BELOW *Anisodontea hypomandarum*
BELOW RIGHT *Annona cherimolia*

pinch them out when small to encourage better foliage growth. Propagate by severing rooted rhizome sections in summer. Plants can only be grown successfully in the home if kept in glass or plastic cases (terrariums) which provide the necessary humidity. *Anoectochilus* derives from the Greek *anoiktos*, open and *cheilos*, lip or labellum, referring to the flat or spreading labellum.

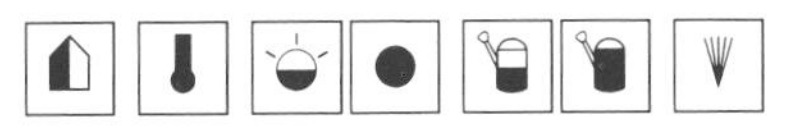

Species cultivated

A. regalis King plant S. India, Ceylon
Leaves 6cm(2½in) long, broadly elliptic and pointed, dark velvety green with a network of golden veins. Flowers 2cm(¾in) long, greenish to white, long-spurred and fringed, several on a 30cm(1ft) stem in spring.

A. roxburghii Himalaya
Leaves 6cm(2½in) long, elliptic and velvety green with the centre of the leaf zoned with gold and the margins suffused red.

A. setaceus Java
Leaves 6cm(2½in) long, broadly elliptic, dark, almost blackish green, with a patterning of silvery white veins.

A. sikkimensis Sikkim
Leaves to 7cm(2¾in) long, green with a red suffusion and gold veins. Flowers olive-green and white, 2cm(¾in) long, opening in autumn.

ANOMATHECA See LAPEIROUSIA

ANOPTERUS

Escalloniaceae

Origin: Tasmania and Eastern Australia. A genus of two species of evergreen shrubs with alternate, leathery leaves and tubular, white flowers in terminal racemes. Propagate by seed or by cuttings in summer. A handsome tub plant for the larger conservatory. *Anopterus* derives from the Greek *ana*, upwards and *pteron*, a wing, referring to the winged seeds.

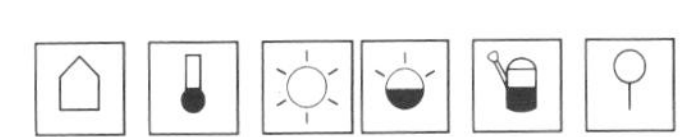

Species cultivated

A. glandulosus Tasmania
Shrub to 2m(6½ft) or so in containers (a small tree sometimes to 12m[40ft] in the wild). Leaves glossy, obovate to oblanceolate, 7–15cm(2¾–6in) long, in dense clusters at the shoot tips. Flowers white, bell-shaped 1.2cm(½in) wide, borne in 15cm(6in) racemes and opening in spring.

ANSELLIA

Orchidaceae

Origin: Tropical and southern Africa. A genus of one very variable species, by some botanists split into two similar species. It is a handsome epiphytic orchid with clusters of leafy, stem-like pseudobulbs. Propagate by division after flowering. *Ansellia* was named for John Ansell who died in 1847, an English gardener who accompanied J.R.T. Vogel and Capt. H.D. Trotter on an expedition to the river Niger, W. Africa in 1841–2.

Anoectochilus sp.

Species cultivated

A. africana (*A. confusa*, *A. gigantea*) Leopard orchid
Robust with leafy pseudobulbs; 60–120cm(2–4ft) tall with arching, lanceolate leaves to 30cm(1ft) in length. Flowers from 7cm(2¾in) wide, long lasting, greenish to bright yellow, usually heavily spotted and barred with chocolate-brown; they are fragrant and borne in many-flowered spikes on simple or branched stems 30–90cm(1–3ft) high during spring and early summer.

ANTEGIBBAEUM See GIBBAEUM
ANTHERICUM See CHLOROPHYTUM
ANTHOLYZA See CHASMANTHE

ANTHURIUM

Araceae

Origin: Humid rain forests of tropical America and the West Indies. A genus of evergreen perennials containing 550 species of which a number are grown for their decorative foliage and flowers. Some are low-growing forming tufts, others climb by aerial roots. The leathery leaves are usually lustrous dark green and can be prettily patterned. The flowers are small, grouped together in a spadix and arise from a flattened spathe which in some species is brightly coloured. This large genus contains many good house plants. Propagate by division or seed in spring. *Anthurium* derives from *anthos*, a flower and *oura*, tail, from the shape of the spadix.

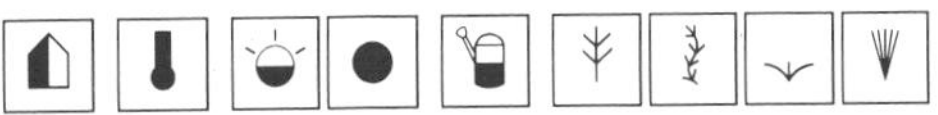

ABOVE *Anthurium crystallinum*
ABOVE RIGHT *Anthurium scherzerianum*
RIGHT *Anthurium andreanum*

Species cultivated

A. andreanum Tailflower Colombia
Erect with a short stem. Leaves long-stalked, oblong-cordate, 15–20cm(6–8in) long, leathery and dark green. Spathe orange-red to scarlet, 10cm(4in) or more long with a rippled surface; spadix yellow. *A.a.* 'Album' has white spathes stained with purple at the base. *A.a.* 'Giganteum' has salmon-red spathes which are much larger than the type. *A.a.* 'Rubrum', has deep red spathes and white spadices with yellow tips.

A. caribaeum See *A. cordifolium*.

A. clarinervium Mexico
Similar to *A. crystallinum*, but with smaller, red-flushed spathes.

A. clavigerum Nicaragua to Brazil and Bolivia
Epiphytic climber to about 2m(6½ft). Leaves digitate or palmatisect, rounded in outline, 1–2m(3–6½ft) wide, each of the seven to 13 segments or leaflets with prominent rounded lobes. Spathes pendent, violet-purple, 30–65cm(12–26in) long, followed by purple berries. An imposing species.

A. cordifolium (*A. caribaeum*) Colombia
Stem short. Leaves ovate-cordate, up to 60cm(2ft) long, lustrous. Spathes narrow, green, about 15cm(6in) long.

A. crassinervium Mexico to Honduras and Venezuela
Rosette-forming epiphyte; stem short and thick. Leaves oblanceolate to obovate, 45–90cm(1½–3ft) or more long, with an elaborate network of prominently raised veins. Spathes 10–15cm(4–6in) long, green; spadix longer and red-purple. Berries red.

A. crystallinum Crystal anthurium Peru, Colombia
Leaves ovate-cordate, 25–38cm(10–15in) long, lustrous deep green, marked with a pattern of silver-white veins. Spathe green, linear-oblong, 9cm(3½in) long, not decorative. Plants with leaves smaller than given above may well be hybrids with the smaller but similar *A. forgetii*.

A. digitatum (*A. pentaphyllum digitatum*) Venezuela
Epiphytic climber to 2m(6½ft), but slow growing. Leaves digitate or palmatisect, rounded in outline, 20–45cm(8–18in) wide, each of the seven to 13 deep green leaflets oblanceolate. Spathes to 7cm(2¾in), narrow and green; spadix longer, purple. Berries red or purple.

A. fissum See *A. palmatum*.

A. forgetii Colombia
Stems short. Leaves ovate-peltate, narrow-pointed, velvety olive-green, the main veins much paler. Spathe very narrow, to 15cm(6in) long, green. Berries white, tinted violet. A small relative of *A. crystallinum* with which it readily hybridizes.

A. grande Bolivia
Another close ally of *A. crystallinum*, but with somewhat longer, narrower leaves.

A. hoffmannii Costa Rica, Panama
Much like *A. regale*, but leaves olive-green and veins more silvery.
A. magnificum Costa Rica
This species differs from *A. crystallinum* in the duller, olive-green colour of its leaves which are on 4-angled stalks, and in its dull purple spathe.
A. palmatum (*A. fissum*) West Indies
Climbing epiphyte with a stem to 4.5m(15ft) in length. Leaves palmate, deeply lobed (palmatisect), 50–80cm(20–32in) wide, the five to 11 lobes or segments oblanceolate and thick-textured. Spathes 6–18cm(2½–7in) long, lanceolate, green. Berries red.
A. pentaphyllum Mexico to Panama and Trinidad to Brazil
Much like *A. digitatum* but leaves composed of only five to 9 leaflets which are usually larger.
A. podophyllum Mexico
Stem short and thick. Leaves palmatisect or digitate, to 90cm(3ft) across, the seven to 11 leaflets are usually pinnately lobed though sometimes with a wavy margin only. Spathes to 8cm(3in) long, green. Berries red.
A. regale Peru
Stem fairly short. Leaves ovate-cordate, 30–45cm(1–1½ft) or more long, dark green with a satiny sheen and pale green veins. Spathes 8–18cm(3–7in) long, lanceolate, green. Spadix longer and whitish.
A. sanguineum Colombia
Much like *A. andraeanum* but with the leaves bluish-green and narrower. Spathe rosy-red; spadix green.
A. scherzerianum Flamingo flower Costa Rica
Leaves erect, short-stalked, deep green, oblong-lanceolate and up to 15cm(6in) or more in length. They set off the broadly ovate 8cm(3in) long spathes which are waxy and bright scarlet contrasting with the yellow, spirally twisted spadix. Cultivars with darker red, white, pink and spotted spathes are grown.
A. veitchii King anthurium Colombia
Leaves cordate, oblong-ovate pendent from a short stalk, deeply quilted, shining dark green, 45–90cm(1½–3ft) in length. Spathe green to ivory-white, 6–8cm(2½–3in) long.
A. warocqueanum Queen anthurium Colombia
In shape the leaves are similar to those of *A. veitchii* but are smooth, with ivory-white veins. Spathe 10cm(4in), green or yellowish, linear-lanceolate; the spadix up to 30cm(1ft) long.
A. watermaliense Costa Rica to Colombia
Stem to 25cm(10in) tall. Leaves 30–60cm(1–2ft) long, ovate-triangular to sagittate, the two basal lobes being longer than broad, thick-textured and a lustrous deep green. Spathes 15–20cm(6–8in) long, triangular-lanceolate, deep purple to blackish, rarely green; spadix to 10cm(4in) long, white, yellow-purple or green. Berries yellow to orange. It is usually represented in cultivation by the black-spathed clone, a striking plant.

ANTIGONON

Polygonaceae

Origin: Tropical America. A genus of eight species of climbing plants, one of which is widely cultivated in the tropics and sub-tropics and deserves to be grown more often in the conservatory. Propagate by cuttings in summer or by seeds in spring. *Antigonon* derives from the Greek *anti*, in place of and *polygonon*, knotweed, denoting its relationship to the genus *Polygonum*.

Species cultivated
A. leptopus Corallita, Coral vine Mexico
A slender, evergreen, semi-woody climber to 6m(20ft) or more, arising from a tuberous (and edible) root. Leaves heart-shaped 2.5–8cm(1–3in) or more long. Inflorescence an axillary raceme or panicle, the stem tips of which bear small, hooked tendrils. Flowers 2cm(¾in) wide, petalless, but with five petal-like sepals in shades of pink or red. *A. l.*

Antigonon leptopus

'Album' is white. Inclined to be shy-flowering under cool conditions, but in tropical temperatures flowers are produced all year.

ANTIRRHINUM

Scrophulariaceae

Origin: Western Mediterranean and western North America. A genus of 30 species of annuals, perennials and sub-shrubs of erect or prostrate habit. They have lanceolate to ovate leaves and racemes of tubular flowers which are pinched to form a mouth opening to five flared lobes. The one species below makes a good pot plant to brighten up the summer conservatory. Propagate by seed in late winter or by cuttings in spring. *Antirrhinum* derives from the Greek *anti*, like and *rhin*, a nose or snout, alluding to the likeness of the flower to the mouth of a dragon.

 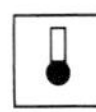 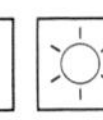 

Species cultivated

A. asarina See *Asarina procumbens*.

A. majus Snapdragon S.W. Europe, but widely naturalized

An erect, woody-based, short-lived perennial usually grown as an annual with ovate to lanceolate leaves to 7cm($2\frac{3}{4}$in) long. Flowers 3–4.5cm($1\frac{1}{4}$–$1\frac{3}{4}$in) long, pink to red-purple. Hardy. Many cultivars are available in colours ranging from white, cream and yellow to pink, red and bronze. The tall, large-flowered sorts such as 'Tetra Snaps' are good for growing as cut flowers. The smaller 'Tom Thumb', 'Magic Carpet' and similar strains are ideal for indoor pot growth – these are often no more than 10–15cm(4–6in) in height. *A.m.* 'Taff's White' with cream-margined leaves makes a decorative foliage plant.

A. maurandioides See *Asarina antirrhiniflora*.

APHELANDRA

Acanthaceae

Origin: Tropical and sub-tropical America. A genus of 20 species of evergreen shrubs including several well-tried house plants. They have opposite pairs of elliptic to ovate, leathery leaves, often veined or mottled grey-silver or white and dense terminal spikes of tubular, 2-lipped flowers, sometimes in the axils of coloured bracts. Propagate by cuttings of young shoots in spring or summer, or by seed in spring. *Aphelandra* derives from the Greek *apheles*, simple and *aner*, male, the anthers having one cell only.

    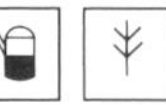

Species cultivated

A. aurantiaca (*A. fascinator*) Mexico

To 90cm(3ft). Leaves broadly ovate, 10–15cm(4–6in) long, dark green with silver-grey veining. Flowers bright orange-scarlet, opening in winter.

A.a. roezlii

More compact with twisted leaves suffused with silver-grey.

A. chamissoniana Brazil

Slender plant to 1.2m(4ft). Leaves elliptic, 7–10cm($2\frac{3}{4}$–4in) long, with a broad, silver-white vein pattern. Flowers and bracts bright yellow in a pagoda-like spike in late autumn and winter.

A. fascinator See *A. aurantiaca*.

A. fuscopunctata Northern South America

To 60cm(2ft) or more with green stems. Leaves to 10cm(4in) or more long, ovate, hairy, dark green above and paler beneath. Flowers coffee-coloured, emerging from sticky-glandular red bracts.

A. ignea See *Chamaeranthemum igneum*.

A. sinclairiana Central America

To 90cm(3ft) or more. Leaves thin-textured, glossy green, prominently veined, 10cm(4in) or more long. Flowers rose pink from between orange bracts; flower spikes clustered at the stem tips.

A. squarrosa Zebra plant Brazil

To 1.2m(4ft) or more. Leaves 15–25cm(6–10in) long, ovate, deep glossy green with contrasting white veins. Flowers and bracts bright yellow, the latter sometimes red-edged, opening from late summer to winter. *A.s.* 'Louisae' is more compact with smaller leaves; a good pot plant. Several cultivars are available, varying in vigour, flower shade and patterning.

A. tetragona West Indies, northern South America

Eventually to 2m($6\frac{1}{2}$ft), of open habit when mature. Leaves broadly ovate, slender-pointed to 15cm(6in) or more long. Flowers 5–8cm(2–3in) long, bright red in long, dense, terminal and axillary spikes. A bold and showy plant.

APONOGETON

Aponogetonaceae

Origin: Africa, Malagasy, Asia and Australia. A genus of 30 species of aquatic perennials with thickened rhizomes, floating or, more usually, submerged leaves and spikes of small flowers held above the water. Propagate by division at flowering time. These are outstanding plants for the home aquarium or conservatory pool. *Aponogeton* derives

BELOW *Aphelandra squarrosa*
BELOW RIGHT *Aphelandra sinclairiana*

from the Latin name *Aqui Aponi*, healing springs, today called Bagni d'Abano, and *geiton*, neighbour; it was originally used for a water plant found there, and later for this genus.

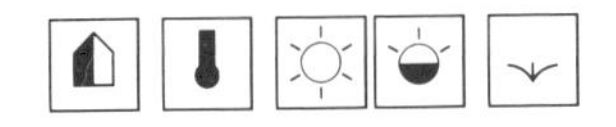

Species cultivated

A. crispus S.E. Asia
Rosette-forming, the whole plant being submerged. Leaves up to 30cm(1ft) long on an unusually short stalk, variable in width – lanceolate to elliptical – usually with wavy or crimped margins, pale to dark green or reddish-tinted. Flowers tiny, white, in a solitary fluffy spike above the water.

A. distachyos Cape pondweed South Africa
Leaves all floating, oblong-elliptic, glossy deep green, 8–15cm(3–6in) long. Flowers glistening white, fragrant in dense forked spikes 5–10cm(2–4in) long in spring to late autumn.

A. ulvaceus Malagasy
Rhizome globular, corm-like. Leaves submerged, 20–35cm(8–14in) long, narrowly oblong, strongly waved and often twisted into spirals, bright translucent lettuce-green. Flower spikes forked, 10–15cm(4–6in) long, bearing numerous minute, off-white flowers.

× APOREPIPHYLLUM
(× APOROPHYLLUM)

Cactaceae

Origin: Hybrid cultivars derived from crossing *Aporocactus* with *Epiphyllum* (or the various hybrids which parade under that name – epicacti). They are succulents, in appearance resembling *Aporocactus*, but bearing larger flowers like those of *Epiphyllum*, making striking plants for hanging baskets. Propagate by cuttings in summer.

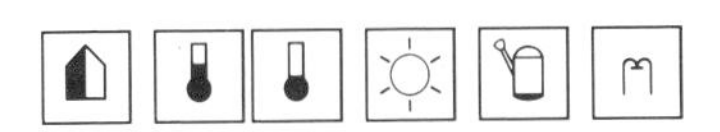

Hybrids cultivated

× A. 'Moonlight', stems vigorous and arching, fresh green with large flowers in soft shades of pink and peach-pink during late spring and early summer.

× A. 'Temple Fire', similar habit, but flowers vivid red.

× APOROPHYLLUM See × APOREPIPHYLLUM

APOROCACTUS

Cactaceae

Origin: Mexico. A genus of six species of epiphytic cacti. They are trailing or pendent plants growing

× *Aporepiphyllum* hybrid

on trees or rocks. The slender stems are cylindrical and bear aerial roots. Flowers are slightly zygomorphic and trumpet-shaped with many tepals. Propagate by cuttings of stem tips in summer or by seed in spring. One of the best genera for growing in hanging baskets. *Aporocactus* derives from the Greek *aporus*, having no way through (impenetrable) and *cactus*, perhaps from their dense masses of hanging stems.

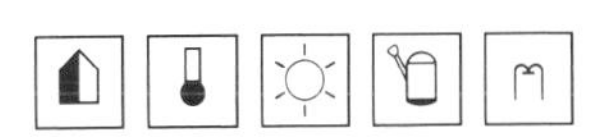

Species cultivated

A. flagelliformis Rat's tail cactus Mexico
Stems pendent to 2m(6½ft) long, ribs ten to 14, aeroles numerous, minute, bearing 15 to 20 spines in radiating clusters. Flowers crimson-pink, 8cm(3in) long, opening in spring.

A. mallisonii (of gardens)
A hybrid between A. *flagelliformis* and *Heliocereus* and correctly known as × *Heliaporus smithii*. Much like a robust, stiffer version of A. *flagelliformis* with larger, fiery red blooms.

A. martianus Mexico
Stems erect or sprawling to 1.5m(5ft) long, dull bluish-green, ribs eight, aeroles 1cm(⅜in) apart, bearing eight radial and three to four thicker, darker, erect spines. Flowers 10cm(4in) or more long, petals scarlet, purple-margined, opening in spring.

Araucaria heterophylla

ARACHIS

Leguminosae

Origin: Brazil, Paraguay. A genus of 15 species of tropical annuals and herbaceous perennials. They have pinnate leaves and prominent stipules at the base of the leaf stalks. The flowers are typical of the pea family. Seed pods, as they ripen, are buried into the ground by the lengthening of their stalks. Propagate by seed. *Arachis* probably derives from the Greek *arachidna*, the old name for a clover with a similar habit of bringing its seed heads to the ground.

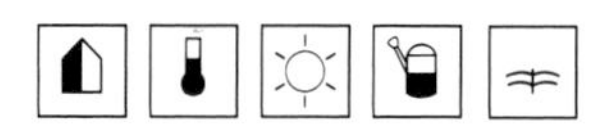

Species cultivated

A. hypogaea Peanut, Groundnut
A bushy, spreading to erect annual 15–60cm(6–24in) tall, grown on a large scale commercially for its oily seeds. Leaves pinnate, of four broadly obovate leaflets 3–7cm(1$\frac{1}{4}$–2$\frac{3}{4}$in) long. Flowers yellow, standard petal rounded 1.5cm($\frac{5}{8}$in) wide. After fertilization, the young fruits (1- to 4-seeded pods) are forced into the soil where they mature.

ARACHNOIDES See POLYSTICHUM
ARALIA See DIZYGOTHECA, FATSIA & TETRAPANAX

ARAUCARIA

Araucariaceae

Origin: South America, Australia, Norfolk Is., New Guinea, New Caledonia, and New Hebrides. A genus of 18 species of evergreen, coniferous trees which make very handsome pot specimens when young. They are very erect with tiers or whorls of horizontal branches and spirally arranged and overlapping ovate or awl-shaped leaves. Propagate by seed in spring. *Araucaria* is named for the Arauco province in Chile, home of the Araucani Indians and of the monkey puzzle tree (*A. araucana*), the first species to be described.

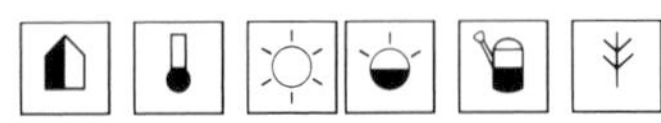

Species cultivated

A. angustifolia Parana pine, Candelabra tree
Brazil to Argentina
To 36m(120ft) or so in the wild. When young, similar to *A. araucana*, but leaves narrower, softer and more loosely arranged.

A.a. saviana
Leaves tinted grey-blue; makes an interesting pot or tub specimen of architectural appearance.

A. columnaris New Caledonia, Polynesia
To 60m(200ft) in the wild. Leaves on young plants triangular or lanceolate to 11cm(4$\frac{1}{2}$in) long.

A. cunninghamii Moreton Bay pine, Hoop pine
Eastern Australia, New Guinea
To 45m(150ft) or more in the wild. Bark peeling off in horizontal bands (hoops), hence the common name. Leaves on young plants are spirally arranged, up to 2cm($\frac{3}{4}$in) long, lanceolate to narrowly triangular. *A.c.* 'Glauca' has silvery-grey leaves and makes a decorative pot or tub specimen.

A. excelsa See *A. heterophylla*.

A. heterophylla (*A. excelsa*) Norfolk Island pine
Norfolk Is. (N.W. of New Zealand)
45–60m(150–200ft) in the wild. Leaves of young plants soft, awl-shaped, incurved and bright green to 11cm(4$\frac{1}{4}$in) long. Very decorative when small.

ARAUJIA

Asclepiadaceae

Origin: South America. A genus of two or three species of woody climbers. They have opposite leaves and small clusters of spreading, bell-shaped flowers. *Araujia* is the Brazilian vernacular name.

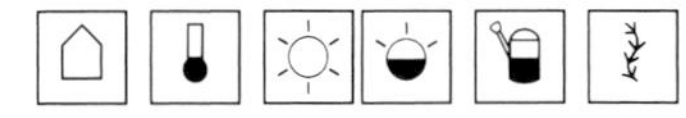

Species cultivated

A. sericofera Cruel plant
Twiner to 7m(23ft) or more. Leaves in opposite pairs, ovate-oblong, 5–10cm(2–4in) long, thinly felted on the undersides. Flowers white, tubular, inflated at the base and expanding to five petal lobes at the top, 2–3cm($\frac{3}{4}$–1$\frac{1}{4}$in) wide, opening in late summer. In the wild, moths visiting these

flowers are sometimes held temporarily by their tongues to the glutinous pollen masses, hence its popular name. Fruit, a grooved pod to 11cm(4½in) long, seeds having long, silky hairs.

ARBUTUS

Ericaceae

Origin: Western Asia, Mediterranean, Western Europe, North and Central America. A genus of about 20 species of evergreen shrubs and trees. Those described below are mainly thought of as specimen plants for the open garden where intense cold is not experienced. All, however, make handsome tub plants, flowering and fruiting young when their roots are confined. They have pleasing, more or less ovate foliage and small trusses of nodding bell-shaped flowers followed by spherical, orange to red fruits. Propagate by seed when ripe or in spring, by cuttings of stem tips in summer and by layering in autumn or spring. *Arbutus* is the old Latin name for the strawberry tree.

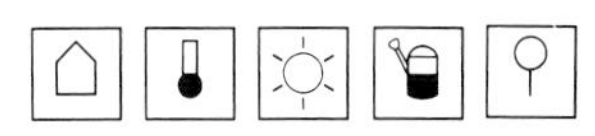

Species cultivated

A. andrachne S.E. Europe, Turkey
Normally a large shrub or small tree, but slow to reach 2m(6½ft) in containers. Leaves oblong to ovate, sometimes toothed, 5–10cm(2–4in) long. Flowers white, 12mm(½in) wide, in compact panicles. Hardy.

A. × andrachnoides (*A. andrachne × A. unedo*)
Midway between the parents but with white flowers in autumn and winter, and cinnamon-red bark on the main stem and branches.

A. unedo Strawberry tree Mediterranean, W. Europe (including S.W. Ireland)
Normally a small tree, but fairly slow to reach 2m(6½ft) in containers. Leaves elliptic to obovate, toothed, 3–10cm(1¼–4in) long. Flowers in compact panicles, white to pink-flushed, 6mm(¼in) long, autumn to early winter. Fruits granular, orange-red, to 2cm(¾in) wide, ripening as the next season's flowers open. Hardy.

ARCHONTOPHOENIX

(PTYCHOSPERMA)

Palmae

King palms

Origin: Australia. A genus of two species of palms which make excellent pot and tub specimens when young. They have arching pinnate leaves composed of many narrow leaflets. The white or lilac flowers and red fruits are not produced on containerized specimens. Propagate by seed in spring at not less than 18°C(65°F). Seedling growth is among the most rapid of the palms. *Archontophoenix* derives from the Greek *archontos*, a chieftain and *phoenix*, the date palm, drawing attention to the majestic appearance of the mature trees – sometimes exceeding 30m(100ft).

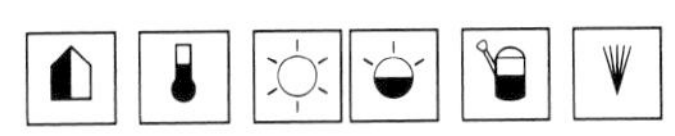

Species cultivated

A. alexandrae Northern bangalow palm, Alexandra palm Queensland
To 3m(10ft) or more in large containers. Leaves to 90cm(3ft) or more long, pale greyish beneath.

A. cunninghamiana Piccabeen palm, Piccabeen bangalow palm, Illawarra palm Queensland and New South Wales
To 3m(10ft) or more in large containers. Leaves to 90cm(3ft) or more long, green beneath. Occasionally grown under the invalid name *Seaforthia elegans*.

ARDISIA

Myrsinaceae

Origin: Tropics and sub-tropics; mostly from Asia and America, some in Australia. A genus of 400 evergreen shrubs and trees. They usually have alternate elliptic to lanceolate leaves and small flowers in terminal or axillary clusters. Propagate by seed or heel cuttings in spring and summer. *Ardisia* derives from the Greek *ardis*, a point – the anthers are spear-pointed.

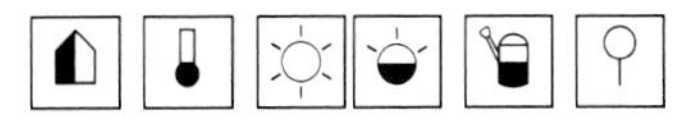

Species cultivated

A. crispa (*A. crenata, A. crenulata*) Marlberry, Coralberry, Spiceberry S.E. Asia
Shrub to 1.5m(5ft). Leaves oblong-lanceolate, 5–

TOP LEFT *Archontophoenix cunninghamiana*
TOP RIGHT *Arbutus unedo*
ABOVE *Ardisia crispa*

10cm(2–4in) in length with wavy margins. Flowers 5-petalled, starry, fragrant, white sometimes flushed with red, clustered at the end of the branches. Fruits bright red, rounded, 6–8mm($\frac{1}{4}$–$\frac{1}{3}$in) in diameter.

ARECA See CHRYSALIDOCARPUS

ARECASTRUM

Palmae
Queen palm

Origin: Brazil. A genus of one species of tall palm which, when young, makes an elegant container plant. Propagate by seed in spring at about 24°C(75°F). *Arecastrum* derives from the S.W. Indian vernacular name *areca* for the betelnut palm, and *astrum*, resembling.

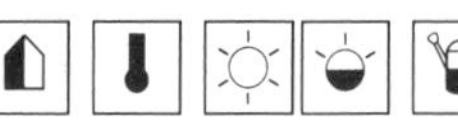

Species cultivated

A. romanzoffianum (*Cocos romanzoffiana, C. plumosa*)
To 3m(10ft) or more in large containers. Leaves pinnate, to 1.5m(5ft) or more long, composed of many very slender, drooping leaflets. The yellow flowers and fruits are not produced on containerized plants.

AREGELIA See NEOREGELIA
ARGYRANTHEMUM See CHRYSANTHEMUM

Argyroderma testiculare

ARGYRODERMA

Aizoaceae

Origin: South Africa. A genus of 50 species of extreme succulents. The shoots which make up each of these intriguing and attractive plants may be solitary or in small tufts or groups. Each shoot is composed of one or two pairs of much swollen leaves. In some two-leaved species they are closely pressed together appearing almost as one with a deep slit at the top. The sessile flowers are large and daisy-like. Propagate by seed in spring or by careful division in early summer (tufted kinds only). *Argyroderma* derives from the Greek *argyros*, silver and *derma*, skin, referring to the whitish or pale grey-green appearance of these plants.

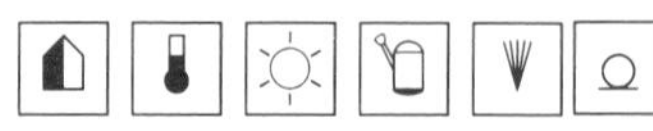

Species cultivated

A. aureum Namaqualand
Shoots solitary; leaves two, 4cm(1$\frac{1}{2}$in) long, connate for two-thirds of their length, then gaping apart. Flowers bright yellow, 3–4cm(1$\frac{1}{4}$–1$\frac{1}{2}$in) wide, borne in late summer.

A. brevipes Namaqualand
Shoots forming loose clumps; leaves two, 5–10cm(2–4in) long, semi-cylindrical, green, sometimes with a reddish tinge. Flowers light purple-red, about 4cm(1$\frac{1}{2}$in) across, borne in summer.

A. delaetii See *A. roseum*.

A. fissum Cape
Closely allied and similar to *A. brevipes*, but leaves grey-white with the upper leaf surface flattened and the tip bluntly keeled.

A. framesii Cape
Shoots tufted; leaves two, about 2cm($\frac{3}{4}$in) long, connate for half this, then opening out to form a gape 6–8mm($\frac{1}{4}$–$\frac{1}{3}$in) wide. Flowers 2cm($\frac{3}{4}$in) wide, rosy-purple, borne in summer. Various forms exist in the wild and are sometimes available commercially; *A.f. minus* is smaller.

A. octophyllum Silver skin Cape
Shoots usually solitary; leaves two or four, 3cm(1$\frac{1}{4}$in) long, almost as wide, barely connate, flat on the upper surface, strongly convex below, gaping fairly widely. Flowers yellow, to 4cm(1$\frac{1}{2}$in) wide, the petals often spiralled. In the past, at least, distributed as *A. testiculare*.

A. patens Cape
Shoots forming wide clumps; leaves two, 2.5–3cm(1–1$\frac{1}{4}$in), connate for at least one-third of this and enclosed in a sheath, the tips widely gaping, and bluntly keeled. Flowers yellow, to 3cm(1$\frac{1}{4}$in) wide, borne in summer.

A. pearsonii (*A. testiculare pearsonii*) Cape
Almost identical with *A. testiculare*, but outer petals purple-red and inner ones brownish-yellow with red stripes, summer.

A. roseum (*A. delaetii roseum*) Namaqualand
Shoots one or two, rarely more. Leaves two or four, 3.5cm(1$\frac{3}{8}$in) long and slightly wider, connate for half their length, upper surfaces flat, lower strongly convex. Flowers red-violet, 8cm(3in) or more wide, the petals lax and drooping over the whole plant, summer.

A. schlechteri Cape
Shoots usually solitary; leaves two, about 2cm($\frac{3}{4}$in) long, connate to half this but erect and gaping very

little, lower surface slightly concave, upper convex. Flowers rose-red, 3cm(1¼in) wide, summer.
A. testiculare Cape
Shoots one to several; leaves two, 2cm(¾in) long, each one hemispherical in outline. Flowers white or cream, 4.5cm(1¾in) wide, summer. At least some plants under this name are referable to A. *octophyllum* (*q.v.*).

ARIOCARPUS

Cactaceae
Living rock cacti

Origin: Southern Texas and Mexico. A genus of six species of highly distinctive slow-growing cacti. They are solitary, or eventually form small groups, and have flattened, globose stems (plant bodies) completely obscured by large, thickened, triangular tubercles. The effect is that of a very short-leaved aloe or agave. Sizeable flowers are borne in the centres of the plants. These cacti must be kept totally dry and cool in winter and watered sparingly in summer. Propagate by seed in spring. *Ariocarpus* derives from the Greek *aria*, the whitebeam tree and *karpos*, a fruit, alluding to the resemblance of the cactus fruits to those of the whitebeam.

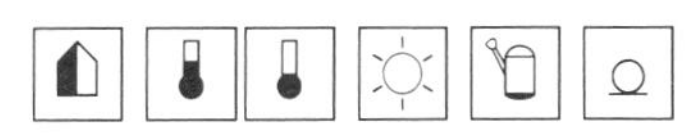

Species cultivated
A. agavoides (*Neogomesia agavoides*) Northern Mexico
Stem solitary, eventually to about 8cm(3in) wide. Tubercles few, grey-green, the outer ones 2–4cm(¾–1½in) long, inner ones smaller, remarkably leaf-like, but with grey-woolly areole on the upper surface. Flowers pink or magenta-pink, 1.5cm(⅝ in) or more across, borne in autumn.
A. fissuratus (*Roseocactus fissuratus*) Star cactus S.W. Texas and Mexico Stems solitary or forming small clusters, eventually 10–15cm(4–6in) wide. Tubercles many, 2–3cm(¾–1¼in) long, greyish to reddish-green, deeply fissured and wrinkled, with a central groove on the upper surface, woolly at the base, spreading to the areole. Flowers pink, 4cm(1½in) or more wide, borne in autumn.
A.f. lloydii
Larger with much less fissured tubercles.
A. kotschubeyanus Central Mexico
Stems forming groups, to 7.5cm(3in) wide, in the wild the almost flat top level with the soil surface. Tubercles many, spreading, triangular, to 12mm(½in) long, dark olive-green, slightly roughened, with the areole in a central woolly groove. Flowers pink to light magenta, 2.5cm(1in) or more wide, borne in autumn.
A.k. albiflorus
Stems smaller; flowers white.
A.k. macdowellii
Stems smaller; flowers magenta.
A. retusus Seven stars Northern Mexico
Stems more or less globose, eventually to 15cm(6in) wide. Tubercles many, divergent, grey to blue-green, 1.5–2.5cm(⅝–1in) or more long, sharply triangular with horny points. Areole near the tip. Flowers about 4cm(1½in) wide, the white petals with a red mid-vein, borne in autumn.
A. scapharostrus Northern Mexico
Stems more or less globose, up to 10cm(4in) wide. Tubercles few, erect to divergent, to 4.5cm(1¾in) long, boat-shaped and grey-green. Areoles minute or lacking. Flowers purple-red, 4cm(1½in) wide, borne in autumn.
A. strobiliformis See *Pelecyphora strobiliformis*
A. trigonus Northern Mexico
Stems flattened globose, eventually to 15cm(6in) wide. Tubercles many, divergent to incurved, 5–8cm(2–3in) long, tapering to a narrow hard point, bright green but woolly at the base; areoles terminal and inconspicuous. Flowers cream to pale yellow, about 5cm(2in) wide, borne in autumn.

ARISAEMA

Araceae

Origin: Mostly E. Africa and Asia; a few in North America. A genus of 150 species of tuberous- or rhizomatous-rooted perennials. Each plant generally has just one or two long-stalked leaves, the leaf blade being pedately divided into three to 15 leaflets or lobes. The flowers are tiny and petalless, the males having two to five stamens; they are borne on a spadix and surrounded by a spathe. Though best grown in a conservatory, these plants

BELOW *Arisaema sikokianum*
BOTTOM *Arisaema candidissimum*

can be flowered in the home. Propagate by offsets at potting time or by seed (sown when ripe if possible). *Arisaema* derives from the Greek *aris*, arum and *haima*, blood, used in the sense of a blood relationship with arum.

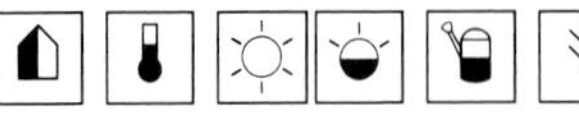

Species cultivated

A. candidissimum China
Leaves solitary on a stalk to 30cm(1ft), trifoliate with 8–20cm(3–8in) long, broadly ovate leaflets. Spathes hooded, 7–10cm($2\frac{3}{4}$–4in) long, white striped with green on the back and beautifully tinged with pink inside; they appear in early summer before the leaf unfurls.

A. consanguineum E. Asia
Leaves of ten to 21 lanceolate-acuminate leaflets borne on a stalk 45–90cm($1\frac{1}{2}$–3ft) high. Spathes, which can reach 18cm(7in) in length, are green with striping which is white on the back and purple-brown within. The length of the spathe is mostly in its tail-like elongated tip; it appears in summer.

A. fimbriatum Malasia
Leaves solitary or in pairs, trifoliate, each leaflet ovate, 10–15cm(4–6in) long. Leaf stalks 30–45cm(1–$1\frac{1}{2}$ft) tall. Spathe 15cm(6in) long or more, brownish-purple with white stripes. Spadix slender, bearing many fine, purple-tinted threads at the top.

A. sikokianum Japan
Leaves in pairs, trifoliate, leaflets broadly ovate, to 15cm(6in) long on 45cm($1\frac{1}{2}$ft) stalks. Spathes waisted, 18–25cm(7–10in) long, brownish-purple, the back with green, purple-centred stripes, the inside with white bands; spadix white, bulbous, opening in spring.

A. speciosum Himalaya
Tuber rhizome-like. Leaves solitary, trifoliate, leaflets oblong-lanceolate 20–40cm(8–16in) long, red-margined, on a brown mottled stalk 60–90cm(2–3ft) high. Spathes to 15cm(6in) or more, brownish-purple on the back, green striped inside, the lower part striped white or pale purple; spadix white, the glossy purple tip extended to a slender tail to 50cm(20in) long, opening in spring.

Arisarum proboscideum

A. tortuosum India
Leaves two or three, composed of 13 to 23 leaflets or lobes, on 60cm(2ft) long stalks. Spathes 8–18cm(3–7in) long, green and purple with paler stripes, opening in summer.

ARISARUM

Araceae

Origin: Mediterranean region. A genus of three species of rhizomatous-rooted perennials having an arum-like spathe around a flower-bearing spadix. Plants are best in a conservatory, but can be grown in a cool room. Propagate by division, offsets or seed (sown as soon as ripe if possible). *Arisarum* derives from the Greek name *arisaron* for one of the species.

Species cultivated

A. proboscideum Mouse plant Spain, Italy
Tubers rhizome-like, creeping and forming colonies. Leaves dark, glossy green, hastate, the blades 8–10cm(3–4in) long. Spathes swollen, joined at their margins into a tube, maroon above, whitish at the base, 2–3cm($\frac{3}{4}$–$1\frac{1}{4}$in) long; they fancifully resemble small mice crouching beneath the leaves, the spathe tips elongated into a 15cm(6in) slender tail. Hardy.

ARISTEA

Iridaceae

Origin: Africa. A genus of 50–60 species of evergreen perennials, the two described below being grown for their attractive blue flowers. They are tufted to clump-forming with terminal panicles of starry, 6-petalled flowers carried well above narrowly sword-shaped leaves. Propagate by seed in spring or by division in early autumn or after flowering. *Aristea* derives from the Greek *aristos*, best, apparently used by its author in the sense of pleasing.

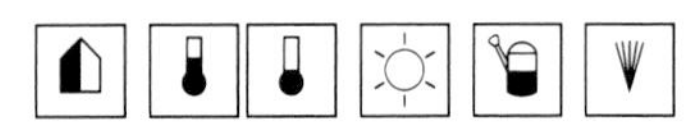

Species cultivated

A. ecklonii Southern to tropical Africa
Tufted to small clump-forming, to about 45cm($1\frac{1}{2}$ft) tall. Leaves linear, generally flexuous. Flowers about 1cm($\frac{3}{8}$in) long, blue, in a loose, spike-like panicle up to 15cm(6in) long. An unusual pot plant.

A. major (*A. thyrsiflora*) South Africa
Clump-forming, to 1.5m(5ft) or more tall. Leaves linear, erect and rigid. Flowers about 2.5cm(1in) wide, rich blue, in spike-like panicles 30–45cm(1–$1\frac{1}{2}$ft) or so long in summer. Makes a statuesque tub plant for the conservatory.

A. thyrsiflora See *A. major*.

ARISTOLOCHIA

Aristolochiaceae
Birthworts

Origin: Tropical and temperate regions of both hemispheres, but more common in the former. A genus of 350 species of mostly woody climbers or scramblers, both evergreen and deciduous, and a few herbaceous perennials. They usually have rounded to ovate or triangular cordate leaves. The curious flowers are formed of a straight or curved tubular perianth that may have a hooded, expanded funnel or dish-like mouth. Small flies enter the mouth, attracted by the odour, and down-pointing hairs prevent their return until the pollen is shed over them; the hairs then wither and the fly escapes to visit another flower and hence effect pollination. Propagate by seed in spring or by cuttings in a propagating frame. *Aristolochia* derives from the Greek *aristos*, best and *locheia*, delivery, alluding to its supposed value in childbirth.

Species cultivated

A. elegans Calico flower, Birth worts Brazil
Vigorous, slender climber capable of attaining 6m(20ft) in a conservatory, but easily kept smaller in containers for the home. Leaves heart- to kidney-shaped, about 8cm(3in) wide. Flowers solitary from the upper leaf axils, pendent on long stalks, shallowly heart-shaped bowls of maroon with white marbling. Blooms in summer and autumn and lacks the unpleasant smell typical of many other aristolochias.

ARTHROPODIUM

Liliaceae

Origin: Australasia and Malagasy. A genus of nine species of perennials, having narrowly lanceolate to linear leaves borne in arching tufts and racemes or panicles of starry flowers, each with six tepals and bearded stamen filaments. Propagate by division or by seed in spring. *Arthropodium* derives from the Greek *arthron*, a joint and *podion*, a small foot, alluding to the jointed flowering pedicels.

 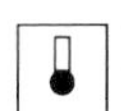 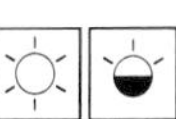  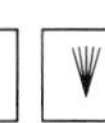

Species cultivated

A. cirrhatum New Zealand
Leaves evergreen, pale to grey-green, up to 60cm(2ft) or more, narrowly lanceolate, from a semi-woody rhizome. Stems 30–90cm(1–3ft) bearing a broad, branched panicle of 2–4cm($\frac{3}{4}$–$1\frac{1}{2}$in) wide, white flowers with slightly reflexed tepals and stamens marked purple with yellow hairs, opening late spring. An unusual and attractive plant.

ARUM See DRACUNCULUS

LEFT *Arundinaria viridistriata*
BELOW *Arthropodium cirrhatum*

ARUNDINARIA

Gramineae
Bamboos

Origin: Widespread in warm areas, but mainly of eastern and southern Asia. A genus of 150 species of bamboos, some of which make elegant specimen tub plants for the conservatory. Some spread by underground rhizomes, others are clump-forming. All those described have slender canes with short, almost horizontal branches and narrowly oblong leaves. Mature, leafy stems create a plumey effect. The grass-type flowers are insignificant and are seldom produced on containerized specimens. Propagate by division in late spring. *Arundinaria* derives from the Latin *arundo*, a reed or cane.

 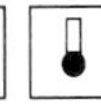 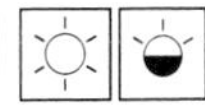  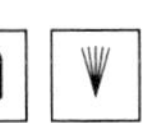

Species cultivated

A. amabilis Tonkin bamboo China
2–3m($6\frac{1}{2}$–10ft) tall in tubs, the canes mid-green. Leaves 10–30cm(4–12in) long, somewhat lustrous bright green above.

A. anceps North-western Himalaya
2m($6\frac{1}{2}$ft) tall or more in containers, the canes glossy deep green. Leaves 10–15cm(4–6in) long, lustrous rich green.

A. auricoma See *A. viridistriata.*

A. fortunei See *A. variegata.*

A. japonica (*Pseudosasa japonica, Bambusa metake*) Japan
2–3m($6\frac{1}{2}$–10ft) in height in containers, the canes olive-green. Leaves 13–25cm(5–10in) or more long and relatively broad, darkish, somewhat glossy green above, greyish-tinted beneath. One of the

easiest and hardiest bamboos, but less elegant than some.

A. murielae (*A. spathacea, Sinarundinaria nitida*) China
2–3m(6½–10ft) tall in containers, canes green with a whitish, waxy patina when young. Leaves to 10cm(4in) long, bright pea green above, duller beneath. One of the most ornamental species for containers.

A. nitida (*Sinarundinaria nitida*) China
Much like *A. murielae* and equally decorative, but canes deep purple or variably purple-flushed.

A. simonii (*Pleioblastus simonii*) Japan
2m(6½ft) or more in height, canes olive-green with a whitish, waxy patina when young. Leaves 8–20cm(3–8in) or more long, green above, greyish beneath.

A. spathacea See *A. murielae*.

A. variegata (*A. fortunei, Pleioblastus variegatus*) Japan
60–90cm(2–3ft) or more, canes pale green, somewhat zig-zag. Leaves 5–15cm(2–6in) long, dark green with white stripes, the latter ageing pale green.

A. viridistriata (*A. auricoma, Pleioblastus viridistriatus*) Japan
60–120cm(2–4ft) or so, canes purplish green. Leaves 5–15cm(2–6in) long, green with conspicuous yellow stripes. Can be cut back hard each year in late winter to maintain a low habit and promote plenty of brightly variegated leaves.

ABOVE *Asarina barclaiana*
RIGHT *Arundo donax* 'Versicolor'

ARUNDO

Graminae

Origin: Tropical and temperate regions of the Old World. A genus of 12 species of giant grasses with broad, flat leaves and plumes of tiny flowers. Propagate by division in spring or by cuttings of side shoots rooted in a propagating case. *Arundo* is the original Latin name for this plant.

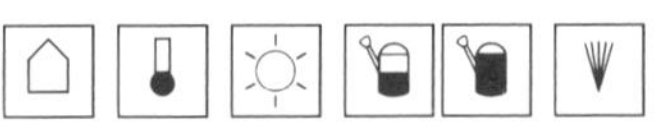

Species cultivated

A. donax Giant reed Mediterranean
Clump-forming, eventually forming a woody base, with stems to 4m(13ft) in the open ground, but less in a pot. Leaves alternate, arching, grey-green to 60cm(2ft) long and 7cm(2¾in) wide. Flowering spikes reddish to whitish in a dense 30–60cm(1–2ft) terminal panicle in autumn. *A.d.* 'Versicolor' ('Variegata') is the most useful for pot culture, being a smaller plant with leaves longitudinally striped with white.

ASARINA

Scrophulariaceae

Origin: West USA, Mexico and the West Indies; one species in Europe. A genus of 16 species of trailing and climbing perennials with alternate, mainly triangular to hastate leaves and slender stalks which act as tendrils. The flowers are 2-lipped and foxglove-like. Plants are best in the conservatory, but can be grown in the home. Propagate by seed or cuttings in spring. *Asarina* derives from the Spanish vernacular name for *Antirrhinum* which is closely related.

Species cultivated

A. antirrhiniflora (*Antirrhinum maurandioides, Maurandya antirrhinifolia*) S.W. USA
Slender climber or trailer to 1–2m(3–6½ft). Leaves triangular 1–3cm(⅜–1¼in) long. Solitary flowers 2–3cm(¾–1¼in) long, purple to rose-pink with a yellowish palate.

A. barclaiana (*Maurandya barclaiana*) Mexico
Woody based climber to 2–5m(6½–16ft). Leaves roughly triangular and shallowly lobed, about 2.5cm(1in) across on twining stalks. Flowers 4–7cm(1½–2¾in) long, open mouthed, deep purple, occasionally pink or white. Sepals gladular hairy.

A. erubescens (*Maurandya erubescens*) Mexico
Climber to 2–5m(6½–16ft). Leaves toothed, triangular and densely greyish pubescent. Flowers to 7cm(2¾in) long, open-mouthed, rose-pink. Seeds winged.

A. procumbens (*Antirrhinum asarina, Maurandya asarina*) Pyrenees
Trailing stems to 60cm(2ft) long. Leaves in opposite pairs to 6cm(2½in) wide, kidney-shaped,

crenate, grey-green with sticky hairs. Flowers 3–4cm(1¼–1½in) long with a white tube streaked pale purple, lips yellow.

ASCLEPIAS

Asclepiadaceae

Origin: North and South America. A genus of 120 species of perennials and sub-shrubs some of which are tuberous-rooted. They have opposite pairs or whorls of lanceolate to ovate leaves and 5-petalled flowers in umbels. Each flower has five stamens which are joined in a tube and crowned by three hooded processes, each one sometimes bearing an incurved horn. Within this tube, the anthers are attached to the stigma. The fruits are pod-like follicles which contain seeds crowned by long silky hairs for wind dispersal. Many species exude a poisonous milky latex when cut or damaged. Propagate by seed in spring. *Asclepias* derives from the name of the Greek god of medicine, *Asklepios*, in a Latinized form.

  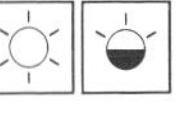  

Species cultivated

A. curassavica Blood flower Tropical America
Sub-shrub to 90cm(3ft) or so in height. Leaves carried in pairs, lanceolate, 5–15cm(2–6in) long. Flowers to 2cm(¾in) wide, orange-red with a yellow staminal crown; they are borne in umbels of five to ten, opening in summer and autumn. A good long term house plant, but even better in the conservatory.

ASCOCENTRUM

Orchidaceae

Origin: Eastern Asia to Philippines, Borneo and Java. A genus of five species of orchids allied to *Vanda*. They have an erect habit with unbranched leafy stems. The flowers, which are borne in lateral, upright racemes, are smallish but showy; they have five tepals, the lower two sometimes larger than the others and the topmost one hooded, and a small labellum with a long narrow nectary spur. Propagate by removing stem tips with aerial roots after flowering or in spring. Plants are best in a conservatory, but can be brought indoors when in bloom – in fact, they are worth trying in the home full-time.

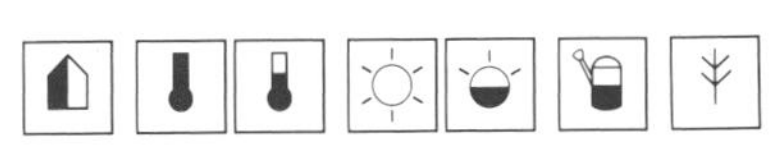

Species cultivated

A. ampullaceum (*Aerides ampullaceum, Gastrochilus ampullaceum, Saccolabium ampullaceum*) Himalaya to Burma
Stems to 25cm(10in) tall, usually less. Leaves linear, irregularly toothed at the tips, leathery, about 13cm(5in) long. Flowers 10–15mm(⅜–⅝in) wide, rose-red to rose-purple, the labellum sometimes a tone paler with a white column, borne in spring to summer.

A.a. houlmeinense
A selected clone with more brightly coloured, larger flowers.

A. miniatum (*Gastrochilus miniatum, Saccolabium miniatum*) Himalaya to Borneo and Malaysia
Stems about 10cm(4in) tall, comparatively thick and woody. Leaves linear, fleshy and firm-textured, 8–20cm(3–8in) long. Flowers 2cm(¾in) wide, usually bright orange-red, but variable in the wild from orange-yellow to vermilion, borne in spring to early summer.

ABOVE *Asclepias curassavica*
LEFT *Ascocentrum ampullaceum*

ASPARAGUS

Liliaceae

Origin: Europe, Africa, Asia and Australia. A genus of 30 species of climbing, sprawling and erect plants, some herbaceous, some shrubby. They are distinguished by the absence of true leaves, their place being taken by needle-like, awl-shaped or broader phylloclades; the true leaves are reduced to tiny scales. The flowers are small, either starry or bell-shaped, with six tepals and are followed by

ABOVE *Asparagus densiflorus* 'Myers'
ABOVE RIGHT *Asparagus falcatus*

berries. Propagate by division or by seed in spring. All the species below make good long-term house or conservatory plants. *Asparagus* derives from the ancient Greek name for the genus.

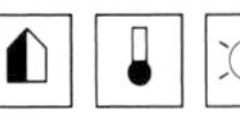 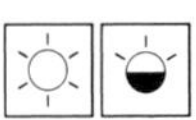  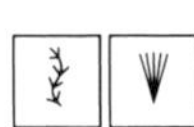

Species cultivated

A. asparagoides (*A. medeoloides, Smilax asparagoides*) Smilax South Africa
Twining, evergreen climber to 3m(10ft) or so. Phylloclades very leaf-like, ovate, glossy bright green, to 3cm($1\frac{1}{4}$in) long. Flowers tiny, greenish white, sometimes followed by red berries. This foliage plant is the so-called Smilax of florists but must not be confused with true members of the genus *Smilax* (*q.v.*).

A.a. myrtifolius Baby smilax
Phyllocades shorter and the plant smaller; the best sort for pot culture.

A. densiflorus South Africa
Tuberous-rooted, tufted; stems to 1.2m(4ft) or more long, arching or pendulous, much branched. Phylloclades usually solitary, linear, 1–2cm($\frac{3}{8}$–$\frac{3}{4}$in) long, bright green. Flowers white, pink tinted, in short racemes; fruits bright red. *A.d.* 'Myers' (*A. myersii*) has stems to 60cm(2ft); erect, with short branched, narrowly plume-like phylloclades in clusters of one to four, about 1cm($\frac{3}{8}$in) long by 1mm($\frac{1}{25}$in) wide. *A.d.* 'Sprengeri' (*A. sprengeri*) is similar to the type, but phylloclades in clusters of one to six (usually three), 1.5–3cm($\frac{5}{8}$–$1\frac{1}{4}$in) long by 1–2.5mm($\frac{1}{25}$–$\frac{2}{25}$in) wide.

A. falcatus Sicklethorn South Africa, Sri Lanka
Twining evergreen climber to 4m(13ft) or so. Phylloclades 4–5cm($1\frac{1}{2}$–2in) long, lanceolate-falcate, bright green, borne in clusters of three or more along the pale stems. Flowers white, tiny, fragrant, in axillary racemes 5cm(2in) long.

A. medeoloides See *A. asparagoides*.

A. myersii See *A. densiflorus* 'Myers'.

A. plumosa See *A. setaceus*.

A. scandens South Africa
Tuberous-rooted herbaceous perennial with scrambling or climbing stems to 2m($6\frac{1}{2}$ft). Phylloclades lanceolate-falcate, mostly 6–12mm($\frac{1}{4}$–$\frac{1}{2}$in) long, borne in clusters of three to five, one of which is longer than the rest (to 2cm[$\frac{3}{4}$in]). Flowers tiny, whitish or pinkish, singly or in pairs from the leaf axils. Berries 6mm($\frac{1}{4}$in) wide, red.

A. setaceus (*A. plumosus*) Asparagus fern South Africa
A twining climber when mature, to 3m(10ft) or more; branches flattened and frond-like, phylloclades bristle-like to 6mm($\frac{1}{4}$in) long, in clusters of eight to 20. Flowers white in small clusters; fruit purple-black. *A.s.* 'Compactus' and *A.s.* 'Nanus' are forms which remain in a juvenile, non-climbing state and are the best for culture in pots where space is restricted.

A. sprengeri See *A. densiflorus* 'Sprengeri'.

ASPASIA

Orchidaceae

Origin: Tropical forests of Central and South America. A genus of ten species of handsome epiphytic orchids growing from clusters of pseudobulbs which are ovoid to ellipsoid, 2-edged and often stalked. Their leaves are lanceolate. Short, erect racemes carry flowers which open widely, comprising narrow, spreading sepals and petals with a prominent, showy lip. Propagate by division when re-potting. *Aspasia* is variously described as being derived from the Greek *aspazomai*, embrace, referring to the way the base of the labellum is enclosed by the column; or for Aspasia, mistress of Pericles (*d.* 429 BC).

 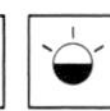 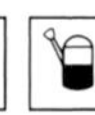 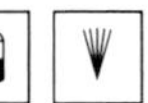

Species cultivated

A. epidendroides Central America, Colombia

Stalked ellipsoid pseudobulbs to 15cm(6in) long. Leaves in pairs, leathery, broadly lanceolate, to 30cm(1ft) by 5cm(2in) wide. Flowers 4cm(1½in) long, lightly fragrant, the sepals greenish with brown-purple cross-banding, the petals lavender to greenish-brown with a white lip, crenate and waved, and marked with purple or yellow, borne in summer.

A. lunata (*Odontoglossum lunatum*) Brazil
Ovoid, flattened pseudobulbs 5cm(2in) long. Leaves 20cm(8in) long, carried either singly or in pairs. Flowers borne in ones or twos, to 4cm(1½in) long with pale green sepals, and petals spotted and barred brown; the lip is white with a central purplish marking.

A. principissima Costa Rica, Panama
Very like *A. epidendroides* but less vigorous. Flowers about 7cm(2¾in) long, borne in spring and summer, sepals and petals yellow-green, longitudinally striped with light brown; the broad lip is cream to pale yellow with a strongly waved margin.

A. variegatum Trinidad to Brazil
Similar to *A. epidendroides* but smaller. Leaves to 15cm(6in) long. Flowers to 6cm(2½in) long, green, cross-banded with purplish-brown, the lip white, dotted with violet, borne in winter to spring.

ASPIDISTRA

Liliaceae

Origin: East Asia. A genus of eight species of evergreen perennials, having horizontal rhizomes and ovate, parallel-veined green leaves held upright on long stalks. Their flowers are unusual in that they have their parts arranged in fours, as opposed to threes which is normal for members of the lily family. Propagate by division. Aspidistras were very popular foliage house plants in Victorian days, being especially tolerant of dust, fumes and general neglect; there was even a song written about them. They have stood the test of time and still make very durable plants for the home. *Aspidistra* derives from the Greek *aspidion*, a small round shield, referring to the rounded end of the large stigma.

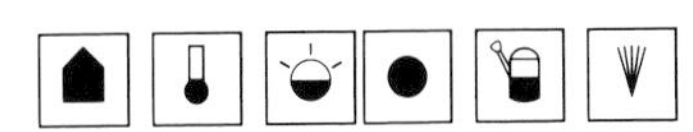

Species cultivated

A. elatior (*A. lurida*) Cast iron plant China
Clump-forming. Leaves long-stalked, oblong-lanceolate, dark lustrous green, 45–75cm(1½–2½ft) long. Flowers solitary, cup-shaped, purple and borne at ground level. *A.e.* 'Variegata' has leaves longitudinally white-striped.

ASPIDIUM See CYRTOMIUM

ASPLENIUM

Aspleniaceae
Spleenworts

Origin: World-wide from tropical to cool temperate zones. A genus of 650 species of perennial ferns, generally of tufted habit or shortly rhizomatous. The elegant leaves are simple, and entire to tripinnatifid; their sori are usually linear, though sometimes oval, and are borne along the veins on the undersides of the pinnae. Propagate by division or by spores in spring, or with some species by removing plantlets in summer. The species described are among the best ferns for the home. *Asplenium* derives from the Greek *a*, not and *splen*, the spleen, an allusion to its supposed medicinal properties in curing diseases of the spleen.

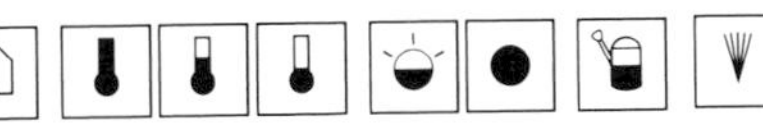

BELOW LEFT *Aspidistra elatior* 'Variegatum'
BELOW *Asplenium scolopendrium* 'Undulatum'

Species cultivated

A. bulbiferum Mother spleenwort
Australasia, Malaysia
Fronds arching to 90cm(3ft) in length, bi- or tri-pinnate, the blade oblong-triangular with lanceolate-triangular pinnules. Sori oblong. Bulbils, and later small plantlets, are borne on the upper side of mature fronds. Often confused with *A. daucifolium*, *q.v.* Temperate.

A. daucifolium (*A. viviparum*) Mother fern
Reunion, Mauritius
Much like *A. bulbiferum* and often confused with that species, but somewhat smaller, the fronds less elaborately cut and more graceful, the final segments of the deeply lobed pinnae being smaller and often forked. Plantlets are borne on the fronds as in *A. bulbiferum*, but are much less prolific. Tropical.

A. nidus (*A. nidus-avis*) Bird's nest fern
Tropical Asia and Australia
Fronds to 90cm(3ft) or more in length, lanceolate, semi-erect, forming a broad, spreading funnel, rich shining green with a black mid-rib. Sori linear. In the wild it is epiphytic, and will tolerate dry conditions for a short while. Tropical.

A. platyneuron Ebony spleenwort
Evergreen, tufted, fronds pinnate to 30cm(1ft) or more long, stalks lustrous brown; pinnae about 2.5cm(1in) long, narrowly oblong with one basal lobe uppermost. A pleasing smallish fern for unheated rooms and conservatories. Cool.

A. scolopendrium (*Phyllitis scolopendrium*)
Hart's tongue fern Europe
Fronds 10–60cm(4–24in) long, rich glossy green, entire, strap-shaped, cordate at the base and variably wavy margined. Sori linear. A very shade tolerant and pleasing fern for any unheated area. Hardy. Several cultivars are known, having waved, lobed or crested fronds: *A.s.* 'Crispum' has strongly crimped frond margins, making it the most decorative cultivar for containers; *A.s.* 'Undulatum' is similar.

A. viviparum See *A. daucifolium*.

ASTER See FELICIA

ABOVE *Astilbe* × *arendsii*
RIGHT *Asplenium nidus*

ASTERANTHERA

Gesneriaceae

Origin: Temperate forests of Chile. A genus of one single species, a self-clinging evergreen climber, found in the wild on tree trunks. Propagate by cuttings. *Asteranthera* derives from the Latin *aster*, a star and *anthera*, anthers, alluding to the anthers of its four stamens which are joined and fancifully resemble a shooting star.

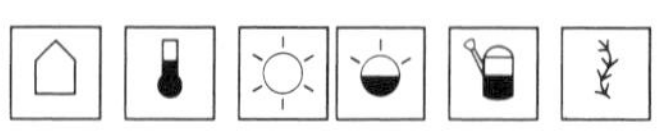

Species cultivated

A. ovata (*Columnea ovata*) Chile
Slender stems to 3m(10ft) or more. Leaves broadly ovate, toothed, 6–20mm(¼–¾in) long carried in opposite pairs. Flowers rich red, borne in ones or twos from the axils; they are tubular, expanding at the mouth to five oblong lobes, the upper two forming a straight hood over the stamens; the flowers are about 2.5cm(1in) long and 4–4.5cm(1½–1¾in) across the spreading lower lobes, opening in summer.

ASTILBE

Saxifragaceae

Perennial spireas

Origin: Eastern Asia, eastern North America. A genus of 25 species of frost hardy perennials usually grown in the open garden but making showy, elegant, short-term pot plants. Those below are clump-forming with dissected, somewhat ferny leaves and dense plume-like panicles of tiny 4- to 5-petalled flowers. Propagate by division when dormant. Plants are potted in autumn or winter and plunged outside or kept in a cold frame until late winter. At this time they are brought into the conservatory to encourage early growth and blooming. After flowering they must be placed outside and either kept watered and fed or planted out. *Astilbe* derives from the Greek *a*, without and *stilbe*, sheen or brightness, alluding to the comparative dullness of the foliage as compared with the earlier described *Aruncus dioicus* (goat's-beard), when both were classified in the genus *Spiraea*. However, it is not a valid comment for *A.* × *arendsii*.

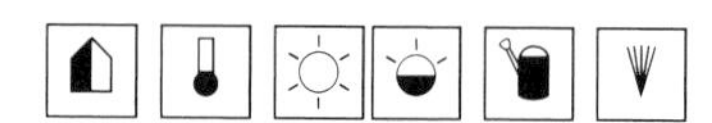

Species cultivated

A.* × *arendsii (*Perennial spireas*) (*A. davidii* × *A. astilboides* × *A. japonica* × *A. thunbergii*)
A complex group of hybrid cultivars raised by the German nurseryman Herr Arends. Leaves coarsely ferny, often coppery or purple-flushed. Floral plumes in shades of white, pink, purple and red. Among many suitable cultivars the following can be recommended (under glass they will start to bloom in late spring): 'Amethyst', rose-lilac, 60–90cm(2–3ft); 'Cologne', bright pink, 40–

50cm(16–20in); 'Fanal', deep red, 30–40cm(12–16in); 'Federsee', rose-red, 50–60cm(20–24in); 'Irrlicht', white with dark foliage, 40–50cm (16–20in).

ASTROPHYTUM

Cactaceae

Origin: Mexico. A genus of six species of globular to shortly cylindrical cacti of distinctive form, some having prominent ribs, variously studded with scale-like branched, white hairs. The areoles are small and woolly, the upper ones bearing the funnel-shaped flowers. Propagate by seed. All species are attractive and successful house plants. *Astrophytum* derives from the Greek *astron*, a star and *phyton*, plant, referring to their more or less star shape when seen from above.

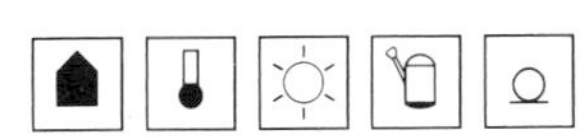

Species cultivated

A. asterias Sea urchin, Sand dollar cactus
Stem without spines, globular, to 8cm(3in) wide but somewhat flattened, ribs eight, broad and rounded. Areoles white between eight deep grooves which divide it into sections. Flowers 3cm($1\frac{1}{4}$in) long, yellow flushed with red at the centre, opening in summer.

A. capricorne Goat's horn cactus
Stem globular when young, becoming ovoid as it ages, 10–20cm(4–8in) high by 10cm(4in) wide,

FAR LEFT *Astrophytum ornatum*
LEFT *Astrophytum asterias*

ribs eight, sharp and covered with small white scales. Areoles brownish, bearing reddish, black or grey, recurved wavy spines 3–10cm($1\frac{1}{4}$–4in) long. Flowers reddish outside, yellow with a reddish base inside, 6–7cm($2\frac{1}{2}$–$2\frac{3}{4}$in) long, opening in summer.

A. myriostigma Bishop's cap cactus
Stem basically spherical, becoming elongated as it ages, to 10–20cm(4–8in) wide, covered densely with white scales and formed into four to eight (usually five) large, triangular ribs. Areoles on the angle of the ribs, brownish and spineless. Flowers 4–6cm($1\frac{1}{2}$–$2\frac{1}{2}$in) long, shiny yellow with a red centre.

A. ornatum
Stem shortly cylindrical, with age reaching 30cm(1ft) in height and 12–15cm($4\frac{3}{4}$–6in) in diameter. Ribs eight, sharp and deep, bearing along their edges areoles with five to 11 awl-shaped, amber to brown spines 3–4cm($1\frac{1}{4}$–$1\frac{1}{2}$in) long. Flowers 7–9cm($2\frac{3}{4}$–$3\frac{1}{2}$in) wide, clear yellow, not occurring on young plants.

A.o. mirbellii
Similar but with golden-yellow spines.

ASYSTASIA See MACKAYA

ATHEROSPERMA

Atherospermataceae

Origin: S.E. Australia. A genus of two species of evergreen trees with opposite pairs of leaves. When young they make decorative plants for a conservatory. Propagate by cuttings with a heel, or by seed in spring. *Atherosperma* derives from the Greek *atheros*, a barb or spine and *sperma*, a seed, referring to the slender pointed seeds.

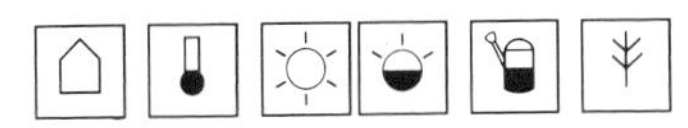

Species cultivated

A. moschatum Victoria, Tasmania
Leaves lanceolate, 5–10cm(2–4in) long, toothed, lustrous rich green above, white downy beneath. Flowers dioecious, 2–3cm($\frac{3}{4}$–$1\frac{1}{4}$in) wide, creamy-

white, borne singly in the leaf axils. The whole plant has a strong aromatic fragrance suggestive of nutmeg.

ATHYRIUM

Athyriaceae

Origin: Cosmopolitan. A genus of 180 species of tufted or clump-forming ferns some also with slender rhizomes. Their fronds are pinnate to tri-pinnate and the sori are borne on the back at the ends of the smaller branched veins. Propagate by division or by spores. The species below is excellent for unheated rooms and conservatories. *Athyrium* derives from the Greek *a*, without and *thyrion*, a small door, referring to the shield-like indusium which covers the spores.

  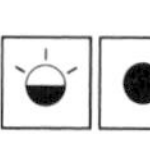  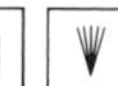

BELOW *Athyrium goeringianum* 'Pictum'
BOTTOM *Aucuba japonica* 'Variegata'

Species cultivated

A. esculentum See *Diplazium esculentum*.

A. goeringianum (*A. iseanum, A. nipponicum*) Japan, Taiwan

Tufted. Fronds deciduous, 20–30cm(8–12in) long, arching, bi-pinnate; pinnules oblong to deltoid-lanceolate, irregularly lobed to 10–17mm($\frac{3}{8}$–$\frac{5}{8}$in) long. Sori small, covered by an indusium. *A.g.* 'Pictum' (Japanese painted fern) is the form usually cultivated, the fronds silvery-grey with purple shading along the midrib; a very decorative fern.

AUCUBA

Aucubaceae (*Cornaceae*)

Origin: Eastern and Central Asia. A genus of three to four species of evergreen shrubs. They have glossy ovate to oblong-lanceolate, leathery leaves borne in opposite pairs and dioecious, 4-petalled but insignificant flowers. Female plants produce showy, glossy, one-seeded fruits, but only if a male plant is also grown. Propagate by seed when ripe, or by cuttings to be sure of the sex of the new plant. *Aucuba* is a Latinized form of the Japanese vernacular name *Aokiba*.

    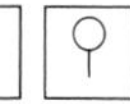

Species cultivated

A. japonica Japan to Himalayas

To 4m(13ft) tall in the open garden, but less in containers. Leaves shining green, ovate to lanceolate, 8–20cm(3–8in) long. Flowers purplish-brown, to 7mm($\frac{1}{4}$in) wide, carried in panicles 7–10cm(2$\frac{3}{4}$–4in) long in spring and followed by ovoid fruits to 1.5–2cm($\frac{5}{8}$–$\frac{3}{4}$in). Many forms with virused leaves producing golden-yellow freckling and spotting are grown (Spotted laurels), notably the male 'Crotonifolia' and 'Speckles', and the female 'Gold Dust' and 'Variegata' ('Maculata'). All hardy.

AVERRHOA

Oxalidaceae

Origin: Probably coastal forests of Brazil (now widely grown and naturalized in the tropics). A genus of two species of large shrubs or small trees grown mainly for their fruit, but which also make interesting tub specimens for the larger conservatory. They have simply pinnate leaves and small panicles of pink or red 5-petalled flowers followed by angularly ovoid berry fruits. Propagate by seed or layering in spring. *Averrhoa* honours the Arabian doctor Averrhoes who lived in Moorish Spain in the 12th century.

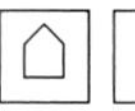  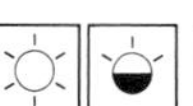  

Species cultivated

A. bilimbi Bilimbi Tropics

2–3m(6$\frac{1}{2}$–10ft) tall in containers. Leaves composed of 21 to 41 oblong, pointed leaflets, each 5–10cm(2–4in) long. Flowers in short narrow panicles direct from the trunk or main branches (cauliflorus), each one 2cm($\frac{3}{4}$in) long, dark red, borne in spring to autumn. Fruits 5–8cm(2–3in) long, cylindrical, obscurely 5-angled, greenish-yellow, very acid and used in making curries, pickles and preserves.

A. carambola Carambola tree Tropics
2–3m(6$\frac{1}{2}$–10ft) tall in containers. Leaves composed of seven to 11 ovate leaflets, each one 3–11cm(1$\frac{1}{4}$–4$\frac{1}{2}$in) long. Flowers in short axillary panicles or arising from the leafless stems immediately below the foliage. Each bloom 8mm($\frac{1}{3}$in) long, pink, borne in summer. Fruits 8–12cm(3–4$\frac{3}{4}$in) long, yellow, ovoid with five sharp, prominent angles (cross-section star-shaped); flesh crisp and juicy, usually sweetly aromatic, but acid forms known.

LEFT *Averrhoa bilimbi*
BELOW *Azolla caroliniana*

AYLOSTERIA See REBUTIA
AZALEA See RHODODENDRON

AZOLLA

Azollaceae (*Salviniaceae*)

Origin: Tropical and warm temperate regions of the Americas. A genus of six species of small floating aquatic plants found in still water where they will cover the surface. They are related to the ferns of tropical and warm temperate climates. Propagate by division. Plants provide a decorative cover for fish in tanks and pools. *Azolla* probably derives from the Greek *azo*, to dry out and *olluo*, to kill, with reference to the rapid death of the plants which follows drying out.

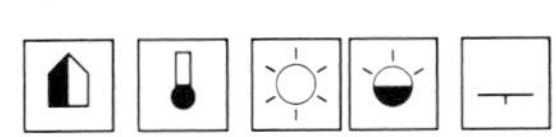

Species cultivated

A. caroliniana Water fern, Mosquito plant
S.E. USA and widely naturalized elsewhere. Moss-like with minute densely overlapping leaves covering the forked stems; pale green, turning red in bright light. When chilled, plants form a continuous carpet over the water, but are easily controlled in an aquarium or pool.

B

BABIANA

Iridaceae

Origin: S. and S.W. Africa and the island of Socotra. A genus of 60 species of cormous plants akin to *Gladiolus*. They have flattened, fan-like clusters of strongly pleated, lanceolate to sword-shaped leaves and short racemes of tubular or funnel-shaped flowers (6-tepalled). Propagate by seed, or by removing offsets or cormlets when re-potting. Plants are best in a conservatory, but can be brought indoors when in bloom. *Babiana* is a Latin form of the Africaans name *babiaan*, baboon, the corms being one of their favourite foods.

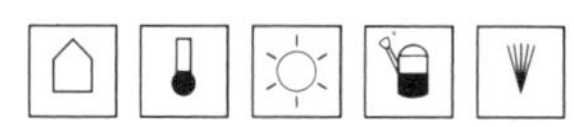

Species cultivated

B. sambucina South Africa
About 15cm(6in) or a little more in height. Leaves lanceolate, slightly pleated. Flowers bluish to deep mauve-purple, scented of elder flowers under warm sunny conditions, opening in late spring.

B. stricta Baboon flower South Africa
To 30cm(12in) or a little more in height. Leaves broadly lanceolate, to 12cm(4¾in) long. Flowers with three outer tepals, white, the three inner blue, opening in late spring. A number of named cultivars, mostly of hybrid origin, are now available with larger flowers in shades of blue, red, violet and white; they are sometimes listed as *B.* × *hybrida*.

ABOVE *Babiana stricta*
ABOVE RIGHT *Bacopa monnieri*

BACOPA

Scrophulariaceae

Origin: Tropical and warm temperate regions mostly of South America. A genus of 100 species of perennial aquatic or partly aquatic species. They have opposite leaves and solitary or paired white or blue flowers. Propagate by division or cuttings. *Bacopa* is a Latinized form of their vernacular South American Indian name.

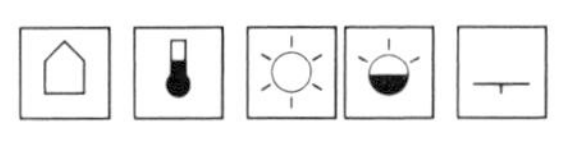

Species cultivated

B. monnieri (*Herpestes monnieri*) Tropics of South & Central America as far north as southern USA
Mat-forming, with creeping or floating stems sometimes shortly erect. Leaves spathulate to obovate, to 2cm(¾in) in length, somewhat fleshy. Flowers bell-shaped, slightly 2-lipped, 8–10mm(⅓–⅜in) long, lilac or blue, borne in summer to autumn.

BAMBUSA See ARUNDINARIA & PHYLLOSTACHYS

BANKSIA

Proteaceae

Origin: Australia. A genus of about 60 evergreen trees and shrubs most of which are native to Western Australian sand heaths. The species described below are all highly distinctive with handsome, saw-edged leathery leaves and terminal, globular to cone-shaped flower spikes in shades of red, orange and yellow guaranteed to catch the eye of even the non-gardener. They are only worthwhile attempting to grow in freely ventilated, sunny conservatories. A compost of equal parts moss peat and grit or coarse sand is essential, with occasional light feedings of magnesium sulphate and nitrogen (in the form of urea) during the growing season. General fertilizers rich in nitrates and phosphates must not be used. Propagate by seed in spring. *Banksia* commemorates Sir Joseph Banks (1743–1820), President of the Royal Society, a generous and wealthy patron of science and, as a young man, one of the first botanists to investigate the flora of Australia.

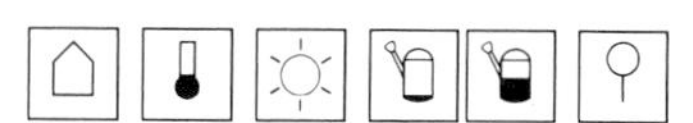

Species cultivated

B. ashbyi Western Australia
Potentially to about 2m(6½ft) in containers (to 5m[15ft] tall in the wild), spreading habit. Leaves linear, to 25cm(10in) long with large, triangular, spine-tipped teeth, somewhat lustrous. Flower spikes 7–10cm(2¾–4in) long by almost half as wide, orange, opening in spring.

B. coccinea Western Australia (Albany area)
About 1.5m(5ft) or more in containers (to 4m[13ft] in the wild), slender, more or less erect habit. Leaves 3–6cm(1¼–2½in) long, obovate with small, prickle-tipped teeth, lustrous. Flower spikes about 6cm(2½in) long and wide, crimson, in spring.

FAR LEFT *Banksia grandis*
LEFT *Banksia hookeriana*
BELOW *Bauera rubioides*

B. grandis Western Australia
About 2m($6\frac{1}{2}$ft) or more in containers (a small to medium sized tree in the wild), erect habit when young, spreading later. Leaves to 30cm(1ft) long, linear with large triangular spine-tipped teeth, lustrous rich green. Flower spikes 15–30cm(6–12in) long, like giant yellow bottle-brushes, opening in spring.
B. hookeriana Western Australia
About 1.2m(4ft) or more in containers (2–3m[$6\frac{1}{2}$–10ft] in the wild), spreading habit. Leaves 10–25cm(4–10in) long, linear, triangular-toothed, semi-glossy. Flower spikes 6–10cm($2\frac{1}{2}$–4in) long by about two-thirds as wide, orange from whitish buds, opening in spring.
B. serratifolia Eastern Australia (Queensland, New South Wales, Victoria)
Potentially to 2m($6\frac{1}{2}$ft) or more in containers (to 6m[20ft] in the wild), erect habit. Leaves linear to oblanceolate, 8–15cm(3–6in) long, saw-toothed, semi-lustrous, mid to deep green. Flower spikes 10cm(4in) or more long, pale greenish cream in bud, opening to orange-yellow and ageing brownish-orange, borne in spring.

BARKERA See EPIDENDRUM

BAROSMA

Rutaceae

Origin: South Africa. A genus of 20 species of small, evergreen shrubs with alternate or paired or, less usually, whorled, obovate to oblong or linear leaves. The 5-petalled, somewhat starry flowers are borne in the upper leaf axils. Propagate by cuttings (best with a heel) or by seed. *Barosma* derives from the Greek *barys*, heavy and *osme*, odour, referring to the strongly aromatic leaves, especially noticeable when they are crushed.

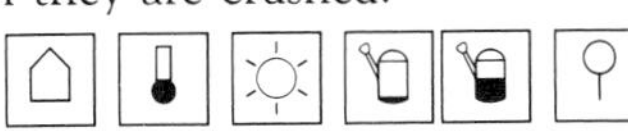

Species cultivated
B. pulchella (*Agathosma pulchella*)
To 1m(3ft) tall, stems slender. Leaves densely borne, leathery and minutely toothed. Flowers rose-purple, to 8mm($\frac{1}{3}$in) wide, freely borne from spring to summer.

BAUERA

Baueraceae (*Saxifragaceae*)

Origin: Australia. A genus of three species of small, evergreen shrubs of rather heath-like appearance from areas of moist soil. They have slender, spreading stems and leaves apparently in whorls of six, though, in fact, made up of a pair of deeply tri-lobed leaves. Flowers are small, white or pink. Propagate by heel cuttings. *Bauera* was named for the Austrian botanical artists Franz (1758–1840) and Ferdinand (1760–1826) Bauer. Franz spent most of his life in England; Ferdinand travelled, taking part in Captain Flinder's pioneer expedition to Australia.

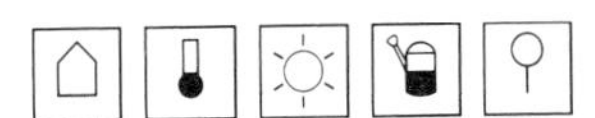

Species cultivated
B. rubioides New South Wales
To 60cm(2ft) in height. Leaf lobes lanceolate, to

12mm(½in). Flowers nodding, borne on thread-like stalks from the upper leaf axils, bowl-like, six to eight red, pink or white petals spreading to 12mm(½in) across at the mouth; they open during spring and summer.

BEAUFORTIA

Myrtaceae

Origin: Western Australia. A genus of 16 species of mainly small to medium-sized wiry-stemmed evergreen shrubs. They are allied to the better known *Callistemon* (bottlebrush), but have shorter flower clusters and much smaller, more crowded leaves. The species described below make attractive pot or tub plants for a sunny, well ventilated conservatory. Propagate by seed in spring or by cuttings in late summer. *Beaufortia* commemorates Mary Somerset, Duchess of Beaufort (*c.*1630–1714), a patroness of botany.

  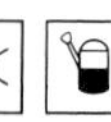 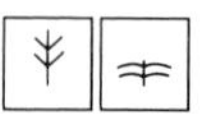

TOP *Beaufortia purpurea*
ABOVE *Beaumontia grandiflora*

Species cultivated

B. eriocephala
To about 60cm(2ft) tall or more, usually of spreading habit. Leaves 5–10mm($\frac{3}{16}$–⅜in) long, linear to lanceolate. Flower spikes terminal, about 2cm(¾in) long and wide, bright red, opening from spring to summer.

B. micrantha
The smallest species, usually under 60cm(2ft) tall, densely twiggy and of spreading habit. Leaves scale-like, to 2mm($\frac{1}{12}$in) long. Flower spikes terminal, about 1cm(⅜in) wide, mauve-pink, opening from spring to summer.

B. orbifolia
Potentially to 2m(6½ft) in containers (to 3m[10ft] in the wild), but easily kept lower by pruning after flowering; erect to spreading habit. Leaves broadly oval to obovate, to 7mm(¼in) long. Flower spikes about 4cm(1½in) long and wide, deep bright red, opening in spring and summer.

B. purpurea
To about 60cm(2ft) or so, densely twiggy and of spreading habit. Leaves lanceolate to linear, 5–10mm($\frac{3}{16}$–⅜in) long. Flower spikes terminal, pinkish to purplish-red, opening in spring and summer.

B. schaureri
Usually about 90cm(3ft) tall, erect to spreading habit. Leaves up to 6mm(¼in) long, linear. Flower spikes 2cm(¾in) long and wide, pink or mauve-pink, opening in spring and summer.

B. sparsa
Up to 2m(6½ft) tall, of fairly open, more or less erect habit with long, slender, elegantly disposed stems. Leaves about 1cm(⅜in) long, ovate to elliptic. Flower spikes 5–7cm(2–2¾in) long and wide, the bright red stamens distinctively long and arching, opening in summer and autumn.

BEAUCARNEA See NOLINA

BEAUMONTIA

Apocynaceae

Origin: Indo-Malaysia, China. A genus of possibly 15 species of woody-stemmed climbers, only one of which is in general cultivation. Although normally high climbing, vigorous species they respond well to container culture and will bloom when fairly small. To initiate flowers they must receive winter night temperatures in the 7–10°C(45–50°F) range, or even a little below this, and, at the same time, must be kept on the dry side. To curtail size, prune immediately after flowering. Propagate by cuttings in late summer. *Beaumontia* commemorates Lady Diana Beaumont (*d.* 1831) of Bretton Hall, Yorkshire.

  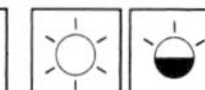  

Species cultivated

B. grandiflora N. India
In containers can be kept to 2–3m(6½–10ft), but capable, when planted out, of attaining twice this and more. Leaves 13–20cm(5–8in) long, ovate-oblong, deep lustrous green above, red-brown hairy beneath, handsome. Flowers 10–13cm(4–5in) wide, trumpet-shaped with five spreading petal lobes, white, fragrant, in small clusters in summer.

B. jerdoniana S. India
Very closely related to *B. grandiflora* and perhaps only a form of it. Generally smaller in all its parts with the leaves smooth beneath and the flowers red or red-flushed in the bud stage.

BEGONIA

Begoniaceae

Origin: Widely distributed in tropical to warm temperate climates, most frequent in South

America, but almost absent from Australasia and the Pacific Islands. A genus of 900 mainly perennial species including some sub-shrubs and climbers. Although very variable, most have characteristic lop-sided, ear-shaped leaves often with beautiful markings. Flowers are dioecious, usually in clusters, with four to five (rarely two) tepals, the ovaries being winged or strongly angled. Some species are tufted and sub-shrubby with a fibrous root system, others are rhizomatous or tuberous. Propagate rhizomatous and tuberous species by division, or by leaf or stem cuttings, or by seed. Most species and hybrids in this huge genus make very good house plants. *Begonia* was named for Michel Bégon (1638–1710), patron of botany and at one time Governor of French Canada.

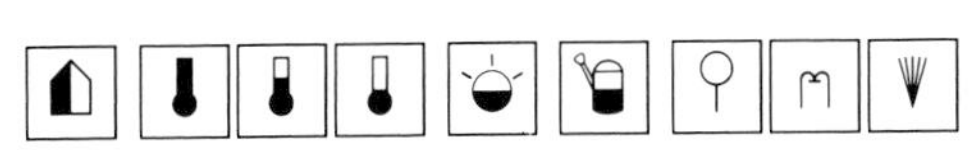

Species cultivated

B. albo-picta Brazil
Shrubby. To 1–2m(3–6½ft) in height, much branched. Leaves elliptic-lanceolate, shining green spotted with silver. Flowers white to greenish-white, 1–1.5cm(⅜–⅝in) wide. Temperate.

B. angularis Brazil
Shrubby. Freely branched to 2m(6½ft). Leaves ovate, slender-pointed, dark glossy green. Flowers white, 8mm(⅓in) wide, borne in large clusters. Temperate.

B. boliviensis Bolivia
Tuberous. Main stems erect 60–90cm(2–3ft) in height, tips and lateral stems sprawling or pendulous. Leaves lanceolate to ovate, to 13cm(5in) long. Flowers scarlet in drooping panicles, somewhat fuchsia-like; males to 5cm(2in) or more long, females less. One of the main parents of the popular race of tuberous begonias (*B.* × *tuberhybrida*). Needs to be kept dry from late autumn to early spring. Temperate.

B. bowerii Miniature eyelash begonia Mexico
Rhizomatous. 10–15cm(4–6in) in height. Leaves broadly ovate, spotted with purple-brown around the margins and edged with white hairs. Flowers white or pink tinged, to 1cm(⅜in) or more wide. Temperate.

B.* × *cheimantha Lorraine/Christmas begonia
Hybrids between *B. socotrana* and *B. dregei*, possibly with other species involved. 'Gloire de Lorraine', raised in 1892, is still the best known cultivar. It is a bushy plant to 45cm(1½ft) having rounded leaves and an abundance of small, icing-pink flowers; the stalks of the flower clusters are also pink. Red and white flowered cultivars are now grown, while 'Mrs Peterson' has bronzed foliage and white, pink-flushed flowers. Pinch the shoots of young plants at least twice to keep them bushy. Some support will be necessary. Tropical.

***B.* × 'Cleopatra'**
A handsome pot plant of hybrid origin. Leaves maple-like, deeply lobed and toothed, yellow-green with a strong suffusion of pinkish-brown. Temperate.

Begonia × 'Cleopatra'

B. coccinea Angel wing begonia Brazil
Shrubby. Stems bamboo-like, swollen at the nodes, 45–90cm(1½–3ft). Leaves obliquely oblong-lanceolate, to 15cm(6in), the upper surface with a waxy bloom, the lower reddish. Flowers coral-red, 1–1.5cm(⅜–⅝in) wide, to 3cm(1¼in) or more long, in pendulous clusters. Temperate. Parent of many hybrids including 'President Carnot' with silver-spotted leaves and paler flowers.

B.* × *corallina (*B. coccinea* × *B. teuscheri*)
Like *B. coccinea*; its 20cm(8in) long, lanceolate leaves white-spotted above, reddish beneath. Flowers pink. Temperate. Sometimes known as *B.* 'Corallina de Lucerna'.

B. cubensis Holly-leaved begonia Cuba
Leaves lustrous deep green, sharply toothed. Flowers white, small. Grown chiefly for its decorative foliage. Temperate.

B. daedalea Mexico
Rhizomatous. To 30cm(1ft) in height. Leaves large, red when young, deepening to mahogany with a network of veins. Flowers white, pink-tinted, borne in a loose cluster. A striking foliage plant. Temperate.

B. dichroa Mexico
Shrubby. Leaves broadly ovate, pointed with wavy margins, glossy green, silver-spotted especially when young. Flowers orange in pendulous clusters. Temperate.

B. dregei Miniature maple leaf begonia
South Africa
Tuberous. Stems to 30cm(1ft) or more. Leaves broadly ovate, toothed, 5–8cm(2–3in) long, bronze above, flushed red beneath. Flowers 2–3cm(¾–1¼in) wide. Temperate. Parent of many notable hybrids.

B.* × *elatior See *B.* × *hiemalis*.

B.* × *erythrophylla (*B.* × *feastii*) (*B. manicata* × *B. hydrocotylifolia*) Beefsteak begonia
20–25cm(8–10in) tall. Leaves kidney-shaped, very thick-textured, to 20cm(8in) wide, lustrous dark green with paler veining above, red beneath with the veining green. Flowers white, 1.2–2cm(½–¾in) wide, in tall panicles. Temperate. *B.* × *e.* 'Bunchii' (Curly kidney begonia) has crested and frilled leaves.

ABOVE *Begonia masoniana*
ABOVE RIGHT *Begonia × corallina*

B. evansiana (*B. grandis*) E. Asia
Tuberous. Erect branched stems to 30–60cm(1–2ft) in height. Leaves ovate-cordate, somewhat lobed, the margins fringed with short hairs, green above, red beneath. Flowers pink, fragrant, 2–3cm(¾–1¼in) wide in loose clusters. Small stem tubers develop. Cool.

B. × feastii See *B. × erythrophylla*.

B. foliosa Fernleaf begonia Colombia
Shrubby, to 90cm(3ft) or more. Much-branched, flattened, arching to pendulous stems, almost frond-like. Leaves ovate-oblong, 1.2–4cm(½–1½in) long, glossy and bronze to dark green. Flowers small, 1.2cm(½in) wide, white with a pink flush. Temperate.

B. fuchsioides Fuchsia begonia Mexico
Shrubby, to 90cm(3ft) or more. Arching, much branched stems and frond-like foliage. Leaves ovate-oblong, 2–5cm(¾–2in) long. Flowers deep pink to rich scarlet, borne in winter to summer. Temperate.

B. glaucophylla Brazil
Rhizomatous. Trailing stems. Leaves 8cm(3in) long, ovate-lanceolate, blue-green, spotted with white above and purple beneath. Flowers small, brick-red, carried in drooping clusters. A good plant for a hanging basket. Temperate.

B. haageana See *B. scharffii*.

B. heracleifolia Star begonia Mexico
Rhizomatous. Leaves long-stalked, 15–25cm(6–10in) wide, palmate, deeply cut into seven to nine sharply toothed lobes, bristly-hairy, each lobe marked with a distinct bronze-green vein. Flowers 2.5–4cm(1–1½in) across, pink, freely borne in panicles 45–60cm(1½–2ft) tall. Temperate.

B. × hiemalis (*B. × elatior*) Winter flowering begonias Garden origin
A group of autumn and winter flowering cultivars produced by hybridizing *B. socotrana* and *B. × tuberhybrida*. They are tuberous or semi-tuberous and are somewhat similar to the Lorraine begonias, *B. × cheimantha*, but have larger leaves and flowers to 5–8cm(2–3in) across, usually semi-double. They are available in a wide colour range, mainly reds and pinks but including apricot and white. Temperate.

B. incarnata Mexico
Fibrous-rooted. To 30–90cm(1–3ft), bushy with erect stems. Leaves ovate-lanceolate, cordate, slender-pointed to 11cm(4½in) long. Flowers pink, to 4cm(1½in) wide in clusters on slender stalks, freely borne for most of the year. Temperate.

B. leptotricha Woolly bear Paraguay
Fibrous-rooted. Hairy stems to about 60cm(2ft). Leaves ovate, wavy-margined, 9cm(3½in) long, silky hairs on the upper surfaces, reddish beneath. Flowers ivory-white, spasmodically throughout the year. Temperate.

B. × 'Lucerna' See *B. corallina*.

B. luxurians Palm-leaf begonia Brazil, Chile
Shrubby. A tall species with bamboo-like stems reaching 2m(6½ft) or more. Leaves digitate, divided into up to 17 arching, short-stalked, lanceolate leaflets each 13–18cm(5–7in) long, bronze-green and finely toothed. The small creamy-white flowers are insignificant. A very handsome foliage plant for a large pot or tub. Tropical.

B. manicata Mexico
Rhizomatous; stout, erect stems. Leaves ovate, glossy green, smooth and somewhat fleshy, 15–20cm(6–8in) long with a prominent ring of downward-pointing red scales at the top of their long stalks. Flowers pale pink, small but freely borne in clusters at the top of long stems. Temperate. *B.m.* 'Aureo-maculata' has leaves blotched with yellow.

B. masoniana Iron cross begonia S.E. Asia
Rhizomatous; low-growing to 15–20cm(6–8in) or more. Leaves very distinctive, broadly ovate to 20cm(8in) or more long with a strongly puckered surface, yellow-green with a bold, dark reddish-brown pattern very like the German Iron Cross. Flowers small, greenish-white and insignificant. A fine foliage plant. Tropical.

B. mazae Mexico
Rhizomatous; slender stems to 30cm(1ft), at first erect, then bending over. Leaves small, broadly ovate, green to bronze with a satiny sheen. Flowers small, pink spotted with red. Temperate.

B. metallica Metal leaf Brazil
Sub-shrubby; to 1m(3ft) or more with succulent stems. Leaves obliquely heart-shaped, 10–15cm(4–6in) long, shallowly lobed and toothed, a shining olive-green with purple veins and red on the undersides. Flowers pinkish-white to 4cm(1½in) across and carried in large clusters. Temperate.

B. olbia Brazil
Fibrous-rooted; branched stems to 15cm(6in). Leaves broadly ovate and thin-textured, bronze-green above, reddish beneath. The small, greenish-white flowers are borne in drooping clusters. Temperate.

B. polyantha Mexico
Fibrous-rooted; slim, erect stems to 60cm(2ft). Leaves elliptic, smooth, dull green. Flowers a clear pink to 2cm(¾in) across, freely borne in winter and spring. A good pot plant. Temperate.

B. × 'President Carnot' See *B. coccinea.*

B. rex Painted leaf/Fan begonia Assam
Rhizomatous. Leaves broadly ovate, the surface crinkly or puckered, dark green with a metallic sheen and a broad, silver-margined band. Flowers pink, in small, erect clusters. Very shade tolerant. Temperate. Most of the named cultivars are of hybrid origin. They have very ornamental and colourful leaves, usually with a dark centre and a margin separated by a whitish or silvery area. Colours can vary from pink, through reds, maroons and browns to all shades of green. These are among the most striking of all foliage begonias.

B. scharffii (*B. haageana*) Elephant's ear Brazil
Shrubby; erect stems to 90cm(3ft). Leaves roughly ovate, 15–25cm(6–10in) long, light olive-green with red veining above, reddish below. Flowers to 3cm(1¼in) or more across, pale pink with red hairs on the petals. A fine pot plant. Temperate.

B. schmidtiana Brazil
Fibrous-rooted, making a low, freely-branched plant. Leaves unevenly heart-shaped to 8cm(3in) long, green above, reddish beneath. Flowers usually pink, sometimes white inside. Temperate.

B. semperflorens Wax begonia Brazil
Fibrous-rooted and bushy; 10–30cm(4–12in) tall. Leaves somewhat fleshy and shining, rounded to broadly ovate, normally green but in some cultivars bronze to purplish-black. Flowers white, pink or red, 2–3cm(¾–1¼in) across, borne freely, sometimes almost hiding the foliage in their profusion. Double-flowered cultivars are also grown. Frequently used for summer bedding, it also makes an undemanding pot plant. Temperate.

B. serratipetala New Guinea
Fibrous-rooted; to 60cm(2ft) tall and wide with erect, then arching and pendent stems. Leaves ovate, slender pointed with shallow lobes and toothing, somewhat pleated, bronze-green with pink to red iridescent spots. Flowers deep pink, 2–3cm(¾–1¼in) across. An attractive pot or hanging basket plant. Temperate.

B. sutherlandii South Africa
Tuberous-rooted; to 15cm(6in) with slender, branched, pendent stems 30–60cm(1–2ft) long. Leaves ovate, to 8–10cm(3–4in) or more in length, shallowly lobed, pale green with red veins. Flowers orange, to 2cm(¾in) across. An excellent species for a hanging basket. Temperate.

B. × tuberhybrida Garden origin
Tuberous. Hybrids derived chiefly from *B. boliviensis*, *B. pearcei* and *B. rosaeflora*. Temperate. A number of groups are recognized, the most important of which are the Camellia-flowered with strong-growing, erect, unbranched stems and large double flowers, and the Pendula group with hanging or arching stems and single or double flowers. The Multiflora or Intermedia group is characterized by abundantly borne smaller flowers on branched stems and a more compact habit. All the hybrids are available in a wide range of colours, from red, pink, orange and yellow to white, some having picotee margins.

Begonia sutherlandii

Begonia semperflorens

BELOPERONE

Acanthaceae

Origin: Tropical and sub-tropical areas of the Americas. A genus of 60 species of evergreen shrubs now often included in *Justicia*. They have pairs of undivided leaves and 2-lipped flowers often with conspicuous bracts. Propagate by cuttings in summer. Pinch the tips from young plants several times to get a bushy specimen. When they become leggy they are best replaced. *Beloperone* derive from the Greek *belos*, an arrow and *perone*, a buckle or rivet, referring to the way the anther lobes are connected.

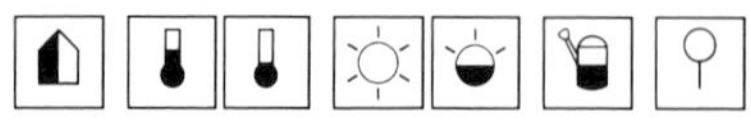

Species cultivated

B. guttata (*Justicia brandegiana*, *Drejerella guttata*) Shrimp plant Mexico
Softly shrubby, to 90cm(3ft) or more, but flowers when small. Leaves 2.5–8cm(1–3in) long, ovate to elliptic, shining. Flowers white, slender, almost hidden beneath the broadly ovate, 2.5cm(1in) long, overlapping shrimp-pink bracts. A popular pot plant. *B.g.* 'Yellow Queen' has bright yellow bracts.

BERBERIDOPSIS

Flacourtiaceae

Origin: Chile. A genus containing one species only, native to warm temperate rain forests. It is a useful cool conservatory tub plant especially for a shady position, and can be grown against a wall or trained across roof space. Propagate by cuttings or seed in spring or by layering a branch into a pot in autumn. *Berberidopsis* derives from *Berberis*, the genus and the Greek *opsis*, like, referring to the general appearance of the plant.

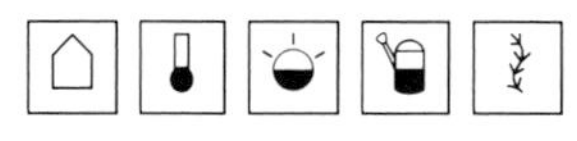

Species cultivated

B. corallina Coral plant
A climber which can reach 6m(20ft). Leaves oblong-cordate to 8cm(3in), prickly toothed, dark green above, glaucous beneath. Flowers rounded or bowl-shaped to 1cm(⅜in) across, crimson-red, borne on long red stalks in pendent racemes in late summer and early autumn.

BERGENIA

Saxifragaceae

Origin: Central and E. Asia. A genus of six species of stout evergreen perennials with semi-woody rhizomes and large, leathery, paddle-shaped leaves. In spring they have showy panicles of pink, red or white bell-shaped flowers made up of five and sometimes more petals. Propagate by division or by seed. Most species are hardy but the more tender *B. ciliata* makes a useful plant for the cool conservatory, flowering in late winter. *Bergenia* was named for Carl August von Bergen (1704–60), German physicist and botanist.

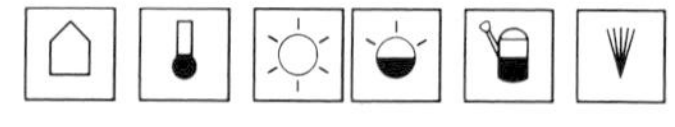

Species cultivated

B. ciliata (*B. ligulata*)
Leaves to 25cm(10in) or more, orbicular to obovate and hairy on both surfaces, dark green. Flower clusters loose and spreading, the flowers white becoming red-flushed as they age, and held within pink sepals; they are slightly fragrant. The form *ligulata* has hairs only on the leaf margins.

BERTOLONIA

Melastomataceae

Origin: Brazil and Venezuela. A genus of ten species of evergreen perennials. They are small plants with decorative ovate to heart-shaped or elliptic, somewhat fleshy leaves often patterned in white or silver or having a metallic, almost crystalline lustre. The flowers, which are carried in terminal clusters, are small and may be removed as soon as the spike appears. They are natives of tropical rain forests where they grow on the ground and are thus adapted to a high humidity and no direct sunlight. In the home these conditions are best provided by a glass case or terrarium. Propagate by stem or leaf cuttings in a case with a minimum

BELOW L–R *Beloperone guttata* 'Yellow Queen', *Berberidopsis corallina*, *Bergenia ciliata ligulata*

temperature of 21–23°C(70–74°F). *Bertolonia* was named for Antonio Bertolini (1775–1869), Professor of Botany at Bologna.

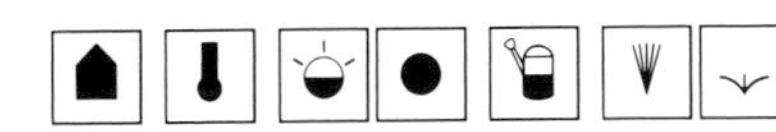

Species cultivated

B. hirsuta Jewel plant
Leaves hairy, vivid green with bold veining and a central reddish zone. Flowers small and white.

B. × houtteana
Leaves ovate to elliptic, 10–18cm(4–7in) long, dull green with cross veins strongly marked with silvery pink. Flowers small and pink.

B. maculata
Leaves usually in rosette-like tufts, ovate-cordate, 10–20cm(4–8in) long, rich olive-green with a narrow, somewhat irregular central silvery stripe. Flowers rose-pink to violet-purple in short, clustered spikes.

B. marmorata
Leaves ovate-oblong, cordate, 11–20cm(4½–8in) long, velvety green with depressed veins giving the leaf a quilted look, and a central silvery stripe; the undersides are purple. Flowers small and red-purple. *B.m.* 'Mosaica' has broader silvery leaf vein marking with pinkish tints.

B.m. aenea
Leaves with a coppery or bronze lustre without the silvery marking.

B. sanguinea
Leaves with a bronze sheen and a central silvery band; reddish beneath.

BESCHORNERIA

Agavaceae

Origin: Mexico. A genus of ten species of evergreen perennials with a strong tuberous rootstock and basal rosettes of thick-textured, narrowly lanceolate leaves. The species mentioned make striking tub plants for the larger conservatory and can be stood outside in summer. Propagate by removing and rooting the sucker-like offsets in late spring, or alternatively by seed in spring. *Beschorneria* was named for Friedrich W.C. Beschorner, a German amateur botanist.

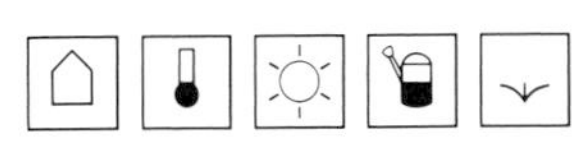

Species cultivated

B. bracteata
Stemless. Leaves 20 to 30, grey-green, 45–60cm(1½–2ft) long and 5cm(2in) wide. Flowering stems branched, up to 2m(6½ft) high with greenish-yellow flowers to 4cm(1½in) long which redden as they age.

B. yuccoides
Stemless. Leaves usually 15 to 20, grey-green, 30–45cm(1–1½ft). Flowering stem branchless, red, to 1.2m(4ft) or more tall; flowers pendent, tubular, bright green, to 5cm(2in) with red bracts.

Bertolonia marmorata

BESSERA

Alliaceae (Liliaceae)

Origin: USA and Mexico. A genus of three species of cormous plants. They have narrowly linear basal foliage and slender flower stalks bearing umbels of nodding flowers with white, pink or red tepals. One plant is in general cultivation making a pleasant pot plant. Propagate by separating offsets or by seed. *Bessera* was named for W.S.J.Th. van Besser (1784–1842), an Austrian Professor of Botany.

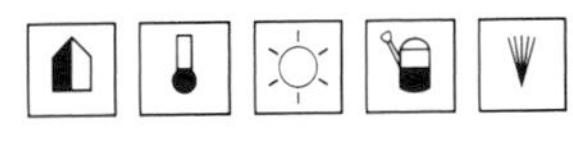

Species cultivated

B. elegans Coral drops Mexico
Leaves 30–60cm(1–2ft) in length, semi-prostrate. Stems to 60cm(2ft), erect, bearing up to 12 nodding, widely bell-shaped flowers each with six scarlet tepals carried in terminal umbels in late summer.

BIFRENARIA

Orchidaceae

Origin: Tropical South America. A genus of ten species of epiphytic orchids forming clusters of ovoid, somewhat angled pseudobulbs each topped by a solitary, oblong-lanceolate to elliptic, leathery leaf. The flowering stems arise from the base of the pseudobulbs and bear one to three spurred flowers each with broadly overlapping tepals and a usually 3-lobed lip (labellum). Propagate by division when the plants become congested. A rewarding orchid for a warm room or conservatory. *Bifrenaria* derives from the Latin *bis*, twice and *frenum*, a bridle, alluding to the way the two pollinia are attached.

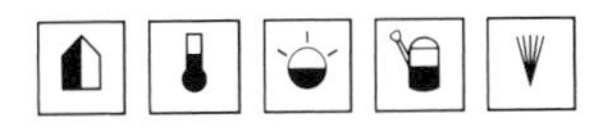

Species cultivated

B. aurantiaca Northern South America, Trinidad
Pseudobulbs to 5cm(2in) tall. Leaves oblong-lanceolate, 10–20cm(4–8in) long, freckled with tiny purple dots. Stems 15–22cm(6–9in) high, bearing seven to 12 flowers opening in autumn,

each bloom about 2.5cm(1in) wide, tepals orange-yellow, variably purple-spotted; labellum somewhat paler.

B. fuerstenbergiana See *B. inodora.*

B. harrisoniae Brazil
Pseudobulbs 5–8cm(2–3in) tall. Leaves oblong-elliptic, leathery and glossy to 30cm(1ft) long. Stems to 15cm(6in), bearing one to two heavily fragrant flowers in summer; each about 7cm($2\frac{3}{4}$in) across, waxy ivory-white to yellowish or greenish, sometimes flushed with red at the tips of the sepals; labellum rich reddish-purple with darker veining.

B. inodora (*B. fuerstenbergiana*) Brazil
Similar in habit to *B. harrisoniae* but the flowers are yellow-green to apple-green and the lip pink, yellow or white; they open from winter to summer and are variably fragrant.

B. tetragona Brazil
Pseudobulbs to 9cm($3\frac{1}{2}$in) tall. Leaves elliptic-oblong to 45cm($1\frac{1}{2}$ft). Flowers carried three to four to a stem in summer, each bloom 4–5cm($1\frac{1}{2}$–2in) across, very fragrant, tepals greenish-yellow streaked purplish-brown; labellum violet-purple flushed.

B. tyrianthina Brazil
Similar in habit to *B. harrisonae* but stronger growing. Flowers to 9cm($3\frac{1}{2}$in) across, red-purple; labellum deep violet; borne from spring to summer.

BIGNONIA

Bignoniaceae

Origin: USA. A genus now reduced to one species – an evergreen woody climber from the south-eastern states of the USA which makes an attractive tub plant for training up a wall or across the roof of a conservatory. With restricted root room and judicious pruning its normally vigorous nature can be held in check. Propagate by cuttings, layering or seed. *Bignonia* was named for Abbé Jean Paul Bignon (1662–1743), Librarian to Louis XIV of France.

Bignonia capreolata

Species cultivated

B. callistegioides See *Clytostoma callistegioides.*

B. capensis See *Tecomaria capensis.*

B. capreolata (*Doxantha capreolata*) Cross vine
Climber to 12m(40ft), or more in the open ground. Leaves made up of pairs of opposite leaflets which are oblong-ovate to lanceolate, 5–13cm(2–5in) long, and three or more branched, twining tendrils. Flowers bell-shaped, tubular, 4–5cm($1\frac{1}{2}$–2in) long, opening to a 5-lobed mouth, reddish-orange shadded to yellow-orange, borne in small axillary clusters in summer.

B. chamberlaynei See *Anemopaegma chamberlaynei.*

B. purpurea See *Clytostoma binatum.*

B. speciosa See *Clytostoma callistegioides.*

B. stans See *Tecoma stans.*

B. unguis-cati See *Doxantha unguis-cati.*

BILLBERGIA

Bromeliaceae

Origin: Tropical and sub-tropical America. A genus of 50 species of evergreen perennials most of which are epiphytic though a few grow naturally on the ground. They have leathery, strap-shaped leaves usually in a rosette which is tubular and water-holding at the base. The rosettes can be solitary but are normally clustered. Flowers are tubular comprising three petals and often colourful, borne within showy bracts atop an arched or sometimes erect stem. Plants are tolerant of poor conditions and neglect but can become shabby and then are slow to recover. With their attractive flowers these bromeliads are rewarding house or conservatory plants. Propagate by division or offsets. *Billbergia* was named for J.G. Billberg (1772–1844), a Swedish botanist.

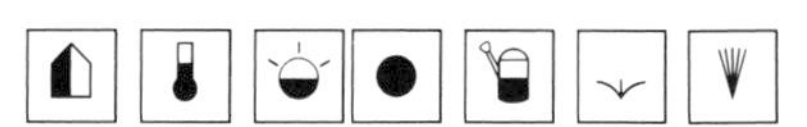

Species cultivated

B. amoena Brazil
Very variable in size. Leaves 20–60cm(8–24in) in length, strap-shaped, slightly spiny, green with silver cross-banding. Bracts rose-pink; flowers green, margined with blue, in 4–6cm($1\frac{1}{2}$–$2\frac{1}{2}$in) long racemes on arching stems, borne in late summer. Several forms are grown with variable cross-banding.

B.a. penduliflora
Narrower, grey-green leaves and orange bracts on more pendent stalks.

B.a. rubra
To 90cm(3ft), pinky-red leaves striped and spotted white.

B.a. viridis
Green leaves with white striping and spotting.

***B.* × 'Albertii'** (*B. distachya* × *B. nutans*)
Leaves linear-oblong, dark green with grey scaling, 30cm(1ft) or more long, in tubular rosettes. Flowers green and blue, bracts rosy-red.

B. × 'Bob Manda' Manda's urn plant
A hybrid of *B. pyramidalis* and favouring that species, but more compact; basic colour coppery olive-green, sometimes with a purple tint, variegated with cream blotches.
B. decora Brazil, Peru, Bolivia
Leaves strap-shaped, pointed, 45–60cm(1½–2ft) long, in erect, tubular rosettes, dark green, grey mottled and cross-banded, the margins minutely spiny toothed. Stem shorter than leaves with a pendent raceme of large, carmine bracts and green flowers with rolled petals.
B. distachia Brazil
A variable plant, but in its basic form much like a taller *B. nutans*, the most obvious difference being the grey-white scurfy, sometimes purple-tinted leaves. *B.d.* 'Maculata' has yellow-blotched leaves.
B.d. concolor
Similar, but flowers green.
B. × 'Fantasia' (*B. saundersii* × *B. pyramidalis*)
In general appearance rather like the *B. saundersii* parent but larger, the coppery-green leaves boldly blotched cream and pink. The pink bracts and blue flowers form a more compact raceme than in *B. saundersii.*
B. forgetii See *Aechmea caudata* 'Variegata'.
B. morelii Brazil
Leaves green, stiff, sword-shaped, to 60cm(2ft) long with spiny margins. Flowers green, blue and red in a pendent spike to 5cm(2in) long with bright red bracts; borne on a stem to 30cm(1ft) in length in late summer.
B. nutans Queen's tears, Friendship plant
Brazil, Argentina and Uruguay
Leaves dark green, to 40cm(16in), borne in tubular rosettes which can build up into dense clumps. Flowering stem erect, arching at the tip; flowers greenish-yellow margined with purple-blue and with pink bracts, opening in early spring. Nectar forms within the flowers and sometimes spills out when the plant is touched giving rise to one of its vernacular names.
B. paniculata See *Aechmea penduliflora.*
B. pyramidalis Brazil
Leaves broadly strap-shaped, to about 30cm(1ft) long, finely toothed, carried in vase-shaped rosettes. Stem erect, bearing red bracts and purple, crimson-tipped flowers in summer. *B.p.* 'Striata' (Striped urn plant) has dark green leaves striped with cream; winter flowering.
B. rhodocyanea See *Aechmea fasciata.*
B. rubro-cyanea See *B. saundersii.*
B. × 'Santa Barbara' Banded urn plant
A hybrid probably involving three species, *B. distachya*, *B. nutans* and *B. pyramidalis*. It is similar to the last named but more vigorous, with grey-green leaves striped with yellow and pink, and forming longer and narrower rosettes.
B. saundersii (*B. rubro-cyanea*)
Leaves strap-shaped, to 40cm(16in), in narrowly tubular rosettes, dark green above, purplish to bronze beneath, both surfaces spotted with white and sometimes also with red. Flowering stems strongly arching, bearing rose-red bracts and pendent yellow-green, blue-edged flowers.

LEFT *Billbergia* × *windii*
BELOW *Billbergia vittata*
BOTTOM *Billbergia zebrina*

B. vittata Brazil
Leaves strap-shaped to lanceolate, 90cm(3ft) or more long, hard-textured, dark grey-green and grey scaly above, boldly grey cross-banded beneath, forming an erect, broadly tubular rosette. Stem shorter than leaves with red bracts to 20cm(8in) long and tipped with a pendent spike of blue, red and greenish-white flowers, the petals of which roll back.
B. × windii (*B. decora* × *B. nutans*)
Leaves linear, grey-green to 30cm(1ft) long, slightly twisted, arching, forming shortly tubular rosettes soon making congested clumps. Flowers greenish yellow with purple-blue margins, together with large rose-red bracts, borne on strongly arching stems from autumn to early summer.
B. zebrina Brazil
Rosette tube-like; leaves strap-shaped, 90cm(3ft) or more long, dark green with grey-white cross-banding. Stem pendent, woolly-scaly, to 40cm(16in) long, bearing large pink bracts and a terminal spike of blossom; each flower up to 8cm(3in) long in bud, greenish-yellow, the petals coiling like a spring on opening.

BIOPHYTUM

Oxalidaceae

Origin: Pan-tropical. A genus of 70 species of perennials, often woody at the base. Two species are grown for their novelty as, like *Mimosa pudica*, the leaflets fold back from their midrib when touched. They can be grown in a pot or pan and make good subjects for a terrarium. Propagate by seed in spring. *Biophytum* derives from the Greek *bios*, life and *phyton*, plant, alluding to the sensitive leaves.

Species cultivated

B. sensitivum Sensitive plant
Leaves pinnate, 5–8cm(2–3in) long, made up of six to 15 pairs of small, oblong leaflets; they are borne in a rosette on a stem which becomes woody and elongates to 10cm(4in) or more, giving the effect of a tiny palm tree. Flowers small, yellow and funnel-shaped, followed by seed capsules which split to their base explosively, flinging out the seed.

B. zenkeri
A very similar plant which is sometimes listed.

Biophytum sensitivum

BLECHNUM

Blechnaceae (*Polypodiaceae*)

Origin: World-wide, but chiefly in the southern hemisphere. A genus of 220 species of handsome, decorative ferns, usually evergreen, but some are partly deciduous. They have pinnate or palmately lobed fronds, in most species forming rosettes but occasionally tufted or rhizomatous, forming mats. The spore clusters (sori) are linear and are found on either side of the midribs of the pinnae, sometimes on separate, fertile erect fronds with narrower pinnae. Those cultivated under cover are chiefly tropical species and grow well in containers if given warmth and shade. They are also more tolerant of drier atmospheres than many ferns. Propagate by division or by spores. *Blechnum* derives from the classical Greek name for an unknown species of fern and was adopted for this genus by Linnaeus.

 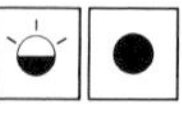 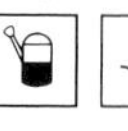 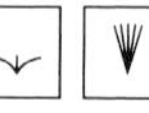

Species cultivated

B. auriculatum (*B. hastatum*) Temperate South America
Fronds up to 60cm(2ft) in length, lanceolate in outline, pinnate, the pinnae lanceolate-falcate, the base of the upper side expanded to an ear-like lobe. Spore-bearing pinnae narrower, with bands of sori midway between the midrib and margins.

B. brasiliense Brazil, Peru
Fronds oblong-lanceolate, pinnatipartite, 60–120cm(2–4ft) long, borne in a wide rosette on a brown, scaly trunk-like rhizome which can be 60–90cm(2–3ft) high with age. Fertile and sterile fronds are the same and the narrow pinnae are finely toothed.

B.b. crispum
Fronds coppery-red when young, smaller and wavy-margined.

B. gibbum (*Lomaria gibba*) New Caledonia and adjacent islands
Fronds 60–90cm(2–3ft) long, lanceolate to narrowly ovate, pinnatipartite, the pinnae very slender and tapering to a point. The narrower, fertile fronds are erect. Rosette wide-spreading, topping a slender, black-scaly trunk to 90cm(3ft) in height. An elegant miniature tree fern.

B.g. platyptera
Somewhat larger and faster growing; it may be of hybrid origin.

B. moorei (*Lomaria ciliata*) New Caledonia
Fronds ovate-oblong, 20–30cm(8–12in) long, pinnate with oblong, lobed or toothed and wavy pinnae. Rosettes neat and spreading; sterile and fertile fronds distinct, the latter narrowly linear, the undersides completely covered with sori. The plant cultivated under this name is probably only a distinct form of *B. gibbum* and not the true species.

B. occidentale Tropical South America
Fronds 35–70cm(14–28in) long, ovate, pinnatipartite, the pinnae narrowly oblong, slightly wavy and closely set. Fertile and sterile fronds similar. Rosette-forming but also spreading by underground stolons.

B. patersonii Strap water fern Australia (Queensland to Tasmania)
Fronds 30–60cm(1–2ft) long, strap-shaped, pointed-tipped, variable in form, from completely entire to deeply lobed; pink when young, maturing deep green. Fertile fronds narrower. Leaf tuft eventually dense, from slowly creeping rhizomes.

BLETIA See BLETILLA & CALANTHE

BLETILLA

Orchidaceae

Origin: Eastern Asia. A genus of nine species of terrestrial orchids with tuber-like pseudobulbs which remain underground. From these arise the pleated leaves and stalked flower clusters. The species described is easy to grow, doing well in a

LEFT *Blechnum gibbum*
BELOW *Bletilla striata*

pot; it is naturally herbaceous and can be kept dormant in winter. Propagate by division in spring. *Bletilla* is a diminutive of *Bletia*, the name of the orchid genus in which *Bletilla* was once included.

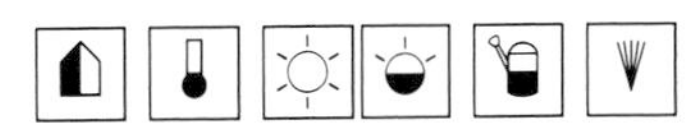

Species cultivated

B. striata (*Bletia hyacinthina* of gardens) China, Japan
Pseudobulbs rounded, to 2cm(¾in) in diameter. Leaves narrowly oblanceolate, pleated, to 30cm(1ft) or more long. Stems 30–60(1–2ft) tall, wiry, bearing a raceme of five to 12 red-purple flowers 3–5cm(1¼–2in) across; the labellum is deeper in colour and markedly keeled. *B.s.* 'Alba' has white flowers, sometimes with a rosy tint, and a pinkish labellum.

BLOOMERIA

Alliaceae

Origin: California, USA. A genus of two species of cormous plants resembling *Brodiaea* and also allied botanically to *Allium* and *Nothoscordum*. Their corms have a fibrous coat and produce only one, or just a few, grassy leaves. The 6–petalled flowers are borne in umbels. The species described makes a good pot plant for the conservatory and can be brought into the home when in bloom. Propagate by seed in spring or by separating offsets at repotting time in autumn. *Bloomeria* commemorates Dr H.G. Bloomer (1821–74), a pioneer botanist of the California flora.

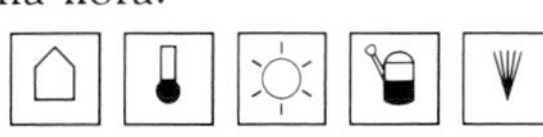

Species cultivated

B. crocea Golden stars
Corm about 1.5cm(⅝in) wide, producing one arching linear leaf up to 30cm(1ft) long. Flowering stem erect, 25–45cm(10–18in) tall, carrying a terminal many-flowered umbel; individual flowers starry, orange-yellow with a medium dark line down each petal, opening in summer.
B.c. aurea
Similar, but with yellow flowers.

BLOSSFELDIA

Cactaceae

Origin: Borders of Argentina and Bolivia (steep mountain slopes). A genus of possibly two or three species of tiny cacti without spines. They have turnip-shaped storage roots and more or less globular, thin-skinned stems which can absorb mist, rain and dew. In the wild the stems may become completely dry and papery without harm, such is their adaptation to drought. They must on no account be over-watered. Except during the hottest weather it is recommended to spray the stems once or twice a week instead of the traditional watering. Propagate by seed in spring or by offsets in summer; plants can also be raised from root cuttings. *Blossfeldia* was named for Harry Blossfeld, German nurseryman and collector of cacti.

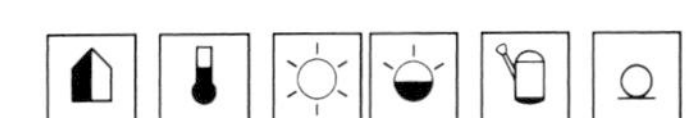

Species cultivated

B. liliputana
Eventually forming clumps or colonies; individual stems 1.2–2.5cm(½–1in) high, varying from

globular to disk-shaped, freely dotted with tiny white woolly areoles. Flowers about 5mm($\frac{3}{16}$in) wide, pale yellow, from the top of the stem in summer.

BOEHMERIA

Urticaceae

Origin: Tropics and sub-tropics, widespread. A genus of about 100 species of trees, shrubs and evergreen perennials, a few of which are grown for their large, handsome leaves. The tiny, petalless greenish flowers are carried insignificantly in the leaf axils. These plants are best used in the conservatory, making good background foliage for more colourful plants. Propagate by cuttings in summer; *B. nivea* also by division or suckers in spring. *Boehmeria* commemorates George Rudolf Boehmer (1723–1803), a German professor of botany.

Species cultivated

B. argentea Mexico

To 2m($6\frac{1}{2}$ft) or more in containers (a small tree to 10m[30ft] in the wild). Leaves about 30cm(1ft) long, the upper surface rugose and glaucous often with silvery or whitish margins. Plants soon get leggy and are best when propagated regularly and discarded after two or three years.

B. nivea Ramie, China grass Tropical Asia

Clump to colony-forming, erect perennial 1–1.5m(3–5ft) tall. Leaves broadly ovate, about 15cm(6in) long, toothed and with a slender pointed tip, medium to deep green above, white-felted beneath. The best foliage is obtained if the plants are cut back to near the base each spring, and propagated regularly. In tropical countries, mainly in southern China and Japan, it is an important crop, grown for its stem fibres which are among the strongest, longest and silkiest known, being superior to cotton but more difficult to process.

ABOVE RIGHT *Boehmeria nivea*
RIGHT *Bolbitis* sp.

BOLBITIS

Lomariopsidaceae (*Aspidiaceae*)

Origin: Pantropical. A genus of 85 species of ferns, the majority of which are epiphytic on trees and mossy rocks by streams. They are characterized by having much reduced spore-bearing fronds, the undersides of which are entirely coated with small sporangia. The species described below are rhizomatous and are best in the shady conservatory, either grown on branches like bromeliads or in pots of loose, peaty compost. Propagate by division or by spores in spring. *Bolbitis* derives from the Greek *bolbo*, a bulb, referring to the curiously swollen leaf veinlets.

Species cultivated

B. cladorrhizans Central America, West Indies

Shortly rhizomatous, eventually forming wide clumps. Fronds arching, 30–60cm(1–2ft) or more long, ovate to triangular-ovate, pinnate at the base, pinnatifid above, the basal pinnae with large rounded teeth.

B. heteroclita Tropical Asia

Rhizomes slender, creeping, eventually forming colonies. Fronds erect, sometimes entire, but usually pinnate, 15–30cm(6–12in) long, divided into three to seven broadly lanceolate, thin-textured, bright green pinnae.

BOMAREA

Alstroemeriaceae

Origin: South America, Mexico and the West Indies. A genus of 150 species, most of which are fleshy- or tuberous-rooted twining climbers. The leaves are usually twisted bringing the undersurface to the top, and the bell-shaped flowers with six tepals are borne in branched, terminal umbels. They grow well in pots, flowering freely, but need sticks for support; alternatively in a conservatory they are effective growing over an evergreen shrub. Propagate by seed or by careful division. *Bomarea* honours Jacques Christophe Valmont de Bomare (1731–1807), a French benefactor of science.

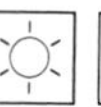

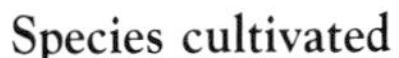

Species cultivated

B. acutifolia Mexico, Guatemala
To 3m(10ft). Leaves oblong-lanceolate. Flowers 2.5–4cm(1–$1\frac{1}{2}$in) long, outer tepals pink or yellow, tipped with green, inner tepals yellow spotted with pink; borne in umbels of ten to 30, each umbel branch with one or two flowers.

B. andimarcana See comments under *B. pubigera*.

B. caldasii Ecuador, Colombia
To 4m(13ft). Leaves ovate-lanceolate, to 15cm(6in) long. Flowers 4cm($1\frac{1}{2}$in) long, bright orange, the outer tepals tipped with green, the inner spotted crimson; borne in umbels of six to 20 in late spring and summer.

B. × cantabrigensis
A hybrid between *B. caldasii* and *B. hirtella*; growth like *B. caldasii*. Flowers tubular, outer tepals pinkish-orange tipped with brownish-green, inner tepals projecting beyond the outer, yellow, freckled with red-purple; borne in a close, umbel-like cluster.

B. carderi Colombia
To 4m(13ft). Leaves oblong-lanceolate to 18cm(7in). Flowers to 7cm($2\frac{3}{4}$in) long, pink, the inner tepals with marginal brown spots, summer.

B. edulis Mexico to Cuba and Peru
To 3m(10ft). Leaves oblong-lanceolate. Flowers 2.5–4cm(1–$1\frac{1}{2}$in) long, the outer tepals pink, inner yellow or green spotted with red; borne in clusters of ten to 30, each umbel-branch with two or more flowers.

B. kalbreyeri Colombia
Often confused with, and rather like, *B. caldasii*. Flowers reddish-orange to brick-red, the inner tepals orange-yellow spotted with red.

B. pubigera Peru
Non-climbing, evergreen. Stems 2–3m($6\frac{1}{2}$–10ft) tall, pushing through shrubs for support. Leaves narrowly lanceolate, 5–12cm(2–$4\frac{3}{4}$in) long with short, twisted white hairs beneath. Flowers tubular, 4.5cm($1\frac{3}{4}$in) long, minutely downy, pale yellow suffused with pink and tipped green. In all but botanical details identical with *B. andimarcana*.

BORONIA

Rutaceae

Origin: Australia. A genus of 70 species of evergreen shrubs. They are generally small to medium-sized, slender-stemmed plants with opposite, simple or pinnate leaves and small 4-petalled, somewhat fleshy-textured flowers in axillary clusters. They are attractive pot plants for a sunny window or conservatory, flowering in spring. With pruning they can be kept compact. Propagate by seed in spring or by cuttings with a heel in late summer. *Boronia* was named for Francesco Borone (1769–1794), an Italian plant collector who worked for Dr Adam Afzelius in Sierra Leone, Professor John Sibthorp in Greece and Sir J.E. Smith in various parts of Europe.

  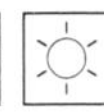  

Species cultivated

B. citriodora Tasmania
Spreading shrub, rarely above 30cm(1ft) tall. Leaves lemon-scented, simply pinnate, composed of three to seven lanceolate to oblanceolate, thick-textured leaflets each up to 1.5cm($\frac{5}{8}$in) long. Flowers star-shaped, soft pink, 1–1.5cm($\frac{3}{8}$–$\frac{5}{8}$in) wide, opening in clusters from the upper leaf axils in spring. This mountain species is the hardiest in cultivation, standing light frost in winter.

B. elatior Western Australia
Shrub to 1m(3ft) or more if not pruned. Leaves imparipinnate, 2.5–4cm(1–$1\frac{1}{2}$in) long, made up of five to 13 linear leaflets. Flowers 6mm($\frac{1}{4}$in) wide, globose, carmine-red, borne in spring.

B. heterophylla Western Australia
Shrub to 1m(3ft), straggling to 2m($6\frac{1}{2}$ft) if not pruned. Leaves 3–5cm($1\frac{1}{4}$–2in) long, undivided, linear or with three to five leaflets. Flowers 8mm($\frac{1}{3}$in) long, ovoid, rose-crimson, in clusters of four to six from the leaf axils in summer.

B. megastigma Western Australia
Low shrub, 30–60cm(1–2ft) high. Leaves with

TOP LEFT *Bomarea pubigera*
TOP RIGHT *Bomarea caldasii*
ABOVE *Boronia citriodora*

three to five narrowly linear leaflets 8–18mm(⅓–¾in) long. Flowers 1–1.2cm(⅜–½in) wide, strongly fragrant, rich purple-brown outside, yellow within, borne singly from the upper leaf axils.

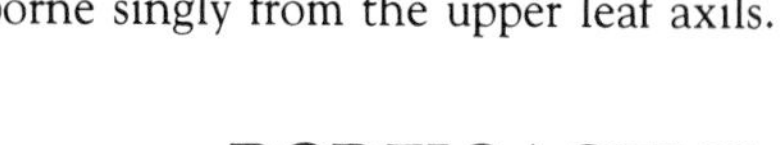

BORZICACTUS

Cactaceae

Origin: South America (Ecuador to N. Argentina). A genus of about ten species of very variably sized cacti, some of which were formerly classified in such genera as *Oreocereus*, *Matucana*, *Submatucana* and others. They are easily grown, the large sorts eventually making statuesque specimen plants. *Borzicactus* is named for the Italian botanist Professor Antonia Borzi (1832–1921), a Director of the Palermo Botanic Garden.

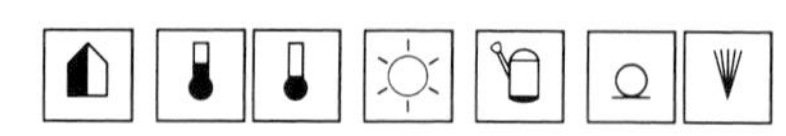

Borzicactus celsianus

Species cultivated

B. aurantiacus (*Submatucana aurantiaca*) Peru
Stems clump-forming, more or less globose, to 15cm(6in) tall, dark green with angular tubercles. Areoles woolly, with 20 to 30 combined radial and central spines, each 2.5–4cm(1–1½in) long. Flowers funnel-shaped, to 9cm(3½in) long, deep red to orange-yellow.

B. celsianus (*Oreocereus neocelsianus*, *Pilocereus celsianus*) Old man of the mountains Bolivia
Stems cylindrical, erect, to 1.2m(4ft), but slow-growing, eventually forming clumps. Ribs ten to 18, rounded and notched, with large oval areoles bearing long, whitish, matted hairs which clothe the entire stem at least when young. Radial spines more or less hidden, central spines one to four, stout, reddish, 5–8cm(2–3in) long. Flowers 9cm(3½in) long, red, usually only on mature plants.

B. decumbens See *Haageocereus decumbens*.

B. haynei (*Matucana haynei*, *M. blanckii*, *M. cereoides*, *M. elongata*, etc) Peru
Stem globose to oblongoid, eventually to 60cm(2ft) high by 10cm(4in) or more thick, but slow-growing. Ribs 25 to 30, not very prominent, with low, rounded tubercles. Areoles numerous, oval, densely yellow-woolly. Radial and central spines similar, 28 to 38, spreading, white, 4cm(1½in) long or more, forming a mesh over the entire stem when young. Flowers about 8cm(3in) long, orange-red, only on mature plants.

Borzicactus trollii

B. strausii See *Cleistocactus strausii*.

B. trollii (*Oreocereus trollii*) Old man of the Andes S. Bolivia to N. Argentina
Stem erect, cylindrical, to 60cm(2ft) or more by 13cm(5in) thick, sometimes solitary but usually forming clumps, slow-growing. Ribs ten to 25 but not very obvious, bearing large oval areoles with abundant white hairs to 5cm(2in) long and clothing the entire stem. Spines eight to 12, yellowish, reddish or white, radials to 2cm(¾in), the one to three centrals to 5cm(2in) long. Flowers carmine to rose-red, about 9cm(3½in) long, but only on mature plants.

BOUGAINVILLEA

Nyctaginaceae

Origin: Tropical and sub-tropical areas of South America. A genus of 18 species of shrubs and scrambling climbers. Those described are climbers having alternate, ovate to elliptic leaves and groups of three small flowers, each made up of three tubular tepals, the whole surrounded by three large, colourful, papery bracts. These in turn are arranged in panicle-like clusters, abundantly borne. They are very colourful climbers suitable for large pots or tubs in the conservatory where they can be trained on to a wall or under the roof. They will also flower when young in small pots. *Bougainvillea* was named for Louis Antoine de Bougainville (1729–1811), a French navigator and sailor who sailed around the world in 1767–9.

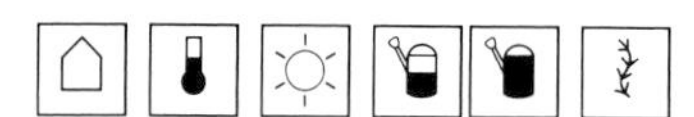

Species cultivated

B. × buttiana
A hybrid between *B. glabra* and *B. peruviana* reaching 6m(20ft) or more in height. Leaves somewhat variable, broadly ovate, to 18cm(7in) long on vigorous stems, usually less than this. Flowers in terminal and lateral clusters. Many cultivars are available, including (colours are for bracts): 'Brilliant', coppery-orange, maturing to cerise; 'Easter Parade', pink; 'Golden Glow', 'Killie Campbell' and 'Orange King', varying shades of orange; 'Jamaica White', pure white; 'Mrs Butt', crimson to magenta; 'Pigeon Blood', deep red; 'Scarlet O'Hara', brilliant orange-red; and 'Surprise', a curious chimaera with bracts that can be rose-purple, white, or a mixture of the two.

B. glabra Paper flower Brazil
To 4m(13ft) or more. Leaves elliptic, to 8cm(3in) long. Bracts to 4cm(1½in) long in shades of cyclamen-purple. Several cultivars are grown including: 'Cypheri', strong-growing, leaves to 11cm(4½in) and bracts to 6cm(2½in); 'Magnifica' (*B. magnifica* 'Traillii'), similar to 'Cypheri' but

freer-flowering and a deeper purple; 'Mrs Leano', freely borne purple-red bracts; 'Sanderiana', especially good for pot culture, flowering freely when small; 'Snow White', the only pure white-bracted cultivar but not often seen; 'Variegata', leaves boldly variegated with white, a useful foliage plant.
B. peruviana Northern South America (Pacific coast area)
Much like *B. spectabilis* and by some botanists considered to be only a form of it. The main differences are a complete lack of hairs (even *B. glabra* has some hairs) and very slender flowers. The bracts, which are bright purplish-pink, tend to have crimped margins. The cultivar names 'Ecuador Pink', 'Lady Hudson' and 'Princess Margaret Rose' all refer to this species.
B. spectabilis Brazil
To 7m(23ft) or more; stems somewhat spiny. Leaves broadly elliptic to ovate. Bracts to 4cm($1\frac{1}{2}$in) or more long, purple to red. *B.s.* 'Mary Palmer' has both carmine and white bracts occurring on the same plant.
B.s. laterita
Bracts smaller, of a pleasing brick-red shade.
B.s. thomasii
Bracts pink, otherwise like the species.

TOP *Bougainvillea spectabilis*
LEFT *Bougainvillea glabra*
ABOVE *Bouvardia laevis*

BOUVARDIA

Rubiaceae

Origin: Tropical and sub-tropical America. A genus of 50 species of shrubs and perennials. They have opposite pairs or whorls of three or more undivided, usually ovate to lanceolate leaves and tubular, 4-lobed fragrant flowers in flattish terminal or axillary clusters or corymbs. They make good pot plants, flowering well when small. Propagate by cuttings in spring in a case with bottom heat. *Bouvardia* was named for Dr Charles Bouvard (1572–1658), at one time in charge of the Jardin du Roi, Paris.

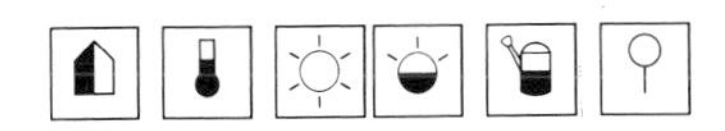

Species cultivated

B. × domestica
Hybrids involving at least three species (*B. ternifolia*, *B. leiantha* and *B. longiflora*). A number of cultivars have been named including: 'President Cleveland', bright scarlet; and 'Rosea', rose-pink. They will flower when 45–60cm ($1\frac{1}{2}$–2ft) tall, starting in summer and continuing into winter.
B. humboldtii See *B. longiflora*.
B. laevis Mexico
To 90cm(3ft) or more tall; stems slender. Leaves lanceolate to ovate, 5–13cm(2–5in) long, tapering to a slender point. Flowers 2.5–4cm(1–$1\frac{1}{2}$in) long, red, borne in autumn and winter. A yellow-flowered form is also known.
B. leiantha Mexico and Central America
To 60cm(2ft) or more. Leaves in whorls of three to five, ovate, tapering to a slender point, 5–8cm(2–3in) long. Flowers to 2cm($\frac{3}{4}$in) long, red to dark red, borne in summer and autumn.
B. longiflora Mexico
To 90cm(3ft) or more. Leaves in pairs, oblong-lanceolate. Flowers singly or in small clusters or corymbs, 5–7cm(2–$2\frac{3}{4}$in) long, white, very fragrant, opening from summer to winter. *B. humboldtii* is probably a hybrid of this, or just a robust form.
B. ternifolia (*B. triphylla*) Scarlet trompetilla
Mexico and Texas
To 60–90cm(2–3ft) tall. Leaves ovate to 5cm(2in) long, in pairs or threes. Flowers fiery scarlet, to 2–3cm($\frac{3}{4}$–$1\frac{1}{4}$in) long, carried in terminal corymbs from summer to winter. Most cultivars listed as belonging to this species are referable to *B. × domestica*.
B. triphylla See *B. ternifolia*.
B. versicolor South America
About 60–90cm(2–3ft) tall; stems slender and arching. Leaves in opposite pairs, lanceolate to ovate, 4cm($1\frac{1}{2}$in) or more long. Flowers 2–3cm($\frac{3}{4}$–$1\frac{1}{4}$in) in length, orange-red, tipped with yellow, borne in summer and autumn.

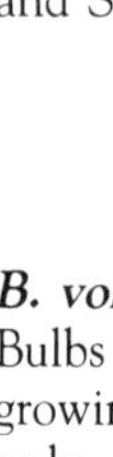

ABOVE *Bowiea volubilis*
RIGHT *Brachychilum horsefieldii*

BOWIEA

Liliaceae

Origin: South and East Africa. A genus of two species of perennial bulbous plants with slender, herbaceous, climbing stems to 2m(6½ft) or more long. The small linear leaves soon fall, the green stems carrying on the work of photosynthesis in their place. *Bowiea* is very easy to grow and being a native of seasonally dry areas is tolerant of irregular watering. Its value is largely as a curiosity, but it is quite decorative in form and the speed with which the young shoots grow from its silvery green bulbs is remarkable. Propagate by seed. *Bowiea* was named for James Bowie (1789–1869), a gardener sent out from Kew to collect plants, working chiefly in Brazil and South Africa.

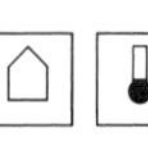 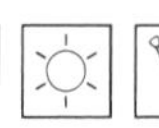  

Species cultivated

B. volubilis Climbing onion South Africa
Bulbs silvery green, rounded, to 20cm(8in) wide, growing largely above soil level, rooting at the base only. Stems green, fleshy. Flowers 6-tepalled, starry, greenish-white to 1cm(⅜in) wide, opening in late summer and autumn.

BRACHYCHILUM

Zingiberaceae

Origin: Islands of S.E. Asia. A genus of two species of tropical perennials forming clumps of hairless, leafy stems. They are epiphytic in the wild but in cultivation grow successfully in pots, needing shade and humidity in summer and becoming almost dormant in winter. Propagate by division when repotting or by seed. *Brachychilum* derives from the Greek *brachys*, short and *cheilos*, lip, referring to the short, lip-like petal.

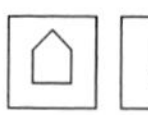     

Species cultivated

B. horsefieldii (sometimes confused with *Alpinia calcarata*)
Clump-forming; stems to 90cm(3ft) or more. Leaves leathery, dark green, lanceolate to narrowly lanceolate, slender-pointed, to 30cm(1ft) long. Flowers white, tubular, with six perianth segments and a shorter, lip-like structure formed by two aborted stamens; they are borne in short, terminal spikes to 8cm(3in) long in summer and are followed by oblong capsules which split open to show the orange interior and three rows of crimson seeds. The fruits which are long-lasting are more attractive than the flowers.

BRASSAIA

Araliaceae

Origin: India to Malaysia, Philippines and N.E. Australia. A genus of perhaps 40 species of trees and shrubs with alternate, digitate leaves which can be large and handsome. The small 5-petalled flowers are borne in racemes or panicles, several of which form terminal clusters. Small berries may follow. The species described below is a popular and durable foliage house plant when young. Propagate by seed when ripe or in spring in warmth, or by air layering in spring. *Brassaia* was named for Samuel von Brassai (*d.* 1897), a Hungarian botanist.

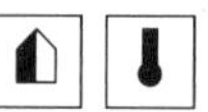    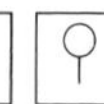

Species cultivated

B. actinophylla (*Schefflera actinophylla*)
Queensland umbrella/Octopus tree Queensland, Australia
To 2–3m(6½–10ft) as a pot or tub plant, but a bushy tree to 12m(40ft) in the wild. Leaves long-stalked, composed of five to 16 ovate-oblong leaflets 15–30cm(16–12in) in length, leathery textured and a polished rich green. Flowers dark reddish in 30–60cm(1–2ft) long racemes, clusters of eight to 20 erupting from mature stem tips, but rarely on containerized plants.

BRASSAVOLA

Orchidaceae

Origin: Mexico, the West Indies to South America. A genus of 15 species of evergreen epiphytic orchids. They are clump-forming with slender, stem-like cylindrical pseudobulbs each terminating in a single, fleshy leaf similar in shape. The flowers, which can be solitary or in racemes, are showy with similar sepals and petals (tepals) and a larger labellum rounded at the base and, like the tepals, with a slender tail-like tip. They grow best in baskets or on slabs of bark or tree fern and are not difficult. Propagate by dividing congested plants in spring. *Brassavola* was named for Antonio Musa Brassavola (1500–1555), an Italian doctor and botanist and Professor at Ferrara.

Species cultivated

B. appendiculata See *B. cucculata.*

B. cucullata (*B. appendiculata, B. odoratissima, Epidendrum cucullatum*) Mexico to South America
Pseudobulbs to 13cm(5in). Leaves slender, to 60cm(2ft) long, arching and somewhat pendulous. Flowers fragrant, usually single, occasionally in twos or threes, slender-tubed with drooping long-tailed tepals to 9cm(3½in) long, white to cream or greenish-white, the tip cream-tinted and fringed; borne mostly in winter.

B. digbyana (*Rhyncholaelia digbyana, Laelia digbyana*) Mexico to Guatemala
Pseudobulbs club-shaped, somewhat flattened, to 15cm(6in) or more. Leaves fleshy, rigid, linear to narrowly elliptic, grey-green, to 20cm(8in) in length. Flowers solitary, up to 18cm(7in) across, creamy-white with a hint of green especially in the throat; fragrant at night; borne in spring to summer.

B. fragrans See *B. perrinii.*

B. glauca (*Rhyncholaelia glauca, Laelia glauca*) Mexico to Panama
Pseudobulbs spindle-shaped, to 10cm(4in) long. Leaves stiff, leathery, oblong-elliptic, grey-green, to 12cm(4¾in) long. Flowers long-lasting and fragrant, to 12cm(4¾in) wide, white to lavender or olive-green, lip marked with a pink spot or lines; borne mainly in spring.

B. grandiflora See *B. nodosa.*

B. nodosa (*B. grandiflora, Epidendrum nodosum*) Mexico to Peru
Stem-like pseudobulbs to 15cm(6in) tall. Leaves erect, linear and very fleshy in texture, to 30cm(1ft) long. Flowers in racemes of one to six, long-lasting, pure white to yellowish-green, to 9cm(3½in) across, fragrant at night; borne chiefly in autumn and winter but spasmodically throughout the year.

B. odoratissima See *B. cucculata.*

Hybrid group

Several *Brassavola* species, notably *B. digbyana*, are among the easiest of epiphytic orchids to cultivate. Their flowers, however, lack colour and substance. To combine the toughness of *Brassavola* with the colourful blooms of less easy orchids it has been hybridized with *Cattleya, Laelia, Epidendrum, Sophronitis* and others. The main hybrids with *Brassavola* as the first parent are × *Brassocattleya*, × *Brassolaelia* and × *Brassolaeliocattleya*, the latter combining all three genera. In general appearance they favour the second parent, particularly as regards flower size, shape and colour, but culturewise they can be treated as *Brassavola*. A surprising number of the many cultivars which have been produced will grow and bloom in the home without difficulty.

BRASSIA

Orchidaceae

Origin: Tropical America. A genus of 50 species of evergreen epiphytic orchids with crowded, relatively large, flattened oblong pseudobulbs borne on a creeping rhizome. Each carries one to three leathery, linear lanceolate to obovate leaves. The long, arching flower stems arise from the base of the pseudobulbs or from the leaf axils and the flowers are carried in few to many-flowered racemes. Each flower has the three outer tepals much larger than the inner, and a distinctive lip. These orchids are best in baskets or on slabs of bark or tree fern, but will grow in pots. Shade from direct sun, but otherwise keep in a light place; they are not very difficult. Propagate by division when re-potting.

Brassia was named for William Brass (*d.* 1783), who collected many plants in West Africa for Sir Joseph Banks.

ABOVE LEFT × *Brassolaeliocattleya* 'St Helier'
ABOVE *Brassia maculata*

Species cultivated

B. brachiata See *B. verrucosa.*

B. longissima (*B. lawrenceana longissima*) Central and South America
Pseudobulbs flattened, oblong-elliptic, to 15cm (6in). Leaves usually solitary, occasionally paired, oblong-elliptic to lanceolate, to 60cm(2ft) long. Flowers yellow to greenish-yellow with very long tepals which can exceed 30cm(1ft) in length, and a yellow to cream lip, all spotted at their bases with red-brown; borne in spring.

B. verrucosa (*B. brachiata*) Mexico to Venezuela
Pseudobulbs narrowly egg-shaped, somewhat furrowed, to 10cm(4in) in length. Leaves oblong-elliptic to lanceolate to 35cm(14in) or more long, in twos. Flowers waxy, pale green, spotted with darker green or with red, lip whitish, with conspicuous black warts at the base, sometimes to 20cm(8in) long, usually smaller; they are borne in racemes of five to eight on stems to 60cm(2ft) or more in spring and summer.

× **BRASSOCATTLEYA** See BRASSAVOLA Hybrid group
× **BRASSOLAELIA** See BRASSAVOLA Hybrid group
× **BRASSOLAELIOCATTLEYA** See BRASSAVOLA Hybrid group

TOP *Breynia nivosa* 'Roseo-Pictus'
ABOVE *Brodiaea volubilis*

BREYNIA

Euphorbiaceae

Origin: South-east Asia to Australia and the Pacific Islands. A genus of 25 species of shrubs or small trees with two ranks of alternate leaves, and insignificant flowers. One species is cultivated making a decorative plant for a warm conservatory. Because it requires plenty of humidity, it is not suitable for the home except in the short term. Propagate by cuttings of mature stems in a case with bottom heat. *Breynia* was named for the German merchant Jacob Breyne (1637–1697) and his son Philip Breyne (1680–1764), a doctor, both authors of works on lesser known plants.

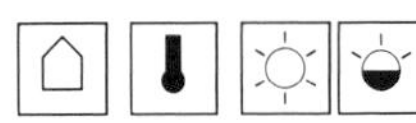

Species cultivated

B. nivosa (*Phyllanthus nivosa*) Snow bush
Pacific Islands
Slender shrub to 1m(3ft) or more with green stems. Leaves short-stalked, broadly ovate, 2.5–5cm(1–2in) long, green with very conspicuous white marbling. Flowers small, petalless, greenish or reddish, borne in the leaf axils. *B.n.* 'Roseo-Pictus' has leaves attractively variegated with pink and white.

BRODIAEA

Alliaceae (*Liliaceae*)

Origin: Western North America. A genus of 40 species of cormous plants, divided by some botanists into several genera including *Hookera*, *Dichelostemma* and *Triteleia*. They have narrowly linear leaves, often channelled, slender erect stems, twining in a few species and topped by rounded, often dense umbels of tubular to starry, 6-tepalled flowers. In some species three of the six stamens have been transformed into staminodes. A dainty flowering plant for a pot or pan in a sunny window or conservatory. Propagate by separating cormlets or offsets at potting time, or by seed in spring. *Brodiaea* was named for James Brodie (1744–1824), a Scottish botanist who specialized in the study of algae.

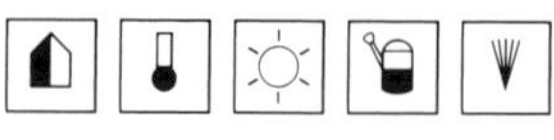

Species cultivated

B. californica (*Hookera californica*)
Leaves to 30cm(1ft) long. Flower stems to 60cm(2ft) tall; flowers in umbels of five to 12, funnel-shaped, 3–4cm($1\frac{1}{4}$–$1\frac{1}{2}$in) long, purple-blue, sometimes pink or white, the petals slightly recurved.

B. coccinea See *B. ida-maia*.

B. coronaria (*B. grandiflora*, *Hookera coronaria*)
Leaves 30cm(1ft) long. Flowers tubular, to 2cm ($\frac{3}{4}$in) long, flared at the mouth into 1.2–2cm ($\frac{1}{2}$–$\frac{3}{4}$in) long lobes, lilac or violet.

B.c. macropoda (*B. terrestris*)
Stem to no more than 5cm(2in) long with the umbels of purple to violet flowers close above the surface of the soil. Good for pots.

B. grandiflora See *B. coronaria*.

B. hyacinthina (*B. lactea*, *Hesperoscordum*, *Hookera* and *Triteleia hyacinthina* have all been used for this species)
Leaves to 40cm(16in) long. Stems 30–60cm(1–2ft) tall; flowers starry, 1.2–2.5cm($\frac{1}{2}$–1in) wide, white with a central green vein on each petal.

B. ida-maia (*B. coccinea*, *Dichelostemma ida-maia*) Fire cracker
Leaves 30–50cm(12–20in). Stems 30–90cm(1–3ft); flowers tubular, 2–4cm($\frac{3}{4}$–$1\frac{1}{2}$in) long, bright scarlet-red, recurved at the mouth with short, pale green lobes.

B. ixioides See *B. lutea*.

B. lactea See *B. hyacinthina*.

B. laxa (*Hookera* and *Triteleia laxa*, *Triteleia candida*) Grass nut, Ithuriel's spear
Leaves 20–40cm(8–16in). Stems to 40cm(16in) or over; flowers tubular, 1–2.5cm($\frac{3}{8}$–1in) long, flared at the mouth with lobes 1–2cm($\frac{3}{8}$–$\frac{3}{4}$in) long, purple-blue.

B. lutea (*B. ixioides*, *Triteleia* and *Hookera ixioides*)
Leaves 20–40cm(8–16in). Stems 30–60cm(1–2ft); flowers starry, 1.5–2cm($\frac{1}{2}$–$\frac{3}{4}$in) across, sometimes to 2.5cm(1in), yellow, each tepal with a dark purple mid-vein.

B. terrestris See *B. coronaria macropoda*.

B. volubilis (*Hookera* and *Dichelostemma volubilis*, *Stropholirion californicum*) Snake lily
Leaves 30–60cm(1–2ft). Stems 50–150cm(20–60in), twining; flowers inflated to 7mm($\frac{1}{4}$in) in length, opening at the mouth to 5–7mm($\frac{3}{16}$–$\frac{1}{4}$in) spreading lobes, pink to rose-red with white staminodes. Support is necessary for the climbing stems.

BROUGHTONIA See COELOGYNE

BROWALLIA

Solanaceae

Origin: Northern South America and the West Indies. A genus of six species of annuals and shrubby perennials. They are bushy plants with ovate, opposite or alternate leaves and loose, terminal racemes, or sometimes solitary flowers. These are blue, violet or white, tubular, and opening at the mouth to five broad, petal-like lobes, the upper one broader than the other four. They make very colourful plants for winter flowering, grown in either pots or hanging baskets, and are usually treated as annuals. Propagate from seed sown in late summer in warmth, for winter flowering. *Browallia* was named for John Browall (1707–1755), a Swedish botanist, Bishop of Abo and a champion of Linnaeus.

   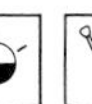 

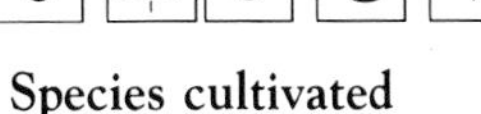

Species cultivated

B. speciosa Bush violet Colombia
Perennial, usually grown as an annual, 45–100cm(18–40in) tall. Leaves ovate, slender pointed, to 10cm(4in) long. Flowers 5cm(2in) across at the mouth, petals not notched, purple above, paler beneath. Varieties include: 'Silver Bells', a white-flowered selection; 'Blue Troll', a dwarf form; 'White Troll', dwarf with white flowers.

B. viscosa South America
Perennial, usually grown as an annual, 30–60cm(1–2ft) tall, sticky-hairy. Leaves ovate, roughly hairy, 2–4cm($\frac{3}{4}$–$1\frac{1}{2}$in) long. Flowers 2.5cm(1in) wide, the petals deeply notched, violet-blue with a white eye. *B. v.* 'Sapphire' is more compact, to 25cm(10in) tall.

BRUGMANSIA See DATURA

BRUNFELSIA

Solanaceae

Origin: Tropical America and West Indies. A genus of 30 species of evergreen shrubs with alternate, elliptic, ovate to lanceolate, leathery leaves and long-tubed flowers, spreading at the mouth to five broad lobes. They are useful flowering shrubs for large pots or containers, needing shade from the hottest sun and a humid atmosphere. Keep leading shoots pinched out for a shapely specimen. Propagate by cuttings, preferably with a heel, in a case with bottom heat. *Brunfelsia* was named for Otto Brunfels (1489–1534), a German physician, botanist and Carthusian monk who produced some of the earliest good drawings of plants.

Browallia speciosa 'White Troll'

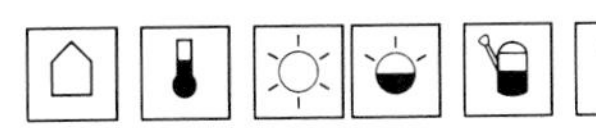

Species cultivated

B. americana (*B. violacea*) Lady of the night
South America
Shrubby, to 3m(10ft) or more in height. Leaves ovate to elliptic, 4–13cm($1\frac{1}{2}$–5in) long. Flowers trumpet-shaped, to 8cm(3in) long, opening yellow, fading to white; they are carried singly and are fragrant at night.

B. calycina (*B. paucifolia calycina*)
Yesterday, today & tomorrow Brazil and Peru
Shrubby, to 60cm(2ft) or more. Leaves oblong-lanceolate, glossy, to 10cm(4in) or more in length. Flowers with wavy petal lobes, to 5cm(2in) across, blue-purple, paling in about three days to white,

FAR LEFT *Brunfelsia americana*
LEFT *Brunfelsia calycina*

giving its vernacular name; they open from winter to summer.

B.c. macrantha (*B. pauciflora macrantha*)
Stronger-growing with larger flowers.

B. latifolia (*Franciscea latifolia*) Tropical Central & South America
Shrubby, 60–90cm(2–3ft) tall. Leaves broadly elliptic to obovate, 6–8cm($2\frac{1}{2}$–3in) long. Flowers 2–3cm($\frac{3}{4}$–$1\frac{1}{4}$in) wide, pale violet with a white eye, fading to all white in a few days; borne in winter to spring.

B. paucifolia See *B. calycina.*

B. undulata White rain tree West Indies
In the open ground eventually a small tree, but in a pot rarely over 1.5m(5ft). Leaves elliptic to lanceolate, 5–15cm(2–6in) long. Flowers fragrant, 3–4cm($1\frac{1}{4}$–$1\frac{1}{2}$in) wide, the lobes strongly waved, white ageing to creamy-yellow and borne in clusters, opening in autumn and spasmodically throughout the year.

B. violacea See *B. americana.*

BRUNSVIGIA

Amaryllidaceae

Origin: South Africa. A genus of 13 species of bulbous plants allied to *Amaryllis*. Very few are cultivated, but the one described below makes a handsome pot plant for the conservatory and can be brought indoors when in bloom. Propagate by seed in spring or by separating the occasionally produced offsets when dormant. *Brunsvigia* commemorates Carl Wilhelm Ferdinand, Duke of Brunswick-Luneberg (1713–1780), a German Prince who patronized the arts and sciences.

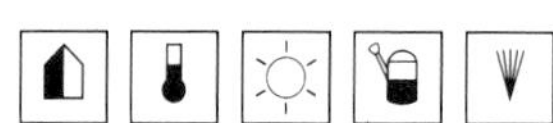

Species cultivated

B. josephinae Josephine's lily Cape, Natal
Mature bulb large, up to 20cm(8in) in diameter, sometimes long-necked and then to 30cm(1ft) long. Leaves deciduous, forming after flowering, tongue-shaped, prostrate to slightly ascending, glaucous, up to 60cm(2ft) long, occasionally more but usually less. Flowering stem stout, 15–30cm(6–12in) high, topped by a loose spherical umbel 30–60cm(1–2ft) wide, composed of 20 to 40 blooms; each flower is 8–9cm(3–$3\frac{1}{2}$in) long, glowing red with small streaks of yellow, borne in late summer.

Buddleia asiatica

BRYOPHYLLUM See KALANCHOE

BUDDLEIA (BUDDLEJA)

Buddleiaceae (Loganiaceae)

Origin: Temperate and sub-tropical areas of Asia, America and South Africa; most frequent in Asia. A genus of 100 species of deciduous or evergreen shrubs or small trees. They have opposite pairs of leaves, occasionally alternate, which are oblong-ovate to lanceolate in shape and small, 4-petalled, tubular flowers in pyramidal panicles. Most species are hardy, but those described make good pot or tub plants for a light room or conservatory. Propagate by cuttings, preferably with a heel, taken in late summer. *Buddleia* was named for the Rev. Adam Buddle (1660–1715), Vicar of Fambridge in Essex and a noted botanist.

Species cultivated

B. asiatica East Indies, China
Slender evergreen shrub to 3m(10ft) or more. Leaves lanceolate, 10–20cm(4–8in) long, sometimes toothed, white-downy beneath. Flowers about 6mm($\frac{1}{4}$in) wide carried in dense, drooping cylindrical panicles to 15cm(6in) long, from late winter to spring.

B. indica See *Nicodemia indica.*

B. madagascariensis See *Nicodemia madagascariensis.*

B. salviifolia South Africa
Erect shrub to 3.5m(12ft) with 4-angled woolly shoots. Leaves lanceolate to elliptic, wrinkled dark green above, grey-hairy beneath. Flowers pale lavender, orange in the throat, carried in dense, lilac-like panicles to 15cm(6in) long in winter.

BULBINE

Liliaceae

Origin: South Africa, Australia. A genus of about 55 species of annuals and perennials, mostly with linear to lanceolate, somewhat fleshy leaves, though a few species in S. Africa have short, broad, highly fleshy ones resembling a *Conophytum*. *Bulbine* derives from the Greek vernacular name for a bulbous plant and transferred to this genus.

 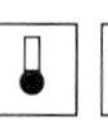 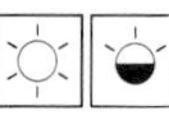  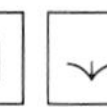 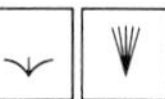

Species cultivated

B. alooides South Africa
Perennial with tuber-like rootstock. Leaves tufted or rosetted, lanceolate to narrowly triangular, up to 10cm(4in) long, fairly fleshy. Flowers 6-petalled, yellow, about 12mm(½in) wide, in racemes to 30cm(1ft) long in summer.
B. annua See under *B. semibarbata.*
B. bulbosa Australia
Perennial with tuber-like rootstock. Leaves in rosettes, linear, 15–25cm(6–10in) long, channelled and fleshy. Flowers yellow, 2cm(¾in) wide, in racemes 25–40cm(10–16in) high, borne in summer.
B. semibarbata Australia
Annual, though sometimes persisting for more than one year, of tufted habit. Leaves linear, somewhat fleshy, to 30cm(1ft) long. Flowers about 2cm(¾in) wide, yellow, in loose racemes up to 45cm(1½ft) high, in summer. Usually parades under the name *B. annua*, a very similar South African species; *B. semibarbata*, however, has only three of its six stamens bearded – in *B. annua* all are bearded.

BULBOPHYLLUM

Orchidaceae

Origin: Pan-tropics and adjacent cooler regions. A large genus of possibly 1,000 or more species of epiphytic orchids. They are mainly small-leaved species with one-leaved short pseudobulbs. The flowers may be solitary or in racemes and are frequently of complex structure, fascinating rather than beautiful. The one floral characteristic most species have in common is a mobile lip which appears as if balanced on a narrow point and rocks in the wind or when touched. The species described below are best in the conservatory, but can be brought into the home when in bloom. Propagate by division in spring or just after flowering. *Bulbophyllum* derives from the Greek *bolbos*, a bulb and *phyllon*, a leaf; each pseudobulb bears a leaf (or two) at its summit, but this is so of many orchids.

Species cultivated

B. barbigerum Tropical West Africa
Pseudobulbs clustered, ovoid, somewhat flattened, to 4cm(1½in) long. Leaves ovate or broadly lanceolate, about 8cm(3in) long, leathery-textured. Racemes of seven to 15 flowers in two parallel ranks. Flowers unpleasantly scented, with three greenish-brown sepals, two almost non-existent petals and a 2.5cm(1in) long, narrow labellum tipped by a crown of very long chocolate-purple hairs, expanding spring to early summer.
B. crassipes Himalaya, Burma, Thailand
Pseudobulbs well-spaced on rhizomes, ellipsoid, to 8cm(3in) long by half as wide. Leaves oblong to lingulate, about 22cm(9in) in length, leathery-textured. Racemes many flowered, horizontal to down-arching. Flowers densely borne, to 12mm(½in) wide, with three yellow, purple-brown spotted sepals, two much reduced petals and a widely tongue-shaped, purplish hairy labellum, opening in spring.

Bulbophyllum lobbii

B. fascinator See *Cirrhopetalum fascinator.*
B. lobbii Thailand, Malaysia, Borneo
Pseudobulbs well-spaced on sturdy rhizomes, 3–5cm(1¼–2in) long, ovoid, somewhat compressed laterally. Leaves to 25cm(10in) in length, narrowly oblong. Flowers solitary, on stems up to 13cm(5in) tall, 8–10cm(3–4in) wide, fragrant, yellow-spotted and suffused red-purple; sepals lanceolate, the two lateral ones falcate; petals like the upper sepal but smaller; labellum small, more or less ovoid, reflexed, expanding in spring and summer.
B. longissimum See *Cirrhopetalum longissimum.*
B. medusae See *Cirrhopetalum medusae.*
B. ornatissimum See *Cirrhopetalum ornatissimum.*
B. purpureorhachis Zaire
Pseudobulbs clustered, oblongoid, about 8cm(3in) long. Leaves in pairs, narrowly tongue-shaped to about 15cm(6in) or more long, rigidly leathery. Flowering stem 20–40cm(8–16in) tall, the upper half ribbon-like, spirally twisted to 4cm(1½in) wide, densely purple-brown spotted. Flowers (borne on this flattened stem) about 12mm(½in) long, brownish-purple; sepals thickly hairy; labellum purple and yellow, small, fleshy and very mobile, opening in autumn.
B. umbellatum See *Cirrhopetalum guttulatum.*

C

CAESALPINIA

Leguminosae

Origin: Widespread throughout the tropics and sub-tropics. A genus of 100 species of trees, shrubs or climbers. They have alternate, bipinnate leaves and racemes or panicles of 5-petalled flowers. With their attractive, ferny foliage and showy flowers, they make good pot or tub plants. Propagate by seed. *Caesalpinia* was named for the Italian Andreas Caesalpini (1524/25–1603), botanist, philosopher, doctor and author of several books, notably *De Plantis* (1583).

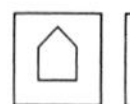   

Species cultivated

C. gilliesii Bird of paradise flower Argentina
Shrub to 3m(10ft), but easily kept smaller by pruning. Leaves to 20cm(8in) long, made up of many 6–12mm($\frac{1}{4}$–$\frac{1}{2}$in) leaflets with black spots on the margins beneath. Flowers 3cm($1\frac{1}{4}$in) wide, bright yellow, with long scarlet stamens to 9cm($3\frac{1}{2}$in); they are carried in 30cm(1ft) erect racemes of 20 to 40 flowers in late summer.

C. pulcherrima (*Poinciana pulcherrima*) Barbados pride Probably West Indies, now pan-tropical
Shrub or small tree to 6m(20ft) in the wild, kept smaller by pruning. Leaves 20cm(8in) or more, with leaflets 12–20mm($\frac{1}{2}$–$\frac{3}{4}$in) long. Flowers 3–4cm($1\frac{1}{4}$–$1\frac{1}{2}$in) wide, orange-yellow, with 5cm(2in) long pendulous, red stamens; carried in 30cm(1ft) erect terminal racemes of 20 to 40 in summer.

C.p. flava
Clear yellow blooms.

BELOW *Caesalpinia gilliesii*
RIGHT *Caladium* × *hortulanum* 'Mme Fritz Koechlin'
BELOW RIGHT *Caladium* × *hortulanum* 'Fire Chief'

CALADIUM

Araceae

Origin: Tropical America and West Indies. A genus of 15 species of perennials. They have underground tubers from which arise beautiful long-stalked, heart-shaped to lanceolate leaves. The flowers are in small, arum-like spathes but are not conspicuous. One of the most colourful of the foliage plants, excellent for the home in a warm position away from direct sun. Dry off and keep dormant for the winter. Propagate by offsets or division of tubers. *Caladium* is derived from the vernacular Malayan name *kaladi*, misused for this genus.

 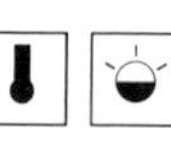  

Species cultivated

C. humboldtii Brazil
Leaves long-stalked, heart-shaped to 6cm($2\frac{1}{2}$in), bright green overlaid with white patterning between the veins. Needs care.

C.* × *hortulanum Angel's wings
The name for the many hybrid cultivars derived

from C. *bicolor* and C. *picturatum*. Leaves broadly triangular to heart-shaped, ranging in colour from white to pinks and reds with variable amounts of green. Those with white and pink leaves are more liable to sun scorch and damage than the darker coloured ones which tend to have firmer leaves. Many named cultivars are grown.

CALANTHE

Orchidaceae

Origin: Tropics and sub-tropics from South Africa eastwards to S.E. Asia and N. Australia; one species in the Americas. A genus of 120 species of mainly terrestrial orchids, both deciduous and evergreen. The former have large, conical to ovoid pseudobulbs sometimes constricted near the middle; the latter have smaller, more globular pseudobulbs almost hidden by the leaf bases. Both have elliptic to lanceolate, pleated leaves. The spurred flowers have similar petals and sepals (tepals) and the labellum is large and lobed. Evergreen species are best in cool conservatories, deciduous ones need warmer conditions and can be flowered in the home. Propagate by division at potting time. *Calanthe* derives from the Greek *kalos*, beautiful and *anthos*, a flower.

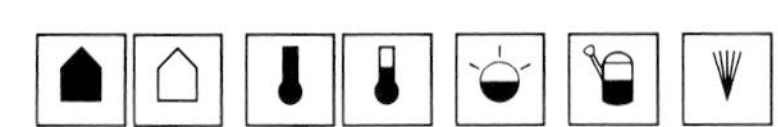

Species cultivated

C. brevicornu Nepal to Assam
Evergreen; leaves oblong to elliptic, to 30cm(1ft) long. Flowers 2–3cm($\frac{3}{4}$–1$\frac{1}{4}$in) wide, purple-brown, shading to yellowish-buff at the centre, the labellum red-purple with a white edge and a yellow keel; they are carried on 60cm(2ft) long racemes in spring. Temperate.

C. masuca (*Bletia masuca*) Himalaya
Evergreen; leaves narrowly elliptic, to 60cm(2ft) long, strongly pleated. Flowers to 4cm(1$\frac{1}{2}$in), rich violet-purple, the labellum darker with a golden-yellow patch; they are carried on 75cm(2$\frac{1}{2}$ft) long racemes in summer and autumn. Temperate.

C. vestita S.E. Asia
Deciduous; pseudobulbs 15–20cm(6–8in) tall, ovoid-conical, constricted at the centre, silvery green; leaves lanceolate, to 90cm(3ft). Flowers to 4cm(1$\frac{1}{2}$in) or more in width, white, in some forms with the labellum flushed pink; they are carried on arching racemes to 90cm(3ft) long which appear before the leaves expand, in winter. Named cultivars are mostly of hybrid origin derived partly from this species. Tropical.

CALATHEA

Marantaceae

Peacock plants

Origin: Tropical America and West Indies. A genus of 150 species of evergreen perennials of tufted or clump-forming habit, some with tubers. The small, asymmetrical flowers are rarely produced on small pot-grown plants. They are grown for their attractively patterned, ovate to lanceolate leaves and make excellent house and conservatory plants needing warmth and shade. Propagate by division at potting time. *Calathea* derives from the Greek *kalathos*, a basket, from the flower clusters which fancifully resemble a basket of flowers as they sit within their bracts.

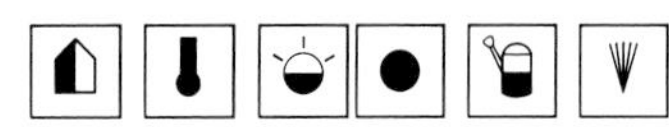

Species cultivated

C. albicans North-eastern South America
This species does not at present appear to be generally available but is worthwhile looking out for. Low-growing, it has broadly elliptic, light to grey-green leaves feathered with a darker grey-green.

C. amabilis See *Stromanthe amabilis*.

C. argyraea Brazil
Leaf-stalks short and winged, blade about 13cm(5in) long, oblong-lanceolate, somewhat lop-sided, grey above with green veins, red-purple beneath. There may be a confusion of identity here, as the plant usually grown under this name has leaf-stalks almost as long as the blade which is also larger.

C. bachemiana Brazil
Tuberous-rooted. Leaf stalks about 15cm(6in) high, blade 15–25cm(6–10in) long, narrowly ovate, slender pointed, silvered-green above with green margins and a pattern of green ovals angling out from the mid-rib, green or suffused purple beneath.

C. bella (C. *kegeliana*) Brazil
Leaf stalks up to 13cm(5in) high, blades about the same length, unequally ovate, cordate, greyish-silver above with dark green lozenge-shaped markings radiating from the midrib; margins and undersurface grey-green.

C. carlina Brazil
Compact habit, leaf blades asymmetrically ovate, about 18cm(7in) long, glossy grey-green, bordered with pure green and bearing a yellow vein pattern.

C. concinna Brazil
Another species little known as yet to gardeners but worth looking for. Leaf blades ovate, yellow-green with a satiny sheen, margined and banded with deep green.

C. crocata Brazil
Plant to about 30cm(1ft) tall. Leaf blades 18–22cm(7–9in) long, narrowly ovate, green and somewhat lustrous. Unlike the majority of calatheas this one is grown as much for its flowers as its foliage. The former are in erect spikes just above the leaves; each flower has orange-red petals and rosy-red sepals and sits in the axil of a glowing orange bract.

C. cylindrica Brazil
Leaf stalks on mature plants up to 90cm(3ft) tall, blades to 45cm(1$\frac{1}{2}$ft) long, oblong-lanceolate to

ABOVE L–R *Calathea picturata, C. lietzei, C. makoyana*

elliptic, thin-textured, light green. Unlike those of most calatheas, the white flowers of this species are borne in attractive 15cm(6in) long spikes, though rather hidden by the leaves.

C. eximia Central America
Tuberous-rooted, forming substantial clumps. Leaf stalks short, the blades asymmetrically ovate to oblong, up to 30cm(1ft) long, grey above with a yellowish midrib and green side veins, purple-red and brown-downy beneath.

C. fasciata See under *C. rotundifolia.*

C. insignis (of gardens) See *C. lancifolia.*

C. kegeliana See *C. bella.*

C. lancifolia (*C. insignis*) Rattlesnake plant Brazil
Leaf stalks 7–30cm($2\frac{3}{4}$–12in) or more long, blades narrowly lanceolate, 10–30cm(4–12in) or more, wavy, marked with dark olive-green oval patterning on either side of the midrib.

C. leopardina Brazil
Leaf stalks about 15cm(6in) tall, blades to 20cm(8in) long, asymmetrically oblong to narrowly ovate, light green above with a pattern of dark green triangles. Flowers yellow, in 5cm(2in) spikes.

C. lietzei Brazil
Leaf stalks to 20cm(8in) tall, blades narrowly ovate to 15cm(6in) long, patterned with alternate deep green and silver cross-bands, purple beneath. Sometimes produces young plantlets on stolons.

C. lindeniana Brazil
Leaf stalk to 35cm(14in), blades 15–30cm(6–12in), elliptic, deep green above with a narrow feathered central zone of yellow-green, purple and grey-green beneath.

C. louisae Brazil
Similar to *C. lietzei* but not producing runners, the leaf pattern more diffuse and feathery.

C. makoyana (*Maranta makoyana*) Peacock plant Brazil
Leaf stalks to 15cm(6in) or more, blades broadly ovate, to 20cm(8in) or more, yellow-green above, patterned with fine lines and alternate large and small ovals in dark green, giving a fanciful likeness to a peacock's tail. The undersurface of the leaf has the same patterning in purple.

C. medio-picta Brazil
Leaf stalks short, the blades 20cm(8in) or more long, spreading and arching, oblong, slender pointed, olive-green above with a central white stripe.

C. metallica Colombia and Brazil
Leaf stalks 10cm(4in) or more, blades to 15cm(6in) long, asymmetrically oblong to elliptic, dark green with an irregular, paler central band above, purplish beneath, at least near the margins.

C. micans Central America to Guyana, Bolivia and Brazil
Leaf stalks and blades of about the same length, the latter up to 13cm(5in) long, oblong-lanceolate, pointed, lustrous green above with a white feathering or irregular band down the middle, pale grey-green beneath. One of the smallest species.

C. musaica Brazil
Plant compact, the leaf blades ovate-cordate to narrowly so, up to 15cm(6in) long, glossy yellow-green above with a network of rich green veins, plain green beneath.

C. ornata Guyana, Colombia, Ecuador
Leaf stalks 20cm(8in), more in larger pots, blades 20–60cm(8–24in), lanceolate to ovate, dark green marked with groups of fine white to pink stripes, purple beneath.

C. picta See *C. warscewiczii.*

C. picturata Brazil, Colombia, Venezuela
Leaf stalks to 20cm(8in), blades to 15cm(6in) or more, elliptic, deep olive-green above marked with silvery lines along the midrib and parallel to the leaf

margins. *C.p.* 'Argentea' has silvery-surfaced leaves with a dark green margin. In *C.p.* 'Vandenheckei' the silvery central line is feathery and the outer lines jagged.

C. roseopicta Brazil
Leaf stalks very short, blades about 20cm(8in) long, somewhat unequal-sided, deep matt green with a pink midrib and a narrow irregular zone of the same colour paralleling the margin.

C. rotundifolia Brazil
Leaf stalks 5–10cm(2–4in) high, blades to 20cm(8in) long, almost orbicular but somewhat lop-sided, not, or barely, patterned. Represented in cultivation by *C.r. fasciata* with olive-green leaves, broadly barred silvery-grey between the veins.

C. rufibarba Brazil
Leaf stalks as long or longer than the blades, the latter 15–25cm(6–10in) long, asymmetrically narrowly oblong to lanceolate, undulate, bright green above, purplish beneath usually with some brownish-red hairs. Stalks suffused red-brown and bearing reddish hairs. Flowers small, yellow, in 5cm(2in) long spikes from purple hairy bracts.

C. undulata Peru
Leaf stalks very short, blades about 10cm(4in) long, asymmetrically oblong, lightly wavy-margined, deep green with a metallic lustre and a central silvery-grey band.

C. veitchiana Ecuador, Peru
Leaf stalks 10–20cm(4–8in) high, blades to 30cm(1ft) or so, unequally oblong to elliptic, lustrous, with a pinnately-lobed and zoned pattern in four shades of green, flushed red beneath. One of the most handsome of all the calatheas.

C. vittata Colombia
Robustly clump-forming, to 90cm(3ft) in height. Leaf blades ovate-oblong, 30–45cm(1–1½ft) long, light green above with streaks of silver radiating outwards and curving up to the margins, yellow-green beneath.

C. warscewiczii (*C. picta*) Costa Rica
In size, vigour and general appearance, much like *C. zebrina*, but the pale midrib zone more irregular and the lateral-vein stripes much narrower.

C. zebrina Zebra plant Brazil
Leaf stalk to 30cm(1ft), blade to 45cm(1½ft), usually smaller on pot plants, oblong-ovate, deep velvety green, with the midrib, veins and margins yellow-green, purple beneath.

CALCEOLARIA

Scrophulariaceae

Origin: Mexico to South America. A genus of over 300 species of annuals, perennials and shrubs, with opposite pairs or sometimes whorls of broadly ovate to linear leaves and showy 2-lipped flowers, the lower lip greatly inflated and pouched, like the toe of a fancy slipper. They are best grown as conservatory plants and brought into the home for flowering; alternatively they can be grown on a cool, sunny windowsill. Propagate by seed, *C. integrifolia* by cuttings in spring. *Calceolaria* derives from the Latin *calceolus*, a little shoe or slipper.

 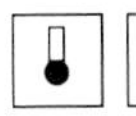 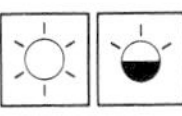  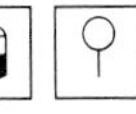

Species cultivated

C. × herbeohybrida Slipper flower, Pocket book plant
Hybrids of garden origin grown as biennials. Plants reach up to 45cm(1½ft) in height. Leaves ovate, somewhat toothed. Flowers to 6cm(2½in) or more wide, in vivid shades of yellow, orange and red, spotted and blotched, carried in large terminal trusses in late spring and summer. Many seed strains are listed in catalogues.

C. integrifolia (*C. rugosa*) Chile
Evergreen shrub to 1m(3ft) or more, usually grown as an annual and propagated by seed or cuttings. Leaves oblong-lanceolate to 5cm(2in) long, the surface above crinkled and dark green, greyish to rusty-hairy beneath. Flowers bright yellow, 12mm(½in) long, borne in dense terminal clusters in summer. Best in full light.

C. rugosa See *C. integrifolia*.

ABOVE LEFT *Calceolaria × herbeohybrida*
ABOVE *Calceolaria integrifolia*

CALENDULA

Compositae

Origin: Mediterranean region. A genus of 20 species of annuals, perennials and sub-shrubs with daisy-like flowers. One species is widely grown in the open garden but also makes a colourful pot plant. Seed sown in February will result in plants blooming from May onwards; an October sowing will start to flower in January provided a minimum night temperature of around 10°C(50°F) can be maintained. *Calendula* derives from the Latin *calendae*, the first day of the month, and *calendarium*, account; the almost perpetual flowering of this plant in its homeland was a constant reminder that interest was owed on the first day of each month.

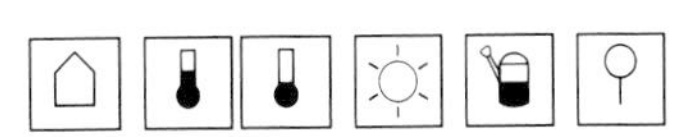

Species cultivated

C. officinalis Pot/Common marigold Southern Europe
A robust, well-branched annual or short-lived

RIGHT *Calendula officinalis* strain
FAR RIGHT *Callistemon citrinus*

perennial, to 50cm(20in) tall. Leaves 7–14cm(2¾–5½in) long, oblanceolate, bright green. Flower heads to 7cm(2¾in) wide, bright orange. Many seed strains and cultivars are available with colours from pale yellow to rich orange as well as single and double-flowered forms. The following are recommended for pot culture: 'Art Shades', 45cm(1½ft), semi-double flowers in cream, apricot and gold. 'Baby Gold', 30–40cm(12–16in), fully double, golden-yellow. 'Baby Orange', 30–40cm(12–16in), double or semi-double, bright orange. 'Orange King', 30cm(1ft), large double orange flowers.

CALLISIA

Commelinaceae

Origin: Tropical America. A genus of 12 species of prostrate evergreen perennials bearing ovate to oblong-lanceolate leaves with clasping bases. The flowers are borne in stalkless clusters usually arising from the axils of leaves or the similar bracts. Useful for both pots and hanging baskets. Because of their tendency to become straggly they are best propagated by cuttings every year, or at least every other year. *Callisia* derives from the Greek *kallos*, beauty.

 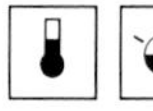 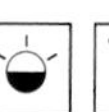  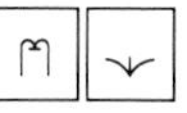

Species cultivated

C. elegans (*Setcreasea striata* of gardens) Wandering Jew Mexico
Trailer with stems to 60cm(2ft) long. Leaves ovate, clasping stems, olive-green with paler stripes and purple beneath, the whole plant shortly velvety-hairy. Flowers 3-petalled, white, to 2cm(¾in) across, opening in autumn and winter.

C. fragrans (*Spironema fragrans, Tradescantia dracaenoides*) Mexico
When young, rosette-forming, the somewhat fleshy, glossy leaves lanceolate, to 25cm(10in) long, creating a dracaena-like appearance. With age the rosette elongates and long runners are formed. Flowers 1.5–2cm(½–¾in) wide, white and fragrant in an elegant terminal panicle. *C.f.* 'Melnickoff' has the leaves variably striped with cream and/or brownish-purple.

C. repens Mexico, Central America, South America south to Brazil
Trailer, the dark stems rooting where they touch soil, eventually to 1m(3ft) or more in length. Leaves crowded in full light, almost overlapping each other, ovate, about 2.5cm(1in) long, ciliate, bright green. Flowers small, white and rather inconspicuous, from the upper leaf axils.

Callisia repens

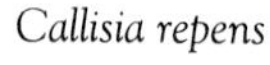

CALLISTEMON

Myrtaceae

Origin: Australia and New Caledonia. A genus of 25 species of evergreen shrubs and small trees. They have alternate linear to lanceolate leathery leaves and dense, showy terminal spikes of red, purplish or yellow flowers, each with long, protruding stamens, giving the effect of a bottlebrush. When the flowers fall the stem continues to grow beyond them and the woody seed capsules are left in a cylinder around the stem where they persist for two years or more. They make attractive tub plants, flowering freely, but they need regular pruning to keep them in check. Propagate by seed in spring or by heel cuttings in summer. *Callistemon* derives from the Greek *kallos*, beauty and *stemon*, a stamen.

  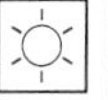  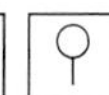

Species cultivated

C. citrinus (*C. lanceolatus*) Lemon bottlebrush
A large shrub to 5m(16ft) in the wild, easily kept smaller in a tub. Leaves linear to lanceolate, 4–9cm(1½–3½in) long. Flower spikes to 10cm(4in) with red stamens. *C.c.* 'Splendens' has brilliant crimson stamens.

C. lanceolatus See *C. citrinus*.

C. speciosus Bottlebrush
To 6m(20ft) or more in the wild, to 3m(10ft) in a tub. Leaves linear, to 13cm(5in) long. Very showy flower spikes, to 13cm(5in) long with deep scarlet stamens.

CALOCEPHALUS

Compositae

Origin: Australia. A genus of 15 species of annuals, perennials and small shrubs from temperate areas of

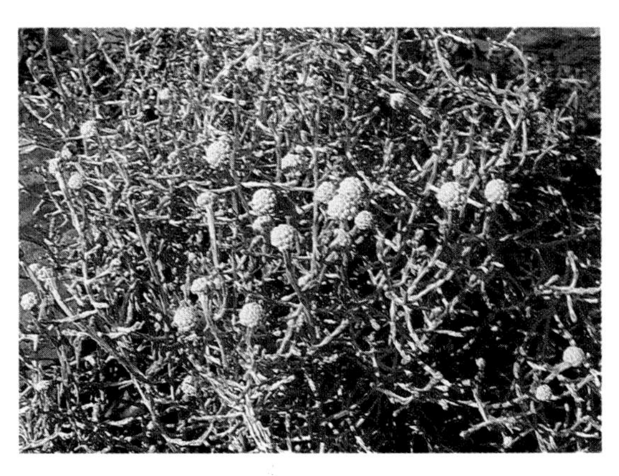

ABOVE LEFT *Calocephalus brownii*
ABOVE *Calothamnus quadrifidus*

Australia, only one of which is in general cultivation, making an attractive foliage pot plant and contrasting well with more colourful subjects. Propagate by cuttings with a heel in late summer. *Calocephalus* derives from the Greek *kallos*, beautiful and *kephale*, head, referring to the flowers of some species.

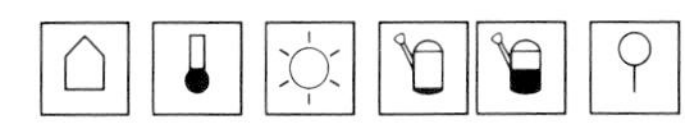

Species cultivated

C. brownii Western Australia
An evergreen shrublet to 30cm(1ft) or more with silvery-woolly, intricately branched stems. Leaves alternate, linear, almost scale-like, 2–3mm($\frac{1}{12}$–$\frac{1}{8}$in) long. Flower heads yellow, tiny, in globular clusters 12mm($\frac{1}{2}$in) wide.

CALONYCTION See IPOMOEA

CALOTHAMNUS

Myrtaceae

Origin: Western Australia. A genus of 25 species of evergreen shrubs allied to the bottlebrushes (*Callistemon*). They have narrowly lanceolate, linear to cylindrical or needle-like leaves and one-sided clusters or spikes of down-curving flowers arising at the bases of newly matured young growths. Petals are tiny and soon fall, their place being taken by bundles of stamens carried on ribbon-like stalks, mainly in shades of red. Propagate by seed in spring or cuttings in summer. The species described below make unusual decorative pot or tub plants for a sunny, airy conservatory. *Calothamnus* derives from the Greek *kalos*, beautiful and *thamnos*, a shrub.

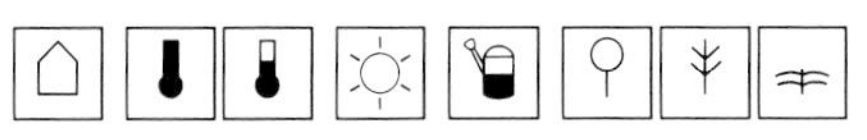

Species cultivated

C. homalophyllus
Erect to spreading habit, 1.2–1.8cm(4–6ft) in height. Leaves linear to oblanceolate, 3–5cm(1$\frac{1}{4}$–2cm) long. Flowers 3cm(1$\frac{1}{4}$in) long, red, in short spikes, borne in summer.

C. quadrifidus
Erect habit, to 2m(6$\frac{1}{2}$ft) or more tall. Leaves needle-like, about 2.5cm(1in) long. Flowers 2.5–4cm(1–1$\frac{1}{2}$in) long, in spikes, borne in summer. The showiest of the cultivated species.

C. rupestris
Spreading habit to 90cm(3ft) or more in height. Leaves needle-like, densely borne, 1.5–2.5cm($\frac{5}{8}$–1in) long, finely downy. Flowers 2.5–4cm(1–1$\frac{1}{2}$in) long, red, in clusters, often on bare stems just below the leaves in summer.

C. villosus
Erect to spreading, well-branched and fairly compact habit, 60–120cm(2–4ft) tall. Leaves needle-like, 2–4cm($\frac{3}{4}$–1$\frac{1}{2}$in) long, thickly grey-downy. Flowers about 2.5cm(1in) long, deep red, in clusters in summer.

CAMELLIA

Theaceae

Origin: India to Indonesia, China and Japan. A genus of about 80 species of evergreen trees and shrubs some of which have handsome foliage and large, showy blooms. The best of these make splendid container plants for the cool or unheated conservatory and can be brought indoors when in flower. The species and cultivars mentioned below are frost hardy except for the swelling flower buds and open blooms. Tending to bloom outside from late winter onwards, they are ideally suited for a frost-free conservatory either full-time or just for their floral display. Unless there is plenty of room it is recommended that the containerized plants are plunged into a bed of peat, sand or weathered ashes for most of the year and brought in from late autumn onwards. Flowering then takes place from mid-winter, depending on how much heat is given. A neutral to acid rooting medium is essential and the plants must never become dry. Propagate by stem or leaf bud cuttings in late summer, by layering in autumn, or by seed when ripe or as soon afterwards as possible. *Camellia* commemorates Georg Josef Kamel (1661–1706), a Czech Jesuit pharmacist who wrote about the flora of the Philippines.

 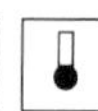 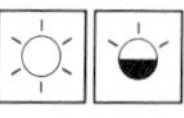  

ABOVE L–R *Camellia granthamiana*, *C. japonica* 'Adolphe Audusson', *C. reticulata* 'Royalty'

Species cultivated

C. granthamiana Hong Kong
Eventually to 3m(10ft) in a pot. Leaves to 20cm(8in) long, broadly ovate, matt dark green with a corrugated surface. Flowers white, to 13cm(5in) across with a large central boss of golden stamens opening from brown, papery sepals in late autumn and early winter.

C. japonica Japan, Korea
Shrub to 3m(10ft) in a pot, a small tree to 9m(30ft) in the open. Leaves 6–12cm($2\frac{1}{2}$–$4\frac{3}{4}$in) long, broadly ovate, rich glossy green. Flowers of the wild species 6–8cm($2\frac{1}{2}$–$3\frac{1}{4}$in) across with five red petals, but in cultivation to 15cm(6in) across, single and double in shades of red, pink or white and bicolour; they flower from winter to spring. The following cultivars give the range of flower shape.
Single-flowered (not more than eight petals): 'Jupiter', bright red; 'Sheridan', pink.
Semi-double (two or more rows of petals and a boss of stamens): 'Adolphe Audusson', red; 'Doncklaeri', red with white blotching, one of the oldest, introduced from Japan in 1834; 'Haku-raku-ten', white, sometimes nearer paeony form; 'Yours Truly', red with white picotee edge and a regular shape.
Anemone Form (one or two rows of petals and a neat central mass of petalodes and stamens): 'Elegans', rose-pink, the central petalodes sometimes pink and white, known since 1831; 'Shiro-Chan', pure white.
Paeony Form (full-petalled with some petaloid stamens and no regular form): 'Debutante', only 1m(3ft) tall, pink; 'Grand Slam', large flowers to 13cm(5in) across, red.
Formal Double (petals in regularly overlapping rows with no stamens showing): 'C.M. Hovey', flowers 10cm(4in) or more across, dusky red; 'Matterhorn', pure white with good foliage.

C. reticulata China
Shrub to 3m(10ft) in a pot, to 6m(20ft) in the open. Leaves to 10cm(4in) long, elliptic, matt green with a network of veins. Flowers to 7cm($2\frac{3}{4}$in) across, funnel-shaped, opening widely, pink. 'Interval' is a semi-double pink with flowers to 15cm(6in) across; 'Royalty' is carmine-pink to red, semi-double, with golden stamens.

C. sasanqua Japan
To 3m(10ft) in a pot, to 6m(20ft) in the open. Leaves 3–7cm($1\frac{1}{4}$–$2\frac{3}{4}$in) long, oblong to elliptic, lustrous dark green. Flowers 4–5cm($1\frac{1}{2}$–$1\frac{3}{4}$in) across, white, pink or pale red, flowering in autumn and early winter. 'Blanchette' is a single white; 'Cleopatra' is semi-double, rose-pink.

C. × williamsii
A group of hybrids between *C. japonica* and *C. saluenensis* flowering in late winter and spring. They include: 'Bow Bells', semi-double, bright rose; 'Debbie', paeony-flowered, pink; 'Francis Hanger', single white; 'J.C. Williams', single pink.

Other hybrids

'Black Lace' (*C. × williamsii × C. reticulata*), formal, red, double flowers to 10cm(4in) across.
'Dr Louis Pollizzi' (*C. saluenensis × C. reticulata*), paeony form, double, rose pink, strong scent.
'Show Girl' (*C. sasanqua × C. reticulata*), large, semi-double, pink.

CAMPANULA

Campanulaceae

Origin: Northern temperate latitudes and into subtropical and tropical zones as a mountain plant. A genus of 300 species of annuals, perennials and subshrubs, the majority being herbaceous perennials. All have 5-lobed flowers which vary from the typical tubular bell-shape to wide open stars. Of the species listed below, *C. fragilis* and *C. isophylla* make charming basket plants for the conservatory or cool room, while *C. pyramidalis* is a pot plant for the conservatory. Propagate by seed in spring or by division. *Campanula* is a diminutive form of the Latin *campana*, a bell.

  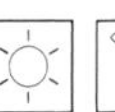  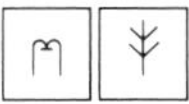

Species cultivated

C. fragilis Italy
Very closely related to *C. isophylla*, but with firmer-textured, almost fleshy leaves, those at the base of the plant more than half as long as the stem leaves. Flowers lilac-blue, with longer, narrower calyx segments than *C. isophylla*.

C. isophylla Italian bellflower N.W. Italy
Tufted; stems trailing to 20cm(8in) or more. Leaves rounded, cordate, toothed, the basal ones soon falling, and less than half the length of the stem leaves. Flowers widely funnel-shaped, almost starry, 1.5–2.5cm($\frac{1}{2}$–1in) across, lilac-blue, borne in late summer. *C. i.* 'Alba' has white flowers; 'Mayi' is a softly-hairy plant; 'Variegata' has white-variegated leaves.

C. pyramidalis Chimney bellflower S.E. Europe
Clump-forming with erect stems 1.5–2m(5–6$\frac{1}{2}$ft) tall in flower. Leaves ovate-oblong to ovate-lanceolate, sub-cordate. Flowers starry, pale blue or white, 3cm(1$\frac{1}{4}$in) across, carried in large pyramidal panicles in summer.

LEFT *Campanula isophylla* 'Alba'
BELOW LEFT *Campelia zanonia* 'Mexican Flag'
BELOW *Canarina canariensis*

CAMPELIA

Commelinaceae

Origin: Mexico to Brazil and West Indies. A genus of one evergreen perennial allied to *Dichorisandra*, *Geogenanthus* and *Tradescantia*. The variegated cultivar makes a striking specimen plant. Propagate by cuttings in summer. The derivation of *Campelia* seems not to have been recorded.

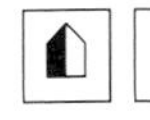 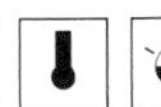 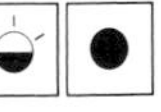  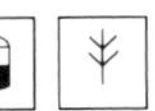

Species cultivated

C. zanonia
Clump-forming with erect habit, the sturdy, sparingly branched stems 60–120cm(2–4ft) in height. Leaves broadly oblanceolate to elliptic, 18–30cm(7–12in) long, rich green and somewhat lustrous. Represented in cultivation mainly by *C. z.* 'Mexican Flag' (*C. z.* 'Albolineata', *Dichorisandra* 'Albomarginata') with leaves boldly white-striped and narrowly red-edged.

CAMPYLONEURUM See POLYPODIUM

CANARINA

Campanulaceae

Origin: Canary Isles and East Africa. A genus of three species of tuberous-rooted perennials with scrambling stems and large pendent bell-flowers. They are allied to *Campanula* but are distinct in having leaves in pairs and fleshy fruits. The species described here is not only beautiful, but flowers in winter and is an especially valuable plant of easy culture. *Canarina* derives from the Canary Isles homeland of the first described species.

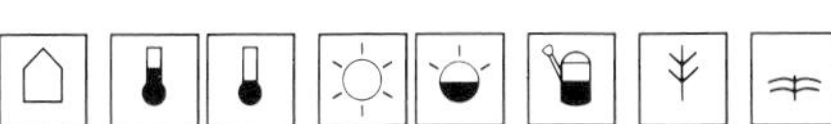

Species cultivated

C. canariensis (*C. campanula*)
Stems clustered at ground level, 2–3m(6$\frac{1}{2}$–10ft) long, needing support. Leaves narrowly triangular, irregularly toothed, 5–8cm(2–3in) long, slightly glaucous above, more so beneath. Flowers widely bell-shaped, about 6cm(2$\frac{1}{2}$in) long, pale orange with darker veins, pendent at the tips of lateral branchlets, appearing in late autumn to early spring.

CANNA

Cannaceae

Origin: Tropical and sub-tropical. A genus of 55 species of clump-forming perennials with stout, tuber-like rhizomes from which arise erect, unbranched stems and bold, alternate, lanceolate to oblong-ovate leaves. The flowers are tubular with three true petals and one to five petaloid stamens of which two or three are coloured and petal-like. The seeds are round and very hard, giving rise to the vernacular name, Indian shot. A showy pot plant. Propagate by division or by seed sown in warmth after a 24-hour soaking. *Canna* derives from the

Canna iridiflora

Greek *kanna*, a reed, referring to the tall, reed-like stems of some species.

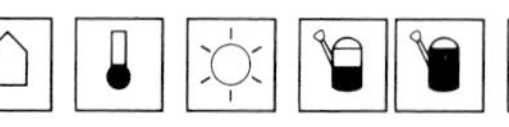

Species cultivated

C.* × *generalis Indian shot
A race of hybrids of garden origin derived largely from *C. flaccida*, *C. coccinea* and *C. indica*. Cultivars vary in height from 60cm(2ft) to 1.5m(5ft). Leaves are ovate-lanceolate, bright green to rich bronze-purple. Flowers 5–8cm(2–3in) wide in bright shades of yellow, orange and red, borne in summer. There is also a cultivar named 'Striped Beauty' which has boldly yellow- to ivory-striped leaves.

C. iridiflora Peru
Leaves oblong, 30–60cm(1–2ft) long. Flowers pink and red, to 5cm(2in) long, borne in summer.

CANTUA

Polemoniaceae

Origin: Warm temperate to sub-tropical areas of South America. A genus of 11 species of evergreen trees and shrubs with alternate or clustered simple leaves and clusters of long, tubular flowers. The species described makes a fine plant for a large tub and can be trained effectively across a conservatory roof on wires. Propagate by cuttings in summer. *Cantua* is a Latin form of the Peruvian vernacular name.

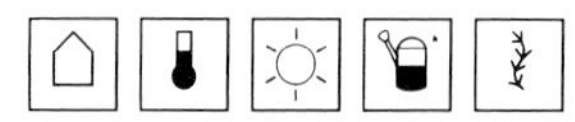

Species cultivated

C. buxifolia (*C. dependens*) Bolivia, Peru, Chile
Slender-stemmed, to 2m($6\frac{1}{2}$ft) or more if trained as a climber. Leaves elliptic to lanceolate, about 2cm($\frac{3}{4}$in) long. Flowers tubular, 7–10cm($2\frac{3}{4}$–4in) in length, expanding to five petal-like lobes at the mouth, pinkish-red striped yellow, borne in pendent terminal clusters in spring.

ABOVE *Cantua buxifolia*
RIGHT *Capsicum annuum* 'Masquerade'

CAPSICUM

Solanaceae

Origin: Warm temperate to sub-tropical areas of the Americas. A genus of 50 species of annuals, shrubs and sub-shrubs. They have alternate leaves which can vary from ovate to lanceolate, and pendulous five-lobed flowers, each with a cone-shaped cluster of five stamens in the centre. These are followed by cylindrical to oblong fruits, well known as 'peppers'. They are usually grown as annuals and make attractive plants when fruiting. Propagate by seed. *Capsicum* derives from the Greek *kapto*, to bite; fruits are hot-tasting.

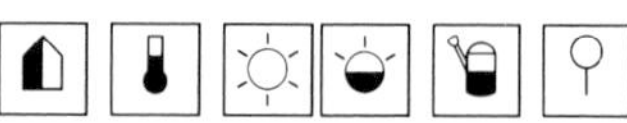

Species cultivated

C. annuum Chillies, Red/Sweet pepper
Probably from Peru but uncertain
Under glass, a bushy annual with oblong-ovate leaves. Flowers white or fawny white, borne singly in the leaf axils and followed by erect or pendent fruits. The many cultivars are grouped into categories according to the shape and size of the fruit. Chillies (*C.a. acuminatum*) have narrowly oblong, pointed, hot-tasting fruits; sweet or bell peppers and paprika (*C.a. grossum*) with oblong to bluntly conical fruits, at first green, then turning red or yellow are grown for culinary uses. For ornamental purposes a number of decorative cultivars have been raised, all with erect fruits, in shades of yellow, orange and deep green.

C. frutescens
Eventually a woody shrub, but usually grown as an annual. Like *C. annuum*, but flowers and fruits borne in pairs. Fruit erect, narrowly conical, 1–2.5cm($\frac{3}{8}$–1in) long, red or yellow. *C.f. baccatum* has globose berries about 1cm($\frac{3}{8}$in) wide.

CARALLUMA

Asclepiadaceae

Origin: Dry regions of Africa and the Mediterranean with an outlier in Burma. A genus of 100 species of succulent plants allied to *Stapelia*. They are tufted or clump-forming and low-growing with cylindrical, 4- to 6-angled fleshy stems bearing rounded tubercles. The leaves are scale-like and soon fall, while the 5-lobed, often striking flowers are starfish- or bell-like, sometimes smelling of carrion. They make intriguing plants for a sunny conservatory or windowsill, but need a little shade from summer sunshine (full light in winter). Propagate by division of old plants or by cuttings in summer, severing single stems at the base and allowing them to dry for two to three days before inserting singly into small pots. *Caralluma* is thought to derive from the mistaken belief that the vernacular name 'car-allum' was used in India.

   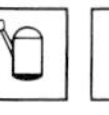 

Species cultivated

C. europaea African and West European shores of the Mediterranean and some islands
Stems branched, spreading to 15cm(6in) long, 4-angled and toothed, grey-green with a few faint, reddish spots. Flowers 13–15mm($\frac{1}{2}$–$\frac{5}{8}$in) wide with almost no smell, greenish-yellow banded with brownish-red, borne in stalkless umbels of up to 13 at any time through the year.

C. joannis Morocco
Stems 6–10cm(2$\frac{1}{2}$–4in) long, 4-angled and toothed, green. Flowers 1.2–2.5cm($\frac{1}{2}$–1in) wide, opening from a bell-shaped base, yellow at the centre with red spots merging to purple; they are velvety hairy and are borne in clusters of two to ten.

C. laterita Botswana, Zimbabwe, Namibia
Stems to 20cm(8in) long with conical teeth, grey-green. Flowers 5–8cm(2–3in) across, brick red and velvety hairy, borne in clusters; they smell strongly of carrion and so are not suitable in the home.

Caralluma europaea

CARDIOSPERMUM

Sapindaceae

Origin: Tropics and sub-tropics, chiefly of the Americas. A genus of 12 species of herbaceous and woody perennial climbers. They have alternate, biternate leaves, tendrils and white flowers. The species described is suitable for a tub in the conservatory where it can be trained against a wall or under the roof. It is essential to provide support in the form of sticks or netting. Propagate by seed sown in spring. *Cardiospermum* derives from the Greek *kardia*, heart and *sperma*, seed, the black seeds having a white, heart-shaped mark.

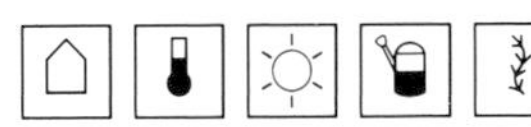

Species cultivated

C. halicacabum Heart-seed, Heart-pea, Balloon vine
This climber is best grown as an annual as it flowers freely in the first summer, but it can be grown as a short-lived perennial. Leaves comprise oblong, slender-pointed leaflets. Flowers tiny, 4-petalled, white, followed by balloon-like, 3-angled capsules 2–3cm($\frac{3}{4}$–1$\frac{1}{4}$in) across.

CAREX

Cyperaceae
Sedges

Origin: Cosmopolitan, mainly in temperate climates. A large genus of 1,500–2,000 species of grassy-leaved perennials, a few of which provide durable, modestly ornamental pot plants. Those described are evergreen and tufted to clump-forming with arching leaves. The catkin-like spikes of tiny petalless flowers are not especially decorative. Propagate by division in spring. *Carex* reputedly derives from the Greek *keiren*, to cut, alluding to the sharp edges of the leaves of some species which can cut a careless finger.

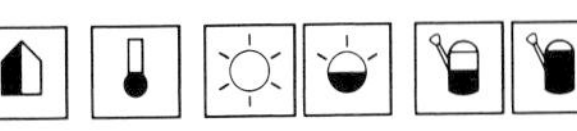

Species cultivated

C. brunnea China, Japan, Taiwan, Philippines
Clump-forming; leaves firm-textured, 25–45cm(10–18in) long, bright- to yellowish-green. *C.b.* 'Variegata' has leaves striped gold and bronze.

C. conica Japan, Korea
Clump-forming; leaves stiff, to 20cm(8in) long, dark green. *C.c.* 'Variegata' has cream-striped leaves.

C. morrowii Japanese sedge Japan
Clump-forming; leaves firm-textured, 20–40cm(8–18in) long, deep semi-lustrous green. Hardy. *C.m.* 'Evergold' ('Aurea') has yellow-striped leaves; there is some doubt as to whether this is a cultivar of true *C. morrowii*, and a hybrid origin has been postulated. *C.m.* 'Variegata' (*C.m. expallida*, *C.m. albomarginata*) has white-striped leaves.

C. ornithopoda Europe to Turkey
Tufted to clump-forming, the soft leaves to 20cm(8in) long, light to mid-green. *C.o.* 'Variegata', has creamy-white striped leaves; best in cool or unheated rooms or conservatories and sometimes only semi-evergreen.

Carex conica 'Variegata'

Carica papaya

CARICA

Caricaceae

Origin: Tropical to warm temperate Americas. A genus of about 40 species of trees, shrubs and climbers. The one species described below is widely grown in tropical countries for its fruit, but as an ornamental it makes an unusual foliage plant. Seeds are sometimes available from seedsmen or can be extracted from shop-bought fruits. Sow in spring at not less than 24–30°C(75–87°F). *Carica* derives from the Greek *karike*, a fig-like plant; used by Linnaeus on account of its fig-like leaves.

  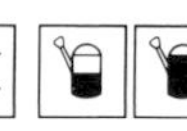 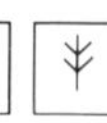

Species cultivated

C. papaya Pawpaw, Papaya Not known truly wild but probably originating in southern Mexico and Costa Rica

Small tree to 3m(10ft) in containers, 5–10m(16–33ft) tall in the open tropical garden. Stem usually pole-like and unbranched, sometimes with a few lateral stems. Leaves at the top in a palm-like cluster, long-stalked; leaf blades palmate, more or less orbicular in outline, to 60cm(2ft) wide or more, divided into seven to 11 deeply cleft or toothed lobes. Flowers trumpet-shaped, about 2.5cm(1in) long, cream or yellow in the leaf axils, fragrant, borne in summer. Plants are dioecious, the males with pendent flower trusses 25–75cm(10–30in) long, the females in small, short-stemmed clusters, or solitary. To obtain fruit, plants of both sexes are needed, though hermaphrodite plants do occur. Under glass, artificial pollination is necessary. The fruits are like small melons, 13–30cm(5–12in) long, green turning yellow or orange when ripe. *C.p.* 'Solo' is a dwarfer-growing hermaphrodite clone; seeds are sometimes offered but cannot be relied on to come 100% true to type.

CARISSA

Apocynaceae

Origin: Warm temperate and sub-tropical regions of Africa, Asia and Australia. A genus of 35 species of evergreen shrubs. They have opposite, leathery leaves and bear thorns which can be branched. The tubular, 5-petalled flowers can be fragrant and are followed by berries which in many species are edible. They make good tub plants. Propagate by seed in spring or by heel cuttings in summer (ideally with bottom heat). *Carissa* is a Latin form of the Indian vernacular name.

Carissa grandiflora

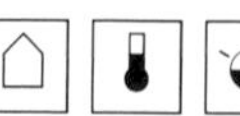 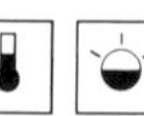  

Species cultivated

C. grandiflora Natal plum South Africa

Evergreen shrub to 5m(16ft) in the wild, but rarely 2m(6½ft) in a tub. Leaves broadly ovate, 2.5–8cm(1–3in) long. Flowers 8–12mm(⅓–½in) long, white, carried singly or in small cymes in late spring; very fragrant. *C.g.* 'Nana' ('Minima') has a dwarf, compact habit and is best for pot culture.

CARNEGIEA

Cactaceae

Origin: S.W. USA and Mexico. A genus of one species of giant cactus, the most familiar of the columnar cacti of the American south-west. It makes an erect-stemmed pot plant and is very slow-growing, taking many years to outgrow its position. *Carnegiea* was named for Andrew Carnegie, the Scottish born American industrialist and philanthropist, benefactor of Dr N.L. Britton and Dr J.N. Rose who monographed *Cactaceae* (1923).

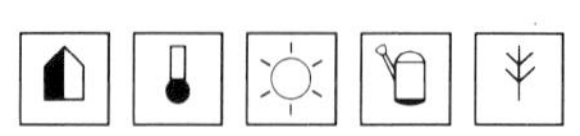

Species cultivated

C. gigantea (*Cereus giganteus*) Saguaro (Sahuaro)

Stems to 18m(60ft) in the wild, columnar and branched, with 12 to 24 ribs. Areoles bear several spines, the central ones to 8cm(3in), the surrounding ones to 2cm(¾in). Flowers white, 11cm(4½in) long and wide, but produced only on old specimens; not occurring on pot plants.

CARPOBROTUS

Aizoaceae

Origin: Mainly South Africa, but also a few in Australia, America and Pacific islands. A genus of 25 to 30 species of prostrate succulents formerly included in *Mesembryanthemum*. They have cylindrical to angular fleshy leaves in pairs and daisy-shaped, often colourful flowers. The species described below have large showy blooms and grow well in a sunny conservatory or home window; they can be especially effective in hanging baskets. Propagate by cuttings from spring to early autumn. *Carpobrotus* derives from the Greek *karpos*, a fruit and *brotos*, edible.

 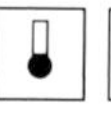 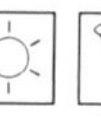  

Species cultivated

C. acinaciformis Hottentot fig Cape
Stems to 90cm(3ft) or more in length. Leaves to 9cm($3\frac{1}{2}$in) long, pale grey-green with keeled edges. Flowers 9–12cm($3\frac{1}{2}$–$4\frac{3}{4}$in) wide, satiny red-purple, borne in summer.

C. chilensis Sea fig Sea coasts of Oregon to Chile
Stems to 60cm(2ft) or more in length. Leaves 3–5cm($1\frac{1}{4}$–2in) long, roughly triangular in cross-section. Flowers 8–9cm(3–$3\frac{1}{2}$in) wide, red-purple, borne in summer.

C. edulis Hottentot fig Cape
Stems to 90cm(3ft) in length. Leaves 8–12cm(3–$4\frac{3}{4}$in) long, triangular in cross-section, the lower edge keeled and minutely toothed, bright green. Flowers 8–10cm(3–4in) wide, purple or yellow, ageing pink, borne in summer.

CARYOTA

Palmae
Fishtail palms

Origin: Sri Lanka, Indo-Malaysia, Solomon Is. and N.E. Australia. A genus of 12 species of large, noble palms of highly distinctive appearance. The leaves are bipinnate in the nature of a fern, with curiously lop-sided leaflets like the components of a fish's tail. Unlike most other palms, they bloom only when full-sized and usually die when the fruits ripen. Young plants in pots provide unusual decoration. Propagate by seed in spring. *Caryota* derives from the Greek *karyon*, a nut, referring to the largish seeds.

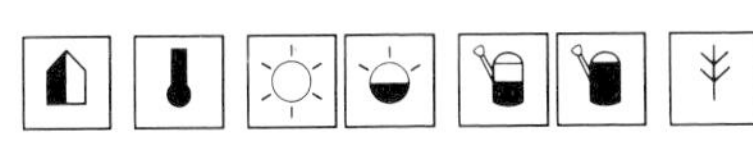

Species cultivated

C. mitis Burma, Malaysia, Philippines
Stem erect, suckering with age to form clumps but only when in large containers, to 3m(10ft) or more (up to 12m[40ft] when planted out). Leaves to 1.5m(5ft), the leaflets lopsidedly wedge-shaped.

C. urens Sago/Wine/Toddy palm India, Sri Lanka, Malaysia
Stem erect, never suckering, up to 5m(16ft) in containers, 24m(80ft) or more when planted out. Leaves to 3m(10ft) long (usually less on containerized plants but twice this on specimens grown in the tropics); leaflets oblique with irregularly lobed and/or toothed tips.

TOP LEFT *Carnegiea gigantea*
TOP RIGHT *Carpobrotus edulis*
ABOVE *Caryota urens*

CASSIA

Leguminosae

Origin: Warm temperate to tropical areas throughout the world. A genus of 500–600 species of trees, shrubs, perennials and annuals. All have alternate, paripinnate leaves and somewhat bowl-shaped, 5-petalled flowers followed by cylindrical or sometimes flattened pods. They are rewarding pot or tub plants, flowering freely when young. Propagate by seed in spring or by cuttings with a heel in late summer with bottom heat. *Cassia*

ABOVE *Cassia didymobotrya*
RIGHT *Casuarina equisetifolia*

derives from the Greek *kasia*, the name for an aromatic species used medicinally and applied to this genus.

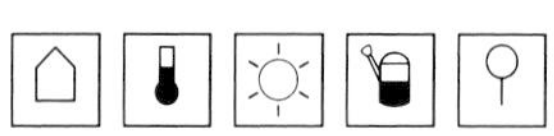

Species cultivated

C. artemisioides Wormwood cassia Australia
Shrub to 2m(6½ft). Leaves divided into six to 14 leaflets, each 6–25mm(¼–1in) long, linear, with a white silky down. Flowers yellow, to 2cm(¾in) wide in dense axillary racemes.

C. corymbosa Argentina
Shrub to 2m(6½ft). Leaves divided into four to six leaflets, each 2–4cm(¾–1½in) long, oblong-ovate, glabrous. Flowers rich yellow, 2–3cm(¾–1¼in) wide, in clusters of three to eight, borne in late summer.

C. didymobotrya (C. *nairobensis*) East Africa
Shrub to 2m(6½ft) or more, but generally less in containers. Evergreen; leaves pinnate, 15–30cm(6–12in) long, each one composed of 12 to 32 ovate to elliptic leaflets up to 5cm(2in) long. Flowers in dense, erect, terminal racemes to 30cm(1ft) long, rich yellow, 4cm(1½in) wide, contrasting strikingly with the glossy, blackish-brown buds; blooming takes place intermittently all year. A handsome tub plant for the conservatory, flowering the first year after sowing.

C. nairobensis See C. *didymobotrya*.

CASUARINA

Casuarinaceae

Origin: Africa to S.E. Asia, Malaysia, Australia and Polynesia. A genus of 30 to 45 species of evergreen trees which superficially resemble conifers. Their leaves are reduced to minute scales – the structures which look like foliage actually being whorls of slender, green stems resembling those of a miniature horsetail. Insignificant and petalless, the minute flowers are arranged in dense spikes, the females later developing into globular or ovoid 'cones' (woody cone-like fruits) bearing winged seeds. The species described below make elegant pot and tub plants with a difference for conservatory or sunny home window. They are easily raised from seed in spring and grow quickly. *Casuarina* derives from the scientific name for cassowary, the drooping sprays of branchlets resembling the curious feathers of that bird.

Species cultivated

C. equisetifolia S.E. Asia to Australia and Pacific islands
Up to 3m(10ft) tall in containers, a tree to 18m(60ft) in the wild; main stem erect with pendent branchlets and grey-green foliage sprays. 'Cones' broadly oblongoid, to 2cm(¾in) long.

C. cunninghamiana Eastern Australia
Much like C. *equisetifolia*, but foliage deeper green and 'cones' almost globular, about 12mm(½in).

C. torulosa Eastern Australia
Up to 3m(10ft) tall in containers, to 20m(65ft) in the wild; main stem erect, branchlets weeping, the foliage sprays very fine, green to bronze-green. 'Cones' globular, slightly flat-topped 2–2.5cm($\frac{3}{4}$–1in) in diameter. Forms with bronzy-purple foliage sprays are cultivated.

CATHARANTHUS

Apocynaceae

Origin: Tropical and sub-tropical Malagasy; one species from India. A genus of five species of evergreen sub-shrubs closely allied to *Vinca* (periwinkle) which it resembles. The species described below is the only one generally cultivated and has become widely naturalized throughout the tropics. It makes a charming pot plant and can be grown in a sunny window or conservatory. Propagate by seed or cuttings in spring. *Catharanthus* derives from the Greek *katharos*, clean or without blemish and *anthos*, a flower.

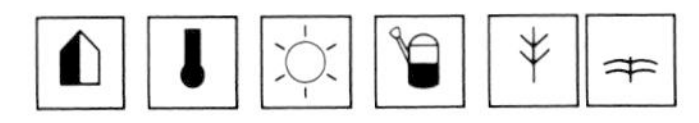

Species cultivated

C. roseus (*Vinca rosea*) Madagascar periwinkle
Erect or spreading to 60cm(2ft). Leaves 3–10cm($1\frac{1}{4}$–4in) long, oblong-ovate, glossy green. Flowers rose-pink to white with a slender tube to 2.5cm(1in) long, opening to five flat, petal-like lobes to about 4cm($1\frac{1}{2}$in) across in spring and summer. *C.r. ocellatus* has white flowers with a red or pink eye; dwarf strains to 20cm(8in) high are available and are excellent for pots.

CATTLEYA

Orchidaceae

Origin: Warmer regions of the Americas from South America to Mexico, including the West Indies. A genus of 60 species of epiphytic orchids. The narrowly leathery leaves arise from erect cylindrical to flattened pseudobulbs and the flowers are often large and showy. All make excellent conservatory plants and thrive well in a sunny window, providing the atmosphere is not too dry. Give some shade from the hottest sun in summer. Propagate by division at potting time. *Cattleya* was named for William Cattley (*d.* 1832), a collector and grower of rare plants.

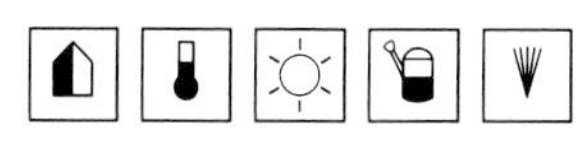

Species cultivated

C. aurantiaca Mexico, Central America
Strong-growing with spindle-shaped pseudobulbs to 35cm(14in) or more tall. Leaves two, usually broadly ovate, fleshy and recurved, to 18cm(7in) long. Flowers to 4cm($1\frac{1}{2}$in) wide, orange to orange-red, in crowded spikes from summer to autumn.

Catharanthus roseus

C. bowringiana Cluster cattleya Central America
Slender to club-shaped pseudobulbs, 30–90cm(1–3ft). Leaves usually two, leathery, oblong, to 20cm(8in) long. Flowers to 7cm($2\frac{3}{4}$in) or more wide, rose-purple, borne on a long-stalked raceme bearing up to 15 blooms in autumn.
C. cinnabarina See *Laelia cinnabarina.*
C. citrina Tulip cattleya Mexico
Rounded to egg-shaped, silvery-green pseudobulbs, about 5cm(2in) tall, bearing two or three strap-shaped pendulous leaves each to 20cm(8in) long. Flowers 7–9cm($2\frac{3}{4}$–$3\frac{1}{2}$in) long, opening to a tulip-like bell, citron-yellow in colour with a golden labellum; they are borne singly or in pairs on a pendent stalk from autumn to spring.
C. dowiana See under C. *labiata.*
C. harrisoniana See C. *loddigesii.*
C. intermedia Cocktail orchid Brazil to Uruguay
Pseudobulbs cylindrical, to 45cm($1\frac{1}{2}$ft) or more tall. Leaves two, leathery, narrowly elliptic, to 15cm(6in) in length. Flowers to 13cm(5in) across, rose-pink to rose-purple, with a purple blotch. Borne in racemes of three to seven blooms in spring and summer. *C.i.* 'Alba' has pure white flowers.
C. labiata Autumn cattleya Brazil, Trinidad
Pseudobulbs club-shaped, somewhat flattened, to 25cm(10in) tall. Leaves solitary, oblong, 15–25cm(6–10in) long. Flowers to 18cm(7in) wide, pale to deep rose-purple, the labellum deep red-purple and frilled, with a yellow blotch, borne in racemes of two to five in autumn. It is a very variable orchid, the following forms being considered by some botanists to be separate species.
C.l. dowiana Queen cattleya Costa Rica
Flowers to 15cm(6in) wide, yellow with a purple labellum, very fragrant, summer to autumn.
C.l. lueddemanniana (C. *lueddemanniana*) Brazil and Venezuela
Pseudobulb almost cylindrical. Flowers to 20cm(8in) or more across, pale purplish-pink suffused with white; the labellum is narrower with a frilled central lobe which is blue-purple; the throat is yellow.
C.l. mossiae (C. *mossiae*) Easter orchid, Spring cattleya
Narrower pseudobulbs and 13–18cm(5–7in) wide

Cattleya citrina

flowers, rose-pink to white, with a much larger, frilled, velvety-red labellum; they are fragrant and open in spring.
C.l. trianaei (*C. trianaei*) Christmas orchid, Winter cattleya
Strong-growing with a darker, yellow or orange blotch on the labellum; opening in winter.
C. loddigesii Brazil, Paraguay
Pseudobulbs cylindrical, to 45cm(1½ft). Leaves two, narrowly elliptic, leathery, to 15cm(6in). Flowers to 11cm(4½in) wide, wavy-textured, bluish pink to lilac-rose, the labellum darker pink, cream or pale yellow at the base, borne in 2- to 6-flowered racemes from late summer to autumn.

ABOVE *Cattleya loddigesii*
ABOVE RIGHT *Cattleya skinneri*
RIGHT *Cautleya spicata* 'Robusta'

C. lueddemanniana See under *C. labiata.*
C. mossiae See *C. labiata mossiae.*
C. perrinii See *Laelia perrinii.*
C. skinneri Mexico south to Colombia, Trinidad
Pseudobulbs club-shaped, somewhat flattened, 30–45cm(1–1½ft) tall. Leaves two, oblong, to 20cm(8in) long. Flowers to 9cm(3½in) wide, rose-purple to darker purple, the labellum white at the base, deeper purple at the tip; they are borne in 4- to 12-flowered racemes in spring or early summer.
C. trianaei See C. labiata trianaei.

Hybrids

Many hybrids and named cultivars have been raised, new names being added every year. Most have larger flowers and all are worth growing.

CAUTLEYA

Zingiberaceae

Origin: Foothills of the Himalaya in warm temperate regions. A genus of five rhizomatous perennials. They form loose clumps of erect stems bearing lanceolate to oblong-ovate leaves and terminate in spikes of rather orchid-like flowers within colourful bracts. The blooms are slenderly tubular with three petal-like lobes and what looks like a labellum but is, in fact, two petaloid sterile stamens. Plants are best in large pots. Propagate by division or by seed in spring. *Cautleya* was named for Major General Sir Proby Thomas Cautley (1802–71), English engineer and naturalist in charge of a number of irrigation projects in India.

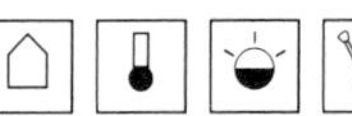 

Species cultivated

C. lutea (*C. gracilis*)
Stems to 45cm(1½ft). Leaves slender, lanceolate, often tinted with red beneath. Flowers yellow with two lobed lips, within purple-red calyces, borne in a spike of three to eight blooms in late summer.
C. gracilis See *C. lutea.*
C. spicata
Stems to 60cm(2ft), sturdy. Leaves oblong-ovate, bright green. Flowers bright yellow within red to green calyces, borne in spikes of up to 15 in late summer. *C.s.* 'Robusta' (*C. robusta* of gardens) has dark maroon bracts contrasting strongly with the flowers.

CELOSIA

Amaranthaceae

Origin: Tropical and sub-tropical regions. A genus of 60 species, mainly annuals. They are erect plants with alternate, lanceolate to ovate leaves and numerous small flowers clustered in dense panicles at the ends of the branches. Celosias make colourful pot plants for the conservatory and can be brought indoors for flowering. Propagate by seed,

moving the young plants to larger containers regularly so that their growth is not checked. *Celosia* derives from the Greek *keleos*, burning, referring to the flame-like shape and colour of the inflorescence of some species.

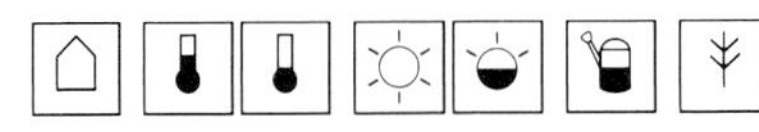

Species cultivated

C. argentea Tropical Asia
Annual, capable of reaching 60–90cm(2–3ft). Leaves linear-lanceolate to ovate, 5–7cm(2–2¾in) long. Flowers silvery-white with bracts of the same colour, borne in panicles which can be erect and plume- or spike-like, or somewhat drooping; they open in summer and autumn.

C.a. pyramidalis (*C. pyramidalis*, *C. plumosa*) Prince of Wales feathers
Flowers and bracts in erect, plume-like panicles in shades of red, yellow and pink. Often listed in catalogues under such cultivar names as 'Golden Feather', yellow, to 25cm(10in) tall; and 'Thompsonii Magnifica', red to 75–90cm(2½–3ft); also in colour mixtures such as 'Fairy Fountains', 25–30cm(10–12in); and 'Argentea Plumosa', 25cm(10in).

C.a. cristata (*C. cristata*) Cockscomb
To 30cm(12in). The flower spikes are curiously crested and flattened into a fan-like or cockscomb shape; this is the result of the flower stems fusing together, a phenomenon sometimes seen in other plants and known as fasciation. They are available in shades of red and yellow, usually in colour mixes such as 'Jewel Box', 15–20cm(6–8in); some single colours are also sold such as 'Fire Chief' (scarlet).

C. plumosa See C. *argenta pyramidalis*.

CELSIA See VERBASCUM

CENTAUREA

Compositae

Origin: Temperate and tropical areas of the Old World, Europe, N. Africa, India and China, with a few in North and South America and one in Australia. A genus of 600 species of annuals, biennials and perennials, sub-shrubs and shrubs.

ABOVE LEFT *Celosia argentea pyramidalis* 'Argentea Plumosa'
ABOVE *Centaurea cineraria*

They are very variable in growth form and leaf shape, but all have similar flowers. These are made up of many tubular florets each opening to a bell-shaped mouth with five lobes. Sometimes all are similar in shape, but in a number of species the outer ring has larger lobes giving the typical cornflower shape. The species described below are good pot plants for the conservatory and the annual species can be brought into the house for flowering. Propagate by seed, or by cuttings of the shrubby sorts taken in late summer. *Centaurea* was named for the half-man, half-horse *centaur* of Greek mythology.

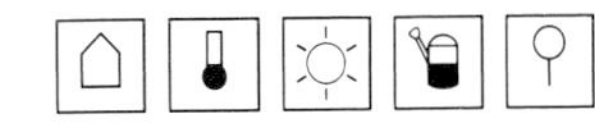

Species cultivated

C. cineraria (*C. gymnocarpa*) Dusty miller
S.E. Europe, Italy
Sub-shrub grown for its foliage. Leaves densely grey-white felty on both surfaces, pinnatisect or bi-pinnatisect, almost fern-like, arching over to 30cm(1ft) or more long. Flowers rarely produced on plants propagated annually; rose-purple, heads 2–3cm(¾–1¼in) across, borne in clusters in late summer.

C. cyanus Cornflower S.E. Europe
Hardy annual, once a familiar cornfield weed, 30–90cm(1–3ft) tall, bushy. Leaves lanceolate, sparingly toothed and sometimes lobed. Flower head solitary, 2.5–5cm(1–2in) across, in shades of blue, purple, red, pink or white in summer. Several cultivars and seed strains are available, the dwarf

ones being best suited for pot culture; they can be bought in colour mixtures such as 'Polka Dot', 30cm(1ft), or single colours such as 'Dwarf Blue', 30cm(1ft).

C. gymnocarpa See C. *cineraria*.

C. moschata Sweet Sultan S.W. Asia

Annual or biennial, to 60cm(2ft). Leaves lanceolate to pinnately lobed, greyish-green. Flowers in solitary heads 4–5cm($1\frac{1}{2}$–2in) across in shades of red, pink, yellow or white; very fragrant. Usually available as a colour mixture.

CEPHALOCEREUS

Cactaceae

Origin: Southern USA to Brazil. A genus of 50 species of cacti, or only one, C. *senilis*, according to the taxonomist Backeberg. They are columnar, sometimes branching and becoming tree-like with age. Their funnel-shaped white, yellow, pink or red flowers are seldom produced on any but the largest potted specimens. Plants are tolerant house or conservatory subjects for a sunny position. Propagate by seed in spring in warmth. *Cephalocereus* derives from the Greek *kephalo*, a head and *Cereus*, a genus of cacti, the flowers being borne on a cephalium-like structure on the sides or top of the stem.

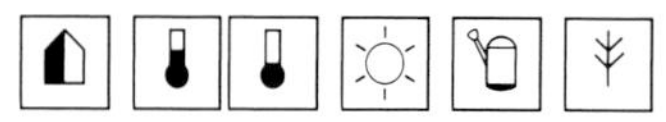

Cephalocereus senilis

Species cultivated

C. chrysacanthus (*Pilocereus chrysacanthus*) Mexico

Erect, eventually branching from the base, to 3m(10ft) or more in height, with ten to 12 prominent ribs. Areoles small, closely set, bearing about 12 golden-yellow or brownish spines 1–2cm($\frac{1}{2}$–$\frac{3}{4}$in) long and a variable amount of hair-like yellow bristles. Flowers pink, 8cm(3in) long, the outside of the tubular base thickly covered with pale yellow or whitish hairs. Quicker growing than most other members of the genus in cultivation. Cool.

C. royenii (*Pilocereus royenii, P. pilocereus floccosus*) West Indies

By some botanists united with C. *nobilis* and described under that name. Stems erect, branched, in the wild to 2m($6\frac{1}{2}$ft) or more, 4–8cm($1\frac{1}{2}$–3in) thick, with six prominent ribs. Four to five central spines to 2cm($\frac{3}{4}$in) or more long arise with ten radial spines and abundant white hairs from a brown areole. Red flowers open from a woolly and spiny cephalium. Temperate.

C. senilis (*Cereus senilis, Pilocereus senilis*) Old man cactus Mexico

Usually unbranched, in the wild reaching more than 6m(20ft), but slow-growing, rarely exceeding 2cm($\frac{3}{4}$in) a year. Young plants have 12 to 15 ribs. The one to five yellow spines are produced from closely set areoles together with 20 to 30 soft, hair-like bristles which are waved and hang down to a length of 6–12cm($2\frac{1}{2}$–$4\frac{1}{2}$in). The red and white flowers are not produced on potted plants. Cool.

CEPHALOPHYLLUM

Aizoaceae

Origin: South Africa. A genus of about 70 species of succulents. They are evergreen perennials of mat-forming or tufted habit, with narrow fleshy leaves in pairs or clusters and showy daisy-like flowers in summer. The species described below are well worth trying in the sunny conservatory or home window. Propagate by seed in spring or by cuttings in summer. *Cephalophyllum* derives from the Greek *kephale*, a head and *phyllon*, a leaf, presumably referring to those species which form dense tufts of long leaves.

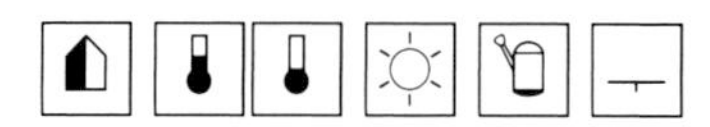

Species cultivated

C. alstonii Cape

Stems prostrate, 30–45cm(1–$1\frac{1}{2}$ft) long. Leaves in erect tufts, semi-cylindrical, grey-green, to about 7cm($2\frac{3}{4}$in) long. Flowers 5–8cm(2–3in) wide, deep glowing red with purple anthers. One of the most beautiful species.

C. cupreum Cape, Namaqualand

Stems prostrate, to 45cm($1\frac{1}{2}$ft) or more long. Leaves in close pairs, pale green, to 6cm($2\frac{1}{2}$in) long, semi-cylindrical, obscurely keeled, somewhat thickened in the upper half (clavate). Flowers to

8cm(3in) wide, yellow and coppery-red. One of the largest flowered species.
C. pillansii Cape, Namaqualand
Stems decumbent, forming large tufts. Leaves in pairs, cylindrical, slender-pointed, rich green, 5–15cm(2–6in) long, but usually around 8cm(3in) long. Flowers satiny-yellow with a red centre, about 6cm(2½in) wide; usually very free-flowering.
C. regale Namaqualand
Stems prostrate, forming mats to 30cm(1ft) across. Leaves 5–9cm(2–3½in) long, tapering to a truncate tip, mid-green, finely dark dotted. Flowers 5cm(2in) across, bright pinkish-purple, on pedicels to 4cm(1½in) high.
C. surrulatum Cape
Stems prostrate, forming mats 30cm(1ft) or more wide. Leaves in pairs, to 7cm(2¾in) long, bright green, more or less triangular in cross-section, the angles finely toothed. Flowers to 5cm(2in) wide, rose-purple.
C. subulatoides Cape
Stems decumbent to prostrate, short and much branched forming thick, sizeable tufts. Leaves in pairs close together, grey-green, semi-triangular in cross-section, 5–7cm(2–2¾in) long. Flowers 4cm(1½in) or more wide, red-purple, freely borne.

CERATONIA

Leguminosae

Origin: Mediterranean region. A genus of one species of evergreen tree, the fruits of which are carob fruits, sometimes known as locusts as eaten by John the Baptist in the wilderness. It is grown for its attractive foliage and for the interest of its biblical associations, making a handsome tub plant. Propagate by seed in spring. *Ceratonia* derives from the Greek *keras*, a horn, referring to the horn-like hardness of the seed pods.

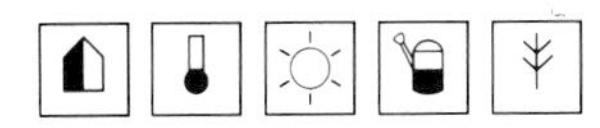

Species cultivated
C. siliqua Carob, Locust, Algaroba, St. John's bread
Evergreen tree to 12m(40ft) or more in the wild. Leaves leathery, paripinnate, with four to ten dark green leaflets, each 2.5–8cm(1–3in) long. Small petalless flowers are followed by 10–20cm(4–8in), tough, leathery pods containing a sweet pulp and hard, shiny oval seeds (these are the original carat weights). Flowers and fruits will be produced only on old, tub specimens.

CERATOPTERIS

Parkeriaceae

Origin: Tropics and sub-tropics. A genus of four species of aquatic ferns valuable for warm water aquaria and for pools in the warm conservatory. They are tufted in habit with bi- to tri-pinnate leaves which, when above water, freely produce plantlets. Fast-growing, they may be grown completely submerged, rooted into sand or compost in the bottom of a tank or pool, floating on the surface, or in wet or barely immersed compost. In the last situation, the fronds are of thicker texture and entirely above the water. Both frond types are edible. *Ceratopteris* derives from the Greek *keras*, a horn and *pteris*, a fern, alluding to the shape of the pinnae of the sterile (non-spore bearing) fronds.

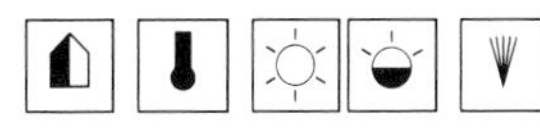

Species cultivated
C. pteridoides Floating/Water fern Florida to Brazil
Sterile fronds widely triangular in outline, to 20cm(8in) or more long, usually free-floating, simply pinnate with triangular pale green pinnae. Fertile fronds up to 40cm(16in) long, usually bi- or tri-pinnate, the pinnules narrowly oblong to linear.
C. cornuta (*C. thalictroides cornuta*) Water fern Tropical Africa, Malagasy to Indonesia and northern Australia
Like C. *thalictroides*, but the sterile fronds lanceolate, the pinnae irregularly lobed.
C. siliquosa See C. *thalictroides*.
C. thalictroides (C. *siliquosa*) Water fern Southern Japan, Pacific islands, S.E. Asia
Sterile fronds pale green, oblong in outline, up to 30cm(1ft) long, submerged or floating but rooted into the substratum. Fertile fronds bi-, tri-, or quadri-pinnate, larger than the sterile ones, the pinnules linear. This is the most widely grown species for aquarium decoration and the one most widely used as a salad vegetable in oriental countries. For C.*t. cornuta* see C. *cornuta* above.

TOP Cephalophyllum regale
ABOVE Ceratopteris thalictroides

Cereus hankeanus

CEREUS

Cactaceae

Origin: South America and the West Indies. A genus of 25 or 50 species according to the classifier. They are columnar, often branched and becoming tree-like. The white, trumpet- to funnel-shaped flowers open at night and are followed by red, juicy and usually edible fruits. They grow well in a sunny position in the home or conservatory, but they are generally fast-growing and so can soon out-grow their headroom. *Cereus* is the Latin name for a wax taper; alluding to the shape of some of the columnar species.

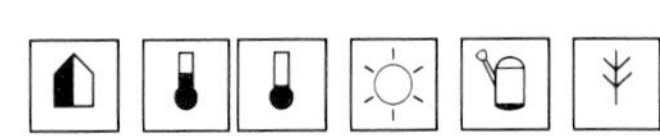

Species cultivated

C. acranthus See *Haageocereus acranthus.*
C. baumannii See *Cleistocactus baumannii.*
C. candicans See *Trichocereus candicans.*
C. chiloensis See *Trichocereus chiloensis.*
C. coerulescens Argentina
Usually unbranched to 1m(3ft) or more; stems 3–4cm(1¼–1½in) thick, blue-green especially when young, with eight blunt ribs. Areoles are almost black, with about four central spines, 2cm(¾in) long and nine to 12 radial ones. The funnel-shaped flowers can reach 20cm(8in) in length, but are produced only on mature plants.
C. decumbens See *Haageocereus decumbens.*
C. forbesii Argentina
Branched stems reaching 4–5m(13–16ft) in the wild, blue-green when young, with four to seven deep, wing-like ribs. Areoles are large, yellowish to brownish and bear one to two central spines to 5cm(2in) long and five to seven awl-shaped radial ones; all have blackish, bulbous bases. Flowers to 25cm(10in) long, white inside, green and purplish-brown outside.
C. geometrizans See *Myrtillocactus geometizans.*
C. giganteus See *Carnegiea gigantea.*
C. hankeanus Argentina
Columnar, unbranched stems, 3–4m(10–13ft), bluish-green when young, becoming darker, about 8cm(3in) thick, with four to five deep, wing-like ribs. Areoles occur in notches on the ribs and carry one central spine, to 3cm(1¼in) and three to four yellowish-brown radial spines. Flowers 12cm(4¾in) long, greenish outside, pinkish-white inside.
C. hexagonus South America (northern region)
Erect and branching to 10m(33ft) tall in the wild, but not fast-growing. Ribs four to five, winged and notched. Areoles 2–4cm(¾–1½in) apart, fairly large, yellowish, ageing greyish, bearing ten or more spines of varying length, the longest to 6cm(2½in). Flowers about 14cm(5½in) long, green and red in bud, opening pure white; usually only produced on large plants.
C. jamacaru South America
Very similar to *C. forbesii*, but the ribs are more deeply notched and the central spines yellow and up to 12cm(4¾in) long. Flowers green on the outside, 20–30cm(8–12in) long.
C. peruvianus Brazil, Argentina
Columnar, eventually to 10m(33ft); stem much branched, 10–20cm(4–8in) wide, with five to eight thick, blunt ribs. The round, brown woolly areoles bear one central, brown, awl-shaped spine to 12mm(½in) long and about seven radial ones. Flowers brownish-green, flushed outside, to 16cm(6¼in) long.
C.p. monstrosa
Under this name are grown a number of crested mutations with flattened and convoluted stems.
C. senilis See *Cephalocereus senilis.*
C. serratus See *Heliocereus serratus.*
C. silvestri See *Chamaecereus silvestri.*
C. smaragdiflorus See *Cleistocactus smaragdiflorus.*
C. spachianus See *Trichocereus spachianus.*
C. speciosus See *Heliocereus speciosus.*
C. strausii See *Cleistocactus strausii.*
C. tetragonus Eastern Brazil
Erect and branching freely, to 3m(10ft) or more, but fairly slow-growing. Ribs four or five, wing-like and slightly notched. Areoles 4–5cm(1½–2in) apart, large, whitish, each one bearing up to six brown to black 1cm(⅜in) long radial, and one or two longer central spines. Flowers 12cm(4¾in) long, red, only appearing on large specimens.
C. versicolor See *Haageocereus versicolor.*

CEROPEGIA

Asclepediaceae

Origin: Tropical and sub-tropical Africa, Asia, Malagasy, Canary Isles and Australia. A genus of

160 species of shrubby or twining perennials, some with fleshy or tuberous roots. Stems sometimes succulent bearing opposite pairs of rounded to linear leaves which in some species soon fall. The remarkable flowers are tubular, often inflated at the base; their five corolla lobes can be reflexed or joined at the tips to form a cage or, in some species where the tips of the lobes are expanded and membraneous, an umbrella-like shape. They grow well in the home or conservatory making intriguing pot or basket plants. Propagate by seed in spring or by cuttings in spring or summer in warmth. *Ceropegia* derives from *keros*, wax, and *pege*, a fountain, the flowers of some species fancifully like a fountain formed from wax.

Species cultivated

C. ampliata S.W. Africa, South Africa (east)
Shortly climbing, relatively thick stems. Leaves small, short-lived, falling before blooming. Flowers in clusters of two to four but only one opening at a time, 5–6cm(2–2½in) long, white with green veins, inflated at the base like a small balloon, then tubular to the mouth which is capped by a cage of narrow red petals.

C. caffrorum Lamp flower S.E. Africa
Twining climber to 1m(3ft) or more. Leaves fleshy, ovate-lanceolate to linear, 1–3cm(⅜–1¼in) long. Flowers hairy, to 5cm(2in) long, with slender lobes, green with purple streaks; they are borne in clusters of three to five.

C. debilis Malawi, Zimbabwe, Zambia
Stems twining or pendent, to 60cm(2ft) or more, with small aerial tubers forming at the nodes. Leaves linear, fleshy, 2–3cm(¾–1¼in) long. Flowers to 2.5cm(1in) long, inflated at the base, expanded at the mouth with narrow lobes, green outside with a reddish tinge, purple-red inside.

C. dichotoma Canary Isles (Tenerife)
Stems erect, branching by forking at the tips, to 90cm(3ft) tall, covered with a white-waxy patina, sometimes purple-tinted. Leaves linear, to 4cm(1½in) long, falling at the end of the growing season. Flowers tubular, yellow, 2cm(¾in) long, in sessile clusters at the nodes.

C. fusca Canary Isles (Tenerife, Gran Canaria)
Much like *C. dichotoma* in overall appearance, but stems more grey-white and flowers dark brownish-red.

C. haygarthii South Africa
Twining or trailing plant, stems to 1m(3ft) long or more. Leaves ovate-cordate, only slightly succulent, 2–5cm(¾–2in) long. Flowers solitary, to about 4cm(1½in) long, the tube bent at the base, flaring widely at the mouth, the five narrow lobes stretched out star-like, cream or pale pink, spotted with maroon; the stigma is maroon and projects from the centre of the flower, terminating in a red, hairy hollow knob about 8mm(⅓in) across.

C. hians Canary Isles (La Palma)
Like *C. dichotoma* in overall appearance, but the yellow flowers are borne on reddish stems.

C. sandersonii Parachute plant, Fountain flower Natal, South Africa
Twining or trailing plant; stems to about 2m(6½ft). Leaves ovate-cordate, fleshy, 3–5cm(1¼–2in) long. Flowers solitary, to 7cm(2¾in) long, the tube widely expanded at the mouth with short, narrow lobes joining to form a parachute-like cover about 5cm(2in) across; they are pale green, lined and mottled with darker shades.

C. woodii String of hearts, Rosary vine
Natal, South Africa
Slender creeping or pendent stems, to 60cm(2ft) or more, often with small, aerial tubers at the nodes. Leaves heart-shaped, fleshy, purple, the upper side strongly marbled with silver, 1.5–2cm(⅝–¾in) long. Flowers about 2cm(¾in) long, slightly curved, inflated at the base and expanded at the mouth, the lobes narrow, red to reddish-brown.

CESTRUM

Solanaceae

Origin: Mexico to Chile, West Indies. A genus of 150 species of shrubs, most of which are evergreen. They have alternate, lanceolate to ovate leaves and tubular, somewhat inflated flowers borne in nodding or arching terminal panicles. Each flower has five pointed, petal lobes at the mouth. They are followed by red to purple-black berries. Cestrums make good tub plants for the conservatory where they are best trained to wires on the back wall and under the roof. Two- and three-year-old stems produce the best flowers, so cut third-year stems

ABOVE LEFT *Ceropegia fusca*
TOP RIGHT *Ceropegia hians*
ABOVE *Ceropegia woodii*

Cestrum aurantiacum

back to the base at the end of their season. Propagate by heel cuttings in late summer in warmth, or by seed in spring. *Cestrum* derives from an ancient Greek name for an unknown species, used by Linnaeus.

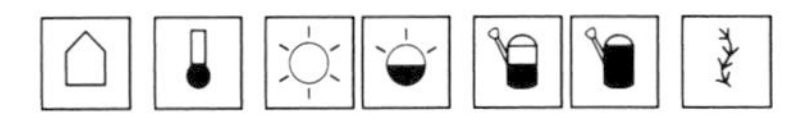

Species cultivated

C. aurantiacum Guatemala
Evergreen semi-climbing shrub to 3m(10ft). Leaves ovate 5–9cm(2–3$\frac{1}{2}$in) long. Flowers 2cm($\frac{3}{4}$in) long with reflexed lobes, bright orange, stalkless, borne in leafy, terminal panicles in summer; their fragrance is most apparent at night.

C. elegans (*C. purpureum*) Mexico
Slender, semi-climbing evergreen shrub to 3m(10ft) or more. Leaves narrowly ovate to lanceolate, 6–12cm(2$\frac{1}{2}$–4$\frac{3}{4}$in) long, on hairy branches. Flowers about 2.5cm(1in) long, red-purple, markedly narrowed at the mouth, the petals reflexed; they are borne in panicles in summer. Globose red fruits, about 1cm($\frac{3}{8}$in) across, follow.

C. fasciculatum Mexico
Semi-climbing evergreen, to 3m(10ft). Leaves ovate-lanceolate, 7–13cm(2$\frac{3}{4}$–5in) long. Flowers 2cm($\frac{3}{4}$in) long, narrowly pitcher-shaped, hairy outside, deep carmine, borne in small rounded clusters making up a leafy panicle, late spring to summer.

***C.* 'Newellii'**
Considered to be a selected form of *C. fasciculatum*, or possibly a hybrid of *C. elegans* and *C. fasciculatum*. It is very similar to the latter species, but its showy flowers are brighter red.

C. purpureum See *C. elegans*.

CHAMAEALOE See ALOE

CHAMAECEREUS

Cactaceae

Origin: Argentina. A genus of only one species of cactus which has been classified as a *Cereus*, and by some botanists is now included in *Lobivia*. It is a very easy plant to grow and will thrive with the minimum of attention. Propagate by detaching stem segments which root readily. *Chamaecereus* derives from the Greek *chamai*, dwarf and *Cereus*, a genus of cacti.

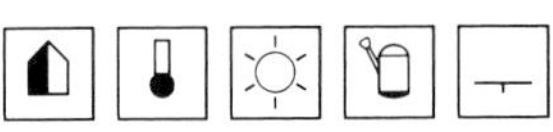

Species cultivated

C. silvestrii Peanut cactus
A small prostrate cactus eventually forming mats to 30cm(1ft) across. Stems cylindrical, branched into short, oblong-ovoid (peanut-shaped) segments, to 6cm(2$\frac{3}{8}$in) long and 1.5cm($\frac{5}{8}$in) wide. They have eight to ten shallow ribs with numerous small, white areoles bearing tiny, bristle-like spines. Flowers orange-scarlet, funnel-shaped, 5–7cm(2–2$\frac{3}{4}$in) long, opening from early spring. Hybrids between this species and members of the genus *Lobivia* are sometimes grown.

RIGHT *Cestrum elegans*
FAR RIGHT *Chamaecereus silvestrii*

CHAMAEDOREA

Palmae

Origin: Mexico to northern South America. A genus of 100 species of small palms, a few solitary but many of slender, suckering habit, with bamboo-like stems and pinnate, rarely entire leaves. The flowers are dioecious in pendent or erect, spike-like inflorescences followed by singly seeded berry-like fruits. They are very successful pot or tub plants and can be grown in the conservatory or home providing the atmosphere is not too dry. Propagate by seed in spring which germinates readily if kept at 20–24°C(68–75°F), or by division of those with suckering stems. *Chamaedorea* derives from the Greek *chamai*, dwarf and *dorea*, a gift, the bright fruits of some species being easily reached when ripe.

Species cultivated

C. costaricana Costa Rica
Clump-forming, suckering species; stems dark green, to 5m(16ft) in the wild, to 1–2m(3–6½ft) in pots. Leaves 60–90cm(2–3ft) long, pinnate, with up to 20 or more pairs of rich green, linear-lanceolate leaflets. An elegant foliage plant.

C. elegans (*Neanthe elegans*) Parlour palm, Dwarf mountain palm Mexico, Guatemala
Slender suckering stems, to 3m(10ft). Leaves 60–120cm(2–4ft) long, pinnate, with broadly lanceolate leaflets, dark green and leathery in texture.

***C.e.* 'Bella'** (*Neanthe bella*)
Similar to C. *elegans*, but with a maximum height of about 1m(3ft) and slower growing. A very tolerant palm for the home which will flower in a pot, producing erect clusters of pale yellow flowers followed by small, globular fruits.

C. erumpens Bamboo palm Honduras
Clump-forming, the slender stems to 3m(10ft) or so. Leaves to 60cm(2ft) or more in length, made up of five to 15 pairs of recurved, ovate leaflets. Relatively tolerant of a dry atmosphere.

C. graminifolia (of gardens) Guatemala
Clump-forming, to 2m(6½ft) or more tall; stems slender, reed-like. Leaves 60cm(2ft) or more long, pinnate, composed of up to 40 slender, 30cm(1ft) long, dark green leaflets, gracefully arching. Flowering stems erect. There is some doubt as to whether this is the true species, as the original botanical description states that the stems are solitary.

C. metallica Miniature fish tail palm Mexico
Solitary-stemmed, to 60cm(2ft) tall. Leaves undivided, to 30cm(1ft) long and 15cm(6in) wide, dark bluish-green, cleft at the apex to give a fish tail effect. A decorative and less usual palm.

C. microspadix Eastern Mexico
Much like C. *erumpens* in appearance, but with larger white flowers and red fruits.

C. oblongata Mexico to Nicaragua
Stem solitary, eventually to 3m(10ft) in height. Leaves pinnate, to 1.2m(4ft) or more, composed of

12 to 18, lanceolate, leathery, deep green leaflets. Flowers green and white; fruit black.

C. seifrizii Reed palm Mexico
Clump-forming, with slender canes. Leaves 60cm(2ft) or more, pinnately divided with 26 to 30 narrow leaflets which are well spaced, giving an effect of light, lacy foliage. Decorative and able to withstand room culture.

CHAMAELAUCIUM

Myrtaceae

Origin: Western Australia. A genus of 12 or more species of wiry-stemmed evergreen shrubs grown for their freely borne flowers. Only the species described here is much grown in the Northern hemisphere, though other species are almost as decorative and worthwhile looking out for, notably C. *ciliatum* and C. *megalopetalum*. Like C. *uncinatum* they will make attractive tub plants. Propagate by seed in spring or by cuttings in summer. *Chamaelaucium* derives from the Greek *chamai*, on the ground (of low growth) and *leukos*, white; the first species to be described was the white form of the dwarf C. *ciliatum*.

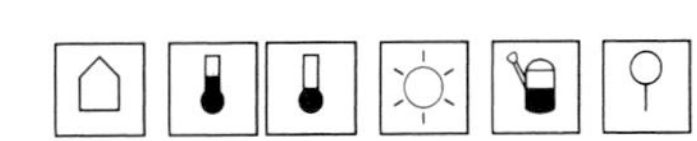

Species cultivated

C. uncinatum Geraldton wax flower
In containers, about 2m(6½ft) high, or less with annual pruning after blooming; in the wild to twice this size. Leaves 2–4cm(¾–1½in) long, linear, triangular in cross-section with a hooked, sharp tip. Flowers about 12mm(½in) wide, varying among individuals from white to pink and deep rose-purple, opening in spring to summer.
The allied C. *ciliatum* is like a dwarf version of C. *uncinatum*, up to 1.2m(4ft) tall and with smaller flowers.

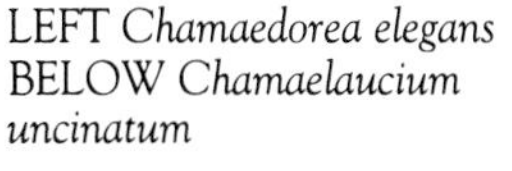

LEFT *Chamaedorea elegans*
BELOW *Chamaelaucium uncinatum*

CHAMAERANTHEMUM
(CHAMERANTHEMUM)

Acanthaceae

Origin: Tropical America. A genus of four or possibly eight species of smallish evergreen perennials and sub-shrubs. They are grown mainly for their handsomely patterned leaves, the small, tubular, white to mauve or yellow flowers being relatively insignificant. Those described below add a touch of class to a collection of foliage plants in the home or warm conservatory. Propagate by division in spring, or by cuttings in summer. *Chamaeranthemum* derives from the Greek *chamai*, on the ground (of low growth) and the related taller genus *Eranthemum*, *q.v.*

Species cultivated

C. gaudichaudii Brazil
Stem prostrate, forming mats to 45cm(1½ft) or more across. Leaves 5–10cm(2–4in) long, elliptic to oblong-ovate, deep green with a bold silvery-grey pattern. Flowers white and lavender. Worth growing as a small hanging basket subject and makes attractive ground cover for a conservatory bed.

C. igneum (*Xantheranthemum igneum*, *Stenandrum igneum*, *Aphelandra igneum*) Peru
Stems prostrate, forming small, tufted mats to 25cm(10in) or more across. Leaves oblanceolate to oblong-elliptic 5–8cm(2–3in) long, deep velvety bronzy-green with a conspicuous deep yellow to reddish vein pattern. Flowers yellow and, though small, more showy than in the other species.

Chamaerops humilis

C. venosum Brazil
Stems prostrate to decumbent, eventually forming mats to 40cm(16in) wide. Leaves broadly ovate, often sub-cordate at the base, up to 8cm(3in) long, almost leathery-textured, downy, greyish-green with silver veins. Flowers lavender and white. The cultivar 'India Plant' has leaves twice as large.

CHAMAEROPS

Palmae

Origin: Western Mediterranean. A genus of two species of fan palm of which one is cultivated. It is considered less decorative than such genera as *Chamaedorea*, but will withstand cooler conditions and is suitable for the conservatory or unheated porch where the more tropical palms will not thrive. Propagate by seed in warmth or by suckers. *Chamaerops* derives from the Greek *chamai*, dwarf and *rhops*, a bush.

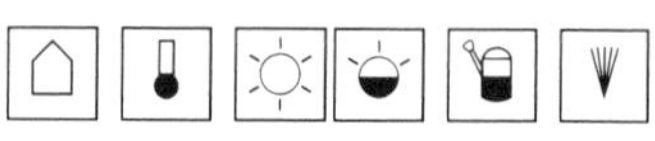

Species cultivated

C. humilis Dwarf fan palm
Suckering and clump-forming, in the wild to 1.5m(5ft), rarely more, and less than this in a pot. Leaves fan-shaped, 60–90cm(2–3ft) across, deeply cut into slender, linear lobes and carried on 1m(3ft) spiny-margined stalks. Flowers small, with six yellowish perianth lobes, borne in dense panicles within a large spathe; they are followed by yellow to brown, small date-like fruits.

CHAMERANTHEMUM See CHAMAERANTHEMUM

CHASMANTHE

Iridaceae

Origin: Tropical to Southern Africa. A genus of about seven species of cormous plants, two of which are grown for their unusually shaped but basically tubular, colourful flowers and upstanding sword-like leaves. They make interesting pot plants for the conservatory and can be brought indoors when in bloom. Propagate by seed in spring or by offsets when dormant. *Chasmanthe* derives from the Greek *chasme*, gaping or yawning and *anthos*, a flower, alluding to the mouth of the flower.

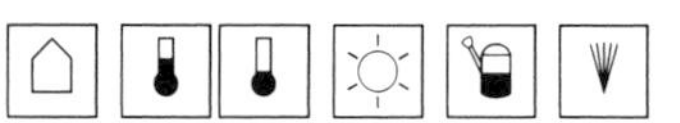

Species cultivated

C. aethiopica (*Antholyza aethiopica*) South Africa
Erect habit, 90–120cm(3–4ft) tall when in bloom. Leaves sword-shaped, shorter than the branched flower spikes. Each bloom orange-red with a 5cm(2in) long curving tube tipped by six narrow

FAR LEFT *Chasmanthe aethiopica*
LEFT *Cheiranthus* × 'Bowles Mauve'

petal lobes; the uppermost lobe is much longer than the rest and thrusts forward carrying on the arch of the tube; the smaller lobes reflex; flowers open in summer.

C. caffra Cape, Natal
Erect, to 45cm(1½ft) tall in bloom. Leaves shorter, linear-lanceolate. Flowers shaped like those of *C. aethiopica*, but a little smaller and bright red; they are carried in unbranched spikes around midsummer.

C. floribunda Cape
Much like *C. aethiopica*, but with leaves up to 4.5cm(1¾in) wide (those of *C. aethiopica* do not exceed 3cm[1¼in]), denser flower spikes and each flower with wider tubes; summer blooming.

CHEIRANTHUS

Cruciferae

Origin: Mediterranean region, Himalayas, Madeira and Canary Isles. A genus of ten species of perennials and sub-shrubs grown for their often showy racemes of fragrant 4-petalled flowers. They make good pot plants for the cool conservatory, those grown as biennials can be brought into the home when in bloom, and though perennials are best discarded after flowering. Propagate by seed in spring for the biennials, by cuttings with a heel in summer, or by seed for the shrubby sorts. *Cheiranthus* is said to derive from the Greek *cheir*, a hand, referring to the practice of carrying flowers in a fragrant bouquet, or by corruption of an Arabic word.

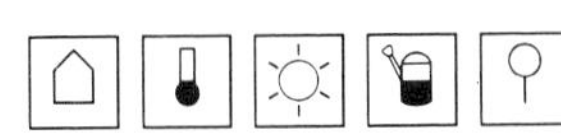

Species cultivated

C.* × *allionii Siberian wallflower
Although now considered probably a hybrid between *Erysimum perovskianum* and *E. decumbens*, this is still usually listed under *Cheiranthus*. It is a short-lived perennial with erect stems to 30cm(1ft) or more tall. Leaves 5–8cm(2–3in) long, narrowly lanceolate. Flowers to 12mm(½in) wide, brilliant orange, in terminal racemes in spring and summer.

***C.* × 'Bowles Mauve'**
A well-branched shrub, 60–90cm(2–3ft) tall, becoming woody with age and then best replaced from cuttings. Leaves narrowly lanceolate to 8cm(3in) long, greyish. Flowers rich mauve-purple, almost throughout the year.

C. cheiri Wallflower Europe
Grown as a biennial, to 45cm(1½ft) tall. Leaves oblong-lanceolate, to 10cm(4in), dull green. Flowers 2–4cm(¾–1½in) across, basically yellow, but selections and hybrids are grown in a wide range of colours. Unless a tall plant is required, the dwarf bushy sorts usually sold as Dwarf Bedding varieties are best. They are available in many beautiful shades of cream, yellow, orange and red. C. 'Harpur Crewe' is a fully double yellow cultivar, to 25cm(10in).

C. mutabilis Canary Isles, Madeira
Sometimes described incorrectly as a synonym of *C. semperflorens*. A well-branched shrub, 60–90cm(2–3ft) tall. Leaves lanceolate, 4–8cm(1½–3in) long. Flowers 1.5–2cm(⅝–¾in) across, opening pale yellow, ageing to lavender-purple giving rise to flower heads of mixed colours in spring.

C. semperflorens Morocco
A sub-shrub to 60cm(2m) tall. Leaves to 8cm(3in) long, greyish green. Flowers 1–1.5cm(⅜–⅝in) wide with white petals from lilac sepals. C. 'Constant Cheer' is a hybrid between this species and *C.* × *allionii*, making a neat dark green bush with dark purplish-red flowers, in 25cm(10in) racemes in summer.

CHEIRIDOPSIS See ALOINOPSIS
CHIAPASIA See DISOCACTUS
CHILENIA See NEOPORTERIA

CHIRITA

Gesneriaceae

Origin: Indo-Malaysia, S.E. Asia and southern China. A genus of 80 species of evergreen perennials and annuals of varied appearance. Some species are rosette-forming, others produce erect leafy stems. A third group produces a short, erect stem topped by one very large leaf and a cluster of smaller ones. All species bear tubular flowers with five, usually rounded, petal lobes. The species described thrive in pots. Propagate by seed in late winter or spring and by cuttings in spring or summer. *Chirita* derives from the Nepalese vernacular name for a gentian; some species have gentian-blue flowers.

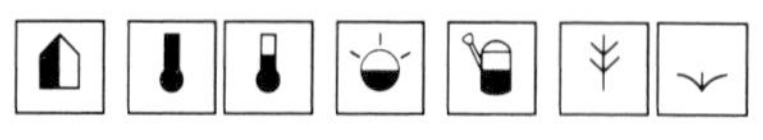

Chirita sinensis

Species cultivated

C. lavandulacea Asia (precise locality unknown)
Annual with fleshy, erect, leafy stems to 60cm(2ft) tall or more if well-grown. Leaves to 20cm(8in) long, elliptic-oblong, cordate, pale green and downy. Flowers 3–4cm($1\frac{1}{4}$–$1\frac{1}{2}$in) long, the tube white with lavender-blue lobes, in clusters from the leaf axils during summer and autumn.

C. micromusa Thailand
Annual to about 30cm(1ft) or a little more. Stem fleshy, erect, topped by one large ovate-cordate leaf up to 25cm(10in) long and several similar but smaller ones. Flowers 12–18mm($\frac{1}{2}$–$\frac{3}{4}$in) long, the tube pale yellow with orange-yellow lobes, from the leaf axils in summer and autumn.

C. sinensis China (including Hong Kong)
Rosette-forming. Leaves stalked, ovate to elliptic, 13–20cm(5–8in) long. The basic species has thick-textured, almost fleshy, rich green corrugated leaves, but silvery-white patterned forms occur in the wild and these are most commonly cultivated. Flowers 3–4cm($1\frac{1}{4}$–$1\frac{1}{2}$in) long, the tube white marked with yellow, the lobes lilac to purple, usually in clusters on stems well above the leaves in summer. Must be grown in sharply drained soil and watered with care.

CHLIDANTHUS

Amaryllidaceae

Origin: South America. A genus of one species, a bulbous plant which makes a delightful conservatory plant and can be brought into the home for flowering. Propagate by separating offsets during the dormant season. *Chlidanthus* derives from the Greek *chlide*, an ornament or luxury and *anthos*, a flower.

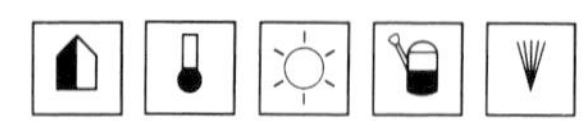

Species cultivated

C. fragrans
Leaves linear, glaucous green, to 30cm(1ft) long. Flowers produced before the leaves, lily-shaped, 10–13cm(4–5in) long, in few-flowered umbels borne on a stout 15–25cm(6–10in) stem in summer.

CHLOROPHYTUM

Liliaceae

Origin: Chiefly from Africa, but also in warmer parts of South America, Australia and Asia. A genus of 215 species of stemless perennials with short rhizomes and fleshy roots. Their leaves are linear or lanceolate-ovate with parallel veins. The flowers are starry with six tepals in loose racemes on slender, branched stems. Those cultivated are non-demanding house or conservatory plants. Propagate those which produce plantlets on their stems by detaching them when they have four to five leaves and potting them singly. All species can also be propagated by seed if available, or by division. *Chlorophytum* derives from the Greek *chlorus*, green and *phytum* a plant, an unremarkable description which could apply to almost every genus.

Species cultivated

C. bichetii See under C. *laxum*.

C. capense (*C. elatum*, *Anthericum elatum*) Spider plant South Africa
Leaves 25–60(10–24in), pale green, on branched stems, to 120cm(4ft). Flowers 1.5–2cm($\frac{1}{2}$–$\frac{3}{4}$in) wide, in small clusters. Confused with and sometimes considered a form of C. *comosum*. C.c. 'Variegatum' has leaves with pale creamy-yellow longitudinal stripes.

C. comosum (*Anthericum comosum*) Spider plant South Africa
Similar to C. *capense*, but leaves 20–45cm(8–18in) long, on branched, 30–60cm(1–2ft) stems. Flowers 2cm($\frac{3}{4}$in) or more wide produced together with small plantlets which weigh down the stems making them pendent. It is this characteristic that makes the spider plant such a good subject for a hanging basket. C. *capense* does not produce plantlets. C.c. 'Mandaianum' has a pale yellow

central variegation. *C.c.* 'Variegatum' has a marginal yellow stripe.
C. elatum See *C. capense.*
C. laxum Tropical Africa
Tufted to clump-forming. Leaves linear to narrowly oblanceolate, 13–20cm(5–8in), arching. Flowers 2cm(¾in) across, white, in slender spike-like panicles. *C.l.* 'Variegatum (*C. bichetii*) has leaves variably cream-striped and margined.

CHOISYA

Rutaceae

Origin: Southern USA, Mexico. A genus of six species of evergreen shrubs with glossy, dark green foliage and very fragrant creamy-white flowers. The species described below makes a splendid tub or pot plant for the conservatory, flowering mainly in spring but producing some blooms on and off throughout the year. Propagate by heel cuttings in summer. *Choisya* was named for Jacques Denis Choisy (1799–1859), Swiss Professor of Philosophy at Geneva and a noted botanist.

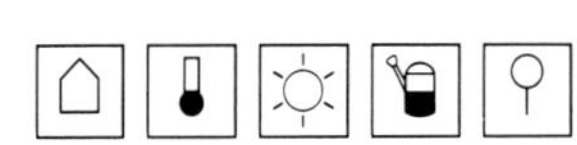

Species cultivated
C. ternata Mexican orange blossom Mexico
A bushy shrub, eventually 2–3m(6½–10ft) tall. Leaves in opposite pairs, trifoliate, the leaflets obovate, 4–7cm(1½–2¾in) with a strong, pungent smell when bruised. Flowers 2.5–3cm(1–1¼in) across, borne in terminal clusters. *C.t.* 'Sundance' is a golden-leaved form.

CHORIZEMA

Leguminosae

Origin: Australia, chiefly in the west. A genus of 15 species of evergreen shrubs and sub-shrubs. They have slender wiry stems and alternate, often spiny-toothed, linear to ovate leathery leaves. Flowers are typical of the pea family in shape, in a mixture of pinks, red and orange. They make attractive pot plants for the conservatory and can be brought into the home for flowering. Propagate by cuttings in early summer and early autumn, or by seed in spring. *Chorizema* is thought to derive from the Greek *choros*, a dance and *zema*, a drink, said to express the joy of a botanist who first found the plant by a spring of fresh water which was desperately needed.

TOP LEFT *Chlorophytum laxum* 'Variegatum'
ABOVE *Chlorophytum comosum* 'Variegatum'
LEFT *Choisya ternata*

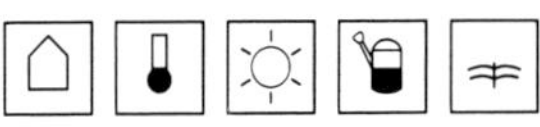

Species cultivated
C. cordatum (*C. ilicifolium*) Western Australia
Wiry, floppy shrub, to 60cm(2ft) tall, more if given support. Leaves 2–5cm(¾–5in) long, ovate to oblong, cordate, with stiff, spine-like teeth. Flowers about 1.5cm(⅝in) wide, the standard petals orange-red with yellow markings at the base, the wing and keel petals purplish; they are borne in 15cm(6in) long axillary and terminal racemes from spring to summer.

Chorizema cordatum

CHRYSALIDOCARPUS

Palmae

Origin: Malagasy, Comoro and Pemba Is. A genus of 20 species of palms one of which is highly ornamental and responds well to container culture. It makes a fine specimen for the conservatory and smaller plants can be used in the home. Propagate by seed in spring kept at 24–27°C(75–80°F). *Chrysalidocarpus* reputedly derives from the Greek *chryos*, gold and *karpos*, a fruit, but perhaps from *chrysalid*, the gold-tinted pupa of the butterfly, the allusion being to the seed of certain species.

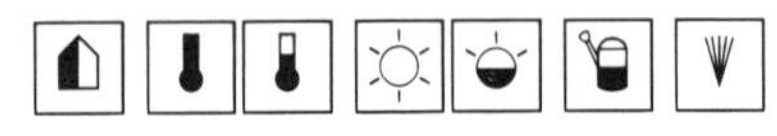

Species cultivated

C. lutescens (*Areca lutescens*) Golden feather palm, Areca palm Malagasy
Stems clustering, 2–3m(6½–10ft) tall in containers, up to 6m(20ft) or more if planted out and like robust canes. Leaves arching, pinnate, 90–180cm(3–6ft) long on containerized plants, with yellowish stalks and many linear leaflets. Specimens in pots are slow to produce stems and are then better suited to home situations. For vigorous growth, large tubs are needed.

Chrysalidocarpus lutescens

CHRYSANTHEMUM

Compositae

Origin: Northern temperate zone. A genus of 200 species of annuals, perennials and shrubs, several of which are mainstay plants in the garden, greenhouse and conservatory. All those described below are typified by showy daisy flowers usually freely borne over a long period. All embellish the conservatory in a colourful way and can be brought into the home when in bloom. Propagate by seed or cuttings in spring; C. *frutescens* also by cuttings in late summer or early autumn. *Chrysanthemum* derives from the Greek *chrysos*, gold and *anthos*, a flower, the first described species having yellow blooms.

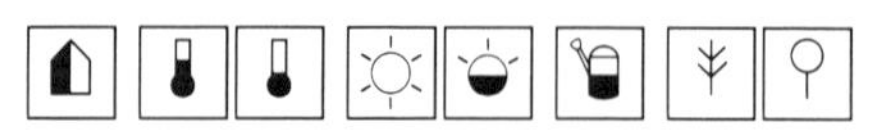

Species cultivated

C. carinatum (C. *tricolor*) North Africa
Hardy annual to 60cm(2ft) tall; stems erect, usually with several branches. Leaves somewhat fleshy, bipinnatifid, bright green, 5–8cm(2–3in) or more long. Flower heads 5cm(2in) or more wide, discs purple, ray florets white with a yellow base, in garden cultivars also variously banded purple, red or maroon; some are double. The allied C. *coronarium* from the Mediterranean region has flower heads with yellow discs and yellow to white ray florets.

C. coronarium See under C. *carinatum*.

C. frutescens (*Argyranthemum frutescens*) Marguerite Canary Isles
Frost-tender shrub of rounded habit, 30–60cm(1–2ft) in height. Leaves evergreen, usually bipinnatisect, 4–8cm(1½–3in) long, sometimes more or less glaucous. Flower heads 2cm(¾in) wide, white, disc yellow, rays white, intermittently all the year. Most of the plants grown as C. *frutescens* are, however, hybrids with other Canary Isles species, e.g. C. *coronopifolium*; popularly known as Paris daisies, they have flower heads 5–8cm(2–3in) wide, some with yellow or pink ray florets, singles and doubles. All are extremely useful conservatory and sunny home window plants flowering as they do in winter, provided a minimum day temperature of 13°C(55°F) can be maintained.

C. morifolium See C. *vestitum*.

C. multicaule Algeria
Hardy annual 15–30cm(6–12in) tall, branching freely from the base. Leaves spathulate in outline, trifid or pinnatifid, 5–10cm(2–4in) long, somewhat fleshy, glaucous. Flower heads about 5cm(2in) wide, golden-yellow throughout, borne in summer and early autumn.

C. × spectabile
Hybrid between C. *carinatum* and C. *coronarium*; at least some of the popular seed strains are of this origin. Sown in late winter; plants can be had in bloom by midsummer.

C. tricolor See C. *carinatum*.

C. vestitum (C. *morifolium*) Florists' chrysanthemums
From this species, probably via hybridization with C. *indicum* and other species, has arisen the extremely variable hybrid cultivar groups, so well known simply as chrysanthemums. They are erect, branched, woody-stemmed, almost hardy perennials, with obovate, prominently lobed leaves which may be deep green or greyish. The daisy

flower heads range from 3–20cm(1¼–8in) or more across, single or fully double. The many hundreds of cultivars have been classified into groups according to their floral characters. The main groups are:
Incurved: globular double blooms of firm-textured, incurving and overlapping florets.
Reflexed: rounded double blooms of florets which curve out and down.
Intermediate: halfway between the two previous groups with some in-curving and some out-curving florets.
Pompon: cushion-shaped to globular double blooms of many short, broad florets.
Spider: sparkler-like double blooms of long slender, quilled florets.
Spoon: like Spider, but with floret tips expanded like paddles (or spoons).
Anemone-centred: daisy-like blooms, but with the disc florets elongated and forming a central cushion surrounded by the longer rays.
Single: daisy-like blooms with a central yellow disc surrounded by one to a few rows of rays.
Charm, *Cascade* and *Korean* are all similar, producing a profusion of small, single to semi-double blooms on freely branching wiry stems.
All groups embrace a wide colour range; shades of red, pink, purple, bronze, yellow and white; some cultivars strikingly bicoloured. Chrysanthemums are known as short day (more accurately long night) plants, as they need nights of at least 9½–10 hours to initiate flower buds. For this reason they bloom naturally from late summer onwards. However, by manipulating the length of darkness with blackout blinds or covers, plants may be had in flower the whole year. Potted plants of this origin have for some time now been part of the florist's stock-in-trade. Such specimens are usually fairly dwarf because young plants are induced to bloom before they are fully grown. If cuttings are rooted from them and grown normally they will attain full height, and flower in autumn. All these florist's chrysanthemums are easily grown from late winter or early spring cuttings, though traditionally, the *Charm*, *Korean* and *Cascade* sorts are raised annually from seed. The young plants need their tips pinching out when about 15cm(6in) tall to induce branching. Late autumn blooming cultivars will need a second pinching out when the first laterals are about 10–15cm(4–6in) long. If allowed to bloom naturally the flower heads are produced in clusters or sprays. This is preferable for general decoration. However, if solitary blooms of large size are required the bud clusters must be thinned (disbudded) as soon as they are large enough to handle. Leave the largest terminal bud in each cluster, then carefully remove the rest by pushing them back or sideways with the finger tip. In addition, remove any side shoots from the leaf axils below the bud clusters. Very many cultivars of florists' chrysanths are available from specialist nurserymen and most garden centres carry stocks of well tried sorts. All are good and their choice is a matter of preferences for size, colour and form.

TOP *Chrysanthemum frutescens* 'Jamaica Primrose'
ABOVE *Chrysanthemum vestitum* cultivar

CHRYSOTHEMIS

Gesneriaceae

Origin: Central and South America. A genus of six species of tuberous-rooted perennials. Those described below have handsome foliage and small but colourful tubular flowers. They are evergreen if kept at a growing temperature and watered regularly, but when they get leggy and untidy with age they can be dried off, cut back, re-potted and started off again. Propagate by basal cuttings in early summer, or by

Chrysothemis pulchella 'Bronze Leaf'

seed in spring. *Chrysothemis* was named for the daughter of Clytaemnestra and Agamemnon, characters from Greek mythology.

 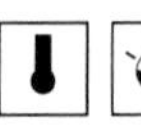  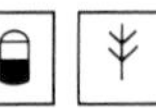

Species cultivated

C. friedrichsthaliana Central America, western Colombia
Stems erect, sparingly branched, 30–60cm(1–2ft) in height, rounded in cross-section. Leaves oblong-ovate to lanceolate, 15–30cm(6–12in) long, hairy, corrugated, bright lustrous green. Flowers about 2.5cm(1in) long, orange-yellow, striped red, each one sitting within a large, angled, bell-shaped, yellow-green calyx, opening in spring and summer.

C. pulchella (*C. tussacia pulchella*) Panama and West Indies to Brazil
Stems erect, usually only sparingly branched, 30–60cm(1–2ft) in height, bluntly square in cross-section. Leaves ovate to widely lanceolate, 15–30cm(6–12in) or more long, hairy, finely corrugated, usually bright glossy green, but in some forms a very dark or olive-green. Flowers about 2.5cm(1in) long, buttercup-yellow with red stripes and markings, each one sitting in a large, angled, bell-shaped bright orange-red calyx; the latter is somewhat fleshy and long persistent, creating the appearance of blossom long after the spring flowers (corolla tubes) have fallen.

Cibotium glaucum

CHYSIS

Orchidaceae

Origin: Tropical America. A genus of six species of deciduous or semi-evergreen epiphytic orchids. They have slenderly club-shaped pseudobulbs to 30cm(1ft) long, which are usually pendent, and pleated lanceolate leaves. The flowers have similar tepals and an often contrasting concave labellum. They are best in hanging baskets, and although most successful in the conservatory, can be grown in the home. *Chysis* is the Greek word meaning melting and refers to the pollinia (pollen masses) which appear to be fused.

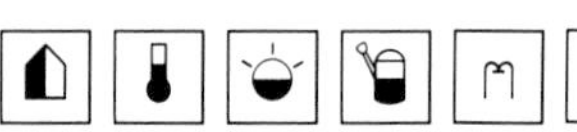

Species cultivated

C. aurea Mexico to Peru
Pseudobulbs 30–45cm(1–1½ft) long. Leaves oblong-lanceolate, to 45cm(1½ft) long, and 5cm(2in) wide, waved and slender-pointed. Flowers to 8cm(3in) in diameter, creamy-yellow, the labellum white, marked with purple to brown; they are borne in 6- to 12-flowered racemes, mainly in summer.

C.a. bractescens (*C. bractescens*) Mexico, Guatemala
Similar, but with bracts on the flowering stems exceeding 2.5cm(1in) – less than 2.5cm(1in) in *C. aurea* and a yellow-blotched labellum.

C. bractescens See *C. aurea bractescens*.

C. laevis Mexico, Costa Rica
Similar to *C. aurea*, but with smaller pseudobulbs and flowers, the latter not exceeding 8cm(3in) in width, the tepals orange-tipped and the wavy-margined labellum blotched and spotted with crimson; spring to summer flowering.

CIBOTIUM

Dicksoniaceae

Origin: Tropical America, Asia and Hawaii. A genus of ten species of mainly large ferns, some of them tree-like. Those listed here make imposing specimens for the warm conservatory. They have large, gracefully arching, usually tripinnate fronds the long stalks of which are densely long-hairy at the base. Propagate by spores in spring. *Cibotium* is from the Greek *kibotos*, a small box, alluding to the shape of the leathery, indusial covering of the sori.

 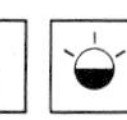   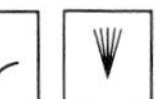

Species cultivated

C. chamissoi Hawaii
A tree fern in its homeland, but slow to form a trunk when grown in containers. Fronds up to 2m(6½ft) in height, the stalks (stipes) with yellow-brown and blackish hairs; blade of frond tripinnate, the lower pinnae broadly lanceolate and up to 45cm(1½ft) long, pinnatisect in the lower part, the ultimate pinnules oblong, blunt. Ferns grown under this name are usually *C. glaucum*.

C. glaucum Hawaii
Very much like *C. chamissoi*, but the fronds with a more obvious glaucous tint and the ultimate pinnules linear and slender-pointed, glabrous.

C. schiedei Mexico and Guatemala
A tree fern in its homeland, only slowly forming a trunk when grown in a container. Fronds to 1.2m(4ft) or more in height, the bases of the stalks densely covered with long, lustrous, brownish-yellow hairs. Blade of frond tripinnate, the lower pinnae oblong-lanceolate, 30–45cm(1–1½ft) or more long, pale green above, glaucous beneath, the ultimate pinnules lanceolate, toothed.

CINERARIA See SENECIO

CIRRHOPETALUM

Orchidaceae

Origin: Tropical Africa to Indo-Malaysia, eastwards to Tahiti. A genus of about 70 species of small, epiphytic orchids closely allied to, and sometimes united with, *Bulbophyllum*. The small, intriguing and sometimes colourful flowers are solitary, or borne in umbels. Like those of *Bulbophyllum*, they have a hinged, mobile lip which rocks in the wind or when touched. The flowers, however, are easily distinguished by the two greatly elongated lateral sepals. In most species of *Cirrhopetalum* these are variously twisted or rolled longitudinally and lie parallel, sometimes united or appearing to be fused together. The species described here are best in the conservatory but may be brought indoors when in flower. Propagate by division in spring or just after flowering. *Cirrhopetalum* appears to derive from the Greek *kirrhos*, yellowish and *petalon*, petalled, alluding to the colour of the first species described. An alternative is *cirrhus*, a curl or tendril, perhaps referring to the long, twisted sepals.

Species cultivated

C. fascinator (*Bulbophyllum fascinator*)
Vietnam, Laos
Pseudobulbs forming clusters, almost spherical, to 2.5cm(1in) high, smooth. Leaves narrowly elliptic, blunt-tipped, leathery, about 5cm(2in) long. Flowers solitary, on stems up to 10cm(4in) tall; lateral sepals 15–20cm(6–8in) or more long, the ends brownish-green and tail-like, the bases green with reddish tubercles; upper sepal and petals fringed with purple hairs; borne in autumn.

C. guttulatum (*Bulbophyllum umbellatum*)
Himalaya
Pseudobulbs forming clusters, ovoid, 2.5–5cm(1–2in) tall. Leaves narrowly oblong, somewhat leathery, 15–25cm(6–10in) long. Umbels of up to nine or sometimes more flowers on stems the same length as the leaves; lateral sepals not greatly elongated, to 2cm(¾in) long, greenish-yellow, purple spotted; borne in autumn.

C. longissimum (*Bulbophyllum longissimum*)
Thailand
Pseudobulbs egg-shaped, sometimes bluntly angular, about 5cm(2in) tall, spaced the same distance apart on creeping rhizomes. Leaves oblong, up to 15cm(6in) long, stiffly leathery. Umbels of five to ten flowers on stems to about 20cm(8in) long; lateral sepals 15–30cm(6–12in) long, united basally, tapering to the tips, buff-pink

Cirrhopetalum longissimum

ABOVE *Cirrhopetalum robustum*
RIGHT *Cissus antarctica*

to pale rose-red with deeper veining; labellum small, yellow-green; borne in winter.

C. medusae (*Bulbophyllum medusae*) Malaysia, Sumatra, Borneo
Pseudobulbs ovoid, more or less 4-angled, about 4cm(1½in) high, well-spaced on creeping rhizomes. Leaves oblong, the blunt tip shallowly notched, stiffly leathery, dark green, to 20cm(8in) long. Umbels densely-flowered, forming rounded heads on stems to 20cm(8in) tall; individual flowers white, cream or straw-yellow, usually with pink or red spots, the lateral sepals 10–15cm(4–6in) long, tapering to thread-like tails; borne in autumn to winter.

C. ornatissimum (*Bulbophyllum ornatissimum*) Himalaya, India
Pseudobulbs ovoid, about 2.5cm(1in) tall, spaced on creeping rhizomes. Leaves oblong-ovate, leathery, to about 15cm(6in) long. Umbels of three to six flowers on a 10cm(4in) tall stem; individual blossoms pale yellow, streaked and/or flushed red to purple, the lateral sepals 8cm(3in) long, lanceolate, tapering to whip-like tails, fragrant; borne in autumn to early winter.

C. robustum (*C. graveolens, Bulbophyllum robustum*) New Guinea
Pseudobulbs clustered, ovoid, to 8cm(3in) tall. Leaves in pairs, elliptic-oblong, slender-pointed, leathery-textured, lustrous deep green, up to 30cm(1ft) or more in length. Flowers six to twelve, arranged in a flat, circular umbel on a strong stem to 10cm(4in) high; each flower 5–8cm(2–3in) long, yellow and reddish-green with a dark red labellum, borne in spring to early summer. A most striking species but with somewhat carrion-scented blooms.

CISSUS

Vitidaceae

Origin: Tropical and sub-tropical regions of the world. A genus of 34 species of climbing and succulent shrubs. They have woody or herbaceous stems with alternate leaves and most have tendrils. Flowers are small, usually greenish and 4-petalled in branched clusters. They are followed by berry-like fruits. All species described are suitable for the conservatory and, with the exception of *C. discolor*, which needs warmth and humidity, are excellent house plants. *Cissus* is derived from the Greek *kissos*, ivy, many species being woody climbers.

Species cultivated

C. antarctica Kangaroo vine Australia
A vigorous woody climber, to 5m(16ft) or more in the wild, far less in a pot. Leaves 7–15cm(2¾–6in), ovate to ovate-oblong, usually somewhat toothed, leathery and glossy, dark green above, paler beneath. Opposite each leaf is a forked tendril. The insignificant flowers are borne in small, axillary clusters. A very tolerant and popular house plant. Temperate.

C. bainesii See *Cyphostemma bainesii*.

C. cactiformis East Africa
Climbing succulent with fleshy stems less than 5cm(2in) thick, constricted at the nodes with four conspicuously winged angles. Leaves small and ovate, occurring only on young growth and soon falling. Temperate.

C. capensis See *Rhoicissus capensis*.

C. discolor Rex begonia vine Cambodia, Java
A slender-stemmed woody climber to about 2m(6½ft). Stems and tendrils dark red. Leaves 10–

15cm(4–6in) long, oblong-ovate, cordate, bristly-toothed, the surface somewhat quilted, velvety-green above with silvery-white patterning between the veins and maroon beneath. Splendid for a hanging basket. Tropical.

C. juttae See *Cyphostemma juttae*.

C. quadrangularis South and tropical Africa
Very similar to *C. cactiformis* with thick, fleshy, 4-angled stems constricted at the nodes, but less than 5cm(2in) thick. Temperate.

C. rhombifolia Grape ivy Mexico to Brazil and West Indies
Woody tendril climber much like, and confused with, the better known *Rhoicissus rhomboidea*. Both can exceed 3m(10ft) in length, with evergreen, coarsely-toothed, lustrous, trifoliate leaves. The easy identification is in the tendrils, those of *C. rhombifolia* being forked at the tips. Temperate.

C. sicyoides Princess vine Probably Brazil
A vigorous, much-branched climber to 3m(10ft) or more. Leaves heart-shaped, about 10cm(4in) long, slender-pointed, somewhat fleshy and a pleasing light green. Under humid conditions, slender aerial roots are produced in profusion, hanging down like a curtain. Temperate. *C.s.* 'Albo-nitens' has leaves with a lustrous silvery patina.

C. voinierianum See *Tetrastigma voinierianum*.

× CITROFORTUNELLA

Rutaceae

Origin: A hybrid arisen in cultivation. A hybrid genus combining species of *Citrus* and *Fortunella*. Three species are known, one of which is frequently met with. They are intermediate between their parents and those described below make attractive pot plants. Propagate by cuttings with a heel in late summer.

Species cultivated

× *C. mitis* (*Citrus mitis*) (*Citrus reticulata* [tangerine] × *Fortunella* [kumquat]) Calamondin
Evergreen shrub to 2m($6\frac{1}{2}$ft), usually less in a pot. Leaves 4–6cm($1\frac{1}{2}$–$2\frac{1}{2}$in) long, ovate to elliptic, paler on the undersides, stalks with very narrow or no wings. Flowers 1–1.5cm($\frac{3}{8}$–$\frac{5}{8}$in) wide, white. Fruits flattened-globose with loose skin, bright orange, to 4cm($1\frac{1}{2}$in) wide, produced on young plants (make good marmalade).

CITRUS

Rutaceae

Origin: S.E. Asia, Indonesia. A genus of 12 species of evergreen shrubs and trees which have been in cultivation for at least 2,000 years. Many are probably of hybrid origin and few can be found truly wild today. They are usually spiny plants with alternate, ovate elliptic-oblong leaves which are

× *Citrofortunella mitis*

aromatic when crushed and borne on usually broadly winged stalks. Flowers are very fragrant and have white fleshy petals. They are borne in small corymbs and followed by large, leathery-skinned fruits (botanically a hesperidium) with juicy, segmented pulp. Citruses make good house and conservatory plants and can be raised from pips as fun plants. For the best fruits, however, named cultivars should be grown. Propagate by seed or by heel cuttings in summer, the latter method for cultivars.

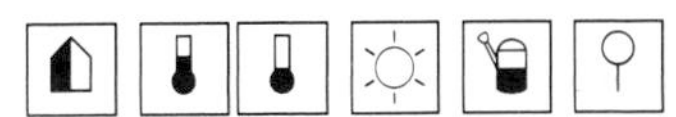

Species cultivated

C. aurantifolia (*C. limetta*, *Limonia aurantifolia*) Lime Probably S.E. Asia, but also wild in N. India
A small tree to 5m(16ft) in open ground, less in pots. Leaves ovate-elliptic, 4–8cm($1\frac{1}{2}$–3in) long with a narrowly winged stalk. Flowers small, the four to five white petals each 8–12mm($\frac{1}{3}$–$\frac{1}{2}$in) long. Fruit oval to globose, 3.5–6cm($1\frac{1}{3}$–$2\frac{1}{2}$in) wide with thick greenish-yellow skins and greenish flesh; very acid.

C. aurantium (*C. bigarardia*) Seville orange S.E. Asia, probably Vietnam
Tree to 10m(33ft), shrubby in a pot. Leaves ovate to elliptic, to 10cm(4in) long, wavy-edged with a broadly-winged stalk. Flowers white, 2–3cm($\frac{3}{4}$–$1\frac{1}{4}$in) wide, very fragrant. Fruit almost globose, somewhat flattened with a thick, coarse, orange-red skin and sour, bitter pulp. The hardiest citrus and the provider of marmalade oranges.

C. bigarardia See *C. aurantium*.

C. decumanus See *C. grandis*.

C. grandis (*C. maxima*, *C. decumanus*) Shaddock, Pummelo Malaysia, Thailand
Tree to 10m(33ft), shrubby in a pot, often without spines. Leaves 5–20cm(2–8in) long, ovate to elliptic, with stems so broadly winged that they look like a second leaf blade. Flowers 5–6cm(2–$2\frac{1}{2}$in) wide,

ABOVE *Citrus limon* 'Variegata'
RIGHT *Citrus medica*

cream with thick-textured petals. Fruits globose to somewhat pear-shaped, solitary, the rind thick and spongy, juice sweetish.

C. limetta See *C. aurantifolia*.

C. limon (*C. limonia*, *C. limonium*) Lemon S.E. Asia

Large shrub or small tree, 3–6m(10–20ft). Leaves elliptic-ovate, toothed, 5–10cm(2–4in) long, stalk with no or only very narrow wings. Flowers white, 4cm(1½in) or more across opening from a pink to purplish bud. Fruit oval, 5–10cm(2–4in) long, thin-skinned with acid juice. *C.l.* 'Variegata' has leaves irregularly margined with yellow and the fruits also striped green and yellow, the markings becoming less obvious as they ripen.

***C.l.* 'Meyer'** (*C. meyeri*) Meyer's lemon

Considered by some to be a lemon-orange hybrid, it is very similar to *C. limon* but has orange-coloured fruits which are lemon-shaped. It will flower and fruit freely when small and makes an excellent pot plant.

***C.l.* 'Ponderosa'** Giant American wonder lemon

A hybrid of *C. limon* and *C. medica*, it has spiny stems and lemon- to almost pear-shaped fruit up to 13cm(5in) long; they are edible, but inferior to true lemons. The plant, however, does fruit when small and is an excellent tub plant for the conservatory.

C. limonia (*C. limonium*) See *C. limon*.

C. maxima See *C. grandis*.

C. medica Citron S.W. Asia

Shrub to 3m(10ft) bearing stout spines. Leaves ovate-oblong, toothed, 5–10cm(2–4in) in length with no wings on the stalks. Flowers white, pink in bud, opening to 3.5cm(1⅓in) or more across. Fruits 5–10cm(2–4in) long, oval, yellow when ripe, the skins thick and rough, the pulp sour with little juice. Grown mainly for ornament.

C. meyeri See under *C. limon*.

C. mitis See × *Citrofortunella mitis*.

C. nobilis See *C. reticulata*.

C. paradisi Grapefruit

Probably a hybrid which arose either in S.E. Asia or in the West Indies where *C. grandis*, one of its possible parents, was grown. Tree to 15m(50ft). Leaves 5–15cm(2–6in) long, ovate to elliptic with a winged stalk. Flowers 4–5cm(1½–2in) across, white. Fruits globular, 8–15cm(3–6in) wide, pale yellow to light orange or greenish, rind thin. Cultivars with pink and red pulp are grown. Can be grown from seed, but less good in pots than most other citrus.

C. reticulata (*C. nobilis*) Mandarin, Satsuma, Tangerine Vietnam, China

Shrub or small tree, 2–8m(6½–26ft) high. Leaves 4–8cm(1½–3in) long, lanceolate to elliptic, with a wingless stalk. Flowers white, 1.5–2.5cm(⅝–1in) wide. Fruits globose, flattened, with thin, loose rind usually orange-red. Hybridizes with grapefruit (Tangelo, 'Ugli') and sweet orange (Tangors, 'Ortanique').

C. sinensis (*C. aurantium sinense*) Sweet orange S. China or Vietnam

Tree to 6–12m(20–40ft) high, usually bearing stout spines. Leaves ovate to elliptic, 5–15cm(2–6in) long, with a narrowly-winged stalk. Flowers white, 2–3cm(¾–1¼in) across. Fruit 4–11cm(1½–4½in), rounded to shortly oval, pale to deep orange. Many cultivars are known.

CLARKIA

Onagraceae

Origin: Western North America and Chile. A genus of 36 species of annuals many of which were formerly classified as *Godetia*. They are slender, erect plants, freely branching with alternate to lanceolate leaves and 4-petalled flowers in terminal spikes. They make fine pot plants for late winter and spring colour in the conservatory and can be brought into the home for flowering. Propagate by seed sown in early autumn; pot into 13cm(5in) pots in October. *Clarkia* was named for Captain William Clark (1770–1838), co-leader with Capt. Lewis of the first expedition to cross the Rockies from east to west to the Columbia River (1804–6).

  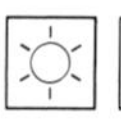 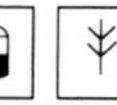

Species cultivated

C. amoena (*Godetia amoena, Oenothera amoena*) California

Bushy, to 60cm(2ft). Leaves lanceolate, 1–6cm($\frac{3}{8}$–$2\frac{1}{2}$in) long. Flowers 4–6cm($1\frac{1}{2}$–$2\frac{1}{2}$in) wide, white, pink, lilac or red, from slender, erect buds.

C.a. whitneyi (*C. grandiflora* of gardens)

Larger flowers to 8cm(3in) in diameter, with a central red blotch. A number of cultivars and seed strains have been raised with single and double flowers; also dwarf strains 20–25cm(8–10in) in height.

C. elegans See *C. unguiculata.*

C. grandiflora See *C. amoena whitneyi.*

C. pulchella Western North America

To 45cm($1\frac{1}{2}$ft) tall. Leaves lanceolate to linear. Flowers to 4cm($1\frac{1}{2}$in) wide with strongly 3-lobed petals, pink to white. Several strains are available, often semi-double, in shades of purple, pink or white.

C. unguiculata (*C. elegans, Oenothera elegans*) Western America

To 1.2m(4ft), sometimes more. The reddish stems bear lanceolate to ovate, 1–6cm($\frac{3}{8}$–$2\frac{1}{2}$in) long leaves. Flowers to 4cm($1\frac{1}{2}$in) across, in a wide variety of pinks, from salmon to lavender and red-purple. A number of single and double cultivars are available; also seed strains in colour mixtures.

CLEISTOCACTUS

Cactaceae

Origin: South America. A genus of 30 species of cacti with slender erect or decumbent, columnar stems having many ribs bearing small, closely set round areoles. The flowers are narrowly tubular and curved to almost straight. Fruits are in the form of small, globular berries. They grow well in the home or conservatory, flowering freely. Propagate by seed in spring or by cuttings of stem tips in summer. *Cleistocactus* derives from the Greek *kleistos*, closed, the flowers barely open.

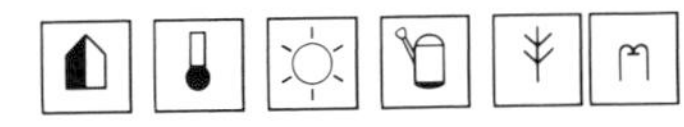

Species cultivated

C. baumannii (*Cereus baumannii*) Scarlet bugler Argentina, Uruguay, Paraguay

Stems to 1m(3ft) or more tall, with about 14 ribs bearing yellow-brown areoles; each has 15 to 20 4cm($1\frac{1}{2}$in) long spines. Flowers orange-scarlet, asymmetric, 6–7cm($2\frac{1}{2}$–$2\frac{3}{4}$in) long, opening in summer and followed by red 1.2cm($\frac{1}{2}$in) fruits.

C.b. flavispinus

Both areoles and spines pale yellow.

C. dependens Bolivia

Stems prostrate or pendulous with ten to 12 shallow

Clarkia amoena whitneyi

Cleistocactus strausii

ribs bearing hairy areoles, the hairs white above, black beneath; each has 11 to 17 spines. Flowers to 4.5cm(1¾in), pink, tipped with pale green.

C. smaragdiflorus (*Cereus smaragdiflorus*) Argentina, Uruguay, Paraguay
Similar to *C. baumannii* and sometimes classified as a variety of it. Spines in clusters of 22 to 28, brown. Flowers reddish-yellow to scarlet, 3.5–5cm(1¼–2in) long with bright emerald-green petal tips.

C. strausii (*Cereus strausii, Borzicactus strausii, Pilocereus strausii*) Bolivia
Stems to 1m(3ft) or more in length, branching from the base and with about 25 ribs; the closely set areoles have 30 hair-like white bristles and three to four pale yellow or white spines giving a white-woolly effect. Flowers 8–9cm(3–3½in) long, red.

CLEMATIS

Ranunculaceae

Origin: Cosmopolitan. A genus of 250 species of woody and sub-shrubby climbers and perennials. The cultivated climbing species, being mainly frost hardy, are grown in the open garden, but a few are tender and can be used very effectively in the cool to temperate conservatory. In addition, many of the showy, large-flowered hybrids thrive under cool conservatory conditions and flower much earlier than outside. They are also particularly useful where a flowering climber is needed for the totally unheated conservatory. Clematis vary greatly in their botanical details. Most climbing species have trifoliate to pinnate leaves and saucer- to bell-shaped flowers. The latter have from four to seven or more petal-like sepals and a central boss of stamens and stigmas. Propagate by stem or leaf bud cuttings in late summer, or by layering in winter or early spring. Seed may be sown of the species, preferably when ripe or in autumn or winter and kept cool. *Clematis* is a Greek name for several sorts of climbing plants, but was used by Linnaeus for this genus alone.

Clematis cirrhosa balearica

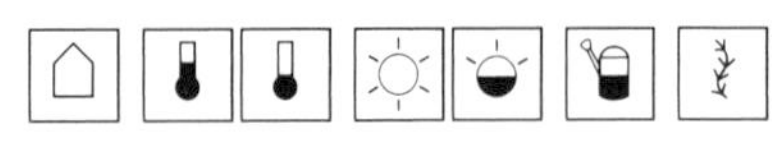

Species cultivated

C. afoliata New Zealand
To 3m(10ft) in height with dark green, rush-like stems and leaves with no proper blade, being reduced to a single leaflet on a wiry mid-rib. Flowers 2cm(¾in) long, nodding, bell-shaped, fragrant, green tinted with yellow.

C. armandii China
Strong-growing evergreen to 6m(20ft) or more. Leaves trifoliate, the leaflets 7–13cm(2¾–5in) long, ovate-lanceolate, bronze when young, becoming a deep glossy green. Flowers 4–6cm(1½–2½in) across, pure white, fragrant, carried in axillary clusters in spring. *C.a.* 'Apple Blossom' has pink-flushed flowers.

C. balearica See *C. cirrhosa balearica.*

C. calycina See *C. cirrhosa balearica.*

C. cirrhosa Southern Europe
Slender evergreen to 4m(13ft) tall. Leaves 3-lobed, ternate or biternate, 2.5–5cm(1–2in) long, dark green. Flowers 4–7cm(1½–2¾in) across, nodding, pale creamy-yellow, sometimes spotted with red inside, opening in winter.

C.c. balearica (*C. balearica, C. calycina*)
Leaves deeply divided, almost fern-like. Flowers creamy-white, always red-spotted within, opening from late autumn through to spring.

C. coccinea See *C. texenis.*

C. florida China
Usually evergreen under glass, to 4m(13ft) tall. Leaves biternate with ovate-lanceolate leaflets 2.5–5cm(1–2in) long. Flowers 6–8cm(2½–3in) across, white with a central boss of black-purple stamens, borne singly in summer. *C.f.* 'Sieboldii' ('Bicolor') has flowers with a central rosette of purple, petaloid stamens.

C. indivisa See *C. paniculata.*

C. napaulensis Northern India, S.W. China
Evergreen, to 10m(33ft), but easily kept to a third of this in containers and with pruning after flowering. Leaves composed of three to five, narrowly ovate leaflets, with or without a few large teeth. Flowers bell-shaped, yellow with purple anthers, in clusters from the leaf axils, opening throughout winter.

C. paniculata (*C. indivisa*) New Zealand
Evergreen, to 10m(33ft), but easily kept smaller. Leaves trifoliate, the leaflets 3–20cm(1¼–8in) long, dark green, leathery texture, broadly ovate, sometimes cordate. Flowers with the sexes separate on the same plant; male flowers 5–10cm(2–4in) across, females smaller, white and fragrant; they are carried in panicles to 30cm(12in) long in spring and summer.

C. phlebantha Western Nepal
A more or less evergreen climber up to 3m(10ft) in height, the whole plant covered with silky white hairs. Leaves pinnate, composed of five to nine broadly ovate leaflets, large-toothed, about 1.5cm($\frac{5}{8}$in) long. Flowers appearing singly from the upper leaf axils, 3–5cm($1\frac{1}{4}$–2in) wide, having five to seven, red-veined white sepals, opening during late summer.

C. texensis (*C. coccinea*) USA (Texas)
A softly shrubby species to 4m(13ft) or more. Leaves pinnate, made up of four to eight leaflets, each 3–7cm($1\frac{1}{4}$–$2\frac{3}{4}$in) long, ovate-cordate, glaucous green. Flowers to 2.5cm(1in) long, pale to deep crimson, nodding, pitcher-shaped.

Hybrids

Large-flowered *Clematis* hybrids of garden origin. Most of the familiar ones grown in the open garden are of hybrid origin with the characters of *C. lanuginosa* and *C. patens* predominating, generally having compound leaves and flattish flowers to 15cm(6in) wide in shades of mauve, purple, blue, red and white. The majority of them grow well in the cool to cold conservatory, but the naturally early (summer-blooming outside) cultivars are best, producing their flowers from mid-spring onwards, depending on warmth. These are best lightly pruned and thinned annually, after the first flush of flowers, to keep them tidy. Among suitable cultivars, the following can be recommended:
'Beauty of Worcester', blue.
'Belle of Woking', mauve.
'Contesse de Bouchard', pink and mauve.
'Henryi', white.
'Lasurstern', lavender blue.
'Niobe', deep red.
'Vyvyan Pennell', deep violet-blue, the first flush double, subsequent ones single.

CLERODENDRUM

Verbenaceae

Origin: Tropics and sub-tropics of the Old World. A genus of 400 species of shrubs, climbers and small trees. They have opposite pairs of usually undivided ovate leaves and small tubular flowers borne in terminal or axillary clusters. Some are rendered showy by a colourful calyx. The berry-like fruits are also decorative in some species. They make attractive pot or tub plants for the conservatory, the climbers needing support. *C. speciosissimum* will flower when only 30cm(1ft) high and can be grown in a bright window in the home when young; it can be grown annually from spring cuttings. Propagate by cuttings in heat in spring or summer, or by seed in spring.

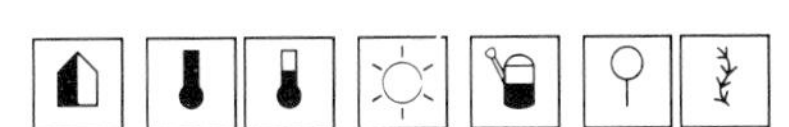

Species cultivated

C. fallax See *C. speciosissimum*.

Clematis 'Niobe'

C. fragrans China, naturalized in many warm countries
Shrubby, to 2m($6\frac{1}{2}$ft) or more, with rigid, angled, white-hairy stems. Leaves to 20cm(8in) or more long, broadly ovate with a truncate or heart-shaped base and coarsely toothed. Flowers 2.5cm(1in) wide, pink or white, fragrant, in dense terminal clusters like those of a hydrangea. Usually represented in cultivation by the double-flowered *C. fragrans pleniflorum*. A good house plant, surprisingly neglected in Britain. Temperate.

C. paniculatum Pagoda flower S.E. Asia
Shrubby, 60–120cm(2–4ft) in height. Leaves broadly ovate to almost circular in outline, 5-lobed, cordate, up to 15cm(6in) long. Flowers scarlet, 12mm($\frac{1}{2}$in) long, grouped into small cymes which in turn are aggregated into 30cm(1ft) long terminal panicles. A showy plant, much like a smaller version of *C. speciosissimum* and worth trying in the home. Temperate.

C. speciosissimum (*C. fallax*) Java glorybean Java
To $3\frac{1}{2}$m(12ft) in the open ground, but about 1m(3ft) in a pot. Leaves broadly heart-shaped, long-stalked, to 30cm(1ft) long, on stout hairy stalks. Flowers 3–4cm($1\frac{1}{4}$–$1\frac{1}{2}$in) long, scarlet, borne in terminal panicles to 25cm(10in) from summer to autumn. Temperate.

Clerodendrum speciosissimum

C. splendens Tropical W. Africa
Woody evergreen twining climber to 3m(10ft) or so, but easily kept smaller in containers. Leaves rich, somewhat lustrous green, oblong to elliptic, 10–15cm(4–6in) long. Flowers bright scarlet, about 2.5cm(1in) wide, in corymbose clusters to 13cm(5in) across, at the stem tips, mainly in summer, sometimes extending into autumn. A striking climber for the warm conservatory; small, potted specimens also in the home. Temperate.

ABOVE *Clerodendrum thomsonae*
ABOVE RIGHT *Clethra arborea*

C. thomsonae Bleeding heart vine, Glory bower Tropical W. Africa
Woody evergreen twining climber, to 4m(13ft) or so tall, but flowering freely when small. Leaves broadly ovate, 7–15cm($2\frac{3}{4}$–6in) long. Flowers to 2cm($\frac{3}{4}$in) wide, crimson, carried within pure white, bell-shaped calyces; they are borne in terminal and axillary clusters in summer. Tropical.

C. ugandense Blue glory bower Tropical Africa
Slender, scrambling evergreen shrub, to 3m(10ft), usually grown as a climber and which blooms when young. Leaves elliptic to narrowly obovate, coarsely toothed. Flowers blue, about 2.5cm(1in) long, in terminal panicles to 15cm(6in) wide, borne in summer or autumn. Best in the conservatory but can be tried in the home. Tolerates temperate conditions.

CLETHRA

Clethraceae

Origin: America, E. Asia and Madeira. A genus of 68 species of deciduous and evergreen shrubs of which one makes a handsome pot or tub specimen for the conservatory, eventually outgrowing its space, but amenable to pruning. Propagate by cuttings with a heel in late summer with bottom heat. *Clethra* derives from *klethra*, the Greek word for an alder tree, the two genera having some similarities in appearance.

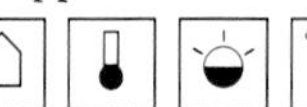   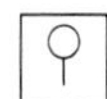

Species cultivated

C. arborea Lily-of-the-valley tree Madeira
Small evergreen tree to 8m(26ft) in the wild, shrubby in a pot. Leaves oblanceolate, finely toothed, 7–15cm($2\frac{3}{4}$–6in) long, bright glossy green above, paler beneath. Flowers white, cup-shaped, fragrant, in pendent racemes to 15cm(6in) long, several together making terminal panicles.

CLIANTHUS

Leguminosae

Origin: Warm temperate areas of New Zealand and Australia with one species in S.E. Asia. A genus of three species of evergreen shrubs. They have alternate, pinnate leaves and large pea flowers with pointed keel and standard petals which fancifully resemble a parrot's beak. *C. formosus* will grow in a sunny window in either home or conservatory; *C. puniceus* is best in a conservatory and needs a winter night temperature down to 10°C(50°F). Propagate by seed in spring; for *C. puniceus* also by cuttings with a heel in late summer. *Clianthus* derives from the Greek *kleos*, glory and *anthos*, a flower, referring to the brightly coloured flowers.

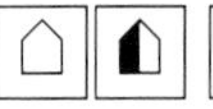 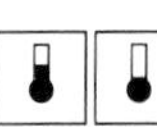 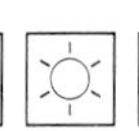 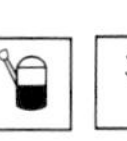 

Species cultivated

C. dampieri See *C. formosus*.

C. formosus (*C. dampieri*) Glory pea, Dampier's pea Australia
Stems white-woolly, prostrate. Leaves have 11 to 21 oval to obovate leaflets, 1–2.5cm($\frac{3}{8}$–1in) long. Flowers 10–14cm(4–$5\frac{1}{2}$in) long, brilliant red with a dark raised blotch at the base of the standard petal, three to six borne together in an umbel-like cluster in summer. A white-flowered form is known. Although technically a short-lived shrub, this species frequently behaves as an annual, dying after flowering. Temperate.

C. puniceus Parrot's bill, Kaka beak, Lobster claw New Zealand
Slender-stemmed spreading shrub to 2m($6\frac{1}{2}$ft).

Leaves having up to 31 narrowly oblong leaflets, 1–2.5cm(⅜–1in) long, with closely pressed silky hairs beneath. Flowers about 10–12cm(4–4¾in) long, red to scarlet, in pendent racemes, opening in late spring and summer. Cool. *C.p.* 'Albus' has creamy-white flowers with a touch of green. *C.p.* 'Roseus' has flowers in shades of scarlet and pink; the plant needs the support of canes or wires.

CLIVIA

Amaryllidaceae

Origin: Warm dry forests of South Africa. A genus containing three species of evergreen perennials. They are clump-forming, with dark green leathery, strap-shaped, arching leaves and 6-tepalled, funnel-shaped flowers borne in terminal umbels. Clivias make good house plants, being very shade tolerant, but must be kept cool for a couple of months in late autumn to early winter to initiate flower buds. The roots are very thick and fleshy and plants are best left undisturbed until too large for their intended use, when they should be carefully divided. Apart from division, propagate by removing offsets when they have three or four leaves, or by seed (sown as soon as ripe). *Clivia* was named for the Duchess of Northumberland, née Charlotte Florentina Clive (granddaughter of Robert Clive), in whose garden it first flowered in Great Britain.

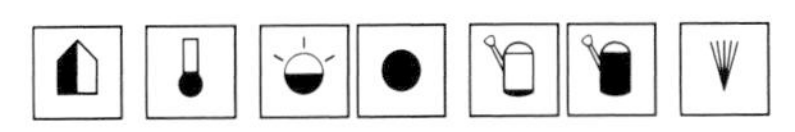

Species cultivated

C.* × *cyrtanthiflora (*C. miniata* × *C. nobilis*)
Midway between its parents, it has narrow-tubed, salmon-red, slightly greenish flowers which expand less at the mouth than *C. miniata*; they are pendent.

C. miniata (*Imantophyllum miniatum*) Kaffir lily
The most attractive species with robust, thick-textured, strap-shaped leaves 40–60cm(16–24in) long and 2.5–5cm(1–2in) wide. Flowers are orange to red, 5–8cm(2–3in) long, broadly funnel-shaped, borne in rounded umbels of up to 20 or more blooms in spring and summer. *C.m.* 'Citrina' has yellow flowers. *C.m.* 'Striata' has variegated leaves.

C. nobilis (*Imantophyllum aitonii*)
Less vigorous than *C. miniata* with leaves 30–50cm(12–20in) long. Flowers tubular, slightly curved and pendent, to 4cm(1½in) long, reddish with green tips, borne in umbels of up to 40, occasionally more, in late spring.

CLUSIA

Guttiferae

Origin: Mostly tropical and sub-tropical America, but also Malagasy and New Caledonia where they inhabit forests. A genus of 145 small trees, shrubs and climbers, often living epiphytically on trees and mossy rocks. They are evergreen, some grown for their foliage, others for both leaves and flowers. The leathery leaves of the species cultivated are oval and obovate, rich green and somewhat lustrous. Their flowers vary in size and are

TOP *Clianthus puniceus*
LEFT *Clivia miniata*

sometimes large and showy, being composed of four to six sepals and four to nine petals and carried in terminal clusters. Propagate by cuttings in summer or by layering in spring. *Clusia* is named for Charles de l'Ecluse (Carolus Clusius of Artois) (1526–1609), the famous Belgian botanist and author of *Rariorum Plantarum Historia* and many other works.

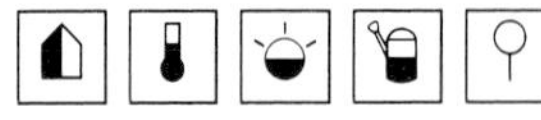

Species cultivated

C. grandiflora Scotch attorney Guyana
Wide-spreading shrub, 3–6m(10–20ft) tall. Leaves elliptic to obovate, dark green above, paler beneath with dark lines, 15–30cm(6–12in) long, clustered towards the tips of robust stems. Flowers about 5cm(2in) wide, pink and white, but not produced on small potted specimens. In the wild this shrub grows epiphytically on large trees; as a pot plant it must have a freely-draining soil.

C. rosea Balsam apple, Fat pork tree, Autograph tree Florida and West Indies to Venezuela
Robust shrub, 2.5–6m(8–20ft) tall. Leaves obovate, deep green, lustrous, up to 20cm(8in) long, borne in opposite pairs. Flowers 5cm(2in) wide, pink and white, rarely produced on small pot plants. In its native countries the 8cm(3in) wide globose, greenish fruits yield a sticky resin used as bird lime. *C.r.* 'Aureo-Variegata' has the leaves splashed with pale green and cream to golden-yellow. The species is epiphytic and both it and 'Aureo-Variegata' need well-drained soil.

CLYTOSTOMA

Bignoniaceae

Origin: Sub-tropical America, mainly where winters are cool and dry. A genus of eight species of evergreen tendril climbers, formerly classified under *Bignonia*. The leaves are composed of one pair of leaflets and a terminal tendril. The flowers are also usually in pairs, either axillary or terminal, foxglove-shaped and large enough to be attractive or even showy. Sizeable, intriguingly spiny seed pods sometimes form. Propagate by cuttings in summer or early autumn, or by layering in spring. *Clytostoma* derives from *klytos*, beautiful and *stoma*, a mouth, referring to the inside of the flowers.

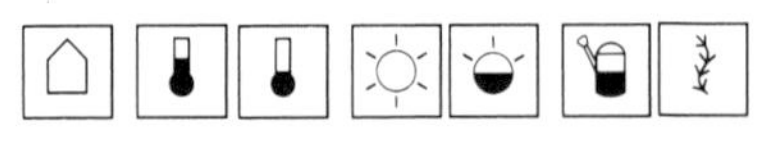

Species cultivated

C. binatum (*C. purpureum*, *Bignonia purpurea*) Love charm Venezuela, Guyana, Paraguay
Wiry stems to 3m(10ft) or more long; leaflets 5–8cm(2–3in) long, elliptic to oblong-lanceolate, bright green above, paler beneath. Flower pairs solitary or in axillary clusters, each bloom about 5.5cm(2¼in) long, mauve with a white, paler throat, borne in spring or summer. Makes an attractive specimen for a conservatory.

C. callistegioides (*Bignonia callistegioides*, *B. speciosa*) Trumpet vine S. Brazil, N. Argentina
Slender wiry stems, 3–5m(10–16ft) long; leaflets 6–10cm(2½–4in), elliptic, somewhat undulate. Flowers 5–9cm(2–3½in) long, pale purple veins, borne in spring or summer. Easily grown in large pots and stands pruning well.

C. purpureum See *C. binatum*.

COBAEA

Cobaeaceae

Origin: Tropical and sub-tropical America. A genus of 18 species of climbing plants with alternate pinnate leaves and branched tendrils. The flowers are bell-shaped and have conspicuous 5-lobed calyces; they are borne in the leaf axils. Only one species is in cultivation – it is a strong-growing climber suitable for the back wall of a conservatory and can be grown as an annual, flowering in its first season. Because of its vigour it is best in a pot.

BELOW *Clytostoma callistegioides*
BELOW RIGHT *Cobaea scandens*

Propagate by seed sown in spring. *Cobaea* was named for the Spanish Jesuit Father Bernardo Cobo (1572–1659), missionary and naturalist in South America.

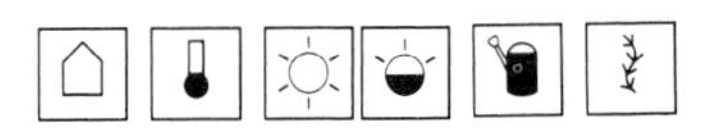

Species cultivated

C. scandens Cup-and-saucer creeper, Cathedral bells Central & South America
Woody climber, eventually to 10m(33ft) or more. Leaves pinnate, made up of four to six ovate or elliptic leaflets, each 10cm(4in) long, the leaf usually terminating in a tendril which has minutely hooked tips. Flowers broadly bell-shaped, 6–8cm(2½–3in) long, opening cream, darkening to purple as they age; they sit on a pale green, saucer-like calyx. C.s. 'Alba' has flowers which do not darken to purple.

COCCOLOBA

Polygonaceae

Origin: Tropical and sub-tropical America. A genus of 150 species of trees, shrubs and woody climbers, two or three of which are grown in warm climate gardens. The species described below makes a handsome and unusual foliage tub plant for the conservatory or a large sunny room. Propagate by seed in spring, by cuttings in summer or by layering in autumn. *Coccoloba* derives from the ancient Greek *kokkolobis* for a kind of grape, used for this genus because of the appearance of the fruits.

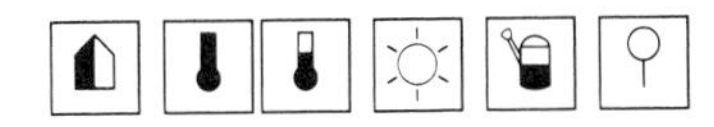

Species cultivated

C. uvifera Seaside grape Florida to Brazil, only on sandy sea shores, much grown by the sea in tropical and warm temperate countries where there is no frost.
Up to 2m(6½ft) or so tall in containers but to 6m(20ft) or more in its own country. Leaves orbicular to broadly heart-shaped, leathery-textured and lustrous, red-veined and somewhat wavy-margined, to 20cm(8in) long. Flowers white, in dense racemes to 25cm(10in) long, followed by 2cm(¾in) long purple berries. Flowering and fruiting not common on potted specimens.

COCHLIODA

Orchidaceae

Origin: Peru, Ecuador, Colombia (Andean region). A genus of four species of epiphytic orchids closely allied to *Odontoglossum* but renowned for their showy flowers in bright shades of red. They have ovoid pseudobulbs, narrow leathery leaves and racemes of flowers composed of five similar tepals and a small labellum. Propagate by division in spring or after flowering. Although best in the conservatory these colourful orchids can, with care, be grown in the home and are well worth trying. *Cochlioda* derives from the Greek *kochlos*, a snail shell, alluding to the shape of the lobes of the labellum of the species first described.

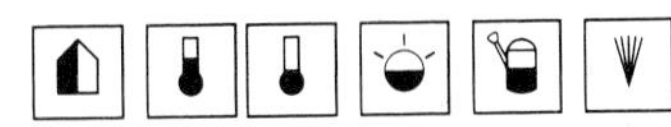

Species cultivated

C. noetzliana (*Odontoglossum noetzlianum*) Peru
Pseudobulbs to 8cm(3in) tall, rather flattened and wrinkled. Leaves usually solitary, linear, about 25cm(10in) or somewhat less in length, soft-textured and blunt-tipped. Racemes arching to pendulous, about 30–45cm(1–1½ft) in length, densely flowered; each flower 2.5–5cm(1–2in) across, bright scarlet, opening during winter and spring, occasionally also in summer.

C. rosea (*Odontoglossum roseum*) Peru

ABOVE *Coccoloba uvifera*
LEFT *Cochlioda vulcanica*

Similar in appearance to C. *noetzliana*, but the leaves usually a little wider, racemes looser and fewer-flowered and the blooms rose-red to rose-carmine, opening in winter.

C. vulcanica Peru
Pseudobulbs and leaves much as in C. *noetzliana*, but the latter rarely above 15cm(6in) in length. Raceme erect, to 35cm(14in) or a little more in height, bearing eight to 20 flowers; each blossom 4–5cm(1½–2in) wide, rose-red, the middle of the labellum a shade paler, expanding from late autumn to early spring.

RIGHT *Cocos nucifera*
BELOW *Codiaeum variegatum pictum*

COCOS

Palmae

Origin: Uncertain and controversial, some authorities favouring tropical western South America, others Melanesia, the latter seeming the most likely. Now widespread in all tropical countries, mainly by the sea. A genus of one species of elegant palm. Formerly it was considered difficult to cultivate in the greenhouse or home, but germinated nuts with a few young leaves are now being offered as house plants. Although attractive and having a romantic aura of tropical islands in the sun, they are not a good buy. Keeping a young coconut palm healthy and growing well must be looked upon as a challenge. Full sun, humidity and an absolute minimum night temperature of 18°C(65°F) is essential, plus a well drained sandy soil, rich in lime and potash. Propagate by seed (nuts) at 26–30°C(79–86°F). *Cocos* derives from *coco*, Portuguese for monkey, presumably because of the likeness of the nuts to the heads of certain monkeys.

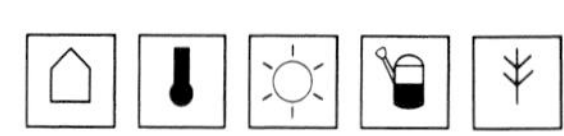

Species cultivated

C. nucifera Coconut palm
Potential height in containers 2m(6½ft) or more, with pinnate, arching leaves of the same length, but usually fading in vigour and dying before this. In tropical countries a stem up to 24m(80ft) in height develops, with leaves to 6m(20ft) long.
C. plumosa (*C. romanzoffiana*) See *Arecastrum romanzoffianum*.
C. weddelliana See *Syagrus weddelliana*.

CODIAEUM

Euphorbiaceae

Origin: Malaysia to the Pacific Islands and northern Australia. A genus of 15 species of evergreen shrubs of which only one is cultivated. It makes an excellent house or conservatory plant. Propagate by tip cuttings in spring or summer. *Codiaeum* derives from the Indonesian (Moluccan) vernacular name *kodiho*.

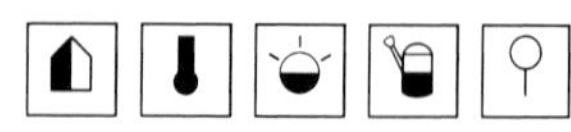

Species cultivated

C. variegatum pictum Croton, Joseph's coat
Malaysia, Polynesia, not known truly wild
A shrub to 1m(3ft) or more. Leaves leathery, glossy, basically ovate but in the many cultivars available varying from narrowly linear to oak- or fiddle-shaped, sometimes waved and twisted; their colour is basically green, then spotted, blotched or suffused with shades of yellow, orange, red and pink, the colouring usually strongest along the veins. Flowers are small and borne in axillary racemes on larger specimens.

Cultivars

Although several cultivars are readily available at garden centres and nurseries they are frequently sold un-named. When names are given they are often confused or wrong. However, plants under the following names can be recommended and are worth looking out for: 'Aucubifolium', 'Bravo', 'Bruxellense', 'Craigii', 'Excellent', 'Fascination', 'Gloriosum Superbum', 'Imperialis', 'Nevada', 'Norma', 'Otto van Oerstedt', 'Punctatum Aureum', 'Reidii', 'Spirale'.

CODONANTHE

Gesneriaceae

Origin: Tropical America. A genus of 15 species of epiphytic evergreen perennials, sub-shrubs and climbers, some of which make tolerant and pleasing house plants. The species described are ideal hanging basket subjects, having ovate to elliptic leaves in pairs and smallish tubular flowers with five petal lobes in their axils. The berry-like fruits which follow may be brightly coloured and add to the plant's season of attraction. Propagate by cuttings in summer, or by seed in spring. *Codonanthe* derives from the Greek *kodon*, a bell and *anthos*, a flower, the tubular base of the blooms being somewhat bell-like at least in some species.

Species cultivated

C. carnosa Brazil
Growth neat and fairly compact, spreading, eventually sub-shrubby, to about 40cm(16in) wide and almost as high. Leaves to 8cm(3in) or so in

LEFT *Codonanthe gracilis*

length, elliptic to obovate, sometimes in whorls of three or four. Flowers about 2cm(¾in) or a little more long, white or pink-tinted with a yellow throat, borne in spring and summer. Fruits red.

C. crassifolia Central America to Peru
Stems prostrate or hanging, rooting at the nodes, to 30cm(1ft) or more long. Leaves to 5cm(2in) long, elliptic to ovate, somewhat fleshy, red-glandular beneath. Flowers to 2.5cm(1in) long, white or sometimes pink-tinted, with a yellow throat. Fruits egg-shaped, 12mm(½in) long, red or pink, showy.

C. gracilis Central America
Stems prostrate or hanging, 30–60cm(1–2ft) long. Leaves 2.5–4cm(1–1½in) long, elliptic to lanceolate, slender-pointed. Flowers up to 2cm(¾in) long, white.

C. macradenia Central America
Much like *C. crassifolia* and sometimes confused with it, but leaves only to 3cm(1¼in) long and relatively broader. Flowers 2.5–3cm(1–1¼in) long with a curved tube, white, red-spotted or mottled in the throat. Fruits egg-shaped, red-purple.

COELOGYNE

Orchidaceae

Origin: W. China and India to Malaysia and the Pacific Islands. A genus of almost 200 species of epiphytic orchids with rounded to flask-shaped pseudobulbs, each usually with two linear to lanceolate leathery leaves. The flowers, which are in racemes or borne singly, have spreading tepals and a prominent labellum which can be lobed and deeply keeled. All grow well in the conservatory and home provided that enough humidity can be given. Those with pendulous flowers are best grown in a basket. Propagate by division. *Coelogyne* derives from the Greek *koilos*, a hollow and *gyne*, female, alluding to the somewhat hollow stigma.

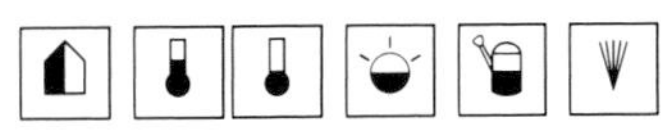

Species cultivated

C. barbata Himalaya
Ovoid, clustered pseudobulbs to 10cm(4in). Leaves oblong-lanceolate, to 45cm(1½ft) long. Flowers 5–8cm(2–3in) wide, white, the labellum crested and fringed with sepia brown; they are fragrant and borne in erect or arching racemes to 45cm(1½ft) long in autumn and winter.

C. corymbosa Himalaya
Ovoid to oblong pseudobulbs, slightly angled, to about 5cm(2in) long. Leaves oblong-lanceolate, to 15cm(6in) long. Flowers 3–5cm(1¼–2in) wide, creamy-white with yellow-brown streaking in the throat, the labellum cream with bright yellow, brown-ringed spots; they are borne in 20cm(8in) racemes of three to five flowers in summer and autumn.

C. cristata Himalaya
Clustered, egg-shaped to globular pseudobulbs becoming wrinkled with age, to 6cm(2½in) long. Leaves linear-lanceolate, to 30cm(1ft) long, two to each pseudobulb. Flowers fragrant, 8–10cm(3–4in) wide, white with strongly waved tepals; the labellum has five rich-yellow keels; they are borne on arching to pendulous racemes, 15–30m(6–12in) long, in winter and spring.

C. dayana Malaysia
Pseudobulbs narrowly conical, strongly ribbed, 13–25cm(5–10in) tall. Leaves carried erect, pleated, to 60cm(2ft) long. Flowers 6–7cm(2½–2¾in) wide, musk-scented, pale yellow-brown; the labellum has margins and keel white with chocolate-brown veining; they are borne in sharply pendent racemes, 45–90cm(1½–3ft) long, in spring and summer.

C. densiflora See *C. massangeana*.

Coelogyne cristata

C. elata Himalaya
Pseudobulbs narrowly ovoid to cylindrical, up to 15cm(6in) in height. Leaves 30–45cm(1–1½ft) long, narrowly lanceolate, thick-textured. Raceme to 50cm(20in) or more, the base erect, the flowering part arching down; flowers white, 5cm(2in) or more wide, the labellum with a divided yellow band and two deep orange keels, fragrant, long lasting, late winter to spring.

C. fimbriata (*Broughtonia linearis*) N. India to China, Thailand, Vietnam
Pseudobulbs ovoid to ellipsoid, to 4cm(1½in) long, spaced on very slender rhizomes. Leaves lanceolate, to 13cm(5in) long, usually rather yellow-green. Flowers solitary or in pairs from tops of most recently formed pseudobulbs, 4cm(1½in) wide, tepals yellowish-green or olive-brown tinted, labellum fringed, yellow striped with brownish-red, fragrant, borne in autumn.

C. flaccida Himalaya
Pseudobulbs spindle-shaped, to 8cm(3in) long. Leaves narrowly lanceolate, to 30cm(1ft) long, glossy green. Flowers fragrant, waxy-white, to 4cm(1¾in) across, the labellum with a golden-yellow centre; they are carried in arching to pendulous racemes, to 25cm(10in), each with seven to 12 blooms.

C. flavida Himalaya
Pseudobulbs ovoid, 3–5cm(1¼–2in) long, well spaced on slender rhizomes. Leaves narrowly lanceolate, slender, pointed, to 15cm(6in) long. Racemes erect, bearing eight to ten yellow flowers, each about 12mm(½in) wide, in summer.

C. fuscescens Himalaya
Pseudobulbs 10–13cm(4–5in) long, more or less cylindrical and clustered. Leaves to 25cm(10in) long, ovate, deep green. Flowers 6cm(2½in) wide, slightly cupped, the tepals translucent, orange-red or sometimes yellowish shaded red-brown; the labellum usually buff, margined with red; they are borne in arching or drooping racemes of five to seven blooms hanging below the leaves in winter.

C. lawrenceana Vietnam
Pseudobulbs broadly ovoid, to 10cm(4in) tall, clustered. Leaves 18–30cm(7–12in) long, lanceolate, somewhat leathery and glossy. Flowers solitary or two together on a stem 15–25cm(6–10in) long; each bloom 6–10cm(2½–4in) wide, tepals tawny-yellow, the labellum white, marked with brown and yellow, fragrant, borne in spring to summer.

C. massangeana (*C. densiflora*, *C. tomentosa*) Malaysia
Pseudobulbs narrowly ovoid, to 10cm(4in) tall. Leaves narrowly elliptic, pleated, to 45cm(1½ft) long. Flowers fragrant, 5cm(2in) wide, pale ochre, the labellum marked with brown, yellow and white; they are borne in pendulous racemes, 15cm(6in) or more long in summer, each bearing up to 20 flowers.

C. nitida (*C. ocellata*) Himalaya, Burma, Vietnam
Pseudobulbs 2.5–5cm(1–2in) long, oblong to ovoid, loosely clustered. Leaves to 25cm(10in) long, lanceolate, bright green. Racemes 3- to 6-flowered, more or less erect, to 20cm(8in) tall; flowers 5cm(2in) wide, white, the labellum marked with brownish-orange and/or yellow spots and veining, borne in late winter or spring.

C. ocellata See *C. nitida*.

C. ochracea Himalaya
Similar to *C. corymbosa*, but pseudobulbs and leaves longer. Flowers white, the labellum yellow with red-flushed keels, borne on an erect raceme in spring and summer.

C. ovalis Himalaya, China, Vietnam
Pseudobulbs 5–8cm(2–3in) long, ovoid to ellipsoid, on slender rhizomes. Leaves to 13cm(5in) long, lanceolate. Flowers usually solitary, sometimes two or more, about 5cm(2in) wide, yellow-brown, the labellum covered and fringed with purple hairs, fragrant, borne in autumn to early winter. Closely allied to *C. fimbriata*.

C. pandurata Black orchid Borneo, Malaysia
Pseudobulbs to 13cm(5in) tall, oblong, strongly

RIGHT *Coelogyne ochracea*
FAR RIGHT *Coelogyne pandurata*

flattened, borne at intervals on a robust creeping rhizome. Leaves narrowly elliptic to lanceolate, 30–45cm(1–1½ft) long, lustrous. Flowers 8–10cm (3–4in) wide, bright pale green, the labellum almost violin-shaped, with black markings and down-curving wrinkled margins, strongly fragrant, borne in summer to autumn.

C. speciosa Java, Sumatra
Pseudobulbs ovoid, clustered, to 5cm(2in) tall. Leaves narrowly elliptic, glossy, to 25cm(10in) long. Flowers musk-scented, 7–8cm(2¾–3in) wide, translucent brownish to olive-green, the labellum white, flushed yellow at the base and marked with red-brown above, fringed and lobed; they are borne in short racemes of two to three, chiefly in spring and summer, occasionally at other times.

C. tomentosa See *C. massangeana.*

COFFEA

Rubiaceae

Origin: Tropics of the Old World, chiefly Africa. A genus of 40 species of evergreen shrubs with opposite pairs of lanceolate to ovate leaves and tubular, white starry flowers in axillary clusters, followed by fleshy, red berry-like fruits. These contain two seeds – the coffee beans of commerce. They are decorative house plants grown for their foliage, though not successful where the atmosphere is too dry. In the conservatory they can be grown in tubs and will flower and fruit. Propagate by tip cuttings in summer with bottom heat, or by seed in spring. *Coffea* derives from the Arabic *kahwah.*

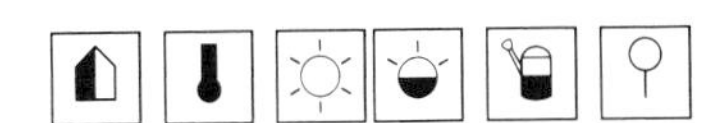

Species cultivated

C. arabica Arabian coffee
Evergreen tree, in the wild to 5m(16ft), a shrub when grown in a pot or tub. Leaves ovate to elliptic, dark glossy green and prominently veined, 5–15cm(2–6in) long. Flowers 1cm(⅜in) across, white with narrow petals, fragrant, in clusters of one to six in late summer. The fruits are to 1.5cm(⅝in) long. *C.a.* 'Nana' is a more compact form and most suitable for growing in the home.

COLCHICUM

Liliaceae

Origin: Europe, Mediterranean region to central Asia and northern India. A genus of 65 species of cormous-rooted plants, the corms rounded to pear-shaped each producing two or more lanceolate to ovate, glossy leaves. The flowers have long stem-like perianth tubes, the ovary lying above the corm, and six tepals opening to form a bowl-shaped bloom. They are usually pinkish-purple and in some species are tesselated or chequered with darker markings and open before the leaves appear.

LEFT *Coffea arabica*
BELOW LEFT *Colchicum speciosum bornmuelleri*

They make good bowl plants for the conservatory or cool room and when potted should be brought immediately into the cool room and not kept cold as for crocus. All those described are autumn-flowering. Propagate by seed when ripe or by separating offsets when dormant. Hardy. *Colchicum* was named for the region of *Colchis* at the eastern end of the Black Sea.

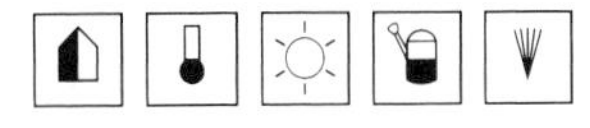

Species cultivated

C. autumnale Autumn crocus, Meadow saffron
Turkey
Leaves lanceolate, 15–30cm(6–12in) long. Tepals 5cm(2in) long, lilac-pink, sometimes very lightly chequered.

C. cilicicum Turkey
Leaves lanceolate, 15–30cm(6–12in) long. Tepals 5cm(2in), dark lilac-pink, sometimes lightly chequered. Produces several flowers to each corm.

C. speciosum Caucasus, Iran, Turkey
Leaves oblong-elliptic, 30–40cm(12–16in) long. Tepals 7–9cm(2¾–3½in), lilac-rose to rose-purple.

C.s. album
Pure white flowers.

C.s. bornmuelleri
Purple-rose with a broad white throat.

Several cultivars are available mostly derived from *C. speciosum*, often as hybrids with another species; they include:
'Lilac Wonder', rose-purple with light chequering and more pointed tepals.
'Water Lily', fully double, rose-lilac; looking, as its name suggests, more like a water-lily than a colchicum.

ABOVE *Coleonema pulchrum*
ABOVE RIGHT *Coleonema* 'Golden Sunset'

COLEONEMA

Rutaceae

Origin: South Africa. A genus of five species of wiry-stemmed, small to medium-sized evergreen shrubs. They are somewhat heath-like in appearance, but the small flowers are 5-petalled and star-like. The species described here make elegant subjects for the conservatory, either planted out or in containers. Propagate by seed in spring or by cuttings in summer. *Coleonema* derives from the Greek *koleos*, a sheath and *nema*, a thread; there are five stamens and five staminodes, the latter concealed in the folded petal bases.

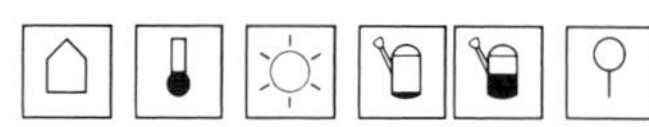

Species cultivated

C. album (*Diosma alba*)
Freely branching habit, 1–2m(3–6½ft) in height, but easily kept smaller by pruning and confining to 15cm(6in) in pots. Leaves densely borne, linear, to about 12mm(½in) long. Flowers in the upper leaf axils, white, about 8mm(⅓in) wide, usually freely produced.

C. pulchrum (*Diosma pulchra*)
Freely branching, spreading habit, 90–120cm(3–4ft) or more tall. Leaves linear 2–4cm(¾–1½in) long. Flowers pink with darker centres or red, up to 2cm(¾in) wide, though with reflexed petals and appearing smaller. 'Golden Sunset' with yellow leaves is generally ascribed to this species, but may be of hybrid origin.

COLEUS

Labiatae

Origin: Tropical Africa and Asia. A genus of 150 species of perennials, annuals and sub-shrubs. They have opposite pairs of ovate, stalked leaves, and small, tubular, 2-lipped flowers which are borne in whorls on a spike-like inflorescence. Usually grown as annuals, they are colourful plants for the home and conservatory. Extra large plants can be obtained by potting on into larger containers each spring. Propagate by seed in spring, or by cuttings in spring and late summer. *Coleus* derives from the Greek *koleos*, a sheath, the stamen filaments being united to enclose the style.

Species cultivated

C. blumei Flame nettle, Painted nettle Java
Shrub to 60cm(2ft), sometimes more. Leaves broadly ovate to narrowly ovate-cordate, coarsely toothed, often hairy beneath. Flowers small, white or blue, borne in a spike-like cluster.

C.b. verschaffeltii is the name for many cultivars and seed strains the leaves of which come beautifully marbled, veined or blotched in a wide variety of colours and shades, mainly of green, pink or red, but also including yellow, purple, brown and creamy-white.

C. fredericii Central Africa
Roughly-hairy perennial to 1.2m(4ft). Leaves broadly ovate to heart-shaped with rounded teeth. Flowers purple-blue in open panicles, to 20cm(8in) or more in length.

C. pumilus (*C. rehneltianus*) Sri Lanka
Similar to *C. blumei*, but of prostrate or trailing growth. Leaves broadly ovate, bright green with

RIGHT *Coleus blumei*
FAR RIGHT *Coleus pumilus*

purplish-brown centres. Several cultivars are grown, some with red or pink coloration, the forms including the name 'Trailing Queen' being most easily available.
C. rehneltianus See C. *pumilus*.
C. thyrsoideus Central Africa
Erect sub-shrub to 90m(3ft) or more. Leaves green, ovate-cordate with tooth-like lobes, hairy, to 15cm(6in) long. Flowers 1cm(⅜in) long, blue, in terminal panicles of three to ten in winter.

COLOCASIA

Araceae

Origin: S.E. Asia to Polynesia. A genus of eight species of rhizomatous or tuberous-rooted perennials grown for their large long-stalked leaves which rise in a cluster from ground level. Plants are best for a warm conservatory, but can be tried in the home where the atmosphere is not too dry. *Colocasia* derives from the Greek *kolocasia*, originally used for the rhizomes of the sacred lotus.

Species cultivated
C. esculenta (including C. *antiquorum*) Taro, Elephant's ear, Cocoyam, Dasheen S.E. Asia
Rarely over 90cm(3ft) in a pot, growing from a tuber-like, starchy corm. Leaves 20–50cm(8–20in), deeply heart-shaped at the base. Spathe pale yellow, 15–25cm(6–10in) long, but only borne on large plants. C.*e.* 'Illustris' has violet-brown leaves which have bright green veins.

COLUMNEA

Gesneriaceae

Origin: Tropical America. A genus of 200 species of evergreen perennials and sub-shrubs, the majority of which are epiphytic. They have opposite pairs of rounded to linear leaves which, in many species, are turned so that all face in the same direction. Stems are climbing or trailing and the showy flowers borne singly or in small groups in the leaf axils, each flower being tubular with a prominently 4-lobed mouth, the upper lobe often longer and forming a hood. They make fine pot plants, the trailing species being particularly decorative in a hanging basket. Propagate by cuttings from non-flowering stem tips in spring or summer, or by seed in spring. *Columnea* was named for Fabio Colonna (1567–1640) using the Latin form of his name, Fabius Columna; an Italian botanist and author of the first work to use copperplate illustrations.

Species cultivated
C. affinis Venezuela
Shrubby, more or less erect, 30–60cm(1–2ft) tall; stems robust, red-purple hairy. Leaf pairs very unequal, the larger one up to 22cm(9in), the smaller one 2.5cm(1in) long, both oblanceolate, dark green above, paler beneath, velvety-purple hairy. Flowers in clusters of two to three, arising from the axils of orange bracts, 3cm(1¼in) long, yellow, covered densely with orange hairs, mainly borne in spring to summer, but intermittently all the year. Berries yellow.
C. × banksii (C. *oerstediana* × *schiediana*)
Stems arching to stiffly pendulous, to 60cm(2ft) or more long. Leaves rich glossy green above, red beneath. Flowers scarlet 6–8cm(2½–3in) long, produced in winter and spring. A very tolerant species, making a good house plant.
C. crassifolia Mexico, Guatemala
Erect and shrubby, to 30cm(1ft). Leaves fleshy in texture, narrowly lanceolate, 5–10cm(2–4in) long, glossy green above. Flowers scarlet with a yellow mouth, 7–8cm(2¾–3in) long, opening in spring and summer.
C. erythrophaea Mexico
Semi-trailing, semi-erect with robust stems. Leaf pairs not quite equal, the leaves 2.5–6cm(1–2½in) long, asymmetrically elliptic to ovate. Flowers strongly 2-lipped, 5–8cm(2–3in) long, orange-red from an orange calyx, each one poised on the end of a 4–5cm(1½–2in) long, red-hairy pedicel below the leaves.
C. gloriosa Goldfish plant Costa Rica
Stem pendulous, to 1m(3ft) long or more. Leaves ovate-oblong, to 3cm(1¼in) long, covered with short brown hairs. Flowers to 8cm(3in) long, very

Colocasia esculenta

TOP *Columnea gloriosa*
ABOVE *Columnea microphylla* and *Selaginella kraussiana*

showy, fiery-red with a yellow throat, the upper lobe very prominent and hooded; they are freely produced from late autumn through the winter to early spring. A fine plant for a hanging basket. *C.g.* 'Superba' ('Purpurea', 'Rubra', 'Splendens') has the leaves densely covered with purple hairs; a striking form.

C. hirta Costa Rica
Stems stiffly trailing, red-hairy. Leaves ovate, 1.5–4cm(⅝–1½in) long, densely hairy. Flowers orange-red, to 7cm(2¾in) long, 2-lipped, the upper lip helmet-shaped, borne in summer.

C. illepida Panama
Shrubby, more or less erect; stems robust, brown-hairy. Leaf pairs unequal, the larger one to 14cm(5½in) long, lanceolate to oblanceolate, hairy, red or red-blotched beneath. Flowers 5–8cm(2–3in) long, yellow with longitudinal maroon stripes from a fringed, green calyx, borne intermittently all the year. Has been listed as 'Butcher's Gold' and 'Panama Gold'.

C. lepidocaula Costa Rica
Shrubby, erect and bushy, to 45cm(1½ft) or so tall. Leaves to 9cm(3½in) long, ovate to elliptic, deep green and lustrous. Flowers to 8cm(3in) long, orange with a yellow throat, strongly 2-lipped, the upper lip helmet-like, borne intermittently all the year.

C. linearis Costa Rica
Erect, shrubby, to 45cm(1½ft) tall. Leaves linear-lanceolate, to 9cm(3½in) long, glossy deep green above. Flowers rose-pink, silky white-hairy, 4.5cm(1¾in) long, produced in spring.

C. magnifica Costa Rica and Panama
Stems slender, but stiff and hairy, more or less trailing. Leaves to 9cm(3½in) long, oblanceolate. Flowers scarlet, strongly 2-lipped, the upper lip helmet-shaped, to 6cm(2½in) long, borne in summer.

C. microphylla Goldfish plant Costa Rica
Rather like *C. gloriosa*. Leaves broadly ovate to almost rounded, to 1cm(⅜in) long, brown-hairy. Flowers 6–8cm(2½–3in) long, very free-flowering. An excellent plant for a hanging basket.

C. minor See *C. teuscheri*.

C. mortonii Panama
Similar to *C. hirta*, but leaves broader and flowers bright scarlet. Has been listed as 'Butcher's Red' and 'Panama Red', etc.

C. nicaraguensis Central America
Shrubby, to 60cm(2ft) or more in height with brown-hairy stems. Leaves in unequal pairs, the larger ones to 13cm(5in) long, lanceolate to ovate, with a silky sheen. Flowers in pairs or threes, 2-lipped, the upper lip helmet-shaped, scarlet and yellow, borne mainly in summer.

C. oerstediana Costa Rica
Closely akin to *C. gloriosa*. Pendent stems to 1m(3ft) or more. Leaves elliptic to ovate, 12mm(½in) long, almost hairless, glossy green. Flowers slender, russet-orange, to 7cm(2¾in) long, borne in spring.

C. ovata See *Asteranthera ovata*.

C. percrassa Panama
Stems trailing, olive-green, 30cm(1ft) or more in length. Leaves to 4cm(1½in) long, elliptic, fleshy and glossy. Flowers tubular, to 3cm(1¼in) long, scarlet and yellow with an arching upper lip, borne in summer.

C. purpureovittata (*Tricantha purpureovittata*) Peru
Shrubby, stems erect to semi-erect, 30–60cm(1–2ft) long. Leaf pairs unequal, the larger one up to 20cm(8in) long, asymmetrically lanceolate to narrowly ovate, quilted above with a lustrous coppery sheen, red-flushed beneath. Flowers about 5cm(2in) or more long, yellow with longitudinal maroon stripes, borne beneath the leaves in summer.

C. sanguinolenta Costa Rica and Panama
Stems erect to arching, 30–60cm(1–2ft) long. Leaf pairs very unequal, the larger ones to 12cm(4¾in) long, oblanceolate, glossy above, sometimes red-spotted beneath. Flowers up to 5cm(2in) long, scarlet, from deeply fringed hairy calyces, opening in summer.

C. schiedeana Mexico
Stems erect, climbing or spreading, to 1m(3ft) long. Leaves in unequal-sized pairs, oblong-lanceolate, 9–13cm(3½–5in) long, softly white-hairy above, red-veined beneath. Flowers solitary or in

pairs, dull yellow with dark red-brown markings, opening in spring and summer.

C. splendens See *Nematanthus longipes.*

C. teuscheri (*Tricantha minor* of gardens) Ecuador

Wiry, creeping or trailing stems. Leaves asymmetrically elliptic to lanceolate, to 6cm($2\frac{1}{2}$in) long, hairy. Flowers narrowly tubular, 4cm($1\frac{1}{2}$in) long, dark maroon with yellow stripes and five small, rounded, yellow-marked lobes; the calyx is bristly feathery. Opening in summer and autumn, this species is very different from most cultivated columneas.

C. tigrina See *C. zebrina.*

C.* × *vedrariensis See hybrid cultivars below.

C. verecunda Costa Rica

Shrubby, erect species, to 60cm(2ft) or more. Leaves lanceolate, 8–10cm(3–4in) long, one of each pair very small, waxy olive-green above, wine-red beneath. Flowers yellow, flushed red-purple beneath, to 4cm($1\frac{1}{2}$in) long, in clusters in the leaf axils in spring and summer.

C. zebrina (*C. tigrina*) Panama

Shrubby, semi-erect, to 60cm(2ft) high and wide, the thick stems constricted at the nodes. Leaf pairs unequal, the larger one up to 18cm(7in) long, asymmetrically lanceolate, smooth above, silvery-hairy beneath. Flowers in axillary clusters, 2-lipped, to 8cm(3in) long, pale yellow with bold longitudinal, maroon stripes, borne in summer.

Hybrid cultivars

'Alpha', well-branched, compact habit; flowers strongly 2-lipped, the upper long and hooded, to 6cm($2\frac{1}{2}$in) in length or more, bright canary-yellow, freely produced intermittently all year.

'Bonfire', a *C. erythrophaea* hybrid forming a well-branched vigorous plant; flowers yellow and crimson, 8cm(3in) long on pendent 5cm(2in) pedicels, intermittently all year.

'Cayugan', one of the Cornell University hybrids; vigorous trailing habit, the red flowers to 8cm(3in) long, borne over a long period.

'Chanticleer', shrubby, well-branched and compact habit with velvety-hairy, pale green leaves; flowers 4cm($1\frac{1}{2}$in) long, light orange, borne over a lengthy period; one of the best columneas for the home.

'Early Bird', spreading habit, but compact with closely set, thick-textured, small glossy leaves; flowers erect, 6cm($2\frac{1}{2}$in) long, orange and yellow with a broad helmet-shaped upper lip, intermittently all year; a good house plant.

'Elvo' (*C. gloriosa* × *C. nicaraguensis*), spreading habit, stems 60cm(2ft) or more with handsome coppery-tinted foliage; flowers bright red, at least 8cm(3in) long, borne in spring; best in the temperate conservatory and very handsome when well-grown.

'Kewensis Variegata' (*C. glabra* × *C. schiedeana*), semi-erect, fairly bushy habit, the smallish elliptic leaves variegated white, flushed pink; flowers at least 5cm(2in) long, bright orange-red.

Columnea 'Stavanger'

'Mary Ann', trailing habit, but well-branched and compact, the narrow leaves rich green; flowers deep pink, borne intermittently throughout the year; one of the few good pink-blossomed columneas and a distinctive house plant.

'Stavanger' (*C. microphylla* × *C. vedrariensis*) Norse fire plant, stems trailing, well-branched to 60cm(2ft) long, with small, rounded, somewhat lustrous leaves; flowers 8cm(3in) long or more, cardinal-red with a prominent helmet-shaped upper lip; one of the best early hybrids and a fine house plant.

C. × *vedrariensis* (*C. magnifica* × *C. schiedeana*), shrubby, erect habit, 60cm(2ft) or more in height with handsome foliage to 10cm(4in) long; flowers scarlet, streaked yellow, 6cm($2\frac{1}{2}$in) or more long; a fine old hybrid still worth growing.

'Yellow Dragon', trailing habit, well-branched and vigorous, the leaves attractively red-flushed beneath; flowers strongly 2-lipped, bright yellow, usually with a touch of red.

COMMELINA

Commelinaceae

Origin: Tropics and to warm temperate regions of the world. A genus of 230 species, most being herbaceous perennials. They can be tufted and erect, or trailing and have alternate to linear leaves. The flowers are borne in small clusters within a boat-shaped bract or spathe, each opening singly and for one day only. They have three petals, one smaller than the other two. They are useful plants for a pot or bed in the conservatory. Propagate by seed or division in spring, or by cuttings in summer. *Commelina* was named for the Dutch botanists Johan Commelijn (1629–1692) and his nephew Caspar Commelijn (1667–1737).

Species cultivated

C. benghalensis Indian day flower Tropical Africa and Asia
Trailing evergreen with stems to 60cm(2ft) long and narrowly ovate to elliptic fleshy leaves, 5–8cm(2–3in) long. Flowers small, sky-blue, the two larger petals kidney-shaped to 1cm(⅜in) long. *C. b.* 'Variegata' has greyish-green leaves with longitudinal white striping.

C. coelestis Day flower Mexico
Erect species growing to 45cm(1½ft) tall. Leaves oblong-lanceolate, 6–10cm(2½–4in) long. Flowers rich-blue, the two larger petals 12mm(½in) long, carried within wavy-edged bracts, opening in summer. *C.c.* 'Alba' has white flowers.

BELOW *Commelina coelestis*
BOTTOM *Conophytum muscosipapillatum*

CONOPHYLLUM See MITROPHYLLUM

CONOPHYTUM

Aizoaceae

Origin: Southern Africa. A genus of 270 species of succulent plants. They are solitary or clump-forming, each plant body or shoot consisting of two fleshy leaves which are completely fused together except for a slit at the top. From this, usually in autumn, daisy-like flowers emerge, often completely hiding the plant beneath. They make good pot or pan plants for the conservatory or window ledge and with a cool rest in winter, preferably in an unheated room, will flower freely. Propagate by seed in spring or by division of congested plants in summer. *Conophytum* derives from the Greek *konos*, cone and *phyton*, plant, referring to the shape of the plant bodies.

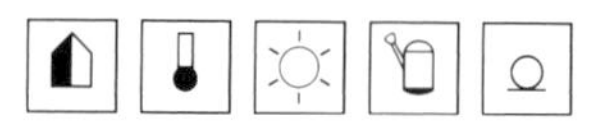

Species cultivated

C. bilobum
Plant body in colonies, 3–5cm(1¼–2in) tall with prominent flattened lobes, grey-green flushed with red. Flowers yellow, 2.5–3cm(1–1¼in) across.

C. calculus
Plant body in colonies, to 2cm(¾in) tall, globose and flattened, pale grey-green. Flowers yellow, 2.5–3cm(1–1¼in) across.

C. framesii
Plant body ovoid, to 12mm(½in) tall, glaucous grey-green with a few dark marks. Flowers cream, 1cm(⅜in) wide.

C. frutescens
Tiny shrublet to 10cm(4in) tall, each stem tipped by a plant body like that of *C. bilobum* in shape, but only 3cm(1¼in) long and dark green. Flowers 2.5cm(1in) wide, rich orange-yellow.

C. gratum
Plant bodies few, loosely clustered, broadly inverted pear-shaped, to 2.5cm(1in) high, glaucous with darker dots. Flowers satiny red-purple, 12mm(½in) wide.

C. meyeri
Almost shrubby, with prostrate, woody branches, 6–10cm(2½–4in) long; plant bodies 2–2.5cm(¾–1in) tall, grey-green, egg-shaped and notched at the apex, shortly hairy. Flowers 12mm(½in) wide, yellow.

C. multipunctatum
Clump-forming, the plant bodies broadly inverted pear-shaped, to 2cm(¾in) high, pale grey-green with dark green dots. Flowers white, opening at dusk, about 12mm(½in) wide.

C. mundum
Clump-forming, the plant bodies broadly obconical, to 12mm(½in) tall, the top flattened and somewhat sunken in the middle, grey-green with transparent dots. Flowers yellow, 12mm(½in) wide.

C. muscosipapillatum
Tiny shrublet to 8cm(3in) high, each stem covered with old brown leaf sheaths; plant bodies terminal, to 4cm(1½in) high with prominent flattened lobes, grey-green to almost whitish, finely velvety-hairy. Flowers 3–4cm(1¼–1½in) wide, rich yellow.

C. nelianum
Plant bodies in colonies, each one 2–4cm(¾–1½in) high with prominent flattened, keeled lobes, grey-green, minutely papillose and dotted with dark green. Flowers to 1.5cm(⅝in) wide, yellow.

C. notabile
Plant bodies to 2.5cm(1in) tall, light blue-green

with prominent, somewhat flattened red-edged lobes. Flowers reddish, 1cm(⅜in) or more wide, in late spring and summer. This species starts into growth earlier than most, often as early as April.

C. obcordellum
Clump-forming, the plant bodies broadly obconical, variable in size, 6–25mm(¼–1in) high and wide, usually flattened at the top with a sunken centre, various shades of green, often suffused pink or red and dotted dark green or purplish. Flowers to 1.5cm(⅝in) wide, white or cream.

C. ornatum
Clump-forming, the bodies inverted conical, to 2.5cm(1in) high, flattened at the top and sunken in the centre, blue-green with a few dark dots. Flowers 2.5cm(1in) wide, yellow; stigmas and stamens red.

C. pearsonii
Clump-forming, very compact, cushion-like when well established. Plant bodies broadly obconical, 8–16mm(⅓–⅔in) high and wide, the top shallowly convex, usually deep glaucous. Flowers up to 2.2cm(⅞in) wide, rosy-magenta.

C. pictum
Plant bodies obconical, up to 1.5cm(⅝in) high by 1cm(⅜in) wide, deep green, usually reddish or purplish on the sides and dotted above with brownish veins. Flowers about 1cm(⅜in) wide, cream.

C. pillansii
Plant bodies solitary or in groups of two to three, obovoid to 2cm(¾in) tall, yellow-green with short, broad lobes bearing translucent spots. Flowers reddish to purple, 2.5cm(1in) across.

C. subrisum
Plant bodies in small groups, obconical, to 1.5cm(⅝in) high, the top flattened and usually slightly depressed in the centre, smoothly white-glaucous. Flowers 2cm(¾in) wide, rich yellow, the petal tips red.

C. tischeri
Loosely clump-forming, old plants producing short woody stems. Plant bodies broadly obcordate, 12mm(½in) tall and almost as wide, slightly laterally flattened with a central cleft to 4mm($\frac{3}{16}$in) deep, grey-green with dark green dots. Flowers 1.5cm(⅝in) wide, pale lilac.

C. truncatum (*C. truncatellum*)
Plant bodies to 2cm(¾in) tall, obconical with a shallow goove along the top, greyish to bluish-green with darker brownish-green spots. Flowers 1.5cm(⅝in) wide, straw-coloured to white.

C. uvaeforme (*C. uviforme*)
Clump-forming, very compact and forming small hummocks with age. Plant bodies almost globular, to 12mm(½in) high and wide, yellowish to grey-green, sometimes flushed red or purple, with darker dots. Flowers 1–1.5cm(⅜–⅝in) wide, creamy yellow.

C. wettsteinii
Broadly clump-forming, eventually widening into mats. Plant bodies very broadly obconical, 1.5cm(⅝in) high by 2–3cm(¾–1¼in) wide, the top flattish or slightly convex, often bulging over the sides, grey-green, minutely white-dotted. Flowers about 3cm(1¼in) wide, violet-purple.

Convallaria majalis

CONVALLARIA

Liliaceae

Origin: Northern temperate zones. A genus of one species of rhizomatous, herbaceous perennial grown as a pot plant for its strongly fragrant white flowers. It can be forced into bloom from December onwards. The traditional method is to dig up crowns in the autumn, ideally ones specially grown for this purpose and kept well manured for two years to give strong plants. Choose the fattest budded with 10–15cm(4–6in) of rhizome. Place these about 5cm(2in) apart upright in 10–15cm(4–6in) pots of a peat compost and keep in a warm place at 16–24°C(60–75°F), ideally in a propagating frame with bottom heat. Batches treated in this way can be brought in at regular intervals of 1–2 weeks to provide a succession. An alternative and easier method is to purchase specially prepared (low temperature treated) crowns. These will force into growth and bloom at much lower temperatures. Even at a 10–13°C(50–55°F) night minimum, flowers can be had in 4–6 weeks. *Convallaria* derives from the Latin *convallis*, a valley, indicating that in the wild they often grow in valley or lowland woods.

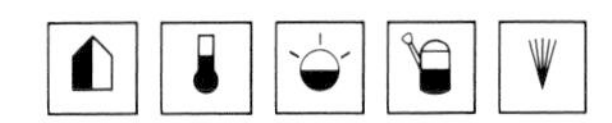

Species cultivated

C. majalis (including *C. montana*)
Lily-of-the-valley
Rhizomatous. Leaves in pairs or threes, deep green, elliptic to lanceolate, 15–20cm(6–8in) long. Flowers waxy-white, strongly fragrant, round bell-shaped, nodding; they are borne in one-sided racemes equalling or shorter than the leaves. The fruits are orange-red berries. *C.m.* 'Aureo Variegata' has yellow longitudinal stripes on the leaves. *C.m.* 'Major', 'Fortin's Giant' and 'Berlin Giant' have larger flowers on long stems. *C.m.* 'Prolificans', ('Flore Pleno') has double flowers. *C.m.* 'Rosea' has pink flowers and is less strong growing.

COPIAPOA

Cactaceae

Origin: Chile. A genus of 15 to 30 species depending upon the classifier. They are normally solitary and have mainly globular to somewhat conical or columnar stems, usually woolly at the top where the flowers arise. These are broadly funnel- to bell-shaped in shades of yellow and red. Copiapoa are good plants for a sunny windowsill or conservatory, but need a cool, dry spell in winter. Propagate by seed in spring or by offsets in summer. *Copiapoa* takes its name from Copiapo, a province of Chile where many of these cacti are found.

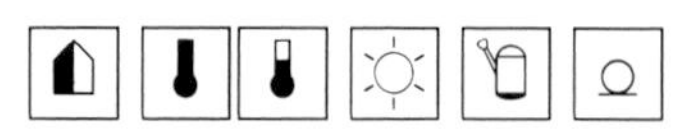

Species cultivated

C. cinerea (*Echinocactus cinerea*)
Stems solitary at first, later building clumps of cylindrical, grey-white glistening stems, to 60cm(2ft) or more with age. They have 18 broad, rounded ribs closely set with grey-woolly areoles bearing five to six spines when young; as they age these fall leaving a single, very sharp spine to 2cm($\frac{3}{4}$in). Flowers yellow, 3–5cm($1\frac{1}{4}$–2in).

C. coquimbana Chile (Coquimbo area)
Stem globular to conical, 13–20cm(5–8in) tall, usually solitary in cultivation, but in its homeland old specimens form clumps. Ribs ten to 17, blunt, warty, bearing large, round, woolly areoles. Radial spines eight to ten, straight or curved, about 1cm($\frac{3}{8}$in) long, central spines one or two, to 2.5cm(1in) long; all spines black, ageing to grey. Flowers about 3cm($1\frac{1}{4}$in) long, bell-shaped, bright yellow, emerging from the woolly top of the stem in summer.

C. echinoides (*Echinocactus echinoides*)
Stems usually solitary, globular to cylindrical, to 13cm($4\frac{1}{2}$in) tall, with about ten prominent ribs bearing large, well-spaced areoles. These have five to seven spreading radial spines and one straight central one, to 3cm($1\frac{1}{4}$in) long. Flowers greenish-yellow, 4cm($1\frac{1}{2}$in) long.

C. gigantea
Stems clustered, globular when young, later becoming columnar, to 90cm(3ft) tall with 14 to 22 ribs, swollen at the areoles. These bear eight to nine dark-tipped, yellow spines, 9–14mm($\frac{1}{3}$–$\frac{1}{2}$in) long. The flowers are yellow and the top of the stem where they grow is covered with a brownish-red wool.

C. humilis
Stem shortly cylindrical, clump-forming, 2.5–5cm(1–2in) tall. Ribs indistinct, divided into prominent tubercles, each one topped by a woolly areole and tipped by ten or more short radial spines. Flowers yellow, 2–2.5cm($\frac{3}{4}$–1in) wide, borne in summer.

C. marginata (*Echinocactus marginata*)
Stems globular to cylindrical, yellow-woolly at the top, with 12 to 15 rounded ribs. Areoles woolly, large, well spaced with seven to nine radial spines and one to two longer central ones, to 2.5cm(1in) in length. Flowers yellow, to 2cm($\frac{3}{4}$in) or more long.

COPROSMA

Rubiaceae

Origin: Malaysia, Australasia, Polynesia and Chile. A genus of 90 species of trees and shrubs with linear to almost rounded leaves borne in opposite pairs, and small dioecious tubular flowers which are inconspicuous. If plants of both sexes are grown, showy berries follow. They make decorative plants for the conservatory and *C. repens* in particular is a good house plant for a cool room. Propagate by seed or cuttings with a heel in summer. *Coprosma* derives from the Greek *copros*, dung and *osme*, a smell, the foliage of some species having a foetid odour when crushed. This is, however, barely noticeable and the plant does not deserve such a name.

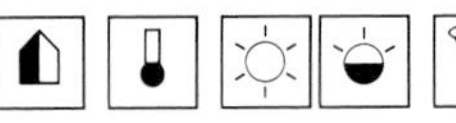

RIGHT *Copiapoa cinerea*
FAR RIGHT *Copiapoa humilis*

ABOVE LEFT *Coprosma repens* 'Marble Queen'
ABOVE *Cordyline terminalis*

Species cultivated

C. lucida New Zealand
Evergreen shrub to 3m(10ft) in the open. Leaves narrowly oblong to elliptic, 7–12cm(2¾–4¾in) long, deep glossy green above, paler and net-veined beneath. Fruits oblong, orange-red to 1cm(⅜in) long.

C. repens New Zealand
Evergreen shrub, spreading at first, eventually erect, to 2m(6½ft). Leaves broadly oblong to ovate-oblong, 6–8cm(2½–3in) long, fleshy and very glossy, rich green above. Fruits ovoid, orange-red, to 1cm(⅜in) long. *C.r.* 'Argentea' has silver variegation. *C.r.* 'Marginata' has leaves margined with yellow. *C.r.* 'Picturata' ('Variegata') has leaves with a central, creamy-yellow blotch. A number of other cultivars are grown with varying variegation.

C. robusta
Potentially a large, fairly erect shrub, but easily maintained at 2m(6½ft) or just below. Leaves elliptic-oblong to broadly lanceolate, 6–13cm(2½-5in) long, dark green and semi-lustrous. *C.r.* 'Williamsii Variegata' is a shrub to 1.2m(4ft) with a semi-pendulous habit and narrower, broadly creamy-white margined leaves.

CORDYLINE

Agavaceae (*Liliaceae*)

Origin: Tropical and warm temperate areas of Australasia and Malaysia with one species in South America. A genus of 15 species of evergreen trees and shrubs often confused in cultivation with *Dracaena*, *q.v.* They have erect, often solitary stems each one ending in a palm-like tuft of long, sword-shaped leaves. The flowers are small, but carried in large terminal panicles, followed by coloured berries. Cordylines make attractive foliage plants for large pots or tubs for the conservatory; *C. terminalis* is also suitable for the home. Propagate by seed in spring or by suckers detached in late spring, or by 6cm(2½in) long stem sections in early summer, treating both these as cuttings until well rooted. *Cordyline* derives from *cordyle* a club, probably alluding to the swollen stem bases of some species which give a rather club-like appearance.

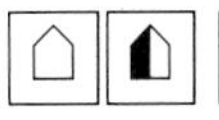  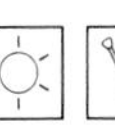 

Species cultivated

C. australis Cabbage palm/tree New Zealand
Tree to 20m(65ft) in the wild, but seldom over 2–3m(6½–10ft) in a container. Leaves linear, 30–90cm(1–3ft) long, erect at first then arching over. Flowers small, white, in broad panicles 60–120cm(2–4ft) long. Fruits white, 4mm(3/16in) across. *C.a.* 'Purpurea' has leaves with bronze to purple suffusion. *C.a.* 'Rubra' has red-bronze leaves.

C. banksii New Zealand
Smaller than *C. australis*. Leaves narrowly lanceolate, drooping at the tips, to 1m(3ft) or more long. Individual flowers about 1cm(⅜in) long. Fruits 4–5mm(3/16in) wide, white or bluish.

C. indivisa (*Dracaena indivisa*) New Zealand
Small tree in the wild, but an elegant pot plant. Leaves lanceolate, 1–2m(3–6½ft) long, often glaucous beneath and purple-flushed above, erect when young, drooping with age. Flowers white, 7–8mm(¼–⅓in) long in compact panicles 60–120cm(2–4ft) long. Fruits 6mm(¼in), bluish.

C. terminalis (*Dracaena terminalis*) Ti tree
Tropical Asia, Polynesia
Stems erect, 1–4m(3–13ft) tall. Leaves broadly lanceolate, 30–60cm(1–2ft) long and up to 15cm(6in) wide, carried on distinct stalks 5–15cm(2–6in) long. Flowers about 6mm(¼in) long, white, sometimes reddish or purplish. Fruits 8mm(⅓in) wide, red. A number of colourful foliage forms are available: *C.t.* 'Baptistii' has deep green to bronze leaves striped with creamy-yellow and pink; *C.t.* 'Tricolor' has broader leaves striped with cream, pink and red.

ABOVE *Coronilla valentina glauca*
ABOVE RIGHT *Correa reflexa pulchella*

CORONILLA

Leguminosae

Origin: Europe, especially the Mediterranean region, Western Asia and the Canary Islands. A genus of 20 species of perennials, shrubs and sub-shrubs with alternate pinnate leaves. They have umbels of pea flowers followed by slender pods which are constricted between the seed (lomenta). Propagate by seed in spring or by cuttings with a heel in late summer. *Coronilla* is a diminutive form of the Latin *corona*, a crown; the umbels of flowers are often circular.

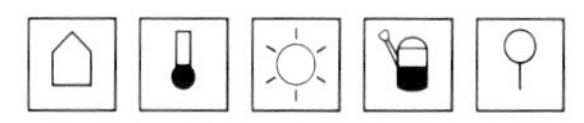

Species cultivated

C. emerus Scorpion senna Europe
Deciduous shrub, 1–2m(3–$6\frac{1}{2}$ft) tall. Leaves glaucous, made up of five to nine obovate leaflets, 1–2cm($\frac{3}{8}$–$\frac{3}{4}$in) long. Flowers 1.5–2cm($\frac{5}{8}$–$\frac{3}{4}$in) long, pale yellow, in 2- to 3-flowered umbels in spring. Fruits curved and constricted, appearing like a scorpion's tail.

C. glauca See *C. valentina glauca.*

C. valentina Mediterranean region, S. Portugal
Evergreen shrub, 1–1.5m(3–5ft) tall. Leaves bright green above, glaucous beneath, made up of seven to 15 notched leaflets, each to 2cm($\frac{3}{4}$in) long. Flowers 7–12mm($\frac{1}{4}$–$\frac{1}{2}$in) long, yellow, with a fragrance reminiscent of ripe peaches, borne in 5- to 12-flowered umbels in late spring and summer.

C.v. glauca (*C. glauca*)
Differs in having leaves which are glaucous on both surfaces. It will flower intermittently through the year.

CORREA

Rutaceae

Origin: Australia. A genus of 11 species of evergreen shrubs of compact habit to 1m(3ft) or more with opposite pairs of ovate to oblong, simple, untoothed leaves bearing stellate hairs. Their flowers are pendent, tubular to bell-shaped, 4-petalled and borne in spring and summer. They make very decorative pot plants. Propagate by heel cuttings in late summer or by seed in spring.

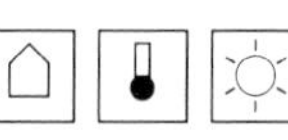  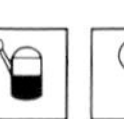 

Species cultivated

C. backhousiana (*C. speciosa backhousiana*)
Tasmania
Well-branched shrub to 2m($6\frac{1}{2}$ft) with slender stems. Leaves ovate-elliptic, 1.5–4cm($\frac{5}{8}$–$1\frac{1}{2}$in) long, dark green above, densely hairy beneath. Flowers to 2.5cm(1in), often less, bell-shaped, flared at the tips, waxy-white to pale yellow-green; they are borne in clusters of one to three or occasionally four.

C. × harrisii
Slender stems, 1–2m(3–$6\frac{1}{2}$ft) tall. Leaves ovate, 1.5–4.5cm($\frac{5}{8}$–$1\frac{3}{4}$in) long, shortly hairy beneath. Flowers scarlet, tubular, to 2.5cm(1in), with protruding yellow stamens.

C. pulchella See *C. reflexa pulchella.*

C. reflexa (*C. speciosa*) E. Australia
Slender-stemmed shrub, 1–2m(3–$6\frac{1}{2}$ft) tall. Leaves leathery, ovate to ovate-oblong, flannelly-hairy beneath, 1.5–4.5cm($\frac{5}{8}$–$1\frac{3}{4}$in) long. Flowers solitary or in twos or threes, tubular, 2–3cm($\frac{3}{4}$–$1\frac{1}{4}$in) long, crimson to scarlet tipped with greenish-white.

C.r. pulchella (*C. pulchella*)
Leaves greenish beneath, flowers rosy red, pinker at the mouth.

C. speciosa See *C. reflexa.*

CORYNOCARPUS

Corynocarpaceae

Origin: N.W. Australia, New Zealand, New Guinea, New Hebrides, New Caledonia. A genus of four or possibly five species of evergreen trees, one of which is sometimes cultivated, making a handsome foliage specimen for a large pot or tub. The flowers, which are small and greenish, are carried in terminal panicles. They are followed by sizeable fleshy fruits with a large, solitary, nut-like seed. It is best in the conservatory but will grow indoors. Propagate by seed when ripe or in spring,

or by cuttings in late summer. *Corynocarpus* derives from the Greek *koryne*, a club and *karpos*, a fruit, alluding to the shape of the fruits.

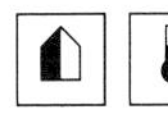 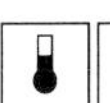 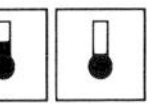 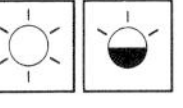  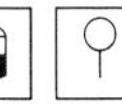

Species cultivated

C. laevigatus Karaka/New Zealand laurel
New Zealand, naturalized in Hawaii
Grown in large containers, 2–3m(6½–10ft) in height, but up to 15m(50ft) in the wild; erect at first then bushy and somewhat spreading. Leaves obovate to elliptic-oblong, 10–20cm(4–8in) long, leathery-textured, dark green and lustrous. Flowers to 1cm(⅜in) wide, greenish-yellow, in stiff panicles, from spring to summer. Fruits narrowly ovoid, to 4cm(1½in) long, orange; flesh edible, but the seeds poisonous until cooked and retted in water (formerly much used by the Maoris as a basic food source). *C.l.* 'Alba Variegatus' has the leaves narrowly white-margined. *C.l.* 'Variegatus' ('Aurea Variegatus') is broadly yellow-margined.

CORYPHA

Palmae

Origin: S.E. Asia, Indo-Malaysia and Sri Lanka. A genus of eight species of large statuesquely handsome fan palms which, when small, make highly ornamental pot or tub foliage specimens. These palms are unique in being monocarpic, that is they grow to maturity in 20 to 80 years, flower and fruit once with enormous prodigality, then die. Propagate by seed. *Corypha* is from the Greek *koryphe*, a summit or hilltop, the allusion being to the great terminal fountain-like flower clusters.

  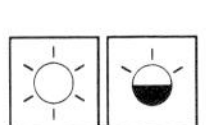 

Species cultivated

C. elata (*C. gembanga*) Gebang/Philippines sugar palm Philippines, East Indies; also northern India and Burma, but perhaps not truly native there
In containers, 1–2m(3–6½ft) tall, (18m[60ft] or more in the tropics). Leaves long-stalked, fan-shaped, to 1m(3ft) wide on container-grown specimens (twice this on well grown trees), deeply cut into up to 80 narrow lobes.
C. gembanga See *C. elata*.
C. umbraculifera Talipot palm Sri Lanka and adjacent coast of India
In containers, up to 2m(6½ft) or more tall. Leaves long-stalked, up to 2m(6½ft) wide on containerized specimens (to 5m[16ft] on well grown trees), deeply cleft into up to 100 lobes. Mature palms produce small cream flowers in an enormous panicle to 6m(20ft) high and wide – the largest single inflorescence in the entire plant kingdom.

ABOVE LEFT *Corynocarpus laevigatus* 'Variegatus'
ABOVE *Corypha elata*

CORYPHANTHA

Cactaceae

Origin: North America to Cuba. A genus of 64 species of cacti, formerly classified under *Mammillaria* and by some botanists considered to include only 20 to 30 species. They have rounded to cylindrical stems, 2.5–15cm(1–6in) long, either solitary or building up wide mounds with age. Tubercles are prominent, each with a marked groove along the upper side and tipped with a spine-bearing areole. The flowers are comparatively large and widely funnel-shaped. They grow well in a conservatory or the home, needing a sunny position and a cool rest in winter. Propagate by seed in spring or by offsets, if available, in summer. *Coryphantha* derives from the Greek *korypha*, summit and *anthos*, a flower, the flowers being borne at the top of the plant body.

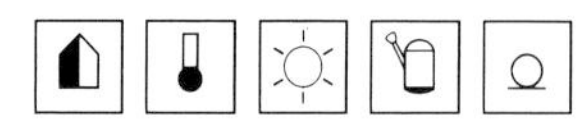

Species cultivated

C. andreae Mexico
Stem globular to slightly elongated, glossy deep green, to 10cm(4in) tall. Tubercles rounded, 2cm($\frac{3}{4}$in) long, the grooves woolly. Radial spines usually ten; central spines five to seven, stouter, up to 2.5cm(1in) long. Flowers 5–6cm(2–2$\frac{1}{2}$in) wide, pale yellow.

C. bergeriana Mexico
Stem solitary, club-shaped, to 13cm(5in) tall. Tubercles conical, grooved to 1–1.5cm($\frac{3}{8}$–$\frac{5}{8}$in). Radial spines 18 to 20, often recurved; central spines about four, recurved, yellowish, to 2cm($\frac{3}{4}$in) long. Flowers to 7cm(2$\frac{3}{4}$in) wide, yellow.

C. bumamma Mexico
Stem solitary, globular, reaching 12cm(4$\frac{3}{4}$in) in time. Tubercles rounded, flattened above, the groove woolly. Radial spines six to eight; central spines not present. Flowers 8–10cm(3–4in) wide, yellow to reddish. Considered by some botanists to be a form of the closely allied *C. elephantidens* which has red to pink flowers.

RIGHT *Coryphantha bumamma*
FAR RIGHT *Coryphantha pallida*

C. clava Central Mexico
Stems generally solitary, cylindrical, bluish to grey-green, eventually to 30cm(12in) tall. Tubercles obliquely conical, pointed, upswept, 1.5cm($\frac{5}{8}$in) or more long, the small areoles woolly when young. Radial spines six to 11, yellowish, about 12mm($\frac{1}{2}$in) long; central spines one to four, thicker, to 2.5cm(1in) long. Flowers to 8cm(3in) wide, pale yellow, opening in summer.

C. cornifera (*C. radians*) Mexico
Stem solitary, globular to elongated, to 10cm(4in) or more tall. Tubercles woolly at the base, obliquely conical. Radial spines 20 or more, each to 3cm(1$\frac{1}{4}$in) long; central spines absent. Flowers 6–7cm(2$\frac{1}{2}$–2$\frac{3}{4}$in) wide, yellow, red at the base.

C.c. echinus (*C. radians pectinata*)
Shorter spines; base of the flowers green.

C. dasyacantha (*Escobaria dasyacantha*)
Texas, New Mexico, northern Mexico
Stems usually solitary, sometimes producing small groups, globular, then shortly cylindrical, to 15cm(6in) in height. Tubercles cylindrical, pointed, about 8mm($\frac{1}{3}$in) long, the areoles small and round, white-woolly at first, later bare. Radial spines 25 to 35, white, bristle-like, the longest to 1.5cm($\frac{5}{8}$in) long; central spines seven to 13, thicker, brown to black, to 2cm($\frac{3}{4}$in). Flowers 4.5cm(1$\frac{3}{4}$in) wide, pink and white, borne in summer.

C. elephantidens See under *C. bumamma*.

C. erecta Central Mexico
Stems usually remaining solitary in cultivation, but clustering in the wild, cylindrical, to 30cm(1ft) tall. Tubercles conical, blunt-tipped, 8mm($\frac{1}{3}$in) long, areoles woolly when young. Radial spines eight to 14, yellow to brown or grey, 12mm($\frac{1}{2}$in) long; central spines two to four, up to 2cm($\frac{3}{4}$in) long, the lowest one reflexed. Flowers 6–8cm(2$\frac{1}{2}$–3in) wide, yellow with red-tipped stamens, opening in summer.

C. macromeris New Mexico, Texas, northern Mexico
Clump-forming, stems cylindrical, about 15cm(6in) tall, green. Tubercles cylindrical, 1.5–3cm($\frac{5}{8}$–1$\frac{1}{4}$in) long, areoles woolly when young. Radial spines ten to 17, awl-shaped, 1–4cm($\frac{5}{8}$–1$\frac{1}{2}$in) long, reddish to whitish; central spines two to four, thicker, brown or black, to 5cm(2in) long.

Flowers to 8cm(3in) wide, deep pink to red-purple, opening in summer.

C.m. runyonii (*C. runyonii*)
Smaller in all its parts and grey-green.

C. minima Texas
Stems solitary or with a few basal branches, to 5cm(2in) tall. Tubercles small, about 3mm($\frac{1}{8}$in) long. Radial spines 13 to 18, bristle-like, white; central spines two to four, thicker and a little

longer. Flowers about 2.8cm(1⅛in) wide, deep pink to purple, opening in summer.
C. pallida Southern Mexico
Stems solitary or forming small clumps, more or less spherical, to 13cm(5in) tall, bluish-green. Tubercles dense, each one bearing about 20 white radial spines to 8mm(⅓in) long; central spines three, awl-shaped, black or black-tipped, to 4cm(1½in) long, the lower one often down-curved. Flowers 7cm(2¾in) wide, satiny pale yellow, opening in summer.
C. pectinata See *C. cornifera echinus.*
C. poselgeriana (*Echinocactus poselgerianus*) Northern Mexico
Stem usually solitary, globular, to 20cm(8in) high or more, bluish-green. Tubercles rhomboid, about 2cm(¾in) long by twice as wide; areoles bare. Radial spines five to seven, bulbous at their bases, the upper ones erect and fascicled, yellow to grey or brown; one central spine, straight or arching, to 5cm(2in) long. Flowers 4–5cm(1½–2in) wide, pink, opening in summer.
C. radians See *C. cornifera.*
C. runyonii See *C. macromeris runyonii.*
C. sulcata Texas
Stems clustered, globular to slightly elongated, 8–11cm(3–4½in) tall, grey-green. Tubercles rounded and flattened, about 1cm(⅜in) long. Radial spines to 15, white; central spines one to three. Flowers 5cm(2in) wide, yellow banded with red, opening in summer.
C. sulcolanata Central Mexico
Stem solitary until long-established, then forming small clumps, globular, depressed at the top, 5cm(2in) or more high by almost twice as wide, dark green and white-woolly. Tubercles conical, somewhat angular and blunt-tipped, to 2cm(¾in) long. Areoles woolly when young. The eight to ten spines all radial, fairly robust, yellow tipped red to brown, the longest to 4cm(1½in) long. Flowers to 8cm(3in) wide, bright yellow, opening in summer.
C. vivipara Pincushion cactus Alberta to Texas
Stems clustered, globular to elongated, to 15cm(6in) tall, grey-green. Tubercles cylindrical. Radial spines 12 to 20 or so, usually white; central spines four to six, brownish to 2.5cm(1in) long, somewhat bulbous at the base. Flowers 5cm(2in) wide, reddish inside, greenish outside, opening for a short time each day, in summer. A variable species.

COSTUS

Costaceae (Zingiberaceae)

Origin: Circum-tropical. A genus of about 150 species of mainly clump-forming, rhizomatous perennials sometimes known colloquially as spiral flag or ginger. This apt name refers to those species with widely spiralling stems which seem to bear leaves only on the outer side of each spiral. The flowering spikes are terminal, composed of crowded bracts which create a cone-like appearance. Usually only one to a few of the comparatively large tubular, orchid-like flowers emerge at any one time, but the repeat performance is prolonged. Propagate by division in spring, or by cuttings of mature stem sections, 2.5–5cm(1–2in) long, in spring or early summer. The species described provide something different in the way of foliage and flowering pot plants and deserve to be seen more often. *Costus* perhaps derives from the Arabic *koost*. The name was used by Pliny for a plant, the identity of which is not now known, with an aromatic root or rhizome.

ABOVE LEFT *Costus igneus*
ABOVE *Costus malortianus*

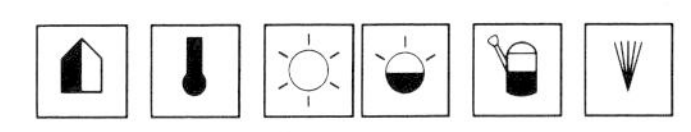

Species cultivated
C. elegans See *C. malortianus.*
C. igneus Spiral ginger Brazil
Up to 60cm(2ft) in height, but often less. Leaves spiralling, up to 15cm(6in) long, oblong-lanceolate, somewhat glossy above, reddish beneath. Flowers up to 6cm(2½in) long with an orange tube and a deep yellow lip, borne in summer.
C. malortianus (*C. elegans, C. zebrinus*) Central America
Usually 60–90cm(2–3ft) tall, but can attain twice this in large containers. Stems and leaves spiralling, the latter obovate, to 30cm(1ft) or more long, bright velvety green with darker zones above, somewhat glaucous beneath. Flowers almost 5cm(2in) long, rich yellow, the large lip marked with reddish zones, borne in summer.
C. pulverulentus See *C. sanguineus.*
C. sanguineus (*C. pulverulentus*) Mexico to western South America
Up to 90cm(3ft) tall, but capable of twice this or more in a tub or planted out. Stems and leaves spiralling, the latter narrowly elliptic to obovate, 15–30cm(6–12in) long, somewhat fleshy-textured, velvety bluish-green above with a central silvery band and yellow-green zones, deep purple-red beneath. Flowers about 4cm(1½in) long wth a red tube and a red or yellow lip, borne in summer. Worth growing for its foliage alone and decorative when small.
C. speciosus Crepe/Malay ginger Indonesia, tropical Asia
To 1.2m(4ft) or more. Stems and leaves spiralling,

the latter oblanceolate, to 20cm(8in) or more long, bright green, slender-pointed. Floral bracts red or red-flushed; flowers 5cm(2in) long, expanding to almost twice this, white, with a yellow-centred lip. Forms with orange-red-centred lips are known.
C. zebrinus See *C. malortianus*.

COTYLEDON

Crassulaceae

Origin: South Africa, one from East Africa. A genus of 41 species of succulent plants which can be divided into two main groups: evergreen, shrubby species with opposite pairs of fleshy leaves; and deciduous species with thickened, short, fleshy stems and alternate leaves. The leaves can be flattened or cylindrical and the flowers 5-lobed and tubular, carried in terminal, umbel-like clusters. Most species have poisonous foliage. They make good pot plants, the evergreen species needing warmer and less dry conditions in winter than the deciduous ones which need a cool, almost dry, winter rest. Propagate by seed in spring or by leaf or stem cuttings in summer. *Cotyledon* derives from the Greek *kotyledon*, meaning a cup or hollow, from the cupped leaves of navelwort first classified in this genus, now *Umbilicus rupestris*.

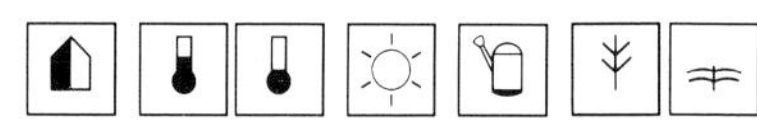

Species cultivated

C. barbeyi East Africa, Eritrea, Arabia
Evergreen shrub to 60cm(2ft) tall. Leaves obovate, glossy light grey-green, 6–14cm(2½–5½in) long. Flowers 2.5cm(1in) or more long, orange-red, borne in panicles held on long stalks well above the leaves, from autumn to winter. Temperate.
C. jacobseniana Cape (Namaqualand)
Semi-prostrate, freely branching evergreen sub-shrub up to 30cm(1ft) wide. Leaves fairly densely arranged, 2–3cm(¾–1¼in) long, narrowly ovoid but tapering to each end, green or with a white, waxy patina. Flowers 6–10mm(¼–⅜in) long, pendent, greenish-red, three to ten in an umbel-like cluster, carried 5–13cm(2–5in) above the leaves in summer. Temperate.

Cotyledon undulata

C. ladysmithensis South Africa
Evergreen shrub, low and spreading to 20cm(8in) tall; stems becoming woody with age. Leaves 3–8cm(1¼–3in) long, obovate, thick fleshy and coarsely white-hairy, with two to four teeth at the apex. Flowers 1–2cm(⅜–¾in) long, brownish-red, in clusters on 15cm(6in) stalks. Temperate.
C. maculata See *Adromischus maculatus*.
C. orbiculata South Africa
Evergreen shrub to 60cm(2ft) or more tall. Leaves 4–14cm(1½–5½in) long, broadly obovate, somewhat convex below, blunt, intensely waxy white and often narrowly red-edged. Flowers 2cm(¾in) long, orange-yellow in panicles on long stalks in late summer. Temperate. A popular house plant.
C.o. dinteri
Dwarf with almost cylindrical leaves.
C.o. oophylla
Compact with egg-shaped leaves bearing a dark red curved mark near their tips.
C. paniculata South Africa
Deciduous shrub to 2m(6½ft) in the wild, but slow-growing, taking many years to exceed 30cm(1ft). Leaves 5–10cm(2–4in) long, lanceolate to ovate, light grey-green, clustered together at the tips of the stems. Flowers about 2cm(¾in) long, dark red with green stripes, opening in summer. Cool.
C. papillaris South Africa (Namaqualand)
Prostrate evergreen shrub, to 30cm(1ft) wide, the branches mostly in opposite pairs, and glandular downy. Leaves at the stem tips, usually rosetted, obovate to wedge-shaped, 1.5–4.5cm(⅝–1¾in) long, 3–6mm(⅛–¼in) thick, dark green, red-tipped or margined. Flowers 8mm(⅓in) long, in shades of red, green and yellow merging together, three to 15 in an umbel-like cluster, carried 5–25cm(2–10in) above the leaves in summer. Temperate.
C.p. glutinosa
Leaves densely downy and flowering stems and calyces sticky hairy.
C. reticulata South Africa
Somewhat like *C. paniculata* but smaller and branching lower to form a small deciduous shrub, 15–30cm(6–12in) high. Leaves almost cylindrical, 1.5–2cm(⅝–¾in) long, fleshy, hairless. Flowers 8–12mm(⅓–½in), erect, greenish-yellow, borne in autumn and winter. After the flowers have fallen, its stalks become woody and spiny adding a distinctive note to the plant. Cool.
C. undulata South Africa
Decorative evergreen shrub to 50cm(20in). Leaves 6–12cm(2½–4¾in) long, widely obovate to almost fan-shaped, the broad ends strongly waved; they are densely covered with a waxy white, floury patina which is strongest on the young leaves and stems. Flowers 2–3cm(¾–1¼in) long, orange-yellow, borne in spring and summer. Temperate.
C. wallichii South Africa
Deciduous shrub to 30cm(1ft) with fleshy stems.

Leaves semi-cylindrical, grey-green, clustered at the ends of the stem branches and falling after the short growing season. Flowers 2cm(¾in) long, green, red-spotted at the tips; carried on 30cm(1ft) long stalks. Cool.

CRASSULA

Crassulaceae

Origin: Widespread throughout the world, but the majority of species, including all those described below, come from southern Africa. A genus of about 300 species of annuals, evergreen perennials and shrubs most of which are succulent. They have opposite, rounded to linear leaves and panicles of more or less starry flowers usually with five (sometimes three to nine) petals. Crassulas make decorative foliage and flowering plants. Propagate by seed in spring, or by cuttings or division in spring or summer.

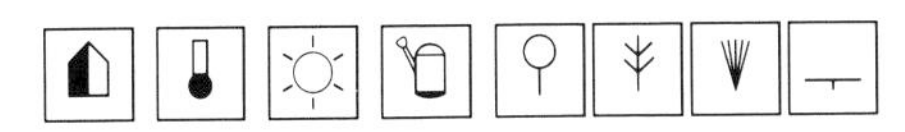

Species cultivated

C. arborescens (*C. cotyledon*) Namaqualand to Natal
Shrub to 1m(3ft) or more with robust stems. Leaves 4–7cm(1½–2¾in) long, broadly ovate, grey-green with a distinctive red edge; needs good light to bring out the colour of its leaves. Flowers white, fading to pink, but seldom produced on pot plants.

C. argentea See *C. portulacea*.

C. barbata South Africa
Eventually clump-forming. Leaves broadly ovate, to 4cm(1½in) long, glossy green, margined with long white hairs, crowded together in angular rosettes. Flowers small, white, borne in erect spikes to 30cm(1ft) or more, from winter to spring. Flowering rosettes die when seed is formed, but offsets replace them.

C. barklyi (*C. teres*) Rattlesnake tail Cape (Namaqualand), Namibia
Stem usually solitary, erect, to 8cm(3in) tall. Leaves orbicular, cupped, green, with ciliate margins, in alternate opposite pairs which tightly overlap each other creating a column about 2cm(¾in) in width. Flowers white or yellow, borne in terminal clusters in autumn.

C. coccinea See *Rochea coccinea*.

C. columnaris Cape (Namaqualand, Karroo)
Much like a more robust *C. barklyi*, the dense column of leaves up to 10cm(4in) high and 4cm(1½in) in width, dark green when actively growing, brownish-green when dormant. Flowers small, orange-yellow or white, fragrant, opening in autumn.

C. cooperi Transvaal
Low mat-forming perennial. Leaves lanceolate to spathulate, rounded-keeled, 1–1.5cm(⅜–⅝in) long, light green, red-flushed above, the surface pitted, the margins finely hairy. Flowers pale pink, opening in spring. Useful for a cool room or porch.

LEFT *Crassula arborescens*
BELOW *Crassula falcata*

C. corymbulosa Cape (coastal regions)
Solitary to tufted, sparingly branched from the base, erect, the stems completely covered by the pairs of alternate leaves; they are narrowly triangular, to 4cm(1½in) long, pale grey-green with darker dots. Flowers small, white, borne in clusters at the top of the 30cm(1ft) stems..

C. cotyledon See *C. arborescens*.

C. deceptrix Cape (Namaqualand)
Small, erect perennial to 5cm(2in) or so tall. Stems at first solitary, then branching at the base to form small clumps. Leave thickly fleshy, rounded triangular, keeled, papillose, whitish-grey, closely packed. Flowers small, white, carried just above the foliage in autumn.

C. deltoidea (*C. rhomboidea*) Silver beads Cape (Namaqualand)
Tiny sub-shrub to 10cm(4in) tall. Leaf pairs perfoliate, thickly fleshy, rhombic-ovate, to 1.5cm(⅝in) long, bluntly keeled, the upper surface ridged, covered with grey farina. Flowers small, off-white to pinkish, borne in terminal cymes in autumn.

C. falcata Natal
Erect-stemmed shrub to 1m(3ft) or so, sparingly branched, the stems hidden by the fleshy, rich grey-green, sickle-shaped leaves, 7–10cm(2¾–4in) long

which are turned so that the blades lie almost vertically. Flowers scarlet, carried in crowded, flattish clusters at the top of grey-white stems in summer. Needs pruning or re-propagating regularly to keep it compact and attractive.
C. gracilis See *C. schmidtii.*
C. hookeri See *C. schmidtii.*
C. lactea Natal and Transvaal
Shrubby, to 30cm(1ft) or more tall. Leaves narrowly obovate, white-dotted. Flowers small, white, fragrant, in dense panicles borne on long stalks in winter.
C. longifolia See *C. perfoliata.*
C. lycopodioides Rat tail plant Namaqualand to Namibia
Small, twiggy sub-shrub to 30cm(1ft) high, the stems slender and erect. Leaves tiny, triangular-ovate, yellowish-green, densely overlapping in four rows so as to hide the stems completely. Flowers tiny, yellowish, arising in the leaf axils in spring and summer.
C. maculata See *Adromischus maculatus.*
C. marnieriana Cape
Semi-erect, sparingly branched, sprawling sub-shrub to about 15cm(6in) wide. Leaves thickly fleshy, almost orbicular, about 8mm(⅓in) across, glaucous with red edges. Flowers small, white, in compact terminal cymes, borne in spring.
C. mesembryanthemopsis Cape (Namaqualand)
Low hummock-forming perennial, to 10cm(4in) or more wide, composed of crowded rosettes. Leaves whitish-green, sometimes red-tinted, thickly fleshy, oblong-cuneate with thickened tips, to 3cm(1¼in) long. Flowers tiny, white, borne in small sessile cymes in autumn.
C. multicava Natal (Transvaal)
Fleshy, erect to spreading shrub to 30cm(1ft) or more tall. Leaves obovate to oval, subcordate, 2.5–8cm(1–3in) long, the surface shallowly pitted, brightish green with a red flush in bright light. Flowers white from pink-tinted buds in conspicuous terminal panicles from late winter to spring. Well worth growing for the early floral display.
C. nealeana (*C. rupestris minor*, *C. perfossa minor*) Cape
Tiny decumbent shrub to 15cm(6in) or more wide. Leaf pairs crowded, perfoliate, broadly oval, shortly pointed, to about 2cm(¾in) long, glaucous or purple-flushed, red-edged. Flowers small, yellow in a panicle-like cluster, 4–5cm(1½–2in) high, in spring.
C. orbicularis Cape, Natal
Mat-forming perennial, to 15cm(6in) wide, composed of loosely arranged, flat rosettes. Leaves spathulate-obovate, 4–5cm(1½–2in) long, glossy green, white-ciliate. Flowers small, white, in a spike-like inflorescence to 20cm(8in) tall, opening in spring.
C. perfoliata (*C. longifolia*) Cape, Natal
Sparingly branched, erect shrub to 60cm(2ft). Leaves lanceolate, somewhat channelled above, tapering to a point, 10–15cm(4–6in) long, grey. Flowers small, red or white in a dense, terminal corymbose cluster to 10cm(4in) wide, borne in summer. Closely allied to *C. falcata*, but with straight leaves which stand at right angles to the stem.
C. portulacea (*C. argentea*) Jade plant
Namaqualand to Transvaal
Dense, freely branching shrub 30–90cm(1–3ft) tall with a robust, woody stem. Leaves to 4cm(1½in) long, thick, fleshy, obovate to spathulate, glossy green with a red edge. Flowers pink, opening in spring. Confused with the similar *C. obliqua* which has obliquely-ovate, narrower silvery-green leaves marked with darker dots.
C. rhomboidea See *C. deltoidea.*
C. rubicunda See *C. schmidtii.*
C. rupestris Bead vine, Rosary vine
Cape, Namaqualand, Karroo
Small decumbent shrub to 25cm(10in) or more wide, the stems repeatedly forking. Leaves in perfoliate pairs, thickly fleshy, broadly ovate, 1–2.5cm(⅜–1in) long and almost as wide, grey-green, reddish-brown edged. Flowers small, yellowish, in terminal clusters, opening in spring and summer.

BELOW *Crassula nealeana*
BELOW RIGHT *Crassula portulacea*

Makes an unusual subject for a small hanging basket.

C. sarcocaulis Cape (mountains)
A gnarled shrublet to 20cm(8in) or more tall. Leaves to 1cm(⅜in) long, lanceolate, green, flushed red (the colour when the plant is kept dry). Flowers bright red to pink, borne in small heads in summer and autumn. Useful for an unheated room or porch.

C. schmidtii Namibia, Transvaal, Natal
C. gracilis, *C. hookeri*, *C. rubicunda* and *C. schmidtiana* are all names under which this plant is erroneously grown in gardens. Low perennial of tufted to mat-forming habit, to 10cm(4in) high. Leaves narrowly lanceolate, 3–4cm(1¼–1½in) long, green with a darker pitted upper surface, convex and reddish beneath. Flowers bright rose to carmine-red, carried in showy terminal clusters in autumn and winter.

C. socialis Cape
Perennial of hummocky to mat-forming habit. Leaves alternate, ovate-triangular, to 6–7mm(¼in) long, in dense rosettes. Flowers white, small, freely borne in clusters on stalks 2.5–4cm(1–1½in) long, in late winter and spring.

C. teres See *C. barklyi.*

C. tetragona Cape
Erect, sparingly branched shrub eventually to 60cm(2ft) or more in height. Leaf pairs well spaced to fairly close together, united at their bases, awl-shaped, 2–4cm(¾–1½in) long, green. Flowers small, white, in terminal, small dense panicles in spring and summer.

× **CRINDONNA** See × AMARCRINUM
× **CRINODONNA** See × AMARCRINUM

CRINUM

Amaryllidaceae

Origin: Tropics and sub-tropics, world-wide. A genus of 100 or more species of bulbous-rooted perennials with large, long-necked bulbs, stalkless, rather fleshy, strap-shaped leaves and large, funnel-shaped lily-like flowers carried in umbels which emerge from spathe-like bracts. They are fine plants for large pots or tubs. When potting, the upper third of the bulb should be above the surface of the soil. Propagate by division, removal of offsets or by seed which must be sown as soon as ripe (if allowed to become dry they lose their viability and will not germinate). *Crinum* derives from the Greek *krinon* which means a lily.

ABOVE LEFT *Crassula schmidtii*
ABOVE *Crassula socialis*

Species cultivated

C. asiaticum Tropical Asian shores
Leaves 90–120cm(3–4ft) long, 8–13cm(3–5in) wide, in arching rosettes. Flowers white, occasionally pinkish, 14–18cm(5½–7in) long, in umbels of 20, fragrant, borne on 2-edged stems, to 60cm(2ft) high, blooming spasmodically through the year if kept in warmth (at tropical temperatures).

C.a. procerum
Larger in all its parts, with leaves to 1.5m(5ft) long by 15cm(6in) wide. Flowers 13cm(5in) or more long, tinged with red.

C. augustum Mauritius, Seychelles
Bulb large, up to 15cm(6in) in diameter with a neck to 30cm(1ft) long. Leaves numerous, arching, 60–90cm(2–3ft) long by up to 10cm(4in) broad. Flowering stem erect, somewhat flattened, 60–90cm(2–3ft) tall; flowers 20 to 30, each one 10cm(4in) long, wine-red outside, paler inside, fragrant, opening in summer.

C. bulbispermum (*C. capense*, *C. longifolium*) South Africa
Bulb flask-shaped with a long neck. Leaves 60–90cm(2–3ft) long by 8–10cm(3–4in) wide, deciduous, with rough margins, slender-pointed and arching over. Flowering stem erect, 30cm(1ft) or more; flowers to 15cm(6in) or more long, white

ABOVE *Crinum bulbispermum*
RIGHT *Crinum* × *powellii* 'Album'

to pink, each petal with a rose-pink line down the centre; they are borne in umbels of six to 12 in autumn. Will stand cool conditions.
C. capense See *C. bulbispermum*.
C. insigne See *C. latifolium*.
C. latifolium (*C. insigne*, *C. speciosa*) Tropical Asia
Bulb large, to 15cm(6in) or more in diameter with a short thick neck. Leaves arching, numerous, 60–90cm(2–3ft) long by 8–10cm(3–4in) wide. Flowering stem to 60cm(2ft) tall; flowers ten to 20, each one 15–20cm(6–8in) long with white, red-keeled petal lobes in summer.
C.l. zeylanicum
Larger bulbs; fewer, wavy-edged leaves; petal lobes with deep purple keels. Formerly given specific rank as *C. zeylanicum*.
C. longifolium See *C. bulbispermum*.
C. macowanii South Africa
Bulbs rounded to 20cm(8in) or more in diameter, tapering to a long neck. Leaves 60–90cm(2–3ft) long, bright green, often wavy, in arching rosettes. Flower stem 60–90cm(2–3ft) tall; flowers to 15cm(6in) or more long, trumpet- to bell-shaped, white or pink, each petal with a rose-purple marking, they are borne in umbels often to 15 in late autumn.
C. moorei South Africa
Similar to *C. bulbispermum*, with the 60–90cm(2–3ft) deciduous leaves having smooth margins. Flowering stems 45–60cm($1\frac{1}{2}$–2ft) tall; flowers very fragrant, pale pink or white, blooming from summer onwards.
C. × powellii (*C. bulbispermum* × *C. moorei*) Swamp lily
A hybrid very similar to the latter parent, with the flowers rose-pink; it is vigorous and flowers in late summer ar.d autumn. *C.* × *p.* 'Album' has pure white flowers.
C. procerum See under *C. asiaticum*.
C. speciosa See *C. latifolium*.
C. zeylanicum See under *C. latifolium*.

CROCOSMIA See TRITONIA

CROCUS

Iridaceae

Origin: Europe, Western Mediterranean to Central Asia, Pakistan. A genus of about 75 species of cormous-rooted perennials most of which have narrow, grassy leaves with a central silvery-white line, and 6-tepalled flowers which open nearly flat in the sun. Each flower has three stamens and a long-stalked tube leading to the ovary which is underground just above the corm. Winter- and spring-flowering species bloom with the leaves, the autumn ones before the leaves develop. Crocus corms are small, rounded and sometimes flattened and are covered by a tunic, often of fibres, which is a character much used for identification. Leaves remain short at flowering time, elongating considerably afterwards. They are colourful plants for a pot in the conservatory or cool room and are best brought into the home as they begin to flower. Propagate by seed when ripe or by offsets when dormant. Hardy. *Crocus* is from the Greek *krokos*, saffron, derived from the Semitic *karkom*, a yellow dye obtained from the stigmas of the saffron crocus (*C. sativus* and allied species).

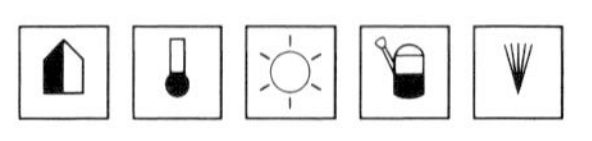

Species cultivated

C. aureus See *C. flavus*.
C. biflorus Italy, Caucasus and Iran
Leaves four to five to each corm, greyish-green. Tepals 3cm($1\frac{1}{4}$in) long, variable in colour, white or blue with purple veining and often with a yellow throat, opening in early spring. *C.b.* 'Parkinsonii' is white to light buff with darker veining. *C.b. weldenii* is white with purple blue freckling or suffusion. *C.b.* 'Fairy' is white, tinted grey on the outside.
C. chrysanthus Yugoslavia to Turkey
Leaves five to seven. Tepals broad, 1.5–3.5cm($\frac{5}{8}$–$1\frac{1}{2}$in) long, orange, feathered outside; winter-flowering. Many cultivars are offered, most of them of hybrid origin with forms of *C. biflorus*. The following can be recommended: 'Cream Beauty',

LEFT *Crocus chrysanthus* 'Snow Bunting'
BELOW *Crocus vernus* cultivar

pale creamy-yellow; 'E.P.Bowles', deep buttery-yellow with prominent feathering; 'Ladykiller', white inside, rich purple outside with white margins; 'Snow Bunting', white and cream with dark lilac feathering; 'Saturnus', orange-yellow with bronze feathering.

C. flavus (*C. aureus*) Yugoslavia to Turkey
Leaves five to six. Tepals 2–4cm(¾–1½in) long, lemon to orange-yellow, flowering late winter to early spring. *C.f.* 'Dutch Yellow' is a deep orange clone.

C. imperati Italy
Leaves five. Tepals 3–4.5cm(1¼–1¾in) long, satiny lilac inside, light buff and usually, but not always, feathered with darker markings on the outside; winter-flowering.

C. sieberi Greece, Crete
Leaves five to seven, green. Tepals 2–3cm(¾–1¼in) long, white to lavender-purple with a large yellow marking on the throat, late winter flowering. *C.s.* 'Hubert Edelsten' has its outer tepals marked and tipped with purple. *C.s.* 'Violet Queen' has wholly purple flowers.

C. speciosus W. Turkey to Caspian Sea
Leaves three to four, after flowers. Tepals 3.5–8cm(1½–3in) long, lilac to purple-blue, veined darker; autumn-flowering. *C.s.* 'Alba' has wholly white flowers. *C.s.* 'Oxonion' is deep purple-blue.

C. vernus Central Italy to Yugoslavia
Leaves three to four. Tepals 2.5–5.5cm(1–2¼in) long, opening in late winter. From this the Dutch crocuses have arisen, chiefly by hybridization with other species. There are many named cultivars of which the following are well tried: 'Joan of Arc', pure white flowers; 'Pickwick', white with contrasting feathering; 'Victor Hugo', rich glossy purple; 'Yellow Mammoth', golden-yellow.

CROSSANDRA

Acanthaceae

Origin: Tropics of the Old World in Africa, Malagasy, Arabia and India. A genus of 50 species of evergreen perennials and shrubs with whorls or opposite pairs of lanceolate leaves. The 4-angled, cone-like flower spikes are terminal or axillary and have overlapping bracts and five broad petal lobes. They can be yellow, orange, orange-red or white. Crossandras are suitable for the conservatory or the home, particularly *C. infundibuliformis*. Propagate by cuttings in spring or summer with bottom heat, or by seed in spring. *Crossandra* derives from the Greek *knossos*, fringe and *aner*, male, referring to the fringed anthers.

Crossandra nilotica

Species cultivated

C. infundibuliformis (*C. undulifolia*) . Firecracker plant India, Sri Lanka
Shrub or sub-shrub to 90cm(3ft) tall. Leaves paired, ovate to lanceolate, pointed, with wavy margins, 7–13cm(2¾–5in) long, deep lustrous green. Flowers salmon-red, spreading, to 3cm(1¼in), borne in spikes to 10cm(4in) long from spring to autumn. A good house plant, flowering freely when small.

C. nilotica East Africa
Shrub to 60cm(3ft) tall. Leaves elliptic, to 10cm(4in) long, often borne in whorls. Flowers pale brick-red to apricot, carried in dense 2.5cm–6cm(1–2½in) spikes from spring to autumn.

C. pungens East Africa
Bushy shrub to 60cm(2ft) tall. Leaves in pairs or whorls, oblong-lanceolate to elliptic, olive-green with a bold creamy-white vein pattern. Flowers pale orange, borne in 10cm(4in) spikes, with keeled and arched bracts, from summer to autumn.

C. undulifolia See *C. infundibuliformis*.

CROTALARIA

Leguminosae

Origin: Tropics and sub-tropics; widespread. A genus of at least 600 species of shrubs, perennials and annuals, the full ornamental potential of which has so far not been exploited. The genus is a varied one with leaves which may be simple, trifoliate or pinnate. The flowers are typical of the pea family with a rounded standard petal and a prominent, curved and pointed keel. Propagate by seed in spring or by cuttings in summer. The species described below add interest and beauty to a collection of conservatory plants. *Crotalaria* derives from the Greek *krotalon*, a rattle or clapper, referring to the somewhat inflated pods when the seeds are ripe.

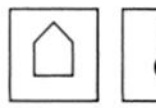 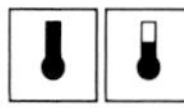 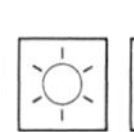 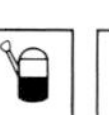 

Species cultivated

C. agatiflora Canary bird bush Kenya and Uganda to Zimbabwe
Shrub of open habit to 3m(10ft), but easily kept to 90–120cm(3–4ft) by regular pruning. Leaves trifoliate, grey-green, each leaflet ovate, 5–8cm(2–3in) long. Flowers in terminal racemes, each one a bright, glowing greenish-yellow, the standard petal to 4cm(1½in) long; intermittently all the year if the bush is trimmed every two to three months.

C. juncea Sunn hemp Probably India, but not now known truly wild
Erect, sparingly branched annual to 1m(3ft) or more in height. Leaves simple, narrowly oblong, 4–10cm(1½–4in) long. Flowers in terminal racemes, each one a rich yellow, the standard about 2.5cm(1in) long, borne in summer. In the tropics this plant is used for green manuring and for its fibres. In India it is second only in importance to jute, having stronger but coarser fibres much used for twine, cord, nets, matting, sacking, etc.

CRYPTANTHUS

Bromeliaceae
Earth stars

Origin: Brazil. A genus of 22 species of evergreen perennials which are rosette-forming and in the wild live largely on the ground, unlike many other bromeliads. They have narrow, evergreen, leathery leaves, usually with toothed edges and are patterned or coloured. The flowers are small, 5-lobed, usually white, and borne in spikes. They are good plants for the conservatory or home, often being used in dish gardens. Propagate by removing suckers or offsets and treating them as for cuttings. *Cryptanthus* derives from the Greek *krypto*, to hide and *anthos*, a flower, the flower buds being concealed by bracts.

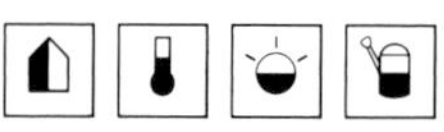

Species cultivated

C. acaulis Green earth star
Flattened rosettes 15–30cm(6–12in) wide. Leaves lanceolate to narrowly triangular, wavy- and spiny-edged, green covered with white scales. *C.a.* 'Ruber' has smaller leaves margined and centred with purplish-bronze and fawn-coloured scales.

C. bivittatus
Spreading rosettes 25–37cm(10–15in) wide. Leaves strap-shaped to 22cm(9in) long, wavy-

BELOW *Crotalaria agatiflora*
BELOW RIGHT *Cryptanthus acaulis*

edged, somewhat fleshy, dull green with two coppery-fawn to buff, longitudinal bands.

C. bromelioides
Spreading rosettes to 35cm(14in) wide, not so flat as in the previous species. Leaves strap-shaped, arching to erect, wavy and finely toothed. *C.b.* 'Tricolor' (Rainbow star) has leaves longitudinally striped and edged ivory-white, and suffused at the base and margins with carmine.

C. fosteranus
Flattened rosettes to 60cm(2ft) or more wide. Leaves thick-textured, stiff, coppery-green to purple-brown, cross-banded with grey to buff lines.

C. zonatus
Flattened rosettes to 50cm(20in) across. Leaves strap-shaped, wavy with finely toothed margins, strongly cross-banded with grey-buff on greenish-brown. *C.z.* 'Zebrinus' has bold, silvery banding.

× CRYPTBERGIA

Bromeliaceae

Origin: Garden hybrids between members of the genera *Cryptanthus* and *Billbergia*, *q.v.* This hybrid genus contains a few ornamental members all of which make durable house plants. In appearance they blend, more or less equally, characteristics of both parents.

Hybrids cultivated

× ***C. meadii*** (*B. nutans* × *C. beuckeri*)
Clump-forming; leaves more or less erect, 20–30cm(8–12in) long, green with pink mottling.

× ***C. rubra*** (*B. nutans* × *C. bahianus*)
Clump-forming; leaves erect, relatively broad, bronze-red. Small white flowers occasionally.

CRYPTOCORYNE

Araceae

Origin: Indo-Malaysia. A genus of 50 species of small, aquatic or wetland evergreen perennials, some of which are desirable aquarium plants and add interest to a warm conservatory pool. They are tufted, rhizomatous plants eventually forming colonies of long-stalked, semi-translucent leaves often flushed or marked with red, purple or yellow. They vary considerably in size and colouring, this depending on light intensity, temperature and water depth. The slender, trumpet-shaped flowering spathes are intriguing, but not big enough to be showy and their appearance is uncertain in cultivation. Propagate by division in spring or summer; they do best when rooted in a soil base. *Cryptocoryne* derives from the Greek *krypto* to hide and *koryne*, club. The club-shaped spadix is hidden within the spathe.

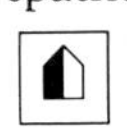    

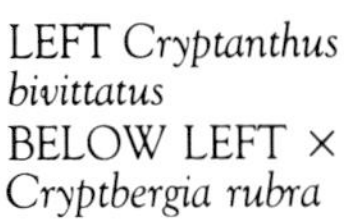

LEFT *Cryptanthus bivittatus*
BELOW LEFT × *Cryptbergia rubra*

Species cultivated

C. beckettii Sri Lanka
Leaf blades narrowly ovate, 8cm(3in) or more long, bright to olive-green above, deep red-purple beneath, usually wavy-margined. A variable and easy species which can be grown submerged or in wet soil only. Grown in the latter way it is the most free-flowering of all cryptocorynes.

C. ciliata India to Malaysia and Indonesia
Leaves usually lanceolate, to 25cm(10in) long on a stalk of almost equal length, pale to mid-green, wavy-margined. Grown as a bog plant it has broader, shorter leaves, flowers freely and produces viable seed. Will stand limy or slightly brackish water. In the wild, it is a variable plant as to leaf shape and size, with several distinct races.

C. cordata Malaysia and Thailand
Leaf blades about 10cm(4in) long, ovate, usually cordate, dark green above, reddish-purple beneath. Grows most vigorously in wet soil only; but once well-established like this it can be gradually submerged and then treated as an aquatic.

C. lingua Borneo
Leaf blades 4–5cm(1½–2in) long, ovate with rounded or bluntly pointed tips, somewhat fleshy, bright mid-green. It can be grown submerged or in wet soil.

C. lucens Sri Lanka
Leaf blades lanceolate to narrowly oblong, to 8cm(3in) long, mid to deep lustrous green. At one time considered to be a form of *C. nevillii*, *q.v.* Easily grown above and below the water surface; vigorous in good soil.

C. nevillii Sri Lanka
Much like *C. lucens* but of lower stature and shorter, somewhat broader leaf blades lacking the bright shining green quality. Nevertheless a pleasing and easy species either submerged or as a bog plant.

C. petchii Sri Lanka
Similar in stature and general appearance to *C. beckettii*, but the leaf blades generally narrower and more wavy, the upper surface olive- to brownish-green with dark green lines; purple flushed beneath.

C. undulata See under *C. willisii*.

C. willisii Sri Lanka
Leaf blades up to 13cm(5in) long, lanceolate, usually strongly wavy-margined, pale to olive-green above, paler beneath and sometimes flushed reddish-purple. A variable and easy, vigorous species which thrives equally well submerged or as a bog plant. Sometimes confused with the similar, or possibly identical, *C. undulata*.

CTENANTHE

Marantaceae

Origin: Tropical South and Central America. A genus of 15 species of evergreen perennials classified by some botanists under *Myrosma*, some of which are grown for their handsome foliage. They have usually narrowly oblong leaves which are undivided and smooth-edged and are similar in general appearance to *Calathea*. The dense flower spikes bear closely overlapping, persistent bracts. Propagate by division in spring or summer. *Ctenanthe* is derived from the Greek *kteis* or *ktenos*, a comb and *anthos*, a flower, referring to the arrangement of the floral bracts.

    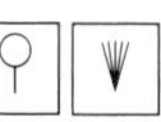

Species cultivated

C. lubbersiana (*Maranta lubbersiana*) Brazil
Tufted with forking stems. Leaves oblong, shortly-pointed, to 20cm(8in) long, deep green mottled and flushed with yellow, paler on the reverse. Flowers white.

C. oppenheimiana Brazil
Shrub to 2m(6½ft) when mature, bushy. Leaves lanceolate, leathery, to 30cm(1ft) or more long, dark green above with silvery-white bands following the lines of the veins on either side of the midribs, red beneath. *C.o.* 'Tricolor' (Never never plant) has the silvery variegation overlain with splashes of creamy-white and a lighter red beneath.

CUNONIA

Cunoniaceae

Origin: New Caledonia, one species in South Africa. A genus of 17 species of evergreen trees and shrubs with pinnate leaves and small white or cream flowers in dense cylindrical spikes. Only one species is generally cultivated and is suitable for conservatory or home when young, making a good foliage pot plant. Propagate by seed in spring or by cuttings with a heel in the summer. Cuttings produce a more shapely plant if taken from erect-growing stems. *Cunonia* was named for John Christian Cuno (1708–1780), a Dutchman who wrote a poem describing his garden.

  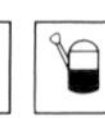 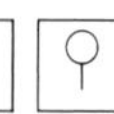

Species cultivated

C. capensis Red alder South Africa
In the wild a tree, but usually 1m(3ft) or so in a pot. Leaves leathery, pinnate with red stalks, each leaf made up of five to nine lanceolate, toothed segments, 5–10cm(2–4in) long. Flowers small, creamy-white, 5-petalled, borne in feathery spikes, to 15cm(6in) long, in the axils of the leaves in late summer.

BELOW L–R *Ctenanthe oppenheimiana* 'Tricolor', *Ctenanthe lubbersiana*, *Cunonia capensis*

CUPHEA

Lythraceae

Origin: Tropics and sub-tropics of the Americas. A genus of 250 species of annuals, perennials and sub-shrubs with opposite pairs of leaves which can be lanceolate to almost rounded. The flowers are usually small but have a tubular, calyx-like receptacle which can be coloured and very showy. They make pretty conservatory and house plants. Propagate by seed in spring, or by cuttings in spring or summer. *Cuphea* derives from the Greek *kyphos* curved, referring to the curved seed capsules.

  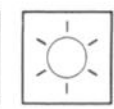  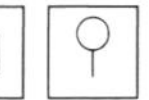

Species cultivated

C. cyanea Mexico
Stickily hairy sub-shrub to 45cm($1\frac{1}{2}$ft) or more tall. Leaves oblong-ovate. Flowers about 2.5cm(1in) long, the tube scarlet and yellow from which protrude tiny violet-blue petals; they are borne in racemes in summer.

C. hyssopifolia False heather Guatemala, Mexico
Freely branched and shrubby, 30–60cm(1–2ft) tall with wiry stems often making a bush wider than tall. Leaves leathery, linear, very heather-like, 6–10mm($\frac{1}{4}$–$\frac{3}{8}$in)long. Flowers to 1cm($\frac{3}{8}$in) long, the tube green and the six petals rose-purple to lilac or white; they are carried in raceme-like clusters in summer and autumn.

C. ignea (*C. platycentra*) Cigar flower Mexico
Wide-spreading, bushy sub-shrub, to 30cm(1ft) or more tall. Leaves lanceolate, 2.5–5cm(1–2in) long. Flowers 2–3cm($\frac{3}{4}$–$1\frac{1}{4}$in) long, tubular, dark red with a dark band and white ring at the end, fancifully like a glowing cigar; they are solitary in the leaf axils and open from spring to summer.

C. miniata (*C. llavea miniata*) Mexico
Shrubby to 60cm(2ft) tall. Leaves pointed-ovate, 2.5–8cm(1–3in) long, hairy beneath. Petals scarlet-red to pale vermilion, opening from a calyx tube to 3cm($1\frac{1}{4}$in) long; they open from summer to autumn. The cultivars sold under this name are mostly of hybrid origin (*C. x purpurea*).

ABOVE LEFT *Cuphea hyssopifolia*
TOP RIGHT *Cuphea ignea*
ABOVE *Cuphea miniata*

CURCULIGO

Hypoxidaceae

Origin: Tropics; widespread. A genus of about ten species of evergreen perennials with thickened, tuber-like rhizomes. The crowded narrow leaves are pleated in some species and resemble those of a young palm. Small, starry, 6-petalled flowers appear near the leaf bases, but the plants are grown mainly for their foliage. Propagate in spring or early summer. *Curculigo* derives from the Latin *curculio*, a weevil, the beak of the ovary being likened to the snout of a weevil.

  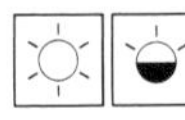 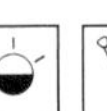 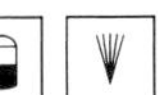

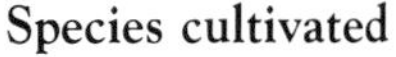

Species cultivated

C. capitulata See *C. recurvata.*

C. orchioides East Indies, Taiwan, China, Japan
Clump-forming, leaves linear to lanceolate, 30cm(1ft) or more in length, widely arching, the outer ones prostrate or hanging over the container edge. Flowers 2cm($\frac{3}{4}$in) wide, yellow, borne in summer. Not a palm-like species, more nearly resembling a *Chlorophytum*.

C. recurvata (*C. capitulata*) Palm grass Tropical Asia, N.E. Australia
Clump-forming, leaves 60–90cm(2–3ft) long, broadly lanceolate to narrowly elliptic, pleated, firm-textured and a glossy rich green. Flowers yellow, borne in dense clusters at soil level in summer. The most handsome species, much like a young palm in the unifoliate stage.

Curculigo recurvata

Curcuma roscoeana

CURCUMA

Zingiberaceae

Origin: Tropical Asia and northern Australia. A genus of about 65 species of deciduous, rhizomatous perennials, a few of which make handsome pot plants. They have simple, more or less erect leaves on shoots direct from the rhizomes and dense flower spikes. The latter are composed of conspicuous bracts, sometimes coloured differently from the flowers which arise in their axils. The curiously-formed tubular flowers, with one hooded petal lobe, have an orchid-like appearance. The plants must be dried off in early autumn and kept warm and dry until the following spring when they are re-potted and watered again. Propagate by division in spring. *Curcuma* is the Latinized version of an Arabic name for the yellow dye plant turmeric (*C. domestica*).

  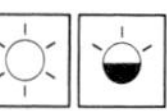 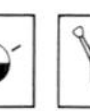 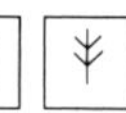

Species cultivated

C. domestica (*C. longa*) Turmeric Not known truly wild; perhaps India, but widely cultivated Rhizomes aromatic with yellow flesh, short, thick and tuber-like. Shoots erect, to 90cm(3ft) or more. Leaf blades 30–45cm(1–1½ft) long, oblong to elliptic, prominently veined, usually a rich green. Flower spikes cylindrical, about 13cm(5in) long; lower bracts green with pale yellow flowers 5cm(2in) long; upper bracts flowerless, white or pink-tinted. The plant is sterile and fruits are not produced. Turmeric is an important crop in India and S.E. Asia, the ground and dried rhizomes being used to flavour and colour curry powder.

C. longa See *C. domestica.*

C. roscoeana Malaysia
Rhizomes white-fleshed. Shoots 45–75cm(1½–2½ft) tall, erect. Leaf blades 15–30cm(6–12in) long, narrowly ovate to broadly lanceolate, smooth, medium green with darker prominent veins. Flower spikes 15cm(6in) or more long, cylindrical, composed of overlapping bracts with recurved tips, green, then bright reddish-orange. Flowers bright yellow, only their tips protruding from the bracts, borne in summer.

CURTONUS

Iridaceae

Origin: South Africa. A genus of only one species – a cormous-rooted perennial rather similar in appearance to Montbretia. It is best in the conservatory, but can be brought into the home for flowering. Propagate by division or by separating offsets when dormant. *Curtonus* derives from the Greek *kyrtos*, bent, referring to the zig-zag bending of the flower spikes.

Species cultivated

C. paniculatus
Clump-forming. Leaves to 90cm(3ft) long, sword-shaped, pleated, carried in fan-like formation. Flowers trumpet-shaped with six lobes, the upper prolonged and hood-like; they are borne in zig-zag spikes making up stiff panicles at the top of the stems to 120cm(4ft) high, in late summer.

CUSSONIA

Araliaceae

Origin: Tropical and South Africa, Malagasy and Mascarene Is. A genus of 25 species of evergreen trees and shrubs. The species described below are grown primarily for their handsome, digitate, juvenile foliage, but when old enough they add a bonus of flowers, which are small but carried in dense spikes or umbels. When older, the leaves become more simplified in design. Propagate by seed or air layering in spring. *Cussonia* commemorates Dr Pierre Cusson (1727–85), a French doctor and Professor of Botany at Montpelier.

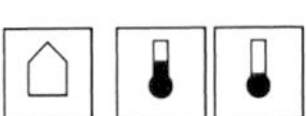  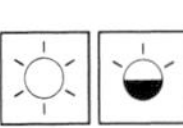 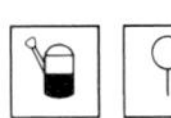

Species cultivated

C. paniculata South Africa
Height in containers to 2m(6½ft) or more; to 5m(16ft) in the wild. Leaves composed of seven to 12 radiating leaflets, the latter 15–30cm(6–12in) long, pinnately lobed, each lobe spine-tipped and sometimes toothed. Flowers white or yellow, in bottlebrush-shaped spikes, up to 8cm(3in) long; these in turn are arranged in panicles up to 30cm(1ft) long.

C. spicata Zambia to South Africa
Height in containers to 2m(6½ft) or more; to 6m(20ft) in the wild. Leaves greyish-green, composed of six to eight obovate, radiating leaflets up to

13cm(5in) long; each leaflet is pinnate, the upper one to three pinnules being completely separate, the lower ones fused to create the shape of the tail of a fish. The whole leaf design is reminiscent of that of a magnified snow crystal. Flowers in small spikes gathered into panicles up to 25cm(10in) long.

CYANOTIS

Commelinaceae

Origin: Tropics and sub-tropics of both hemispheres. A genus of 50 species of perennials, most of which are evergreen. They are weak-stemmed, trailing or tufted plants somewhat similar to *Tradescantia* with small, linear to ovate alternate leaves, the bases of which shield the stems. The 3-petalled flowers are usually purple-blue. Propagate by cuttings in summer. *Cyanotis* derives from *kyanos*, blue and *anthos*, flower.

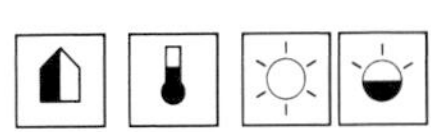

Species cultivated

C. kewensis Teddy bear vine Southern India
Prostrate with trailing, branching stems to 30cm(1ft) long. Leaves ovate, 2–4cm($\frac{3}{4}$–$1\frac{1}{2}$in) long, overlapping, olive-green, with purple beneath, clothed (together with the stems) in rusty-brown hairs. Flowers 1cm($\frac{3}{8}$in) wide, rose-purple, from winter to spring.

C. sillamontana See *Tradescantia sillamontana.*

C. somaliensis Pussy ears East Africa
Stiffly trailing stems to 20cm(8in) or more long. Leaves linear-lanceolate to ovate-lanceolate, to 4cm($1\frac{1}{2}$in) long, shining green, clothed with long white hairs. Flowers bright violet-blue, from winter to spring.

C. veldthoutiana See *Tradescantia sillamontana.*

ABOVE *Cyanotis somaliensis*
LEFT *Cyathea medullaris*

CYATHEA

Cyatheaceae
Tree ferns

Origin: World-wide tropics, mostly in the mountains; also some temperate areas of the southern hemisphere. A genus of 600 species of tree ferns developing erect, palm-like stems closely woven with aerial roots, giving the appearance of brown bark, and terminating in a rosette of large bi- or tri-pinnatifid fronds. When small they make decorative pot plants, needing moisture and humidity. Propagate by spores in spring. *Cyathea* derives from the Greek *kyatheion*, a little cup, alluding to the shape of the sori which cover the spores.

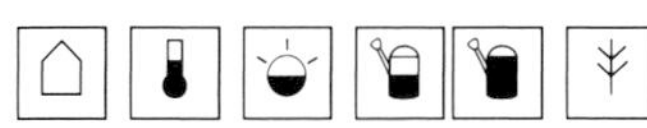

Species cultivated

C. arborea West Indian tree fern Puerto Rico, Cuba, Jamaica
Fronds tripinnate, to 2m($6\frac{1}{2}$ft) long when mature, light, almost yellow-green on both sides.

C. dealbata (*Alsophila tricolor*) Silver tree fern, Ponga New Zealand, Lord Howe Island
Stem 2–3m($6\frac{1}{2}$–10ft) in containers, to 2m(30ft) in the wild. Leaves 1–2m(3–$6\frac{1}{2}$ft) long or more, tripinnate, the ultimate pinnules 1.5mm($\frac{5}{8}$in), dark green above with a contrasting silvery white patina beneath.

C. medullaris Sago fern, Black tree fern New Zealand, Tasmania, Victoria
Leaves in the wild can reach 6m(20ft); they are bi- to tri-pinnate, rather leathery and dark green above, paler beneath. Slow-growing.

C. smithii (*Alsophila smithii, Hemitelia smithii*) Soft tree fern New Zealand
Stem 2m($6\frac{1}{2}$ft) or more high in containers, to 8m(25ft) in the wild. Leaves 1–1.5m(3–5ft) long or more, bi- to tri-pinnate, the secondary pinnules 3–6cm($1\frac{1}{4}$–$2\frac{1}{2}$in) long, bright green above, paler beneath.

Cycas revoluta

CYCAS

Cycadaceae

Malagasy, E. and S.E. Asia, Indo-Malaysia, Australasia and Polynesia. A genus of 20 species of primitive seed-bearing plants of palm-like appearance. They have stout woody trunks, sometimes to several metres tall, eventually with a few branches or suckering from the base. The leathery or hard-textured leaves are pinnate and carried in a terminal rosette. Female 'flowers' have the appearance of much reduced woolly leaves bearing scattered globules, but those of the males are cone-like. The species described below make statuesque container specimens and are generally more tolerant than palms. Propagate by seed in spring; well developed suckers can be removed at the same time and treated as cuttings. *Cycas* is an ancient Greek name for a palm, taken up by the botanist Linnaeus for this genus.

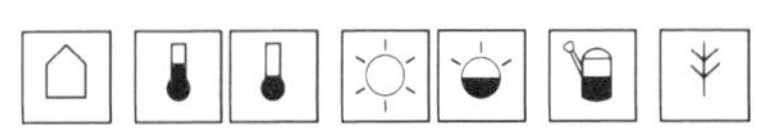

Species cultivated

C. circinalis Fern palm, Sago fern palm
Old World tropics
Stems to 2m(6½ft) in containers but very slow-growing; in warm climate gardens they can attain 6m(20ft) in height. Leaves to 1.5m(5ft) or more long, the crowded leaflets to 30cm(1ft) long, linear-lanceolate, usually glossy rich green, but sometimes glaucous-tinted.

C. revoluta Japanese sago palm Japan, Ryukyu Is.
Stem to 1.5m(5ft) or more tall in containers; 2–3m(6½–10ft) in the wild. Leaves to 75cm(2½ft) long, the crowded leaflets about 15cm(6in) or more long with revolute margins and a spiny tip, usually a lustrous rich green. Seeds to 4cm(1½in) long, ovoid, flattened, bright red, but only produced if plants of both sexes are grown together and hand pollination is carried out.

CYCLAMEN

Primulaceae

Origin: Europe, Mediterranean region to Iran. A genus of about 20 species of perennials with rounded, somewhat woody corms. From these arise long-stalked, rounded to ovate-cordate leaves, sometimes lobed and spotted or marked with silver. The flowers are solitary and pendent with five petals which reflex to give a characteristic shuttlecock appearance to the flower. In all species except C. *persicum*, after the flower is fertilized its stem coils down, pulling the rounded seed capsules to ground level. They are decorative plants for a cool conservatory, the one described below also making a fine house plant for a cool room, but it is very short-lived in warmth. Propagate by seed sown in late summer or late winter. Dry off the corm in late spring and keep dry until late summer. Repot into fresh compost and start watering again when growth begins. *Cyclamen* derives from the Greek *kyklos*, circular, from the way the stems spiral down.

 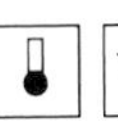 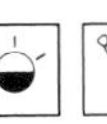 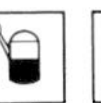 

Species cultivated

C. persicum Southern and eastern regions of the Mediterranean from Algeria to Lebanon; also some of the islands of Greece
Leaves rounded to broadly ovate, toothed, dark green patterned with silver. Flowers rose to pale pink and white with elegantly twisted, narrow petals with a darker basal blotch, fragrant, opening in winter and spring. The florists' cyclamen which are normally sold under this name are the result of generations of careful selection and breeding. They are much larger plants with fleshier, more or less rounded to kidney-shaped leaves and larger flowers which can be 5–7cm(2–2¾in) long. They are available in shades of pink, red, scarlet, mauve, purple and white, some with crested petals, but have very little, if any fragrance.

LEFT & ABOVE *Cyclamen persicum* cultivars

CYCNOCHES

Orchidaceae

Origin: Tropical America. A genus of 12 species of deciduous orchids with fleshy, cylindrical, stem-like pseudobulbs and three to four ribbed leaves. The flowers are unisexual, an unusual condition in orchids, sometimes mixed in the same racemes, sometimes separate. There is also a strong difference in the flowers of each sex. They are large and waxy and because of a twist in their stalks are borne upside down. Their sepals and petals are similar, being oblong-elliptic in shape, while the labellum is shield-shaped and the column slender and down-curving. These orchids are suitable for the conservatory, but can be brought into the home for flowering. Propagate by division. *Cycnoches* derives from *kyknos*, swan, alluding to the curved column like the neck of a swan.

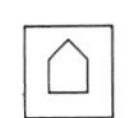    

Species cultivated

C. egertonianum (*C. ventricosum egertonianum*) Mexico to Brazil
Pseudobulbs slender, 8–15cm(3–6in) long. Male flowers to 4cm(1½in) or more wide, tepals green sometimes with a bronze to purple suffusion, or maroon to brown spotting; labellum white to green, flushed with purple, the column long and slender; they are borne on racemes to 45cm(1½ft) long in autumn and winter. Female flowers are similar, but larger and fleshier.

C. ventricosum Swan orchid Mexico, Guatemala, Honduras
Pseudobulbs slender, both sexes similar. Flowers 10–12cm(4–4¾in) wide, tepals light green to yellow and a white lip with a long claw; the column curves forward; flowers are borne in racemes in summer and autumn and are fragrant.

CYMBALARIA

Scrophulariaceae

Origin: Temperate regions of the Old World. A genus of 15 species of small perennials which have been classified in *Linaria*. They are trailers or creepers with rounded to kidney-shaped, sometimes lobed leaves which are palmately veined. Their flowers are small, tubular with a rear nectary spur and a 5-lobed mouth forming two lips, the lower having three lobes, the upper two. The species described make pretty pot or hanging basket plants. Propagate by division, cuttings or seeds, all in spring. Hardy. *Cymbalaria* derives from the Greek *kymbalon*, a cymbal, alluding to the almost circular leaves of some species.

    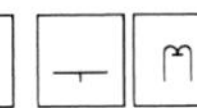

Species cultivated

C. muralis (*Linaria cymbalaria*) Ivy-leaved toadflax, Kenilworth ivy Southern Europe
Low, evergreen perennial with prostrate or trailing

Cymbalaria muralis 'Alba'

Cymbalaria pallida

stems 30–60cm(1–2ft) long. Leaves 1.5cm($\frac{5}{8}$in) or more long, long-stalked, kidney-shaped with three to seven lobes, lustrous green above, purplish beneath. Flowers lilac-blue, the spur 3mm($\frac{1}{8}$in) long and the throat yellow; they are borne singly in the leaf axils from spring to late autumn. *C.m.* 'Alba' has white flowers. *C.m.* 'Globosa Rosea' forms small, congested hummocks to 15cm(6in) wide and is suitable for a pot or pan.

C. pallida Italy
Similar in general appearance to *C. muralis*, the leaves paler green, hairy beneath and on the leaf stalks. Flowers 1.5–2.5cm($\frac{5}{8}$–1in) long, pale lilac-blue with a yellow throat and a 6–9mm($\frac{1}{4}$–$\frac{1}{3}$in) spur; they open in summer and autumn.

CYMBIDIUM

Orchidaceae

Origin: Tropical Asia and Australia, often in montane forestlands. A genus of 40 species of epiphytic and terrestrial orchids which are clump-forming with usually oval to conical pseudobulbs. The bases of the linear, arching leaves sheathe the pseudobulbs. Flowers are rarely solitary, more usually borne in arching to pendent racemes. They have spreading, elliptic tepals and a prominent 3-lobed lip in a contrasting colour. Cymbidiums are suitable for the warm conservatory or home and several are tolerant of short spells at lower temperatures. For home cultivation, however, the species have been largely superceded by their hybrids which are numbered in thousands, new cultivars being produced every year. Among them is a strain of dwarf plants with erect flower spikes suitable for growing on a windowsill and these are now the main target of orchid breeders. Only the species are described here, as annually produced orchid catalogues can be referred to for the latest cultivars. Propagate by division. *Cymbidium* derives from the Greek *kymbe*, a boat, referring to the hollow, fancifully boat-like shape in the labellum.

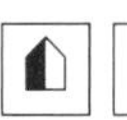   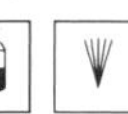

BELOW *Cymbidium* 'Loch Lomond'
BOTTOM LEFT *Cymbidium* 'Annan'
BOTTOM RIGHT *Cymbidium lowianum*

Species cultivated

C. aloifolium India, Sri Lanka
Pseudobulbs very small, hidden by the bases of the leaves; these are arching to pendulous, fleshy and strap-shaped, 30–45cm(1–1$\frac{1}{2}$ft) long. Flowers in vertically pendulous racemes, about 30–60cm(1–2ft) long, the individual blooms well spaced, to 4.5cm(1$\frac{3}{4}$in) across; tepals fawn to yellow, the labellum red-brown with a central yellow marking and a white margin.

C. dayanum Sikkim to Malaysia and Philippines
Almost no pseudobulbs. Leaves very slender, arching to 75cm(2$\frac{1}{2}$ft) long. Flowers in pendulous racemes, about 30cm(1ft) long, the blooms well spaced, to 7cm(2$\frac{3}{4}$in) wide; tepals white, marked with a central maroon to purplish streak, the labellum red-purple with white and yellow markings; they are fragrant and open in summer.

C. devonianum Himalaya, Khasia Hills
Pseudobulbs clump-forming, usually hidden by the expanded bases of three to five leaves. Each leaf 20–35cm(8–14in) long, leathery, broadly oblanceolate and tapering to a long petiole. Racemes pendulous, to 30cm(1ft) or more, bearing many flowers; each bloom about 4cm(1$\frac{1}{2}$in) wide, waxy-textured, the tepals olive-green to buff-yellow streaked or spotted brownish-purple, the labellum dark to light rose-purple; they open in spring and summer.

C. eburneum Himalaya from Sikkim to Burma
A terrestrial species with small pseudobulbs hidden by the leaf bases. Leaves narrowly sword-shaped to linear, bright green, to 60cm(2ft) long and 1.5cm($\frac{1}{2}$in) wide. Racemes erect, to 30cm(1ft) long carrying one to three flowers; each bloom is 7–10cm(2$\frac{3}{4}$–4in) wide, white to ivory, the labellum marked with yellow; they are fragrant and open in late winter and spring.

C.e. parishii (*C. parishii*)
Shorter, broader leaves, to about 45cm(1$\frac{1}{2}$ft) long by 2.5cm(1in) wide, which are twice as wide as those of the type species, and stems with three to seven flowers.

C. ensifolium China and Japan to Sumatra and Java
Leaves arching, slender to 30cm(1ft) long. Racemes erect, to 30cm(1ft) long, with seven or more flowers each 5cm(2in) wide; tepals light green, marked with red-brown lines, the labellum pale green, spotted and lined with red; they are fragrant and open in summer.

C. erythrostylum Vietnam
Pseudobulbs clump-forming, ovoid to oblongoid, about 5cm(2in) long. Leaves several, arching, linear, to about 38cm(15in) long, thin-textured. Racemes erect to arching, to 45cm(1$\frac{1}{2}$ft) or more long, bearing four to seven flowers; each bloom about 10cm(4in) wide with white tepals and a cream lip, heavily lined and/or spotted crimson-purple, opening in autumn.

C. finlaysonianum Malaysia to Indonesia and Philippine Is.
Pseudobulbs short, stem-like. Leaves rigidly leathery, sword-shaped, 60–90cm(2–3ft) long by

5cm(2in) wide. Racemes pendent, 60–120cm(2–4ft) long, bearing well-spaced flowers each 4–6cm($1\frac{1}{2}$–$2\frac{1}{2}$in) wide; tepals yellowish to olive-green, the labellum white, the centre lobe tipped with red-purple, the side lobes pink streaked with red, opening in summer.

C. giganteum Himalaya, Khasia Hills
Pseudobulbs ovoid, 10–15cm(4–6in) long, somewhat compressed. Leaves narrowly sword-shaped to linear, up to 90cm(3ft) long, keeled. Racemes arching, to 90cm(3ft) long, bearing seven to 15 well-spaced flowers; tepals yellow-green cross-lined with red, the labellum spotted or blotched with red; they open in autumn.

C. lowianum Burma, Khasia Hills
Very similar to C. *giganteum*, but the flowers are more greenish with the tepals cross-lined only faintly and the labellum buff to yellow with a crimson central lobe.

C. mastersii Himalaya
Pseudobulbs not produced. Leaves arching, narrowly sword-shaped to linear, 75cm($2\frac{1}{2}$ft) long, 2cm($\frac{3}{4}$in) wide. Racemes, erect, to 30cm(1ft) long, bearing four to ten flowers; tepals ivory-white with faint red spots and an orange marking on the lip; remain rather bell-shaped.

C. parishii See C. *eburneum parishii*.

C. tigrinum Burma
Pseudobulbs ovoid, to 4cm($1\frac{1}{2}$in) long. Leaves short, thick-textured, oblong-lanceolate, to 15cm(6in) long. Racemes held almost at right angles to the plant, sometimes arching, bearing two to five flowers, 5–9cm(2–$3\frac{1}{2}$in) across; tepals olive-green, lined and spotted with red, labellum with a white central lobe, outer lobes yellow, all marked with red-purple stripes; they open in summer.

C. tracyanum Burma
Pseudobulbs ovoid, 10–13cm(4–5in) long, somewhat compressed. Leaves narrowly sword-shaped. Racemes arching, to 60cm(2ft) or more, bearing up to 20 flowers; tepals yellow-green, spotted and streaked with dark red, sometimes with a red suffusion; labellum cream, spotted and striped with red; they are fragrant and open in late autumn.

CYPELLA

Iridaceae

Origin: Mexico to Argentina. A genus of 15 species of bulbous-rooted plants with tufts of narrowly lanceolate to linear basal leaves and somewhat iris-like blooms borne singly or in clusters of up to six in the axils of bracts at the top of the slender, wiry stems. Each bloom lasts for one day only. Plants are suitable for the conservatory and can be brought into a cool room for flowering. Propagate by separating offsets, or by seeds sown in warmth as soon as ripe if possible. *Cypella* derives from the Greek *kypellon*, a goblet, referring to the bowl-shaped base of the blooms.

  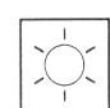  

Cymbidium 'Western Rose'

Species cultivated

C. drummondii See *Herbertia drummondii*.

C. herbertii Argentina, Uruguay, Brazil
Leaves linear-lanceolate, 30–50cm(12–20in) tall; stems about the same height, branched at the top. Flowers 4–7cm($1\frac{1}{2}$–$2\frac{3}{4}$in) wide, chrome-yellow to copper, the outer tepals with a purple central line near the base, inner tepals thickly purple-spotted; they open in late summer.

C. plumbea Blue tiger lily Brazil, Uruguay
Leaves sword-shaped, to 60cm(2ft) or more long; stems to 75cm($2\frac{1}{2}$ft), not branched. Flowers 6–8cm($2\frac{1}{2}$–3in) wide, mauve to dull blue with a yellow throat, sometimes flushed and spotted with brown and opening in late summer and autumn.

CYPERUS

Cyperaceae

Origin: Tropical, sub-tropical and warm temperate regions throughout the world. A genus of 550 species of evergreen perennials with a few annuals. Those cultivated are all perennials, a few grow from tubers, but mostly rhizomatous or stoloniferous. They have tapering grass-like leaves and tiny, petalless flowers each made up of a single ovary and one to three stamens enclosed in bractlets and arranged in flattened, grass-like spikelets. These make up umbels or heads of flowers. All may be grown in pots in a conservatory and the smaller species are good in the home, especially in cool rooms. Propagate by division or seed in spring. *Cyperus* is the Greek word for a sedge.

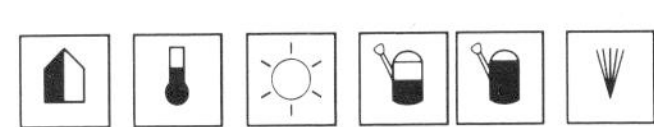

Species cultivated

C. albostriatus See C. *diffusus*.

C. alternifolius Umbrella grass/plant
Malagasy, Mauritius
Clump-forming, the stems 60–120cm(2–4ft) tall, stiff and channelled. Leaves borne around the top

of the stem like the spokes of an umbrella, slender, 20–40cm(4–8in) long. Flowers borne in stalked spikelets at the centre of the leaf rosette, green to brown. In the wild, a bog plant, but in a pot it is best if stood in a saucer or tray of water. *C.a.* 'Gracilis' is much smaller, only 30–45cm(12–18in) tall. *C.a.* 'Variegatus' has leaves striped longitudinally with white, sometimes all white.
C. diffusus of gardens (*C. albostriatus*)
South Africa
Suckering stems 30–60cm(1–2ft) tall with basal, narrowly strap-shaped leaves and an umbrella-like rosette of five to ten similar leaves. Flowers in small branched spikelets at the centre of the leaf rosette. *C.d.* 'Variegatus' has leaves streaked with white.
C. esculentus Chufa/Tiger nut Tropical Asia, Africa and America
Tufted and stoloniferous, 10–60cm(4–24in) high; narrow, arching basal leaves to 30cm(1ft) and small yellowish clusters of flowers. Not particularly decorative but grown as a fun plant from its tubers which can be purchased as tiger or chufa nuts.
C. papyrus Egyptian paper reed, Papyrus
Tropical Africa
Aquatic and clump-forming, the stems triangular, dark green, 1.5–4m(5–13ft) tall. No basal leaves. Inflorescence rounded, made up of many tiny spikelets on graceful, arching, pendulous hair-like stalks, 20–45cm(8–18in) long. Can be grown in a pot standing in a container of water.

BELOW *Cyperus alternifolius*
BOTTOM *Cyphomandra betacea*

CYPHOMANDRA

Solanaceae

Origin: South America, West Indies. A genus of 30 species of shrubs and trees. They have large, alternate leaves and racemes or clusters of bell-shaped flowers. One species is cultivated and makes an ornamental pot plant for the conservatory producing edible fruits. Propagate by seed sown in spring and pot on regularly. In this way it can be treated as a biennial and replaced before getting too large. *Cyphomandra* derives from the Greek *kyphos*, a tumour or swelling and *aner*, male; the tissue between the two anther lobes is thickened or swollen.

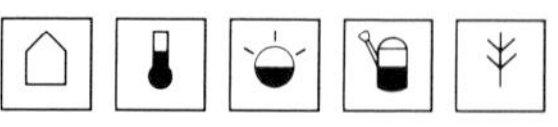

Species cultivated
C. betacea Tree tomato Peru
An erect, tree-like evergreen shrub with soft stems 2.5–6m(8–20ft) tall, usually rather short-lived. Leaves to 30cm(1ft) long, alternate, ovate-cordate, softly hairy, with prominent veins, sometimes purplish when young. Flowers 2.5–3cm(1–1¼in) wide, pale pink or white, opening from a purplish-flushed bud, with five slender-pointed, somewhat reflexed petals and a central cone of yellow stamens; opening chiefly in spring and summer, but also spasmodically throughout the year if sufficiently warm. Fruits are about 5cm(2in) long, egg-shaped, yellowish to brick-red with a distinct sweetish, yet slightly acid taste.

CYPHOSTEMMA

Vitidaceae

Origin: Southern Africa. A genus of at least four species of succulent shrubs formerly included in *Cissus*. They have a swollen caudex from which arise many branches, the soft stems having a yellowish papery bark. The leaves are deciduous and fleshy. They are intriguing plants for the sunny conservatory or room. Propagate by seed in spring. *Cyphostemma* derives from the Greek *kyphos*, a tumour or swelling and *stemma*, a crown; the main stem or crown of the plant is greatly distended with water storage tissue.

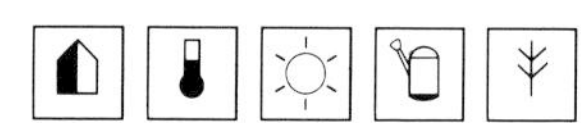

Species cultivated
C. bainesii (*Vitis bainesii, Cissus bainesii*)
West Africa
Stem bottle-shaped, to 60cm(2ft) tall, more in the wild. Leaves shortly stalked, usually trifoliate, the leaflets to 11cm(4½in) long, occasionally single, toothed and woolly, falling in late summer. Flowers small and green in terminal corymbs, followed by round, red fruits.
C. juttae (*Cissus juttae*) Namibia
Stem barrel-shaped, reaching a great size in the wild, forming trunks to 6m(20ft) tall, far less than this in cultivation. The yellowish-grey papery bark peels off in strips as the plant ages. Leaves pointed-ovate, to 15cm(6in) long, waxy glaucous-green. The insignificant flowers are followed by yellow or red fruits.
C. macropus Namibia
Similar to *C. bainesii*, but with smaller leaves of one leaflet only. The plants are more freely branched and when mature have a shrubby appearance.

CYRTANTHUS

Amaryllidaceae

Origin: Tropical and Southern Africa. A genus of 47 species of bulbous plants with narrowly strap-shaped to linear leaves and umbels of tubular, down-curving flowers on leafless stalks. They are attractive pot plants for the conservatory and can be brought into the home for flowering. Propagate by removing offsets or by seed in spring. *Cyrtanthus* derives from the Greek *kyrtos*, arched and *anthos*, a flower, alluding to the curved flowers.

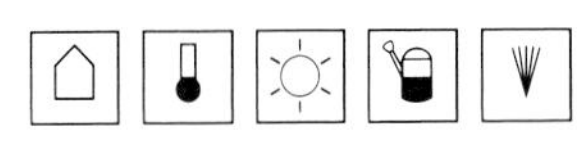

Species cultivated

C. mackenii Ifafa lily South Africa
Each bulb produces up to six linear leaves 20–30cm(8–12in) long. Flowers 5cm(2in) long, reflexed at the mouth. Normally white to cream but occasionally pinkish, fragrant, borne in umbels of four to ten on 15–30cm(6–12in) stems in spring.

C. o'brienii South Africa
Very similar to *C. mackenii*, but flowers bright red on stems to 20cm(8in) and scentless.

C. parviflorus South Africa
Each bulb produces three to six linear leaves which, when mature, attain 20–30cm(8–12in) in length. Flowers red, about 2.5cm(1in) long, with six short, straight, rounded lobes, carried in umbels of five to ten on erect stems standing 15–20cm(6–10in) in height.

CYRTOMIUM

Aspidiaceae

Origin: S.E. Asia and Polynesia. A genus of about ten species included by some botanists in the genus *Phanerophlebia*. They have densely scaly rhizomes and once-pinnate fronds with broad, asymmetrical leaflets which may or may not be toothed. Brown spore capsules with centrally fixed indusia are borne abundantly on the undersides of the fronds. They are very tolerant pot plants for the home or conservatory, succeeding well in a cold porch or a heated room. Propagate by division of multi-crowned plants or by spores. *Cyrtomium* derives from the Greek *kyrtos*, arched, with reference to the arching fronds.

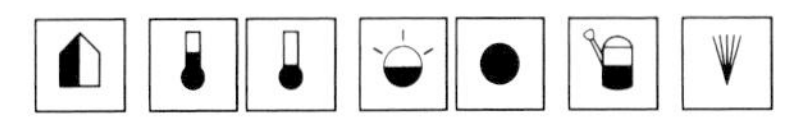

Species cultivated

C. falcatum (*Aspidium falcatum*, *Polystichum falcatum*) Holly fern Asia from India to Korea and Japan, Hawaii and South Africa
Tufted growth with 35–65cm(14–26in) long arching fronds; these are pinnate, with up to 23 oblong-ovate, wavy-edged pinnae ending in a tail-like point, shining deep green above, the undersides softly brown-hairy with abundant sori on fertile fronds.

Cyrtanthus sp.

C.f. rochfordianum
Stronger growing, with broader fronds having the margins distinctly toothed as well as wavy.

C. fortunei (*Aspidium falcatum fortunei*) Japan, Korea and China
Basically like *C. falcatum*, but somewhat smaller in all its parts, with each frond composed of up to twice as many, less taper-pointed, and less glossy leaflets. More frost hardy than *C. falcatum* and suitable for unheated rooms.

Cyrtomium fortunei

CYRTOSTACHYS

Palmae

Origin: Malaysia, especially New Guinea, Solomon Is. A genus of ten species of slender, clump-forming feather palms. One only is sometimes grown in greenhouses, conservatories and homes but other species are worth seeking out.

Propagate by seed sown at not less than 27°C(80°F). *Cyrtostachys* derives from the Greek *kyrtos*, curved and *stachys*, a spike, alluding to the flowering spikes.

  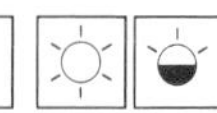  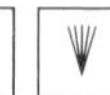

Species cultivated

C. lakka Sealing wax palm Borneo
Stems slender, erect, like those of a robust bamboo, eventually 2–3m($6\frac{1}{2}$–10ft) in large containers, but slow growing. Leaves 60–120cm(2–4ft) long, pinnate, erect to slightly arching, composed of 20 to 50 slender leaflets. The most distinctive feature is the sealing-wax-red leaf sheafs, stalks and midribs which seem to glow even in poor light. The allied C. *renda* from Sumatra is similar, but the red stalks are less intensely so, sometimes almost brownish.

RIGHT *Cyrtostachys lakka*
FAR RIGHT *Cytisus* × *spachianus*

CYTISUS

Leguminosae

Origin: Europe, Mediterranean region and North Atlantic Islands. A genus of 30 species of deciduous and evergreen shrubs with alternate leaves, simple or trifoliate which may be short-lived leaving green stems to carry out their functions. The flowers are typical of the pea family and the two species described below are good flowering shrubs for the conservatory or windowsill and flower freely when small. Propagate by seed in spring or by cuttings with a heel in late summer, preferably with bottom heat. *Cytisus* derives from the Greek *kytisos*, a name for several woody plants of the pea family.

 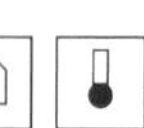 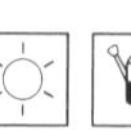  

Species cultivated

C. canariensis (*Genista canariensis*, *Teline canariensis*) Florists' genista Canary Islands
Evergreen shrub to 3m(10ft). Leaves trifoliate, very short-stalked or stalkless, the leaflets obovate to elliptic, 6–12mm($\frac{1}{4}$–$\frac{1}{2}$in) long, flowers yellow fragrant, in short, lax racemes in spring and summer.

C. fragrans (C. *racemosus*) See C. × *spachianus*.

C.* × *spachianus (C. *fragrans*, C. *racemosus*, *Teline* × *spachianus*)
A natural hybrid between C. *stenopetalus* and C. *canariensis* and often confused in gardens with the latter species. Leaves trifoliate on stalks 6mm($\frac{1}{4}$in) long; leaflets obovate, 1–2cm($\frac{3}{8}$–$\frac{3}{4}$in) long, green above, silky-hairy beneath. Flowers rich yellow, to 12mm($\frac{1}{2}$in) long, very fragrant, borne in slender racemes 5–10cm(2–4in) long in winter and spring.

D

LEFT *Darlingtonia californica*
BELOW LEFT *Dasylirion acrotriche*

DAEDALACANTHUS See ERANTHEMUM

DARLINGTONIA

Sarraceniaceae
Pitcher plants

Origin: USA (California and Oregon). A genus of carnivorous plants containing one species which makes an intriguing pot or pan plant in a cool conservatory. It must be grown in equal parts of peat and sphagnum moss and kept moist by standing the pot in a shallow saucer of water. Contrary to popular belief, they do not need insects fed to them. Propagate by division or by seed in spring, keeping the seeds very moist. *Darlingtonia* was named for Dr William Darlington (1782–1863), an American botanist.

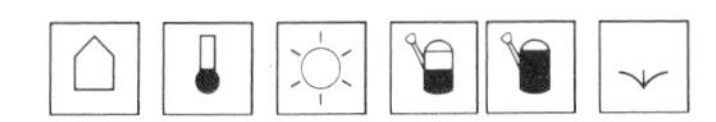

Species cultivated

D. californica Cobra plant, California pitcher plant
Rosette-forming herbaceous perennial. Leaves 10–45cm(4–18in) tall, erect, yellow-green with conspicuous net-veining; they are folded and fused to form a slenderly trumpet-shaped container, the top forming a broad hood with a 2-lobed green and brown appendage, the whole fancifully like the head of a cobra. The pitcher contains a digestive fluid into which unwary insects can fall. Flowers solitary, nodding and bell-shaped with five greenish sepals and five purple petals.

DASYLIRION

Agavaceae

Origin: Mexico and South-western USA. A genus of about 15 species of xerophytic, usually woody-based evergreen perennials or small palm-like trees or shrubs with a similar appearance to a narrow-leaved yucca or agave. The linear leaves are arranged in very dense rosettes often of almost spherical outline. Small, whitish, 6-tepalled flowers are carried in racemes or narrow panicles, but only rarely on containerized specimens. Propagate by suckers removed in spring, or by seed sown at the same time. The species described below make fine architectural foliage plants. Although not fully succulent they are plants of arid regions and thrive in dry atmospheres. *Dasylirion* is derived from the Greek *dasys*, thick and *lirion*, a lily; the genus was formerly classified in the *Liliaceae* and the plants are dense or thick.

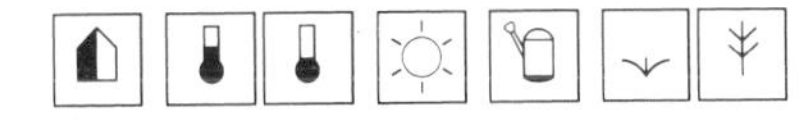

Species cultivated

D. acrotriche (*D. acrotrichum*) Mexico
As a containerized specimen this species behaves as an evergreen perennial, but when old, a woody trunk 60–90cm(2–3ft) high develops. Leaves to about 90cm(3ft) long by 1cm($\frac{3}{8}$in) wide, margins thickly set with minute teeth and larger spines terminating in brush-like fibres.
D. bigelowii See *Nolina bigelowii*.
D. glaucophyllum Mexico
In general appearance like *D. acrotriche*, but with glaucous leaves lacking the brush-like tip.
D. longissimum Mexico
Leaves very slender, to 1.2m(4ft) or more in length by about 6mm($\frac{1}{4}$in) wide and almost as thick, roughly 4-angled in cross-section with a slender, tapered point. Old specimens form a woody trunk 1–2m(3–6$\frac{1}{2}$ft) in height.

DATURA

Solanaceae

Origin: Warm temperature regions, chiefly in Central America. A genus of 24 species of annuals, shrubs and trees, the shrubby members of which are sometimes included in *Brugmansia*. They are mainly poisonous, having alternate leaves and very large trumpet-shaped flowers opening to five spreading lobes. Daturas make free-flowering and beautifully fragrant shrubs for the conservatory. Propagate by cuttings in late spring and summer. *Datura* is a Latin version of the Hindustani *dhatura*, or possibly of the Arabic *tatorah*.

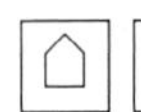 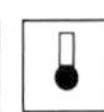 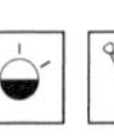  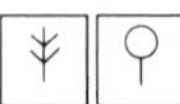

Species cultivated

D. arborea Peru, Chile
Large shrub to 3m(10ft). Leaves hairy, ovate-lanceolate to elliptic-ovate. Flowers white, 18–20cm(7–8in) long with a spathe-like green calyx, very fragrant, opening summer to autumn. Much confused in cultivation with the hybrid *D.* × *candida*, which is often grown for it.

D. aurea Columbia, Ecuador
Shrub to 11m(35ft) in the wild. Leaves toothed on young plants, ovate and smooth-edged when mature. Flowers pendulous, white or yellow-gold, 15–25cm(6–10in) long with recurved teeth, opening from a 2- to 5-lobed calyx in summer and autumn.

D.* × *candida Angel's trumpet
Hybrid between *D. aurea* and *D. versicolor*. Tall shrub to 6m(20ft). Leaves downy, ovate to oblong-elliptic. Flowers usually white, occasionally creamy-yellow or pink, 20–30cm(8–12in) long, the five lobes recurved and extended to 5–10cm(2–4in) long, powerfully fragrant. Fruits rarely observed. *D.* × *c.* 'Plena' (*D* × *c.* 'Knightii', *D. knightii*) has double flowers, one inside the other.

D. cornigera
Confused with and in cultivation probably represented by *D.* × *candida*.

D. knightii See *D.* × *candida* 'Plena'.

D. sanguinea Columbia, Chile, Peru
Shrub to 11m(35ft) in the wild. Leaves 18cm(7in) long, ovate-oblong, downy, toothed on young plants, entire when mature. Flowers pendent, 18–25cm(7–10in) long, orange-red, with the base of the corolla tube yellow, opening from late summer to winter.

D. suaveolens Brazil to Mexico
Hairless shrub 3–4m(10–13ft) tall. Leaves 15–30cm(6–12in) long, ovate-oblong to elliptic. Flowers bell-shaped, more or less pendent, 20cm(8in) or more long, opening to pointed lobes, 4cm(1½in) long; they are white and open from a 5-toothed, angled, inflated calyx in summer.

ABOVE *Datura* × *candida* 'Plena'
RIGHT *Datura sanguinea*

DAVALLIA

Davalliaceae

Origin: S.W. Europe, Canary Is., Malagasy and tropical and sub-tropical Pacific Islands and Asia as far north as Korea and Japan. A genus of 40 species of mainly small to medium-sized epiphytic ferns with creeping, scaly rhizomes and 3- to 4-pinnate, somewhat leathery fronds, triangular to ovate in outline. They are excellent plants for hanging baskets or can be grown on tree bark, moss sticks or in pans, being suitable for the conservatory or home but it is necessary to provide extra humidity in warm weather. All species become dormant for at least a short time between autumn and late spring. Species from hot dry places, such as *D. canariensis*, are dormant for four to six weeks in summer; the tropical species can grow their new fronds as the old die away. Propagate by division, by rhizome sections taken as cuttings, or by spores. *Davallia* commemorates Edmond Davall (1763–98), a Swiss botanist and friend of James Edward Smith (1759–1828), founder of the Linnean Society of London, to whom he bequeathed his herbarium.

Species cultivated

D. canariensis Hare's foot fern Canary Is., Spain, N. Africa
Rhizomes stout, appearing furry, with a dense covering of narrow, chaffy brown scales. Fronds 25–45cm(10–18in), triangular, quadripinnatifid.

D. dissecta See *D. trichomanoides*.

D. fijiensis (*D. solida fijiensis*) Rabbit's foot fern Fiji

LEFT *Delosperma nubigenum*
FAR LEFT *Davallia canariensis*

Rhizomes stout, bearing scales with long marginal hairs. Fronds long-stalked, the blades 30–60cm(1–2ft) long, broadly triangular, quadripinnatifid, bright green. A vigorous and elegant species.
D. mariesii Squirrel's foot fern, Ball fern
Japan, Korea, Taiwan, China
Rhizomes stout, densely covered with light brown scales. Fronds 15–20cm(6–8in) long, ovate to triangular in outline, quadripinnatifid, the segments oblong-lanceolate, sharp-toothed, 1–2.5mm ($\frac{1}{25}$–$\frac{1}{10}$in) long.
D. pyxidata Squirrel's foot fern Australia
Rhizomes fairly slender, densely covered with hairy-margined brownish-red scales. Fronds 25–45cm(10–18in) long, triangular, tri- to quadripinnatifid, bright glossy green.
D. solida Malaysia to N.E. Australia
Much like *D. fijiensis* but usually tripinnatifid and therefore of somewhat coarser, but still handsome appearance.
D. trichomanoides (*D. dissecta*) Squirrel's foot fern Malaysia
Much like *D. mariesii*, but rhizome scales whitish or pale yellowish-brown and fronds larger, 30–45cm(1–1½ft) long. Almost as hardy as *D. mariesii*, standing light frost for several days in succession.

DEAMIA See SELENICEREUS

DELOSPERMA

Aizoaceae

Origin: South Africa. A genus of about 150 species of mainly woody-based perennials or dwarf shrubs with succulent leaves in opposite pairs and daisy-like, often showy flowers. The species described below can be grown in a sunny window or conservatory, the prostrate-growing sorts making effective hanging basket subjects. Propagate by cuttings in summer or early autumn, or by seed in spring. *Delosperma* derives from the Greek *delos*, evident and *sperma*, a seed; an apt name as the capsules are not fully closed at the tops and the seeds can be seen at all stages of development.

  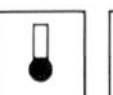 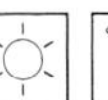  

Species cultivated

D. echinatum (*D. pruinosum*) Cape
Densely bushy species, eventually 15–30cm(6–12in) in height and width. Leaves crowded, ovoid to hemispherical, bright green, entirely covered with prominently bristle-tipped papillae. Flowers white or yellow, about 1.5cm($\frac{5}{8}$in) wide, developing intermittently throughout the year but mainly in summer. A highly distinctive and rare species worth growing for its foliage alone and well worth seeking out.
D. lehmannii Cape (Karroo, etc.)
Mat-forming with branches to 30cm(1ft) in length. Leaves of the main stems connate at the base, up to 2.5cm(1in) long, 3-angled with convex sides, grey-green sometimes intensely so; leaves of the short side shoots very much smaller and crowded. Flowers 4cm(1½in) wide, pale yellow, opening at midday in the summer.
D. lydenburgense Transvaal
Mat-forming with stems to 20cm(8in) or more in length. Leaves linear, soft-textured, pointed, 4.5–5.5cm(1¾–2¼in) long. Flowers in small loose clusters, 2.5cm(1in) wide, purple, opening in summer.
D. nubigenum Cape
Mat-forming with well-branched stems to 15cm(6in) or so in length. Leaves ovate to elliptic or almost linear, thick and pointed, 6–12mm(¼–½in) or more long, pale green, minutely papillose and often red-tinted. Flowers about 2cm(¾in) wide, orange-red to bright yellow, mainly opening in late spring and summer. In the wild this species inhabits rocky slopes and crevices high in the mountains. In cultivation it can be grown in an unheated conservatory or room and must be kept cool in winter if it is to flower well.
D. pruinosum See *D. echinatum*.

DENDROBIUM

Orchidaceae

Origin: Asia to Australasia and the Pacific Islands. A genus of 1,400 species of epiphytic orchids most of which have stem-like or club-like pseudobulbs, lanceolate leaves which are flat, or rolled under (never pleated) and the flowers in racemes. The blooms have spreading tepals, the outer three often

ABOVE *Dendrobium densiflorum*
ABOVE RIGHT *Dendrobium nobile*

narrower, and a lobed or entire labellum sometimes elaborately fringed. They are good container plants for the conservatory and many will do well in the home; the evergreen sorts need only a short resting period, but the deciduous kinds must be given a cool, almost dry spell after flowering is completed. Propagate by division when re-potting, or by removing stem pseudobulbs and using as cuttings, potting plantlets when well formed. Long pseudobulbs can be cut into 2-noded sections. *Dendrobium* derives from the Greek *dendron*, a tree and *bios*, life, the plants being epiphytic.

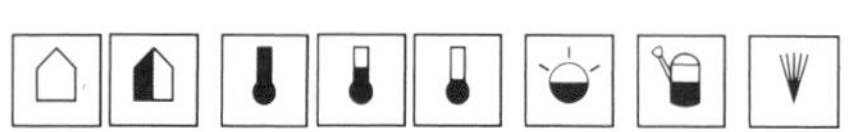

Species cultivated

D. aggregatum Burma, Thailand, Laos, S. China
Evergreen. Pseudobulbs to 7cm(2¾in), erect and clustered, oblong-ovoid, somewhat 4-angled. Leaves solitary from each pseudobulb, leathery and stiff, to 7cm(2¾in) long. Flowers about 4cm(1½in) wide, tepals deep golden-yellow with a broad lip; honey-scented, opening in spring. Temperate.

D. aphyllum See *D. pierardii.*

D. aureum See *D. heterocarpum.*

D. bigibbum Queensland (Australia), New Guinea
Evergreen. Pseudobulbs to 45cm(1½ft) tall, cylindrical, stem-like. Leaves to 10cm(4in) long, oblong-lanceolate, up to six, carried towards the top of each pseudobulb. Flowers to 5cm(2in) wide, the tepals usually rosy-mauve; labellum red-purple with a white crest in the throat. Tropical.

D.b. phalaenopsis (*D. phalaenopsis*)
Similar but more robust, with pseudobulbs to 1m(3ft); leaves to 15cm(6in); flowers to 9cm(3½in) wide, the tepals rosy-mauve with a red-purple labellum, but sometimes pure white.

D. chrysotoxum Himalaya to S. China
Deciduous. Pseudobulbs 20–30cm(8–12in) tall, club- to spindle-shaped. Leaves to 13cm(5in) long, oblong, leathery. Flowers 4–5cm(1½–2in) wide, the tepals golden-yellow, the labellum marked deep orange to brown, hairy above, fringed, in arching terminal racemes to 30cm(1ft), from autumn to spring. Temperate.

D. densiflorum Himalaya to Burma
Evergreen. Pseudobulbs to 30–45cm(1–1½ft) tall, club-shaped, 4-angled. Leaves 15cm(6in), elliptic, deep glossy green, carried near the top of the pseudobulbs. Flowers 5cm(2in) wide, tepals golden-yellow, with a velvety darker orange lip, not fringed; borne in long, dense, pendent racemes, to 22cm(9in) long, in spring. Temperate.

D. devonianum Himalaya, S.E. Asia to China
Deciduous. Pseudobulbs to 90cm(3ft) or more long, stem-like, often branched, pendent. Leaves to 14cm(5½in) long, linear-lanceolate. Flowers 5–7cm(2–2¾in) wide, creamy-white with a pink flush and inner tepals tipped with red-purple; labellum white with a purple border or tip, blotched with orange and lacily fringed; fragrant and borne in ones and twos from the stem nodes in spring and summer. Temperate.

D. fimbriatum Himalaya to Malaysia
Semi-deciduous. Pseudobulbs to 2m(6½ft) tall, stem-like, erect or arching. Leaves to 15cm(6in) long, lanceolate, pointed. Flowers 5–7cm(2–2¾in) wide, tepals deep orange to bright yellow, the labellum deeper orange, velvety and fringed; borne in 8- to 15-flowered racemes in spring. Temperate.

D.f. oculatum Burma
Similar but more robust, with two velvety, black-brown blotches on the labellum; the form most seen in cultivation. Temperate to tropical.

D. heterocarpum (*D. aureum*) S.E. Asia from India to the Philippines and Moluccas
Deciduous. Pseudobulbs to 60cm(2ft) long, erect to pendent, stem-like. Leaves to 13cm(5in) long, oblong-lanceolate, leathery. Flowers 6–8cm(2½–3in) wide, tepals cream to yellow, sometimes suffused with brown or green; labellum white with a central golden disk or entirely yellow; fragrant and carried in twos and threes from the stem nodes chiefly in late winter and spring, but often at other times. Temperate.

D. kingianum Australia
Evergreen. Pseudobulbs 20–30cm(8–12in), clustered, club-shaped, often thick below, thinner above, often branching. Leaves to 10cm(4in) long, lanceolate, three to six from each pseudobulb. Flowers 2.5cm(1in) wide, tepals pale to purplish-pink, the lip usually with a darker mark, occasionally all white; borne in spikes of two to nine from winter to spring; fragrant. Temperate.

D. loddigesii China
Deciduous. Pseudobulbs 20cm(8in) long, stem-like, prostrate or pendent, often branched. Leaves 7–9cm(2¾–3½in) long, broadly lanceolate, shining green. Flowers 4–5cm(1½–2in) wide, tepals pale pinkish-mauve to rosy-purple; labellum fringed, white with a large orange central blotch, sometimes tipped with purple; carried singly on the leafless pseudobulbs in spring. Cool to temperate.

D. moschatum Himalaya to Thailand
Semi-deciduous. Pseudobulbs 1–2m(3–6½ft) long, stem-like, arching to pendent, brownish striped. Leaves to 18cm(7in) long, oblong-ovate, pendent, leathery. Flowers 7–10cm(2¾–4in) wide, tepals fawny-cream suffused with pink; labellum slipper-shaped, pale yellow with two dark purple markings; musk-scented and carried in pendent 5- to 10-flowered racemes in spring and summer. Temperate to tropical.

D. nobile S.E. Asia from Himalaya to Taiwan
Semi-deciduous. Pseudobulbs to 60cm(2ft) tall, stem-like to narrowly club-shaped, erect. Leaves to 10cm(4in) long, oblong with notched tips. Flowers to 8cm(3in) wide, not opening out fully, normally with white tepals tipped with pink; labellum white, red in the throat, but in some forms with richer coloured flowers; fragrant and borne in clusters of one to three from the nodes of leafless pseudobulbs. Temperate.

D.n. virginale
A pure white form.

D. phalaenopsis See *D. gibbum phalaenopsis*.

D. pierardii (*D. aphyllum*) S.E. Asia, Himalayas to Malaysia
Semi-deciduous. Pseudobulbs 1–2m(3–6½ft) long, slender, pendent, stem-like. Leaves to 10cm(4in) long, pointed lanceolate. Flowers 5cm(2in) wide, tepals white, lightly flushed with pink, often semi-transparent and fragile; labellum yellow with purple streaking at the base; borne in clusters of one to three from the nodes of leafless pseudobulbs from late winter to late spring. Temperate.

D. primulinum S.E. Asia, Himalayas to Malaysia
Deciduous. Pseudobulbs to 30cm(1ft) long, slender, arching to pendent. Leaves 4cm(1½in) long, oblong, leathery. Flowers 5–6cm(2–2½in) wide, white, the tepals flushed pink; labellum pale to brighter yellow, streaked with red-purple; borne singly from the nodes of leafless pseudobulbs in spring. Cool.

D. speciosum Australia, New Guinea
Pseudobulbs to 90cm(3ft) tall, stem-like, sturdy, somewhat ribbed. Leaves to 25cm(10in), oblong-lanceolate, leathery. Flowers 5–7cm(2–2¾in) wide, not opening fully, tepals narrow, white to pale yellow; labellum 3-lobed, cream to yellow, streaked or spotted with red-purple; borne in dense, arching racemes to 45cm(1½ft) long. Temperate

D. wardianum S.E. Asia, Himalaya to Thailand
Deciduous. Pseudobulbs to 1m(3ft) or more tall, sturdy, stem-like, erect or arching. Leaves to 10cm(4in) long, narrowly lanceolate, shining green. Flowers to 9cm(3½in) wide, tepals white, tipped with reddish-purple; labellum hooded, marked with orange-yellow inside and two purple spots near the base; borne in clusters of one to three from the leafless pseudobulbs in spring. Temperate to tropical.

LEFT *Denbrobium wardianum*
FAR LEFT *Dendrobium pierardii*

DENDROCHILUM

Orchidaceae

Origin: Indo-Malaysia to the Philippines and New Guinea. A genus of 100 to 150 species of epiphytic orchids with small, usually clustered pseudobulbs. Unlike most cultivated orchids, they produce dense, raceme-like spikes of tiny flowers which, in the most popular species, hang down like large, elegant catkins. Although best in the conservatory, they can be grown in the home and are well worth trying there. They are best kept in small pots or shallow pans. Propagate by division in spring. *Dendrochilum* derives from the Greek *dendron*, a tree, and *cheilos*, a lip; the blossoms of this tree-dwelling genus have prominent lips.

 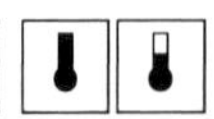  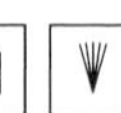

Species cultivated

D. cobbianum (*Platyclinis cobbiana*) Philippines
Pseudobulbs densely clustered, narrowly ovoid to sub-conical, usually furrowed, dark green, 2.5–5cm(1–2in) long. Leaves solitary, long-stalked, elliptic to narrowly lanceolate, softly leathery-textured, somewhat wavy, up to 15cm(6in) in length. Inflorescence stem about 20–30cm(8–12in) tall from the top of which hangs a raceme of almost equal length; flowers up to 2cm(¾in) wide, white to sulphur or straw-yellow, usually with a darker or orange labellum, strongly fragrant. Autumn is the usual flowering season.

Dendrochilum filiforme

D. filiforme (*Platyclinis filiformis*) Philippines
Pseudobulbs clustered, 2.5cm(1in) long, ovoid to conical, becoming grooved after maturity. Leaves solitary, longish stalked, linear-lanceolate, erect to ascending and firm-textured, 13–20cm(5–8in) long. Inflorescence stem to 25cm(10in) tall with a pendent raceme almost as long; flowers numerous (to 100), 6mm(¼in) wide, arranged in two ranks, pale to greenish-yellow, fragrant, opening in summer.

D. glumaceum (*Platyclinis glumacea*)
Philippines
Pseudobulbs clustered, ovoid, 3–5cm(1¼–2in) long, when young with reddish sheathing bracts. Leaves longish-stalked, narrowly lanceolate, 20–30cm(8–12in) long, often wavy-margined. Inflorescence stem to 20cm(8in) in length, almost thread-like, arching, the 15–20cm(6–8in) long raceme continuing the curve; flowers 1.2cm(½in) or more in width, white to pale yellow with a green, yellow or white labellum, coumarin (sweet hay) scented. Spring is the usual flowering period.

DENNSTAEDTIA

Dennstaedtiaceae (*Polypodiaceae*)
Cup ferns

Origin: Pantropical plus temperate South America, North America, Australasia and Japan. A genus of 70 species of sizeable ferns, some of which make handsome specimen foliage plants and ground cover for the shady conservatory. They have creeping, hairy rhizomes and lacy, several- to many times pinnate fronds bearing sori sheltered by cup-shaped indusia. Propagate by division or spores in spring. *Dennstaedtia* commemorates August Wilhelm Dennstedt (1776–1826), a German botanist.

  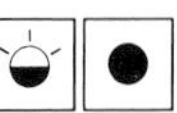  

Species cultivated

D. adiantoides See *D. bipinnata*.

D. bipinnata (*D. adiantoides*) Mexico to South America, West Indies
Fronds (including the stalks) about 1.2m(4ft) tall, twice this when planted out in moist rich soil, bi- to tripinnate, the ultimate lobes rhomboidal, toothed, glossy green. Grown in a tub, this fern makes an eye-catching specimen foliage plant for the large, tropical conservatory.

D. davallioides Lacy ground fern Australia, Queensland to Victoria
Rhizomes far-creeping, forming extensive colonies. Fronds (including the stalks) 90–120cm(3–4ft) in height, tripinnate, the ultimate segments oblong and deeply pinnatifid. Provides splendid ground cover for the temperate conservatory bed, but may need curbing from time to time.

D. obtusifolia Central America, West Indies, Trinidad, south to Paraguay
Fronds about 90cm(3ft) tall, twice this in rich moist soil, triangular in outline, quadri-, or more

times, pinnate, thinly hairy and lacily elegant. Requires tropical conditions.

DERMATOBOTRYS

Scrophulariaceae

Origin: South Africa, Natal, Zululand. A genus of one species of deciduous, epiphytic shrub suitable for the sunny room or conservatory. Propagate by seed in spring or by cuttings in summer. *Dermatobotrys* derives from the Greek *derma*, skin (or bark) and *botrys*, a cluster, perhaps because the first flowers arise from leafless stems.

Species cultivated

D. saundersii

Sparingly branched shrub 60cm(2ft) or more in height. Leaves in opposite pairs, ovate, up to 10cm(4in) or more long, coarsely toothed and slightly fleshy, but firm-textured. Flowers in small clusters at the tips of the previous season's stems, about 5cm(2in) long, tubular, with five short lobes, light carmine-red outside, yellow inside. The first flowers usually appear on bare stems to be almost immediately followed by young leaves; flowers occasionally appear in spring and summer. For a suitable compost see introductory section on Bromeliads. When dormant in summer the plants must be kept dry then barely moist until foliage appears, when water may be given more freely, but never in excess. An unusual hanging basket subject.

DIANELLA

Liliaceae

Origin: Tropical Asia, Australia and Polynesia. A genus of 30 species of tufted to clump-forming evergreen perennials. They have linear, grassy or strap-shaped leaves in fan-shaped clusters, and panicles of small, usually blue or white flowers are followed by purple to blue berries. Plants are suitable for the conservatory and can be brought into the home. Propagate by seed or by division, both in spring. *Dianella* is a diminutive form of Diana, the goddess of hunting. The reasoning behind the name is obscure.

Species cultivated

D. caerulea Eastern Australia (New South Wales)

Shrubby, stems clustered, erect to 60cm(2ft) tall, but taking several years to attain it. Leaves 30–45cm(1–$1\frac{1}{2}$ft) long, leathery, deep green, arching. Panicles usually above the leaves, about 30cm(1ft) long; flowers 8mm($\frac{1}{3}$in) wide, blue, borne in spring. Fruits rounded to shortly oval, blue to blue-purple.

ABOVE LEFT *Dermatobotrys saundersii*
ABOVE *Dianella tasmanica*

It seems likely that *D. caerulea* is confused in cultivation with *D. tasmanica* and perhaps also *D. intermedia* and other species.

D. ensifolia East tropical Africa, Malagasy, S.E. Asia, Australia, Hawaii

Shrubby, with woody branched stems 1–1.5m(3–5ft) in height. Foliage comparatively dense on mature specimens; leaves strap-shaped, arching, to about 30cm(1ft) long, deep green. Flowers about 8mm($\frac{1}{3}$in) wide, blue to bluish-white, in smallish panicles borne in summer. Fruits to 12mm($\frac{1}{2}$in) long, deep blue.

D. intermedia

The true species from Norfolk Is., Fiji and adjacent islands is not in cultivation, but the name is misapplied to *D. nigra*, *q.v.*

D. nigra (*D. intermedia* of gardens) New Zealand

Shortly rhizomatous sub-shrub, stems clustered, erect to 15cm(6in) or so. Leaves 25–60cm(10–24in) long and 1–1.5cm($\frac{3}{8}$–$\frac{5}{8}$in) wide, dark green. Flowers 7mm($\frac{1}{4}$in) wide, greenish-white, in broad lax panicles which overtop the leaves. Fruits rounded to oblong, shining violet-blue, sometimes greyish to pale blue, 7–17mm($\frac{1}{3}$–$\frac{2}{3}$in) long.

D. tasmanica S.E. Australia, Tasmania

Erect shrubby plant to 1.5m(5ft) high. Leaves 30–60cm(1–2ft) long by 1–3cm($\frac{3}{8}$–$1\frac{1}{4}$in) wide, linear, arching. Flowers 1.2–2cm($\frac{1}{2}$–$\frac{3}{4}$in) wide, pale blue, in narrow panicles. Fruits 1–2cm($\frac{3}{8}$–$\frac{3}{4}$in) long, glossy deep blue to violet.

DIANTHUS

Caryophyllaceae

Origin: Temperate areas of the northern hemisphere, rarely in the southern hemisphere, commonest in southern Europe. A genus of 300 species of annuals, evergreen perennials and sub-shrubs with linear, often grey-green leaves in opposite pairs and 5-petalled flowers opening widely from a tubular calyx, the base of which is enclosed by two to three pairs of bracts. This character separates

Dianthus from closely related genera. *Dianthus* derives from the Greek *di* of Jove or Zeus, and *anthos*, a flower, literally flower of the gods.

Of the two species and their forms described below, *D. chinensis* makes an attractive short-term pot plant which, although it is perennial, is usually grown as an annual. It is best in a conservatory or a sunny window in a cool room, but can be brought into warmer conditions for flowering. It is raised from seed. Carnations derived chiefly from *D. caryophyllus* are conservatory plants, usually being brought into the home as cut flowers. If grown as outlined below they will bloom during the winter, the time when flowers are most appreciated. Insert cuttings in late winter or obtain young plants in spring and pot on as required until a final container size of 15cm(6in) is reached. During this period, firstly insert a 90cm(3ft) cane for support, then pinch out the growing tips at the tenth leaf pair. Several lateral stems will result and these must be pinched out beyond their fifth to seventh leaf pair. Sub-laterals will grow out in response to this treatment and should be allowed to grow unhindered until early to mid July. At this time all young growths must be pinched out. Subsequent shoots must be left to produce flower buds and, providing a minimum temperature of 7–10°C(45–50°F) is maintained, they will open in winter. If this final pinching is left until August, flowering is delayed until spring. On the other hand, if the sub-laterals are not pinched, flowering will start in the first summer and continue into autumn. Well-branched plants may need several canes for support. Specially made carnation rings make the securing of the plants to stakes quick and easy. By moving on into larger containers and adding longer canes each spring, the plants can be kept from two to three years. They do get very tall and ungainly, however, and there is much to be said for rooting cuttings annually and discarding one-year plants. To get the best out of the plants, once the first young flower buds are seen, apply liquid feed at seven day intervals and continue until flowers cease or the plants are potted on. For quality blooms, such as one would purchase from a florist, disbudding is necessary. Flower buds are produced in small clusters and, to disbud, the smaller side buds are snapped out, leaving only the large terminal one to develop.

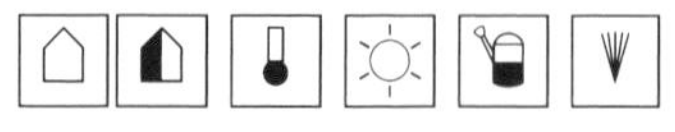

Dianthus chinensis heddewigii

Species cultivated

D. caryophyllus Carnation Central Mediterranean region

Perennial of loosely tufted habit becoming woody at the base with age. Leaves linear, to 8cm(3in) long, grey-glaucous. Flowers up to 6cm(2½in) across, mainly white, pink and red. For culture under glass, it is best to grow named cultivars and dozens of these are available from specialist nurserymen in a wide colour range. All are double-flowered; some are of one colour only and are known as Selfs, others have two or more colours or shades in the same bloom. If the two-colour patterning is irregular they are known as Fancies; if the second colour is restricted to the margins of the petals they are Picotees.

D. chinensis Indian pink E. Asia

Biennial or short-lived perennial of tufted habit. Leaves lanceolate to linear, dark green. Flowers to 4cm(1½in) across, red to white, often darker at the centre, borne on stems 30–45cm(12–18in) high, branching near the top.

D.c. heddewigii

The form usually grown, represented in gardens by many seed strains and cultivars only 15–25cm(6–10in) tall; more compact and in a wide range of shades of red, pink and white, variously marked.

DIASCIA

Scrophulariaceae

Origin: South Africa. A genus of 42 species of erect to prostrate annuals and perennials. They have ovate leaves and 5-petalled, usually pink flowers which have two spurs. All the species make good pot plants for the conservatory and sunny window sill and those with pendent stems are very attractive in hanging baskets. Propagate perennials by cuttings in spring or late summer, or all by seed sown in autumn if minimum temperatures are over 10°C(50°F), or in spring. *Diascia* derives from the Greek *di*, two, and *askos*, a sac, referring to the two nectary spurs.

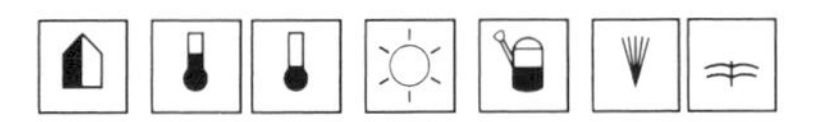

Species cultivated

D. barberae

Erect, bushy annual 30–45cm(1–1½ft) tall. Leaves ovate, toothed. Flowers 1.5–2cm(⅝–¾in) across, rose-pink spotted with green in the throat. Propagate by seed in autumn for spring flowering, otherwise in summer.

LEFT *Diascia rigescens*
BELOW LEFT *Diascia vigilis*

D. rigescens Natal
Evergreen, clump-forming perennial, erect at first, but old plants becoming decumbent; stems square in cross-section, up to 45cm($1\frac{1}{2}$ft) tall including the flowering portion. Leaves sessile, broadly ovate, dark green, 1.2–2.5cm($\frac{1}{2}$–1in) long, sharply toothed. Flowers about 2cm($\frac{3}{4}$in) wide, dusky crimson-pink, in long tapering racemes from summer to late autumn.
D. vigilis Natal
Evergreen, rhizomatous perennial, spreading to decumbent; stems angular, to 30cm(1ft) or more tall including the flowering part. Leaves shortly stalked, ovate to narrowly so, sub-cordate, toothed, light green, 1.5–2.5cm($\frac{5}{8}$–1in) long. Flowers about 2cm($\frac{3}{4}$in) wide, soft luminous pink, in elegant racemes in summer to autumn. It makes a most attractive hanging basket plant.

DIASTEMA

Gesneriaceae

Origin: Central and tropical South America. A genus of about 40 species of deciduous perennials, a few of which make attractive house plants. They are allied to *Isotoma* and *Achimenes* and have similar scaly rhizomes. Spreading to erect and bearing stalked leaves in pairs, the stem tips produce stalked clusters of tubular flowers with five rounded petal lobes. Propagate by separation of the rhizomes when re-potting in spring or by cuttings of young stems. *Diastema* derives from the Greek *di*, two, and *stemon*, a stamen; the four stamens are in pairs.

Species cultivated
D. pictum See *D. vexans*.

D. quinquevulnerum Venezuela
Stems about 15cm(6in) in height, branching from the base. Leaves up to 8cm(3in) long, ovate to elliptic, sharply toothed, pale green with impressed veins. Flowers in racemes of ten or more, 2cm($\frac{3}{4}$in) long, white with yellow throats, each petal lobe bearing a violet spot, opening in summer.
D. vexans (*D. pictum*) Colombia
Small tufted plant with stems to 10cm(4in) tall. Leaves crowded, 8cm(3in) long, ovate to lanceolate, toothed, hairy, carried on long stalks. Flowers one to several from the uppermost leaf axils, or terminating the stem, 16mm($\frac{5}{8}$in) long, white with a purple-brown spot at the base of each petal lobe, borne in summer.

DICENTRA (DIELYTRA)

Fumariaceae

Origin: W. Himalaya to Siberia, North America. A genus of 20 species of clump-forming or rhizomatous perennials. The leaves are deeply dissected, rather fern-like and glabrous, being quite

Dicentra spectabilis

decorative in their own right, but it is the curious flowers which attract most attention. These are pendent and 4-petalled, the outer two petals pouched and spurred, the inner ones smaller – the overall shape prompting various apt common names. The frost-hardy species described below makes an attractive short-term conservatory plant and can be brought into the home for flowering. Pot roots in autumn and keep above 10°C(50°F) from early winter onwards. Propagate by division or root cuttings in a cold frame in winter. *Dicentra* derives from the Greek *dis*, twice and *kentron*, a spur; the two nectaries are spur-like in shape in some species.

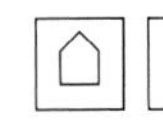  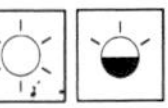  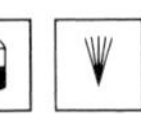

Species cultivated

D. spectabilis Bleeding heart, Lady in the bath, Dutchman's breeches Japan, Korea, China
Clump-forming with decorative foliage; stems to 45cm($1\frac{1}{2}$ft) or more, branched. Leaves long-stalked, biternate; leaflets glaucous, 3- to 5-lobed. Flowers heart-shaped, outer petals rose-crimson, the protruding inner ones white; they are borne in arching racemes in spring. Place outside or in a cold frame when foliage yellows and repot annually.

BELOW *Dichorisandra thyrsiflora*
BOTTOM *Dichorisandra reginae*

DICHELOSTEMMA See BRODIAEA

DICHORISANDRA

Commelinaceae

Origin: Tropical America. A genus of about 35 species of perennials, those described below having handsome lanceolate foliage and terminal clusters of attractive flowers. They look effective when planted out in the warm conservatory border and can be grown in containers in the home. Propagate in early spring by division, or by cuttings in summer. *Dichorisandra* derives from the Greek *dis*, twice, *chorizo*, to part and *aner*, male, referring to the way in which two of the six stamens spread apart from the others.

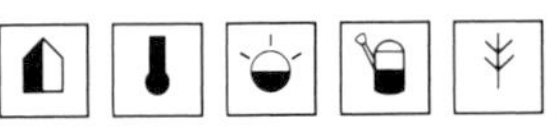

Species cultivated

D. reginae (*Tradescantia reginae*) Peru
Stems more or less erect, sparingly branched, slowly forming clumps, eventually to 60cm(2ft) or more in height. Leaves in two ranks, lanceolate to elliptic, up to 18cm(7in) long, red-purple below, dark purplish, glossy green above with two submarginal silvery bands and lesser streaks which tend to fade with age. Flowers about 2.5cm(1in) wide, each of the three blue-purple petals with a white base; they are produced in compact panicles in summer, but only on long-established specimens and then not freely – however, this is essentially a foliage plant and any flowers are a bonus.

D. thyrsiflora Brazil
Stems erect, robust, sparingly branched and almost cane-like on mature specimens, up to 1.2m(4ft) in height when in large containers or planted out. Leaves 15–30cm(6–12in) long, broadly lanceolate, lustrous deep green above, purple-flushed beneath. Flowers 3-petalled, blue to deep blue-purple, about 2–2.5cm($\frac{3}{4}$–1in) wide, borne in dense panicles up to 15cm(6in) or more in length during the summer and autumn. *D.t.* 'Variegata' has leaves with two longitudinal silvery bands and a red midrib.

DICKSONIA

Dicksoniaceae

Origin: Malaysia to Australasia, St. Helena, tropical America. A genus of 30 species of large ferns, most of which build up trunks with age. Their fronds are arching and bi- or tri-pinnate giving an elegantly lacy effect. They make very attractive foliage plants for shady positions. Propagate by

LEFT *Dicksonia antarctica*
BELOW *Didymochlaena truncculata*

spores in spring at about 15°C(61°F). *Dicksonia* was named for James Dickson (1738–1832), a British nurseryman and botanist.

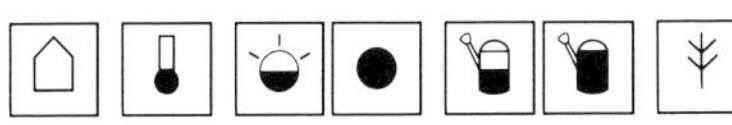

Species cultivated

D. antarctica Soft tree fern Eastern Australia from Queensland to Tasmania
Stem eventually to 5m(16ft) tall in the open garden, covered with a rusty brown bark-like surface of matted roots. When young the fronds grow from ground level; they are borne in terminal rosettes, each frond being 1.3–3m(4½–9ft) long, tripinnate, the sori at the ends of the veins near the margins of the leaflets which reflex to cover them.

DIDISCUS See TRACHYMENE

DIDYMOCHLAENA

Aspidiaceae

Origin: Circum-tropical, Natal. A genus of one species of dwarf evergreen tree fern which makes a handsome specimen for the tropical, shaded conservatory. Propagate by spores or division in spring. *Didymochlaena* derives from the Greek *didymos*, twin and *chlaina*, a cloak, referring to the indusium which covers and protects the sporangia.

Species cultivated

D. truncculata (*D. lunulata*, *D. sinuosa*)
Stem erect, eventually to 1.2m(4ft) or more in height. Fronds bipinnate, in a dense, terminal rosette, 90–120cm(3–4ft) long, potentially to twice this when planted out, ultimate pinnules more or less oblong, wavy margined, 2–2.5cm(¾–1in) long, with a similar texture to that of a maidenhair fern (*Adiantum*).

DIEFFENBACHIA

Araceae

Dumb canes

Origin: Tropical America and West Indies. A genus of 30 species of shrubby, evergreen perennials with robust, erect stems, sometimes fleshy, carrying terminal tufts of large oblong to ovate leaves. The tiny flowers are carried like those of arum, on a spadix within a narrow, arum-like spathe. The plant is very poisonous and care must be taken that sap does not get into the mouth or eyes. One particular symptom of poisoning is speechlessness, hence the common name 'dumb cane'. They are attractive foliage plants for the conservatory or home. Propagate by cuttings of stem tips or by stem sections 6–8cm(2½–3in) long. These are taken from the lower, leafless stems and inserted horizontally into the rooting medium. *Dieffenbachia* was named for J. P. Dieffenbach (1790–1863), the Administrator of the Royal Palace Gardens at Schonbrunn in Vienna.

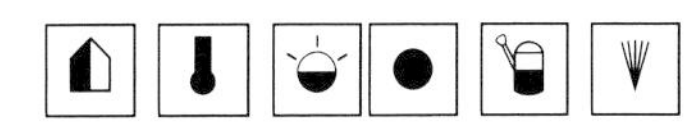

Species cultivated

D. amoena Giant dumb cane Tropical America
Robust, stems reaching 1.5m(5ft) or more. Leaves 30–40cm(12–16in), oblong and glossy dark green,

irregularly marked with white lines and spots along the side veins.

D.* × *bausei (*D. maculata* × *D. weirii*)
Garden hybrid. Fairly robust, stem eventually to 90cm(3ft) or more in height. Leaves to 30cm(12in) long, broadly lanceolate, yellow-green with a dark green margin and irregular small blotches plus a rash of white speckles.

D. bowmannii Eastern Brazil
Robust, to 1.2m(4ft) or even more in time. Leaves ovate-elliptic, usually narrowly so, 45–60cm(1½–2ft) long, basically dark green with zones of light green along the lateral veins.

***D.* 'Exotica'**
Possibly a form or hybrid of *D. maculata*. One of the smaller and less robust dieffenbachias, but still a good pot plant. Stems to 45cm(1½ft) or so. Leaves ovate, to 25cm(10in) long, slender, pointed, dark green with a variable but bold marking of white or greenish white.

ABOVE *Dieffenbachia maculata*
ABOVE RIGHT *Dieffenbachia* × *bausei*
RIGHT *Dieffenbachia seguine*

***D.* 'Hoffmannii'** Costa Rica
A decorative plant much like a larger leaved, more robust *D.* 'Exotica', perhaps of hybrid origin.

D. imperialis Peru
Robust, stems eventually to 1.5m(5ft) tall. Leaves ovate to elliptic, up to 60cm(2ft) long, leathery-textured, dark green with greyish feathering along the midrib and a scattering of yellow to pale green blotches.

D. leopoldii Costa Rica
Moderately robust, stems to 60cm(2ft) or so. Leaves 25–35cm(10–14in) long, rich velvety green with a prominent white to cream midrib.

D. longispatha Panama and Colombia
Moderately robust, stems to 60cm(2ft) or so. Leaves narrowly oblong-lanceolate, 45–55cm(18–22in) long, leathery-textured, plain green.

D. maculata (*D. picta*) Common dumb cane
Brazil
To 1m(3ft) or so tall. Leaves to 25cm(10in) long, oblong-elliptic to ovate, green spotted and blotched with creamy-white. Many forms and cultivars are grown with various amounts of cream coloration. *D.m.* 'Rudolph Roehrs' has the leaves almost completely coloured pale yellow-green, showing the normal dark green only on the narrow margins and veins.

D.* × *memoria-corsii (*D. picta memoria*)
This hybrid of *D. maculata* and *D. wallisii* tends to favour the first parent in overall appearance, but the green leaves are irregularly marked with light grey especially along the midrib, plus a scattering of white spots.

D. oerstedii Mexico and Costa Rica
Robust, stems to 90cm(3ft) or more. Leaves ovate to oblong-lanceolate, often somewhat asymmetrically so, deep green. *D.o.* 'Variegata' has ivory-white midribs.

D. picta See *D. maculata*.

D. picta memoria See *D.* × *memoria-corsii*.

D. seguine Tropical America
One of the largest species, stems eventually attaining 2m(6½ft) or more in height. Leaves variable, from ovate to elliptic and lanceolate, to 45cm(1½ft) long, glossy deep green, somewhat fleshy. Several variegated forms are known: 'Irrorata', very distinct in having a thinner-textured yellow-green leaf with dark green edges and blotches and whitish stalks; 'Lineata', green leaf but white-striped leaf stalks; 'Liturata', with a cream zone on either side of the midrib; 'Nobilis', spotted bright pale green.

DIELYTRA See DICENTRA

DIETES

Iridaceae

Origin: South Africa, Australia, Lord Howe Island. A genus of four species of rhizomatous perennials formerly included in *Moraea*. They differ from *Moraea* only in having rhizomes and evergreen

LEFT *Dimorphotheca sinuata*
FAR LEFT *Dietes bicolor*

leaves, the latter being sword-shaped as in *Iris*. The species described here are best in the conservatory, especially when planted out in a bed or border where they develop into clumps and generally flower freely. They also respond to pot culture and can be brought into the home when in bloom. Propagate by seed in spring or by division after flowering or in late winter.

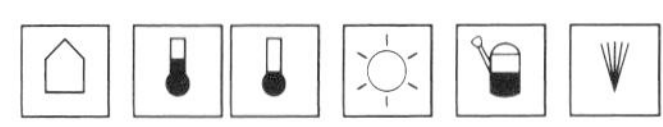

Species cultivated

D. bicolor (*Moraea bicolor*) South Africa
Clump-forming, the narrow leaves to 75cm(2½ft) in length, often arching. Flowering stems up to 60cm(2ft) in height, usually branching; flowers 5cm(2in) wide, lemon-yellow with a purple-black basal blotch on each of the outer tepals, borne in summer.

D. vegeta (*Moraea catenulata, M. iridoides*)
African iris South Africa
Clump-forming, the narrow deep green leaves usually exceeding 60cm(2ft) in length. Flowering stems up to 60cm(2ft) tall, branched; flowers 6cm(2½in) wide, white, with the three outer tepals yellow or brown spotted and the style crests blue, borne in summer.

DIGITALIS See ISOPLEXIS

DIMORPHOTHECA

Compositae

Origin: South Africa. A genus of about seven species of annuals, evergreen perennials and subshrubs. Most are of spreading habit, with narrow leaves and brightly coloured or white daisy-like flower heads. They make colourful summer-flowering annuals for the conservatory and home. Propagate by seed; sown in warmth in January, they will flower from May to August. *Dimorphotheca* derives from the Greek *dis*, twice, *morpho*, form or shape and *theca*, fruit, referring to the two different shapes of seed in each head.

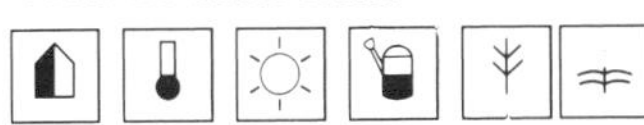

Species cultivated

D. annua See *D. pluvialis*.
D. aurantiaca See *D. sinuata*.
D. barberiae See *Osteospermum barberiae*.
D. calendulacea See *D. sinuata*.
D. ecklonis See *Osteospermum ecklonis*.
D. jucunda See *Osteospermum jucundum*.

D. pluvialis (*D. annua*)
Erect, branching annual to 30cm(1ft) tall. Leaves hairy, narrowly oblong-ovate, 2.5–8cm(1–3in) long with blunt teeth. Flowers 5–6cm(2–2½in) wide, ray florets white or creamy-white above, purple beneath, discs golden-brown; they open from summer to autumn.

D. sinuata (*D. aurantiaca* of gardens, *D. calendulacea*)
Spreading annual to 30cm(1ft) tall. Leaves 8cm(3in) long, oblong to spathulate, coarsely toothed. Flowers to 5cm(2in) wide, ray florets bright orange-yellow, sometimes purplish at their bases, disc florets yellow; they open from early summer to autumn.

DINTERANTHUS

Aizoaceae
Living stones

Origin: Southern Africa. A genus of about six species of highly succulent perennials. They are closely allied to and resemble *Lithops*, but each shoot may have one to three pairs of leaves. The daisy-like flowers appear in late summer/autumn and, in the case of *D. pole-evansii*, in spring. As with *Lithops*, the species of this genus listed below can be successfully grown in a sunny home window, or conservatory. Propagate by seed in spring. *Dinteranthus* honours Professor Kurt Dinter (1868–1945), a German botanist who studied the flora of South-west Africa.

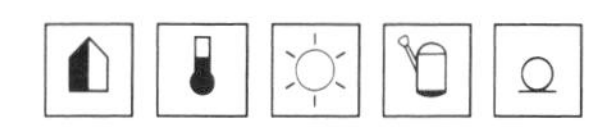

Species cultivated

D. inexpectatus S.W. Africa, Namaqualand
Leaves in a single pair, solitary or in small clusters, broadly ovoid with a transverse slit, grey with translucent green dots; each leaf top, from the

Dinteranthus punctatus

central slit, 2cm(¾in) or slightly more long. Flowers yellow, 2.5cm(1in) wide, opening in early autumn.
D. microspermus S.W. Africa, Namaqualand
Like a larger version of *D. inexpectatus*, but having a deeper slit between each leaf pair, and yellow petals with reddish tips.
D. pole-evansii Cape
Leaves in a single pair, usually solitary, oblongoid, with a deep, narrow, transverse slit, grey, often tinted red or yellow; from the centre slit each leaf top is 1.4cm(½in) long. Flowers rich, satiny-yellow, 4cm(1½in) wide, opening in late spring.
D. puberulus See *D. punctatus*.
D. punctatus (*D. puberulus*) Cape
Leaves usually in two pairs, the shoots which bear them forming small clusters; each leaf pair is united for up to half their length, obovoid and slightly angular, 2.5–3cm(1–1¼in) long, coppery grey-green with dark green dots and a coating of microscopic, velvety hairs. Flowers yellow, 2.5cm(1in) or a little more wide, expanding in autumn.
D. wilmotianus Cape
Leaves in a single pair, solitary, very broadly boat-shaped, grey, often tinted pink and dotted deep purple; each leaf from the central suture 3.5cm(1⅓in) or more long, somewhat keeled at the tip. Flowers deep yellow, 3cm(1¼in) wide, opening in autumn.

DIONAEA

Droseraceae

Origin: North and South Carolina, USA. A genus of only one species, a carnivorous herbaceous perennial. It makes a fascinating pot plant and is not difficult to grow in the conservatory or home, being best in a pot of a 50:50 mix of peat and sphagnum moss stood in a shallow tray or pan of water. Keep out of direct sun in summer, and in a dry atmosphere a glass cover will keep the necessary humidity. Propagate by dividing old plants or by seed in spring. *Dionaea* derives from *Dione*, the Greek name for Venus.

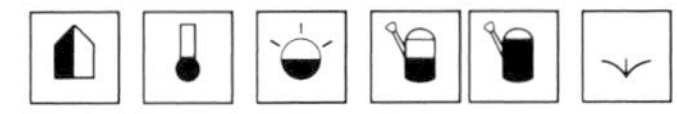

Species cultivated

D. muscipula Venus fly trap
A rosette forming plant with up to eight spreading leaves, 8–15cm(3–6in) long, which are remarkably adapted for trapping small insects. Each leaf has a broadly winged stalk and an oblong blade which is divided centrally into two lobes which are margined with stiff bristles and bear three highly sensitive trigger hairs on their surface. When two or

Dionaea muscipula

more of these hairs are touched the leaf snaps shut like a closing book, the bristles interlocking and forming a perfect trap. The unfortunate insect which has triggered off the mechanism is digested by juices secreted by glands on the surface. When digestion is completed the trap opens and is reset for the next arrival. Flowers white, 5-petalled, 2cm($\frac{3}{4}$in) wide and borne in umbel-like clusters on stems 8–15cm(3–6in) tall in summer.

DIOON

Zamiaceae

Origin: Mexico and Central America. A genus of three to five species of primitive, palm-like plants superficially resembling the better known *Cycas*, and another member of the plant group popularly known as cycads. They form erect, usually shortish trunks topped by a rosette of long, pinnate leaves of firm and leathery texture. The 'flowers', which are only produced on sizeable plants, are borne in cone-like structures in the centre of the rosette. In their native countries, the large, edible, starchy seeds are ground into a coarse flour or fine meal, while the pith within the stem yields a high quality sago. Propagate by seed sown in spring or summer. The species described below make interesting and decorative container plants for the warm room or conservatory. They are slow growing. *Dioon* derives from the Greek *dis*, twice and *oon*, an egg; the seeds are carried in pairs.

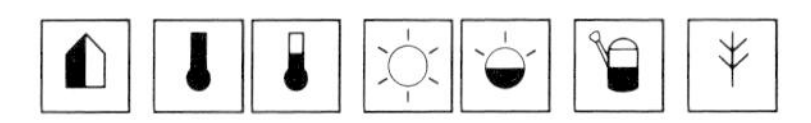

Species cultivated

D. edule Mexican fern palm, Chestnut dioon/dion Mexico

In its native country, a dwarf palm-like tree with a trunk to 2m(6$\frac{1}{2}$ft) in height; in containers trunks of even 60cm(2ft) will take many years to form. Leaves 60–120cm(2–4ft) or more long, composed of many narrow, spine-tipped leaflets, deep green with a waxy bluish patina; the stalks of young leaves are white-woolly.

D. spinulosum Mexico, Yucatan

Similar to *D. edule*, but in the wild eventually much taller and more elegant. Leaflets deep green and with spiny-toothed margins. Equally as slow-growing as *D. edule*, possibly even more so.

DIOSCOREA

Dioscoreaceae

True yams

Origin: Tropics and sub-tropics – widespread. A genus of at least 500 species of mainly herbaceous, tuberous or rhizomatous perennials with twining stems. Among their number are several species which are an important food crop in tropical countries. A much lesser number have ornamental foliage and are well worth trying in a warm room or conservatory. Propagate by dividing the rootstock, by cuttings of young basal shoots or by seed in spring. *Dioscorea* honours Pedanios Dioscorides, the first century Greek doctor, herbalist and author of the original *Materia Medica*.

Dioon edule

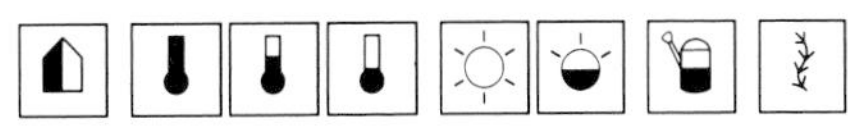

Species cultivated

D. amarantoides See under *D. dodecaneura*.

D. batatus (*D. opposita*) Chinese/cinnamon yam Eastern Asia

Stems up to 3m(10ft), bearing small axillary tubers. Leaves in opposite or almost opposite pairs, ovate, cordate and slender pointed, up to 8cm(3in) long. Flowers tiny, insignificant, in small axillary spikes, notable only for their cinnamon scent. Temperate to cool.

D. bulbifera Air potato, Aerial yam E. Asia
Stems to 6m(20ft) bearing tubers of up to 1km(2lb) or more in weight. Leaves alternate to sub-opposite, ovate, cordate, 15–25cm(6–10in) long. The abundantly produced aerial, edible tubers give this plant curiosity value. Temperate to tropical.
D. discolor Ornamental yam Surinam
Stems on mature plants exceeding 2m(6½ft) in height. Leaves alternate, ovate, cordate, 10–15cm(4–6in) long, velvety, olive-green with a pattern of silvery or silvery-pink veins above, purple beneath. One of the most handsome of the variegated-leaved yams. Tropical.
D. dodecaneura Brazil
Stems to 2m(6½ft) or more. Leaves narrowly ovate, cordate, up to 15cm(6in) long, bronzy-green with a somewhat metallic sheen, the veins picked out in grey with a wider zone along the midrib, purplish beneath. Tropical. This appears to be the correct name for plants listed as *D. amarantoides* 'Eldorado' and *D. multicolor* 'Eldorado'.
D. elephantipes See *Testudinaria elephantipes*.
***D. multicolor* 'Eldorado'** See *D. dodecaneura*.
D. opposita See *D. batatus*.

RIGHT *Dioscorea discolor*
BELOW *Diplazium proliferum*

DIOSMA See AGATHOSMA & COLEONEMA
DIPLACUS See MIMULUS
DIPLADENIA See MANDEVILLA

DIPLAZIUM

Athyriaceae (Aspidiaceae)

Origin: Tropics and northern hemisphere – widespread. A genus of about 400 species of ferns of medium to large size. The genus is a variable one with affinities to *Athyrium* (lady ferns) and *Asplenium* (spleenworts). Some are rhizomatous, others of tufted habit and a few are low tree ferns. Their fronds range from simple to bi- or tri-pinnate. The evergreen species make pleasing foliage specimens. Propagate by spores, detaching well grown plantlets from the leaves and where possible, by division, all in spring. *Diplazium* originates from the Greek *diplasios*, double, alluding to the sporangia which occur in pairs on either side of a leaf veinlet.

Species cultivated
D. esculentum (*Athyrium esculentum*)
Vegetable fern, Paco India to Polynesia
Rhizome erect and trunk-like, to 60cm(2ft) in height. Leaves in a terminal rosette, up to 1.2m(4ft) or even longer on well grown mature specimens, simply pinnate when young, later bipinnate, bright green. Best in partial shade or even with some full light.
D. proliferum Tropical Africa, Asia and Polynesia
Tufted habit, the simply pinnate fronds 60–120cm(2–4ft) or more in length, arching, the pinnae narrowly triangular, sessile, leathery and bright green. Mature fronds bear plantlets in the axils of the pinnae with the main leaf stalk.

DISCHIDIA

Asclepiadaceae

Origin: Indo-Malaysia to Polynesia and Australia. A genus of 80 species of epiphytic climbing plants allied to *Hoya* which cling to their supports by climbing stems and/or aerial roots. In certain species some of the leaves are modified into pitchers, the function of which is not fully understood

LEFT *Disocactus nelsonii*
FAR LEFT *Dischidia bengalensis*

though they seem to be primarily rainwater and humus collectors. Frequently the roots of the plant grow into the pitchers, taking up water and food substances from the humus. In the wild, ants may inhabit the pitchers and may keep away other insects which would feed upon the plant. The structure of the flowers resembles that of *Hoya*, but the outer segments curve down to form an urn or bell-shape, obscuring the central corona. Although interesting, these flowers are not large enough to be showy and the plants are mainly grown for their leaves and pitchers. The species described below are best grown in the conservatory, but are also worth trying in the home as companion plants to epiphytic orchids. They are best trained on moss poles or cylinders. Propagate by cuttings in summer or by seed in spring. *Dischidia* derives from the Greek *dischides*, twice parted or cleft, referring to the segments of the corona.

Species cultivated

D. bengalensis India
Stems 1–2m(3–6½ft) or more in length, but can be kept to the height of the moss pole. Leaves in pairs, fleshy, glaucous, 2.5–4cm(1–1½in) long, obovate to elliptic. Flowers 5mm(⅕in) long, in axillary umbels.

D. rafflesiana Malayan urn vine Malaysia, New Guinea
Stems eventually to several metres (yards) in length but easily kept in check. Leaves in pairs, fleshy, grey-green to deep green, about 2.5cm(1in) long, broadly ovate to circular. Pitchers pear-shaped, 5–13cm(2–5in) long, purplish within. Flowers 6mm(¼in) long, cream with a purple throat.

DISOCACTUS

Cactaceae

Origin: Northern South America, Cental America, Mexico and West Indies. A genus of about seven species of epiphytic cacti allied to *Epiphyllum*. They have slender, cylindrical main stems which bear flattened, leaf-like branches. The trumpet-shaped flowers appear from notches on the edges of these flattened stems. Suitable for the home and conservatory, these graceful cacti make good hanging basket subjects; cultivate as for bromeliads. Propagate by seed in spring or by cuttings in summer, using mature whole lateral branches or their tips. *Disocactus* derives from the Greek *dis*, twice, *isos* equal and *cactus*, referring to the two equal-numbered rows or rings of petals.

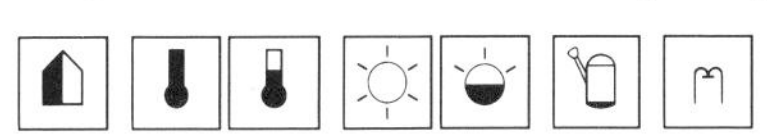

Species cultivated

D. macranthus (*Pseudorhipsalis macrantha*) Mexico
Stems arching to pendent, up to 90cm(3ft) in length, the flattened side branches up to 5cm(2in) wide. Flowers 8cm(3in) wide by 5cm(2in) long, yellow, borne in autumn and winter.

D. nelsonii (*Chiapasia nelsonii, Epiphyllum nelsonii*) South Mexico, Guatemala
Stems at first erect, eventually arching to pendent, 60–120cm(2–4ft) in length; flattened side branches up to 25cm(10in) long by 3–4cm(1¼–1½in) wide. Flowers to 10cm(4in) long by 6cm(2½in) wide, rose-purple, the petal tips recurved, reputedly violet-scented. A showy and free-flowering species.

DIZYGOTHECA

Araliaceae

Origin: Australasia. A genus of 17 species of shrubs and small trees with alternate digitate leaves having seven to 11 leaflets which are often wavy-edged or lobed. Only the first species described is commonly cultivated, making a decorative foliage plant for the home or the warm conservatory, but best with additional humidity in a dry room. Propagate by imported seed in warmth, by air-layering in spring and summer or by taking stem sections as cuttings. *Dizygotheca* derives from the Greek *dis*, twice, *zygos*, a yoke and *thake*, a case, there being twice as many anther lobes as would be expected from the number of stamens.

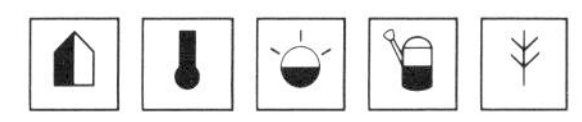

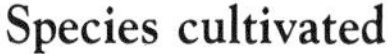

Species cultivated

D. elegantissima (*Aralia elegantissima*)
Erect, little branched shrub to 2m($6\frac{1}{2}$ft), making a small tree in the wild. Leaves long-stalked, with seven to ten linear-lanceolate, deeply toothed leaflets to 15cm(6in) long by 1.5cm($\frac{5}{8}$in) wide; when young they are a shining coppery-red becoming deep green as they mature. Flowers are not produced on pot specimens. Large, flowering-sized specimens have much broader, stiffer leaflets.

D. kerchoveana (*Aralia kerchoveana*) Polynesia
Juvenile plants similar to those of *D. elegantissima*, but leaflets usually nine to 11, up to 1.5cm($\frac{5}{8}$in) wide, lustrous deep green with a pale midrib.

D. veitchii (*Aralia veitchii*) New Caledonia
Juvenile plants with slender, erect stems to 2m($6\frac{1}{2}$ft) or more. Leaves digitate, composed of seven to 11, narrowly elliptic, wavy-edged leaflets to 13cm(5in) long, coppery-green above, deep red beneath.

RIGHT *Dizygotheca elegantissima*
BELOW *Dodonaea viscosa* 'Purpurea'
BOTTOM *Dodonaea viscosa*

DODONAEA

Sapindaceae

Origin: Tropics and sub-tropics, chiefly Australia. A genus of 60 species of evergreen trees and shrubs with alternate leaves and small, insignificant flowers. One is grown as an effective foliage plant. Propagate by seed in spring in warmth, or by cuttings. *Dodonaea* was named for Rembert Dodoens (1575–1585), Professor of Medicine at Leiden, Royal physician and herbalist, and author of works on plants.

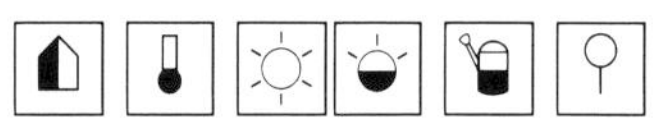

Species cultivated

D. viscosa Hop bush, Akeake
Shrub to 2m($6\frac{1}{2}$ft) or more in a container, making a 6m(20ft) tree in the wild. Leaves 6–10cm($2\frac{1}{2}$–4in) long, about 2.5cm(1in) wide, narrowly ovate to elliptic, firm-textured. Flowers petalless, 4mm ($\frac{3}{16}$in) long, greenish-yellow, in small panicles, followed by seed capsules 1.5cm($\frac{5}{8}$in) long and wide, 2- or 3-winged, reddish to purple. *D.v.* 'Purpurea', with purple or coppery foliage, is the form most frequently cultivated.

DOLICHOS

Leguminosae

Origin: Tropics and sub-tropics. A genus of 70 species of perennials, annuals, sub-shrubs and climbers all of which have trifoliate leaves and white to purple flowers typical of the pea family which can be axillary or terminal. The fruits are pods, often curved and flattened, and in the species which is usually grown, edible. In spite of this it is grown as an ornamental conservatory plant producing its flowers in the first season from seed. Propagate by seed sown in spring. *Dolichos* was the ancient Greek name for a long-podded bean, used for this genus by Linnaeus.

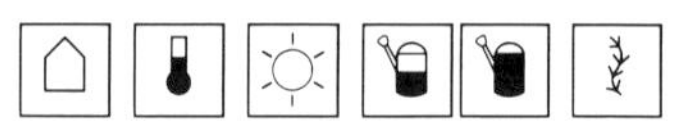

Species cultivated

D. lablab Hyacinth bean, Indian bean, Lablab bean Tropical Asia
Short-lived perennial climber, 2–5m($6\frac{1}{2}$–16ft) tall; to 9m(30ft) in the wild; often grown as an annual. Leaves with ovate leaflets, 8–15cm(3–6in) long, the lateral pair obliquely ovate. Flowers 1.5–2cm($\frac{5}{8}$–$\frac{3}{4}$in) long, purple or white, borne in axillary racemes to 30cm(1ft) long, opening in summer and autumn.

D.l. lignosus (*D. lignosus*) Australian pea Asia
Semi-erect, bushy climber, usually grown as an annual. Leaves triangular-ovate, to 4cm($1\frac{1}{2}$in) long. Flowers rose-purple to white, clustered at the end of the stalks in summer and autumn.

DOLICHOTHELE

Cactaceae

Origin: Texas and Mexico. A genus of about 12 species of small cacti, by some authorities included in the very closely related *Mammillaria*. They have globular, clustering stems with large cylindrical tubercles and relatively large, widely funnel-shaped flowers. Like *Mammillaria*, the species described below are free-flowering and thrive in the home and conservatory. Propagate by seed in spring, or by removing offsets in spring and summer. *Dolichothele* derives from the Greek *dolichos*, long and *thele*, a nipple, an allusion to the very prominent tubercles.

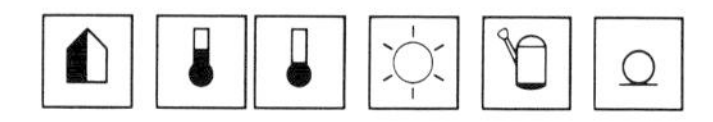

Species cultivated

D. longimamma (*Mammillaria longimamma*) Central Mexico
Stems 8–13cm(3–5in) high and wide, clustering and eventually forming large hummocks. Tubercles to 5cm(2in) or more long, tipped by a rosette of 1–2cm($\frac{3}{8}$–1in) long radial, and one to three central spines. Flowers 6cm($2\frac{1}{2}$in) wide, canary-yellow, freely produced. An easily grown species, tolerating partial shade.

D. sphaerica (*Mammillaria sphaerica*)
N. Mexico, S. Texas
Stems about 5cm(2in) high and wide, clustering. Tubercles up to 1.5cm($\frac{5}{8}$in) long, pale green, tipped by a rosette of nine to 15 slender, yellow to white spines, each one up to 1cm($\frac{3}{8}$in) long. Flowers 6cm($2\frac{1}{2}$in) or more wide, pale yellow.

DOMBEYA

Byttneriaceae (*Sterculiaceae*)

Origin: Africa, Malagasy and Mascarene Islands. A genus variously estimated at between 50 and 350 species of shrubs and trees. Those most commonly cultivated have palmate, lobed leaves and mallow-like 5-petalled flowers arranged in dense, umbel-like cymes from the leaf axils. They are best in the conservatory, but small flowering specimens can be brought into the home. *Dombeya* commemorates Joseph Dombey (1742–94), a French botanist who collected plants in South America and died in a Spanish prison.

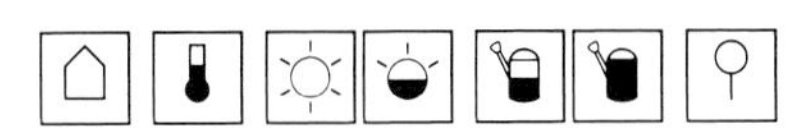

Species cultivated

D. burgessiae South Africa to Kenya
About 2m($6\frac{1}{2}$ft) tall in containers, but capable of twice this when planted out, well branched. Leaves up to 20cm(8in) long and wide, usually sharply 3-angled or lobed, softly downy. Flowers about 2.5cm(1in) long, white with pink to red veins, fragrant, opening in late summer and autumn.

D. × cayeuxii (*D. burgessiae × wallichii*)
Garden hybrid; to 5m(15ft) when planted out, but easily kept to 2m($6\frac{1}{2}$ft) in a container. Leaves 20–30cm(8–12in) long and wide, cordate, toothed, hairy. Flowers pink, 2.5–4cm(1–$1\frac{1}{2}$in) long, many together in almost spherical umbels 10–13cm(4–5in) wide, pendent on long stalks, opening in late autumn to early spring.

D. tanganyikensis East Africa
A rather smaller species than those described above, up to 2m($6\frac{1}{2}$ft) and easily kept to half this. Leaves 8–13cm(3–5in) long and wide, shallowly 3- to 5-lobed, toothed and densely hairy. Flowers 2.5–4cm(1–$1\frac{1}{2}$in) long, pink, in long-stalked umbels to 8cm(3in) wide.

D. wallichii Malagasy
In the wild a small tree to 10m(30ft) in height, but easily kept to about 2m($6\frac{1}{2}$ft) in a tub. Leaves rounded to broadly ovate, cordate, toothed, up to 30cm(12in) long, sometimes 3-winged. Flowers red or pink, about 4cm($1\frac{1}{2}$in) long, in dense, almost spherical umbels, pendent on long stalks, opening in winter.

DOROTHEANTHUS

Aizoaceae

Origin: South Africa. A genus of ten species of succulent annuals formerly classified in *Mesembryanthemum* and still included in that genus in many gardens. They have linear or spathulate to almost cylindrical leaves and stalked, daisy-like flowers which only open in sun. They make attractive pot plants for the sunny conservatory and will flower in the home. Propagate by seed sown in spring. *Dorotheanthus* was named for Dorothea, mother of Professor C. Schwantes, a specialist in the study of *Mesembryanthemum* and its allies, with the Greek *anthos*, a flower.

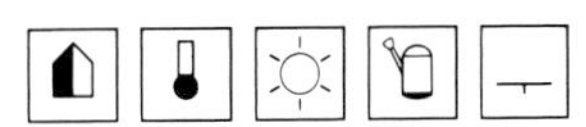

Species cultivated

D. bellidiformis (*Mesembryanthemum criniflorum*) Livingstone daisy S.W. Cape
Prostrate, mat-forming annual to 30cm(1ft) or more wide. Leaves obovate to almost cylindrical, 8cm(3in) long, covered with tiny glistening spots like sugar grains. Flowers 3–4in($1\frac{1}{4}$–$1\frac{1}{2}$in) wide with petals in shades of red, pink, orange or white, with darker centres, sometimes bicoloured, the stamens being blackish-purple; borne singly in summer.

D. tricolor (*Mesembryanthemum tricolor*)
Spreading annual forming mats 30cm(1ft) across and 8cm(3in) high. Leaves linear to spathulate, 5–8cm(2–3in) long. Flowers 4cm($1\frac{1}{2}$in) wide, the petals red, white or pink, often with a contrasting centre, and black-purple stamens; they are borne singly in summer.

LEFT *Dombeya × cayeuxii*
FAR LEFT *Dolichothele longimamma*

Dorstenia foetida

DORSTENIA

Moraceae

Origin: Mainly tropical America and Africa with at least one species in India. A genus of about 170 species of perennials and sub-shrubs, some evergreen, others deciduous; some are semi-succulent or otherwise adapted to arid climates. Although not showy the inflorescence of this genus is intriguing and unique. The tip of each flowering stem is expanded into a flat or hollowed, somewhat fleshy structure in which are embedded numerous tiny green florets. The equally small, one-seeded fruits which follow are ejected to a considerable distance by pressure from the turgid tissue surrounding. The mechanism is comparable to that of squeezing an orange pip between thumb and finger. Only the species *D. contrajerva* is commonly seen in botanical collections, but the others described below are sometimes available and worth looking out for. All are best in a conservatory, but would probably grow also in the home. *Dorstenia* honours Theodore Dorsten (1492–1552), a German botanist and Professor of Medicine at Marburg.

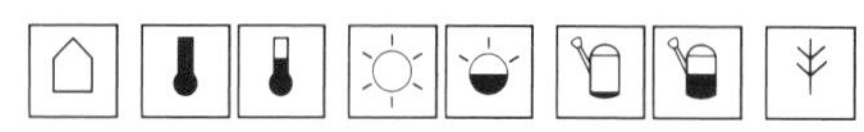

Species cultivated

D. argentata (*D. argentea*) Brazil
Erect sub-shrub to 30cm(1ft) or so in height. Leaves broadly lanceolate, up to 13cm(5in) long, basically deep green but with a broad, irregular, silvery zone down the centre, in some examples covering most of the leaf surface. Inflorescence concave, margined with purplish tubercles.

D. argentea See *D. argentata*.

D. barteri West Africa
Shrubby, eventually to 60cm(2ft) tall, the erect stems sparingly branched. Leaves 13–18cm(5–7in) long, elliptic, usually hairy. Inflorescence more or less circular, flat, 2.5–5cm(1–2in) wide, the greyish flower cluster surrounded by a membranous green margin from which radiate slender, ray-like projections.

D. contrajerva Northern South America to Mexico and the West Indies
Evergreen perennial, 30cm(1ft) or more tall, with slowly creeping rhizomes. Leaves long-stalked, the blades 10–20cm(4–8in) long, roughly triangular in outline, 3-lobed, the central one largest and pointed-tipped, glossy dark green. Inflorescence 2.5–5cm(1–2in) wide, flattish, somewhat 4-angled and irregularly lobed.

D. foetida Southern Arabia
In the arid wilds of its homeland this shrubby species forms low mounds of thickened and contorted stems arising from a flattened, swollen stem (caudex) partially imbedded in the soil. In cultivation the stems are erect, more succulent and taller, up to 15cm(6in) or so. Leaves at stem tips only, lanceolate to narrowly ovate, finely crimped at the margins and with a pale vein pattern. Inflorescence about 2cm($\frac{3}{4}$in) or more wide, circular, flattened, with several slender marginal rays creating a starry appearance. In bright light the inflorescences take on a reddish tinge; they are freely produced throughout summer.

D. gigas Southern Arabia, Socotra
The most convincingly shrub-like species, with a swollen, sparingly branched stem eventually to 1.2m(4ft), but very slow growing. Leaves 10–15cm(4–6in) long, oblanceolate, finely bullate with recurved margins, glossy deep to olive-green. Inflorescences similar to those of *D. foetida*, but with fewer, shorter rays and less profuse.

DOXANTHA

Bignoniaceae

Origin: South America. A genus of an indeterminate number of species, some being classified in *Bignonia*, others in *Macfadenya* with little agreement among botanists as to their correct naming. They are attractive evergreen, woody climbers with some leaflets modified into tendrils and somewhat foxglove-like flowers, best grown in the conservatory where they can be trained up the back wall and under the roof. Propagate by cuttings in summer or by seed in spring. *Doxantha* derives from the Greek *doxa*, glory and *anthos*, a flower.

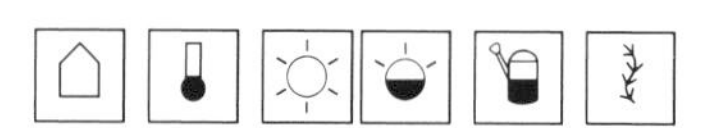

Species cultivated

D. capreolata See *Bignonia capreolata*.

D. unguis-cati Cats' claw vine Argentina
Slender climber to 10m(32ft). Leaves with two slender, lanceolate leaflets 4–5cm($1\frac{1}{2}$–2in) long and three straight, unbranched tendrils ending in sharply hooked tips which will cling to any rough surface. Flowers to 10cm(4in) across, tubular, opening to five broad lobes, yellow, borne singly in the leaf axils in late spring and summer.

DRACAENA

Agavaceae (Liliaceae)

Origin: Tropical and sub-tropical Africa and Asia. A genus of 150 species of trees and shrubs with

FAR LEFT *Doxantha unguis-cati*
LEFT *Dracaena angustifolia*
BELOW *Dracaena deremensis* 'Bausei'
BOTTOM *Dracaena goldieana*

lanceolate, leathery leaves growing either in tufts at the tips of the stems or more normally along the stem lengths, in the former case giving the plant a rather palm-like appearance. The flowers are small and rather insignificant and are carried in panicles or occasionally in heads or spikes. They are followed by often colourful berries. Dracaenas make attractive foliage plants for house or conservatory. Propagate by cuttings, either basal, tip or stem sections in spring or summer in warmth, also by seed in warmth. *Dracaena* derives from the Greek *drakaina*, a dragon, from the dragon tree (*D. draco*) of the Canary Islands.

Species cultivated

D. angustifolia Solomon Islands
To 2m(6½ft) or more in a large container, but easily kept to half this. Stems slender, suckering and branching. Leaves 15–25cm(6–10in) long, narrowly lanceolate, recurved, leathery-textured, glossy rich green. *D.a.* 'Honoriae' has creamy-yellow leaf margins. A graceful species deserving of wider recognition.

D. deremensis Tropical Africa
To 3m(10ft) or more in a pot, slow-growing. Leaves to 45cm(1½ft) long by 5cm(2in) wide, sword-shaped, glossy green with faint longitudinal striping, usually erect, occasionally arching over. It sometimes produces its reddish flowers when small. *D.d.* 'Bausei' has two broad creamy-white bands separated by a narrow greyish line; 'Warneckei' has green leaves with pale grey-green bands edged with yellow.

D. draco Dragon tree Canary Is.
To 3m(10ft) or so in a pot, capable of reaching 15m(50ft) or so in the open, but very slow-growing. In a tub it will remain single-stemmed, but when mature it flowers, the stems then branching and building up a wide head. Leaves 40–60cm(16–24in) long, glaucous green, spine-tipped. Flowers greenish, followed by bright orange fruits.

D. fragrans Corn plant Tropical Africa
Stems to 3m(10ft) in a pot, twice this in the open. Leaves to 60cm(2ft) or more long and 10cm(4in) wide, strap-shaped, gracefully arching over and down. The fragrant yellow flowers, when produced, are followed by orange-red fruits. *D.f.* 'Massangeana' has broad yellow and pale green central bands; 'Victoria' has a silvery grey-green central band bordered with creamy-yellow.

D. goldieana Tropical Africa
Stems erect, fairly slender to 1.2m(4ft) or more but slow-growing. Leaves ovate with distinct stalks, to 22cm(9in) long, basically a lustrous deep green, but closely cross-banded with dense zones of grey spots. Flowers opening at night are white, fragrant, in globular heads, followed by round red fruits 1.2cm(½in) in diameter.

D. godseffiana See *D. surculosa*.

D. hookeriana South Africa
Stems erect and stout, to 2m(6½ft) or more in height. Leaves 45–75cm(1½–2½ft) long, linear, firm-textured, glossy-green, in substantial terminal rosettes. *D.h.* 'Gertrude Manda' is almost identical,

ABOVE *Dracaena surculosa*
TOP *Dracaena marginata* 'Tricolor'

but with heavier textured leaves up to 10cm(4in) wide. It is reputedly a hybrid between *D. hookeriana* and the rarely seen *D. grandis*. *D.h.* 'Latifolia' has a very different appearance with leaves up to 9cm(3½in) wide; 'Variegata' has white-striped leaves.

D. indivisa See *Cordyline indivisa*.

D. marginata Madagascar dragon tree Malagasy
To 1–2m(3–6½ft) high in a pot, considerably more in the open, but always slow-growing. Leaves 30–40cm(12–16in) long, narrowly linear, green edged with red, with the central vein of the same colour. Very closely allied to *D. concinna* and *D. cincta* which may sometimes replace it in cultivation. *D.m.* 'Variegata' has cream-striped leaves; 'Tricolor' is similar, but with additional red-edging to its leaves.

D.* × *masseffiana (*D.* 'Pennock') (*D. fragrans* 'Massangeana' × *D. surculosa*)
Stems fairly slender, erect to leaning, usually unbranched. Leaves broadly lanceolate to narrowly ovate, about 15cm(6in) or more long, strongly arching, deep lustrous green with a sparse rash of yellow speckles. Mature plants may bear conspicuous spike-like panicles of pale yellow flowers in summer.

D. phrynioides Fernando Po.
Stem fairly robust and short, to about 30cm(1ft), rarely more. Leaves crowded, with a distinct stalk, 15–20cm(6–8in) long, ovate, taper-pointed, dark green with yellow spots. Flowers pure white, in dense head-like sessile clusters in late summer, but largely hidden by the leaves.

D. reflexa See *Pleomele reflexa*.

D. sanderiana Ribbon plant Central Africa
Stems cane-like, to 2m(6½ft) high. Leaves 15–25cm(6–10in) long, narrowly lanceolate, pale to grey-green with bold creamy-white margins.

Dracunculus vulgaris

D. surculosa (*D. godseffiana*) Gold dust dracaena Central Africa
A very distinct species making a shrub to 1m(3ft) high, with slender-branched stems. Leaves 8–13cm(3–5in) long, broadly elliptic to obovate, dark green dusted with creamy-white spots; they are carried in twos and threes and are widely spaced along the stems. Racemes of small, greenish-yellow flowers are followed by red fruits to 2cm(¾in) or more. *D.s.* 'Florida Beauty' has the white spotting merging to produce a leaf more white than green.

D. terminalis See *Cordyline terminalis*.

D. thalioides Tropical Africa, Sri Lanka
Stems robust, eventually to 60cm(2ft) or more, but fairly slow-growing. Leaves lanceolate to 30cm(1ft) or more in length, leathery-textured, deep green, each one carried on a longish, grey-spotted stalk. Mature plants bear erect racemes of sepia-tinted white flowers in summer.

D. umbraculifera Mauritius, Java
Stems robust, usually unbranched, to about 2m(6½ft) in containers; attaining 3m(10ft) or more planted out. Leaves strap-shaped, arching, to about 90cm(3ft) long, rich, somewhat lustrous green. Well established plants bear conspicuous terminal panicles of red-tinted white flowers in summer.

DRACUNCULUS

Araceae

Origin: Mediterranean region and the Canary Islands. A genus of two species of tuberous-rooted herbaceous perennials with rounded rhizomes from which arise erect arrow-shaped leaves, their bases sheathing each other. Tiny flowers are borne on an erect spadix within an arum-like spathe. The species described below makes an interesting foliage plant with the flowers as a bonus. Unfortunately, the flowers smell foul and at that stage plants cannot be kept in the home. Propagate by removing offsets when dormant, or by seed sown, if possible, as soon as ripe. *Dracunculus* is Latin for a little dragon, originally referring to a plant no longer identifiable and taken up for this genus.

Species cultivated

D. vulgaris (*Arum dracunculus*) Dragon plant
Mediterranean
Up to 1m(3ft) tall, the apparent stem being composed of the rolled, flattened, whitish leaf stalks which are strongly mottled with green. Leaves deeply cut into 11 to 15 lanceolate-oblong leaflets up to 15–20cm(6–8in) long. Spathes 25–35cm(10–14in) long, wavy-margined, soft-textured and usually folding back or spreading horizontally, chocolate to reddish-purple; spadix to 30cm(1ft) or more, dark brown, foul-smelling. Flowering takes place in summer. Fairly hardy.

FAR LEFT *Dregea corrugata*

DREGEA

Asclepiadaceae

Origin: Tropical and South Africa to China. A genus of 12 species of climbing plants very closely related to *Asclepias*. Only one species is cultivated and this makes a very decorative plant for the conservatory. Propagate by seed in spring or by cuttings in late summer. *Dregea* commemorates Johan Franz Drege (1794–1881), a German botanist who collected plants in South Africa.

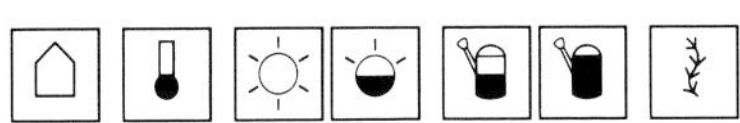

Species cultivated

D. corrugata (*Wattakaka sinensis*) China
Stems woody, twining, 2–3m($6\frac{1}{2}$–10ft) in length. Leaves more or less evergreen, or deciduous after long-term hard frost, in opposite pairs, ovate-cordate, to 10cm(4in) long, densely grey velvety downy beneath. Flowers 5-petalled with a small corona, white or cream with red speckling, 1.5cm($\frac{5}{8}$in) wide, fragrant, borne in axillary, stalked umbels during the summer. In overall floral effect there is a strong resemblance to *Hoya*. Occasionally the intriguing seed pods form; these are paired, incurving, spindle-shaped follicles, either smooth or corrugated and 5–7cm(2–$2\frac{3}{4}$in) long. Fairly hardy.

DREJERELLA See BELOPERONE

DRIMIOPSIS

Liliaceae

Origin: Tropical and South Africa. A genus of 22 species of bulbous plants related to *Scilla*. Unlike *Scilla*, the value of the species cultivated is more in their broad, spotted leaves than in the small white to greenish flowers. Propagate by seed in spring or by separating offsets when dormant in late summer. *Drimiopsis* derives from the Greek *Drimia*, a closely allied genus of bulbous plants and *opsis*, like or resembling; *Drimia* is from the Greek *drimys*, acrid – the irritant sap.

   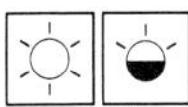  

LEFT *Drimiopsis maculata*

Species cultivated

D. kirkii Tanzania
Bulb globose, about 4cm($1\frac{1}{2}$in) or more thick. Leaves lanceolate, narrowed to a stalk-like base, up to 30cm(12in) long, somewhat fleshy, bluish-green with dark spots. Flowers about 6mm($\frac{1}{4}$in) long, white, in a dense short spike above the leaves, appearing in spring or early summer.

D. maculata South Africa
Bulb globose, about 2.5–4cm(1–$1\frac{1}{2}$in) thick. Leaves ovate, more or less cordate, with a definite stalk, wavy-margined, deep glossy green with dark spots. Flowers about 5mm($\frac{1}{5}$in) long, white ageing greenish, in short dense spikes above the leaves in spring.

DRIMYS

Winteraceae

Origin: South America, Australasia to Borneo. A genus of 65 species of evergreen trees and shrubs with alternate leaves and 5- to 20-petalled flowers in umbel-like clusters. They are followed by fleshy, berry-like fruits and make splendid foliage plants for the cool conservatory with a bonus of attractive flowers when the plant is sufficiently mature. Propagate by cuttings with a heel in summer or by seed sown, if possible, as soon as ripe. *Drimys* is the Greek word for acrid, referring to the taste of the bark of *D. winteri* which can be used medicinally.

 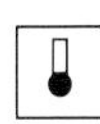 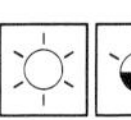  

Species cultivated

D. aromatica See *D. lanceolata.*

D. colorata (*Pseudowintera colorata*) Pepper tree New Zealand
Aromatic shrub to 2m($6\frac{1}{2}$ft) tall. Leaves to 6cm($2\frac{1}{2}$in) long, elliptic to obovate, green flushed with bronzy-red, glaucous beneath. Flowers to 1cm($\frac{3}{8}$in) wide, 5- to 6-petalled, greenish-yellow, in axillary clusters of two to five in summer. Fruits are almost dry, red to black berries.

D. lanceolata (*D. aromatica, Winterania latifolia, Tasmannia latifolia*) Mountain pepper
S.E. Australia, Tasmania
A dense shrub in a container, but making a tree to 5m(16ft) in the wild. Young stems bright red, becoming brown. Leaves about 5–8cm(2–3in) long, but very variable, elliptic to oblanceolate, shining dark green. Flowers 1.2–2cm($\frac{1}{2}$–$\frac{3}{4}$in) wide, dioecious, the male flowers with up to eight creamy-white petals, the female with usually no more than four. They are borne from the terminal leaf axils, each on a single stalk, but appearing as an umbel.

RIGHT *Drimys winteri*
FAR RIGHT *Drosanthemum hispidum*

D. winteri (*Wintera aromatica*) Winter's bark
The name of a very varied species, in its broadest sense ranging from Mexico south to Tierra del Fuego, but now considered to be correct only for plants from central Chile and adjacent Argentina. It makes a shrub in a container, a tree to 5m(16ft) or more in the open garden in a mild climate. Leaves oblong-elliptic to 20cm(8in) long, glossy dark green, glaucous beneath. Flowers 3–4cm($1\frac{1}{4}$–$1\frac{1}{2}$in) wide, with five to 20 creamy-white petals, borne in branched clusters in early summer; they are fragrant.

D.w. chilensis
Branches freely from the base, the flowers having six to 14 petals, borne in clusters of four to seven; *D.w. andina* is smaller and bushier.

DROSANTHEMUM

Aizoaceae

Origin: South Africa. A genus of 70 to 95 species of flowering succulents formerly included in *Mesembryanthemum*. Technically dwarf shrubs, they have pairs of cylindrical to angled fleshy leaves which bear translucent papillae, suggesting rather the carnivorous plant genus *Drosera*. The daisy-like flowers are showy and freely borne, but open only in the afternoons. They are suitable for pots or hanging baskets. Propagate by seed in spring or by cuttings in summer. *Drosanthemum* derives from the Greek *drosos*, dew and *anthos*, a flower, which is somewhat confusing as it is the dew-like leaf papillae which are referred to.

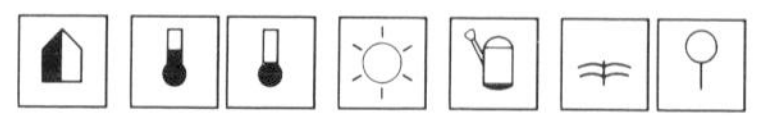

Species cultivated

D. floribundum Cape
Stems slender, creeping, freely branching, eventually forming low hummocks to 30cm(1ft) or more across. Leaves cylindrical to narrowly obovoid, 10–15mm($\frac{3}{8}$–$\frac{5}{8}$in) long, light green. Flowers pale glowing pink, almost 2cm($\frac{3}{4}$in) wide, opening in abundance during the summer and autumn.

D. hispidum Namibia to Namaqualand
Well-branched shrub, eventually to 60cm(2ft) tall by 90cm(3ft) or more wide, but easily kept to half

this. Leaves in well-spaced pairs, cylindrical, glossy light green, taking on reddish tints in bright light, 1.5–2.5cm($\frac{5}{8}$–1in) long. Flowers about 3cm($1\frac{1}{4}$in) wide, satiny, deep purple, opening in abundance from spring to autumn.

D. speciosum Dewflower Cape
More or less erect to spreading shrub, 30–60cm(1–2ft) high and wide, composed of slender, branched stems. Leaves semi-cylindrical, to 1.5cm($\frac{5}{8}$in) long, bright green, the pairs well spaced apart. Flowers about 5cm(2in) wide, bright orange-red with a green-tinted centre, opening mainly in summer, sometimes earlier and later.

DROSERA

Droseraceae
Sundews

Origin: Widespread in the tropical and temperate zones of both hemispheres. A genus of 100 species of carnivorous perennials and annuals. Among the perennials, some are deciduous and some evergreen. In the former group certain species are tuberous-rooted. There are two main groups: sessile, rosette-forming either solitary or in tufts or clumps, and stem-forming; the latter may be branched or unbranched, with alternate leaves. Some resemble small shrubs, others are climbers. All have 5- to 8-petalled flowers which, in some species, are comparatively large and showy. The leaves of all species are covered with prominent, red-stalked hairs (tentacles). These secrete a glistening sticky fluid which adheres as a terminal globule. Small creatures, mainly flies, mistake the fluid for nectar and are stuck to it. The struggling causes the surrounding tentacles to bend towards it thus smothering and pushing it down to the leaf surface. At this point a digestive enzyme breaks down the animal protein ready to be absorbed by the plant. Propagate by division or seed in spring. The perennial species described below make intriguing subjects for conservatory or home terrarium (plant case). *Drosera* derives from the Greek *droseros*, dewy, the leaf hairs bearing dew-like droplets.

   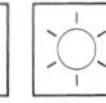  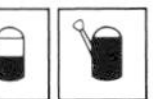 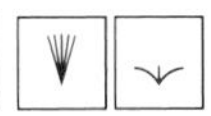

Species cultivated

D. anglica Great sundew Europe to northern Asia
Tufted to small clump-forming. Deciduous; leaves more or less erect, long-stalked, the blade narrowly obovate to oblong, up to 3cm($1\frac{1}{4}$in) long. Flowers in racemes, white, 5mm($\frac{1}{5}$in) wide, with five to eight petals, opening in summer.

D. binata (*D. dichotoma*) New Zealand, S.E. Australia including Tasmania
Tufted to small clump-forming. Evergreen; leaves long-stalked, erect, the blade forking into two linear lobes 8–15cm(3–6in) long, resembling the prongs of a hay fork. Occasionally leaves with four or more lobes are formed. Flowers 1.5–2.5cm($\frac{5}{8}$–1in) wide, in corymbose racemes 30–50cm(12–20in) tall, opening in summer.

D. capensis South Africa
Clump-forming. Evergreen; leaves 7cm($2\frac{3}{4}$in) or more long, narrowly oblong. Flowers in racemes 2cm($\frac{3}{4}$in) wide, purple, formed of five petals and opening in summer.

D. dichotoma See *D. binata*.

D. menziesii Pink rainbow Western Australia
Stems erect, 10–30cm(4–12in) tall, from a small tuber. Deciduous; leaves in small scattered clusters, up to 1.2cm($\frac{1}{2}$in) wide including the long, spreading tentacle hairs, red flushed. Flowers 2.5cm(1in) wide, pink to red, opening in summer.

D. rotundifolia Common sundew Temperate zone, widespread
Rosettes solitary or tufted. Deciduous; leaves long-stalked, on the soil or inclined above it, orbicular, to 1cm($\frac{3}{8}$in) wide, expanding in summer.

D. spathulata S.E. Australia, New Zealand
Tufted to clump-forming. Evergreen; leaves spreading, the blade 5mm($\frac{1}{5}$in) long, but tapering into a broad, reddish stalk and appearing larger. Flowers in short racemes, white, or rarely pink, about 1cm($\frac{3}{8}$in) wide, opening in summer.

LEFT *Drosera binata*
BELOW *Drosera rotundifolia*

DRYNARIA

Polypodiaceae

Origin: Old World tropics and Australia. A genus of about 20 species of mainly epiphytic, rhizomatous ferns, those described here making very effective hanging basket specimens. Like the better known stag's horn fern (*Platycerium*), they produce two very distinct types of frond. The sterile fronds are stalkless and press fairly closely to their tree branch host, acting as collectors of dead leaves and other organic debris which then breaks down into humus and is invaded by the fern's roots. The fertile fronds are long-stalked and simply pinnate to pinnatifid. Grown on a slab of rough bark, section of tree fern stem or in a basket, the ferns form compact mounds of overlapping sterile and gracefully arching fertile

ABOVE *Dryopteris filix-mas* 'Cristata'
ABOVE RIGHT *Drynaria quercifolia*

fronds, together creating a very pleasing composition. Propagate by spores or division in spring. *Drynaria* derives from the Greek *drys*, oak; the sterile leaves are rather like those of certain oaks.

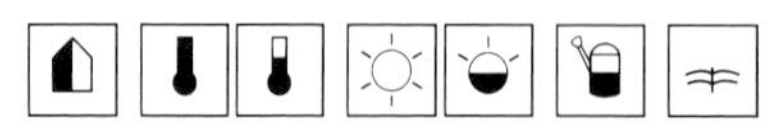

Species cultivated

D. quercifolia Oak fern India, Malaysia, Australia
Rhizomes thick, short and woody. Sterile fronds 20–30cm(8–12in) long, obovate, lobed. Fertile fronds 60–90cm(2–3ft) long including the stalk, by 30cm(1ft) or more wide, deeply pinnatifid. Sporangia in two rows between the lateral veins of the pinnae.

D. sparsisora
Almost identical to *D. quercifolia* above, but the sporangia are scattered irregularly.

DRYOPTERIS

Aspidiaceae

Buckler ferns, Wood ferns

Origin: World-wide from tropical to cool temperate regions. A genus of 150 species of evergreen and deciduous ferns with thick, short rhizomes and arching rosettes of fronds which are usually bi- or tri-pinnate. The following species are hardy, but make decorative pot plants for the cool conservatory or an unheated porch or room. Propagate by division of plants with two or more crowns, or by spores in spring. *Dryopteris* derives from the Greek *drys*, oak and *pteris*, a fern, referring to the supposed association of the fern with oak woods.

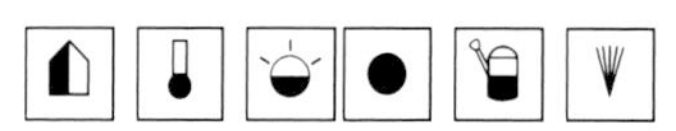

Species cultivated

D. borreri See *D. pseudomas*.

D. erythrosora Japanese buckler/shield fern China, Japan, Philippine Is.
Evergreen; fronds 30–60cm(1–2ft) or more long, broadly ovate in outline, bipinnate, the pinnules coarsely toothed, the whole frond flushed coppery-red when young. Sori bright red at first, darkening as the spores mature.

D. filix-mas Male fern North temperate zones and mountains in the southern hemisphere south to Peru, Malagasy, Java and Hawaii
Deciduous to semi-evergreen; fronds 30–120cm(1–4ft) long, oblong to elliptic-lanceolate, pinnate, the pinnae very deeply lobed and toothed so as to appear almost bipinnate. Many forms have arisen with varying frond-shapes. These include: 'Cristata', pinnae tips crested; 'Decompositum', fronds more finely cut; 'Linearis', slender, more elegant fronds; 'Polydactyla', frond and pinnae tips crested.

D. pseudomas (*D. borreri*) Golden scaled male fern Europe, S.W. Asia
Similar to *D. filix-mas*, usually evergreen with thicker textured, more erect fronds which are narrower in outline, the pinnules having only a few blunt teeth near their apex. The stalks and midribs of the fronds have shaggy tan to brown scales.

DUDLEYA

Crassulaceae

Origin: South-western USA, Mexico. A genus of between 40 and 60 species of succulents very closely related to *Echeveria*. They are mostly clump-forming evergreen perennials, though many species eventually produce stems and then appear semi-shrubby. The fleshy leaves are borne in terminal tufts or rosettes and are often grey to white-waxy farinose. The 5-petalled flowers may be bell- or star-shaped depending on how wide they expand. They are borne in cymose clusters on stems well above the leaves. Dudleyas thrive in containers, the white-leaved sorts making striking specimen plants. Propagate by seed in spring or by offsets removed in early summer. *Dudleya* honours William Russel Dudley (1849–1911), an early Professor of Botany at Stanford University, California.

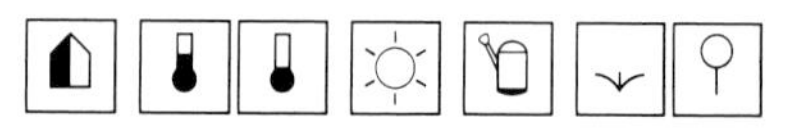

Species cultivated

D. attenuata (*Echeveria attenuata, E. edulis*) Baja California – sea coasts
Semi-shrubby, 15–30cm(6–12in) tall. Leaves erect, cylindrical to narrowly obovoid, glaucous, up to 10cm(4in) long. Flowering stems up to 25cm(10in) in height, with several to many branches; flowers starry, the petals to 8mm(⅓in) long, yellow with red lines, opening in summer.

D.a. orcuttii Southern California (USA) and northern Baja California (Mexico)
White, rose-flushed petals.

D. densiflora (*Echeveria nudicaulis*) California
Semi-shrubby, the thick, sparingly branched, erect stems to 10cm(4in) tall. Leaves cylindrical, pointed-tipped, 6–15cm(2½–6in) long, waxy white. Flowering stems 15–30cm(6–12in) in height, with several to many branches; flowers starry, the petals to 1cm(⅜in) long, white or pink-tinted, summer.

D. farinosa (*Echeveria compacta, E. eastwoodiae, E. septentrionalis*) Central and northern California – sea coasts
Semi-shrubby, the usually well-branched stems 10–30cm(4–12in) tall. Leaves ovate-oblong, pointed, 3–6cm(1¼–2½in) long, green or grey-white, sometimes red-flushed in strong light. Flowering stems 15–30cm(6–12in) or more in height with three to five branches, some of which can be forked above; flowers bell-shaped, the petals 1cm(⅜in) or more long, lemon-yellow, opening in summer.
D. hassei (*Echeveria hassei*) Santa Catalina and Guadalupe Islands
Semi-shrubby, the usually much-branched stem 10–30cm(4–12in) tall. Leaves linear-lanceolate, obtuse, 4–10cm(1½–4in) in length, white farinose. Flowering stems up to 30cm(1ft) tall with two to four branches which may be simple or forked; flowers starry, the petals up to 1cm(⅜in) long, white, opening in summer.

DURANTA

Verbenaceae

Origin: Mexico, Central and South America, Florida and West Indies. A genus of 36 species of trees and shrubs, one of which is widely grown in warm countries and as a conservatory pot plant in colder climes. Propagate by seed or layering in spring, or by cuttings in summer. *Duranta* commemorates Castore Durantes (c. 1529–1590), a Papal doctor and botanist in Rome.

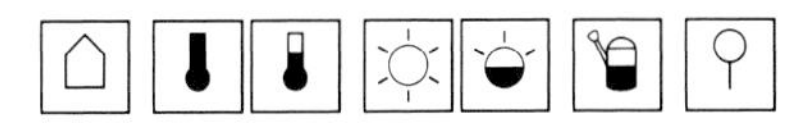

Species cultivated

D. repens (*D. plumieri*) Golden dewdrop, Pigeon berry, Skyflower Florida to Brazil, naturalized elsewhere
Shrub or small tree to 3m(10ft) or more planted out, but easily kept to half this or less in a container. Stems freely branching, erect to spreading, sometimes spiny. Leaves more or less evergreen, in opposite pairs, ovate to obovate, sparingly to coarsely toothed, 1.2–5cm(½–2in) long, or up to twice this on extra strong shoots. Flowers shortly tubular with five rounded petals, lilac-blue, about 1cm(⅜in) across, in simple or branched racemes to 15cm(6in) long, appearing intermittently from spring to autumn, but mainly in summer. Fruits berry-like, globular, yellow, to about 1cm(⅜in) wide, but sparingly produced on indoor pot plants. *D.r.* 'Alba' has pure white flowers; 'Variegata' has the leaves white margined.

DYCKIA

Bromeliaceae

Origin: South America. A genus of perhaps 80 species of evergreen, clump-forming perennials with narrow leaves in rosettes. Unlike most other members of the bromeliad family they are soil dwelling and adapted to arid conditions. The rigid leaves are usually semi-succulent within and margined with sharp spiny teeth. The 3-petalled flowers are a shade of yellow and carried in long spikes well above the leaves. Those described below thrive equally well in the home or conservatory and make striking specimens. However, because of the hook-like marginal spines of some species they must be sited in the home with care. Propagate by seed or separating offsets in spring. *Dyckia* is named for Prince Joseph Salm-Reifferscheid-Dyck (1773–1861), a German botanist who wrote several noteworthy books on succulent plants.

ABOVE LEFT *Dudleya hassei*
ABOVE *Dyckia fosteriana*

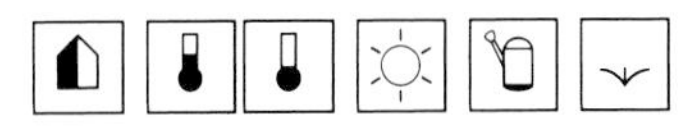

Species cultivated

D. brevifolia (*D. sulphurea, D. princeps*) Brazil
Leaves 20–30cm(8–12in) long, white-scaly beneath, arching, with 2mm($\frac{1}{16}$in) marginal spines. Flowers up to 2cm(¾in) long, yellow and green, in spikes about 30cm(1ft) tall.
D. fosteriana Brazil
A choice miniature species with spreading rosettes 10–13cm(4–5in) across. Leaves strongly recurved, green to silvery-grey, with comparatively large, hooked, marginal spines. Flowers deep yellow to orange, about 12mm(½in) long, in a loose, spiralling raceme, to 20cm(8in) tall.
D. princeps See *D. brevifolia*.
D. rariflora Brazil
Leaves 10–15cm(4–6in) in length, recurved, green and grey scaly with soft, blackish marginal spines. Flowers orange, 12mm(½in) long, up to 12, well spaced on a stem to 60cm(2ft) in height. This is the true plant described by J.H. Schultes (1804–40). The name is also an invalid synonym of the next species and the two are sometimes confused in cultivation.
D. remotiflora (*D. rariflora* of some authors, not J.H. Schultes) Uruguay, southern Brazil, northern Argentina
Leaves linear, up to 30cm(1ft) long, strongly recurved, bearing pale, smallish, hard straight spines. Flowers orange-yellow, 2–2.5cm(¾–1in) long, eight to 12 on a stem 30–45cm(1–1½ft) in height.
D. sulphurea See *D. brevifolia*.

E

ECBALLIUM

Cucurbitaceae

Origin: Mediterranean region. A genus of just one species of tuberous-rooted perennial usually grown as an annual, cultivated in pots mostly as a fun plant though it has decorative leaves. Propagate by seed. *Ecballium* is derived from the Greek *ekballein*, to cast out, as the fruits eject the seeds.

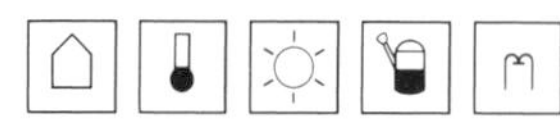

Species cultivated

E. elaterium Squirting cucumber

Prostrate, weak-stemmed trailer, with hispid stems to 60cm(2ft) in length. Leaves 4–10cm(1½–4in) long, triangular to cordate, rough-hairy, light greyish-green. Flowers yellow, unisexual, the females solitary with a prominent ovary, males in axillary racemes. Fruits 4–5cm(1¾–2in) long, ovoid, green and bristly hairy, when touched ejecting their seeds in a stream of watery fluid which comes out through the stalk hole.

ECCREMOCARPUS

Bignoniaceae

Origin: Chile and Peru. A genus of five species of evergreen climbers, becoming woody but usually grown as annuals or short-lived perennials. They have opposite pairs of pinnate or bipinnate leaves ending in branched tendrils and usually racemes of tubular flowers from a bell-shaped calyx. The species grown is a fine quick-growing pot plant capable of flowering when small. It can be grown in a conservatory or in the home around or over a wire support. Propagate by seed in early spring. *Eccremocarpus* derives from the Greek *ekkremes*, hanging and *karpos*, fruit.

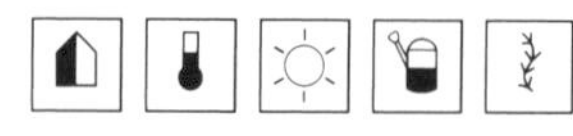

Species cultivated

E. scaber Chilean glory flower Chile

Climber with 4-angled stems, making 5m(16ft) or more in a season. Leaves bipinnate, ending in a pair of tendrils, the leaflets 1–2cm(⅜–¾in) long, asymmetrically ovate-cordate. Flowers 2–3cm(¾–1¼in) long, bottle-shaped, fuller below than above, opening slightly at the mouth; they are orange-red and borne in racemes 10–15cm(4–6in) long opposite the leaves. Fruits are inflated seed pods 4cm(1½in) long, the flattened seeds having a circular, membranous wing.

E.s. aureus

Golden-yellow flowers.

E.s. carmineus

Carmine-red flowers. Often incorrectly called *E.s. ruber*.

Ecballium elaterium

ECHEVERIA

Crassulaceae

Origin: The Americas from S.W. USA to Mexico and Argentina. A genus of 150 species of evergreen succulent perennials which are usually stemless with rosettes of fleshy, often attractive leaves. The flowers are tubular to bell-shaped with five fleshy petals. They are decorative plants for the conservatory and sunny windowsill. Propagate by detaching rosettes or by leaf cuttings, both in summer, or by division or seed in spring. *Echeveria* was named for the Spanish botanical artist Athanasio Echeverria Godoy who accompanied an expedition to Mexico in 1787–1797.

Species cultivated

E. affinis Black echeveria

Rosettes stemless or almost so, to 10cm(4in) wide, dense, solitary or in small groups. Leaves oblanceolate with a short pointed tip, brownish-olive to greenish-black in full light and with sparing watering; deep to bright green in poorer light and with regular watering. Inflorescence 20–35cm(8–14in), composed of several simple or forked cymes; flowers up to 1.2cm(½in) long, scarlet, expanding in late summer and early autumn.

E. agavoides (*Urbinia agavoides*)

Rosettes stemless, 10–25cm(4–10in) wide, solitary or in small clumps. Leaves thick, ovate, pale green with a deep red to brownish spine-like tip. Inflorescence 20–45cm(8–18in) tall, composed of one, two or more simple cymes; flowers about

1.2cm(½in) long, the red petals with yellow tips. An easily grown plant and one of the largest rosette species, but slow growing.

E.a. corderoyi
More robust, with leafier rosettes and more flowers per inflorescence. It also offsets more freely, forming large clumps and stands some frost.

E. atropurpurea (*E. sanguinea*)
Stem robust, usually unbranched, eventually to 20cm(8in) in height, topped by a rosette 20–30cm(8–12in) wide. Leaves spathulate to obovate, somewhat boat-shaped, coppery-red to purple with a glaucous patina. Inflorescence 30–60cm(1–2ft) high, racemose; flowers red, about 1.2cm(½in) long, appearing in summer.

E. attenuata See *Dudleya attenuata.*
E. compacta See *Dudleya farinosa.*
E. crenulata
Much like *E. gibbiflora*, but the leaves green, usually with wavy red margins.

E. derenbergii Painted lady
Clump- or mat-forming with stemless or shortly-stemmed rosettes to 6cm(2½in) or more wide. Leaves to 6cm(2½in) long, broadly spathulate, with a grey-white glaucous bloom and reddish margins and tips. Flowers 1–1.5cm(⅜–⅝in) long, orange, in arching racemes to 8cm(3in).

E. desmetiana See *E. peacockii.*
***E.* × 'Doris Taylor'** See under *E. pulvinata.*
E. eastwoodiae See *Dudleya farinosa.*
E. edulis See *Dudleya attenuata.*
E. elegans (of gardens) See *E. harmsii.*
E. gibbiflora
Shortly-stemmed when young, eventually extending to 60cm(2ft), sometimes branched. Leaves 15–25cm(6–10in) long, obovate to oblong-spathulate, light grey-green tinged with purple-red. Flowers to

TOP LEFT *Eccremocarpus scaber*
ABOVE *Echeveria glauca*
LEFT *Echeveria harmsii*

2.5cm(1in) long, light red, in panicles, on stems 60cm(2ft) or more tall, borne in autumn and winter. *E.g.* 'Carunculata' has looser rosettes of somewhat longer leaves; these have conspicuous warty outgrowths along the leaf blades, largest near their bases.

E. glauca (*E. secunda glauca*) Blue echeveria
Clump-forming, usually stemless with rosettes to 10cm(4in) across. Leaves 2.5–8cm(1–3in) long, broadly obovate, very glaucous with a blue-grey patina and sometimes with a narrow reddish margin. Flowers 8–10mm(⅓–⅜in) long, bright red, in racemes 20–30cm(8–12in) tall, erect at first, arching at the tips, opening in spring and summer.

E. harmsii (*Oliveranthus elegans*)
Branched sub-shrub, with stems eventually to 30cm(1ft) tall. Leaves 2–3cm(¾–1¼in) long, clustered at the stem tips, but not in a true rosette; they are lanceolate-spathulate, keeled beneath and

Echeveria leucotricha

green. Flowers 2–3cm(¾–1¼in) long, narrowly urn-shaped, scarlet, the petals tipped with yellow; they are borne singly or in twos or threes on stalks 10–20cm(4–8in) long in late spring.

E. hassei See *Dudleya hassei*.

E. leucotricha Chenille plant
Stems to 30cm(1ft) or more in height, branching, each one tipped by a loose rosette up to 15cm(6in) wide. Leaves lanceolate, shortly but densely white-hairy except for a brownish tip superficially resembling those of *Kalanchoe tomentosa*. Inflorescence 30cm(1ft) or more tall forming a thyrsoid spike; flowers about 2cm(¾in) long, cinnabar-red, opening from late winter to spring.

E. multicaulis Copper roses
Shrubby, eventually to 90cm(3ft) or more, each stem tipped by a rosette up to 10cm(4in) wide. Leaves obovate-spathulate, deep lustrous green with reddish margins. Inflorescence 25–30cm(10–20in) tall, racemose; flowers 1.2cm(½in) long, red outside, yellow within, opening in winter.

E. nodulosa
Stems to 45cm(1½ft) tall, sparingly branched, each one leafy towards the tips and rosetted, the latter 8–15cm(3–6in) wide. Leaves obovate to spathulate, somewhat keeled, covered with minute, whitish papillae and strongly flushed red along the margins and keel. Inflorescence up to 60cm(2ft) tall, but usually less, racemose. Flowers about 1.2cm(½in) long, tan-yellow with a touch of red, summer.

E. nudicaulis See *Dudleya densiflora*.

E. paraguayense See *Graptopetalum paraguayense*.

E. peacockii (*E. desmetiana*, *E. subsessilis*)
Rosettes stemless or almost so, usually solitary, up to 13cm(5in) wide. Leaves oblong to oblong-obovate, blue-green with a dense, waxy-white patina. Inflorescence 20–35cm(8–14in) tall, terminating in a simple cyme; flowers 1.2cm(½in) long, glowing rich red, maturing in early summer. A most attractive species, but must not be over-watered or become chilled.

***E.* × 'Perle von Nurnberg'** (*E. gibbiflora metallica* × *E. potosina*)
Rosettes up to 15cm(6in) wide, eventually topping a slow-growing stem. Leaves obovate with a metallic, grey-purple patina. A delightful hybrid well worth growing entirely for its foliage.

E. pulvinata Plush plant
Stems to 20cm(8in) in height, eventually with a few branches, tipped by loose rosettes up to 10cm(4in) wide. Leaves obovate, thick, softly and shortly white-hairy. Inflorescence 20–30cm(8–12in) tall, terminating in a thyrsoid raceme; flowers 1.2–2cm(½–¾in) long, red, or yellow and red, appearing in late winter to spring. *E.p.* 'Frosty' has extra white-hairy leaves; 'Ruby' has narrower leaves with red hairs at the tips and along the margins. The hybrid 'Doris Taylor' (*E. pulvinata* × *E. setosa*) blends the characters of both parents and has comparatively large orange blooms.

***E.* × 'Pulv-Oliver'** (*E. harmsii* × *E. pulvinata*)
Stems to 30cm(1ft) in height, branching, with scattered leaves and terminal rosettes to 10cm(4in) wide. Leaves oblanceolate to rhomboid, reddish-tipped, densely hairy. Inflorescence 15–20cm(6–8in) tall, racemose; flowers coral-red, 2.5cm(1in) long, appearing in summer.

E. rosea Desert rose
Stems to 30cm(1ft) tall, branching, the tips bear-

ing rosettes 12–18cm(4¾–7in) wide. Leaves oblanceolate to spathulate, semi-cylindrical near their bases, pale to grey-green to reddish, the latter especially so in winter. Inflorescence 15–35cm(6–14in), racemose; flowers 1.2cm(½in) long, yellow with red sepals and borne in the axils of red bracts, opening from autumn to spring, sometimes in other seasons.

E. runyonii
Stemless or short-stemmed rosettes, eventually forming clumps; rosettes to 10cm(4in) across. Leaves to 8cm(3in) long, spathulate, up-curving, glaucous grey-blue, appearing almost waxy. Flowers 2cm(¾in) long, pink, on forking racemes 15–20cm(6–8in) long.

E. sanguinea See *E. atropurpurea.*

E. secunda
Much like *E. glauca*, but the leaves light green or only slightly glaucous with reddish tips.

E. septentrionalis See *Dudleya farinosa.*

E. × 'Set-Oliver' (*E. setosa × E. harmsii*)
This shrubby hybrid resembles the latter parent, but has broader, brighter green leaves with mahogany margins. The flowers too are almost as big and similarly coloured, but carried in an inflorescence 30–40cm(12–16in) tall.

E. setosa Firecracker plant
Stemless rosettes 10–15cm(4–6in) across, producing many offsets. Leaves to 5cm(2in) long, spathulate to club-shaped, densely covered with short, white bristly hairs. Flowers yellow, suffused with red, in panicles on stems 20–30cm(8–12in) tall in summer.

E. shaviana
Rosettes stemless, 10–20cm(4–8in) wide, usually solitary. Leaves spathulate to obovate, glaucous, the margins more or less crimped and pink-tinged. Inflorescence to 30cm(1ft) in height, bearing one or two simple cymes; flowers 16mm(⅝in) long, deep pink, opening in summer.

E. subsessilis See *E. peacockii.*

ECHINOCACTUS

Cactaceae

Origin: Southern North America. A genus of ten species of globular to cylindrical cacti. They have prominently ribbed stems and large areoles, woolly at the top of the plant. The funnel-shaped flowers are 3–7cm(1¼–2¾in) long and usually yellow, but generally borne only on large plants. They are attractive subjects for a sunny window sill or conservatory and although eventually large are slow-growing. Propagate by seed in spring in warmth. *Echinocactus* derives from the Greek *echinos*, a hedgehog and *cactus*; an obvious allusion.

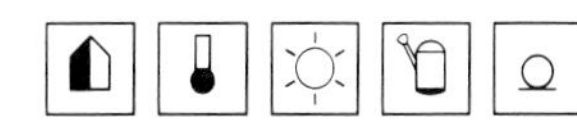

Species cultivated

E. cinerea See *Copiapoa cinerea.*

E. echinoides See *Copiapoa echinoides.*

E. eyriesii See *Echinopsis eyriesii.*

E. grusonii Golden barrel, Golden ball
Central Mexico
Globular until old then elongating, eventually reaching 1m(3ft) in the wild, though for cultivated specimens 30cm(1ft) is large. Ribs 20 or more, glossy green, bearing yellow to grey-woolly areoles, eight to ten radial spines to 2cm(¾in) long and three to five thicker central ones, to 5cm(2in) long; they are golden-yellow, sometimes reddish at the base, paling with age. Flowers yellow.

E. horizonthalonius Northern Mexico, Southern USA
Globular, flattened on top, eventually reaching 25cm(10in) high and more wide, slow-growing. Ribs seven to 13, broad, bearing many woolly areoles, six to nine radial spines, straight or recurved, 2–4cm(¾–1½in) long and a single central one not always present but if so, to 5cm(2in) long, yellowish to pink or reddish becoming grey as they age. Flowers, which are sometimes produced on relatively small plants, pink, 5–7cm(2–2¾in) long.

E. ingens Mexico
Stem globular, depressed, becoming cylindrical with age and over 1m(3ft) high. Ribs five to eight when small, increasing with age to as many as 50, with well-spaced, yellowish woolly areoles, eight radial and one central spine, stiff, brown, 2–3cm(¾–1¼in) long. Flowers reddish-yellow.

E. marginata See *Copiapoa marginata.*

E. platyacanthus Mexico
Broadly globular, depressed, eventually to 60cm(2ft) across and high; slow-growing. Ribs to 30, obtuse, densely woolly at the top of the plant; four radial spines 1.5cm(⅝in) long and three to four central ones, 2.5cm(1in) long, grey-brown. Flowers yellow.

E. polselgerianus See *Coryphantha poselgeriana.*

E. pulcherrima See *Frailea pulcherrima.*

E. rinconensis See *Thelocactus rinconensis.*

E. setispinus See *Thelocactus setispinus.*

E. texensis (*Homalocephela texensis*) Texas, Mexico
Globular and flattened to 15cm(6in) high and 30cm(12in) across, densely woolly at the top. Ribs 13 to 27, acute, with woolly, well-spaced areoles,

Echinocactus grusonii

six to seven reddish radial spines of two lengths, the larger two to 4cm($1\frac{1}{2}$in), the others half as long, one central one pointing downwards. Flowers bell-shaped, to 6cm($2\frac{1}{2}$in) long, pink with red in the throat, fragrant, freely borne. Fruits red.

ECHINOCEREUS

Cactaceae

Origin: Southern North America. A genus of 75 species of cactus, generally clump-forming with oval to cylindrical stems usually less than 40cm(16in) long, of erect, sprawling or pendent habit, most with very spiny areoles and widely funnel-shaped showy flowers in summer. Their fruits are spiny. Plants are suitable for a sunny window sill or conservatory and are easy to grow and flower. Propagate by seed or by cuttings in summer. *Echinocereus* derives from the Greek *echinos*, a hedgehog and *cereus*, a closely allied cactus genus differing in having spineless fruits.

Species cultivated

E. baileyi See *E. reichenbachii albispinus*.

E. blanckii Mexico
Stems cylindrical to 35cm(14in) long, 3cm($1\frac{1}{4}$in) thick, at first erect, later becoming prostrate; they are blue-green and have five to six ribs which carry the brownish areoles. Radial spines usually nine, white, central spines black. Flowers large, to 9cm($3\frac{1}{2}$in) long, reddish-violet, sometimes with a pale centre, freely produced.

E. caespitosus See *E. reichenbachii*.

E. chloranthus USA (New Mexico, Texas), Mexico
Cylindrical, to 20cm(8in) long, occasionally branching from near the base, erect. Ribs 18 to 20, low and rather rounded, with closely set white areoles. Radial spines 12 to 20, 1cm($\frac{3}{8}$in) long, centrals three to six, the longest over 2.5cm(1in) long. Flowers borne along the stems are 2.5cm(1in) long, funnel-shaped, yellow-green suffused brown.

E. cinerascens Central Mexico
Cylindrical, to 20cm(8in) long, freely branching from the base, forming clumps of erect or semi-sprawling stems. Ribs six to eight, with yellow or whitish areoles. Radial spines eight to ten, central ones one to four, all are about 2cm($\frac{3}{4}$in) long, reddish. Flowers to 7cm($2\frac{3}{4}$in) long, red-purple, opening from brownish-violet buds.

E. dubius USA (S.E.Texas)
Cylindrical, 10–20cm(4–8in) long, freely branching from the base, erect at first, soon semi-prostrate, yellowish-green. Ribs seven to nine with rounded, yellowish-green areoles. Radial spines five to eight, spreading, 1–3cm($\frac{3}{8}$–$1\frac{1}{4}$in) long, central spines one to four, sometimes curving, 4–7cm($1\frac{1}{2}$–$2\frac{3}{4}$in) long. Flowers to 6cm($2\frac{1}{2}$in) long, broadly funnel-shaped, pink.

E. enneacanthus Strawberry cactus USA (New Mexico, Texas), Mexico
Cylindrical, to 30cm(1ft) long, freely branching from the base, more or less erect, rather soft in texture. Ribs eight to 13 with round, whitish areoles. Radial spines usually eight or nine, sometimes up to 15, white, to 1.5cm($\frac{5}{8}$in) or more, central ones one to four, brownish, to 4cm($1\frac{1}{2}$in). Flowers 5–6cm(2–$2\frac{1}{2}$in) wide, red, freely produced.

E. knippelianus Mexico
Rounded to shortly columnar, solitary, to 5cm(2in) tall, dark green. Ribs five, shallow and widely spaced, with small rounded areoles. Spines lacking when the plant is young, later one to three, recurved, white and bristle-like, to 1.5cm($\frac{5}{8}$in) long. Flowers 4cm($1\frac{1}{2}$in) wide, funnel-shaped, purplish-violet, darker on the outside.

E. paucispinus triglochidiatus See *E. triglochidiatus*.

E. pectinatus Hedgehog cactus Central Mexico
Oval to cylindrical, to 20cm(8in) or more long, branching from the base. Ribs about 20, with closely set white areoles, hairy when young. Radial spines 20 to 25, to 9mm($\frac{1}{3}$in) long, pinkish at first, then white, central ones two to six, very short. Flowers 6–8cm($2\frac{1}{2}$–3in) long, funnel-shaped, bright pink, white-hairy and spiny outside, freely produced.

RIGHT *Echinocereus blanckii*
FAR RIGHT *Echinocereus reichenbachii perbellus*

E. pectinatus rigidissimus See *E. rigidissimus.*
E. pentalophus Mexico
Cylindrical, to 13cm(5in) long, branching from the base, prostrate, light green. Ribs five, often spiralling, bearing dense white areoles. Radial spines only, three to five whitish or yellowish, later becoming grey. Flowers to 13cm(5in) long, opening almost flat, lilac to pinkish-violet.
E. perbellus See *E. reichenbachii perbellus.*
E. poselgeri See *Wilcoxia poselgeri.*
E. reichenbachii (*E. caespitosus*) Lace cactus
USA (Texas), Mexico
Usually solitary, spherical at first, later cylindrical, erect, up to 20cm(8in) high. Ribs ten to 19 with closely set areoles. Radial spines 12 to 36, comb-like, of varying lengths, white or yellowish with darker tips, central ones sometimes one or two, rarely to seven, but often absent. Flowers 6–7cm($2\frac{1}{2}$–$2\frac{3}{4}$in) long, bright pink.
E.r. albispinus (*E. baileyi*) USA (Oklahoma)
Similar, but with stems to 10cm(4in) high, branched from the base, the 16 radial spines to 5cm(2in) long, straight, white; no centrals. Flowers to 5cm(2in), pink, with toothed petals.
E.r. perbellus
Stems branching from the base, 5–10cm(2–4in) high, with closely set areoles. Radial spines 12 to 15, about 1.5cm($\frac{5}{8}$in) long, brownish or reddish. Flowers 4–6cm($1\frac{1}{2}$–$2\frac{1}{2}$in) long, purple.
E. rigidissimus (*E. pectinatus rigidissimus*)
Rainbow cactus USA (Arizona) and Mexico
Cylindrical, to 30cm(1ft) long, erect. Ribs 16 to 23 with closely set areoles. Radial spines 16 to 20, comb-like and spreading, to 1.5cm($\frac{5}{8}$in) long, whitish, pale pink, brownish or reddish, arranged in distinct bands of separate colours, giving rise to its popular name. Flowers 6–7cm($2\frac{1}{2}$–$2\frac{3}{4}$in) long, pink, with white hair and spines outside.
E. triglochidiatus (*E. paucispinus triglochidiatus*)
Texas, Arizona and Colorado
Oval to shortly columnar, 20cm(8in) in height, branching from the base, some stems erect, others more or less prostrate. Ribs five to ten bearing white-woolly areoles. Radial spines three to six, grey, to 2.5cm(1in) long, one central to 4cm($1\frac{1}{2}$in). Flowers 5cm(2in) long, bright red.
E. tuberous See *Wilcoxia poselgeri.*

ECHINOFOSSULOCACTUS

Cactaceae

Origin: Mexico. A genus of 32 species of variable cacti, included by some botanists in the genus *Stenocactus*. Most are globular or shortly cylindrical with attractive spination and often showy blooms. They are suitable for a sunny window sill or conservatory, most species being free-flowering. Propagate by seed in spring in warmth. *Echinofossulocactus* derives from the Greek *echinos*, a hedgehog, the Latin *fossulo*, a ditch and *cactus*, the hollows between the spiny ribs being very marked.

  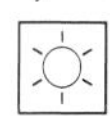  

Echinofossulocactus albatus

Species cultivated

E. albatus
Stem normally solitary, globular and depressed at the top, eventually becoming cylindrical to 13cm(5in) in diameter with age, bluish-green. Ribs numerous, wavy, white-woolly above. Radial spines about ten, bristle-like, to 1.5cm($\frac{5}{8}$in) long, central spines four, thicker and longer, to 11cm($4\frac{1}{2}$in), all yellowing to white. Flowers 2cm($\frac{3}{4}$in) long, funnel-shaped, pure white.
E. coptonogonus
Spherical, flattened at the top, to 10cm(4in) in diameter, with up to 14, rather broad ribs with well spaced areoles. Spines three to five, all similar, curved upwards, the longest 3cm($1\frac{1}{4}$in), soft and red at first, then brown to grey and horny. Freely borne flowers are 2.5cm(1in) long, opening flat to 4cm($1\frac{1}{2}$in) at the mouth, white with a central pink to purplish line; stamens are reddish and the anthers violet.
E. zacatecasensis Brain cactus
Found near Zacatecas, Mexico, hence the tongue-twisting name. Globular, pale green, to 10cm(4in) in diameter, having about 55 thin, wavy ribs, with the upper areoles white-woolly. Radial spines ten to 12, slender and spreading, white, to 1cm($\frac{3}{8}$in) long, central ones three, the middle one thick, flattened and straight to 4cm($1\frac{1}{2}$in) long, the outer two shorter and hooked. Flowers 3–4cm($1\frac{1}{4}$–$1\frac{1}{2}$in) across, petals white, tipped with pink.

Echinofossulocactus zacatecasensis

Echinopsis werdermanniana

ECHINOPSIS

Cactaceae
Sea urchin cacti

Origin: South America. A genus of 35 species of globular or cylindrical cacti with prominently ribbed, solitary or clump-forming stems and trumpet-shaped flowers often with very long tubes. They are very tolerant house or conservatory plants, best in a sunny window. Propagate by offsets or by seed. *Echinopsis* derives from the Greek *echinus*, a hedgehog and *opsis*, like.

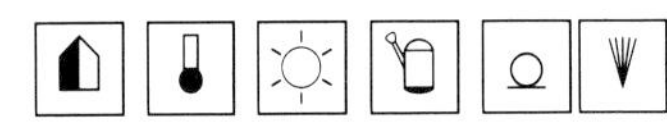

Species cultivated

E. aurea See *Lobivia aurea.*

E. eyriesii (*Echinocactus eyriesii*) Argentina, southern Brazil
Stem globular, branching from the base, becoming elongated with age, to 15cm(6in) in diameter. Ribs 11 to 18 with round, well-spaced woolly areoles, ten radial spines and four to eight central ones all similar, dark brown. Flowers 22cm(10in) long from base of tube, opening widely to 12cm(4¾in) across, pure white.

E. forbesii See *E. rhodotricha.*

E. multiplex (*E. oxygona*) Brazil
Globular, elongating with age to 15cm(6in), branching from base and sides to form a mounding clump, pale to yellow-green. Ribs 12 to 14, sharp, bearing well-spaced areoles. Radial spines eight to nine, awl-shaped; central spines two to five, to 4cm(1¾in) in length, all brown but tipped almost black. Flowers 18–20cm(7–8in) long including the tube, pink; they are freely produced and fragrant.

E. oxygona See *E. multiplex.*

E. pentlandii See *Lobivia pentlandii.*

E. rhodotricha (*E. forbesii*) Argentina, Paraguay
Stem oval to cylindrical, branching from the base to 30–80cm(12–32in) tall. Ribs eight to 13, bearing well-spaced yellowish to grey areoles. Radial spines four to seven, thick, to 2cm(¾in) long; central spine one, sometimes none, to 3.5cm(1½in) long, all yellowish to brown. Flowers 15cm(6in) long to base of tube, 8cm(3in) across, white.

E. werdermanniana (*Trichocereus werdermanniana*) Bolivia
Cylindrical stems, eventually tree-sized, green to grey reaching 60cm(2ft) in diameter at the base. In its young state, however, it is very decorative, with six ribs (later to 12), bearing white, closely set areoles. Radial spines 12 to 18, to 4cm(1½in) long, fawny-yellow becoming grey; central spines six to nine, reddish-brown becoming horny. Flowers at the top, pink when young, becoming more red with age.

ECHIUM

Boraginaceae
Viper's bugloss

Origin: Europe to W. Asia, Northern and Southern Africa, Canary Is. A genus of 40 species of annuals, biennials and shrubs with spirally arranged entire leaves. The flowers are obliquely funnel-shaped and borne in simple or forked cymes, many together making spike- or panicle-like clusters. The shrubby species are ornamental flowering plants for pots and tubs in the conservatory, while the annual described below is a short-term pot plant for conservatory or home. Propagate by seed or by cuttings of the shrubby sorts. *Echium* derives from the Greek *echion*, the name for viper's bugloss.

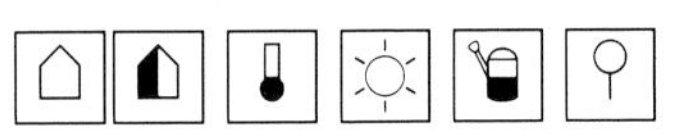

Species cultivated

E. fastuosum Canary Is.
Evergreen shrub to 1.2m(4ft) in height. Leaves oblong-lanceolate, tapering to fine points, the whole covered in soft, greyish hairs. Flowers deep blue, crowded into erect panicles in spring and summer.

E. lycopsis (*E. plantagineum*) Purple viper's bugloss S. & W. Europe
Annual, sometimes a biennial, branching from the base to form bushy plants. Leaves ovate, narrowing up to the stalks to become oblong-lanceolate, 5–14cm(2–5½in) long, stiffly but softly white-hairy. Flowers 2.5cm(1in) long, pink, purple and blue in short, one-sided clusters in summer and autumn. Several named seed strains are available, the dwarf ones being the best for pots.

E. plantagineum See *E. lycopsis.*

BELOW *Echium fastuosum*
BELOW RIGHT *Echium lycopsis*

EICHHORNIA

Pontederiaceae

Origin: Southern USA to Argentina and West Indies. A genus of seven species of aquatic perennials, often weed-pests in sub-tropical waterways, but very attractive in tubs and glass tanks. They are usually free-floating, though they will root into mud in shallow water. Their entire, glossy leaves are decorative in themselves, but their racemes of lavender to white flowers are their chief attraction, gaining them an alternate name, water hyacinth. Propagate by division or by separation of rosettes. *Eichhornia* commemorates J.A.F.Eichhorn (1779–1856), Minister of Education in Prussia.

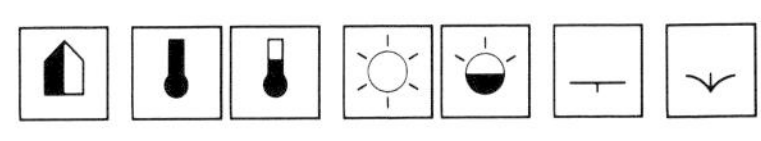

Species cultivated

E. crassipes (*E. speciosa*) Water Hyacinth Sub-tropical regions of the Americas
Evergreen plant spreading from rhizomes. Leaves with bulbous stalks filled with spongy air tissue, leaf blades rounded to kidney-shaped, 5–11cm(2–4½in) across, glossy green. Flowers funnel-shaped, opening almost flat, 4–5cm(1¾–2in) wide, lilac-blue, with six tepals, the uppermost with an orange-yellow blotch; they are borne in erect, dense racemes on long stalks above the leaves in summer.

ELAPHOGLOSSUM

Lomariopsidaceae (Aspidiaceae)

Origin: Tropics and sub-tropics, but mainly South America. A genus of perhaps 400 species of evergreen, mainly epiphytic ferns. They are decidedly un-fernlike in appearance, with simple, often leathery or slightly fleshy fronds, the fertile ones being completely covered beneath with sporangia. Propagate by spores or division in spring. Cultivate as for Bromeliads (see Introductory section). *Elaphoglossum* derives from the Greek *elaphos*, a stag and *glossa*, a tongue, alluding to the shape of the fronds of certain species.

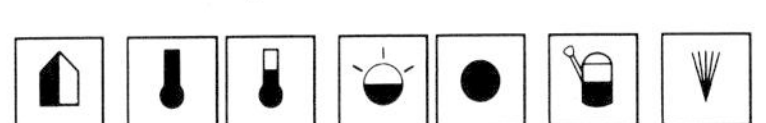

Species cultivated

E. crinitum (*Acrostichum crinitum*) Elephant's ear fern West Indies, Mexico, Central America
Clump-forming, though often staying as a solitary rosette for several years. Fronds 25–45cm(10–18in) or more long, oval-oblong, thick-textured, slightly wavy, ciliate and with scattered hairs, borne on long, densely scaly-hairy stalks. Spore-bearing fronds are smaller and relatively longer stalked.

E. hirtum (*E. acrostichum paleaceum*) Tropics
Rhizomatous, eventually forming small colonies. Leaves 15–30cm(6–12in) long, lanceolate, tapering gradually to base and tip, densely scaly-hairy on both surfaces, those on the margins reddish. Spore-bearing fronds narrower.

ABOVE *Eichhornia crassipes*
LEFT *Elaphoglossum crinitum*

E. longifolium West Indies, tropical America
Rhizomes short, thick and fairly slow-growing. Fronds eventually form clumps 30–45cm(1–1½ft) in length, narrowly lanceolate, slightly wavy, glabrous.

ENCEPHALARTOS

Zamiaceae

Origin: Tropical and South Africa. A genus of 20 to 30 species of dioecious cycads. They are primitive, palm-like plants with rigidly leathery, pinnate

RIGHT *Encephalartos lehmannii*
FAR RIGHT *Epacris impressa*

leaves in rosettes with flowering and fruiting spikes resembling cones. The species described below eventually develop palm-like trunks, but other species produce their leaves at or near ground level from a below-ground, tuber-like stem. Although best in the conservatory, they tolerate conditions indoors and make intriguing specimen pot plants. Propagate by seed in spring or summer. *Encephalartos* is derived from the Greek *en*, within, *kephale*, a head and *artos*, bread; the seeds of certain species yield a starchy flour or meal, formerly used by African natives as a food source.

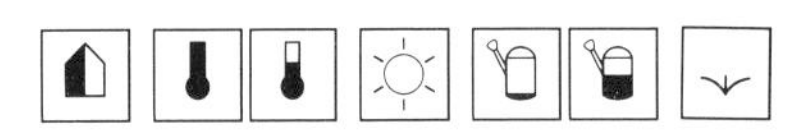

Species cultivated

E. altensteinii Bread tree cycad South Africa
In the wild this is one of the largest species, producing a trunk 15m(45ft) in height; in containers, many years elapse before a stem forms. Leaves to about 1.2m(4ft) in length, twice this on mature specimens; leaflets 10–15cm(4–6in) long, narrowly oblong with several spiny teeth. Flowering cones only on fully grown specimens, 30–45cm(1–1½ft) long, yellow-green.

E. ferox South Africa
Trunk eventually to 60cm(2ft) or more in height, but it is many years before any signs are seen of it in containerized specimens. Leaves 1–2m(3–6½ft) in length, glaucous, the lowest leaflets modified to spines, the upper ones ovate-oblong, to 15cm(6in) long, toothed and spine-tipped. Flowering cones 25–50cm(10–20in) in length, the plumper female cones shaded pink to red, but these are only formed on mature plants of a good age.

E. lehmannii South Africa
Trunk eventually to 2m(6½ft) or more, but slow to form on containerized specimens. Leaves 90–150cm(3–5ft) long, richly glaucous; leaflets linear, 13–20cm(5–8in) long, with a few small teeth. Flowering cones seldom seen on pot plants, 25–40cm(10–16in) long; the female ones are more plump and a reddish-brown.

ENCEPHALOCARPUS See PELECYPHORA
ENSETE See MUSA

EPACRIS

Epacridaceae
Australian heaths

Origin: Australasia. A genus of 40 species of evergreen shrubs with wiry stems, small leathery leaves and 5-lobed tubular flowers. The species described make splendidly elegant and colourful pot plants for the sunny, airy conservatory. They can be brought indoors when in bloom. Propagate by cuttings and layering in summer or by seed in spring. If it is possible, the plants benefit by being stood outside during the summer. A lime-free compost is essential for success. *Epacris* derives from the Greek *epi*, upon and *acris*, a summit, alluding to the hilltop habitat of certain species.

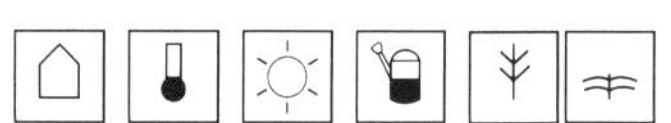

Species cultivated

E. impressa South Australia, New South Wales, Victoria, Tasmania
A variable species, but usually erect, to 60cm(2ft) or more in height, of fairly open habit. Leaves sessile, linear to broadly lanceolate, rigidly pointed-tipped. Flowers about 1.5cm(⅝in) long, white, pink or red, usually abundantly borne from the leaf axils along the upper part of the stems, mainly in late winter and spring.

E. longiflora New South Wales, Sydney area
Erect shrub to 90cm(3ft) or more when young, then arching and sprawling unless pruned annually. Leaves broadly lanceolate to ovate, rounded to cordate at the base and prickly-pointed, up to 1.2cm(½in) long. Flowers 2–2.5cm(¾–1in) long, the tube crimson with a white tip and petal lobes, mainly in spring, occasionally a few at other times especially on old plants. Closely related to *E. impressa* and just as free-blooming, but less hardy.

EPIDENDRUM

Orchidaceae

Origin: Tropical America, one species in West Africa. A genus of about 400 species of very variable epiphytic orchids. Some have tall, reed-like stems, others more obvious pseudobulbs which may be rounded, ovoid, cylindrical or conical. Their leaves range from elliptic to strap-shaped and cylindrical and the flowers have three sepals and two petals, both similar (tepals), with a simple to 3-lobed lip. They are best in the conservatory, but can be grown in the home in pots or orchid baskets. Propagate by division, those with stem-like pseudobulbs by cuttings from these. *Epidendrum* derives from the Greek *epi*, upon and *dendron*, a tree, from their epiphytic habit.

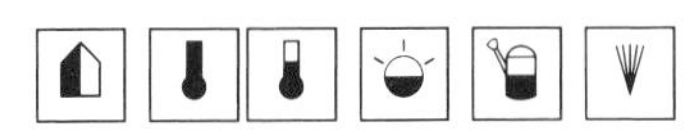

Species cultivated

E. alatum Mexico and Central America
Pseudobulbs more or less pear-shaped, to 13cm(5in) tall. Leaves narrowly lanceolate, stalked, stiffly leathery, up to 45cm(1½ft) in length. Inflorescence a simple or branched raceme, 60–90cm(2–3ft) tall; flowers about 5cm(2in) wide, variable in colour, but the narrow tepals usually a pale greenish-yellow and brownish-purple, the 3-lobed labellum yellow with reddish or purplish markings; spicily fragrant, opening between spring and autumn.

E. brassavolae Mexico to Panama
Pseudobulbs narrowly ovoid to 15cm(6in) or more high, growing in a cluster from a stout rhizome. Leaves to 40cm(16in) in length, narrowly ovate to oblanceolate, leathery in texture, often a rather yellowish-green. Flowers 7–10cm(2¾–4in) wide, rather fleshy with yellow to greenish-brown sepals and a straw coloured to cream-coloured labellum stained at the tip with red-purple, six to nine on each 45cm(1½ft) raceme, opening from summer to autumn.

E. brevicaule See *E. schlecterianum*.

E. ciliare Mexico and West Indies to Brazil
Pseudobulbs flattened, oblong-cylindrical to spindle-shaped, to 15cm(6in) tall. Leaves to 30cm(1ft) long, glossy dark green, elliptic-oblong and blunt ended. Flowers 8–18cm(3–7in) across, rather spidery, with very narrow palest yellow-green sepals and a 3-lobed creamy-white lip, the outer lobes deeply fringed; the 4- to 7-flowered racemes are up to 30cm(12in) tall.

E. cochleatum Cockle-shell orchid Mexico to Brazil
Pseudobulbs ovoid, clustered, to 10cm(4in) and sometimes 20cm(8in) high. Leaves 15cm(6in) or more in length, oblong to linear-lanceolate. Flowers 7–10cm(2¾–4in) across, tepals greenish-yellow, to 4cm(2½in) long, swept away from the rounded, cockle-shell-like lip, which is black-purple below with a white centre and pale, radiating lines. The whole bloom is held upside down, in 4- to 7-flowered racemes, 30–45cm(1–1½ft) long.

E. cucculatum See *Brassavola cucculata*.

E. difforme Florida and Mexico to Brazil
Slender, zig-zag stems to 30cm(1ft) or more long. Leaves about 10cm(4in) long, oval to oblong-lanceolate, glossy green. Flowers 3cm(1¼in) or more wide, the tepals green to yellow-green, the lip more or less kidney-shaped; they are fragrant and borne in umbel-like clusters chiefly in autumn, but appearing spasmodically throughout the year.

E. endresii Costa Rica, Panama
Pseudobulbs stem-like, to 22cm(9in) tall, clustered. Leaves elliptic to ovate-cordate, about 2.5–4cm(1–1½in) long, fleshy and hard-textured. Racemes erect, up to 13cm(5in) tall; flowers 2.5cm(1in) wide, white, spotted or blotched with violet-purple, fragrant, expanding during winter and spring.

E. fragrans Mexico and West Indies to Brazil
Pseudobulbs clustered, ellipsoid to 10cm(4in) long. Leaves 10cm(4in) or more long, oval to oblong, borne singly. Flowers 4cm(1½in) across, the sepals yellow to greenish-white with a white, shell-like labellum marked with faint purple lines radiating outwards from the centre; they are carried upside down, in 2- to 8-flowered racemes to 13cm(5in) long, in winter and spring and are fragrant.

E. ibaguense (*E. radicans*) Mexico to Peru and Brazil
Stems slender, climbing by aerial roots and capable of reaching several metres (yards) in length. Leaves to 10cm(4in) long, ovate-oblong to elliptic, stiffly fleshy. Flowers 3cm(1¼in) across, in shades of orange, red and yellow, the fringed labellum with a yellow blotch; they are borne in short, dense racemes on long wiry stems, spasmodically throughout the year.

LEFT *Epidendrum cochleatum*
ABOVE *Epidendrum ibaguense*

BELOW *Epidendrum pseudoepidendrum*
BELOW RIGHT *Epidendrum lindleyana*
BOTTOM *Epidendrum vespa*

E. lindleyana (*E. spectabile, Barkeria spectabilis*) Mexico, Guatemala, Honduras, Costa Rica
Pseudobulbs narrowly spindle-shaped to stem-like, up to 15cm(6in) tall. Leaves oblong to narrowly lanceolate, shortly pointed to acuminate, somewhat fleshy and often purple-flushed or streaked. Raceme fairly compact on a wiry stem 30–60cm(1–2ft) in length; flowers 5–8cm(2–3in) wide, with three lanceolate sepals, two ovate petals and a superficially similar labellum, all of one colour, usually rose-purple but varying from white to deep purple, appearing in winter and spring.
E. nocturnum Southern Florida and Mexico, south to Brazil; naturalized in tropical West Africa
Pseudobulbs stem-like, 30–75cm(1–2½ft) tall. Leaves variable, linear to oblong-lanceolate or elliptic, leathery-textured, 15–20cm(6–8in) in length. Raceme simple or sometimes branched, short, 2- to 5-flowered, but only one opening at a time; each bloom 8–13cm(3–5in) wide, ochre-yellow to greenish-white, strongly fragrant especially at night, opening intermittently all the year, but mainly in summer and autumn.
E. nodosum See *Brassavola nodosa*.
E. polybulbon Mexico, Central America, Cuba, Jamaica
Pseudobulbs ovoid, to 3cm(1¼in) in height, spaced on slender, creeping, branching rhizomes. Leaves ovate to elliptic, 5–8cm(2–3in) long, leathery-textured, glossy green. Flowers usually solitary, borne at the tips of the pseudobulbs, about 3cm(1¼in) wide, tepals yellowish or brownish, sometimes with red streaks, labellum usually white, sometimes marked or flushed yellow, fragrant, long-lived, expanding from early winter to spring.
E. pseudoepidendrum Costa Rica, Panama
Pseudobulbs stem-like, 60–90cm(2–3ft) tall, clustered. Leaves narrowly lanceolate to oblong, about 15cm(6in) long, leathery-textured, deep green. Racemes few-flowered, erect, to 15cm(6in) tall; each flower 6cm(2½in) across, tepals bright apple-green, labellum relatively large, bright orange to orange-red, maturing in summer.

E. radiatum Mexico to Venezuela
Pseudobulbs clustered, ovoid to ellipsoid, strongly ribbed, to 13cm(5in) tall. Leaves to 30cm(1ft) long, lanceolate to strap-shaped, rigid and leathery, two or three to each pseudobulb. Flowers 2–3cm(¾–1¼in) across, creamy-white to greenish-cream, the white labellum with purple radiating lines; they are borne upside down, in racemes on arching stems, in spring and summer.
E. radicans See *E. ibaguense*.
E. schlecterianum (*E. brevicaule*) Mexico to Central America south to Brazil, West Indies
Pseudobulbs totally stem-like, concealed by overlapping stem bases, to 6–10cm(2½–4in) tall, clustered and branched. Leaves 3cm(1¼in) long, very fleshy, with reddish or transparent margins. Flowers solitary or in pairs terminating the pseudobulbs, 2–3cm(¾–1¼in) wide, but not expanding fully, usually green to yellow-green, sometimes tinted pink, red or purple, appearing in summer.
E. spectabile See *E. lindleyana*.
E. stamfordianum Mexico to Colombia
Pseudobulbs clustered, spindle-shaped, to 25cm(10in) long. Leaves to 25cm(10in) long, linear to narrowly oblong, leathery. Flowers 4cm(1½in) or more long, greenish-yellow, spotted with red, the labellum fringed, yellow; they are borne in branched racemes to 60cm(2ft) tall, from winter to spring.
E. variegatum See *E. vespa*.
E. vespa (*E. variegatum*) Cuba and Costa Rica to Brazil
Pseudobulbs rounded to almost spindle-shaped and up to 20cm(8in) tall. Leaves 15–25cm(6–10in) long, oblong-lanceolate. Flowers 2cm(¾in) long, yellowish-green with red or purplish-brown spotting; they are borne upside down on racemes to 30cm(1ft) tall.
E. vitellinum Mexico, Guatemala
Pseudobulbs clustered, oval to conical, 2.5–6cm(1–2½in) tall. Leaves 20–30cm(8–12in) long, linear-lanceolate, blue-green. Flowers 4cm(1½in) wide, bright orange-red with a largely yellow labellum; they are borne in branched inflorescences to 45cm(1½ft) tall.

EPIPHYLLOPSIS See RHIPSALIDOPSIS

EPIPHYLLUM

Cactaceae

Origin: Tropical South America to Mexico. A genus of 21 species of epiphytic cacti which are woody at the base when mature, the stems flattened and broadened until they resemble leaves. The flowers, which are trumpet-shaped and have many tepals, are large and showy. Some species open their blooms at night, others by day. They are excellent plants for the conservatory and home, in pots or in hanging baskets, but they must have a cool, dry rest in winter, though unlike most cacti, they must never become totally dry. Propagate by cuttings, taking stem sections ('leaves') in summer, or by seed in spring. *Epiphyllum* derives from the Greek *epi*, upon and *phyllon*, a leaf, alluding to the way the flowers appear to grow from the leaf edges which are in fact modified stems.

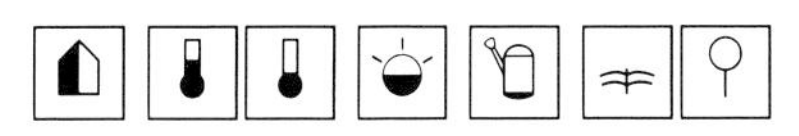

Species cultivated

E. ackermannii (*Nopalxochia ackermannii*) Orchid cactus Mexico
Stems 30–60cm(1–2ft), bearing tiny areoles with a few spines on their crenate margins. Flowers to 15cm(6in) long, crimson. The true species is very rare in cultivation, its place being taken by a wide range of hybrids, for which see below.
E. anguliger S. Mexico
Erect, bushy plant, its leaf-like branches lanceolate, with deep, rounded lobes. Flowers to 15cm(6in) long, white within, yellowish-brown outside, opening to 10–13cm(4–5in) in diameter at the mouth; they are fragrant.
E. nelsonii See *Disocactus nelsonii*.

Hybrids

These are hybrids involving *E. ackermannii* and *E. crenatum* with species of *Heliocereus* and *Selenicereus*. There are now hundreds of named cultivars often still called Phyllocacti or Orchid cacti and frequently listed under *E. ackermannii*. The following are popular or can be recommended, but a wide choice is available from specialist nurserymen; most bloom in spring:
'Autumn', coral-pink.
'Cambodia', purplish-red with ruffled tepals.
'King Midas', golden-yellow with darker centre and outer tepals.
'London Glory', deep red to purple, very free-flowering.
'Pacesetter', inner tepals lavender with a paler serrated edge and orange centre, outer tepals tangerine with a purplish edge; very striking.
'Professor Ebert', orchid- to fuchsia-pink.
'Reward', yellow flowers; inner tepals paler cream with yellow striping, outer tepals pure yellow.
'Sun Burst', wide-open flowers of burnt orange, the outer tepals narrow and flaring, the inner with a red-bronze eye.
'Truce', inner tepals pure white, outer pale green.

EPIPREMNUM

Araceae

Origin: Tropical Asia. A genus of about ten species of climbers with aerial roots, having alternate leaves markedly different in juvenile and adult stages, and arum-like flowers which are, however, not produced on pot plants. The species described is grown for its decorative foliage particularly in its variegated forms. Propagate by cuttings and layering in warmth in summer. *Epipremnum* derives from the Greek *epi*, upon and *premnon*, a trunk, alluding to its habit in the wild of climbing up tree trunks.

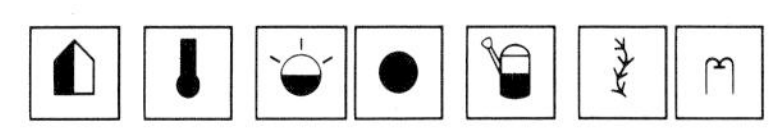

Species cultivated

E. aureum (*Pothos aureus*, *Scindapsus aureus*, *Raphidophora aurea*) Devil's ivy, Golden Pothos, Taro vine Solomon Is.
Still usually sold as *Scindapsus*, this plant owes its variety of Latin names to the fact that until 1956 it was not seen in bloom, so accurate naming was not possible. In the wild it reaches 12m(40ft) or more in height, in a pot it will seldom exceed 2m(6½ft). Leaves ovate-cordate, to 15cm(6in) long on pot plants, dark green with yellow marbling. *E.a.* 'Marble Queen' is more frequently met with, being less robust than the true species and having dark green leaves heavily marbled with creamy white. *E.a.* 'Tricolor' has mid-green leaves marbled with pale green, cream and yellow.

ABOVE *Epipremnum aureum*
ABOVE CENTRE *Epiphyllum ackermannii* hybrid
TOP *Epiphyllum anguliger* hybrid

ABOVE *Episcia cupreata*
RIGHT *Episcia dianthiflora*

EPISCIA

Gesneriaceae

Origin: Tropical America and West Indies. A genus of 40 species of prostrate to trailing perennials with pairs of ovate to elliptic leaves often patterned with white and gold and often with a metallic sheen. They have tubular flowers opening to five flared, rounded lobes. They make fine hanging basket or pan plants for a warm conservatory or room if sufficient humidity can be provided. Propagate by division, by detaching plantlets growing upon the runners, or by cuttings, all in warmth in summer. *Episcia* derives from the Greek *episkios*, shaded, referring to their natural habitat.

Species cultivated

E. cupreata Flame violet Colombia, Venezuela
Creeping or trailing, with stems to 40cm(16in) or more in length. Leaves elliptic, slightly toothed, to 6–8cm(2½–3in) long, their surfaces puckered or quilted, normally green with a coppery flush and often having the veins picked out in silver, white or pink. Flowers bright red, 2–2.5cm(¾–1in) wide, spotted with yellow within. Several named forms are available with varying silver variegation or a deeper coppery flush to the leaves.

E. dianthiflora Lace flower Mexico
Tufted at first, sending out creeping, hairy runners bearing small plantlets. Leaves oval, crenate, to 4cm(1½in) long, thickish, dull green, with a purple midrib, softly and shortly downy above. Flowers 2.5cm(1in), pure white, the petals deeply fringed.

E. fulgida See *E. reptans*.

E. lilacina Central America
Tufted at first, producing creeping runners bearing small plantlets. Leaves ovate-lanceolate, 6–8cm(2½–3in) long, hairy, green, often with a bronze or reddish suffusion, with the centre veins picked out in paler green above, pink beneath. Flowers slender-tubed, white, opening at the mouth with a yellow patch in the throat.

E. reptans Colombia to Brazil
Stems trailing, to 60cm(2ft) or more long. Leaves ovate, 5–12cm(2–4¾in) long, quilted, dark green flushed with mahogany, the veins silver. Flowers deep flame-red, to 4cm(1½in) long, toothed or fringed at the apex, the throat unspotted.

Hybrids

A number of named cultivars of hybrid origin have been raised, some with cream to yellow flowers, and they are all very attractive but not as yet easy to obtain. Well worth looking out for.

EPITHELANTHA

Cactaceae

Origin: South-western USA, Mexico. A genus of three species of cacti one of which is cultivated widely in both home and conservatory. They superficially resemble members of the genus *Mammillaria*, but differ in producing their flowers from the tips of the tubercles instead of from their bases. Propagate by seed in spring or by offsets in summer. *Epithelantha* derives from the Greek *epi*, upon and *thele*, a nipple, in reference to the mode of flowering.

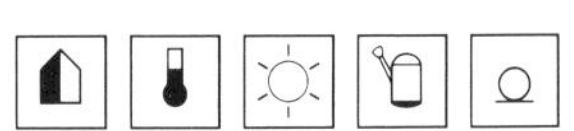

Species cultivated

E. micromeris Button cactus Arizona, New Mexico, Texas, Mexico
Plant body (stem) solitary at first, later sometimes forming small clumps more or less spherical, but flattened above and tapered below, 2.5–5cm(1–2in) long and wide, sometimes much more in cultivation. Tubercles very small, closely packed in rows, each one topped by a rosette of 20 to 40 tiny white spines up to 3mm(⅛in) long, the overall effect being that of a close, firm white web over the surface of the stem. Flowers pale pink to white, up to 1.2cm(½in) long, funnel-shaped, lasting less than one day. Fruits cylindrical, red, showy, like those of *Mammillaria*. The stems, like those of *Lophophora*, contain hallucinatory alkaloids.

E.m. greggi
Stems up to twice as large and regularly forms quite large clumps.
E.m. rufispina
Reddish-brown tipped spines.
E.m. unguispina
Develops blackish central spines with age.
E. pachyrhiza
More robust, elongated stems which form clumps.

ABOVE *Epithelantha micromeris*
LEFT *Eranthemum pulchellum*

ERANTHEMUM

Acanthaceae

Origin: Tropical Asia. A genus of 30 species of shrubs and perennials, one of which has long been grown for its pure blue winter flowers. This makes a splendid pot plant ideally for the conservatory, but which can be flowered in the home. Propagate by cuttings in late spring or early summer. Young plants should be pinched twice as they develop to promote bushiness. Flowered specimens should be spurred back after flowering for the same reason and also to provide growth suitable for cuttings. *Eranthemum* is an ancient Greek name used for this genus by Linnaeus; it means 'lovely flower'.

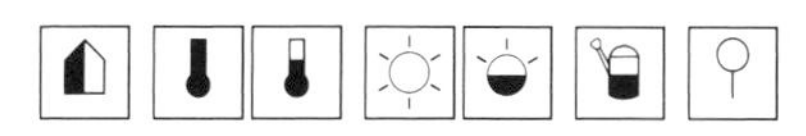

Species cultivated
E. atropurpureum See *Pseuderanthemum atropurpureum.*
E. cooperi See *Pseuderanthemum cooperi.*
E. pulchellum (*E. nervosum, Daedalacanthus nervosum*) Blue sage India, naturalized elsewhere
Shrub to 1.2m(4ft), but easily kept to half this if regularly pruned. Leaves evergreen, in opposite pairs, ovate to elliptic, prominently veined, deep green, 8–15cm(3–6in) in length. Flowers rich blue, about 3cm($1\frac{1}{4}$in) long, tubular with five rounded petals, carried in terminal and axillary spikes up to 8cm(3in) in length.
E. reticulatum See *Pseuderanthemum reticulatum.*

ERICA

Ericaceae
Heaths

Origin: Chiefly South Africa, but some from North Atlantic Islands, Europe and North Africa to Turkey and Syria. A genus of about 600 species of wiry-stemmed evergreen shrubs with narrow linear to oblong leaves, usually with rolled under margins. The flowers are bell- or urn-shaped and very freely borne. Most ericas need an acid, peaty compost as they live in association with a fungus which lives within the entire plant, but particularly in the roots (endotrophic mycorrhiza), and helps to provide essential minerals. These fungi mostly require acid conditions. The hardy species do not thrive under glass with the exception of *E. carnea* which can be grown in pots kept outside until autumn, then brought into the conservatory for flowering. Apart from this species those described below are best in the conservatory, but can be brought into the home for flowering. Propagate by heel cuttings in spring (*E. carnea* in late summer); species also by seed in spring. *Erica* derives from the Greek *ereike* or the Latin *erice*, ancient names for the tree heath, *Erica arborea*, which is common in the Mediterranean region.

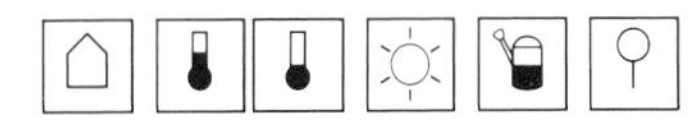

Species cultivated
E. caffra See *E. subdivaricata.*
E. canaliculata South Africa
Erect, to 1m(3ft) or more in a pot; 3–5m(10–16ft) in the open, with white-hairy stems. Leaves 4–6mm($\frac{1}{6}$–$\frac{1}{4}$in) long, recurved, carried in whorls of three. Flowers about 3mm($\frac{1}{8}$in) long, bell-shaped, white to pale pink, carried in leafy panicles from spring to early summer. Temperate.
E. carnea (*E. herbacea*) Winter heath
Central and S. Europe
Neat shrub to 25cm(10in). Leaves 5–8mm($\frac{1}{5}$–$\frac{1}{3}$in) long, linear. Flowers 5–6mm($\frac{1}{5}$–$\frac{1}{4}$in) long, narrowly urn-shaped, rosy-red, borne in dense racemes from autumn to spring. Tolerant of a limy compost. Many cultivars are available with coloured foliage

Eriobotrya japonica

BELOW *Erica versicolor*
BOTTOM LEFT *Erica cerinthoides*
BOTTOM RIGHT *Erica ventricosa*

and flowers, from pale pink to red, also white. Cool.

E. cerinthoides South Africa
Erect and bushy, 60–90cm(2–3ft) tall. Leaves linear, to 1.2cm($\frac{1}{2}$in) long, in whorls of four to six. Flowers in terminal clusters, 2–3cm($\frac{3}{4}$–1$\frac{1}{4}$in) long, tubular, slightly inflated in the middle and constricted at the mouth, crimson, appearing from early summer until autumn. Temperate.

E. gracilis South Africa
Freely branching shrub to 45cm(1$\frac{1}{2}$ft) high. Leaves 3–5mm($\frac{1}{8}$–$\frac{1}{5}$in) long, linear, borne in whorls of four or more. Flowers 3–4mm($\frac{1}{8}$–$\frac{1}{6}$in) long, rounded, rose-purple, opening in autumn and winter. Temperate.

E. herbacea See *E. carnea.*

E. × hyemalis South African hybrid
Erect shrub to 60cm(2ft) with 5–10mm($\frac{1}{5}$–$\frac{3}{8}$in) linear leaves. Flowers 1.5–2cm($\frac{5}{8}$–$\frac{3}{4}$in) long, tubular, rose-pink shading to white at the mouth and carried in long racemes in autumn and winter. Temperate.

E. subdivaricata (*E. caffra*) South Africa
Freely branching, erect at least when young, up to 90cm(3ft) in height. Leaves 4mm($\frac{1}{6}$in) long, linear, in whorls of four. Flowers 3mm($\frac{1}{8}$in) long, white, bell-shaped, fragrant, many together in leafy, raceme-like clusters up to 10cm(4in) long, opening in winter. Temperate.

E. ventricosa South Africa
Bushy, erect shrub, eventually up to 2m(6$\frac{1}{2}$ft) in large containers. Leaves crowded, linear to awl-shaped, ciliate, 1.2–1.7cm($\frac{1}{2}$–$\frac{2}{3}$in) long, in whorls of four. Flowers in dense terminal umbels, 1.2–2cm($\frac{1}{2}$–$\frac{3}{4}$in) long, ovoid, inflated at the base and constricted at the mouth, lustrous red, pink or white, opening during the summer and autumn. Temperate.

E. versicolor South Africa
Freely branching, more or less erect, 60–120cm(2–4ft) in height. Leaves to 4mm($\frac{1}{6}$in) long, linear, in whorls of three. Flowers in clusters of three topping all the short lateral stems, 2–2.5cm($\frac{3}{4}$–1in) long, tubular, red, with the mouth and lobes bright greenish-yellow, appearing from autumn to spring. Temperate.

ERIOBOTRYA

Rosaceae

Origin: Southern Asia. A genus of 30 species of evergreen trees with large, entire leaves and creamy-white, 5-petalled flowers in terminal panicles. They are followed by ovoid fruits. Propagate by seed when ripe or by cuttings with a heel in late summer, with bottom heat. *Eriobotrya* derives from the Greek *erion* wool, and *botrys*, a cluster of grapes, referring to the woolliness of the flower clusters. A handsome foliage plant.

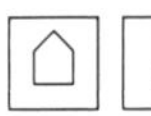  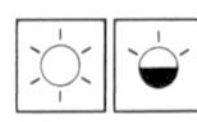  

Species cultivated

E. japonica Loquat China, Japan
Large shrub or small tree with woolly young stems, 4–10m(13–33ft) tall in the open, but can be kept smaller in a tub or large pot. Leaves 15–30cm(6–12in) long, oblong-lanceolate, with conspicuous parallel veins, dark green above, rusty-hairy beneath. Flowers 2cm($\frac{3}{4}$in) wide, creamy-white and fragrant, borne in pyramidal panicles in autumn and winter. Fruits ovoid, 4cm(1$\frac{1}{2}$in) long, downy, pale golden-yellow with white or yellowish flesh, edible.

ERVATAMIA See TABERNAEMONTANA
ERYSIMUM See CHEIRANTHUS

ERYTHRINA

Leguminosae

Origin: Tropics and sub-tropics of both hemispheres. A genus of 100 species of mainly trees and

ABOVE LEFT *Erythrina crista-galli*
ABOVE *Eschscholzia californica*

shrubs, most with thick, somewhat spiny stems. They have trifoliate leaves, usually deciduous, sometimes evergreen, and pea flowers in racemes followed by pods of hard, often red seeds. The species described below is a fine tub or pot plant for the conservatory and although a tree, is grown as a perennial, by cutting it to the ground annually or every other year, flowering when small. Keep cool and almost dry in winter. Propagate by seed or by soft cuttings taken with a heel, in spring, using bottom heat. *Erythrina* derives from the Greek *erythros*, red, referring to the flower colour.

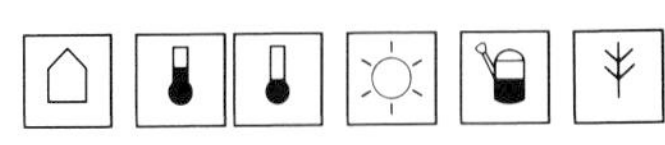

Species cultivated

E. crista-galli Cockspur coral tree Brazil
As cultivated, a thorny, erect shrub, 1–2m(3–6½ft) tall. Leaves with broadly elliptic leaflets to 10cm(4in) long, somewhat glaucous and thick-textured. Flowers 5cm(2in) long, brilliant red with a large, reflexed standard petal to 5cm(2in) long, a narrow keel and very small wings; they are borne in dense, terminal racemes and open in summer.

ERYTHRORHIPSALIS See RHIPSALIS

ESCHSCHOLZIA

Papaveraceae

Origin: Western North America. A genus of about ten species of annuals and perennials with alternate, finely ternately divided leaves and 4-petalled poppy-like flowers, rolling in dull weather to protect the style and the normally 16 stamens. Fruits are slender, pod-like capsules opening explosively when ripe. Hardy. Propagate by seed sown in autumn for winter to spring flowering plants, in spring for late summer blooms. *Eschscholzia* was named for Dr Johann Freidrich Eschscholz (1793–1831), a Russian doctor and naturalist of German extraction who accompanied Russian scientific expeditions including one to the Pacific coast of North America.

Species cultivated

E. caespitosa (*E. tenuifolia*) California, USA
Leaves three times ternate, glaucous. Flowers 2.5–4cm(1–1½in) wide, bright yellow, on stems usually 15–20cm(6–8in) tall in cultivation. 'Miniature Primrose' with primrose-yellow flowers on stems 12–15cm(4¾–6in) tall is probably derived from this species.

E. californica Californian poppy California to Washington, USA
Leaves dissected into almost thread-like segments, glaucous blue-grey. Flowers 4–6cm(1½–2½in) across, orange-yellow, on stems 20–40cm(8–16in) tall. Many colour strains are available, some with double flowers: 'Ballerina' has semi-double flowers, the petals fluted, in shades of yellow, orange, pink, carmine and crimson-scarlet.

E. tenuifolia See *E. caespitosa*.

ESCOBARIA See CORYPHANTHA

ESPOSTOA

Cactaceae

Origin: South America. A genus of six to 11 species of columnar, often large, branching cacti which make fine specimen plants for the sunny conservatory or house window. Some species are thickly hairy, others have the normal spiny armature and resemble *Cereus*. Flowers only appear on mature plants of many years. They are borne on a lateral cephalium – an elongated section of the stem tip modified from several ribs. Propagate in spring by seed or cuttings of branch tips. *Espostoa* commemorates Nicholas E. Esposto, a botanist in Lima, Peru.

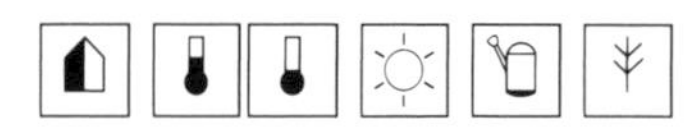

Species cultivated

E. blossfeldiorum (*Thrixanthocereus blossfeldiorum*) Northern Peru
Stem up to 90cm(3ft) or so, simple or with a few branches. Ribs 18 to 25, bearing white-woolly areoles which soon become glabrous. Radial spines 20 to 25, glossy-textured to 1cm($\frac{3}{8}$in) long; central spines one, awl-shaped, black-tipped, to 3cm($1\frac{1}{4}$in) long. Flowers narrowly funnel-shaped, 5cm(2in) in length, white.

E. lanata Snowball cactus Northern Peru, Southern Ecuador
Stems of mature plants in the wild up to 5m(16ft) or more, but taking many years to attain half this in a container. Ribs about 20, low and rounded, bearing areoles with persistent white or yellowish hair and spines. Radial spines glassy-yellowish or reddish, 12 or more, 6mm($\frac{1}{4}$in) long. Flowers white, but not on small, potted specimens. This is the most decorative species and the best for the home as it likes warmth and tolerates a certain amount of shade.

Espostoa lanata

EUCALYPTUS

Myrtaceae

Gum trees, Eucalypts

Origin: Australia and Indo-Malaysia. A genus of about 500 species of evergreen trees and shrubs many of which are colloquially known as gums. They have aromatic, leathery leaves and multi-stamened petalless flowers which may be white, cream, yellow, pink or red. In most species the leaves of young plants (juvenile leaves) are borne in pairs and are a different shape from the adult ones. They are often more brightly hued, e.g. in the familiar blue gum, *E. globulus*, the juvenile leaves are ovate-cordate and blue-white, whereas the adult ones are lanceolate and more green than grey. The species described below make good tub or pot plants for conservatory and home when young. To retain juvenile foliage they must be cut back annually, or not less than every other year, in spring. Propagate by seed in spring. *Eucalyptus* derives from the Greek *eu*, well and *kalypto*, to cover; the united sepals and petals form a cap over the numerous showy stamens which is shed when the flower opens.

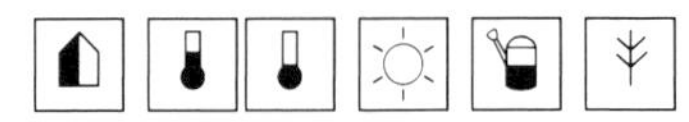

Species cultivated

E. citriodora Lemon-scented gum Queensland
Tree to 45m(150ft) tall in the wild, but easily maintained at under 2m($6\frac{1}{2}$ft) with annual pruning. Juvenile leaves oblong-lanceolate, 8–15cm(3–6in) long, rough-textured, hairy; adult leaves similar, but smooth; both are fragrant of oil of citronella when bruised. Flowers white, but not borne on cut back specimens.

E. ficifolia Red-flowering gum Western Australia (Albany area)
Small tree to 8m(25ft) or more in the wild or planted out, but easily kept to a third of this in a container. The leaves of juvenile and adult plants are similar, broadly lanceolate, to 10cm(4in) long, deep green. Flowers 3–4cm($1\frac{1}{4}$–$1\frac{1}{2}$in) wide, pale to deep red, usually produced on large, root-bound, tubbed specimens.

E. globulus (Tasmanian) Blue gum Tasmania and Victoria
Large tree in the wild or planted out, but easily maintained at about 2m($6\frac{1}{2}$ft) in containers. Juvenile leaves ovate-cordate, 8–15cm(3–6in) long, brightly glaucous; adult leaves lanceolate-falcate, 13–30cm(5–12in) long, greyish-green. Flowers creamy white, but not borne on cut back specimens. The easiest eucalypt for containers, the juvenile phase being far the most decorative.

E. macrocarpa Mottlecah Western Australia
Spreading shrub to 3m(10ft) or more in the wild,

ABOVE LEFT *Eucalyptus nitens*
ABOVE *Eucharis korsakovii*

but easily kept to half this in a container. Leaves of the juvenile type only, in opposite pairs, oblong-ovate to broadly lanceolate, to 10cm(4in) or more long, intensely silvery blue-grey. Flowers are the largest in the genus, to 8cm(3in) wide, red. Fruits the largest (widest) in the genus, like ornate 5cm(2in)-wide buttons. Perhaps the most intensely glaucous-leaved of all the eucalypts and well worth growing for that alone.

E. nitens Shining gum Victoria

Large tree in the wild, but easily kept to 2m(6½ft) or so in a container. Young stems square in cross-section, the angles with prominent, red-edged, crimped wings. Juvenile leaves to about 13cm(5in) long, oblong-ovate-cordate, glaucous with a red mid-vein; adult leaves to 20cm(8in) in length, lanceolate, lustrous green. Flowers white, but not borne on cut back containerized plants. Although not well known at present, this species rivals *E. globulus* as a decorative foliage plant when young and is worthwhile seeking out. It will stand several degrees of frost.

EUCHARIS

Amaryllidaceae

Origin: Tropical South America. A genus of about ten species of bulbous plants allied to *Hymenocallis*. Like that genus they have 6-tepalled flowers with a central cup or corona formed of outgrowths from the bases of the stamen filaments. The species described below thrive in containers and add a touch of class to a collection of flowering pot plants. Propagate by seed and offsets in spring. *Eucharis* derives from the Greek for charming, or pleasing.

Species cultivated

E. grandiflora (*E. amazonica*) Amazon lily Colombia, Peru

Bulbs eventually forming clumps. Leaves evergreen, long-stalked, the blades 20cm(8in) long, oblong to ovate or elliptic, heavily textured, lustrous deep green. Flowering stems erect, 30–60cm(1–2ft) tall, topped by an umbel of three to six flowers; each bloom narcissus-like, but with a slender curving tube, 8cm(3in) wide, white, fragrant. Summer is the usual flowering time but if after each new flush of leaves has matured the plant is kept on the dry side for about six weeks, spring and autumn or winter flushes can be triggered off.

E. korsakovii

Like a half-sized version of *E. grandiflora* which tolerates temperate conditions.

EUCOMIS

Liliaceae

Pineapple lilies

Origin: Tropical and southern Africa. A genus of 14 species of bulbous perennials, the bulbs with widened, almost stem-like base plates and fleshy-textured strap-shaped leaves. Their flowers are 6-tepalled, star-shaped and usually in shades of green or white, carried in cylindrical racemes and topped by a tuft of green leaves. They make good pot plants for the cool conservatory and are best repotted annually in early spring. Propagate by removing offsets, or by seed, both in spring. *Eucomis* derives from the Greek *eu*, good and *kome*, hair, the terminal tufts of leaf-like bracts fancifully like a good head of hair.

  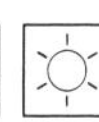 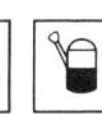 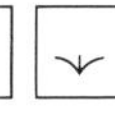

BELOW *Eucomis bicolor*
RIGHT *Eucomis autumnalis*
FAR RIGHT *Eugenia brasiliensis*

Species cultivated

E. autumnalis (*E. undulata*) South Africa
Leaves to 60cm(2ft) long, ovate-lanceolate, with wavy margins and grooved at the mid-vein. Flowers 1–1.5cm($\frac{3}{8}$–$\frac{5}{8}$in) wide, greenish or white in 45cm($1\frac{1}{2}$ft) long cylindrical racemes, from late summer to autumn.

E. bicolor South Africa (Natal)
Very similar to *E. autumnalis* but with pale green flowers having purple-margined tepals and longer stalks; they are borne in racemes 35–40cm(14–16in) long.

E. comosa (*E. punctata*) Pineapple flower
South Africa
Leaves broadly lanceolate without wavy edges, to 60cm(2ft) long. Flowers to 1.5cm($\frac{5}{8}$in) wide, yellow-green to whitish, sometimes flushed with purple, with a dark purple ovary; they are borne in racemes on purple-spotted stalks, the whole 30–60cm(1–2ft) in height.

E. pole-evansii Giant pineapple flower
Southern Africa (Transvaal, Swaziland)
The largest species, with deeply channelled leaves, 60cm(2ft) or more long. Flowers greenish-white, on long stalks, borne in racemes on stems, giving an overall height of 90–180cm(3–6ft).

E. punctata See *E. comosa*.

E. undulata See *E. autumnalis*.

E. zambesiaca East Africa
Leaves broadly lanceolate, slightly wavy-edged, 45–60cm($1\frac{1}{2}$–2ft) long. Flowers white, with short stalks giving a dense raceme; they are borne on a green stalk, to 60cm(2ft) high.

EUGENIA

Myrtaceae

Origin: Tropics and sub-tropics, widespread. A genus of approximately 1,000 species of evergreen trees and shrubs, some of which make useful large pot or tub plants. They have opposite pairs of simple leaves and solitary or clustered, many-stamened, 5-petalled, white flowers followed by berries in shades of yellow, purple, red and blue-black. Propagate by seed in spring or by cuttings in summer. *Eugenia* commemorates Prince Eugène of Savoy, France (1663–1736), reputedly a Patron of botany.

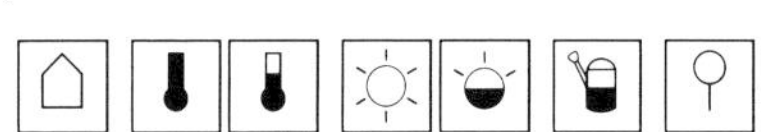

Species cultivated

E. australis See *Syzygium paniculatum*.

E. brasiliensis (*E. dombeyi*) Brazil
Sizeable, round-headed tree in the wild, but easily kept to 2m($6\frac{1}{2}$ft) or so. Leaves obovate to oblong-ovate, 10–13cm(4–5in) long, leathery-textured, deep lustrous green. Flowers solitary. Fruits about the size of a cherry, long-stalked, red to purple-black, edible and sweet.

E. dombeyi See *E. brasiliensis*.

E. jambos See *Syzygium jambos*.

E. myriophylla (*Myricaria myriophylla*) Brazil
Normally a large, well branched shrub, but easily restricted to 1.2m(4ft) or under. Leaves densely crowded, linear to narrowly lanceolate, up to 4cm($1\frac{1}{2}$in) long, but usually less. Flowers solitary.

Grown for its frothy foliage on slender, arching to pendulous stems.

E. myrtifolia See *Syzygium paniculatum.*

E. paniculata See *Syzygium paniculatum.*

E. smithii (*Acmena smithii, A. floribunda*) Lillypilly Australia

Large tree in the wild, but easily kept to 2–3m(6½–10ft) in a tub. Leaves ovate to oblong-lanceolate, 4–9cm(1½–3½in) long, lustrous rich green, aromatic when bruised. Flowers in terminal panicles 5–10cm(2–4in) across. Fruits purple or purple and white, the size of a pea.

E. uniflora Pitanga, Surinam cherry Brazil

In its native land a large shrub or small tree 8m(25ft) in height, but amenable to pruning down to 2m(6½ft) or so. Leaves ovate to elliptic, 3–6cm(1¼–2½in) long, purple-red when young, then a rich green. Flowers fragrant, solitary or in few-flowered clusters. Fruits edible, globose, deeply fluted, up to 3cm(1¼in) in diameter, glossy pale to deep red or yellow, sweet and juicy. Much grown in tropical countries for its fruit and ornamental appearance.

EUONYMUS

Celastraceae

Origin: Almost cosmopolitan, but mainly Asia. A genus of about 170 species of deciduous and evergreen shrubs, trees and climbers. Among the evergreen species those described below provide useful pot or tub plants for unheated rooms and conservatories. They have opposite pairs of simple leaves and small axillary cymes of generally insignificant, greenish, whitish or purplish flowers. The fruits are often red or orange, but only in the deciduous species are they conspicuously colourful. Propagate by seed when ripe (dried seeds can take one or two years to germinate) or by cuttings in late summer. *Euonymus* derives from the Greek *euonymon*, meaning 'of good name'. Reputedly this is used in an ironic sense as many of the species are poisonous to livestock.

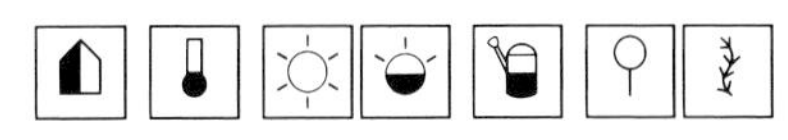

Species cultivated

E. fimbriatus See *E. lucidus.*

E. fortunei China, Japan, Korea

Evergreen self-clinger with aerial roots. Leaves to 6cm(2½in) long, elliptic, dark green. Flowers small, greenish-white, not produced on small plants. Hardy. 'Emerald n' Gold' has yellow-margined leaves; 'Silver Queen' tends to remain low, but will eventually produce climbing shoots, leaves with creamy-white margins.

E.f. kewensis

A dwarf mutant with rounded to oblong leaves 7–10mm(⅓–⅜in) long, with paler veins.

E.f. radicans

Leaves to 4cm(1½in) long. Both it and its variegated forms can be used to cover the back wall of an unheated conservatory. The dwarf non-climbing sorts are useful pot plants.

E. japonicus Japan

Shrub to 2m(6½ft) or more in a pot, to 7m(23ft) in the open. Leaves to 8cm(3in) long, obovate to elliptic, glossy green. Flowers 5mm(⅕in) across, whitish-green followed by rounded, 7mm(⅓in) wide greenish or pinkish fruits. Hardy. 'Albomarginatus' has a central gold blotch; 'Microphyllus Pulchellus' is dwarf, to 30cm(1ft) high, with leaves 1–2.5cm(⅜–1in) long, variegated with gold; 'Microphyllus Variegatus' is similar with silver variegation.

E. lucidus (*E. fimbriatus*) Himalaya

Large shrub or small tree when planted out, but in containers easily kept to 1.2–1.8m(4–6ft) in height. Leaves up to 13cm(5in) long, lanceolate to narrowly oval, finely toothed, leathery-textured, lustrous rich green when mature, red when young. Flowers small, green. Fruits 4-winged, about 1.2cm(½in) wide, splitting to disclose orange-coloured seeds. Will not stand severe frost and best grown at cool temperatures.

Euonymus fortunei 'Silver Queen'

EUPATORIUM

Compositae

Origin: Mainly the Americas, but some also in Europe, Africa and Asia. A genus of about 1,200 species of shrubs and perennials grown for their abundantly borne clusters of tiny groundsel-like flower heads. Among the frost-tender shrubs are several worthy of greater recognition as pot plants. Propagate by seed in spring or by cuttings in summer. *Eupatorium* honours Mithridates Eupator, King of Pontus, an ancient district of modern Turkey, who reputedly discovered that one species acted as an antidote to poison.

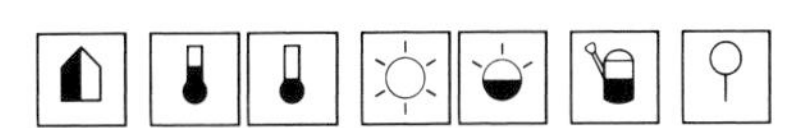

Species cultivated

E. atrorubens Mexico

Softly woody-stemmed shrub 60–90cm(2–3ft) in height. Stems robust, bearing opposite pairs of ovate leaves 15–30cm(6–12in) long, deep green above, purple-hairy beneath, toothed. Flowers lavender to reddish-purple in terminal panicles up to 30cm(1ft) across, mainly in winter. The finest species in both foliage and flower. Temperate.

E. ianthinum See *E. sordidum.*

E. ligustrinum See *E. micranthum.*

E. micranthum (*E. ligustrinum, E. weinmannianum*) Mexico

As a container plant, this bushy shrub rarely exceeds 90–120cm(3–4ft), but when planted out, twice or three times this height can be expected. Leaves in opposite pairs, elliptic to lanceolate, 4–9cm(1½–3½in) long, smooth, lightish green. Flowers white, sometimes with a hint of pink, fragrant, in flattish corymbs 10–20cm(4–8in) wide,

Eupatorium sordidum

in autumn. Hardier than the other two species mentioned here, thriving under cool temperatures and surviving light frost.

E. sordidum (*E. ianthinum*) Mexico
Shrub to 90cm(3ft) or more with fairly robust stems. Leaves in opposite pairs, ovate to broadly oblong, toothed, up to 10cm(4in) long with longish stalks, deep green. Flowers violet-purple, fragrant, in dense corymbose clusters to 9cm(3½in) wide, mainly in winter. Temperate.

E. weinmannianum See *E. micranthum*.

EUPHORBIA

Euphorbiaceae

Origin: Cosmopolitan, though most frequent in sub-tropics and warm temperate regions. A genus of some 2,000 species of annuals, perennials and shrubs, many being succulent. They differ vegetatively in almost every respect, but florally they are remarkably alike. All have a cyathium – a small cup-shaped whorl of bracts which are fused together. This contains several male flowers each reduced to a single stamen, and one female flower which is no more than a 3-lobed ovary. The bracts often bear crescent-shaped nectar glands, but sometimes have larger, petal-like structures. The cyathia are borne in dichasial cymes the stalks of which are known as rays and grow in umbel-like clusters (pseudumbels) from the tips of the stems. At the base of the cyathia are separate pairs of bracts called raylet leaves, and a ring of larger bracts called pseudumbel leaves at the base of each pseudumbel. These can be brightly coloured as in poinsettia. The fruits, which are 3-lobed capsules, open explosively. All the species described here are suitable for the conservatory and home, especially when small. Succulents need a dry resting season. *E. fulgens* and *E. pulcherrima* are short-day plants needing 9–10 hours of darkness before they initiate flower buds. This needs to be done for about three months before flowers are required and can easily be carried out by covering with a box or blinds sufficiently early each evening to give the required amount of darkness. The milky latex of most species of *Euphorbia* is highly irritant to eyes, mouth and tender skin. Propagate by seed, division where possible or by cuttings in summer; for the succulent species washing the ends of the cuttings or dipping them in powdered charcoal and allowing them to dry for several days before insertion. *Euphorbia* was reputedly named for Euphorbos, doctor to the king of Mauritania.

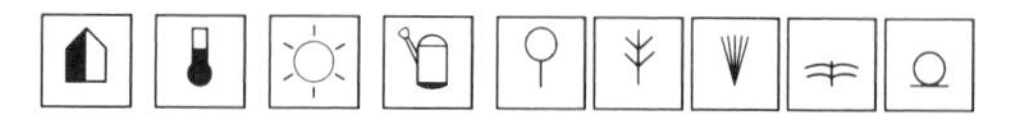

Species cultivated

E. abyssinica Abyssinia
Sizeable succulent in the wild, but 2–3m(6½–10ft) in large containers. Stems robust, erect, with eight almost wing-like, often wavy angles, each angle with a brownish-grey horny margin and bearing closely set, paired, 1cm(⅜in) long thorns. Leaves linear-lanceolate, to 5cm(2in) long, on the young stem tips only and soon falling.

E. aphylla Canary Is.
Bushy, mound-forming shrub 45cm(1½ft) or more in height, the stems fleshy, erect, bright greyish to yellowish-green, leafless.

E. atropurpurea Canary Is. (Tenerife only)
Sparingly branched shrub to 1m(3ft) or more, with somewhat succulent stems. Leaves oblanceolate, glaucous, to 10cm(4in) long, in deep rosettes or

tufts at the stem tips. Cyathia and raylet leaves red-purple, forming a panicle-like cluster from the leaf rosettes, developing in winter and spring.

E. bubalina South Africa
Shrub to 1.5m(5ft), the succulent, erect stems thickened towards the tips, green ageing grey-green. Leaves lanceolate, 6–10cm($2\frac{1}{2}$–4in) long, light green, on the young stems only, falling in autumn. Cyathia surrounded by red-edged or entirely red raylet leaves, appearing in summer.

E. bupleurifolia South Africa
Stem erect, usually unbranched, ovoid to shortly cylindrical, succulent, up to 20cm(8in) in height by 8cm(3in) thick, boldly tubercled by the persistent leaf stalk bases. Leaves lanceolate, 10–15cm(4–6in) long, light green, from the top of the plant in summer, then deciduous. Cyathia solitary on 5cm(2in) stalks with two green to red raylet leaves, developing in summer.

E. canariensis Canary Islands
Succulent leafless shrub, to 90m(3ft) or so in containers (three times this in the wild where it forms wide, mound-like masses). Stems mainly branching at the base, sometimes also forking above, erect, prominently 4- or 5-angled, green with a waxy-white patina; each angle with a horny margin and bearing a row of paired, 1.5cm($\frac{5}{8}$in) long thorns. Cyathia reddish-green, solitary on short stalks at the tips of the stems, but rarely on containerized specimens. A decorative species, but very slow growing.

E. caput-medusae Medusa's head South Africa
Stem thick, succulent, rarely above 15cm(6in) tall, but occasionally more on old specimens. Branches fleshy, crowded, radiating from the top, prostrate to decumbent, rather like a multi-armed octopus, greyish-green, eventually to 45cm($1\frac{1}{2}$ft) or more in length. Leaves linear-lanceolate, 1.5–2.5cm($\frac{5}{8}$–1in) long, at the branch tips only and short-lived.

E. cereiformis (*E. erosa*, *E. leviana*) South Africa
Erect, succulent shrub to 90cm(3ft) in height. Stems dark green, branched at the base and higher up furrowed into nine to 15 ribs, the ridges of which are tubercled and spiny. Leaves minute, short-lived. Cyathia solitary at the stem tip, the raylet leaves purple. An easy species to cultivate.

E. erosa See *E. cereiformis*.

E. fulgens Scarlet plume Mexico
Elegant deciduous shrub to 1.2m(4ft) tall. Stems slender, arching, wand-like, usually sparingly branched. Leaves elliptic to lanceolate, 5–10cm(2–4in) long, bright green. Cyathia with petal-like appendages, in small clusters from all the upper leaf axils, creating bright scarlet sprays in winter.

E. gorgonis South Africa (Cape)
Stem fleshy, more or less globose, up to 10cm(4in) high and wide, tubercled with persistent swollen leaf bases. Branches succulent, crowded, radiating stiffly from the top, up to 5cm(2in) long, green, sometimes red-tinted. Leaves minute, soon falling. Cyathia solitary, from the main stem only, with brownish to bright crimson bands. Much like a short-branched *E. caput-medusae*.

E. grandicornis Cow's horn Southern Africa (Natal, north to Kenya)
Succulent shrub, eventually to 2m($6\frac{1}{2}$ft). Stems much branched, constricted into rounded or ovoid segments with three wing-like angles; each angle wavy, horny-margined and bearing wide-spreading pairs of sharp, 2–5cm($\frac{3}{4}$–2in) long spines.

E. grandidens South Africa (Cape)
Eventually a succulent tree to 15m(50ft) in the wild, but easily kept at 2m($6\frac{1}{2}$ft) or so in containers. Stems erect with branches in whorls, the latter with three to four often spirally twisted angles; each angle with a horny, toothed margin bearing pairs of 4–6mm($\frac{1}{6}$–$\frac{1}{4}$in) long spines. Leaves minute and only rather short-lived. A statuesque and easily grown species.

E. horrida South Africa, Cape to Karroo
Succulent, cactus-like shrub to 90cm(3ft) in height, mainly branched from the base and forming clumps. Stems erect, with up to 14 furrows, the ridges of which are toothed and bear single or clustered spines. The latter are derived from the

BELOW *Euphorbia grandicornis*
BOTTOM LEFT *Euphorbia fulgens*
BOTTOM RIGHT *Euphorbia atropurpurea*

hardened flowering peduncles and are up to 4cm(1½in) long. Cyathia solitary, green.

E. ingens Southern Africa, north to Zimbabwe and Zambia
In the wild, a succulent tree to 10m(30ft) in height, but in containers easily kept to 1.5m(5ft) or less. Young specimens and the ultimate branches of older ones are dark green when mature and constricted somewhat at the end of each growing season. They have three to five wing-like angles which are ribbed, wavy and sometimes bear tiny spines; leafless.

E. leviana See *E. cereiformis*.

E. mammillaris South Africa (Cape)
Small, succulent, cactus-like, suckering species to 20cm(8in) tall by eventually twice as wide. Stems semi-erect to prostrate, cylindrical, to 5cm(2in) thick or more, with seven to 17 tuberculate angles. Persistent, woody, flowering peduncles provide a scattering of 1.2cm(½in) long spines. Leaves scale-like, short-lived. Cyathia solitary, with yellowish or purple glands.

E. mellifera Madeira, Canary Is.
Shrub to 2m(6½ft) in a pot, but in the wild 15m(48ft). Leaves narrowly lanceolate, mid to dark green with a paler central vein; raylet leaves pale to reddish-brown. Cyathia richly honey-scented, flowering in early spring.

E. meloformis South Africa
Cactoid succulent with a globose stem to 10cm(4in) tall, rarely more, with eight to ten prominent ribs. Between the ribs the skin is grey-green, or reddish with a paler ribbed pattern. Leaves tiny, linear, short-lived. Persistent, woody, flowering peduncles to 2cm(¾in) or more long form scattered blunt-tipped spines.

E. milii (*E. splendens*) Crown of thorns
Malagasy
Evergreen shrub, spreading to 90cm(3ft) and reaching 30–60cm(1–2ft) in height. The densely spiny stems are grooved. Leaves 1.5–10cm(⅝–4in) long, obovate to almost rounded. Cyathia borne in small umbels and have showy, bright red raylet leaves. In the form *lutea* they are lemon-yellow.

E. obesa Gingham golf ball South Africa
A remarkable small succulent reduced to a ball-shaped, later pear-shaped stem, 8–12cm(3–4¾in) tall, with five longitudinal ribs; in colour it is a light greenish-grey, patterned with a chequered network of reddish lines. The plants are dioecious, the cyathia small and greenish-yellow.

E. pseudocactus Cactus spurge South Africa (Natal)
Succulent, cactoid shrub 1–2m(3–6½ft) in height. Stems erect or inclined, branched, constricted into ovoid segments with three to five angles; each angle has a horny, toothed edge and bears paired spines about 1.2cm(½in) long. Between the ribs the skin is a medium to deep green with a contrasting, yellowish, arching V-shaped pattern.

E. pulcherrima Poinsettia Mexico
A familiar pot plant, usually seen below 1m(3ft) in height, but in the wild to 4m(13ft). Leaves to 15cm(6in) or more long, ovate to lanceolate, toothed and sometimes lobed. Pseudumbel leaves normally bright red, in clusters to 15–30cm(6–12in) across. Cultivars in shades of pink and white are also available.

E. splendens See *E. milii*.

E. tirucalli Rubber spurge, Finger tree, Milk bush Tropical and Southern Africa
Succulent shrub to 2m(6½ft) or more with freely

RIGHT *Euphorbia mellifera*
BELOW *Euphorbia meloformis*
BOTTOM *Euphorbia pulcherrima*

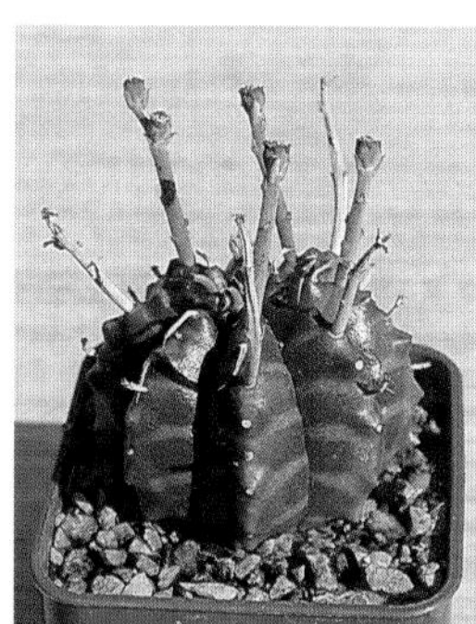

branching, smooth, green, jointed, finger-like stems to 6mm($\frac{1}{4}$in) or more thick. Leaves very small, linear, soon falling. A fast-growing easy species, though not of the first order.

EURYA

Theaceae

Origin: Eastern Asia to the Pacific region. A genus of 130 species of evergreen trees and shrubs allied to *Camellia*, but with small flowers and berry-like fruits. The one species described here can be used as a handsome foliage plant. Propagate by seed when ripe or in spring, or by cuttings in late summer. *Eurya* reputedly derives from the Greek *euru*, broad, but in what sense is not recorded.

  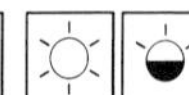  

Species cultivated

E. japonica China, Korea, Japan, Taiwan and Malaysia
As a tub specimen, up to 1.2m(4ft) or so and of dense habit. Leaves alternate, 4–8cm($1\frac{1}{2}$–3in) long, leathery-textured, toothed, elliptic to oblong-lanceolate, lustrous deep green. Flowers white to greenish-yellow, 6mm($\frac{1}{4}$in) wide, singly or in small axillary clusters. Berries glossy black, the same size as the flowers.

EUSTOMA

Gentianaceae

Origin: Mexico, Southern USA, northern South America. A genus of three species of perennials, annuals or biennials, one of which is now making a comeback as a highly decorative flowering pot plant. It is best grown in the conservatory, but can be brought into the home when in bloom. Propagate by division in autumn or by seed in early spring. *Eustoma* derives from the Greek *eu*, good and *stoma*, mouth, but is used in the most literal sense for 'a pretty face', e.g. the lovely flowers.

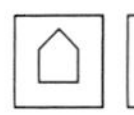 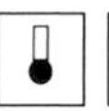   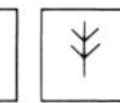

Species cultivated

E. grandiflorum (*E. russellianum*, *Lisianthus russellianus*) Prairie gentian Colorado and Nebraska to Texas and northern Mexico
Erect perennial grown as an annual or biennial, 60–90cm(2–3ft) in height. Leaves simple, in opposite pairs, to 8cm(3in) long, oblong to ovate. Flowers 5-petalled, opening widely bowl-shaped to 5cm(2in) in width, pale satiny purple in loose, terminal, corymbose clusters in summer. The original species is not too easy to grow, but plant breeders have been working on it and there are new compact seed strains 35–45cm(14–18in) tall in shades of blue, pink, purple and white. Recommended are: 'Yodel Blue', 'Yodel Pink' and 'Yodel White'.

EXACUM

Gentianaceae

Origin: Tropical Asia and India, islands of the Indian Ocean. A genus of 40 species of annuals, biennals and perennials with opposite, entire leaves and rounded flowers in clusters from the leaf axils. They are splendid flowering pot plants. Propagate by seed in spring, or for larger plants sow in autumn and these will flower earlier than spring-sown plants. *Exacum* derives from the Gaelic name *exacon*, a vernacular for *Centaurium* and used by Linnaeus for this genus.

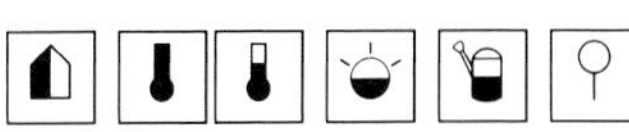

Species cultivated

E. affine Socotra (Indian Ocean)
Bushy annual or biennial, 15–25cm(6–10in) tall. Leaves broadly ovate to elliptic, 1.5–4cm($\frac{5}{8}$–$1\frac{1}{2}$in) long, shiny green. Flowers 1.5–2cm($\frac{5}{8}$–$\frac{3}{4}$in) across, rotate, 5-petalled, lavender-blue to purplish-blue, darker at the base of the petals, stamens yellow; they open from summer to autumn.

ABOVE *Exacum affine*
TOP *Eustoma grandiflorum*

F

FARFUGIUM See LIGULARIA

FASCICULARIA

Bromeliaceae

Origin: Chile. A genus of five species of evergreen perennials related to *Billbergia* and *Cryptanthus*. The two species described are ground dwelling (terrestrial), but some others are epiphytes. They form dense clumps or mounds of narrow-leaved rosettes. The tubular, 3-petalled flowers are borne in clusters in the hearts of mature rosettes. When of a good size, these species make unusual specimen plants, especially when planted out in the conservatory border. They can also be grown in the home. Propagate by removing offsets in spring. *Fascicularia* derives from the Latin *fasciculus*, a bundle, referring to the densely clustered leaf rosettes.

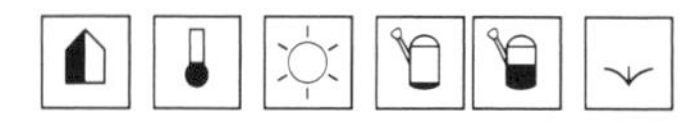

Species cultivated

F. bicolor

Leaves to 45cm(1½ft) long by 1.5cm(⅝in) wide, arching, stiffly leathery, spiny-toothed and sharply pointed. The smaller inner leaves of mature, flowering-sized rosettes are bright red or partially so. Flowers pale blue, to 4cm(1½in) long, in dense, rounded clusters surrounded by bracts which just exceed them.

F. pitcairnifolia

Almost identical to *F. bicolor*, but the floral bracts are shorter and the blue flowers, slightly darker.

× FATSHEDERA

Araliaceae

Tree ivies

A bigeneric hybrid raised in 1910 by the French nursery firm Lizé Frères of Nantes, involving the genera *Fatsia* and *Hedera*. From *Hedera* it has inherited a partially climbing habit. It is a good pot plant, best grown either with support or regularly pinched to keep bushy and is very tolerant of heat and cold, and poor light. Propagate by cuttings in summer and autumn. × *Fatshedera* is a combination of the two generic names.

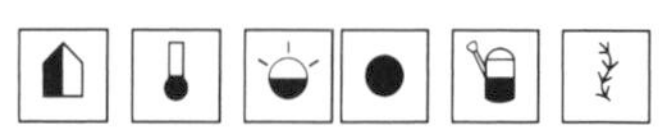

Species cultivated

× ***F. lizei*** (*Fatsia japonica* 'Moseri' × *Hedera helix hibernica*) *Tree ivies*

Semi-climber or shrub, to 3m(10ft). Leaves ivy-like, 20–40cm(8–16in) wide, palmately lobed with five to nine divisions, the margins wavy. Flowers pale green, 5mm(⅕in) across, in rounded umbels which are carried in terminal panicles. × *F.l.* 'Variegata' is irregularly cream-margined and is not so strong growing.

FATSIA

Araliaceae

Origin: Japan, Korea and Taiwan. A genus of one or two species of evergreen shrubs with alternate palmate leaves which are dark green and conspicuously white-veined. The small whitish flowers are borne in umbels which in turn make up loose panicles. They make good house and conservatory plants though are not quite as tolerant as × *Fatshedera*. Propagate by seed in spring, by cuttings in summer or by air-layering in spring. *Fatsia* is a

BELOW *Fascicularia bicolor*
BELOW RIGHT × *Fatshedera lizei* 'Variegata'

Latinized version of *fatsi*, said to be an old Japanese name for the plants.

 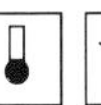 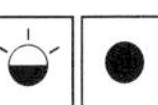  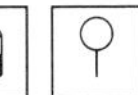

Species cultivated

F. japonica (*Aralia japonica, Aralia sieboldii*) False castor-oil plant Japan, S. Korea
Large shrub or small tree in the wild, to 2m(6½ft) or so in a pot, the stems covered in conspicuous leaf-scars. Leaves 20–40cm(8–16in) wide, with five to nine wavy-margined lobes. Flowers 5mm(⅕in) wide, milky-white, followed by glossy black, rounded berries. *F.j.* 'Moseri' is more compact, with larger leaves; *F.j.* 'Variegata' has white-margined leaves.

F. papyrifera See *Tetrapanax papyriferus*.

Fatsia japonica

FAUCARIA

Aizoaceae

Origin: Cape, South Africa. A genus of 37 species of dwarf succulents easily grown in the home or conservatory. The almost stemless tufts of highly fleshy leaves have flat upper surfaces margined with upstanding, prominent whitish teeth, hence the colloquial name of 'tiger jaws' sometimes applied to members of this genus. The showy flowers are daisy-like and large for the size of the plant. Propagate by seed in spring or by cuttings in summer. *Faucaria* derives from the Latin *faux*, a throat, the paired leaves (described above) resemble a gaping mouth, fancifully like that of a tiger.

  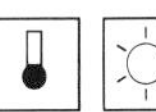 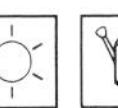 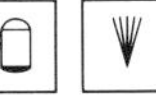

Species cultivated

F. felina Cat's jaws
Leaves to 4.5cm(1¾in) in length, boat-shaped, with a prominently keeled tip and three to five teeth on each of the upper edges, bright green, dotted white, sometimes with a reddish flush when past maturity. Flowers about 5cm(2in) wide, rich yellow, opening only in the afternoon from late summer to late autumn.

F. tigrina Tiger jaws
Much like a somewhat longer version of *F. felina*, the leaves broader and up to 5cm(2in) in length, with each margin bearing eight to ten teeth. Flowers rich yellow, 5cm(2in) across, autumn.

F. tuberculosa
Leaves about 2cm(¾in) long, roughly rhomboidal with a prominent, bluntly rounded, keeled tip, each margin bearing three large and a few smaller teeth, plus blunt to tooth-like tubercles on the upper surface; deep green. Flowers 4cm(1½in) across, yellow, borne from late summer to autumn.

FEIJOA (ACCA)

Myrtaceae

Origin: South America. A genus of two species of evergreen shrubs with opposite, pinnate-veined

ABOVE *Faucaria felina*
LEFT *Faucaria tigrina*

Feijoa sellowiana

leaves and long-stalked, solitary flowers made conspicuous by a central brush-like tuft of red stamens. One species is in cultivation and is suitable for growing in a pot or tub. Propagate by heel cuttings in summer, with bottom heat. *Feijoa* was named for Don J. da Silva Feijoa, Director of the Natural History Museum of San Sebastian, Spain, who botanized in Brazil.

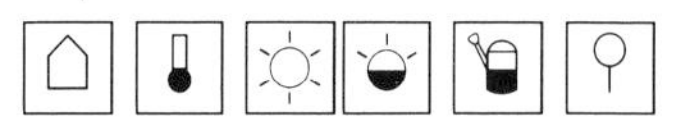

Species cultivated

F. sellowiana (*Acca sellowiana*) Pineapple guava Brazil, Uruguay
Small tree or shrub in the open garden, to 2m($6\frac{1}{2}$ft) or so in a pot. Leaves oval to ovate, 3–8cm($1\frac{1}{4}$–3in) long, dark shining green above, white-felted beneath. Flowers 3–4cm($1\frac{1}{4}$–$1\frac{1}{2}$in) wide, with four petals, reddish-purple at the centre paling to almost white at the margins and white-woolly beneath; they reflex from the long, dark red stamens. Egg-shaped fruits, technically though not obviously berries, follow; they are up to 5cm(2in) long and have a pleasant flavour.

FELICIA

Compositae

Origin: Tropical and Southern Africa. A genus of 60 species of dwarf sub-shrubs and annuals with entire linear to ovate leaves and abundant daisy-like flower heads which are usually in a shade of blue, mauve or white. They make attractive summer pot or hanging basket plants for the conservatory and can be brought into the home for flowering on a sunny windowsill. Propagate annuals by seed, perennials by cuttings or seed. *Felicia* was named for Herr Felix (*d.* 1846), a German official of Regensburg.

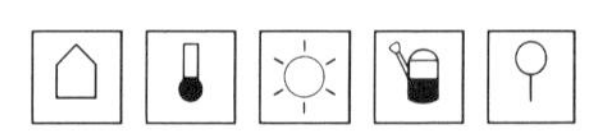

Species cultivated

F. amelloides (*Agathaea coelestis*) Blue marguerite South Africa
An erect to bushy perennial sub-shrub, 30–45cm(1–$1\frac{1}{2}$ft) tall. Leaves opposite, broad to roundish ovate, about 2.5cm(1in) long. Flowers 3–4cm($1\frac{1}{4}$–$1\frac{1}{2}$in) across on long stalks, borne very freely in summer, but opening only in sun.

F. bergeriana (*Aster bergeriana*) Kingfisher daisy South Africa
Annual, 10–15cm(4–6in) tall, bushy. Leaves lanceolate to narrowly oblong, toothed, 2.5–4cm(1–$1\frac{1}{2}$in) long. Flowers 2–3cm($\frac{3}{4}$–$1\frac{1}{4}$in) across, bright blue with yellow to almost black centres; they are borne abundantly on long, hairy stalks throughout the summer.

FENESTRARIA

Aizoaceae

Origin: Southern Africa. A genus of two species of flowering succulents worthy of a place in any collection of such plants. In the wild, the leaves are buried up to the tips in desert sand, the intense sunlight making contact with the chlorophyll through the transparent 'windowed' tip. In the home or conservatory the leaves should be entirely uncovered to take advantage of our less intense sunlight. The daisy-like flowers are disproportionately large and showy. Propagate by seed or very careful division in spring. *Fenestraria* derives from the Latin *fenestre*, a window, alluding to the leaf character described above.

  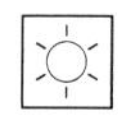 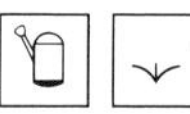

RIGHT *Felicia bergeriana*
FAR RIGHT *Fenestraria aurantiaca*

Species cultivated

F. aurantiaca Cape
Tufted growth habit, eventually forming small hummocks. Leaves 2–3cm($\frac{3}{4}$–$1\frac{1}{4}$in) long, narrowly obovoid with a slightly flattened tip, greyish on the sides. Flowers 3–5cm($1\frac{1}{4}$–2in) wide, bright yellow, borne from late summer to autumn.
F. rhopalophylla Namaqualand, Namibia
In foliage almost the same as *F. aurantiaca*, but individual leaves a little shorter and thicker. Flowers 1.8–3cm($\frac{3}{4}$–$1\frac{1}{4}$in) wide, white, autumn.

FEROCACTUS

Cactaceae

Origin: South-west USA, Mexico. A genus of 35 species of large, unbranched and usually very spiny cacti with funnel- to bell-shaped flowers. They make handsome pot plants for the conservatory and sunny windowsill, never reaching great size in a pot. Propagate by seed in spring, in warmth. *Ferocactus* comes from the Latin *ferox*, fierce and *cactus*; an apt name for the genus.

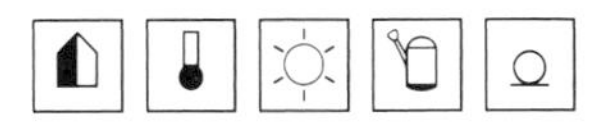

Species cultivated

F. acanthodes S. California
Stem oval, eventually becoming cylindrical, to 3m(10ft) tall with age. Ribs 13 to 23, blunt, with areoles sunken along their edges. Radial spines nine to 20, curved, each 4–6cm($1\frac{1}{2}$–$2\frac{1}{2}$in) long, meshing to form a basket-like covering to the stem, usually red or pink, sometimes white or yellowish; central spines flexible, somewhat flattened, to 12cm($4\frac{3}{4}$in) long. Flowers to 5cm(2in) long, yellow to orange, followed by yellow, red-tinged fruits.
E. haematacanthus (*F. stainesii haematacanthus*) Mexico
Stem globular to shortly columnar, bright green, rarely exceeding 50cm(20in), woolly at the top. Ribs 12 to 20, notched, straight and broad. Radial spines six, to 4cm($1\frac{1}{2}$in), deep red tipped with yellow; central spines about four, to 6cm($2\frac{1}{2}$in) long, thick. Flowers 6cm($2\frac{1}{2}$in) long, purplish-red.
F. horridus California
Stem globular and depressed, later oval then cylindrical, to 2m($6\frac{1}{2}$ft). Ribs 13, broad and blunt with widely spaced areoles. Radial spines eight to 12, 3–4cm($1\frac{1}{4}$–$1\frac{1}{2}$) long, white, spreading; central spines six to eight, 5–7cm(2–$2\frac{3}{4}$in) long, very stiff, spreading, one flattened and hooked, to 10cm(4in) long on large specimens.
F. latispinus Fish hook cactus Mexico
Stem broadly globular, depressed, reaching only 30cm(1ft) in height. Ribs eight, increasing to 20 with age, narrow. Radial spines six to 12, 2–2.5cm($\frac{3}{4}$–1in) long, red or white; central spines about four, stouter and to 3.5cm($1\frac{1}{2}$in), one much flattened and hooked, to 7cm($2\frac{3}{4}$in) long, reddish, brightest when young. Flowers 3.5cm($1\frac{1}{2}$in) long, white or reddish.
F. setispinus See *Thelocactus setispinus*.
F. wislizenii Texas to Arizona
Stem globular, depressed becoming oval, then cylindrical, to 1–2m(3–$6\frac{1}{2}$ft). Ribs 15 to 25, acute, with large, well-spaced areoles. Radial spines about 20, bristle- or awl-like, to 5cm(2in) long; central spines four, to 5–6cm(2–$2\frac{1}{4}$in) long, very thick and flattened with a hooked tip. Flowers 5–6cm(2–$2\frac{1}{4}$in) long, reddish-yellow, green on the outside.

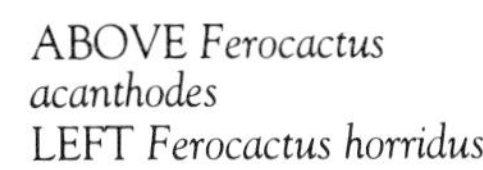

ABOVE *Ferocactus acanthodes*
LEFT *Ferocactus horridus*

FERRARIA

Iridaceae

Origin: Tropical to southern Africa. A genus of two species of cormous plants allied to both *Iris* and *Tigridia* and in appearance rather like a fusion of the two. The one species in general cultivation provides something different in the way of a pot plant. The corms are best left undisturbed for several years

Ferraria undulata

before re-potting. After the leaves yellow and start to die they must be kept dry and warm until autumn when watering is resumed. Propagate by offsets removed when dormant or by seed in spring. *Ferraria* commemorates Giovanni Battista Ferrari (1584–1655) of Siena, an Italian Jesuit botanist and author of *Hesperides*.

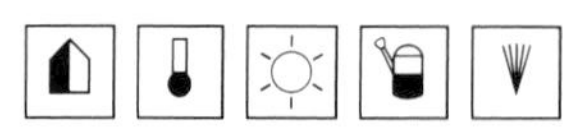

Species cultivated

F. undulata (*F. crispa*) South Africa
Leaves sword-shaped, folded, up to 30cm(1ft) or more in length, arranged in flattened shoots like those of an iris, the outer ones arching or flopping when full-sized. Flowering stems 30cm(1ft) or more in height, arising in spring, branching at the top. Individual flowers from the axils of large green bracts, about 5cm(2in) across, 6-tepalled, their bases overlapping to form a bowl-shaped base, their tips reflexing; each tepal coffee-purple, marked with white at the base and centre and elaborately waved and crimped at the margins like the finest moss-curled parsley. Regrettably, these fascinating blooms are marred by an unpleasant, though not strong scent.

FICUS

Moraceae

Figs

Origin: Tropical and sub-tropical regions of the world, commonest in India and S.E. Asia. A genus of 800 species of chiefly evergreen trees, shrubs and climbers, with alternate leaves many of which are grown as house and conservatory plants for their decorative foliage. In pots their size will be restricted by the small space for roots, but in the wild many are large trees; others start life as epiphytes, putting down roots which thicken and eventually completely surround and kill their host plant; some are straightforward climbers. Heights given below are for container-grown specimens. Most evergreen figs rarely flower or fruit in pots, but when they do, the flowers are remarkable though not decorative, being held within the hollowed-out swollen receptacle. This receptacle can be rounded, ovoid or pear-shaped and opens by means of a tiny hole through which, in the wild, a tiny gall wasp enters to lay its eggs and thus pollinate the flowers. The fruits are small, egg-shaped to rounded. Propagate by seed in spring or by cuttings or layering in summer. *Ficus* is the Latin name for the edible fig.

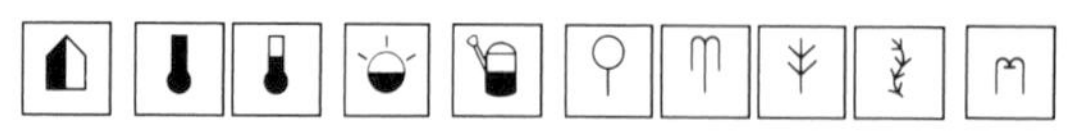

Species cultivated

F. aspera Pacific Islands
Shrub to 1.2m(4ft) or more in containers, but attaining small tree size if planted out. Leaves oblong to ovate, slender-pointed, thin-textured, to 20cm(8in) long; they are hairy beneath and dark green above with a white mottling or spotting. Fruits rounded, green striped white and pink, about 2.5cm(1in) long.

F. auriculata (*F. roxburghii*) Himalaya
Shrub to 1.2m(4ft) or more in containers, a small tree in its homeland. Leaves up to 40cm(16in) long, broadly ovate-cordate, varying from rounded-tipped to slender-pointed, handsome. Fruits pear-shaped, 6cm($2\frac{1}{2}$in) long, silky-hairy, spotted white or red-brown, borne directly from leafless main stems on mature plants.

F. benghalensis Banyan India and Pakistan
Large tree with trunk-like prop roots from the horizontal branches in tropical countries, but easily kept under 2m($6\frac{1}{2}$ft) in containers. Leaves up to 20cm(8in) long, elliptic to broadly ovate, leathery-textured, rich green. Fruits globose, orange-red, 1.2cm($\frac{1}{2}$in) wide, in axillary pairs, but rarely borne on containerized plants.

F. benjamina Weeping fig Tropical Asia
Large tree in the wild, easily kept to 2m($6\frac{1}{2}$ft) or so in a pot. Leaves 5–10cm(2–4in) long, ovate, tapering to a slender point, shining dark green on arching to pendent stems. Fruits globose, 1–1.5cm($\frac{3}{8}$–$\frac{5}{8}$in) wide, black. Very decorative. For plants sold under the name 'Hawaii', see *F. microcarpa* 'Hawaii'.

F. cyathistipula Tropical Africa
Although a tree in its native land this species is easily kept under 60cm(2ft) as a pot plant. Stems erect to spreading, intriguingly clothed with large inflated stipules. Leaves lustrous deep green, to 15cm(6in) or more in length, narrowly obovate to oblanceolate, prominently veined. Fruits not observed on pot grown specimens.

F. deltoidea (*F. diversifolia*) Mistletoe fig Malaysia
Usually grown as a pot shrub 30–60cm(1–2ft) in height, but capable of attaining twice this when planted out, or in a large tub. Leaves 3–8cm($1\frac{1}{4}$–3in) long, broadly obovate, usually bright green above and red-brown-tinted beneath. Fruits 9mm($\frac{3}{8}$in) wide, globular, yellow, on long stalks from the leaf axils even on young plants.

F. diversifolia See *F. deltoidea*.

F. elastica Rubber plant India, Malaysia
Growing in pots 2–3m($6\frac{1}{2}$–10ft) tall, but in the wild making a tree to 30m(100ft) tall or more; relatively slow-growing. Leaves 15–30cm(6–12in) long, oblong to elliptic, rich glossy green with a paler central vein. Fruits not produced on pot plants. 'Decora' is the commonly grown rubber plant with broader leaves than the type, tinted pinky-bronze when first opening. 'Doescheri' has leaves variegated with grey-green, creamy-yellow and white, the darkest colour near the pale veins; they also have pink stalks. 'Schrijveriana' is similar with somewhat broader leaves and red leaf stalks. 'Zulu Shield' has similar variegation with dark-red leaf stalks.

F. lyrata (*F. pandurata*) Fiddle-leaf fig Tropical West Africa
Easily kept to 1m(3ft) in a pot, but an epiphytic

climber to 12m(40ft) or more in the wild. Leaves to 30cm(12in) or more long, fiddle-shaped, dark glossy green. Fruits 4cm($1\frac{1}{2}$in) across, rounded.

F. macrophylla Australian banyan, Moreton Bay fig Eastern Australia
As a pot plant, much like *F. elastica*, but leaves a little smaller and thinner-textured; reputedly not so durable in the home. It is a large banyan-like tree in its homeland.

F. microcarpa (*F. nitida, F. retusa*) Laurel fig S.E. Asia and eastern Australia
Large shrub or small tree with the overall effect of *F. benjamina* as a pot specimen, but with blunt-pointed, obovate leaves. Makes an effective house plant 45–120cm($1\frac{1}{2}$–4ft) in height. 'Hawaii' is margined and splashed with creamy-white; almost always seen erroneously named under *F. benjamina*.

F. microphylla
No botanical validity but sometimes used by gardeners for *F. rubiginosa*, or a form of it, *q.v.*

F. nitida
A confusing name; officially a synonym of *F. benjamina* (*q.v.*), but also wrongly used for *F. microcarpa*.

F. palmeri Baja California
Young plants of this white-stemmed, tree-sized fig species make pleasing pot plants to 1.5m(5ft) or so. Leaves ovate-cordate with pointed tips, up to 18cm(7in) long. Fruits globular, 1.2cm($\frac{1}{2}$in) wide, white, carried in pairs from the leaf axils.

F. pandurata See *F. lyrata*.

F. petiolaris Mexico
Although a sizeable tree in the wild this is another species adaptable for container growth, easily kept 90–120cm(3–4ft) tall. Leaves about 8cm(3in) across, orbicular-cordate with short points, rich green above, tufts of white hairs in the vein axils beneath. Fruits 1.2cm($\frac{1}{2}$in) wide, globular, villous hairy when young, almost smooth later.

F. pumila (*F. repens*) Creeping fig, Climbing fig E. Asia, Australia
Self-clinging climber producing, when mature, non-climbing flowering shoots with elliptic to oblong leaves 5–10cm(2–4in) long. On pot grown specimens foliage is normally juvenile in character, the leaves 1–2.5cm($\frac{3}{8}$–1in) long, ovate-cordate, with prominate veins. Fruits 5cm(2in) long, pear-shaped, yellowish-green.

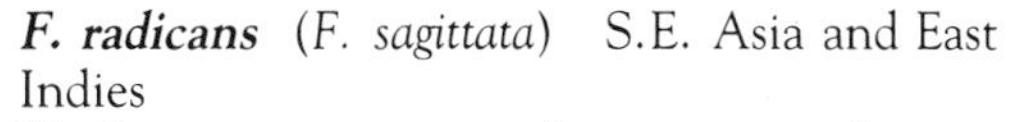

F. radicans (*F. sagittata*) S.E. Asia and East Indies
Trailing stems erect at first in a pot. Leaves 5–10cm(2–4in) long, lanceolate to ovate-lanceolate, dark green with prominent veins. Fruits rarely produced on pot plants. *F.r.* 'Variegata' has leaves with an irregular white margin, sometimes creamy-white to the mid-vein.

F. religiosa Pepul, Bo tree, Sacred fig tree India to S.E. Asia
In its native country this is a large tree, deciduous or partially so in the dry season. As a pot plant it can be kept to 2m($6\frac{1}{2}$ft) and is usually evergreen. The handsome dark green leaves are broadly ovate, up to 15cm(6in) long with a slender tail-like tip. To the Buddhists and Hindus the bo tree is the most sacred of several different so-called sacred figs. Shrines are built beneath them and devout worshippers will not injure nor prune even the smallest seedling.

F. repens See *F. pumila*.

F. retusa See *F. benjamina*; also sometimes erroneously used for *F. microcarpa*, *q.v.*

F. roxburghii See *F. auriculata*.

F. rubiginosa Port Jackson fig Eastern Australia
As a container plant this can be kept to 90cm(3ft) or under; planted out it attains large shrub or small tree status with rooting branches. Leaves up to 10cm(4in) long, elliptic-oblong to ovate, dark

TOP LEFT *Ficus benghalensis*
TOP CENTRE *Ficus microcarpa* 'Hawaii'
TOP RIGHT *Ficus elastica* 'Schrijveriana'
ABOVE LEFT *Ficus religiosa*
ABOVE *Ficus lyrata*

green and lustrous above, brownish-red and pubescent beneath. Fruits in pairs from the leaf axils, globular, 1.2cm($\frac{1}{2}$in) wide, reddish pubescent and warted.

F. sagittata See *F. radicans*.

F. sycamorus Sycamore/Mulberry fig (The sycamore of the Bible) Southern Africa north to Egypt, cultivated widely in the eastern Mediterranean region

In its homeland this fig grows to 18m(60ft) and is partially deciduous; as a pot plant it can be kept to 1.2m(4ft) or so and is usually evergreen. Leaves broadly ovate to orbicular, cordate, up to 15cm(6in) long, rugose and somewhat bluish-green. Fruits obovoid to rounded, about 2.5cm(1in) long, edible, in panicles from the trunk, branches and leafless stems, but rarely borne on containerized specimens.

F. villosa Malaysia

Climbing species which clings to its support by aerial roots in the same way as ivy; stems to 3m(10ft) or more in length, but easily pruned to size. Leaves to 8cm(3in) long, broadly ovate-cordate with a slender pointed tip, dark green with brown ciliate margins. A useful plant for covering conservatory walls and moss-sticks and it makes an unusual hanging basket subject.

FITTONIA

Acanthaceae

Origin: Peru. A genus of two species of creeping, evergreen perennials with broad, decorative leaves having the veins marked in a contrasting colour. Their small flower spikes add little to the display. These are plants for the warm conservatory or home, where they need additional humidity, growing well in terrariums, and can also be used for hanging baskets. Propagate by careful division or by cuttings in warmth in summer. *Fittonia* was named for Elizabeth and Sarah Mary Fitton who wrote *Conversations on Botany* in 1817.

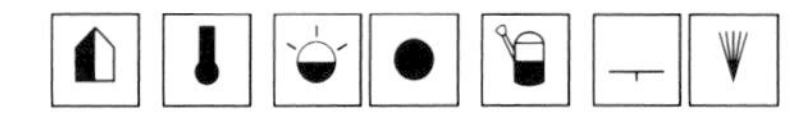

Species cultivated

F. argyroneura See *F. verschaffeltii argyroneura*.

BELOW *Fittonia gigantea*
BELOW RIGHT *Fittonia verschaffeltii argyroneura* 'Minima'

F. gigantea

Creeping to erect perennial which can exceed 30cm(1ft) in height. Leaves broadly oval, 8–12cm(3–4$\frac{3}{4}$in) long, shining green with slightly depressed fine red veins. Flowers yellowish, in spikes up to 15cm(6in) long and although not colourful add to the effectiveness of the plant.

F. verschaffeltii Painted net leaf, Nerve plant

Creeping to tufted, to 15cm(6in) tall. Leaves oval, 7–10cm(2$\frac{3}{4}$–4in) long, dark green with the veins picked out boldly in red, forming an elaborate network over the leaf surface. Flowers small, reddish to yellowish, borne in short spikes.

F.v. argyroneura Silver net leaf

The most commonly seen form with the veining clearly marked in white, a very decorative plant. *F.v.a.* 'Minima' is similar, but much smaller, seldom exceeding 10cm(4in) in height, with leaves 2–3.5cm($\frac{3}{4}$–1$\frac{1}{2}$in) long.

F.v. pearcei

Brighter green leaves with pink veins.

FORTUNELLA

Rutaceae

Kumquats

Origin: Eastern Asia and the Malay Peninsular. A genus of four to six species of evergreen shrubs and small trees. They have simple leaves, white flowers with five or more petals and orange-yellow to red fruits like those of small to miniature oranges. All the species mentioned here provide handsome, long-term pot plants attractive at all times of the year. Propagate by seed when ripe or in spring, or by cuttings in summer. *Fortunella* honours Robert Fortune (1812–1880), a Scottish horticulturist, plant collector, traveller and author responsible for the establishment of the tea industry in India (from China) and the introducer to Britain of many now popular garden plants.

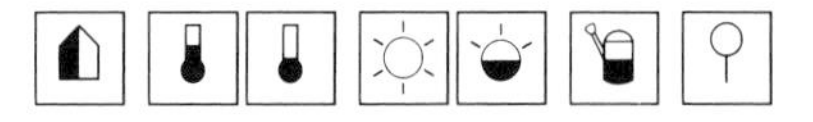

Species cultivated

F. crassifolia Meiwa kumquat

Unknown provenance probably of hybrid origin and originating in China. A well branched shrub, slightly thorny, 60–120cm(2–4ft) in height as a container specimen. Leaves 4–8cm(1$\frac{1}{2}$–3in) in length, ovate, sometimes broadly so, thick-textured, deep green. Fruits globular, 2.5–4cm(1–1$\frac{1}{2}$in) in diameter, orange, the thick rind sweet and edible. *F.c.* 'Variegata' has the leaves irregularly and heavily blotched grey-green and cream.

F. hindsii Hong Kong kumquat Southern China

Erect, almost fastigiate spiny shrub or small tree easily kept at 90–180cm(3–6ft) in height in containers. Leaves to 4cm(1$\frac{1}{2}$in) or more long, elliptic to obovate, deep green. Fruits 1.2cm($\frac{1}{2}$in) or a little more in diameter, bright orange-red, usually freely borne. An unusual and very ornamental fruiting

FAR LEFT *Fortunella crassifolia* 'Variegata'
LEFT *Fortunella hindsii*

pot plant deserving wider recognition. It has been used as a container plant in China for centuries and the spicy fruits are bottled and used for flavouring.

F. margarita Nagami/Oval kumquat Southern China
Fairly bushy erect shrub to 3m(10ft) tall in its native country, but easily kept to half this in containers. Leaves 4–8cm(1½–3in) long, broadly lanceolate to ovate, lustrous deep green. Flowers fragrant. Fruits broadly oblongoid, about 4cm (1½in) long, rich orange-yellow with acid but flavoursome pulp and sweet rind; in China at least it is often candied.

FOURCROYA See FURCRAEA

FRAILEA

Cactaceae

Origin: Andes and sub-tropical South America. A genus of 12 or more species of small, largely rounded cacti with comparatively large flowers. They thrive in containers, their small size making them ideal for narrow window ledges or shelves. Propagate by seed in spring, or by offsets in summer. *Frailea* commemorates Manuel Fraile (*b.* 1850), a Spanish gardener for many years Superintendent of the U.S. Department of Agriculture's Cactus Collection in Washington D.C.

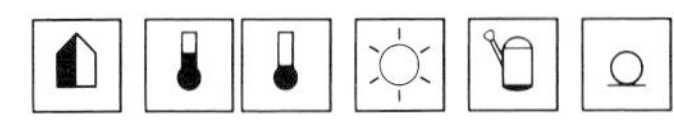

Species cultivated

F. castanea Brazil, Paraguay
Stem 2–4cm(¾–1½in) wide, globular, depressed at the top, with ten to 15 broad, shallow ribs, brownish to olive-green. Areoles minute, each one set with a cluster of about five, tiny, reflexed black spines. Flowers yellow, as big as the plant; during dull weather the flowers fail to open, but self pollination still takes place and large seed pods are set. The curiously shaped, comparatively large, cup-shaped seeds are adapted to water dispersal in the wild, their grassy savanna homeland being subject to brief summer flooding.

F. columbiana Colombia
Stems sub-globose to 4cm(1½in) wide, narrowly depressed at the top, with 15 to 18 bright green shallow ribs. Areoles very small, each one crowned by 20 to 25 small yellowish spines, the largest to 6mm(¼in) long. Flowers yellow, about 2.5cm(1in) long.

F. gracillima Paraguay
Stems starting globular, but becoming cylindrical with age, eventually to 10cm(4in) tall by 2.5cm(1in) thick and forming clusters. Ribs 12 to 14, low and rounded. Areoles small, set with up to 18 bristle-like white spines up to 6mm(¼in) long. Flowers 3cm(1¼in) long, yellow with a reddish throat.

F. pulcherrima (*Echinocactus pulcherrima*) *Uruguay*
Stems forming clusters, shortly ovoid, to 5cm(2in) tall by 2.5cm(1in) thick, dark green. Ribs 18 to 21, low and rounded. Areoles small, crowned with ten

Frailea collection, including *F. castanea* (centre) and *F. columbiana* (right)

to 14 tiny white to pale brown spines. Flowers 2.5cm(1in) long, yellow.

F. pygmaea Uruguay, northern Argentina
Stems globular, depressed at the top, up to 3cm(1¼in) wide, dark green, forming clusters. Ribs 12 to 21, low and rounded. Areoles tiny, set with six to nine bristle-like, recurved to adpressed spines. Flowers 2–2.5cm(¾–1in) long, yellow.

FRANCISCEA See BRUNFELSIA

FRANCOA

Saxifragaceae

Origin: Chile. A genus of one very variable species of evergreen, herbaceous perennial, divided by some botanists into four or five entities. They make good pot plants for the conservatory and can be brought into the home for flowering. Propagate by seed or division in spring, or by cuttings in late spring; plants from seed flower in the second season. *Francoa* was named for Francisco Franco, a 16th-century Spanish doctor and a patron of botany.

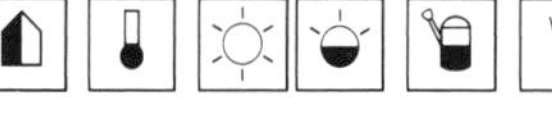

Species cultivated

F. appendiculata Bridal wreath
Rhizomatous, clump-forming. Leaves pinnate, 15–30cm(6–12in) long, the terminal leaflet by far the largest, broadly ovate, somewhat lobed and wavy, forming basal rosettes with a few smaller stem leaves. Flowers 5-petalled, bell-shaped, pale pink, darker spotted, sometimes white or almost red and carried in simple or branched racemes on long stems, 60–90cm(2–3ft) tall, in summer.

F.a. ramosa (*F. ramosa*)
Flowers white, in freely branched panicles.

F.a. sonchifolia (*F. sonchifolia*)
Flowers pink, in loose racemes.

ABOVE *Francoa appendiculata*
RIGHT *Freesia × kewensis*

FREESIA

Iridaceae

Origin: South Africa. A genus of 20 species of cormous, deciduous perennials with slender stems and narrow leaves mostly in basal fans. The beautiful and colourful flowers are borne in one-sided spikes and face upwards. One species, together with the hybrids derived from it, is readily available, making splendid fragrant plants for the conservatory and to be brought into the home for flowering. They are also long-lasting cut flowers and are mostly grown for this purpose. For winter flowering, pot in September and for a succession of bloom, pot batches fortnightly until December, the last will open in spring. Water freely when growing and from the first appearance of the flowers until they finish give a fortnightly liquid feed. It is usually necessary to support the slender stems with canes and twine. Dry off when the leaves yellow in summer. Corms specially treated for the plant to flower in late summer can now be purchased. Propagate by separating offsets when re-potting, or by sowing seed in spring. *Freesia* was named for Frederick Freese (*d.*1876), German doctor and a friend of the botanist who named the genus.

  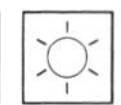  

Species cultivated

F. × kewensis (*F. × hybrida*)
A group of modern cultivars derived from crossing *F. armstrongii* with *F. refracta*. They are similar to *F. refracta*, but the flowers are to 5cm(2in) long and fuller, less fragrant, in a wide range of shades of red, mauve, pink, yellow, cream and white; there are also some double-flowered forms.

F. refracta
Leaves sword-shaped, to 30cm(1ft) long, mainly in basal fans. Flowers to 3cm(1¼in) long, opening to six broad lobes from a funnel-shaped tube, white and yellow, opening in spring; they are very fragrant.

FREMONTODENDRON
(FREMONTIA)

Sterculiaceae

Origin: South-western USA and Mexico. A genus of two species of evergreen shrubs. Their leaves are alternate, undivided, with three to seven lobes and dull green in colour with a covering of tiny, starry hairs. The flowers are solitary and bowl-shaped, made up of five sepal-like petals which are bright golden-yellow. They are splendid flowering plants for a cool conservatory and are best trained flat against the back wall. Propagate by cuttings or seed in warmth in spring. *Fremontodendron* was named for Major General Fremont (1813–1890), explorer of western North America and amateur botanist, his name used together with the Greek *dendron*, a tree.

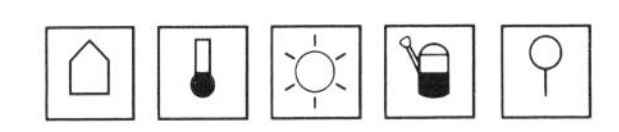

Species cultivated

F. californicum California, Arizona
Well branched shrub from 2m(6½ft), to 9m(30ft) in the wild, with densely hairy shoots. Leaves 5–10cm(2–4in) long, sometimes entire, more often 3-lobed, with a pale fawn felty covering beneath, dull green above. Flowers cup-shaped, opening almost flat to 5cm(2in) or more across, golden-yellow, opening from early spring to late autumn with a main flush in spring. *F.* 'California Glory' is a hybrid between *E. californicum* and *F. mexicanum*, having slightly larger, more richly coloured blooms; it is very floriferous.

F. mexicanum S. California (USA), Baja California (Mexico)
Similar to *F. californicum*, but leaves usually thicker, with five lobes. Flowers to 7cm(2¾in) across with more pointed petals which give a starry effect, orange-yellow with a reddish tinge at the base, opening from spring to autumn.

ABOVE *Frithia pulchra*
LEFT *Fremontodendron californicum*

FRITHIA

Aizoaceae

Origin: South Africa. A genus of one species of flowering succulent the foliage of which is almost identical to that of *Fenestraria*. As in that genus, the leaves are almost buried in its desert homeland with only the transparent windowed tips showing. In the home or conservatory the plant must be grown with its leaves exposed to our less intense sunlight. A desirable succulent for the home, but not quite as easy to grow as *Fenestraria*. Propagate by seed or by very careful division. *Frithia* commemorates Frederick Frith, a South African enthusiast and collector of succulents.

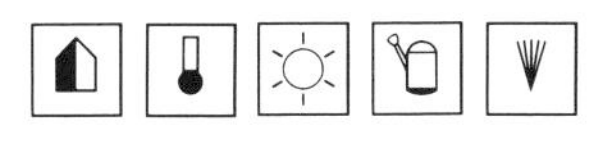

Species cultivated

F. pulchra Transvaal
Of tufted habit, eventually forming hummocks. Leaves 2cm(¾in) long, usually obovoid with a shallowly rounded tip, grey-green on the sides. Flowers carmine with a white centre or entirely white, about 2cm(¾in) wide. Unlike *Fenestraria*, this species rests in the summer, when it must be kept dry or almost so, and grows during the winter when it must be watered regularly and with care. It also needs maximum winter sunlight.

FRITILLARIA

Liliaceae
Fritillaries

Origin: North temperate zone, especially in S.E. Europe and W. Asia. A genus of 85 species of bulbous plants with grassy leaves and 6-tepalled nodding bell-shaped flowers, often distinctly shouldered and frequently chequered, flowering in late winter and spring. They grow well in pots in a cool conservatory providing interest from late winter to spring. Pot in autumn, watering once growth is visible, but allowing the soil to almost dry out between each watering. Dry off when leaves yellow and keep barely moist until autumn. Propagate by removing offsets and bulblets in autumn when potting, or by seed in spring. *Fritillaria* derives from the Latin *fritillus*, a chequerboard or dice box, from the patterning on the flowers of some species.

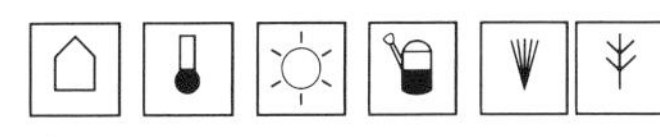

Species cultivated

F. acmopetala Cyprus, Syria, Turkey
30–40cm(12–16in) tall. Leaves linear, grey-green.

ABOVE *Fritillaria acmopetala*
ABOVE RIGHT *Fritillaria pallidiflora*
TOP *Fritillaria bucharica*

Flowers one to three on a stem, long bell-shaped to 3–4cm(1¼–1½in), narrowing at the mouth then flared with recurved sepals, jade-green to yellowish, waxy, with purple markings, in some forms with a large purple blotch at the top of the inner sepals.
F. bucharica (*Rhinopetalum bucharica*)
Afghanistan and adjacent USSR
15–25cm(6–10in) tall. Leaves broadly lanceolate at the base, narrowing rapidly up the stem. Flowers 1.5–2.5cm(⅝–1in) across, opening almost flat, with white petals and dark nectaries, one to 15 borne in the axils of the upper leaves.
F. camtschatcensis Black sarana E. Asia from USSR to Japan, N.W. America
30–40cm(12–16in) tall. Leaves narrowly lanceolate, usually glossy green, borne in whorls. Flowers to 3cm(1¼in) long, the bells opening widely, dark chocolate or maroon-purple, one to four on each stem.
F. imperialis Crown imperial Iran, W. Himalaya
Robust plant; stems 75–120cm(2½–4ft) high. Leaves alternate, bright glossy green, making a dense arching spread of foliage at the base of the stems. Flowers 5–6.5cm(2–2½in) long, yellow, orange or red, with large drops of sweet nectar on the nectaries inside the large bells; they have a musky odour which some people find unpleasant. The bulbs also have the same but stronger smell. *F.i.* 'Aurora' is red-orange; 'Lutea Maxima' has large yellow flowers; 'Rubra Maxima' is in shades of dark red.
F. meleagris Snake's head fritillary North-west and central Europe
To 30cm(12in) tall. Leaves narrowly lanceolate, grey-green, at intervals along the stems. Flowers to about 4cm(1½in) long, with wide shoulders and almost straight bells, 3.5cm(1⅓in) wide; they are a light to dark red-purple, with darker chequering, the shades very variable. *F.m. alba* is white and occurs in wild populations; 'Artemis' has greyish-purple flowers; 'Pomona' has purple chequering on a pale background.
F. michaelowskyi Turkey to USSR (Armenia)
10–15cm(4–6in) tall. Leaves lanceolate, 5–8cm(2–3in) long, becoming narrower up the stem. Flowers broadly bell-shaped, 2cm(¾in) long and wide, red-purple with a broad yellow band at the mouth, pendent, on slender arching stems. A strikingly coloured bloom.
F. pallidiflora USSR
30–100cm(12–40in) tall. Leaves broadly lanceolate, glaucous blue-green, mostly at the base of the stem. Flowers 4cm(1½in) long, broadly bell-shaped, to 3cm(1¼in) wide, pale yellow, one to four to a stem.
F. persica Cyprus, Turkey, Iran
30–100cm(12–40in) tall. Leaves grey-green, lanceolate, densely carried in the basal half of the stem. Flowers 1–2cm(⅜–¾in), brownish-red to deep purple, shallowly bell-shaped, borne in a long raceme of up to 30 blooms. *F.p.* 'Adiyaman' has greener leaves and rich plum-purple flowers.

FUCHSIA

Onagraceae

Origin: Mexico to South America, New Zealand and Pacific Is. A genus of about 100 species of mainly shrubs, with entire leaves in opposite pairs or whorls, and distinctive pendent flowers made up of a perianth tube with four spreading, coloured sepals and four down-pointing petals often of contrasting colours. Those described below are all excellent and colourful house and conservatory plants and can be grown in pots or hanging baskets. They should be potted annually in spring, cutting back the previous year's canes to 2–3cm(1in). Propagate by cuttings in spring or late summer, growing the latter on through winter. Once the cuttings are rooted, put into 7.5–10cm(2½–4in) pots, and if growing a bush or pyramid-trained specimen, pinch out the tip at about 10–15cm(4–6in). For a bush, pinch again when the new lateral growths are about 10cm(4in) and a second time if needed, to make a really well-balanced plant. For a pyramid, tie the strongest of the new growths to a cane, remove the weakest and pinch out the rest at four or five leaf pairs. When the leading shoot has made another 8–10cm(3–4in), repeat until the desired height is reached. Standards are best trained from a strongly growing cutting. Tie the stem to a cane and pinch out all side growths until the main stem is three to four leaf pairs taller than the required height. Pinch out the tip and any subsequent shoots arising from the top buds. For a hanging basket, the shoots usually need two early pinchings, as for a bush specimen, to encourage sufficient stems to form. *Fuchsia* was named for Leonhart Fuchs (1501–1566), German doctor and herbalist.

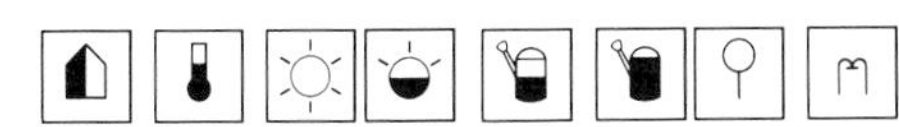

Species cultivated

F. austromontana (*F. serratipetala*) Peru
Bushy shrub, eventually to 3m(10ft) or more. Leaves 3–7cm($1\frac{1}{4}$–$2\frac{3}{4}$in) long, narrowly oblong, toothed, borne in pairs or whorls. Flowers 4–5cm($1\frac{1}{2}$–2in) long, carmine-red, opening in summer.

F. boliviana Peru to Argentina
Lax shrub or, in the wild, a small tree, but easily maintained at 2m($6\frac{1}{2}$ft) in a large pot. Leaves elliptic-ovate, 10–18cm(4–7in) or more in length, pubescent, with a red stalk and mid-vein. Flowers crimson, each one having a very slender, trumpet-shaped tube to 4cm($1\frac{1}{2}$in) long, sepal lobes slender, spreading then recurved, petals arching, somewhat shorter, borne in dense, terminal, drooping racemes or narrow panicles in summer to autumn.

F.b. luxurians El Salvador to Ecuador
Flowers with tubes to 6cm($2\frac{1}{2}$in) in length. This variety is much confused with *F. corymbiflora*, *q.v.* Fruits black-red, edible, with an almost fig-like flavour.

F. corymbiflora Ecuador, Peru
Scrambling, lax shrub to 3m(10ft) or more high. Leaves 7–13cm($2\frac{3}{4}$–5in) long, oblong-lanceolate, hairy. Flowers with long, slender tubes, to 6cm($2\frac{1}{2}$in) long, deep red, borne in terminal drooping clusters in summer. Probably fairly uncommon in cultivation, plants under this name being *F. boliviana*, or more usually, its variety *luxurians*. True *corymbiflora* has flowers with petals longer than the non-reflexing sepals. The two species have long been confused and are closely allied.

F. fulgens Mexico
Shrub to 2m($6\frac{1}{2}$ft) tall. Leaves to 8cm(3in) or so long, ovate-cordate. Flowers with long slender tubes, to 8cm(3in) long, scarlet, the tepals tipped with green; they are borne in leafy clusters at the end of the shoots in summer.

F. magellanica Chile, Argentina
Erect, hairless shrub to 4m(13ft) tall. Leaves 2.5cm(1in) long, lanceolate to ovate, in pairs or whorls of three, occasionally of four. Flowers 4–5cm($1\frac{1}{2}$–2in) long with a short tube, calyx and sepals red, petals purple, opening from summer onwards to late autumn. 'Aurea' has golden foliage and is not so strong-growing as the type; 'Pumila' is smaller in all its parts; 'Riccartonii' is the hardiest, with darker sepals and broader petals; 'Variegata' has pink leaf margins with a cream flush; 'Versicolor' has greyish leaves tinted with pink when young, becoming white when mature.

F.m. macrostemma (*gracilis*)
Slender habit with narrow leaves and flowers.

F.m. molinae (*alba*)
Very pale pink flowers.

F. procumbens New Zealand
Prostrate or trailing habit. Leaves 6–20mm($\frac{1}{4}$–$\frac{3}{4}$in) long, rounded to broadly-ovate. Flowers 1.2–2cm($\frac{1}{2}$–$\frac{3}{4}$in) long, the narrow tube pale orange-yellow, the sepals green, tipped with purple, sharply reflexed; they open in summer and are followed by pinkish-purple to dull red fruits, to 2cm($\frac{3}{4}$in) long.

F. serratifolia See *F. austromontana*.

F. triphylla Haiti, San Domingo
Shrub or sub-shrub to 60cm(2ft) tall, densely hairy. Leaves 4–10cm($1\frac{1}{2}$–4in) long, lanceolate to lanceolate-ovate, in whorls of three, more rarely of two or four. Flowers 4cm($1\frac{1}{2}$in) long with a slender, tapering tube, orange-scarlet, carried in terminal racemes. *F.t.* 'Thalia' is a free-flowering cultivar derived from this species and almost indistinguishable from it. With minimum temperatures to 15°C(60°F) in winter, it will flower all year.

Hybrid cultivars

Hundreds are now available in shades of white, pink, red and purple, often in two contrasting colours with the petals single, semi-double or fully double. The following can be recommended:
'Aintree', single, tube and sepals ivory, corolla red.
'Constellation', double, white throughout with some green marking.
'Display', single, tube and sepals dark pink, corolla widely flared, cerise-pink.
'Gay Fandango', semi-double, tube and sepals carmine-pink, corolla magenta-red.

ABOVE *Fuchsia* 'Display'
LEFT *Fuchsia boliviana*

ABOVE *Fuchsia* 'Golden Marinka'
RIGHT *Fuchsia procumbens* fruits
BELOW *Furcraea foetida* 'Mediopicta'

'Golden Marinka', single, tube and sepals bright red, foliage yellow and cream variegated, often giving an all-over gold effect, splendid for a hanging basket.
'Orange Crush', single, tube and sepals salmon-orange, corolla orange-red.
'Pink Dessert', single, pink throughout, the corolla flared.
'Royal Velvet', double, very large flowers, tube and sepals red, corolla royal purple.
'Swingtime', double, tube and sepals bright red, corolla creamy-white with feint pink veins, good in a pot or basket.
'Ting-a-Ling', single, pure white throughout with a widely flared corolla.

FURCRAEA (FOURCROYA)

Agavaceae

Origin: Tropical America. A genus of 20 species, mainly of semi-desert plants superficially resembling *Agave* in foliage. Some species have short to tall trunk-like stems with a dense rosette of leaves at the top, others are stemless. The narrow, often spiny-toothed leaves are more or less fleshy within and supported by tough fibres which can be used as a hemp substitute. As a rule only large rosettes many years old, such as develop when planted out, will bloom. Like those of *Agave*, the flowering stems are very tall and much-branched, but loose and fountain-like with 6-tepalled white and starry flowers. After the seed has set the rosette dies leaving only small offsets or none at all. When young, the species described below make handsome container plants. *Furcraea* honours Antoine Francois Fourcroy (1755–1809), a French chemist and naturalist much involved in the French Revolution.

Species cultivated

F. foetida (*F. gigantea*) Mauritius hemp
Northern South America, but widely grown as an ornamental and in Mauritius and St. Helena especially for its fibres. Usually trunkless as a container plant, shortly stemmed with age. Leaves deep green, 1–2.5m(3–8ft) in length, with only a few small, curved, spiny teeth on slightly wavy margins. Flowering stem to 7.5m(25ft) in height, with numerous 4cm(1½in) wide flowers. *F.f.* 'Mediopicta' has the leaves bearing a broad, central, longitudinal silvery-white band, sometimes broken down into stripes and lesser bands with green in between.

F. selloa Colombia
Trunk to about 90cm(3ft) in height, but only on plants of good age. Leaves 90cm(3ft) or more long, glossy deep green with well spaced, hooked brown teeth. Flowering stem to 6m(20ft) in height with pendent green and white flowers. *F.s.* 'Marginata' has the leaves margined white, ageing creamy-yellow.

G

GALANTHUS

Amaryllidaceae
Snowdrops

Origin: Europe to W. Asia. A genus of about 20 species of bulbous perennials, each bulb having two (occasionally three) narrow leaves. Snowdrop flowers are borne singly, pure white in colour with green markings and consist of six tepals, the outer three being long and spreading and the centre three forming a short cup or tube. They are useful winter or early spring flowers for a pot in a cold or cool conservatory or unheated room or porch. Hardy. Propagate by offsets or by seed as soon as ripe. *Galanthus* derives from the Greek *gala*, milk and *anthos*, flower, from its colour.

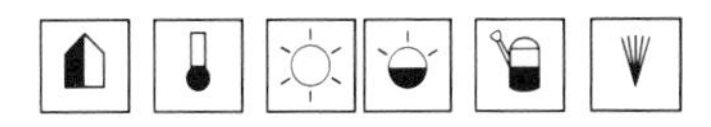

Species cultivated

G. elwesii Giant snowdrop Turkey
Leaves glaucous, erect, with hooded tips. Flowers 3.5–4.5cm(1½–1¾in) wide with broadly ovate outer tepals, the inner with both apical and basal green markings, the presence of the former at once distinguishing it from robust forms of G. *nivalis*.

G. nivalis Common snowdrop Europe
Leaves somewhat glaucous, flat. Flowers 2–3cm(¾–1¼in) wide, the inner tepals notched and marked at the base with green. G.*n.* 'Plena' has double flowers. There are several other cultivars and a number of named hybrids, all of which are similar.

GALTONIA

Liliaceae

Origin: South Africa. A genus of four species of bulbous plants with erect to arching, strap-shaped leaves borne in rosettes. The pendent, 6-lobed flowers are bell-shaped and are carried in racemes at the top of leafless stems. They are statuesque pot plants for the cool conservatory and can be brought into the home for flowering. Propagate from offsets or by seed in spring. *Galtonia* was named for Sir Francis Galton (1822–1911), anthropologist, meteorologist and distinguished all-round scientist.

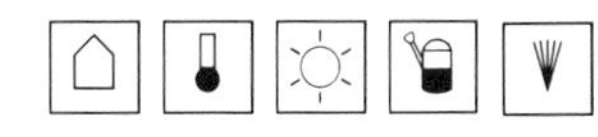

Species cultivated

G. candicans Summer hyacinth
Bulbs large and round, carrying four to six shining-green leaves, to 60cm(2ft) or so long in summer.

Galanthus nivalis

Flowers 2.5–3.5cm(1–1½in) long, narrowly bell-shaped and white, fragrant, borne in racemes of 15 to 30 or more on a stem 60–90cm(2–3ft) high.

G. viridissima
Leaves broader, more yellow-green, narrowing suddenly to an acute tip. Flowers broadly bell-shaped, 2–5cm(¾–2in) long, pale green.

GARDENIA

Rubiaceae

Origin: Warmer areas of Asia and Africa. A genus of 250 species of evergreen shrubs and trees with entire leaves in pairs or sometimes whorls of three or more. The flowers have five to 11 waxy petals, usually white or cream in colour. They are good conservatory plants and can be brought into the house at flowering time, but need good light and a humid atmosphere for success. Propagate by spring or summer cuttings with bottom heat. *Gardenia* was named for Dr Alexander Garden (1730–1791), a physician and naturalist of Scottish origin who lived in Charleston, S. Carolina, USA, and who corresponded with Linnaeus.

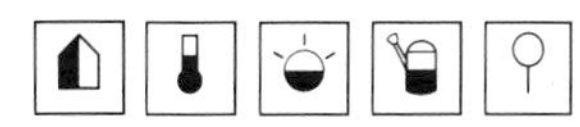

Species cultivated

G. citriodora See *Mitriostigma axillare*.

G. jasminoides (G. *florida*, G. *grandiflora*) Common gardenia, Cape jasmine S. China
Shrub to 60cm(2ft) or more as a pot plant, to

Gardenia jasminoides

2m(6½ft) in the open. Leaves to 10cm(4in) long, lanceolate to obovate, deep shining green, in whorls of three. Flowers to 8cm(3in) across, white and beautifully fragrant, borne in summer and autumn. The rarely seen wild form has five petals. G.j. *fortuniana* has larger, double flowers and is the sort usually in cultivation; 'Veitchiana' is winter-flowering, needing higher temperatures to flower (tropical); 'Prostrata' is a low-growing form with smaller leaves, sometimes listed as G. *radicans*.

G. taitensis Society and adjacent Pacific islands
Much like G. *jasminoides*, but with single, scented blooms composed of five to nine radiating petals.

G. thunbergia South Africa
This evergreen shrub with whitish stems can eventually exceed 2m(6½ft) in height, but is easily kept down to half this. Leaves elliptic, sometimes broadly so, 10–15cm(4–6in) long, lustrous rich green. Flowers usually 8-petalled, fragrant, white, 8cm(3in) or more wide, opening from winter to spring.

GASTERIA

Liliaceae

Origin: Southern Africa; all species listed below are native to Cape Province of South Africa unless stated. A genus of 50 to 70 species of succulent, clump-forming perennials with fleshy, almost stemless leaves arranged in two ranks. The tubular, swollen-based flowers are in tall, wiry racemes. They are suitable for conservatory or home, their leaves being decorative throughout the year. Keep almost dry in winter and always allow the compost to become almost dry before watering. Propagate by division or by seed in spring. *Gasteria* derives from the Greek *gasta*, a belly, an allusion to the swollen flower tube.

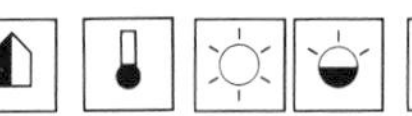 

Species cultivated

G. armstrongii
Young leaves inclined, older ones prostrate, tongue-shaped, up to 5cm(2in) long, arranged in two parallel opposite ranks (distichous), deep green with some white tubercles. Leaf tips broadly rounded with a small, abrupt point in the middle. Old plants produce somewhat longer, triangular-ovate leaves in rosettes.

G. batesiana (G. *carinata*, G. *subverrucosa marginata*) Natal
Leaves triangular-lanceolate, up to 15cm(6in) long, stiff, keeled, dark olive-green, roughened by dark green and white tubercles arranged in spiralling rosettes with only a few offsets. Flowers about 4cm(1½in) long.

G. brevifolia
Leaves arranged distichously, the young ones erect, older ones spreading, 8–15cm(3–6in) long, tongue-shaped, the rounded tip with a small abrupt point, tuberculate, dark green with confluent white spots arranged in transverse bands, margins toothed. Flowers 2cm(¾in) long, in racemes to 60cm(2ft) tall.

G. carinata See G. *batesiana*.

G. humilis
Leaves in spiralling rosettes, triangular, 8–10cm(3–4in) long, upper surface concave, lower surface sharply keeled, lustrous deep green with whitish tubercles and white spots, the latter sometimes confluent and either scattered or arranged into irregular transverse lines.

G. liliputana
Leaves in spiralling rosettes, 5–10cm(2–4in) long, lanceolate, dark green marbled and spotted with white, the upper surface hollowed, the lower rounded and keeled. Flowers 1.5cm(⅝in) long, red, carried in 10cm(4in) racemes.

G. lutzii
Rosettes distichous when young, then spiralling. Leaves up to 25cm(10in) long though usually somewhat less, strap-shaped, rounded-tipped and more or less acuminate, smoothly green or with a reddish tinge or suffusion, with a few large, paler green spots. Flowers green and red flushed, 2.5cm(1in) long, in panicles to 90cm(3ft) or more tall.

G. maculata
Rosettes distichous or slightly spiralled. Leaves 16–20cm($6\frac{1}{2}$–8in) long, tongue-shaped, flattened above, hollowed below, dark green with bands of white spots sometimes merging. Flowers 1.5–2cm($\frac{5}{8}$–$\frac{3}{4}$in) long, in two shades of red, in racemes to 120cm(4ft).

G. marmorata
Rosettes distichous. Leaves to 15cm(6in) long, lanceolate, dark green with pale green confluent spots. Flowers 2cm($\frac{3}{4}$in) long, reddish, on racemes to 75cm($2\frac{1}{2}$ft).

G. obtusa
Leaves in spiralling rosettes, young ones erect, older ones spreading, 10–15cm(4–6in) long, triangular-lanceolate, blunt-tipped, keeled or 3-angled beneath, glossy rich green above, with transverse bands composed of many small, rather indistinct spots.

G. obtusifolia
Rosettes distichous, composed of up to 14 leaves when mature. Individual leaves 15–18cm(6–7in) long, tongue-shaped, green with whitish-green spots arranged in transverse bands. Flowers 2.5cm(1in) long, in racemes to 75cm($2\frac{1}{2}$ft) in height. *G. o.* 'Variegata' has the leaves marked with creamy-white.

G. picta
Clump-forming, the rosettes elongating with age. Leaves distichous or arranged spirally, strap-shaped to lanceolate with a blunt tip, dark green, almost glossy, with irregularly disposed cross-bands of confluent white dots. Flowers reddish-pink in usually fairly dense racemes.

G. pseudonigricans
Leaves triangular-lanceolate to almost strap-shaped, in distichous rosettes, 15–20cm(6–8in) long, erect at first then spreading or recurved, lustrous green with a vague white spotting.

G. pulchra
Leaves arranged more or less distichously on stems 15–30cm(6–12in) in height, narrowly sword-shaped, 3-angled, 20–30cm(8–12in) long, glossy, but a rather murky green, with oblong, confluent white spots arranged in transverse bands or lines. Flowers 2cm($\frac{3}{4}$in) long, in panicles to 90cm(3ft) tall.

G. subverrucosa marginata See G. *batesiana*.

G. verrucosa Ox tongue
Rosettes distichous. Leaves to 23cm($9\frac{1}{4}$in) long, lanceolate with triangular apexes, concave above, rounded beneath, dull green with small white warts sometimes merging. Flowers 2–2.5cm($\frac{3}{4}$–1in) long, on racemes to 60cm(2ft).

GASTROCHILUS See ASCOCENTRUM

ABOVE LEFT *Gasteria batesiana*
ABOVE *Gasteria picta*

× GASTROLEA

Liliaceae

Origin: Garden hybrid. A hybrid genus of about 15 succulent 'species' derived from crossing together various species of *Gasteria* and *Aloe*. In appearance they blend the characters of the parents with usually a nod in the direction of *Gasteria*. When produced, the flowers are tubular, usually reddish with green tips arranged in racemes above the leaves. They are useful foliage pot plants.

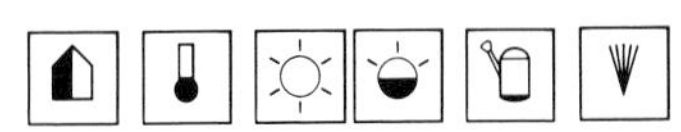

Hybrid species cultivated

× ***G. beguinii*** (*Aloe aristata* × *Gasteria verrucosa*)
Leaves erect, up to 15cm(6in) long, triangular ovate, taper-pointed, deep green, reddish towards the tip in bright light, banded transversely with white spots.

× ***G. smaragdina*** (*Aloe variegata* × *Gasteria ? candicans*)
Leaves more or less spreading, to 20cm(8in) in length, stiff and fleshy, narrowly triangular to triangular-lanceolate, channelled and keeled towards the apex, lightish green patterned with greenish-white spots in irregular transverse bands.

× *Gastrolea smaragdina*

GAZANIA

Compositae

Origin: Tropical and Southern Africa. A genus of 40 species of evergreen perennials, prostrate to semi-prostrate in growth with brightly coloured, daisy-like flowers. The flowers open only in sun, pot plants giving their best display in the sunny conservatory. They are best re-propagated annually or every second year before plants become too large and straggly, and they need a cool rest in winter. Propagate by cuttings taken in late summer, or by seed. *Gazania* was named for Theodore of Gaza (1398–1478), who translated the botanical works of Theophrastus into Latin.

  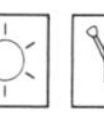 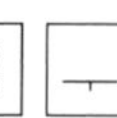

Species cultivated

G. ringens South Africa
Leaves linear to spathulate, green above, silky-white beneath. Flowers 8cm(3in) or more in diameter, orange to cream, pink or bronze with a darker spot at the base of each petal.

Hybrids

Many hybrids have been raised, with *G. bracteata*, *G. pavonia*, *G. pinnata*, *G. ringens* and *G. uniflora* as parents. A few named cultivars are available as plants raised by cuttings, but easiest to obtain are the seed strains. The following are recommended:
'Chansonette', 22–30cm(9–12in); flowers 8cm (3in) across, of good texture, in a mixture of yellows, apricot, orange, bronze, mauve and red; they are borne on stems to 30cm(12in) or more high.
'Mini-Star', 20cm(8in); flowers 5cm(2in) or more across, available both in a colour mix or as a single colour, e.g. 'Mini-Star Yellow' and 'Mini-Star Tangerine'; leaves deep green above.

Gazania 'Mini-Star Tangerine'

GELSEMIUM

Loganiaceae

Origin: China, Borneo, Sumatra, S.E. USA, Mexico to Guatemala. A genus of two species of evergreen twining climbers, one of which makes a charming adornment for the conservatory. Propagate by cuttings in summer. *Gelsemium* derives from the Italian name, *gelsomino*, for jasmine.

 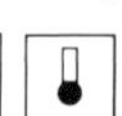 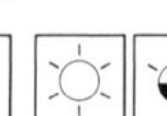  

Species cultivated

G. sempervirens False/Carolina yellow jasmine
S.E. USA, Mexico, Guatemala
Slender climber, 3–5m(10–15ft) in height; stems twist anti-clockwise. Leaves in opposite pairs, lanceolate, 4–10cm(1$\frac{1}{2}$–4in) or more long, lustrous green. Flowers funnel-shaped, with five spreading oblong lobes, 2.5–4cm(1–1$\frac{1}{2}$in) long by almost as wide, bright pale to deep yellow with darker, almost orange throats, fragrant, borne in terminal clusters in spring to summer or later. The flattened, 1.2cm($\frac{1}{2}$in) long seed capsules contain winged seeds.

GENISTA See CYTISUS

GEOGENANTHUS

Commelinaceae

Origin: Tropical South America. A genus of three or four evergreen perennials, one of which provides a striking indoor foliage plant. In appearance it superficially resembles the allied *Dichorisandra*, but differs radically in producing its flowers from the leafless nodes near the base of the stems, often quite close to the ground. *Geogenanthus* derives from the Greek *ge*, earth, *genea*, birthplace and *anthos*, a flower, referring to the flowering habit described above.

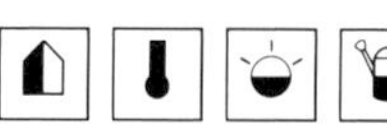

Species cultivated

G. undatus Seersucker plant Brazil, Peru
Stems in small groups from a thick rhizome, erect, or reclining with age, 15–25cm(6–10in) tall, unbranched. Leaves in a terminal rosette of three or four, broadly ovate, 8–11cm(3–4$\frac{1}{2}$in) long, almost fleshy and leathery-textured, strongly puckered, dark green with silvery-green vein stripes above,

purple red beneath. Flowers rarely produced and only on mature plants of several years, to 2cm(¾in) wide, purple with three fringed petals lasting less than one day.

GERANIUM See PELARGONIUM

GERBERA

Compositae

Origin: Africa, and Asia east to Bali. A genus of 70 species of herbaceous perennials with leaves in rosettes and solitary flowers on long stalks. The flowers of wild species have one or two rows of ray florets. They are attractive flowering plants for the sunny conservatory or window sill and are best re-propagated every two or three years to keep them compact. Gerberas are also very long-lasting as cut flowers. Propagate by seed, by division or by cuttings of non-flowering shoots, all in spring. *Gerbera* was named for Traugott Gerber (*d.* 1743), a German naturalist and traveller.

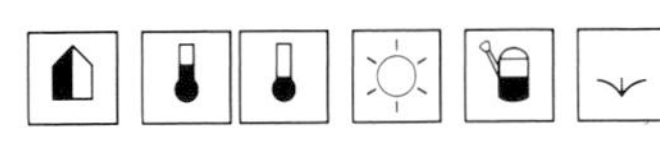

Gerbera hybrid

Species cultivated

G. jamesonii Barberton daisy South Africa
Clump-forming, with 25–40cm(10–16in) long, deeply pinnately lobed, oblong-spathulate leaves, dull green above, white-woolly beneath. Flowers to 10cm(4in) wide, yellow to orange-red, on leafless stems to 45cm(1½ft) high.

Hybrids and seed strains

These come in a wide range of colours including red, pink, orange, yellow and white, many are double or semi-double.
'Happipot' is a dwarf seed strain 20–30cm(8–12in) high, bred especially for growing in pots. The flowers are semi-double to double and up to 8.5cm(3½in) across.

GESNERIA

Gesneriaceae

Origin: Tropical America, West Indies. A genus of about 50 species of evergreen perennials and shrubs some of which are cultivated for their pleasing foliage and attractive flowers. They have simple, often hairy leaves in pairs and tubular to funnel-shaped flowers from the upper axils. The species described below perform well as pot plants. Propagate by seed in spring or by cuttings of non-flowering shoots in spring and summer. *Gesneria* honours Conrad von Gesner (1516–1565), a famous German naturalist of his time, author of the vast *Historiae Animalium*, the originator of bibliography and much else.

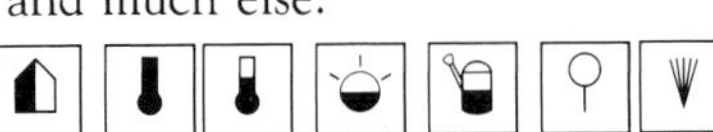

Species cultivated

G. cardinalis See *Sinningia cardinalis*.
G. cuneifolia Puerto Rico, Cuba, Hispaniola
Tufted to clump-forming perennial, with age producing short erect stems and then attaining about 30cm(1ft) in height. Leaves oblanceolate, cuneate, up to 15cm(6in) long, smooth and dark green above. Flowers tubular, slightly curved, solitary from the leaf axils, red or yellow, about 3cm(1¼in) long.
G. ventricosa West Indies
Shrub, 60–90cm(2–3ft) tall as a container plant, but blooming when half this; in its homeland, specimens of 5m(16ft) in height have been recorded. Leaves elliptic, toothed, 10–15cm(4–6in) long. Flowers 4cm(1½in) long, tubular, expanding towards the mouth, orange-red with prominently exserted stamens.

GIBASIS See TRADESCANTIA

Gesneria hybrid

ABOVE *Gibbaeum velutinum*
TOP *Gibbaeum fissoides*

GIBBAEUM

Aizoaceae

Origin: South Africa. A genus of 30 species of variable flowering succulents, including species which resemble living stones (*q.v. Lithops* and *Dinteranthus*) and others with more elongated leaves and a looser growth habit. The daisy-like flowers may appear from late summer to autumn or spring. Propagate by seed or by careful division in spring. *Gibbaeum* derives from the Latin *gibba*, humped or swollen, referring to the very fleshy leaves of certain species.

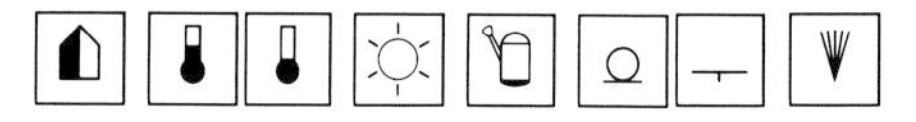

Species cultivated

G. album Cape, Karroo
Small clump-forming, each shoot composed of two fused, irregularly hemispherical leaves, together about 2.5cm(1in) wide, microscopically white-downy. Flowers pure white, 2.5cm(1in) wide.

G. fissoides (*G. nelii, Antegibbaeum fissoides*) Cape
Tufted to small clump-forming, each shoot of one or two pairs of basally-fused leaves. Individual leaves about 5cm(2in) long, semi-erect, almost cylindrical, but flattened on the inner face, rounded-tipped, smooth, greyish-green. Flowers 2.5–5cm(1–2in) wide, rose-purple.

Gilia capitata

G. gibbosum Cape, Karroo
Tufted compact habit from a somewhat woody rootstock, each shoot of one very unequal pair of leaves. Largest leaf up to 6cm(2½in) long, slightly incurving, semi-cylindrical with barely flattened inner surface; smaller leaf of similar shape, but only one third as large. Flowers to 3cm(1¼in) wide, pink to magenta-purple.

G. muirii See *G. schwantesii.*

G. nelii See *G. fissoides.*

G. petrense Cape, Karroo
Very small species, eventually forming small mats or low hummocks. Shoots of one or two pairs of leaves which are connate for one third of their length and lie parallel; each leaf about 1cm(⅜in) long and half as wide and thick, flattened on the inner face, rounded and sharply keeled-tipped on the back, whitish grey-green. Flowers 1–1.5cm(⅜–⅝in) wide, red.

G. schwantesii (*G. muirii*) Cape, Karroo
Habit and unequal leaf pairs like those of *G. gibbosum*, but slightly larger and broader-based, with keeled, hooded tips. Smaller leaf almost half the length of the larger one; leaf surface dark green sometimes with grey or brown overtones, microscopically velvety pubescent. Flowers to 5cm(2in) wide, pure white.

G. velutinum Cape, Karroo
Tufted; old plants with short, prostrate, somewhat woody stems clothed in dry remains of previous leaves; each shoot with one or two leaf pairs each year. Leaves in unequal pairs, triangular in outline, the larger to 6cm(2½in) long by 3cm(1¼in) wide at the base, the tip keeled and incurving like a hook, grey pubescent. Smaller leaf similar, but two-thirds as long, united basally with the larger. Flowers to 5cm(2in) wide, pink or red, borne in summer.

GILIA

Polemoniaceae

Origin: North and South America. A genus of 20 to 120 species, some classified by botanists under other genera including *Collomia*, *Eriastrum* and *Ipomopsis*. They are mostly annuals and biennials with some perennials, all having deeply dissected leaves. Propagate those described below by seed, sown in autumn for a late winter to early spring display from the annuals, a summer one for the biennials. They are very decorative pot plants for the cool conservatory (temperate for winter flowering), and can be brought into the home when in bloom. *Gilia* was named for Fillipo Luigi Gilii (1756–1821), an Italian astronomer.

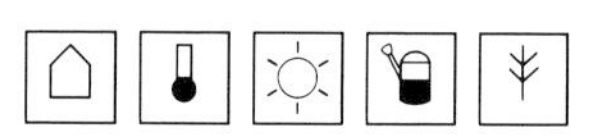

Species cultivated

G. androsaceus See *Linanthus androsaceus.*

G. capitata Blue thimble flower N.W. USA
Erect, slender-stemmed annual, to 50cm(20in) or more. Leaves 10cm(4in) long, bi- or tri-pinnate with linear segments. Flowers 6–8mm(¼–⅓in) long, light blue to lavender, 50 or more crowded into rounded terminal heads, 2.5cm(1in) across.

G. rubra (*G. coronipifolia, Ipomopsis rubra*) Standing cypress, Skyrocket S.E. USA
Biennial or short-lived perennial, with filiform segments. Flowers 3–5cm(1¼–2in) long, narrowly trumpet-shaped, pink, red or white, borne in spike-like panicles on leafless stalks, 90cm(3ft) or more high.

G. tricolor Bird's eyes California, USA
Erect and bushy annual, to 50cm(20in) tall. Leaves bipinnate, the segments narrow. Flowers 1.2–

2cm($\frac{1}{2}$–$\frac{3}{4}$in) across, lavender-purple, the deep yellow tube marked with purple spots, borne on leafy stems.

GLADIOLUS

Iridaceae
Gladioli, Sword lilies

Origin: Europe and the Mediterranean region, S.W. Asia, tropical and southern Africa. A genus of 300 species of perennials with rounded to flattened corms and fan-like tufts of narrow to broadly sword-shaped leaves. Their usually colourful flowers are widely funnel-shaped with a short tube and three large upper petal lobes and three smaller lower ones which are often patterned. There are a number of species suitable for a conservatory, but not hardy enough for the open garden though all the garden hybrids raised from them can be grown in pots. They are derived from such South African species as G. *cardinalis*, G. *carneus*, G. *natalensis*, G. *oppositiflorus*, G. *papilio*, G. *primulinus* and G. *saundersii*, and all are available in almost any colour except blue. For pot culture, the smaller sorts classified as *Miniature*, including *Butterfly* and *Primulinus* are the most suitable. Planted in autumn and with temperatures in the cool range they will flower in early summer; kept at temperate level they will flower in spring. The species mostly flower earlier especially at the warmer temperatures indicated. Propagate by offsets or spawn when potting, the species also by seed in spring. *Gladiolus* derives from the Latin for a small sword, referring to the shape of the leaf.

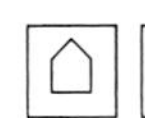    

Species cultivated

G. alatus South Africa
About 30cm(1ft) tall, sometimes a little more. Flowers about 6cm($2\frac{1}{2}$in) wide, strongly zygomorphic, all tepals tapering to a short stalk (claw), the upper three brick-red to salmon-pink, the lower three lime-green tipped with red or pink, fragrant, opening in spring to early summer. Leaves linear to lanceolate with prominent longitudinal ribs.

G. blandus See G. *carneus*.

G. byzantinus Central and eastern Mediterranean region
To 75cm($2\frac{1}{2}$ft) tall. Flowers 5cm(2in) long, the tepals touching or overlapping, rich purplish-red with a narrow red line along the centre of the lower tepal, opening late spring to early summer. 'Albus' is a very beautiful white form.

G. callianthus (*Acidanthera bicolor*, *A.b. murielae*) Ethiopia, Malawi, Tanzania
To 90cm(3ft) tall. Flowers four to six or more to a stem, each 6–8cm($2\frac{1}{2}$–3in) wide with a long tube opening widely to six white tepals, blotched maroon at the centre and very fragrant; they open in autumn.

G. carinatus South Africa
Of variable height, but usually about 45cm($1\frac{1}{2}$ft). Leaves linear, almost grasslike. Flowers more or less funnel-shaped, but variable, some forms opening more widely about 3.5–4.5cm($1\frac{1}{2}$–$1\frac{3}{4}$in) long, pale to deep violet-blue, less commonly pink, red or yellow. The commonest form in cultivation is pale mauve with darker veining, the lower tepals arching down and banded primrose-yellow, strongly scented violet, expanding mid to late spring.

G. carmineus South Africa
About 25cm(10in) tall, sometimes more. Leaves one to three per corm, more or less prostrate, to 45cm($1\frac{1}{2}$ft) long. Flowers fairly widely funnel-shaped, almost regular 5–8cm(2–3in) long, carmine to rose-pink. A very distinct and un-gladiolus like species with spikes of one to three upward-facing blooms in late summer or autumn before the leaves, the latter developing in winter.

G. carneus (G. *blandus*) Painted lady South Africa
30–60cm(1–2ft) tall. Flowers 5–6cm(2–$2\frac{1}{2}$in) across, the tepals opening wide, basically pink, mauve-lilac or cream, the lower three tepals with arrow-head or V-shaped markings in red, yellow or purple. They flower in early spring or late winter given warmth.

G. citrinus (G. *symmetranthus*, *Ixia spathacea*) South Africa
Height 15–20cm(6–8in), occasionally more or less, very slender. Flowers solitary or in pairs, rarely three, erect, ixia-like, 3–5cm($1\frac{1}{4}$–2in) long, yellow, maroon at the base, opening in late spring. Leaves narrowly rush-like. The symmetrical flowers and very slender foliage makes this the most atypical of all gladioli and a nice talking point for the gardener.

Gladiolus citrinus

Gladiolus illyricus

G.* × *colvillei (G. *nanus* of gardens)
Hybrid between G. *tristis* and G. *cardinalis* to 45cm(1½ft) tall. Flowers almost erect, the tepals pointed, scarlet, the lower ones with yellow markings inside. 'The Bride' ('Albus') has pure white flowers and is the parent of the Nanus group.
G. gracilis South Africa
Variable in height, but usually 20–35cm(8–14in). Flowers 4–5cm(1½–2in) long, basically funnel-shaped but with the lower tepals arching down, the upper three thrusting forwards to form a wide hood; they are mauve-blue, or rarely pink, the lower tepals with a pale yellow zone marked with broken maroon lines; strongly violet-scented and expanding in spring. Leaves deeply longitudinally grooved, rush-like.
G. illyricus South and west Europe
To 30–50cm(12–20in) tall, occasionally more, of slender habit. Leaves narrowly sword-shaped. Flowers red to red-purple, up to 4cm(1½in) long, the two lower tepals darker, with white markings, borne in spring to early summer.
G. italicus See G. *segetum*.
G. nanus See under G. × *colvillei*.
G. orchidiflorus South Africa
Usually about 45cm(1½ft) in height, but sometimes more. Flowers strongly zygomorphic and superficially orchid-like, 3–5cm(1¼–2in) long, the upper tepal firm and strap-shaped, arching over and down so that its apex (and those of the stamens and stigma) is opposite, or below the centre of the flower. Colouring variable, from pale chartreuse to shades of grey-green or biscuit, the lateral upper tepals with a dark median line, the lower lateral tepals bearing a yellow, purple-edged blaze or a purple blotch; fragrant, opening in late spring. Leaves very variable from short and falcate to long and linear, the latter type being the commonest.
G. papilio (G. *purpureo-auratus*) South Africa
Stems 50–90cm(20–36in) in height. Leaves grey-green. Flowers in a spike of one to six, sometimes more, drooping, each is 4–4.5cm(1½–1¾in) long, strongly hooded, in shades of yellow, tinged with mauve to purple, particularly the lower three tepals.
G. primulinus Maid of the mist S.E. tropical Africa
Up to 60cm(2ft) or more in height. Flowers 5–10cm(2–4in) long, the upper three tepals large and forming a large hood, the lower three much smaller with recurved tips, primrose-yellow throughout, expanding in summer. Leaves erect, sword-shaped. Some authorities make this a colour form of the widespread mainly red G. *natalensis*.
G. purpureo-auratus See G. *papilio*.
G. segetum (G. *italicus*) Mediterranean
60–75cm(2–2½ft) tall. Flowers 5cm(2in) long, the tepals tapered and widely expanded, rose-purple, the lower three with paler markings; opening in early summer.
G. symmetranthus See G. *citrinus*.
G. tristis South Africa
To 45cm(1½ft) tall. Leaves narrow, more or less cylindrical. Flowers to 7cm(2¾in) wide, funnel-shaped, erect, creamy-yellow striped and shaded reddish-brown, fragrant, opening in late winter and early spring. G.*t. concolor* has slightly larger, plain yellow to cream flowers.

GLECHOMA

Labiatae

Origin: Europe and Asia. A genus of ten to 12 species of creeping perennials with kidney-shaped leaves and small, mauve 2-lipped flowers. Although the species described is usually thought of as a weed, its variegated form is surprisingly decorative, especially so when it is grown in a hanging basket. It is not as hardy as the green-leaved type. Propagate by division in spring, or by cuttings in spring to autumn. *Glechoma* is from the Greek *glechon*, a name for another member of the mint family, probably pennyroyal, but chosen for this genus by Linnaeus.

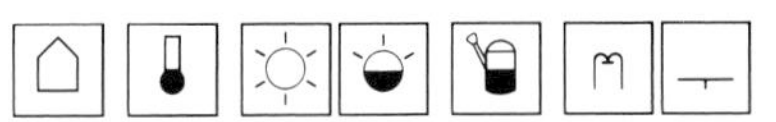

Species cultivated
G. hederacea Ground ivy
Prostrate, rooting at the nodes. Leaves 1–3cm(⅜–1¼in) wide, rounded to broadly ovate-cordate, softly hairy, the margins crenate. Flowers 1.5–2cm(⅝–¾in) long, mauve-blue, borne in whorls in the axils of the opposite pairs of leaves. G.*h.* 'Variegata' has the leaves splashed with white. Hardy.

GLOBBA

Zingiberaceae

Origin: Southern China, Indo-Malaysia. A genus of 50 species of evergreen rhizomatous perennials which combine grace of habit with pleasing foliage and intriguing rather than beautiful flowers. They are clump-forming, sometimes widely so if planted out in the conservatory border, with lance-shaped, stalked leaves. The small flowers are in arching or pendent racemes or panicles and are combined with prominent pale green or coloured bracts. In some species the lowest floral bracts contain bulbils instead of flowers. Each bloom has a slender corolla tube with three irregularly sized petal lobes and a larger staminodal lip. The seeds are surrounded by a fleshy aril which, in some species at least, are colourful enough to be showy when the capsules open. Propagate by division or seed in spring or by bulbils when mature (fall easily when touched). Globbas are best in the conservatory, but can be brought indoors when in bloom. *Globba* is the Latin for the Indonesian name galoba.

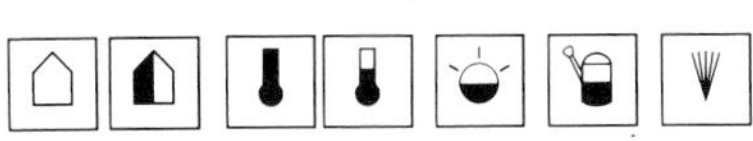

Species cultivated

G. atrosanguinea Borneo, Thailand
Stems to 60cm(2ft) tall or more. Leaves to 20cm(8in) in length, slender-pointed, deep green. Panicles inclined to arching, with red bracts and 2.5cm(1in) long yellow flowers; blooms intermittently all the year if warm enough.

G. schomburgkii Thailand
Stems about 45cm(1½ft) tall. Leaves about 13cm(5in) long, slender-tipped. Panicles arching to pendent with pale green bracts, the lowest bearing bulbils; flowers 2cm(¾in) long, yellow, the lip with red spots, borne in late summer.

G. winitii Thailand
Stems up to 90cm(3ft) but usually less. Leaves to 20cm(8in) in length, ovate-oblong, slightly wavy and with a long slender tip. Panicles pendent, with large rose-purple to magenta bracts and yellow flowers having curved tubes 1.2cm(½in) long, borne in autumn.

GLORIOSA

Liliaceae

Creeping/Climbing lilies

Origin: Africa and Asia. A genus o five species of perennial climbers with narrow, rather brittle tubers. They bear slender stems and oblong-lanceolate leaves often elongated at the tip into small tendrils. The long-stalked showy blooms are borne singly from the axils of the upper leaves and are normally red or yellow or a combination of the two. The six tepals are reflexed sharply and are sometimes waved; the six stamens spread widely beneath them. They are suitable for large pots in a conservatory and can be allowed to climb through a shrub, or supported by twiggy sticks and if there is enough room they can be brought into the house for flowering. Pot the tubers in spring and dry off after they have flowered when the leaves begin to yellow, storing dry through the winter. Propagate by separating offsets or by seed in spring. *Gloriosa* is the Latin name for glorious.

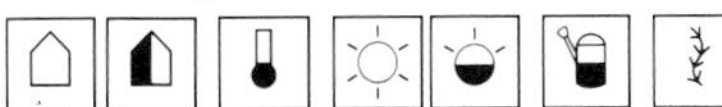

Glechoma hederacea 'Variegata'

Species cultivated

G. carsonii East Africa
To 1m(3ft) tall. Leaves to 13cm(5in) long. Flowers purple-red, edged and centred with yellow; tepals strongly reflexed to 6cm(2½in) long.

G. rothschildiana Glory lily Tropical Africa
To 2m(6½ft) tall. Leaves broadly ovate-lanceolate, to 18cm(7in) long and up to 5cm(2in) wide. Flowers red with yellow margins, the yellow changing to red as it ages; tepals strongly reflexed and wavy edged, 5–8cm(2–3in) long.

G. simplex Tropical Africa
To 1.2m(4ft) tall. Leaves to 15cm(6in) long. Flowers orange-yellow; tepals broadening towards their tips and only slightly waved.

G. superba Tropical Africa
To 1.5m(5ft) or more tall. Leaves 10–15cm(4–6in) long. Flowers yellow at first, changing to red; petals to 7cm(2¾in) long, not so sharply reflexed, strongly waved and crimped along the margins. G.s. 'Lutea' has yellow flowers.

GLOTTIPHYLLUM

Aizoaceae

Origin: South Africa. A genus of 50 species of easily grown flowering succulents. They are mainly tufted perennials with long, thick, somewhat soft-textured, very juicy leaves. The daisy-like flowers are large and showy opening from autumn to early spring. Propagate by seed in spring or by cuttings in summer. *Glottiphyllum* derives from the Greek *glotta*, a tongue and *phyllon*, a leaf, alluding to the texture and shape of the leaves.

Species cultivated

G. fragrans Cape, Karroo
Leaves somewhat obliquely tongue-shaped, 8cm(3in) or more in length by 2.5cm(1in) wide. Flowers 8–10cm(3–4in) wide, satiny rich yellow, fragrant.

G. linguiforme Cape, Oudtshoorn
Leaves distichous, to 6cm(2½in) long, glossy bright green. Flowers 5–7cm(2–2¾in) wide, rich yellow. One of the best known species by name, but not infrequently represented in cultivation by hybrids.

G. regium Cape, Oudtshoorn
More restricted in growth form than most species, with open tufts of shoots composed of one to two pairs of leaves only. Leaves semi-erect, 5-10cm(2–4in) long, sometimes less, comparatively narrow, 10–15mm(⅜–⅝in) wide and thick, smooth light green. Flowers 4cm(1½in) wide, yellow.

GLOXINIA

Gesneriaceae

Origin: Central and tropical South America. A genus of six species of plants allied and similar to *Achimenes* and having the same sort of scaly rhizomes. Formerly this genus including several other species, including the popular florist's gloxinia now classified as *Sinningia speciosa* (*q.v.*). The species described here make attractive plants for the home and conservatory. Propagate by separating the scaly rhizomes during their winter dormancy, or when re-potting, or by seeds in early spring. *Gloxinia* honours Benjamin Peter Gloxin, a German botanist and author of *Observationes Botanicae* (1785).

Species cultivated

G. gymnostoma (*Achimenes gymnostoma*)
Argentina
Stems 40–60cm(16–24in) in height, erect. Leaves ovate, to 8cm(3in) long, toothed and hairy, fairly long-stalked, prominently veined. Flowers solitary from the upper leaf axils, 3.5cm(1⅓in) long, funnel-shaped with five petal lobes, the lower three larger, rose-pink, red-spotted, borne in summer.

G. maculata See G. *perennis*.

G. perennis (G. *maculata*) Colombia to Peru
Stems to 60cm(2ft) or more, erect. Leaves long-stalked, broadly ovate, 13–18cm(5–7in) long, coarsely toothed, lustrous deep green, sometimes red-tinted beneath. Flowers solitary, from the smaller upper leaf axils, forming a raceme; each flower 2.5–4cm(1–1½in) long, with five petal lobes, the central lower one toothed and incurving, soft lavender-blue, the throat deep purple, borne in summer.

GODETIA See CLARKIA
GOETHEA See PAVONIA

ABOVE *Gloxinia gymnostoma*
ABOVE CENTRE *Glottiphyllum linguiforme*
TOP *Gloriosa rothschildiana*

GOMPHRENA

Amaranthaceae

Origin: Tropics of Central and South America with outliers in S.E. Asia and Australia. A genus of 100 species of annuals, biennials and perennials with entire, opposite leaves and usually rounded heads of tiny flowers. The species described below can be grown in a pot in the conservatory or on a sunny windowsill and has a very long season of flowering. Propagate by seed sown in early spring. *Gomphrena* is the Latin name for an unknown sort of amaranth.

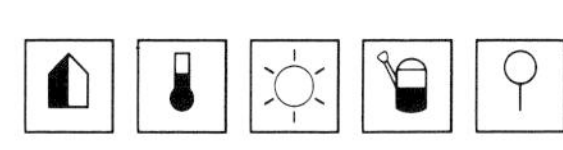

Species cultivated

G. globosa Globe amaranth Tropical Asia
Erect, much branched annual to 45cm(1½ft) tall. Leaves to 10cm(4in) long, oblong to elliptic, in opposite pairs. Flowers tiny, almost hidden within the pink, purple, white, yellow or orange papery bracts, which together make up rounded to ovoid heads 2.5–4cm(1–1½in) long. 'Nana' or Dwarf Form is only 15cm(6in) tall.

GOSSYPIUM

Malvaceae

Origin: Tropics and sub-tropics. A genus of 16 to 67 species, depending on the botanical authority consulted. It includes annuals, shrubs and small trees with superficial resemblances to *Abutilon* and *Hibiscus*. Best known are the species which yield the cotton of commerce, two of which are described below. Apart from their commercial interest they are attractive plants in bloom and well worth trying as pot plants. Propagate by seed in spring. *Gossypium* derives ultimately from the ancient Arabic name *kutum* for cotton.

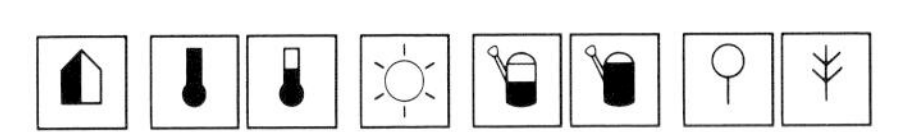

Species cultivated

G. arboreum Tree cotton Exact provenance unknown, probably tropical Asia
Basically an evergreen shrub or small tree to 3m(10ft) in height, but can be grown as an annual. Leaves palmate, deeply and narrowly 3- to 7-lobed, to 12cm(4¾in) across, long-stalked. Flowers borne singly from the upper leaf axils, hibiscus-like, 5-petalled, yellow to red-purple, about 5cm(2in) wide subtended by a ring of three or more large ovate, leafy, long-toothed bracts. The oblongoid seed pods (bolls) are up to 5cm(2in) long, copiously pitted with oil glands; they contain seeds bearing long, woolly whitish hairs (lint or raw cotton).

G. herbaceum Levant/Turkish cotton Probably Africa, Turkey, India
Usually an annual 60–120cm(2–4ft) tall, but sometimes persisting for more than one year if kept warm in winter. Leaves palmate, 10cm(4in) or more long and wide, cleft to about halfway into three to five broad, wavy lobes. Flowers borne singly from upper leaf axils, hibiscus-like, 5-petalled, yellow with a purple centre, about 6cm(2½in) wide, subtended by three large, broadly ovate, toothed bracts. The seed pods (bolls) are rounded, up to 5cm(2in) long, with a few oil glands; the seeds within bear copious, long, woolly grey-white hairs (lint) which, when spun, becomes cotton thread.

G. sturtianum (*G. sturtii*) Sturt's desert rose Central Australia
Evergreen shrub to 1.5m(5ft) tall, sometimes much more, usually rather sparingly branched. Leaves broadly ovate, entire or shallowly 3-lobed, grey-green, about 5cm(2in) long. Flowers 6–8cm(2½–3in) wide, 5-petalled, clear mauve with a red-purple centre. Seeds shortly fuzzy-hairy only.

Gomphrena globosa

GRAPTOPETALUM

Crassulaceae

Origin: S.W. USA and Mexico. A genus of ten to 12 species of succulent, evergreen perennials. They have entire, fleshy leaves and umbel-like clusters of flowers in the axils, and are grown for their ornamental foliage, making handsome plants for a sunny conservatory or window sill. Propagate by leaf and stem cuttings. *Graptopetalum* derives from the Greek *graptos*, to write or paint, with reference to the lines on the petals of some species.

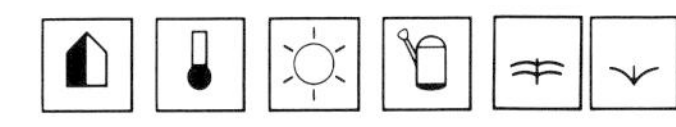

Species cultivated

G. amethystinum (*Pachyphytum amethystinum*) Mexico
Stems robust, more or less erect at first, but becoming decumbent with age and branching from the base. Leaves oblanceolate to obovate-oblong, 3–6cm(1¼–2½in) long, very thick, blue-grey with a lavender patina. Flowers about 16mm(⅝in) wide, light greenish-yellow with reddish transverse bands, appearing in spring and early summer.

Gossypium sturtianum

Graptopetalum paraguayense

G. pachyphyllum Mexico
Stems prostrate to decumbent, about 20cm(8in) long, branching mainly near their bases. Leaves club-shaped, to 2cm($\frac{3}{4}$in) long, round in cross-section, blue-grey, red-tipped in good light. Flowers about 1.2cm($\frac{1}{2}$in) wide, white with red dots, in small terminal panicles during early summer.
G. paraguayense (*Echeveria paraguayense, Sedum weinbergii*) Mother of pearl plant, Ghost plant Mexico
Spreading, loosely tufted stems, to 30cm(1ft) long, ending in rosettes of leaves, 5–7cm(2–2$\frac{3}{4}$in) long, obovate-spathulate, fleshy; flat above, keeled beneath, glaucous blue-grey with a mother-of-pearl lustre. Flowers 1.5–2cm($\frac{5}{8}$–$\frac{3}{4}$in) wide, white, spotted with red, borne in small panicles in spring.

GRAPTOPHYLLUM

Acanthaceae

Origin: Australia and Pacific islands, perhaps also West Africa, but also widely cultivated. A genus of ten species of evergreen shrubs with opposite, usually entire and colourful leaves. They have tubular flowers opening widely at the mouth, generally in shades of red and purple. The species described below is a popular house and conservatory plant. Propagate by cuttings in late spring and summer. *Graptophyllum* derives from the Greek *graptos*, to write or paint and *phyllum*, a leaf, from the patterning on the leaves.

Species cultivated
G. pictum Caricature plant New Guinea
To 2m(6$\frac{1}{2}$ft) in the wild, less than 1m(3ft) in a pot. Leaves elliptic-ovate, to 15cm(6in) long, green or purplish, marked with yellow splashes and pinkish along the main vein and stalk. Flowers tubular, to 4cm(1$\frac{1}{2}$in), red to purple, in short terminal spikes.

GREENOVIA

Crassulaceae

Origin: Canary Islands. A genus of four species of succulent plants allied to and resembling the houseleeks (*Sempervivum*). The species described here have showy, yellow, starry flowers and thrive in containers. The main growing season is autumn to spring, with an almost dry summer resting period. Propagate by seed or offsets treated as cuttings in late summer. *Greenovia* commemorates the English geologist George Bellas Greenough (1778–1855).

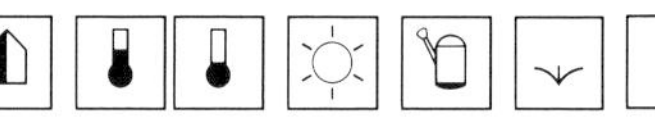

Species cultivated
G. aizoon Tenerife
Rosettes 5–8cm(2–3in) wide, several to many, forming clumps or hummocks. Leaves very broadly spathulate, finely and softly hairy, light green. Flowering stem 10–15cm(4–6in) tall, topped by a panicle of 1–2cm($\frac{3}{8}$–$\frac{3}{4}$in) wide yellow flowers composed of about 20 linear petals.
G. aurea
Rosettes 13–25cm(5–10in) wide, sometimes even larger, in the wild clustering and forming small

BELOW *Graptophyllum pictum*

BELOW RIGHT *Greenovia aizoon*

clumps, but often solitary in cultivation. Leaves cuneate to spathulate, bluish-green, densely overlapping. Flowering stem 30–45cm(1–1½ft) in height, topped by a broad panicle of 2.5cm(1in) wide, bright yellow flowers composed of 30 to 35 petals.

GREVILLEA

Proteaceae

Origin: Australia, New Caledonia to E. Malaysia. A genus of 190 species of evergreen trees and shrubs with alternate leaves, often pinnately divided, and racemes of petalless flowers, their bright colours coming from the calyx (or perianth) tube from which the long style protrudes. G. *robusta* is a large tree in the wild and is grown as a pot plant for its foliage only. The other species mentioned below are shrubby and are grown for their showy flowers. All are good for a cool conservatory and as foliage plants in the home. Propagate by seed sown in spring or by cuttings taken with a heel in summer with bottom heat. *Grevillea* was named for Charles Francis Greville (1749–1809), a founder of the Horticultural Society of London.

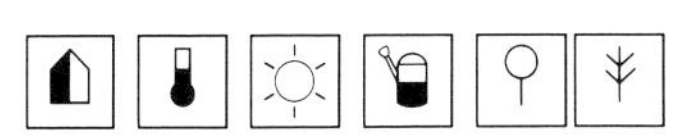

Species cultivated

G. banksii Australia
Tall shrub, in the wild to 5–6m(16–20ft), sometimes a small tree. Leaves deeply pinnate or pinnatifid, 15–30cm(6–12in) with up to 11 narrow linear leaflets, silky-hairy beneath. Flowers (including the long style) red, in crowded terminal racemes.

G. juniperina See G. *sulphurea*.

G. robusta Silky oak Queensland, New South Wales (Australia)
Tree to 30m(100ft) in the wild, but in a pot rarely over 2–3m(6½–10ft). Leaves ferny, bipinnately lobed, to 45cm(1½ft) long on young plants, smaller

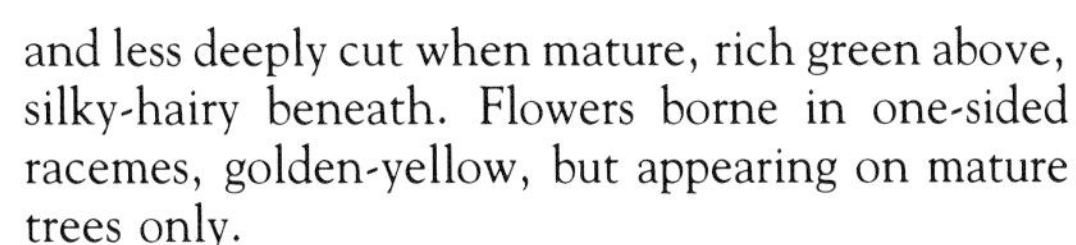

and less deeply cut when mature, rich green above, silky-hairy beneath. Flowers borne in one-sided racemes, golden-yellow, but appearing on mature trees only.

G. rosmarinifolia S.E. Australia
Freely branched shrub, to 2m(6½ft). Leaves 3–4cm(1¼–1½in) long, linear, with rolled under margins, the upper surface dark green, silvery silky-hairy beneath. Flowers 2–3cm(¾–1¼in) long, red, but occasionally pink or white, in short crowded racemes.

G. sulphurea (G. *juniperina*) New South Wales (Australia)
Shrub to 2m(6½ft). Leaves 1–2.5cm(⅜–1in) long, linear, ending in a sharp point, dark green above, paler and downy beneath. Flowers 1–2.5cm(⅜–1in) long, pale yellow, in terminal clusters of 12 or more.

GRISELINIA

Griseliniaceae (Cornaceae)

Origin: New Zealand, Chile and Brazil. A genus of six species of evergreen shrubs with alternate, ovate or obovate glossy leaves, thick in texture. The flowers, which are small and insignificant, are not produced on small plants. The species described are handsome foliage plants for a cool conservatory and a cool or unheated room. Propagate by cuttings, best taken in late summer with a heel, or by seed in spring. *Griselinia* was named for Francesco Griselini (1717–83), an Italian naturalist.

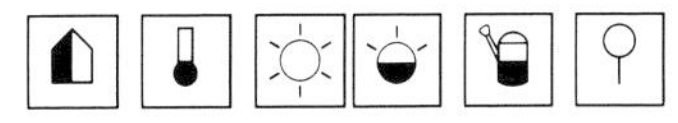

Species cultivated

G. littoralis New Zealand
Shrub or sometimes a tree in the wild, to 10m(33ft) or more. Leaves broadly ovate to oblong, 3–10cm(1¼–4in) long, glossy green, often rather yellowish green. G.*l*. 'Variegata' has white-variegated foliage.

FAR LEFT *Grevillea banksii*
LEFT CENTRE *Grevillea robusta*
BELOW *Griselinia littoralis*

G. lucida New Zealand
Shrubby, usually 2–3m(6½–10ft) tall, sometimes a tree in the wild. Leaves 10–18cm(4–7in) long, broadly ovate, unequal at the base giving a lop-sided effect. Variegated forms available.

GUZMANIA

Bromeliaceae

Origin: Tropical America, West Indies. A genus of 110 species of evergreen rosette-forming perennials the majority of which are epiphytic. The leaves are strap-shaped and, as with many other bromeliads, overlap at their bases to form water-holding reservoirs. The 3-petalled tubular flowers are yellow or white and carried in spikes or panicles often held within showy bracts. They are suitable for conservatory or home as long as some humidity is given. Propagate by offsets removed in late spring and summer. *Guzmania* commemorates Anastasio Guzman, an 18th-century Spanish naturalist.

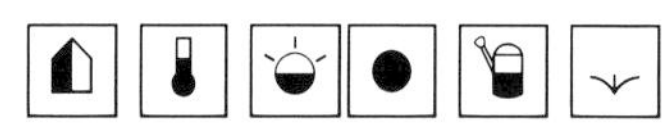

Species cultivated

G. berteroana Flaming torch Puerto Rica
Flowering stem robust, erect, up to 40cm(16in) tall, bearing overlapping, broadly oval, glossy cinnabar-red bracts. Flowers 5–6cm(2–2½in) long, rich yellow, the petal tips flared. Leaves 30cm(1ft) or more in length, broadly strap-shaped to elliptic, mid-green, or flushed red in sites open to direct sunlight.

G. lingulata Scarlet star West Indies to Brazil
Leaves to 45cm(1½ft) long by 3cm(1¼in) wide, metallic green. Flowers about 4cm(1½in) long, white, but almost hidden by the crimson-red bracts which make a star-shaped 'flower', and are carried on stems 18–30cm(6–12in) tall.

G. monostachya (*G. tricolor*) S. Florida, West Indies to Brazil
Leaves to 45cm(1½ft) long by 2cm(¾in) wide, green. Flowers 2.5cm(1in) long, white, within salmon-red to green bracts striped with brownish-red and brightest at the top of the spike; they are carried on stems 30–40cm(12–16in) long.

G. musaica Panama, Colombia
Flowering stems up to 30cm(1ft) or so tall, erect, clad with overlapping, elliptic, pink bracts and topped by a short, head-like spike of cupped, glossy pink bracts and 5cm(2in) long rich yellow flowers. Leaves 60–75cm(2–2½ft) long, broadly strap-shaped, rounded-tipped with a short spike, pale green with an elaborate zoned pattern of dark green to red-brown lines.

G. picta See *Nidularium fulgens*.

G. sanguinea Costa Rica to Colombia and Ecuador
Leaves to 30cm(1ft) long, the outer ones green, the

Guzmania sanguinea

inner of each rosette bright red with yellow-green centres. Flowers 4cm($1\frac{1}{2}$in) wide, almost stemless.
G. tricolor See G. *monostachya*.
G. vittata Colombia to Brazil
Flowering stems erect, to about 50cm(20in) in height bearing partially overlapping, tapered, lanceolate bracts terminated by a congested, head-like panicle of bracts and 2cm($\frac{3}{4}$in) long white flowers. Leaves up to 60cm(2ft) long, forming an erect rosette of narrow, vase-like outline, dark, glossy green with a cross-banding of silvery scales.
G. zahnii Costa Rica
Flowering stem to 50cm(20in) tall, concealed in the lower part by reddish, overlapping leaf-like bracts topped by smaller redder bracts and a dense ovoid panicle to 15cm(6in) long composed of 3cm($1\frac{1}{4}$in) long, yellow flowers and smaller floral bracts of the same colour. Leaves up to 60cm(2ft) long, strap-shaped, pointed, pale to olive-green with longitudinal dark red lines.

Hybrids
A number of colourful hybrids have been raised, some having deep red, glossy leaves.

GYMNOCALYCIUM

Cactaceae

Origin: South America. A genus of 60 species of globular to shortly cylindrical cacti with prominent ribs bearing areoles above 'chin-like' projections. They have relatively few spines, which are stout and usually recurved and generally conspicuous, funnel- to trumpet-shaped flowers. They make good plants for the conservatory or home, needing a sunny position, and a cool rest in winter. Propagate by seed in spring or by removing offsets in summer. *Gymnocalycium* derives from the Greek *gymnos*, naked and *kalyx*, a bud, the flower buds of this genus lacking spines, bristles or hair.

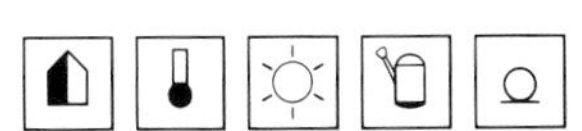

Species cultivated
G. andreae Argentina
Stems globular, bluish-green, somewhat flattened, to 4cm($1\frac{1}{2}$in) across. Ribs eight, areoles whitish, bearing about seven white radial spines to 1cm($\frac{3}{8}$in) long and one to three central ones which curve upwards. Flowers 3cm($1\frac{1}{4}$in) across, bright yellow, funnel-shaped.
G. baldianum (G. *venturianum*) Argentina
Stems solitary, to 4.5cm($1\frac{3}{4}$in) high by 7cm($2\frac{3}{4}$in) wide, depressed at the top, dark grey-green, with small areoles on nine to 11 broadly rounded ribs. Radial spines five, about 6mm($\frac{1}{4}$in) long, yellowish to grey; central spines absent. Flowers about 4cm($1\frac{1}{2}$in) long, funnel-shaped, red.
G. bruchii See G. *lafaldense*.
G. denudatum Spider cactus S. Brazil to N. Argentina
Stems more or less globular, eventually to

Guzmania 'Glory of Ghent'

10cm(4in) tall, rich green, with five to eight shallow, broad ribs bearing slightly woolly areoles. Spines radial only, five to eight, about 10mm($\frac{3}{8}$in) long, usually yellowish and up-curving. Flowers widely funnel-shaped, 5–7cm(2–$2\frac{3}{4}$in) long, white or pale pink.
G.d. paraguayense
More prominent, grey-green ribs and white flowers.
G. gibbosum Argentina
Stem at first globular, later becoming cylindrical, to 20cm(8in) tall and 9cm($3\frac{1}{2}$in) wide. Ribs 12 to 19, notched, with grey-woolly areoles and prominent 'chins'. Radial spines seven to ten, pale brown, to 3.5cm($1\frac{1}{2}$in) long; central spines one or two or absent. Flowers 6cm($2\frac{1}{2}$in) long, white to reddish.
G.g. nigrum
Handsome, with black spines.
G. loricatum See G. *spegazzinii*.
G. lafaldense (G. *bruchii*) Argentina
Stems clump-forming, globular, up to 4cm($1\frac{1}{2}$in) tall, with about 12 low ribs bearing elongated, somewhat white-woolly areoles. Radial spines 12 to 15, very slender, white, at least at the tips, about 5–10mm($\frac{1}{5}$–$\frac{3}{8}$in) long; central spines slightly longer, usually one to three, sometimes absent. Flowers 3–5cm($1\frac{1}{4}$–2in) long, purplish-pink.
G. mihanovichii Paraguay
Stem globular, somewhat flattened, greyish-green,

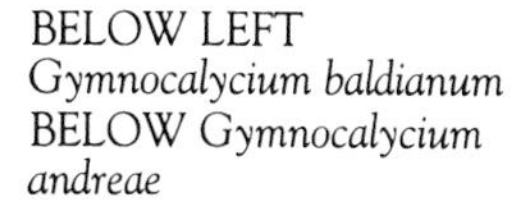

BELOW LEFT *Gymnocalycium baldianum*
BELOW *Gymnocalycium andreae*

to 5cm(2in) wide. Ribs about eight, broadly triangular, with small closely set areoles. Radial spines seven to eight, yellowish to pale brown, 4cm(1½in) long; no central spines. Flowers 4–5cm(1½–2in) long, yellow-green to reddish.

A number of mutant forms have arisen lacking chlorophyll, and yellow or red in colour. Being unable to produce their own food they can be maintained only by grafting on to a green stock, cuttings of a *Hylocereus* being frequently used. These mutants are widely available under such names as 'Red Head', 'Red Cap', 'Yellow Cap', 'Hibotan' (yellow) and 'Blondie'.

G. multiflorum S. Brazil to Argentina
Stems solitary or forming clusters, rounded, to 9cm(3½in) tall by up to 13cm(5in) broad, with five to ten prominent ribs, bearing large, elliptic areoles to 1cm(⅜in) long. Radial spines only, seven to ten, sturdy, awl-shaped, yellowish to pinkish, arranged in a comb-like formation, the longest to 3cm(1¼in) long. Flowers funnel-shaped, about 4cm(1½in) long, pink or white.

G. platense N. Argentina
Stems globular, eventually to 9cm(3½in) high and somewhat wider, with ten to 14, blunt, grey-green ribs, bearing grey-woolly, elongated areoles. Spines five to seven, usually all radials, low arching, white and reddish, the longest to 1.5cm(⅝in). Flowers about 5cm(2in) long, slender-tubed, white.

G. quehlianum Argentina
Stem globose, flattened, to 4cm(1½in) high and 15cm(6in) wide, bluish-green. Ribs eight to 13, with prominent 'chins' beneath the grey-woolly areoles. Radial spines five, pale yellow, to 1cm(⅜in) long; no central spines. Flowers 5–6cm(2–2½in) long, white inside, green outside with a red centre and edging underneath.

G. saglione N. Argentina
Stems usually solitary, rounded, eventually to 30cm(1ft) thick and almost as tall, with up to 30 or more rounded, grey to blue-green ribs bearing large, oval, woolly areoles. Radial spines seven to ten or more, awl-shaped, recurved, 2.5–4cm(1–1½in) long, reddish to greyish or off-white; central spines somewhat shorter but straight. Flowers about 4cm(1½in) long, white or pale pink. This is a comparatively fast-growing species and appreciates light shade during the hottest summer weather.

G. schickendantzii Argentina
Similar to G. *saglione*, but not quite as big, with radial spines only and 5cm(2in) long flowers which may vary from white to reddish and yellowish.

G. spegazzinii (G. *loricatum*) N. Argentina
Stems usually solitary, globose to shortly cylindrical, eventually up to 15cm(6in) in height, with ten to 13 flattish ribs bearing round, brownish areoles. Radial spines only, five to seven, sturdy, recurved, awl-shaped, brown to grey, up to 3cm(1¼in) or more long. Flowers to 8cm(3in) long, pink to white.

G. venturianum See G. *baldianum*.

Gynura 'Purple Passion'

GYNURA

Compositae

Origin: Africa to E. Asia and Malaysia. A genus of 100 species of evergreen perennials, shrubs and climbers with rather fleshy, alternate, entire leaves and flower heads rather akin to groundsel. They make good pot plants, grown for their decorative foliage. Propagate by cuttings in late spring or summer. *Gynura* is derived from the Greek *gyne*, female and *oura*, a tail, referring to the long stigmas.

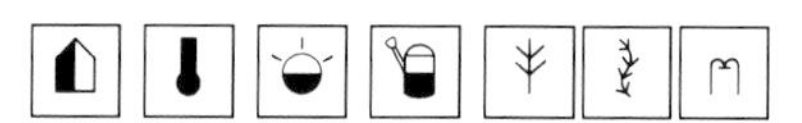

Species cultivated

G. aurantiaca Velvet plant Java
Erect, soft shrub, to 90cm(3ft), becoming partially climbing with age, and then to 3m(10ft). Stems and leaves densely clothed with velvety violet-purple hairs. Leaves 10–20cm(4–8in) long, ovate to elliptic, coarsely double-toothed. Flowers 1.5–2cm(⅝–¾in) wide, orange-yellow, ageing purple; they are carried in loose corymbs in winter.

G. bicolor Moluccas
Erect sub-shrub, to 9cm(3ft) tall. Leaves ovate-lanceolate, coarsely lobed and toothed, downy, pale greenish-purple above, rich purple beneath. Flowers 1.5–2cm(⅝–¾in) wide, orange-yellow, borne in winter.

G. procumbens See G. *sarmentosa*.

G. 'Purple Passion'
A hybrid, probably between G. *aurantiaca* and G. *sarmentosa*, with semi-twining stems. Leaves to 11cm(4½in) long, lanceolate to oblong-lanceolate, purple-hairy, with lobed and toothed margins, purplish-green above, deep red-purple beneath.

G. sarmentosa (G. *procumbens*) Malaya to Philippines
Erect sub-shrub, later trailing or climbing. Leaves to 13cm(5in) long, ovate, lanceolate, toothed, green with a purple mid-vein. Flowers orange.

H

HAAGEOCEREUS

Cactaceae

Origin: Peru. A genus of about 40 species of cylindrical-stemmed densely small-spiny cacti with funnel-shaped nocturnal flowers. Most species grow erect, but a few recline or grow completely prostrate, an unusual type of growth for this sort of cactus. The species described below are suitable for the home and conservatory, the prostrate sorts making unusual hanging basket specimens. Propagate by seed in spring or by cuttings in summer. *Haageocereus* couples the name of the German nursery firm of Haage with that of *Cereus*.

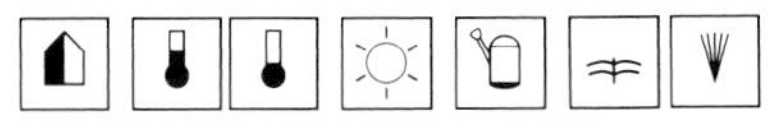

Species cultivate

H. acranthus (*Cereus acranthus*)
Stems branching from the base to form clumps, erect, to 90cm(3ft) in height by 8cm(3in) thick. Ribs ten to 14, prominent, slightly notched on the large areoles. Radial spines yellow, 20 to 30, each one 1cm($\frac{3}{8}$in) long; central spines usually two, awl-shaped, 2cm($\frac{3}{4}$in) long. Flowers 6–9cm($2\frac{1}{2}$–$3\frac{1}{2}$in) long, white with green shading; a pink form is recorded.

H. decumbens (*Cereus decumbens*, *Borzicactus decumbens*)
Stems branching from the base, prostrate, to 90cm(3ft) in length by up to 10cm(4in) thick. Ribs about 20, not prominent, notched between the small, yellowish woolly areoles. Radial spines about 30, each one 5mm($\frac{1}{5}$in) long, white when young; central spines two to five, black to dark brown, 2–4cm($\frac{3}{4}$–$1\frac{1}{2}$in) long. Flowers 7cm($2\frac{3}{4}$in) long, white, fragrant.

H. versicolor (*Cereus versicolor*)
Stems erect, branching from the base to form clumps up to 1.5m(5ft) in height by 8cm(3in) thick. Ribs 12 to 22, not prominent, closely set with hairy areoles. Radial spines 25 to 30, yellow, red or bicoloured, 5mm($\frac{1}{5}$in) long; central spines one or two, to 4cm($1\frac{1}{2}$in) long, at least some of them pointing downwards. Flowers 10cm(4in) long, white. *H.v.* 'Aureospinus' has entirely yellow spines.

HABRANTHUS

Amaryllidaceae

Origin: Sub-tropical and tropical America. A genus of perhaps 20 species of bulbous plants related to both *Hippeastrum* and *Zephyranthes* and in appearance much like a diminutive version of the former, but with only one flower per stem. Easily grown in pots, they make pleasing short-term flowering plants for home and conservatory. Propagate by seed, or offsets removed when re-potting, both in spring. Plants are best left in their containers until fairly congested before dividing. Keep dry during the winter. *Habranthus* derives from the Greek *habros*, graceful and *anthos*, a flower.

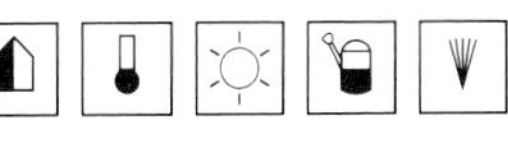

Species cultivated

H. andersonii South America, particularly Uruguay and Argentina
Leaves narrowly strap-shaped, 13–15cm(5–6in) long, more or less arching. Flowering stems erect, about 15cm(6in) in height, topped by a 4cm($1\frac{1}{2}$in) wide, 6-tepalled flower; tepals yellow to coppery-yellow inside, coppery-red striped outside, mainly in early summer. All-yellow and all-coppery coloured forms are known.

H.a. texanus Texas
Orange-yellow and red flowers. Some authorities give it separate specific rank as *H. texanus*.

H. brachyandrus (*Hippeastrum brachyandrum*) S. Brazil
Leaves narrowly strap-shaped, to 30cm(1ft) in length, tapered towards the tips. Flowering stems erect, to 30cm(1ft) tall, topped by a 8cm(3in) wide flower; tepals clear mauve-pink shading to darkest red or purple at the base, opening in late summer.

H. robustus See *H. tubispathus*.

H. texanus See *H. andersonii texanus*.

H. tubispathus (*H. robustus*, *Zephyranthes robusta*) Argentina
Leaves narrowly strap-shaped, spreading to recurved, about 22cm(9in) long, usually developing after blooming. Flowering stems to 22cm(9in) or more in height topped by a 6–8cm($2\frac{1}{2}$–3in) wide flower; tepals rose-red or pink with a hint of purple, shading to white at the base; early autumn.

ABOVE *Habranthus tubispathus*
TOP *Haageocereus acranthus*

HADRODEMAS

Commelinaceae

Origin: Guatemala. A genus of one species of evergreen perennial closely related to *Tradescantia*. Like other tropical members of that genus, it is very amenable to cultivation in the home and conservatory and makes a handsome and unusual specimen plant. Propagate by cuttings or by removing the plantlets which form on the flowering stems. *Hadrodemas* derives from the Greek *hadros*, well developed or of good size and *demas*, a living body, presumably referring to the eventually large size of the plant.

  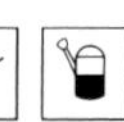  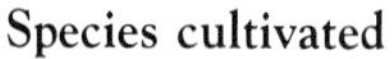

Species cultivated

H. warszewiczianum (*Spironema/Tripogandra/Tradescantia warszewicziana*)
Stem robust, erect, sometimes forking, otherwise unbranched, up to 90cm(3ft) long and then sometimes sprawling. Leaves narrowly ovate, stem-clasping, somewhat fleshy, bright green, to 30cm(1ft) in length, crowded along the stem and arching outwards, creating a Dracaena-like plant. Flowers small with three persistent, purplish to lilac sepals and three rounded rose-purple petals which last but one day, many together in a wide spreading panicle well above the leaves.

Haemanthus magnificus

HAEMANTHUS

Amaryllidaceae

Origin: Tropical southern Africa, Arabia, Socotra. A genus of 50 species of bulbous perennials with broadly strap-shaped, evergreen or deciduous leaves and many-flowered umbels of small, often colourful flowers. When potting, keep the neck of the bulb at or just above soil level. Dry off deciduous species when the leaves yellow. Propagate by removing offsets when potting, or seed sown as soon as gathered. *Haemanthus* is derived from the Greek *haima*, blood and *anthos*, a flower; most species have red flowers.

  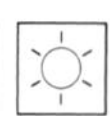  

Species cultivated

H. albiflos Paintbrush South Africa
Leaves 20cm(8in) long, 10cm(4in) wide, dark green, tongue-shaped and fringed with white hairs. Flowers white, in 5cm(2in) wide umbels, the barely protruding anthers yellow, carried on 20cm(8in) stems and followed by bright red fruits. The plant needs an early summer resting period.

H. coccineus South Africa
Leaves deciduous, 30–60cm(1–2ft) long, about 15cm(6in) wide, broadly strap-shaped. Flowers in umbels 5–8cm(2–3in) across, coral-red, with white, yellow-tipped anthers within two scarlet bracts which overtop the umbel; borne on brown spotted stems to 30cm(1ft) high in autumn when the plant is leafless. Dry summer rest essential.

H. katherinae Blood flower Natal, Transvaal (S. Africa), Zimbabwe
Leaves 20–30cm(8–12in) long, oblong. Flowers in umbels, 15–25cm(6–10in) across, bright salmon-red, with prominent stamens and spreading or drooping bracts; they are borne on 30–60cm(1–2ft) tall stems in summer.

H. magnificus Natal, Transvaal
Leaves 30cm(1ft) or more, ovate to oblong-ovate, wavy. Flowers in umbels, 10–15cm(4–6in) wide, deep orange-red, surrounded by maroon to greenish bracts, in summer to early autumn.

H. multiflorus Blood flower Tropical Africa
Leaves 30cm(1ft) long, lanceolate. Flowers markedly stalked, in umbels to 15cm(6in) across, deep red; stems are 30–60cm(1–2ft) tall, very similar to *H. katherinae*, but spring-flowering.

H. natalensis Natal (S. Africa)
Leaves 30cm(1ft) long by 10cm(4in) wide, shining green on stems 60–90cm(2–3ft) tall which overtop the flowers. Umbels of scarlet blooms with protruding stamens are held within a cup of seven to eight deep red bracts on leafless stalks in autumn.

H. pole-evansii Zimbabwe (mountain forests)
Leaves 30–60cm(1–2ft) long, oblanceolate, fresh green, deciduous. Flowers about 4.5cm(1¾in) wide, starry, rich salmon-pink on 90–120cm(3–4ft) flowering stems, from winter to spring. Rather like a giant version of *H. multiflorus*, but with a flower colour unique in its genus.

HAMATOCACTUS See THELOCACTUS

HARDENBERGIA

Leguminosae

Origin: Australia. A genus of two species of woody-stemmed evergreen climbers. They have twining stems, simple or compound alternate leaves and axillary racemes of small pea flowers. Their freedom of flowering and late winter to spring season places them among the most desirable of small climbers for the conservatory. They will also tolerate pot culture and can be grown in the home if regularly pruned after blooming. Propagate by seed in spring or by cuttings in summer. *Hardenbergia* commemorates Countess Franziska von Hardenberg, the sister of Baron Carl A.A. von Hugel (1795–1870), Austrian traveller and Patron of Horticulture.

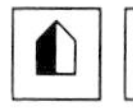  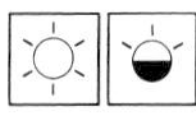  

Species cultivated

H. comptoniana Wild sarsparilla W. Australia
Up to 3m(10ft) in height with slender, wiry, branched stems. Leaves compound, composed of three to five ovate to lanceolate leaflets, the largest 6–9cm($2\frac{1}{2}$–$3\frac{1}{2}$in) long. Racemes 8–13cm(3–5in) long, flowers violet-blue, about 9mm($\frac{1}{3}$in) long.

H. violacea Eastern Australia
To 3m(10ft) in height with slender, wiry well-branched stems. Leaves composed of one lanceolate to ovate 8–13cm(3–5in) long leaflet which is actually broadest at the base, tapering to a blunt tip. Racemes to 10cm(4in) or more in length, flowers blue-purple, about 9mm($\frac{1}{3}$in) long. Pink and white-flowered forms are known.

HARIOTA See RHIPSALIS
HATIORA See RHIPSALIS

HAWORTHIA

Liliaceae

Origin: Southern Africa. A genus of 150 species of mostly clump-forming succulent perennials with fleshy, rather triangular-shaped leaves either in rosettes or crowded on short erect stems. The flowers are tubular and often curved, appearing to be 2-lipped though in fact they are made up of six narrow lobes. They are carried in racemes on long wiry stems and although quite pretty, the main attraction of the plants is the patterning and disposition of their leaves. The flowers of all the species are greenish-white and very similar to each other. Propagate by division or by offsets. *Haworthia* was named for Adrian Hardy Haworth (1768–1833), an English entomologist who was also an authority on succulent plants.

  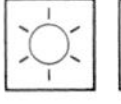  

LEFT *Hardenbergia violacea*
BELOW LEFT *Haworthia attenuata*

Species cultivated

H. armstrongii Cape Province
Stems erect, to 10cm(4in) or more in height. Leaves crowded, pointing upwards, 3–4cm($1\frac{1}{4}$–$1\frac{1}{2}$in) long, lanceolate, often keeled on the backs, deep matt-green, with a slightly waxy-white patina and a slim white point.

H. attenuata Cape Province
Rosette-forming. Leaves 6–8cm($2\frac{1}{2}$–3in) long, oblong triangular, tapering to a slender point, dark green, the surface covered with bands of white tubercles, densest on their backs.

H. batesiana Cape Province (Valley of Desolation)
Leaves in rosettes, eventually forming sizeable hummocks; individual leaves broadly lanceolate to oblong, with incurving, tapering bristle-tips, pale, semi-translucent green with a tesselation of darker veins.

H. coarctata Cape Province
Stems erect, eventually to about 20cm(8in) in height. Leaves triangular-lanceolate, tapering to a point, up to 6cm($2\frac{1}{2}$in) long, deep green with lines of small, greenish-white tubercles.

H. cooperi Cape Province, Karroo, etc.
Rosette-forming, much like a larger form of *H. obtusa*.

H. cuspidata Cape Province, George Cape
Rosette-forming. Leaves 2.5cm(1in) long, thick-textured, ovate-triangular, narrowing to an abrupt point, rounded on the back, pale green with a transparent tip.

H. cymbiformis Cape Province
Rosette-forming. Leaves obovate, 1.5–2.5cm(⅝–1in) long, lower surface keeled at the tip and strongly concave, grey-green with a shortly pointed transparent tip.

H.c. obesa
Slightly longer, thicker, bristle-tipped leaves.

H.c. translucens
Totally translucent leaves with dark longitudinal veins.

H. fasciata Zebra haworthia Cape Province
Rosette-forming. Leaves 3–4cm(1¼–1½in) long, firm, triangular-lanceolate, tapering, strongly convex beneath, glossy green with large white tubercles grouped into regular transverse lines.

H. fasciata major
Leaves 5–8cm(2–3in) long.

H. greenii Cape Province, Grahamstown area
Stems more or less erect to procumbent with age, 10–20cm(4–8in) long. Leaves to 4cm(1½in) long, broadly lanceolate with a keeled, incurving tip, reddish to copper-green patterned with rows of whitish tubercles.

ABOVE *Haworthia maughanii*
RIGHT *Haworthia venosa*

H. limifolia Cape Province, Transvaal
Rosette-forming. Leaves 8–10cm(3–4in) long, rigid, broadly ovate-triangular, sharply tapering, strongly convex on the back, roundish keeled, dark green with whitish transverse ridges.

H. margaritifera See *H. pumila*.

H. maughanii Cape Province, Karroo, etc.
Rosette-forming. Leaves similar to those of *H. truncata*, but more or less cylindrical and arranged in a spiral.

H. obtusa (*H. cymbiformis obtusa*) Cape Province
Rosette-forming. Leaves dense, cuspidate, erect and almost cylindrical in cross-section, ovoid to obovoid, 2.5cm(1in) or more long, translucent for the upper half or one third, with dark, longitudinal veining.

H. planifolia Cape Province
Rosette-forming. Leaves 3.5–5cm(1½–2in) long, narrowly ovate, tapering abruptly to a short point, flat above, convex beneath, light greyish-green.

H. pumila (*H. margaritifera*) Pearl plant
Cape Province
Rosette-forming. Leaves triangular-ovate, tapering, up to 8cm(3in) long, up-curving, keeled at the tip, dark green, with large pearly tubercles. *H.p. maxima* is larger in all its parts; *H.p. minima* is smaller.

H. pygmaea Cape Province
Rosette-forming. Leaves glossy dark green, about 2.5cm(1in) long or a little more, narrowly oval, the tips abruptly truncate and translucent. It is natural for the leaves to shrink considerably during the resting period.

H. reinwardtii Cape Province, widespread
Stems more or less erect, 10–15cm(4–6in) tall. Leaves closely overlapping, incurving, up to 5cm(2in) long, lanceolate, deep green with prominent, whitish tubercles. Variable species regarding leaf and stem length and density of tubercles.

H. reticulata Cape Province
Rosette-forming. Leaves oblong-lanceolate, about 3–5cm(1¼–2in) long, ending in a long bristle tip and with toothed margins, pale green, the upper third almost transparent, with darker veins.

H. retusa Cape Province, Karroo, etc.
Rosette-forming. Leaves almost erect, up to 5cm(2in) long, very fleshy, the upper half bent almost at right angles, its triangular surface flattened and translucent with longitudinal pale veins.

H. setata Cape Province, Karroo, etc.
Rosettes almost rounded in outline, composed of slender tipped, lanceolate leaves, 2.5cm(1in) or more in length, deep green and keeled at the tips, with numerous, marginal, prominent pale bristles. This is a highly distinctive and variable species, well worth seeking.

H.s. gigas
Big rosettes composed of leaves up to 6cm(2½in) long bearing even more prominent marginal bristles.

H. subfasciata Cape Province
Rosette-forming. Much like *H. fasciata*, but very

much larger, the leaves up to 13cm(5in) long.

H. tessellata Star window plant Cape Province to S.W. Africa
Rosette-forming, composed of shortly tapering, triangular, toothed leaves, 3–5cm(1¼–2in) long, with a glossy, semi-transparent upper surface set with a tesselated pattern of pale or reddish veins. This is a somewhat variable species with regard to leaf size, shape and vein colouring.

H. truncata Cape Province
Leaves six to eight together in opposite ranks borne fanwise; each is 2cm(¾in) long, incurved, dark green to brownish, roughly tuberculate, the top looking as if it has been cut off, the exposed surface being all that is above ground in its native habitat.

H. venosa Cape Province
Rosette-forming, but these elongate somewhat with age, forming a short stem. Leaves triangular-lanceolate, spreading to recurved, 5–7cm(2–2¾in) long, with keeled tips and toothed margins, the upper surface green to purple-green, often bearing a tesselated vein pattern.

HEBE

Scrophulariaceae

Origin: Mainly New Zealand, also S.E. Australia, Tasmania, New Guinea, southern South America, Falkland Is. A genus of about 100 species of evergreen shrubs or occasionally small trees which have opposite pairs of entire leaves and racemes or panicles of small tubular flowers which open to four lobes. Those described below make attractive shrubs for the conservatory, the smaller ones make pretty, short-term house plants. Propagate by cuttings of non-flowering shoots. Because of the ease with which they hybridize, seedlings are rarely true to type. *Hebe* was named for the Greek goddess of youth, cup-bearer to the gods and who married Hercules.

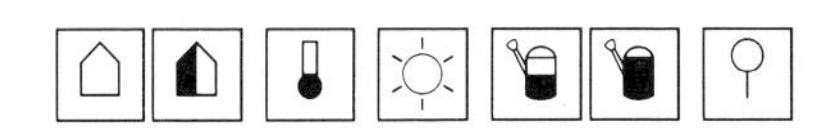

Species cultivated

H. × andersonii
A hybrid between *H. speciosa* and *H. salicifolia* or the allied *H. stricta.* Leaves to 11cm(4½in), oblong-lanceolate, leathery. Flowers violet, fading through mauve to white; they are borne in 10–15cm(4–6in) long racemes in summer. 'Variegata' has smaller, grey-green leaves which are margined with creamy-white.

H. diosmifolia New Zealand (North Island)
Shrub 60–150cm(2–5ft) tall, occasionally more in the wild. Leaves 1.5–2.5cm(⅝–1in) long, lanceolate to obovate, dark green above, paler beneath, on hairy stalks. Flowers lavender-blue to white, fragrant, in dense racemes to 4cm(1½in) long, from the upper leaf axils; very free-flowering. Two forms appear to be available, the small one described here being an excellent pot plant.

***H. × franciscana* 'Blue Gem'**
A hybrid between *H. speciosa* and *H. elliptica* making a dense, rounded shrub to 1–1.5m(3–5ft) or more. Leaves 3–6cm(1¼–2½in) long, elliptic to obovate, shining green and rather fleshy. Flowers violet, in dense racemes 3–6cm(1¼–2½in) long, borne in summer. 'Variegata' has the leaves broadly margined creamy-white.

H. speciosa New Zealand (North Island)
Rounded shrub, 1–2m(3–6½ft) tall. Leaves to 10cm(4in) long, elliptic to obovate-oblong, dark green. Flowers in racemes to 10cm(4in) long, red-purple to violet, borne in the upper leaf axils. Rarely grown, usually being represented by one of its hybrids or cultivars. The following can be recommended:

'Alicia Amherst', elliptic-ovate leaves to 9cm(3½in) long and deep violet-blue flowers.

'Gauntlettii', leaves to 13cm(5in) long and rosy-pink flowers.

'La Seduisante', leaves to 10cm(4in) long, often purplish beneath, with magenta-purplish flowers.

'Simon Delaux', leaves to 10cm(4in), flowers crimson.

ABOVE LEFT *Haworthia venosa*
ABOVE *Hebe diosmifolia*

HECHTIA

Bromeliaceae

Origin: Southern USA and Mexico to Central America. A genus of 35 to 45 species of ground-dwelling, dioecious bromeliads related to and resembling *Dyckia* and like that genus, native to arid regions. They form solitary or clustered dense rosettes of slender leaves usually armed with sharp, hooked teeth. The dioecious 3-petalled flowers are carried in large, diffuse panicles, but are not showy. Propagate by seed or division in spring. *Hechtia* honours Julius Gottfried Konrad Hecht (*d.* 1837), a councellor to the King of Prussia (Germany).

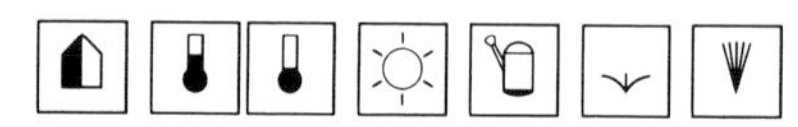

Species cultivated

H. argentea Central Mexico
Rosettes usually solitary. Leaves to 45cm(1½ft) in length, linear, low-arching, with a silvery-white patina on both surfaces. Panicle to 45cm(1½ft) tall, the 1cm(⅜in) long, white or greenish flowers in clusters. A slow-growing species not very easy to obtain, but possibly the most decorative.

H. epigyna Mexico
Rosettes usually solitary. Leaves 30–45cm(1–1½ft) long, linear, slightly channelled, with white hooked teeth, basically bright green, but grey-scaly. Flowering stem up to 45cm(1½ft) or more in height, set with many 5mm(⅕in) long lilac flowers.

H. marnier-lapostollei Mexico
Rosettes eventually clustering and forming loose hummocks. Leaves to 13cm(5in) in length, narrowly triangular, spiny-margined and tipped, recurved, silvery-grey, taking on reddish tints in bright sunlight. Flowers small, white, in panicles above the leaves.

Hechtia argentea

HEDERA

Araliaceae
Ivies

Origin: Europe to the Caucasus, Himalaya and Japan; also Canary Is. and Madeira. A genus of about five species of evergreen climbers; 15 species according to some botanists. They have juvenile stems which climb by short clinging roots, and have attractive, usually ovate, palmately-lobed leaves. The mature phase has erect stems with unlobed leaves and is fertile, bearing small, 5-petalled, greenish-yellow flowers in terminal umbels. These are followed by black berries. They make very good pot or hanging basket plants, the small-leaved cultivars being particularly effective. Propagate by cuttings from spring to autumn, those taken from non-climbing stems producing the bushy, so-called tree ivies. *Hedera* is the Latin name for ivy.

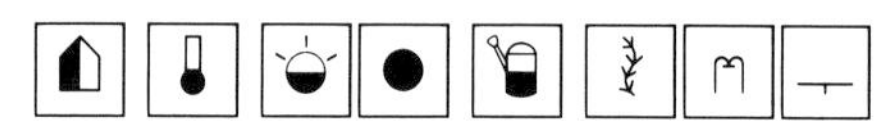

Species cultivated

H. canariensis Canary Island ivy Canary Is., Madeira, N.W. Africa
Stems often purplish-red. Leaves to 15cm(6in) wide, broadly ovate-cordate, with three to five shallow lobes, dark glossy green. 'Gloire de Marengo' has almost un-lobed leaves, grey-green with a creamy-white margin of variable width.

H. colchica Persian ivy Caucasus to N. Iran
Strong-growing. Leaves to 12cm(4¾in) wide, ovate, heart-shaped, un-lobed, dark glossy green. 'Dentata' has larger leaves, to 20cm(8in) long, distantly toothed, more suitable for pot culture. 'Dentata Variegata' has dark green leaves shaded with grey-green and with irregular cream margins.

H. helix Common/English ivy Europe to the Caucasus
Leaves 3–10cm(1¼–4in) wide, broadly ovate, with three to five lobes, dark glossy green on fertile stems. Very variable in foliage characters, producing a wide range of forms among which are some of the best ivies for the home:
'Buttercup', leaves 5–7cm(2–2¾in) across, 6–8cm(2½–3in) long, 3- to 5-lobed, bright yellow to yellow-green in the sun, much less yellow in shade.
'Deltoidea', leaves 6–10cm(2¼–4in) across, 8–10cm(3–4in) long, 3-lobed, the lobes rounded, dark green.
'Eva', leaves 2.5–3.5cm(1–1½in) across and in length, usually 3-lobed, grey-green with an irregular cream margin; stems purplish-red.
'Glacier', leaves 3–6cm(1¼–2½in) across and in length, shallowly 3-lobed, grey-green shading to silvery-grey, often with a very narrow cream edge; stems purplish-green.

LEFT *Hedera helix* 'Goldheart'
FAR LEFT *Hedera helix* 'Eva'
BELOW LEFT *Hedera helix* 'Parsley Crested'

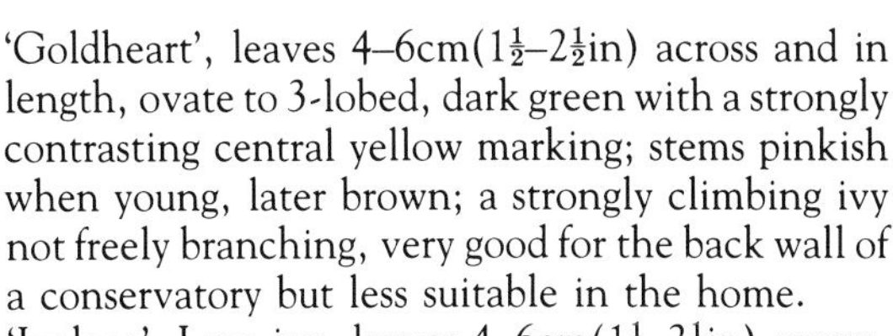

'Goldheart', leaves 4–6cm(1½–2½in) across and in length, ovate to 3-lobed, dark green with a strongly contrasting central yellow marking; stems pinkish when young, later brown; a strongly climbing ivy not freely branching, very good for the back wall of a conservatory but less suitable in the home.
'Ivalace', Lace ivy, leaves 4–6cm(1½–2½in) across, 4–5cm(1½–2in) long, basically 3-lobed, the leaf margins waved and crinkled, glossy green.
'Koniger's Auslese', leaves 3–7cm(1¼–2¾in) across, 5–8cm(2–3in) long, 3- to 5-lobed, the central one twice as long as the side lobes, outer lobes when present pointing backwards; an excellent pot plant.
'Parsley Crested', leaves 4–6cm(1½–2½in) across and in length, the almost un-lobed leaves with crimped margins; very distinctive.
'Sagittifolia', leaves 2–3cm(1¾–2¼in) across and in length, basically of three long and narrow lobes, the central one the longest, sometimes with two backwards pointing basal lobes, grey-green with an irregular creamy-white margin.

HEDYCHIUM

Zingiberaceae

Origin: Malaysia, S.W. China to India; also Malagasy. A genus of 50 species of rhizomatous, clump-forming perennials with erect, unbranched, rather bamboo-like stems. The lanceolate, entire leaves are borne in two ranks and the narrowly tubular flowers are in terminal spikes. Each flower has a conspicuous petal-like lip which is formed by an enlarged stamen. They are best in a conservatory and are suitable only for a large room when in bloom. Propagate by division in spring. *Hedychium* derives from the Greek *hedys*, sweet and *chion*, snow, the first described species having pure white, fragrant flowers.

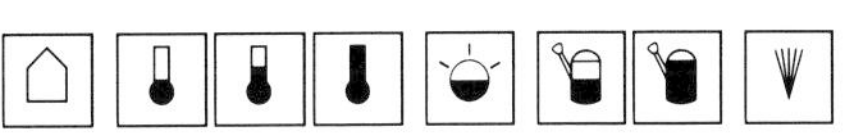

Species cultivated

H. coccineum Red/Scarlet ginger lily India, Burma
To 1.2–2m(4–6½ft). Leaves to 50cm(20in) long, broadly lanceolate. Flowers about 5cm(2in) long, bright red, in spikes 25cm(10in) long and 5cm(2in) wide, opening in late summer and autumn. Cool.

H. coronarium Garland flower, Butterfly ginger lily Tropical Asia, naturalized in tropical America
Height about 2m(6½ft), often less in containers. Leaves oblong-lanceolate, 45–60cm(1½–2ft) in length, slender-pointed, downy beneath. Flower spikes 20–30cm(8–12in) long, composed of four to six comparatively large, white, fragrant blooms; each flower has a 4–8cm(1½–3in) long tube, a broad 5cm(2in) long staminodal lip and three narrower petal-like lobes. The flowering season extends from late spring to autumn. Temperate to tropical.

H. densiflorum India including the Himalayas
To 3m(10ft) tall. Leaves 35cm(14in) long and 10cm(4in) wide, oblong-lanceolate. Flowers orange to coral-red, in 20cm(8in) long spikes, from summer to autumn. Cool.

Hedychium gardnerianum

H. gardnerianum Kahili ginger N. India
To 2m($6\frac{1}{2}$ft) tall. Leaves 40cm(16in) by 13cm(5in) wide. Flowers 5cm(2in) long, pale yellow, in spikes to 35cm(14in) or more in length, opening from summer to autumn. Cool.
H. greenei India
Height from 1–2m(3–$6\frac{1}{2}$ft), the erect stems bearing oblong-lanceolate, acuminate leaves, up to 25cm(10in) long. Flower spikes dense, to 13cm(5in) long; individual flowers have 10cm(4in) long tubes topped by three narrow, bright red lobes and a broader, darker, staminodal lip of equal length. Late summer to autumn is the usual flowering season. Cool.

HEERIA See HETEROCENTRON
HEIMERLIRIODENDRON See PISONIA
× **HELIAPORUS** See APOROCACTUS

HELICHRYSUM

Compositae

Origin: Europe, Africa, S. India, Sri Lanka, Australia. A genus of 300 to 500 species of annuals, perennials, shrubs and sub-shrubs. They have attractive, alternate, linear to ovate leaves which are usually hairy. The flowers have no ray florets being made up entirely of disc florets, but the bracts which surround them (the involucre) are chaffy, petal-like and sometimes coloured. Some species are grown especially for dried flower arranging. Those described are attractive foliage plants for a conservatory. Propagate by seed in spring or from cuttings taken from late spring to late summer. *Helichrysum* derives from the Greek *helios*, sun and *chrysos*, golden, referring to the flower heads of some species.

Species cultivated
H. angustifolium See *H. italicum*.
H. italicum (*H. angustifolium*, *H. serotinum*) Curry plant S. Europe
Sub-shrub to 30cm(1ft) or more. Leaves 1–3cm($\frac{3}{8}$–$1\frac{1}{4}$in) long, linear, silvery-grey, with a thin felting

Helichrysum petiolatum

of hairs. Flower heads 2–3m
low, held in small clusters on l
H. petiolatum Liquorice pla
Trailing, partially erect shrub
1.2m(4ft) or more wide. Leave
broadly ovate, cordate, cove
with a dense white felt. Flo
white to buff, in terminal hea
less vigorous and has pale yellowish-green felting. A pretty hanging basket plant.
H. serotinum See *H. italicum*.
H. serpyllifolium South Africa
Trailing shrub, densely branched and forming mats to 90cm(3ft) or more wide by about 10–15cm(4–6in) deep. Stems densely white-woolly. Leaves 6–16mm($\frac{1}{4}$–$\frac{5}{8}$in) long, slightly folded upwards and wavy, densely white-woolly beneath, green above with a light coating of white down. Flowers insignificant, seldom produced. Makes an unusual hanging basket subject.
H. splendidum South Africa
Spreading shrub to 30cm(1ft) or more. Leaves to 2cm($\frac{3}{4}$in) long, oblong, with three prominent veins, silvery-woolly on both surfaces. Flower heads very small, yellow, borne in rounded terminal clusters on long stems in summer.

HELICONIA

Heliconiaceae (*Musaceae*)
Lobster claws

Origin: Tropical America. A genus of 80 species of mainly large to giant evergreen perennials. In appearance and relationship they come between *Strelitzia* (bird of paradise flower) and *Musa* (banana). They are mostly clump-forming with often long-stalked, paddle-shaped leaves. When mature they bear erect to pendulous flower spikes, the most conspicuous part being colourful, boat-shaped bracts mostly arranged distichously. In most but not all species, only the tips of the small, 6-tepalled flowers protrude from the bracts. Heliconias are seen at their best when planted out in the tropical conservatory, but also respond to pot culture and when young can be brought into the home as foliage plants. Propagate by seed and offsets in late spring. *Heliconia* derives from Mt. Helicon, Greece, home of the mythological Muses. It seems to be a tongue-in-cheek name used by Linnaeus (its author) to indicate the close relationship to *Musa*.

Species cultivated
H. bihai (*H. distans*) Northern South America, Lesser Antilles
Up to 5m(15ft) or more in height, the leaf blades 90cm(3ft) long or more, oblong-oval, pointed. Floral bracts 15–25cm(6–10in) long, dark red with yellow tips, about six together forming the striking, erect inflorescence. In cultivation there seems to be

confusion between this species and the similar *H. caribaea*, *H. humilis* and *H. wagneriana*.
H. distans See *H. bihai*.
H. latispatha S. Mexico to Central America
Up to 3m(10ft) in height, the broadly oblong-oval leaf blades 90cm(3ft) long. Floral bracts about 15cm(6in) in length, usually orange, but sometimes entirely yellow or red, ten to 20 arranged spirally forming the showy, erect inflorescence. Unlike most heliconias, the inflorescence rises above the leaves.
H. schiedeana Mexico
Similar to *H. latispatha*, but the bracts narrower, entirely red with the small yellow flowers visible. Inflorescence below the leaf blades.

HELIOCEREUS

Cactaceae

Origin: Mexico, Central America. A genus of five to seven species of slender, angled or wing-stemmed cacti. Some species are more or less erect and self-supporting, others have trailing or scrambling stems which, in the wild, grow up through desert scrub. Their funnel-shaped flowers are often large and showy. Propagate by seed in spring or by cuttings in summer. *Heliocereus* derives from the Greek *helios*, the sun and *Cereus*, the genus in which these cacti were once classified. *Cereus* have nocturnal flowers whereas those of *Heliocereus* open in the sun.

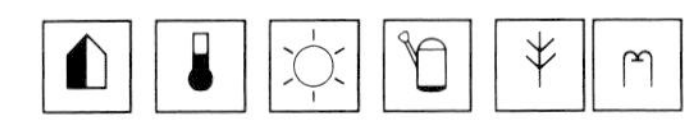

Species cultivated

H. serratus (*Cereus serratus*) Guatemala
Stems erect to 30cm(1ft) or more in height, branched, 4-angled, the angle-margins toothed with some small yellowish spines. Flowers long, purple-red.
H. speciosus (*Cereus speciosus*) Central Mexico
Stems erect to trailing, 90cm(3ft) or more in length, branched, 4-angled, to 5cm(2in) thick, with small whitish or brownish spines. Flowers up to 15cm(6in) long, satiny-scarlet to carmine with a bluish sheen.

HELIOPHILA

Cruciferae

Origin: Southern Africa. A genus of 75 species of annuals, biennials and perennials with 4-petalled flowers in leafless racemes and pod-like fruits. The species described makes a pretty pot plant for the conservatory or the home window sill. Propagate by seed in late autumn and winter for winter and spring flowering, and in spring for summer blooms. *Heliophila* derives from the Greek *helios*, sun and *philos*, I love.

  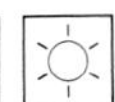  

ABOVE *Heliophila longiflora*
ABOVE LEFT *Heliocereus serratus*
TOP *Heliconia schiedeana*

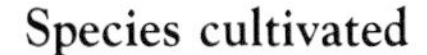

Species cultivated

H. longifolia
Slender, sparingly branched annual, 30–45cm(1–$1\frac{1}{2}$ft) in height. Leaves to 2cm(5in) long, narrowly linear. Flowers 1–2cm($\frac{3}{8}$–$\frac{3}{4}$in) across, bright blue, freely borne in wand-like racemes. A dainty pot plant with the quality of flax. Cool; temperate for winter flowering.

HELIOTROPIUM

Boraginaceae

Origin: Tropics to warm temperate regions. A genus of about 250 species of annuals, shrubs and sub-shrubs with usually alternate, entire leaves and very fragrant flowers in one-sided cymes. The species described below is a good pot plant for the

ABOVE RIGHT *Heliotropium arborescens* 'Marine'
BELOW *Helipterum roseum*
BOTTOM *Hemigraphis repanda*

conservatory and can be brought indoors to a sunny windowsill. Propagate by cuttings from young shoots in spring or mature growth in late summer, also by seeds in late winter. *Heliotropium* derives from the Greek *helios*, sun and *trope*, turning, an allusion to the fallacy that the flowers turn to face the sun.

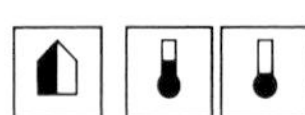 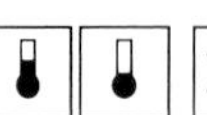 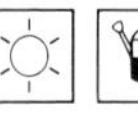 

Species cultivated

H. arborescens (*H. peruvianum*) Common heliotrope Peru
Shrub eventually to 2m($6\frac{1}{2}$ft) or more. Leaves 2.5–8cm(1–3in) long, lanceolate, deep green with very conspicuous veining. Flowers 6mm($\frac{1}{4}$in) long, purple to lavender, also white, in dense cymes, to 7cm($2\frac{3}{4}$in) long. A number of cultivars are grown, some more dwarf in habit, some of varying shades of purple. Not all are strongly fragrant.

HELIPTERUM

Compositae

Origin: South Africa and Australia. A genus of 60 to 90 species of annuals, perennials, shrubs and sub-shrubs with alternate, undivided leaves and rounded heads of disc florets with colourful chaffy bracts closely akin to those of *Helichrysum*. The plant described below makes a pretty pot plant for a sunny conservatory or windowsill, and the 'everlasting' flowers can be dried. Propagate by seed in early spring. *Helipterum* derives from the Greek *helios*, sun and *pteron*, wing; the pappus on the seed is made up of radiating feather-like bristles like a child's drawing of the sun.

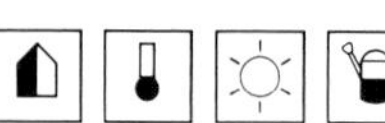

Species cultivated

H. roseum (*Acroclinum roseum*) Western Australia
Annual, 30–60cm(1–2ft) tall, freely branching at the base making an erect bush. Leaves about 6cm($2\frac{1}{2}$in) long, linear to lanceolate. Flowers 2.5–5cm(1–2in) across having bright pink bracts, opening in summer to autumn. *H.r. album* has white bracts; *H.r. plenum* has double flowers.

HELXINE See SOLEIROLIA

HEMIGRAPHIS

Acanthaceae

Origin: Tropical S.E. Asia and Australia. A genus of about 100 species of annuals, perennials and sub-shrubs related to *Ruellia*, only two of which are in general cultivation. These are spreading plants classified as evergreen perennials, but somewhat woody at the base when old. They have opposite pairs of coloured leaves and small, tubular white flowers. Both species are useful foliage plants for the home and conservatory where they provide colour contrast to green and variegated subjects. They can also make good ground cover for the conservatory border and provide something uncommon for a hanging basket. Propagate by cuttings from spring to early autumn. *Hemigraphis* derives from the Greek *hemi*, half and *graphis*, a brush, alluding to the way the hair covers the stamen filaments.

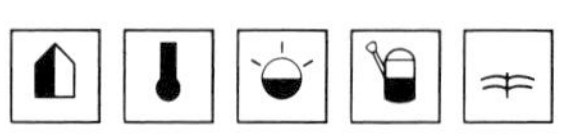

Species cultivated

H. alternata See *H. colorata*.
H. colorata (*H. alternata*) Red/Flame ivy
Probably Malaysia (not India as often stated)
Stems prostrate and rooting, to 30cm(1ft) or more in length. Leaves broadly ovate-cordate, rounded-toothed, somewhat puckered, 6–10cm($2\frac{1}{2}$–4in) long, with metallic blue-purple patina above and strongly red-purple beneath. Flowers 1.2–2cm($\frac{1}{2}$–$\frac{3}{4}$in) long, white with purple lines, in bracted clusters, but relatively inconspicuous. *H.* 'Exotica' has somewhat smaller flowers and a more bushy habit. It was found in New Guinea and may well be a distinct species.
H. repanda Probably Malaysia
Stems prostrate to decumbent, rooting, to 25cm(10in) or more in length. Leaves lanceolate to linear, boldly toothed, up to 6cm($2\frac{1}{2}$in) long, semi-lustrous purple-green above, deep purple beneath. Flowers the same as *H. colorata*.

HEMIONITIS

Hemionitidaceae (*Polypodiaceae*)

Origin: Northern tropical America and tropical Asia. A genus of seven species of evergreen ferns. They are tufted to small clump-forming with neat, comparatively small, rounded to triangular fronds on slender, dark stalks. When young the stalks are comparatively short and the plants compact. When mature, spore bearing or fertile fronds appear which have long slender stalks giving the plant a two-tiered effect. These very distinct ferns are ideal for the shady conservatory and can be grown indoors. *Hemionitis* derives from the Greek *hemionos*, a mule. In ancient times these ferns were worn by women to prevent pregnancy.

Species cultivated

H. arifolia Tropical Asia
Overall height of mature plant about 30cm(1ft). Fronds sharply sagittate, 5–13cm(2–5in) long, smooth, rich green. Stalks black with brown scales.

H. palmata (*Asplenium hemionitis*) Tropical America
Overall height of mature plant 20–30cm(8–12in). Fronds up to 15cm(6in) wide, palmate, deeply cut into five pubescent lobes, each one edged with small, shallow, rounded lobes. Stalks brown.

HEMITELIA See CYATHEA
HEPTAPLEURUM See SCHEFFLERA

HERBERTIA

Iridaceae

Origin: Southern USA, south to Argentina and Chile. A genus of about ten species of cormous perennials allied to *Iris* and *Tigridia*. They have narrow leaves and flowers composed of three large outer tepals, three very much smaller inner tepals and a 3-lobed style, each lobe may be twice or three times divided. The species described here make pretty pot plants for the cool conservatory and can be brought indoors when in bloom. Propagate by seed in spring or by removing offsets when dormant (late autumn to spring). *Herbertia* honours Dr William Herbert (1778–1847), Dean of Manchester and an amateur botanist of high repute with a special knowledge of bulbous plants.

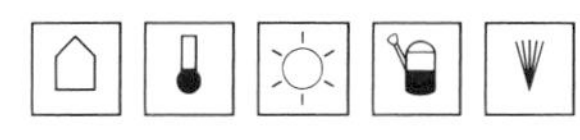

Species cultivated

H. drummondii (*Cypella drummondii*, *Alophia drummondiana*) Texas
Leaves 15–30cm(6–12in) long, lanceolate to ovate, pleated. Flowers about 5cm(2in) wide, the outer tepals white at the base and lavender at the tip with a dark purple zone between; inner tepals small, purple; spring.

H. platensis Argentina
Leaves sword-shaped, glaucous, 60cm(2ft) or so long. Flowers to 10cm(4in) wide, china-blue, the three outer petals arching down, on stems 90–120cm(3–4ft) in height; early summer to autumn.

H. pulchella Southern Brazil
Leaves narrowly sword-shaped, 13–25cm(5–10in) long, strongly ribbed. Flowers to 8cm(3in) wide, the outer tepals light blue-purple with yellow and white bases thickly purple-spotted, on stems just topping the leaves in early summer.

HERMODACTYLUS See IRIS
HERPESTES See BACOPA
HESPEROSCORDUM See BRODIAEA
HESPEROYUCCA See YUCCA

HETEROCENTRON

Melastomataceae

Origin: Mexico and Central America. A genus of 12 to 27 species (depending upon the botanical authority consulted) of shrubs, sub-shrubs and perennials. The only species in general cultivation is described below. It provides colourful ground cover for the conservatory border and looks splendid in a large hanging basket. Propagate by cuttings in spring and summer. *Heterocentron* derives from the Greek *heteros*, variable or diverse and *kentron*, a spur or point, referring to the dimorphic stamens, the larger ones having bristle-like appendages or spurs.

Species cultivated

H. elegans (*Heeria elegans*, *Schizocentron elegans*) Mexico, Guatemala, Honduras
Stems trailing, forming mats to 60cm(2ft) or more wide, woody-based when old. Leaves in opposite pairs, ovate, to 1.2–2.5cm($\frac{1}{2}$–1in) long, smooth and usually rich green. Flowers 2.5cm(1in) wide, composed of four, rounded, spreading petals, bright purple-rose, often produced in profusion during the summer. Classified as a sub-shrub, but best considered as a perennial.

BELOW *Hemionitis arifolia*
BOTTOM *Heterocentron elegans*

BELOW *Hibbertia scandens*
BOTTOM LEFT *Hibiscus rosa-sinensis*
BOTTOM RIGHT *Hibiscus schizopetalus*

HIBBERTIA

Dilleniaceae

Origin: Malagasy, New Guinea, New Caledonia and Fiji, but mainly Australia. A genus of about 100 species of evergreen shrubs and twining climbers. They have alternate, simple leaves and widely expanded, 5-petalled, usually yellow flowers. The species described are best in the conservatory, though the climbing species will grow in the home where room permits. Propagate by seed or layering in spring, or by cuttings in summer. *Hibbertia* honours George Hibbert (1757–1837), founder of a private botanic garden at Clapham and who sent James Niven to collect new plants in South Africa.

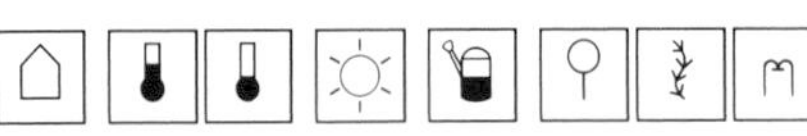

Species cultivated

H. cuneiformis Western Australia
Fairly bushy shrub, easily kept to around 90–100cm(3–4ft) in a container (twice or three times this height in its native eucalyptus forest home). Leaves lanceolate-cuneate, 2–4cm($\frac{3}{4}$–$1\frac{1}{2}$in) long, prominently toothed at the tips, smooth deep green. Flowers to 3cm($1\frac{1}{4}$in) wide, bright mid-yellow.

H. dentata Eastern Victoria to Queensland
Stems trailing or climbing, 2–3m($6\frac{1}{2}$–10ft) or so in length. Leaves ovate to narrowly so, distantly toothed, up to 6cm($2\frac{1}{2}$in) long, often red when young. Flowers 2.5–4cm(1–$1\frac{1}{2}$in) wide, deep to pale yellow.

H. scandens (*H. volubilis*) Guinea flower, Snake vine Queensland, New South Wales
Vigorous climber to 6m(20ft) or more in its native country, but easily kept to one-third of this in a container. Leaves 4–9cm($1\frac{1}{2}$–$3\frac{1}{2}$in) long, elliptic to oblanceolate, almost leathery, lustrous deep green. Flowers about 4cm($1\frac{1}{2}$in) wide, sometimes more, bright yellow, borne mainly in summer, but intermittently throughout the year if warm enough.

H. volubilis See *H. scandens*.

HIBISCUS

Malvaceae

Origin: Tropics and sub-tropics. A genus of 250 to 300 species of annuals, perennials, shrubs and trees with alternate, often palmately lobed leaves and 5-petalled flowers which are usually very showy. The species described below are free-flowering shrubs for the sunny conservatory and while still small can be used in the home in a sunny window. Propagate by cuttings from late spring to late summer with bottom heat. *Hibiscus* is the Greek name for mallow, applied by Linnaeus to this closely related genus.

Species cultivated

H. rosa-sinensis Rose of China Tropical Asia
Evergreen shrub to 2m($6\frac{1}{2}$ft) or so in a pot. Leaves to 15cm(6in) long, ovate, often coarsely toothed,

glossy green. Flowers 10–15cm(4–6in) across, often more in selected cultivars, red in the type species, pink, apricot, orange, yellow and white in the many named cultivars grown chiefly in tropical and sub-tropical regions.

H.r. cooperi
Narrower leaves variegated with pink and white.

H. schizopetalus Japanese lantern East Africa
Evergreen shrub 2–3m(6½–10ft) tall, with slender, somewhat arching branches. Leaves 2–13cm(¾–5in) long, ovate, toothed. Flowers to about 6cm(2½in) across, pink, with deeply fringed reflexed petals and a long stamen column; they are pendent, borne on long stalks in summer.

HIPPEASTRUM

Amaryllidaceae

Origin: Central and South America. A genus of 75 species of bulbous plants with tufts of strap-shaped leaves and bold 6-petalled, funnel-shaped, lily-like flowers on erect, leafless stems. They are suitable for home or conservatory, and need to be dried off when the leaves begin to turn yellow in summer. Propagate by offsets or by seed sown in spring in warmth. If kept without a dry resting period the seedlings will mature and flower more quickly, but must then be rested after flowering. *Hippeastrum* derives from the Greek *equus*, a horse and *hippeus*, rider, because the flower buds within their spathes fancifully suggest a horse and rider.

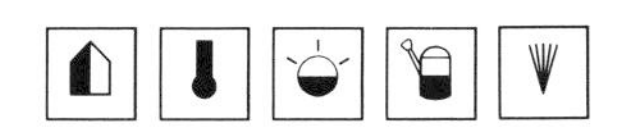

Species cultivated

H. × ackermannii (*H.* .× *acramanii*)
'Amaryllis'
This name covers the popular large-flowered hybrid cultivars derived from *H. aulicum*, *H. elegans*, *H. reginae*, *H. reticulatum* and *H. striatum*, now so freely available in garden centres and multiple stores. They are vigorous plants with thick-textured, dark green leaves to 60cm(2ft) long, robust flowering stems 45–60cm(1½–2ft) or more tall and umbels of two to four huge lily-like flowers in shades of red, pink, orange and white, sometimes striped, opening either just before or with the young leaves in late winter or spring. Both named cultivars and un-named seedlings are commercially available.

H. advenum Chile
Leaves about 30cm(1ft) long, linear, arching, usually somewhat glaucous. Flower stem erect, about as long as the leaves, bearing two to six blooms; each flower, borne in winter, is about 5cm(2in) long, composed of six narrow tepals; red is the usual colour, but a yellow form is known.

H. brachyandrum See *Habranthus brachyandrus*.

H. equestre (*Amaryllis equestre*, *Hippeastrum puniceum*) South America
Leaves to 45cm(1½ft) long by 5cm(2in) wide. Flowering stem strongly erect from 30–60cm(1–2ft) high, carrying two to four blossoms; these are

Hippeastrum × *ackermanii* 'Red Lion'

10cm(4in) or more wide, bright red with green at the base, opening in winter and spring before the leaves develop.

H. leopoldii Peru
Leaves arching, strap-shaped, 45–60cm(1½–2ft) long, rich green. Flowering stems sturdy, erect, to 45cm(1½ft) tall, sometimes more, bearing usually one or two blooms; each flower 15–18cm(6–7in) wide, formed of six broad red, white and green tepals, opening in spring or summer.

H. pratense Chile
Leaves linear, arching, dark green, to 25cm(10in) long. Flowering stem erect, up to 30cm(1ft) in height, bearing two to four blooms; each flower bright red, 5–8cm(2–3in) wide, opening with the young leaves in late spring.

H. procerum (*Worsleya rayneri*) Blue amaryllis Brazil
Leaves strap-shaped, usually about 45–60cm(1½–2ft) in length, but sometimes much more, or less. Flowering stem robust, two-edged, erect, 30–45cm(1–1½ft) tall, bearing two to six, occasionally up to twice as many flowers; each bloom is widely funnel-shaped, to 13cm(5in) or more long, composed of six comparatively narrow, wavy, white-based, lilac to mauve-blue tepals generally with darker spots. August is the usual flowering period, but it can be earlier or later.

H. psittacinum Brazil
Leaves up to 60cm(2ft) long, arching, strap-shaped, with a light glaucous patina. Flowering stem erect, robust, 60–90cm(2–3ft) tall, bearing two to four blooms; each flower is 10–13cm(4–5in) long, funnel-shaped, strikingly striped with green and crimson; they open with the leaves in summer.

Hippeastrum × *striatum* 'Crocatum'

H. reticulatum (*Amaryllis reticulata*) Brazil
Leaves to 30cm(1ft) or more long. Flowering stems to 30cm(1ft) tall bearing five to six flowers; these are 7cm($2\frac{3}{4}$in) wide, pink or purplish-red, sometimes white, but always with the darker veining, opening in autumn.
H. rutilum See *H. striatum*.
H. striatum (*H. rutilum*) Brazil
Leaves broadly strap-shaped, up to 30cm(1ft) in length, bright green. Flowering stem erect, about the same length as the leaves, bearing two to four blooms; each flower about 10cm(4in) long, glowing crimson, tepals with a prominent green keel in the lower half, opening usually in late spring. *H.s.* 'Crocatum' has smaller, pink blooms; 'Citrinum' is a rare yellow and 'Fulgidum' is the best crimson-flowered selection.
H. vittatum Peru
Leaves 60–90cm(2–3ft) long, broadly strap-shaped, arching, rich green. Flowering stems erect, robust, about the same length as the leaves, bearing two to six blooms; flowers widely funnel-shaped, 13–15cm(5–6in) wide, white with bold red stripes, opening in spring.

HIPPOBROMA See ISOTOMA

HOFFMANNIA

Rubiaceae

Origin: Mexico to northern Argentina. A genus of 100 species of shrubs and perennials some of which are grown for their handsome foliage. They have opposite pairs of prominently veined leaves and rather insignificant 4-, rarely 5-petalled flowers in small axillary clusters. Hoffmannias are excellent foliage plants for the warm conservatory and worth trying in the home at least in the short term. Propagate by cuttings in summer. *Hoffmannia* commemorates George Franz Hoffmann (1761–1826), a Dutch Professor of Botany in turn at both Gottingen and Moscow.

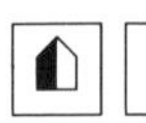 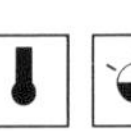   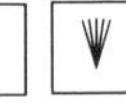

Species cultivated

H. ghiesbreghtii Southern Mexico and Guatemala
Shrub eventually to 1.2m(4ft) in height, but easily grown to half this in pots. Stems 4-angled or winged; leaves oblong-lanceolate to 30cm(1ft) long, deep velvety bronze-green above with silvery or pink veins, red beneath. Flowers yellowish almost hidden by the leaves. *H.g.* 'Variegata' has the leaves irregularly marked pink and cream.
H. refulgens Mexico, Central America
Shrubby perennial to 30cm(1ft) or more tall; stems cylindrical often hairy. Leaves obovate to 15cm(6in) in length, corrugated between the veins, coppery, purplish or olive-green with a metallic lustre above, purple-red beneath. Flowers bright red, but not displayed to advantage. *H.r.* 'Vittata' has the veins and margins of the upper leaf surfaces picked out in grey.
H. roezlii Mexico
Much like *H. refulgens* and possibly only a variant of it, with broader leaves up to 20cm(8in) in length which are lighter in tone and more iridescent. Flowers darker red.

HOMALOCEPHALA See ECHINOCACTUS
HOMALOCLADIUM See MUEHLENBECKIA

RIGHT *Hoffmannia ghiesbreghtii* 'Variegata'
FAR RIGHT *Hoffmannia roezlii*

HOMALOMENA

Araceae

Origin: Tropical Asia and South America. A genus of about 130 species of evergreen perennials, those described below being grown for their ornamental leaves. They are clump-forming with mostly, but not entirely, arrowhead to heart-shaped leaves. The tiny petalless flowers are enclosed in a usually greenish spathe hidden by the leaves. Propagate by division or seed in spring. *Homalomena* reputably derives from a mis-rendered Malayan vernacular name translated into the Greek *homalos*, flat and *mene*, moon, which appears to make little sense.

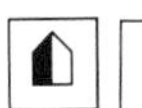    

Species cultivated

H. lindenii New Guinea
To about 75cm($2\frac{1}{2}$ft) in height, the leaf-blades triangular-ovate, with a long, slender point, bright green with all the main veins picked out in yellow.

H. sulcata Borneo
To 45cm($1\frac{1}{2}$ft) or more in height, the leaf blades broadly ovate (almost triangular-ovate), 13–20cm(5–8in) long, green above, coppery beneath.

H. wallisii Venezuela
To about 15cm(6in) in height, but with a much greater spread. Leaf blades elliptic to ovate-oblong, 13–20cm(5–8in) in length, on short stalks, arching or recurving, bright green blotched and marbled yellow. Very distinct from the species above, being much like an *Aglaonema* in general appearance.

HOMERIA

Iridaceae

Origin: South Africa. A genus of about 40 species of cormous plants grown for their elegant and colourful blooms. They have solitary or sparse tufts of sword-shaped leaves and widely expanded, 6-petalled flowers in loose branching clusters. Each flower lasts less than a day, but several to many buds open successionally. Propagate by seed in spring or by offsets when dormant. Re-pot annually in autumn or early spring. Best in the conservatory, but can be brought indoors when in bloom. *Homeria* derives from the Greek *homereo*, to meet, referring to the stamen filaments which are joined together to form a tube. Not commemorating the Greek poet Homer as sometimes stated.

 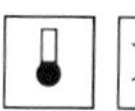 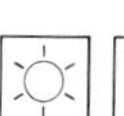  

Species cultivated

H. collina (*H. breyniana*) Cape tulip
Leaves two to four per corm, about 60cm(2ft) in length when mature, linear, arching to recurved. Flowering stems erect, about 45cm($1\frac{1}{2}$ft) in height, the tips bearing slender, pointed, spindle-shaped green bracts which shelter the flower buds; flowers to 5cm(2in) or more wide, fragrant, in shades of pink, yellow or cream, borne in summer. *H.b.* 'Aurantiaca' has pink petals with a narrow basal blotch.

H. lilacina
A half-sized version of *H. collina* with lilac flowers, each petal having purple veins and a yellow-speckled basal purple blotch.

LEFT *Homalomena wallisii*
BELOW LEFT *Homeria collina* 'Aurantiaca'

HOODIA

Asclepiadaceae

Origin: South Africa, Namibia, Angola. A genus of 100 species of succulents allied to *Stapelia*. They form clumps of more or less erect, cylindrical, fleshy stems thickly set with rows of pointed tubercles on low ribs. The comparatively large bowl- to bell-shaped flowers are borne near the tips of the stems.

RIGHT *Howeia forsterana*
BELOW *Hoodia macrantha*

Hoodias add interest and variety to a collection of succulents. Propagate by individual stems taken as cuttings or by careful division in late spring. *Hoodia* honours a Mr. Hood, a keen collector of succulent plants around the mid 19th century, but of whom nothing else seems to have been recorded.

Species cultivated

H. bainii Cape, Namibia
Stems to 20cm(8in) tall by 2.5–4cm(1–$1\frac{1}{2}$in) thick, with 12 to 15 rows of spine-tipped tubercles. Flowers about 6cm($2\frac{1}{2}$in) wide, bell-shaped, matt yellow, reddening as they fade, borne in summer.

H. gordonii Cape, Namaqualand, Namibia
Stems grey-green, to 30cm(1ft) or more in height by 5cm(2in) thick with up to 14 rows of spine-tipped tubercles. Flowers 8–10cm(3–4in) wide, saucer-shaped, pale purple with yellow stripes, borne in summer.

H. macrantha Namibia
In the wild, this is the largest species with stems 90cm(3ft) or so in height, but in cultivation often only half this. Tubercles spine-tipped in about 15 rows. Flowers are the largest in the genus, reputedly to 20cm(8in) wide, but usually less, bowl-shaped, light purple with yellowish veins, the lobes darker purple, hairy, borne mainly in summer.

HOOKERA See BRODIAEA

HOWEIA (HOWEA)

Palmae

Origin: Lord Howe Island, S.E. Pacific. A genus of two species of palm, growing to 20m(65ft) in the wild, having erect stems bearing arching, pinnate leaves. They make very good pot or tub plants when small. Propagate by seed in spring in heat. *Howeia* was named after Lord Howe Island where they are found.

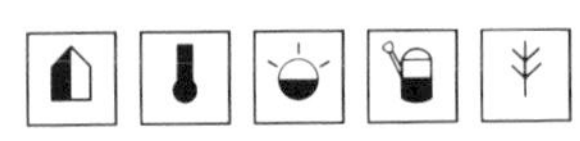

Species cultivated

H. belmoreana (*Kentia belmoreana*) Curly palm, Sentry palm
To 2–3m($6\frac{1}{2}$–10ft) tall in pots. Leaves to 1m(3ft), strongly arching, short-stalked; leaflets linear-lanceolate, pointed outwards.

H. forsterana (*Kentia forsterana*) Kentia palm, Sentry palm, Thatchleaf palm, Paradise palm
To 2–3m($6\frac{1}{2}$–10ft) tall in pots. Leaves less strongly arching, long-stalked, leaflets linear-lanceolate, pointed downwards.

HOYA

Asclepiadaceae

Origin: S.E. Asia and the Pacific Isles. A genus of 200 species of evergreen climbers, some epiphytic, with opposite pairs of fleshy, entire leaves and waxy, 5-petalled flowers in large, pendent umbels. They are good house and conservatory plants, the climbers needing a support of canes or wires, the more shrubby sorts, particularly *H. bella*, being best in a hanging basket. Propagate by cuttings in warmth in summer. *Hoya* was named for Thomas Hoy (*d.* 1809), Head Gardener to the Duke of Northumberland at Syon House, Isleworth.

Species cultivated

H. australis Australia (Queensland, New South Wales), New Guinea
Climber to 5m(16ft) or so, but less in containers. Leaves rounded to broadly obovate, about 8cm(3in) long, firmly fleshy, rich green. Flowers fragrant, about 1.2cm($\frac{1}{2}$in) wide, white with five red-purple spots in the centre, up to 50 in each umbel, opening in summer. Tropical.

H. bella Miniature wax plant India
Epiphytic, branching shrub to 45cm($1\frac{1}{2}$ft) tall, with arching to pendulous stems. Leaves to 3cm($1\frac{1}{4}$in) long, ovate-lanceolate. Flowers 1cm($\frac{3}{8}$in) wide, white with a red centre, in umbels of eight to ten, opening in summer. Tropical.

H. carnosa Wax plant S. China to Northern Australia
Climber to 6m(20ft) or more. Leaves 5–8cm(2–3in), ovate to obovate. Flowers 1.5cm($\frac{5}{8}$in) wide, white ageing to pink with a pink centre; they are

very fragrant, especially at night, and are borne in large umbels from late spring to autumn, the same umbel often producing several flushes of bloom. *H.c.* 'Variegata' has cream-edged leaves. Temperate.

H. cinnamomifolia Java
Stems twining, to 3m(10ft) or so in containers, up to twice this planted out. Leaves 8–13cm(3–5in) long, ovate, slender-pointed, thick-textured, with a very thick stalk. Flowers about 16mm(⅔in) wide, bright yellow-green with a deep magenta centre, carried densely in almost globular umbels and expanding in summer. Tropical.

H. coronaria Java
Slow-growing, thick-stemmed climber, rarely more than 2–3m(6–10ft) in containers. Leaves broadly oblong to oval, leathery-textured, downy beneath, up to 15cm(6in) long. Flowers 2.5–3.5cm(1–1⅜in) across, widely bell-shaped, yellow to white with five red spots in the centre, carried in axillary umbels in summer. One of the larger-flowered hoyas and distinct with its cupped petals. Tropical.

H. imperialis Borneo
Vigorous climber to 6m(20ft) or so, but less in containers. Leaves 10–22cm(4–9in) in length, narrowly oblong to elliptic, somewhat downy, leathery-textured. Flowers up to 8cm(3in) wide – and the largest of the cultivated species – brown-purple to deep magenta with a cream or yellowish centre, eight to 12 in each umbel, opening in summer. Tropical.

H. kerrii Thailand and Laos
Climber to 3m(10ft), less in containers. Leaves dark green, hairy, broadly obovate to orbicular-cordate, with a notched tip, up to 10cm(4in) or more in length. Flowers 1.2cm(½in) wide, white with a rose-purple centre, carried in many-flowered umbels during summer. Tropical.

H. longifolia Central Himalaya
Climber to 3m(10ft) or so, less in containers. Leaves oblanceolate, up to 20cm(8in) long, thickly fleshy but firm, borne almost vertically on thick stalks. Flowers 2–4cm(¾–1½in) across, white or pink flushed with deeper pink or reddish centres, opening in summer. Tropical.

H.l. shepherdii
Thinner-stemmed and smaller growing with shorter, relatively narrower, channelled leaves and 1.2–2cm(½–¾in) wide flowers.

H. multiflora Malacca
Technically a climber, but in containers it tends to form a loose bush 30–90cm(1–3ft) high and wide. Leaves elliptic, to 10cm(4in) long, bright green. Flowers with rolled margins, swept-back petals, about 2.5cm(1in) long, straw-yellow with a white or brownish centre; they are arranged in loose umbels which look just like a flight of diminutive shuttlecocks in late summer. Tropical.

H. purpureo-fusca (*H.* 'Silver Pink') Java
Much like *H. cinnamomifolia*, but the leaves with a raised silvery-pink mottle and the flowers having greyish to reddish-brown, white-hairy petals with a pink centre. Tropical.

Hoya carnosa

LEFT *Hoya australis*
FAR LEFT *Hoya bella*

HUERNIA

Asclepiadaceae

Origin: Southern and eastern Africa, Ethiopia and Arabia. A genus of about 30 species of succulents related to *Stapelia* and *Hoodia*. The stems are much like those of *Hoodia*, but with fewer and more prominent ribs. The flowers also are similar, but have bell-shaped bases and arise at the base or below the middle of the young stems. They make an interesting addition to a succulent collection, but beware of the slight carrion odour given off by the flowers. Propagate by seed, cuttings or by careful division in late spring. *Huernia* honours Justin Heurnius (1587–1652), a Dutch missionary and reputedly the first collector of South African (Cape) plants. His name was mis-spelt by the describing botanist, R. Brown.

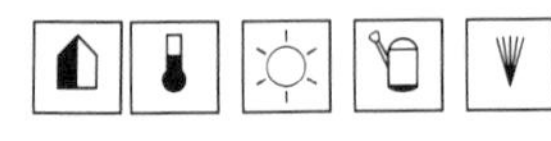

Species cultivated

H. hystrix Natal, Orange Free State, etc.
Stems to 12cm(4¾in) tall, erect to ascending, sometimes decumbent, grey-green. Ribs usually five edged with pointed tubercles. Flowers to 4cm(1½in) wide, broadly bell-shaped, yellow with transverse broken red lines.

H. macrocarpa Ethiopia
Stems to 10cm(4in) long, with five ribs edged with large, pointed tubercles. Flowers bell-shaped, 2cm(¾in) wide, greenish-yellow outside, yellow within bearing fine, brown transverse stripes. *H. m. cerasina* has flowers coloured red inside; *H. m. flavicoronata* is maroon and yellow; and *H. m. penzigii* has larger blooms entirely dark maroon within.

H. pillansii Cape
Stems to 4cm(1½in) tall and almost half as thick, densely tufted. Ribs to 15 or more, shallow but thickly set with pointed, bristle-tipped tubercles, dark green or purple-tinted. Flowers up to 3cm(1¼in) wide, the five long-pointed lobes creating a starfish outline, pale yellow, enlivened with many red papillae.

H. primulina Cape
Stems light grey-green, sometimes red-spotted, up to 8cm(3in) tall by 1.2cm(½in) thick. Ribs four to five, set with teeth-like tubercles, the pointed tips of which are deciduous. Flowers bell-shaped with five broadly triangular lobes, palest primrose to rich yellow, the small basal corona dark purple, borne in autumn to winter. One of the most attractively flowered species.

H. zebrina Owl eyes Transvaal, Natal, Namibia
Stems to 8cm(3in) tall by 2.5cm(1in) thick, 5-angled, each angle bearing spreading teeth-like tubercles. Flowers to 4cm(1½in) or more wide with a thick, convex annulus around the small bell-shaped eye, lobes spreading, sharply triangular, yellow to greenish-yellow banded and spotted brownish-purple.

Huernia sp.

HUMEA

Compositae

Origin: Southern Australia and Malagasy. A genus of seven species of biennials, perennials and shrubs which look remarkably unlike the average member of the composite (daisy) family. The biennial species described below makes a statuesque specimen plant for a conservatory. Where there is sufficient room it can be brought indoors. Propagate by seed in late summer. *Humea* was named for Lady Amelia Hume (1751–1809) of Wormleybury, Hertfordshire, an amateur botanist and gardener.

Species cultivated

H. elegans
Flowering plants can reach 2–2.5m(6½–8ft) in height, but are often less than this; main stem unbranched and woody below. Leaves to 30cm(1ft), oblong to ovate-lanceolate, rather wrinkled, the bases clasping the stem, decreasing in size up the stem. At the top are the tiny, reddish-pink flowers which are borne in profusion in loose, pendent panicles 40–90cm(20–36in) long, giving the plant a very graceful appearance; they are produced in summer and are long-lasting.

HUMULUS

Cannabidaceae (Cannabaceae)

Origin: Northern hemisphere. A genus of two species of twining perennials with rough stems and opposite pairs of broad, lobed leaves. They are dioecious, the small male flowers in loose panicles, the females in short, cone-like spikes with green bracts. Both species have variegated forms which are very decorative foliage plants for the back wall of a conservatory. Propagate by seed or cuttings in spring. They can be grown as annuals or deciduous perennials, needing a cool resting period in winter. Hardy. *Humulus* is the Latin form of an old low German name for the hop.

  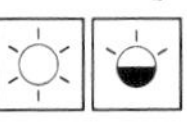  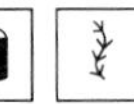

Species cultivated

H. japonicus Japanese hop Japan, China and adjacent islands
To 6m(20ft) or more in the open. Leaves 5–13cm(2–5in) long and wide, 5- to 7-lobed, the upper ones 3- to 5-lobed, carried on stalks more than the length of the leaf. Flowers borne in late summer, the female 'hops' to 1cm(⅜in) long, dark green. *H.j.* 'Variegatus' has leaves splashed and streaked with white.

H. lupulus Common hop North temperate zone
Stems to 9m(30ft) in the open. Leaves to 15cm(6in) long, 3- to 5-lobed and coarsely toothed, carried on stalks shorter than the leaf length. Flowers borne in late summer, the female 'hops' to 2cm(¾in) long, pale green to straw-coloured when mature. *H.l.* 'Aurea' with leaves flushed golden-yellow is very decorative.

LEFT *Humulus japonicus* 'Variegatus'
BELOW *Hyacinthus orientalis* 'Pink Pearl'

HYACINTHELLA See PSEUDOMUSCARI

HYACINTHUS

Liliaceae
Hyacinth

Origin: Central and eastern Mediterranean. A genus now considered to contain only one species, others once classified here now being included in *Brimeura, Bellevalia, Hyacinthella* and *Pseudomuscari.* It is a bulbous species from which many robust cultivars have been bred, with denser spikes of flowers and in a range of colours. They are favourite late-winter blooming plants for the home and conservatory. Pot the bulbs as soon as received, keep in a cool place until a good root system has been built up and the shoots are approaching 5cm(2in) tall, then bring into a warm room, but in a temperature not above about 16°C(61°F) for good growth and flowers. Propagate by removing offsets when dormant. Seed can be sown, but the resulting plants are rarely as good as their parents. *Hyacinthus* derives from an early pre-Greek language in which *hyakinthos* denoted blue, and not, it is now believed, from the youth of that name who was accidentally killed by Apollo with a discus.

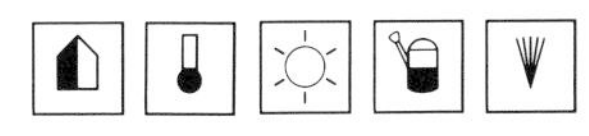

Species cultivated

H. orientalis Common hyacinth
Leaves 25–35cm(10–14in) long, strap-shaped, fleshy, deeply channelled, to 2.5cm(1in) or more in width. Flowers 2–3cm(¾–1¼in) long, bell-shaped, the six tepals widely spreading to reflexed at the mouth, very fragrant, borne in a lax to dense raceme on 20–30cm(8–12in) tall stems in late winter to spring. Represented in cultivation by many named cultivars (Dutch hyacinths) including: 'City of Haarlem', yellow flowers; 'Delft's Blue', dark blue; 'Innocence', the best white; 'Lord Balfour', looser racemes, but of an almost magenta-pink; 'Orange Boven', an unusual shade of apricot; and 'Pink Pearl', bright, clear pink.

H.o. albulus Roman hyacinth
Smaller stems and flowers, several flowering stems rising from each bulb. It is nearer to the original wild type and is found in southern France.

HYDRANGEA

Hydrangeaceae (Saxifragaceae)

Origin: Northern temperate zone, Central and South America. A genus of 23 to 80 species (depending upon the classification followed). They are shrubs, climbers and small trees, some evergreen, others deciduous. All have opposite pairs of entire leaves which can be minutely or deeply toothed. The flowers are made up of normal fertile florets and much larger sterile ones; all have four to five petals and are carried in corymbs or panicles at the ends of the branches. One species is frequently grown as a pot plant for its large trusses of pink, blue or white flowers. Propagate by cuttings in late summer, and over-winter the plants outside in a cold frame if possible. Bring them into warmth in winter and they will leaf and flower. Afterwards return them to the garden or a cool place treating them very much as short-term pot plants. For larger plants, cuttings can be taken in spring, potting on through the summer and pinching out the tips at three to four leaves to make a bushy plant. Overwinter in a frame if possible and bring indoors in late winter as before. *Hydrangea* derives from the Greek *hydro*, water and *aggos*, a jar, referring to the tiny cup-shaped seed capsules.

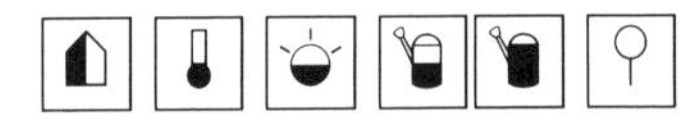

Species cultivated

H. macrophylla (*H. hortensis*) Common/ French hydrangea Japan
Deciduous shrub. Leaves 10–15cm(4–6in) long, broadly ovate to obovate, fresh green. Flowers are produced in two combinations, all sterile, or mixed as in the wild type. *H.m. macrophylla* cultivars have almost all sterile florets and are known as Hortensias (Mop-headed type). *H.m. serrata* cultivars have a flat corymb of small, fertile flowers surrounded by a ring of large-petalled sterile ones; these are the 'lacecaps'. Cultivars listed as blue need an acid compost to produce this colour, those listed as pink will be mauve without the addition of lime. Hardy.

BELOW *Hydrangea macrophylla serrata* cultivar
BOTTOM LEFT *Hydrangea macrophylla macrophylla* cultivar
BOTTOM RIGHT *Hylocereus undatus*

HYLOCEREUS

Cactaceae

Origin: West Indies and Mexico to Peru. A genus of about 20 species of climbing epiphytic or semi-epiphytic cacti. They have comparatively slender, angled or winged rooting stems and large trumpet-shaped flowers which open at dusk. Equally amenable to culture in the home or conservatory, they make spectacular specimens when in bloom, especially if there is room for them to grow large. Propagate by cuttings in summer. A bromeliad compost gives the best results. *Hylocereus* derives from the Greek *hyle*, the wood, and the allied genus *Cereus*; the genus inhabits tropical woodland.

  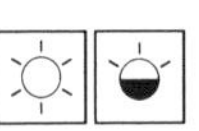  

Species cultivated

H. lemairei Trinidad, Tobago
Stems to 3m(10ft) or more in length if planted out, triangular in cross-section, rooting from one face only, deep green. Spines small, one or two from the well spaced areoles. Flowers about 25cm(10in) in length, white outside, reddish-flushed inside. Fruits ovoid, to 8cm(3in) long, purple and edible.

H. triangularis See under *H. undatus*.

H. undatus Queen of the night Tropical America; exact provenance unknown, much planted and naturalized in other tropical countries
Stems 3–5m(10–15ft) in length, 3- or sometimes only 2-winged with horny, roundly-notched margins. Spines small, singly or in tufts of two to five. Flowers in late summer 25–35cm(10–14in) long, white from yellow-green buds. Fruits ovoid, red, to 10cm(4in) or more long, edible and tasty. The best known species, but sometimes confused with *H. triangularis*, a smaller plant with softly-angled, not hornily-winged, stems and flowers up to 20cm(8in) long.

HYMENOCALLIS

Amaryllidaceae
Spider lilies

Origin: Southern USA, Central and South America. A genus of about 30 species of bulbous plants with strap-shaped leaves and 6-tepalled flowers like yellow or white spidery daffodils. They are borne in terminal umbels on erect, leafless flowering stems. Plants are best in a conservatory and brought into the home when flowering. Deciduous species must be kept barely moist once the leaves begin to yellow and until new leaves are well developed in spring. *Hymenocallis* derives from the Greek *hymen*, a membrane and *kallos*, beauty, referring to the corona which is made up of joined membraneous outgrowths from the stamens.

    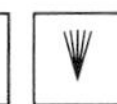

Species cultivated

H. americana See *H. littoralis*.
H. calathina See *H. narcissiflora*.
H. caribaea (*Pancratium caribaeum*) West Indies
Leaves evergreen, 30–60cm(1–2ft) or more in length, more or less erect, broadly strap-shaped, glossy. Flowering stem erect, robust, 2-edged, about 30cm(1ft) in height, carrying six to ten flowers; each bloom white, fragrant, with a tube to 6cm(2½in) long, a funnel-shaped corona to 2.5cm(1in) and six linear petals to 10cm(4in) long. Mid to late summer is the usual flowering time. This species can be, and is, sometimes confused with the allied *H. latifolia*, *q.v.*
H. × festalis (*Ismene × festalis*)
Hybrid between *H. longipetala* and *H. narcissiflora*, very like the latter, but with longer stamens.
H. harrisiana Mexico
Leaves evergreen, broadly strap-shaped, tapering to a short stalk-like base, more or less erect, about 30cm(1ft) long. Flowering stem 25cm(10in) long, bearing several blooms; each flower has a slender, greenish tube to 13cm(5in) long, a funnel-shaped corona of 2cm(¾in) long and six narrow white petals, expanding in summer.
H. latifolia Florida, West Indies
Much like *H. caribaea*, but with arching, evergreen leaves and the flowers having the tube longer than the petals.
H. littoralis (*H. americana*; *H. senegambica* in part) Tropical America, naturalized in West Africa
Leaves evergreen, strap-shaped, widely arching, up to 90cm(3ft) long, bright green. Flowering stems 45–75cm(1½–2½ft) tall, bearing four to eight or more white flowers; each bloom has a green-flushed tube about 15cm(6in) long, a funnel-shaped corona to 5cm(2in) and six linear petals 8–13cm(3–5in) in length. Late spring to summer is the flowering season.
H. × macrostephana
Hybrid of *H. narcissiflora* and *H. speciosa* which is semi-evergreen. Leaves 60cm(2ft) or more long, bright green. Flowering stems to 60cm(2ft), bearing fragrant blooms with staminal cups to 7cm(2¾in) long in summer.
H. narcissiflora (*H. calathina*) Basket flower, Peruvian daffodil Peru, Bolivia
Leaves to 60cm(2ft) long. Flowering stems also to 60cm(2ft) tall, carrying blooms with a 5cm(2in) long corona from which the stamens project a further 2.5cm(1in), opening in spring and summer.
H. speciosa West Indies
Leaves 45–60cm(1½–2ft) long, deep green, tapering at the base to appear stalk-like. Flowering stems to 40cm(16in) tall, carrying umbels of seven to 12 pure white, very fragrant blooms with staminal cups 3–4cm(1¾–1½in) long, opening in summer. Best with tropical temperatures.

HYPOCYRTA See NEMATANTHUS

ABOVE *Hymenocallis narcissiflora*
TOP *Hymenocallis littoralis*

Hypoestes phyllostachya

HYPOESTES

Acanthaceae

Origin: Tropical Africa and Asia, especially Malagasy. A genus of 40 to 150 species (depending upon the botanical authority consulted). It comprises evergreen shrubs, sub-shrubs and perennials with opposite pairs of simple leaves sometimes colourfully variegated, and terminal spike-like clusters of tubular, 2-lipped flowers. The species described are popular pot plants. Propagate by cuttings from spring to late summer or by seed in spring. *Hypoestes* derives from the Greek *hypo*, under and *estia*, a house, a slightly obscure reference to the way the floral calyces are covered by bracts.

  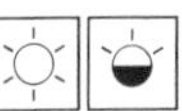  

Species cultivated

H. aristata South Africa
Sub-shrub to 90cm(3ft) or more in height usually grown as a perennial and then rarely above 60cm(2ft). Leaves 5–8cm(2–3in) long, ovate, plain soft green. Flowers about 2.5cm(1in) long, rose-purple, in long, tiered spikes. One of the best of the flowering hypoestes in cultivation, especially useful for its late winter to spring season.

H. phyllostachya Polka dot plant, Freckle face Malagasy
Sub-shrubby perennial to 60cm(2ft) in height, but kept to less if pruned annually. Leaves up to 6cm(2½in) long, ovate, deep green, freckled with small, irregular pink spots. Flowers about 2cm(¾in) long, lavender, borne singly from the leaf axils. *H.p.* 'Splash' is a selected form with larger, brighter leaf spots. *H. phyllostachya* was for long mis-identified as *H. sanguinolenta*, an allied species with leaves veined, not spotted, red.

I

IMANTOPHYLLUM See CLIVIA

IMPATIENS

Balsaminaceae

Origin: Warm temperate and tropical Africa and Asia, also Europe and a few hardy species in North America. A genus of about 600 species of annuals, perennials and soft shrubs, most of which have fleshy, rather succulent stems, undivided, often toothed leaves and distinctive asymmetrical flowers. These have three, rarely five, small sepals, one of which has a petal-like spur, and five petals, the upper one often hooded. The other four unite into pairs which are often deeply lobed. Those described are splendid plants for home and conservatory. Propagate by cuttings taken from spring to late summer, or by seed in spring. *Impatiens* derives from the Latin for impatient; seed capsules open explosively when ripe.

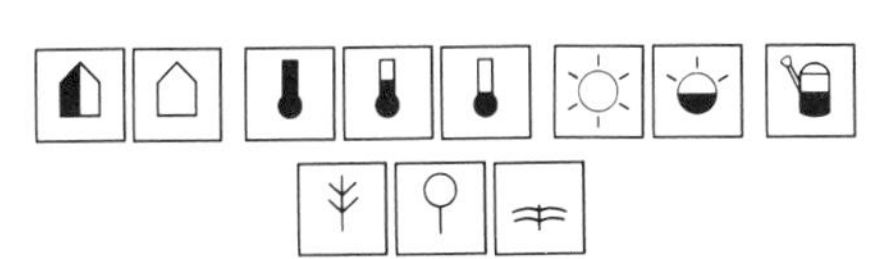

Species cultivated

I. balsamina Rose balsam, Touch-me-not
India, China, Malaysia
Sparingly branched, erect annual to 60cm(2ft) or more tall. Leaves 7–15cm(2¾–6in) long, lanceolate, toothed with a pointed tip. Flowers to 4cm(1½in) across, in shades of red, pink, purple, yellow and white. Best as a pot plant for the conservatory. In cultivation the species is rare, being represented by double cultivars such as 'Camellia-flowered' to 45cm(1½ft) tall and 'Tom Thumb' to 25–30cm(10–12in). Cool.

I. flaccida Southern India, Sri Lanka
Shrubby perennial much like the familiar *I. walleriana*, with red-tinted, branched stems to 70cm(28in) in height. Leaves alternate, ovate, 3–10cm(1¼–4in) long, sometimes larger. Flowers flat, red, mauve or rarely white, to 3.8cm(1½in) wide with a slender spur, 2–3.5cm(¾–1⅜in); they are borne singly or in twos and threes directly from the leaf axils, not on a peduncle as in *I. walleriana*, intermittently throughout the year. Tropical.

I. hawkeri New Guinea
Perennial to 60cm(2ft) tall. Leaves 10cm(4in) or more long, ovate-oblong, in opposite pairs in whorls of three. Flowers 5–7cm(2–2¾in) wide, red or brick-red with a 7cm(2¾in) spur, opening from summer to autumn. Tropical.

I. holstii See *I. walleriana*.

I. magnifica See *I. sodenii*.

I. marianae Assam
Creeping perennial with decumbent stems, sometimes more or less erect. Leaves alternate, broadly oblong-ovate, round-toothed, 2.5–6cm(1–2½in) long, rich green, painted with silvery white between the veins. Flowers pale purple, 2–3cm(¾–1¼in) long, solitary or in pairs from the upper leaf axils. A very attractive foliage plant. Tropical.

Impatiens marianae

***I.* × New Guinea Hybrids** See under *I. schlechteri*.

I. niamniamensis Western and Central tropical Africa
Erect perennial 30–60cm(1–2ft) or so in height, sparingly branched or unbranched, usually red-flushed. Leaves alternate, mainly on the upper part of the stems, broadly ovate to elliptic, 6–20cm(2½–8in) long, coarsely round-toothed. Flowers about 3cm(1¼in) long, formed largely of a pouch-like spur, varying from pure red to red and white, red and pink and red and yellow. In the especially showy cultivar 'Congo Cockatoo' each bloom is pale green, deep yellow and crimson. The flowers are borne on short stalks in groups from the leaf axils and bare stems below in summer and autumn. Tropical.

I. platypetala Java
Somewhat shrubby perennial (sometimes behaving as an annual) with branched, almost succulent reddish stems, to 45cm(1½ft) or more tall. Leaves ovate to lanceolate, often slender-pointed, 8cm(3in) long, prominently veined, carried in whorls. Flowers flat, about 4cm(1½in) wide, with a slender spur, red, purple, pink or white, opening in summer. In *I.p. aurantiaca* ('Tangerine') from the Celebes, the flowers are orange-yellow with a red eye. Tropical.

I. oliveri See *I. sodenii*.

I. petersiana See *I. walleriana petersiana*.

I. pseudoviola Kenya to northern Tanzania
Trailing, usually decumbent perennial, but sometimes with stems to 30cm(1ft) tall. Leaves

BELOW *Impatiens sodenii*
BELOW CENTRE *Impatiens* 'New Guinea Hybrid'
BOTTOM *Impatiens walleriana* 'Variegata'

ovate, crenate, up to 5cm(2in) long. Flowers flattish or slightly cupped, about 2cm($\frac{3}{4}$in) wide, the lower two petals much narrower than the rest, purple to pinkish-purple or palest pink, with a violet stripe down the centre of the four lower petals. A free-flowering species, blooming from spring to autumn and well worth trying in a hanging basket. Tropical.

I. repens India, Sri Lanka
Procumbent annual or short-lived perennial. Leaves to 4cm(1$\frac{1}{2}$in) long, broadly ovate-cordate. Flowers to 3cm(1$\frac{1}{4}$in) long, yellow and hairy, with a very broad spur. Effective in a hanging basket, flowering in summer. Tropical.

I. schlechteri New Guinea
Shrubby perennial to 60cm(2ft) or more, but easily maintained at half this height in containers. Leaves ovate to oblanceolate, 5–10cm(2–4in) long, toothed, glossy rich green, often red-flushed, carried in whorls of three to seven. Flowers bright red, pink or orange, flat, about 4.5cm(1$\frac{3}{4}$in) wide with pale slender spurs of equal length, opening from spring to autumn. This species is not common in cultivation, but is one parent of the new race known as New Guinea Hybrids. These can have green, red or bronze leaves, usually variegated with an irregular, yellow, mid-vein stripe. The glistening flowers range from shades of red to orange, pink, purple and white. Tropical.

I. sodenii (*I. magnifica*, *I. oliveri*) Tropical eastern Africa, southern Kenya to northern Tanzania (mountains)
Shrub-like perennial with robust, erect, sparingly branched stems to 1.5m(5ft), but usually less in containers. Leaves in whorls of six to ten, oblanceolate 8–18cm(3–7in) in length, prominently toothed, rich green with a pale midrib. Flowers flat, but the petals slightly waved, 6cm(2$\frac{1}{2}$in) or more across, apple blossom pink to pure white or white and pink with a white spur to 8cm(3in) or so in length, usually solitary or in pairs from the upper leaf axils, opening from autumn to spring or later. Tropical.

I. sultanii See *I. walleriana*.

I. walleriana (*I. holstii*; now including *I. sultanii*) Busy Lizzie, Patient Lucy Tanzania, Mozambique
Shrubby perennial to 60cm(2ft) or more tall. Leaves 4–10cm(1$\frac{1}{2}$–4in) long, lanceolate to ovate. Flowers 3–5cm(1$\frac{1}{4}$–2in) across, flat, with a 4cm(1$\frac{1}{2}$in) spur, scarlet, orange, purple, magenta, pink or white, opening throughout the year if temperatures are kept above 13°C(56°F). Cool to temperate.

I.w. petersiana
Similar with longer, bronze-tinted leaves.

IOCHROMA

Solanaceae

Origin: Tropical South America. A genus of 25 species of shrubs and trees, those described below being distinctive and colourful plants for the conservatory. They have alternate, simple leaves and nodding or pendent clusters of sizeable tubular flowers with flared mouths. Berries may follow, each one enclosed by an enlarged calyx. Propagate by cuttings in summer or by seed in spring. *Iochroma* derives from the Greek *ion*, violet and *chroma*, colour; the first described species had violet-purple blooms.

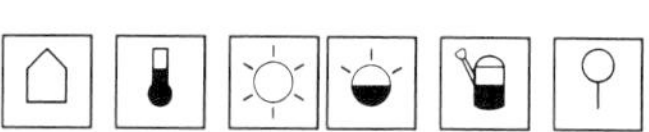

Species cultivated

I. coccineum Central America
Evergreen shrub to 3m(10ft), but easily kept to half this in a tub and pruned regularly after flowering. Leaves ovate to oblong, pointed, 8–13cm(3–5in) long, usually rich green and somewhat lustrous. Flowers 4–5cm(1$\frac{1}{2}$–2in) long, bright red with pale yellow throats, borne in arching clusters of eight or more in summer.

I. cyaneum (*I. lanceolatum*, *I. tubulosum*) N.W. South America
Evergreen shrub 2–3m(6$\frac{1}{2}$–10ft) in height if planted

LEFT *Ipheion uniflorum*
FAR LEFT *Iochroma cyaneum*
BELOW *Ipomoea acuminata*

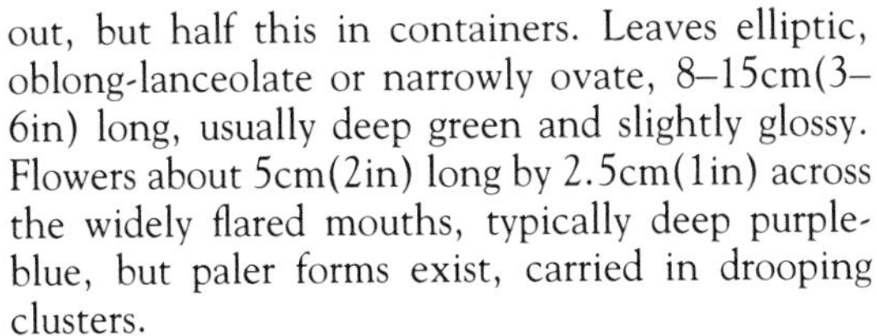

out, but half this in containers. Leaves elliptic, oblong-lanceolate or narrowly ovate, 8–15cm(3–6in) long, usually deep green and slightly glossy. Flowers about 5cm(2in) long by 2.5cm(1in) across the widely flared mouths, typically deep purple-blue, but paler forms exist, carried in drooping clusters.

I. lanceolatum See *I. cyaneum*.
I. tubulosum See *I. cyaneum*.

IPHEION

Liliaceae

Origin: Mexico to South America. A genus of several species of small bulbous plants, the one cultivated being variously classified under *Brodiaea*, *Milla*, *Triteleia* and *Tristagma*, but now thought to belong here. It is a pretty plant for spring blooming in the conservatory or can be brought into the home at flowering time. Propagate by the abundant offsets or by seed in spring. *Ipheion* is of unknown derivation.

Species cultivated

I. uniflorum Spring star flower Argentina, Uruguay
Leaves tufted, to 20cm(8in) long, linear, pale green, all arising from ground level. Flowers tubular, the six sepals opening to pointed, almost triangular lobes, giving a starry effect; they are white to very pale blue and borne singly on a 15cm(6in) stem in spring. *I.u.* 'Wisley Blue' has mauve-blue flowers; 'Froyle Mill' is a darker mauve-blue.

IPOMOEA

Convolvulaceae
Morning glories

Origin: Tropical to warm temperate regions, worldwide. A genus of about 500 species of annuals, perennials and shrubs, many being climbers. They all have tubular to funnel-shaped flowers. The annual and perennial sorts make fine pot plants for

a sunny conservatory. Those listed below are climbers and need the support of canes or wires. Propagate by seed in spring in warmth, pre-soaking them in tepid water for up to 12 hours before sowing. *Ipomoea* derives from the Greek *ips*, worm and *homoios*, similar to, a curious origin thought to be a reference to the twining stem tips, though it may more aptly refer to the curling roots of some weedy bindweeds.

  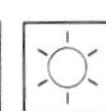  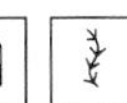

Species cultivated

I. acuminata (*I. learii*, *Pharbitis learii*) Tropical America
Vigorous perennial to 6m(20ft) and more. Leaves 10–20cm(4–8in) long, rounded-ovate to heart-shaped. Flowers 6–10cm($2\frac{1}{2}$–4in) or more wide, funnel-shaped, blue-purple, ageing pinkish-red.

I. batatas Sweet potato Tropical America (exact area unknown and not found growing truly wild)
Cultivated throughout the tropics and sub-tropics and in some warm temperate countries for its edible tuberous roots. Stems prostrate, or weakly climb-

Ipomoea batatus

ing, to 1.2m(4ft) or more in length. Leaves to 15cm(6in) long, very variable even on the same plant, basically broadly ovate-cordate to bluntly sagittate with or without varying degrees of lobing. Extremely lobed forms come into the digitate category; usually soft green but in some cultivars variably purple-flushed. Flowers 3–5cm(1¼–2in) long, purple, often rather hidden by the leaves. An unusual foliage plant for the warm conservatory, responding well to hanging basket treatment. The very dark purple leaved cultivar 'Blackie' is worth seeking out.

I. bona-nox (*I. noctiflora, Calonyction bona-nox*) Moonflower Tropical America
Perennial to 4m(13ft). Leaves to 20cm(8in) long, broadly ovate, sometimes 3-lobed. Flowers 15cm(6in) long and wide, white with a green tube; they are fragrant but open only at night.

I. coccinea Red morning glory, Star ipomoea S.E. USA
Annual climber to 3m(10ft). Leaves to 15cm(6in) wide, ovate-cordate, sometimes toothed, on slender stalks. Flowers to 4cm(1½in) wide, bright scarlet with yellow in the throat and borne two or more together on long stalks. Seed capsules reflexed.

I. hederacea (*Pharbitis hederacea*) Tropical America
Annual climber to 3m(10ft). Leaves to 15cm(6in) wide, broadly ovate-cordate, usually shallowly 3-lobed. Flowers to 5cm(2in) across, purple, red, pink and blue, with broad sepals which contract abruptly to a linear, spreading or recurved tip.

I. learii See *I. acuminata.*

I. lobata (*Mina lobata*) Mexico to South America; widely cultivated
Perennial, reddish-stemmed twiner, 3m(10ft) or more in height. Leaves to 8cm(3in) long, palmate, basically 5-lobed but with the middle three much larger than the rest, the central one constricted where it joins the main blade – a leaf of distinctive and pleasing shape set off by its usually red stalk. Flowers about 2cm(¾in) long, tubular, crimson ageing to yellow, borne in one-sided spikes to 8cm(3in) in length on long red peduncles.

I. nil (*Pharbitis nil*) Tropics
Annual climber to 4m(13ft). Leaves to 15cm(6in) wide, broadly ovate-cordate, shallowly 3-lobed. Flowers to 5cm(2in) across, in shades of red, purple and pink with lanceolate, slender-tipped sepals. 'Scarlett O'Hara' has crimson flowers.

I. noctiflora See *I. bona-nox.*

I. purpurea (*Pharbitis purpurea*) Common morning glory Tropical America
Annual climber to 3m(10ft). Leaves to 13cm(5in) wide, broadly ovate. Flowers 7cm(2¾in) long, white, pink, magenta and purple-blue, with oblong, short pointed sepals. 'Violacea' has double purple flowers.

I. quamoclit Cypress vine Tropical America; now naturalized throughout the tropics
Annual, slender twiner 1–5m(3–16ft) much depending on the fertility of the soil and the temperature. Leaves highly atypical of the genus, being ovate in outline but pinnately cut to the midrib into many well-spaced linear lobes creating a fern-like quality. Flowers scarlet, about 2.5cm(1in) long by almost as wide, tubular for most of their length, expanding at the top into five ovate lobes, singly or in pairs on long peduncles.

I. tricolor Tropical America
Perennial climber to 4m(13ft). Leaves 15–25cm(6–10in) wide, ovate-cordate, slender-tipped. Flowers 10cm(4in) long and wide, purplish blue with a white tube. 'Flying Saucers' is striped blue and white; 'Heavenly Blue' is a rich sky blue.

IPOMOPSIS See GILIA

IRESINE

Amaranthaceae

Origin: Tropical and warm temperate regions. A genus of 70 to 80 species of perennials and subshrubs some of which are climbers. They have colourful ovate to lanceolate leaves and very insignificant flowers best removed if they occur on pot plants. They make good conservatory and window sill plants, growing well in pots and hanging baskets. Propagate by cuttings easily rooted in spring in warmth; pinch the tips of the young plants to encourage bushiness. *Iresine* derives from the Greek *eiros*, wool, the flowers being woolly.

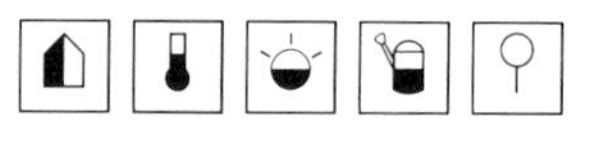

Species cultivated

I. herbstii Beefsteak plant South America
To 60cm(2ft), sometimes more, in a container less than this if propagated annually. Stems red. Leaves to 13cm(5in) long, ovate to rounded, notched at the tip, the surface sometimes puckered, deep purplish-red, with paler to yellowish veins. *I.h.* 'Brilliantissima' has deep purplish-red leaves with the veins picked out in scarlet.

I.h. aureo-reticulata
Bright green leaves with the same vein pattern.

I. lindenii Blood leaf Ecuador
To 60cm(2ft) tall. Leaves 5–8cm(2–3in) long,

Iresine herbstii aureo-reticulata

narrowly lanceolate, sharply pointed, deep red.
I.l. formosa
Broader, green leaves with conspicuous yellow veining.

IRIS

Iridaceae

Origin: North temperate zone. A genus of 300 species of bulbous and rhizomatous perennials. Most have sword-shaped, parallel-veined basal leaves, often in two ranks, the bases overlapping; some have narrowly linear leaves. The flowers are highly characteristic, being made up of six tepals, the outer three of which are usually larger and arch or hang down; these are the falls. The three inner tepals are usually erect and arched inwards, rarely spreading; these are the standards. The three stamens are protected by petal-like style branches which are often 2-lobed or crested at the apex. Propagate by seed when ripe or by offsets. All those mentioned below are bulbous, making attractive flowering plants for the conservatory and can be brought into the home at flowering time. *Iris* is the Greek word for rainbow, from the wide range of colours in the flowers.

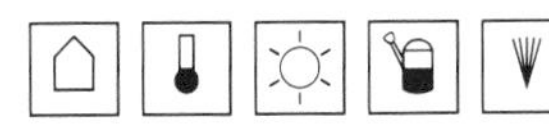

Species cultivated

I. danfordiae Turkey
Leaves to 30cm(1ft) long, 4-angled, cylindrical, 10cm(4in) or less at flowering time. Flowers 5cm(2in) across, bright yellow with greenish spotting on the falls; they appear in winter, the bulbs then splitting up and these can be planted in the garden to grow though they are unlikely to flower the second year.
I. histrio Turkey to Israel
Leaves 30cm(1ft), 4-angled, cylindrical, not above 10cm(4in) at flowering time. Flowers to 5.5cm(2¼in) long, falls pale lilac, veined, tipped and flecked violet, with a yellow crest.
I. histrioides Turkey
Leaves to 30cm(1ft), 4-angled, appearing with or after the blooms. Flowers to 5cm(2in) long, blue, the falls with white markings at the centre and an orange crest; free flowering. 'Major' is richer-coloured and the form usually seen in cultivation.
I. latifolia See *I. xiphioides*.
I. pavonia See *Moraea neopavonea*.
I. reticulata Turkey, Caucasus, Iran, Iraq
Leaves to 40cm(16in) or so long, cylindrical, 4-angled, only 10cm(4in) or so when blooming. Flowers to 7cm(2¾in) long, blue to purple, the falls marked white, with an orange crest. 'Cantab' is pale blue; 'J.S. Dijt' is red-purple; 'Natasha' is almost white, the falls dove-grey, standards ageing a soft blue. A number of hybrid cultivars are also available (*I. histrioides* × *I. reticulata*): 'Clairette', pale blue standards, purple falls; 'Joyce', rich blue flowers; 'Wentworth', purple-blue flowers.
I. tuberosus (*Hermodactylus tuberosus*)
Southern France to Eastern Mediterranean
Rootstock tuberous, finger-like. Leaves cylindrical, 4-angled, 40–60cm(16–24in) when mature, more or less greyish-green. Flowers solitary, on stems 15–30cm(6–12in) tall, about 5cm(2in) long, with greenish-yellow standards and style arms and velvety plum-purple falls, borne in spring. More intriguing than beautiful.

LEFT *Iris danfordiae*
BELOW *Iris reticulata* 'Natasha'

I. xiphioides (*I. latifolia*) English iris
Pyrenees, N.W. Spain
Leaves 45cm(1½ft) long, linear, channelled. Flowers 6–8cm(2½–3in) long, deep purple-blue, the almost rounded falls with a golden central blotch; they are carried on 45–60cm(1½–2ft) stems in summer. White, blue and purple cultivars are available.
I. xiphium Spain to S. France, N. Africa
Rather similar to *I. xiphioides*, but more slender, with smaller, earlier flowers. The familiar Dutch irises are hybrids of *I. xiphium* and are available in a wide range of colours including white, yellow, bronze, purple, mauve and blue and some bicolours.

ISMENE See HYMENOCALLIS
ISOLOMA See KOHLERIA
ISOLEPSIS See SCIRPUS

ISOPLEXIS

Scrophulariaceae

Origin: Canary Islands and Madeira. A genus of four species of evergreen shrubs closely allied to

Digitalis and under which name they were originally classified, being, in effect, shrubby foxgloves with racemes of showy tubular flowers. They are best in a conservatory, but can be brought indoors while in bloom. Propagate by seed in spring or by cuttings in late summer. *Isoplexis* derives from the Greek *isos*, equal and *pleko*, to plait or contrive, said to refer to the 'equally contrived' lower lobes at the mouth of the corolla tube (those of *Digitalis* are unequal).

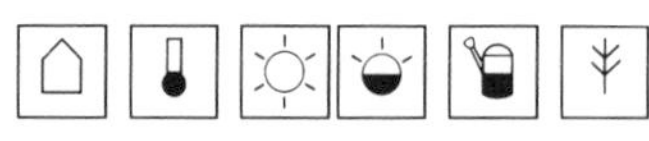

Species cultivated

I. canariensis (*Digitalis canariensis*) Tenerife
Eventually to 1.5m(5ft) in height, but usually less in a container, sparingly branched. Leaves lanceolate to narrowly ovate, toothed, fairly thick-textured, dark glossy green, up to 15cm(6in) long. Flowers bright orange-red to brownish-orange, about 3cm(1¼in) long, in a dense columnar terminal spike to 30cm(1ft) in height, borne in summer.

I. sceptrum (*Digitalis sceptrum*) Madeira
Similar to *I. canariensis*, but with leaves oblong-oval to obovate and up to 25cm(10in) long. Flowers also longer, arranged in a closely packed ovoid raceme up to 15cm(6in) tall in late summer.

ABOVE *Isoplexis canariensis*
RIGHT *Ixia* hybrid

ISOTOMA

Campanulaceae

Origin: Mostly Australia with one species from the West Indies, Central and South America. A genus of 11 annuals and perennials, by some botanists now added to *Laurentia*. Both genera are close to *Lobelia* and have tubular 5-lobed flowers of similar structure. Propagate by seed in spring or by cuttings in late summer. *Isotoma* derives from the Greek *isos*, equal and *toma*, a section; the corolla lobes are more or less equal.

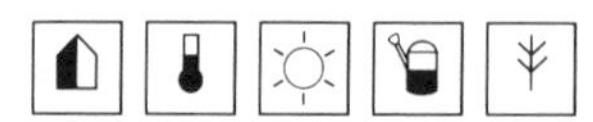

Species cultivated

I. longiflora (*Lobelia/Laurentia/Hippobroma longiflora*) West Indies, Central and South America, naturalized elsewhere in the tropics
Erect, usually short-lived perennial 40–60cm (16–24in) in height, often rather sparingly branched, but variable in this respect. Leaves lanceolate to oblanceolate, 10–20cm(4–8in) long, very coarsely toothed, rich green. Flowers up to 8cm(3in) long, white, slenderly tubular with lobes to 2cm(¾in) long, borne in summer. An unusual and attractive pot plant when well grown, best treated as an annual. The sap of this plant is very poisonous.

IXIA

Iridaceae

Origin: South Africa. A genus of 30 to 40 species of cormous perennials, with narrow fans of sword-shaped leaves and racemes of somewhat crocus-like flowers. They are very pretty plants for the cool conservatory and are excellent cut flowers for the home. Propagate by seed in spring, or by separating corms when dormant. Pot in autumn and with temperatures of medium range they will flower in the early summer, and if they can be kept warmer, they will flower in spring. Dry off when the leaves yellow. *Ixia* is the Greek word for bird lime, these plants having a very sticky sap.

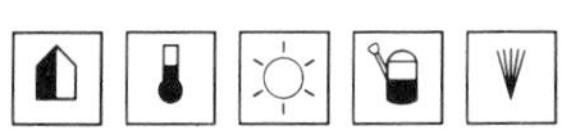

Species cultivated

I. maculata Cape
Leaves 15–30cm(6–12in) tall, linear to lanceolate. Flowers 4cm(1½in) wide, yellow-orange to white with a brown-purple eye; they are borne in short spikes on stems to 45cm(1½ft) high.

I. spathacea See *Gladiolus citrinus*.

I. viridifolia Cape
Leaves to 45cm(1½ft), linear, in tufts of five to seven. Flowers to 4cm(1½in) across, the petals narrower than in *I. maculata* and a remarkable metallic blue-green with a purplish black eye.

Hybrids

Most freely available is a group of hybrids derived

from *I. maculata*, *I. patens* which is pink to light red, the crimson *I. speciosa* and perhaps other species. They vary in colour from white to yellow, orange, red, pink and purple and also in flower shape, some being almost bowl-shaped, others starry.

IXIOLIRION

Amaryllidaceae

Origin: Western and Central Asia. A genus of three species of bulbous plants with long, slender leaves and umbels of white, mauve or purple star-shaped flowers. One species is available and suitable for a cool conservatory, or may be brought into the home when in bloom. Propagate by seed or by separating offsets when dormant. *Ixiolirion* derives from the botanical name *Ixia* used together with the Greek *leiron*, a lily, referring to the similarity of the flowers of the true species.

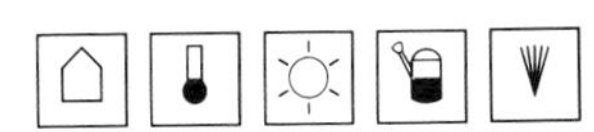

Species cultivated

I. tataricum (*I. ledebourii*, *I. montanum*)
Turkey to Afghanistan and Siberia
Basal leaves to 30cm(1ft) tall; stem leaves shorter. Flowers 4cm(1½in) long, trumpet-shaped, with six tepals, pale to deep blue, sometimes with a purple suffusion; they are borne in umbels of four to eight, the individual blooms with different lengths of stalk, on a 30–45cm(1–1½ft) stem in late spring.
I.t. pallasii (*I. pallasii*)
Flowers rose-purple.

IXORA

Rubiaceae

Origin: Pan-tropical. A genus of about 400 species of shrubs and trees grown primarily for their colourful flowers (but see *I. borbonica*). They have opposite pairs of simple leaves and terminal corymbs of tubular, mostly 4-petalled flowers which, in quite a few species, are fragrant. Propagate by seed in spring or by cuttings in summer. This is a popular genus in the tropics, lending itself to hedging as well as ornamental shrubberies and container specimens. They make excellent tub plants for the conservatory and where there is room they can be brought into the home at least while in bloom. Several good named hybrids have been raised in the tropics and are well worth seeking. (See under hybrids below). *Ixora* derives from the Portuguese rendering of the Sanscrit Israra (the Hindu god Siva).

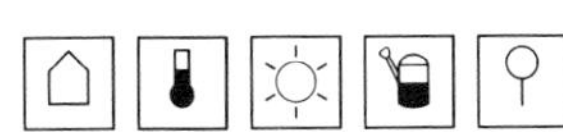

Species cultivated

I. borbonica Reunion Island
Fairly bushy shrub to 2m(6½ft) or so, but easily kept to half this height. Leaves 15–25cm(6–10in) long, lanceolate, leathery-textured, lustrous, mossy-green with a netting of carmine-red veins. Flowers small, off-white. A handsome foliage shrub with something of the appeal of a *Codiaeum* (croton).

Ixora coccinea

I. chinensis China, Malaysia
Much like the better known *I. coccinea*, but smaller in all its parts and more compact in habit, the petals with blunt tips. *I.c.* 'Alba' has white blooms.
I. coccinea Flame-of-the-woods India
Variable in stature, but to about 1.5m(5ft) or so tall. Leaves elliptic to broadly so, or oblong, to 10cm(4in) long, usually deep green and glossy. Flowers about 4cm(1½in) long, with broadly pointed petal tips, bright red, in corymbs to 13cm(5in) wide, but usually less, borne in summer. Cultivars in various shades of red, pink, orange-yellow and yellow are known.
I. javanica Jungle geranium S.E. Asia
In its homeland, a large shrub or small tree to 7.5m(25ft) in height, but easily kept to 2m(6½ft) or below in a container. Leaves ovate-oblong, up to 20cm(8in) long, leathery-textured with a waxy shine, rich green. Flowers 4cm(1½in) long, salmon-red, in clusters 6cm(2½in) or more across in summer.

Hybrids

Garden origin. The origins and parentage of the following *Ixora* cultivars seems not to have been recorded. In general they have the appearance of *I. coccinea*, often more floriferous and with larger leaves:
'Angela Busman', shrimp-pink flowers; compact habit.
'Frances Perry', large trusses of deep yellow flowers.
'Gillettes Yellow', pale yellow.
'Helen Dunaway', deep orange flowers freely produced; tall habit.
'Henry Morat', fragrant pink flowers.
'Herrera's White', a good white-flowered sort.
'Superking', deep red, floriferous; compact habit.

J

JACARANDA

Bignoniaceae

Origin: Tropical America and West Indies. A genus of 50 species of trees and shrubs with pinnate, or more usually bipinnate, fern-like leaves. When mature they have panicles of blue or violet flowers. These do not occur on pot plants which are grown solely for their lacy foliage. They are most decorative and thrive in the home or warm conservatory; they should be cut back or discarded when they get too large or leggy. Propagate by seed sown in warmth in spring, potting the seedlings singly as soon as they have their first true leaves; cuttings can also be taken, preferably with a heel, in early summer from plants which have been cut back. *Jacaranda* is a latinized form of the Brazilian/Indian vernacular name.

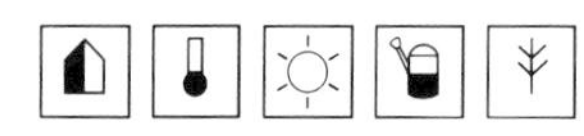

Species cultivated

J. acutifolia Peru
Tree to 10m(35ft) in the wild, but in a pot, to 2m($6\frac{1}{2}$ft). Leaves bipinnate with up to eight pinnae having linear-lanceolate leaflets to 1cm($\frac{3}{8}$in) long.

J. mimosifolia (*J. ovalifolia*) Argentina
Tree to 12m(40ft) in the wild, in a pot to 2m($6\frac{1}{2}$ft). Leaves bipinnate with 12 or more pinnae having oblong-oval leaflets to 1cm($\frac{3}{8}$in) long.

JACOBINIA See JUSTICIA

Jasminum nitidum

JASMINUM

Oleaceae

Jasmines

Origin: Asia, Africa and Australia, mostly from tropical regions. A genus of 200 to 300 species of deciduous and evergreen shrubs. They have opposite pairs of leaves which can be pinnate, trifoliate or simple. The flowers are tubular with six (rarely from four to nine) petal-like lobes opening flat at the mouth and followed by black berries. Plants are suitable for the conservatory or a home window sill where they can be trained on wires or canes. Propagate by cuttings in late summer. *Jasminum* was derived from the Arabic name Yasmin for this genus.

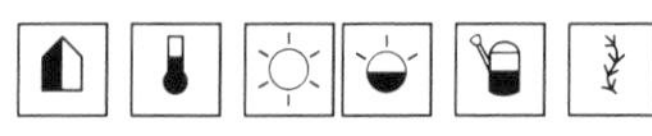

Species cultivated

J. angulare (*J. capense*) South Africa
Evergreen scrambler to semi-twiner, reaching 2m($6\frac{1}{2}$ft) or more with support. Leaves pinnate, composed of three to five ovate leaflets, each one up to 4cm($1\frac{1}{2}$in) long. Flowers in terminal clusters, white, sometimes from pink buds, the tube 2–4cm($\frac{3}{4}$–$1\frac{1}{2}$in) long, topped by five to seven rolled-margin petal lobes about half as long, fragrant. Summer to autumn is the usual flowering season.

J. capense See *J. angulare*.

J. mesneyi (*J. primulinum*) Primrose jasmine Western China
Evergreen scrambler to 3m(10ft), not twining. Leaves 3–7cm($1\frac{1}{4}$–$2\frac{3}{4}$in) long, trifoliate, oblong to lanceolate. Flowers often partly double, opening to six to ten lobes, 4cm($1\frac{1}{2}$in) long, bright yellow, borne in the leaf axils in spring and summer.

J. nitidum Windmill/Star jasmine Admiralty Islands, Southern Pacific
Evergreen semi-twiner up to 3m(10ft) tall, but less in containers. Leaves simple, elliptic-lanceolate, to 7cm($2\frac{3}{4}$in) long, deep lustrous green. Flowers white from purple-red-flushed buds, the tube 2cm($\frac{3}{4}$in) long, topped by six to eleven narrow petal lobes of almost the same length, fragrant. Summer is the main flowering period.

J. odoratissimum Madeira, Canary Islands
Usually a loose, floppy shrub to 1.2m(4ft) or so with support, but can scramble to almost twice this. Leaves evergreen, trifoliate, or sometimes with five leaflets, but also mixed with simple ones at the base of shoots and below the inflorescences; leaflets narrowly ovate, 2.5–5cm(1–2in) long, dark green and glossy. Flowers in clusters terminating the side shoots, clear yellow, usually scented, the tube about 1.5cm($\frac{5}{8}$in) long, topped by four to six 1cm($\frac{3}{8}$in) long petal lobes, opening in summer.

J. officinale Common white/Summer jasmine, Poet's jessamine Caucasus to Western China
Vigorous deciduous twiner to 4m(13ft) in a pot, to 10m(33ft) in the open, with green stems. Leaves trifoliate, the leaflets 1.5–2.5cm($\frac{5}{8}$–1in) long, ovate, the end one tapering to a slender point.

Jasminum polyanthum

Flowers about 2cm(¾in) across, white, very fragrant, carried in terminal clusters from summer into autumn. *J.o. affine* has pink-tinged buds; 'Aureovariegatum' has the leaves and shoots irregularly blotched with yellow.

J. polyanthum Pink jasmine, Chinese jasmine W. China
Evergreen or partly deciduous twiner to 3m(10ft) or more. Leaves trifoliate, the terminal one slender-pointed. Flowers 2cm(¾in) wide, white, pink in bud, brightest in good light, very fragrant and borne in abundance in spring and summer.

J. primulinum See *J. mesneyi.*

J. rex Thailand, Cambodia
Vigorous evergreen twiner with slender stems to 3m(10ft) or more. Leaves simple, broadly ovate, up to 20cm(8in) in length, dark green. Flowers white, in small axillary and terminal clusters, the tube about 5cm(2in) long, topped by six to nine elliptic petal lobes each of the same length. This is the showiest and largest-flowered of the white jasmines, good forms having blooms up to 8cm(3in) across; regrettably there is no fragrance. Under warm conditions flowering is intermittent throughout the year.

J. sambac Arabian jasmine India to Burma and Sri Lanka
Evergreen twiner to 3m(10ft) or so in height. Leaves simple, broadly ovate, about 8cm(3in) long, but sometimes less or up to 13cm(5in), dark green. Flowers white, fragrant, in tight clusters of three or more, mainly terminating lateral shoots, the tube 1.2cm(½in) long topped by four to nine, oblong to rounded petal lobes. *J.s.* 'Maid of Orleans' (*J.s. multiplex*) has semi-double flowers, while those of 'Grand Duke of Tuscany' ('Flore Pleno') are fully double like small gardenias. Under warm conditions flowering takes place at intervals throughout the year. The flowers are used to scent certain kinds of Chinese tea. They are also sacred to the Hindus and used for making garlands.

JATROPHA

Euphorbiaceae

Origin: Mainly tropics and sub-tropics with a few species in temperate North and South America. A genus of 175 species of perennials, shrubs and trees, some of them native to arid regions and having water storage tissue. The species described below are shrubs with alternate leaves and terminal, stalked corymbs of colourful 5-petalled flowers. Rather more curious than beautiful, they nevertheless are worthy of a place in the conservatory or home. Propagate by seed or by cuttings in summer. *Jatropha* derives from the Greek *iatros*, a physician or doctor and *trophe*, food, several species are, or were, used medicinally.

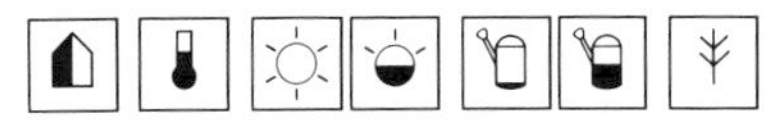

Species cultivated

J. hastata See *J. integerrima.*

J. integerrima (*J. pandurifolia, J. hastata*) Cuba
Slender, fairly sparingly branched shrub 90–150cm(3–5ft) in height. Leaves 8–15cm(3–6in) long, extremely variable in form, from oblong to obovate, sometimes constricted in the middle and with or without basal side lobes. Flowers about 2.5cm(1in) wide, bright red, in clusters of ten or more in summer.

Jatropha integerrima

J. multifida Coral plant, Physic nut Tropical America
Large shrub or even a small tree in the wild, but easily kept to 1.2m(4ft) in a tub. Leaves with long slender stalks, the blades almost circular in outline, to 30cm(1ft) wide, deeply cleft into eight to ten lobes, green above, glaucous beneath. Flowers small, scarlet, in corymbs about 5cm(2in) wide in late spring to late summer.
J. pandurifolia See *J. integerrima.*
J. podagrica Tartogo, Gout plant C. America
Deciduous shrub 45–90cm(1½–3ft) or more in height, the base of the main stem club-shaped, swollen with water storage tissue; when young it is unbranched with two or three stems, when mature, sparingly branched; stems rough, with persistent horny stipules. Leaves in a cluster at the top of the stems, long-stalked, the blades peltate, broadly 3-lobed, 10–20cm(4–8in) wide, smooth, slightly waxy-glaucous. Flowers small, scarlet to coral-red, in clusters up to 5cm(2in) wide, intermittently throughout the year if warm enough. Must be kept dry in winter when most or all the leaves fall.

ABOVE *Juanulloa aurantiaca*
TOP *Jovellana violacea*

JOVELLANA

Scrophulariaceae

Origin: Southern South America and New Zealand. A genus of seven species of perennials and sub-shrubs closely related to *Calceolaria*. They have opposite, entire leaves and 2-lipped flowers within a 4-lobed calyx. They are best in a conservatory and can be brought into the home when flowering. Propagate by cuttings in summer. *Jovellana* was named for Caspari Melchiori de Jovellanos, an 18th-century patron of botany in Peru.

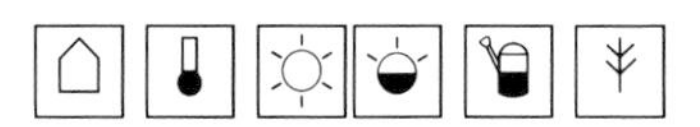

Species cultivated
J. sinclairii New Zealand
Erect sub-shrub to 1.2m(4ft). Leaves 8cm(3in) long, ovate-oblong, coarsely double-toothed on 5cm(2in) stalks. Flowers about 8mm(⅓in) across, bell-shaped, white to very pale lilac spotted with red inside; they are carried in corymbs in summer.
J. violacea Chile
Erect sub-shrub to 2m(6ft), but can be kept much smaller. Leaves 3cm(1¼in) long, coarsely toothed to almost lobed. Flowers 1cm(⅜in) across, pale purple with yellow in the throat and violet-purple spotting within; they are carried in terminal corymbs in summer.

JUANULLOA

Solanaceae

Origin: Mexico to northern South America. A genus of about 12 species of shrubs, only one of which is widely cultivated. It is best in the conservatory, but worth trying in the home at least while in bloom. Propagate by cuttings in summer. *Juanulloa* commemorates two 18th-century travellers in Chile and Peru – George Juan and Antonio Ulloa.

Species cultivated
J. aurantiaca Peru
Evergreen shrub eventually to 2m(6½ft), but easily kept to half this height and flowering when young. Leaves alternate, elliptic to oblong, 8–13cm(3–5in) long, thick-textured, mid to deep green, felted beneath. Flowers orange, tubular to narrowly urn-shaped, 4–5cm(1½–2in) long, in short, pendent, terminal racemes. In the wild this shrub is an epiphyte, but it will thrive in ordinary compost provided it is very well drained; a Bromeliad compost with regular feeding is ideal.

JUBAEA

Palmae

Origin: Chile. A genus of one species of large palm which, when young, makes a handsome foliage pot or tub plant. Propagate by seed in spring or summer. *Jubaea* is named for King Juba of Numidia, Africa (*d.* 46 B.C.).

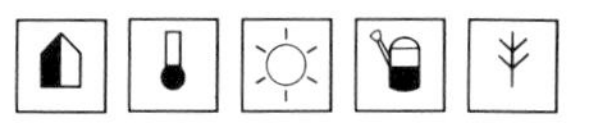

Species cultivated
J. chilensis (*J. spectabilis*) Chilian wine/Honey palm, Coquito Central Chile (coastal region)
In the wild, a tree to 30m(90ft) in height, with a massive elephant-grey trunk. It is slow-growing especially when in containers, where it may never form a trunk. Leaves pinnate, rarely above 1.2m(4ft) in pots, but two or three times this length on mature trees; leaflets lanceolate to linear, 30–45cm(1–1½ft) long, usually a shade of greyish-green. Flowers small, maroon and yellow, in

panicles to 90cm(3ft) or more in length, but only on large specimens. Fruits almost spherical 2.5–4cm(1–1½in) long, yellow when nearly ripe, then brown with a very hard shell and coconut-like flesh, hence the name coquito. These nuts are used in confectionary. In Chile the trees are tapped for their sweet sap which, when boiled down, can be turned into molasses and sugar or wine. To extract the sap the trunk is deeply cut and after about six months often results in the death of the tree. As a result of this exploitation the tree is comparatively rare in the wild. The largest known specimen under glass can be seen in the Temperate House at Kew Gardens where it is about 18m(55ft) tall.

LEFT *Jubaea chilensis*
BELOW *Justicia carnea*

JUSTICIA (JACOBINIA)

Acanthaceae

Origin: Tropics and sub-tropics, also temperate North America. A genus of about 300 species of perennials, shrubs and sub-shrubs with opposite pairs of ovate to elliptic leaves and racemes or spikes of flowers often with conspicuous bracts. The flowers are tubular, 2-lipped and range from reds and purples to yellow and white. They are good pot plants. Propagate by cuttings in spring with bottom heat, potting-on regularly. After flowering, cut back stems to 5cm(2in) and give a barely moist rest, or take the subsequent shoots as cuttings and discard the old plant. *Justicia* was named for James Justice (*d.* 1754), a Scottish gardener.

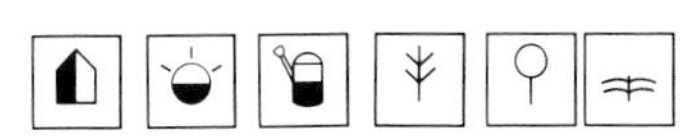

Species cultivated

J. brandegiana See *Beloperone guttata.*
J. carnea (*Jacobinia carnea, J. pohliana, J. velutina*) King's crown Northern South America
Shrub to 2m(6½ft). Leaves to 25cm(10in) long, oblong to ovate, hairless or shortly velvety-hairy. Flowers 5cm(2in) long, pink to rose or purple-pink, in dense spikes 10–13cm(4–5in) long in summer and autumn.
J. coccinea See *Pachystachys coccinea.*
J. floribunda See *J. pauciflora.*
J. pauciflora (*J. rizzinii, J. floribunda, Libonia floribunda*) Brazil
Low, soft shrub 30–60cm(1–2ft) high. Stems softly downy, freely branching. Leaves to 2cm(¾in) long, oblong to broadly obovate, in unequal pairs. Flowers to 2.5cm(1in) long, scarlet, tipped with yellow and borne in small, nodding, axillary clusters from autumn onwards to late spring.
J. pohliana See *J. carnea.*
J. rizzinii See *J. pauciflora.*
J. suberecta Uruguay
Perennial to 60cm(2ft) or so tall with spreading, grey-velvety stems. Leaves 5–7cm(2–2¾in) long. Flowers 3cm(1¼in) long, brick-red, carried in terminal clusters in summer and autumn.
J. velutina See *J. carnea.*

K

KAEMPFERIA

Zingiberaceae

Origin: Tropical Africa and S.E. Asia. A genus of 70 species of rhizomatous and tuberous perennials grown for both foliage and flowers. The thick fleshy rhizomes are aromatic and give forth clumps of simple, arching leaves and somewhat orchid-like flowers. The latter have three true petals and an equally large or larger staminodal lip or labellum. Needing a humid atmosphere, kaempferias thrive best in the warm conservatory, but may be brought into the home for short periods. Propagate by division in late spring. *Kaempferia* honours Engelbert Kaempfer (1651–1716), a German doctor and botanist who travelled widely in the east, particularly Japan where he resided for two years, later writing books on the plants and history of that country.

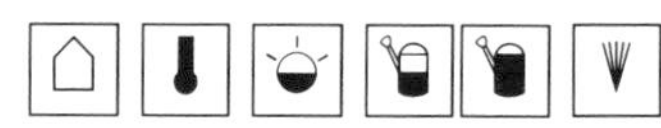

Species cultivated

K. pulchra Thailand, Malaysia
Leaves one or two per shoot, spreading horizontally, broadly ovate, to 13cm(5in) long, quilted, dark bronze-green with a broken grey zone midway between the margins and the midrib. Flowers 4cm(1½in) wide, light purple, the lip spotted white with yellow at the base, borne in summer.

K. roscoana Peacock plant, Dwarf ginger lily
Burma
Much like *K. pulchra*, but the leaves purple-bronze and more iridescent with a pale green zone.

K. rotunda Tropical crocus S.E. Asia
Leaves two per shoot, erect, narrowly elliptic to oblong-lanceolate, up to 30cm(1ft) or more in length, dark and variegated with light green above, purple beneath. Flowers almost 5cm(2in) wide, white with a pale purple lip, in clusters of five to ten in summer.

Kaempferia pulchra

KALANCHOE

Crassulaceae

Origin: Africa and southern Asia, one species in tropical America. A genus of 125 to 200 species of succulent shrubs and perennials some of which have been included in *Bryophyllum* and *Kitchingia*. They have fleshy leaves in opposite pairs and 4-petalled tubular flowers carried in terminal panicles. They are good house and conservatory plants, best on a sunny window sill. Propagate by stem or leaf cuttings from late spring to summer or by seed in warmth in spring. Some species produce plantlets along the leaf edges and these can be detached and potted. *Kalanchoe* is a Latinized form of the Chinese name for one species.

Species cultivated

K. beharensis Velvet leaf Malagasy
Shrubby perennial with erect stems to 1.4m(4½ft), branching only sparingly when grown in a pot. Leaves 15cm(6in) or more long, triangular-lanceolate, wavy, olive-green, finely brown woolly-hairy. Flowers 1.2cm(½in) long, yellowish, but rarely seen on pot-grown specimens.

K. blossfeldiana Flaming Katy Malagasy
Well-branched perennial to 30cm(1ft) tall. Leaves to 7cm(2¾in) long, ovate-oblong, rich shining green with a narrow red edge. Flowers 1cm(⅜in) long, scarlet, borne in dense clusters from winter to early summer. A number of cultivars of this species have been raised with flowers in shades of red, orange and yellow. These are all good house plants, but are short-day plants and normally flower only from late autumn to spring. By giving them extra darkness in summer (thereby reducing the daylight to 12 hours or less) they can be kept in flower all the year.

K. diagremontiana Devil's backbone Malagasy
Somewhat shrubby perennial 60–90cm(2–3ft) tall. Leaves 10–20cm(4–8in), narrowly triangular, spotted or marbled with red-brown, with tiny plantlets growing along the margins. Flowers about 2cm(¾in) long, glaucous-purple, pendent.

K. fedtschenkoana Malagasy
Well-branched perennial to 30–50cm(12–20in) with decumbent, rooting, non-flowering stems. Leaves to 5cm(2in) long, boldly scalloped, blue-green, sometimes with small plantlet buds developing along the margins. Flowers orange-yellow to reddish.

K. flammea Somali Republic
Erect perennial with sparingly branched stems to 40cm(16in) or more tall. Leaves obovate, sinuately toothed or entire, light grey-green, to 8cm(3in) long. Flowers about 2cm(¾in) wide, orange-red or pink, with a yellow tube, carried in corymbose clusters during the winter and spring.

K. gastonis-bonnieri Malagasy
Perennial with erect stems to 60cm(2ft) in height, usually branched only from the base. Leaves with a

whitish patina, narrowly ovate to lanceolate or spathulate, crenate, 13–20cm(5–8in) long, the bases of each leaf pair amplexicaule. Flowers pendulous, in a terminal corymb, to 15cm(6in) wide; each bloom is composed of an inflated, long-ovoid calyx 2–3cm(¾–1¼in) long and a 4cm(1½in) long corolla with recurved, pink, reddish or yellowish petal lobes, usually expanding in winter and spring.

K. grandiflora East Africa, India
Erect perennial 45–60cm(1½–2ft) in height, the stems branching mainly from the base. Leaves obovate, crenate, to 8cm(3in) long, surfaced with a waxy blue-purple to reddish patina. Flowers in fairly dense, corymbose cymes, each blossom 2–2.5cm(¾–1in) wide, yellow. A handsome species in both leaf and flower, but rather shy-blooming in cultivation.

K. longiflora South Africa, Natal
Shrubby perennial to 60cm(2ft) in height, the mainly erect stems 4-angled. Leaves obovate to ovate-oblong, prominently toothed in the upper half, 4–9cm(1½–3½in) long, pale grey-green, taking on yellow to orange tints under strong light or low temperatures. Flowers in terminal corymbose cymes, the individual blooms yellow, about 2cm(¾in) long. *K.l.* 'Coccinea' has leaves which turn pink to scarlet in strong sunlight.

K. manginii Malagasy
Semi-erect to decumbent perennial with wiry stems branched mainly at the base. Leaves obovate-cuneate, 2.5–4.5cm(1–1¾in) long, thick and fleshy. Flowers pendent, in loose terminal clusters of three to nine, narrowly bell-shaped, bright red, about 2.5cm(1in) long, appearing in spring.

K. marmorata Ethiopia, Somali Republic
Shrubby perennial to 90cm(3ft) tall, the decumbent to erect stems branching mainly at the base. Leaves obovate, up to 10cm(4in) long, greyish-green with a blotched or mottled pattern in light purple-brown. Flowers white, in terminal cymes, slenderly tubular, 6–8cm(2½–3in) long, opening in late winter and spring. Formerly known as *K. grandiflora*, but not to be confused with the true species, *q.v.*

K. marnieriana Malagasy
Decumbent to semi-erect, the woody-based stems often with aerial roots. Leaves overlapping, broadly oval, about 3cm(1¼in) long, fairly thick and fleshy, blue-green with purple spots above and bearing marginal plantlets. Flowers pink, 2–3cm(¾–1¼in) long, in terminal clusters usually in the winter to spring period.

K. millotii Malagasy
Well-branched shrub to about 40cm(16in) tall. Leaves 3–4cm(1¼–1½in) long, ovate, softly hairy. Flowers 1cm(⅜in) long, yellow and orange.

K. pilosa (of gardens) See *K. tomentosa*.

K. pinnata Tropics and sub-tropics, now spread widely by man so that its origins are uncertain
Little branched, shrubby perennial to 40–180cm(16–72in) tall. Leaves pinnately divided into three to five leaflets, each being 7–20cm(2¾–8in) long, oblong to rounded, the margins scalloped and bearing marginal plantlets. Flowers to 3.5cm(1½in) long, greenish-white flushed with red, within a bell-shaped calyx.

LEFT *Kalanchoe flammea*
BELOW LEFT *Kalanchoe manginii*
BELOW *Kalanchoe pumila*

K. pumila Malagasy
Branched perennial to 20cm(8in) or more in height. Leaves 2–3cm(¾–1¼in) with a waxy-white patina. Flowers to about 1cm(⅜in) long, red-purple.

K. thyrsiflora South Africa (Cape to Transvaal)
Erect perennial to 60cm(2ft), sometimes twice as tall, glaucous whitish. Leaves obovate to spathulate, up to 15cm(6in) long, especially those of the young basal growths, waxy-white-surfaced. Flowers in dense, terminal, thyrse-like panicles, the individual blooms narrowly urn-shaped, 1.5cm(⅝in) long, yellow, from late winter to spring.

K. tomentosa (*K. pilosa* of gardens) Pussyears, Panda plant Malagasy
Small, slow-growing shrub 20–50cm(8–20in) tall. Leaves 4–8cm(1½–3in) long, oblong-obovate, thick in texture and covered with a dense, silvery felting, marked at the tips with dark brown. Flowers about 1cm(⅜in), yellowish to purplish.

K. tubiflora Chandelier plant Malagasy
Erect, little-branched perennial to 1m(3ft) tall. Leaves 3–12cm(1¼–4¾in) long, cylindrical, grooved, green with red-brown spotting; they are

BELOW *Kalanchoe uniflora*
BELOW CENTRE *Kochia scoparia*
BOTTOM *Kennedia sp.*

carried in ones to threes and have plantlets near their ends. Flowers to 2.5cm(1in) long, pendent and tubular, salmon-red to scarlet.
K. uniflora Malagasy
Trailing, epiphytic perennial, good for a hanging basket. Leaves 1–2cm($\frac{3}{8}$–$\frac{3}{4}$in) long, rounded to ovate, mid-green. Flowers about 2cm($\frac{3}{4}$in) long, urn-shaped, red to purplish.
K. velutina Central Africa
Velvety-hairy perennial with creeping and rooting, robust fleshy stems. Leaves narrowly ovate, to 10cm(4in) long. Flowering stems erect, with narrower leaves and terminal rounded clusters of 1.2cm($\frac{1}{2}$in) long blooms which, though usually yellow, can also be pink.

KENNEDIA (KENNEDYA)

Leguminosae

Origin: Australia. A genus of 16 species of perennial trailing and climbing plants with showy pea flowers. The climbing species are stem twiners and all species have trifoliate leaves arranged alternately. They are best in the conservatory where the climbers can be used effectively twining through a planted or container shrub; they can also be tried in the home. Propagate by seed in spring. Seeds germinate erratically and respond to heat shock treatment. Pour boiling water into a tea cup or vessel of similar size to a depth of 4–5cm(1$\frac{1}{2}$–2in). Drop in the seeds immediately and allow to soak for 12 hours, then sow in the usual way. *Kennedia* is named for John Kennedy (1759–1842), a London nurseryman and able botanist.

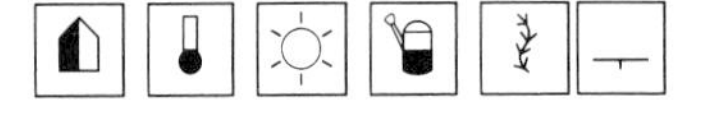

Species cultivated
K. coccinea Western Australia (S.W. coast)
Vigorous climber to 2m(6$\frac{1}{2}$ft) or more. Leaflets 3–7cm(1$\frac{1}{4}$–2$\frac{3}{4}$in) long, broadly oblong, sometimes lobed, mid to deep green. Flowers about 1.2cm($\frac{1}{2}$in) wide in axillary umbels, bright orange-red, borne in spring to summer.
K. nigricans Black bean Western Australia
Strong-growing climber to 2m(6$\frac{1}{2}$ft) or more. Leaflets ovate-cordate, 5–10cm(2–4in) long with a notched tip; sometimes unifoliate. Flowers 3–4cm(1$\frac{1}{4}$–1$\frac{1}{2}$in) long, velvety black-purple, the reflex standard petal with a broad yellow splash, borne in axillary umbels in spring and summer.
K. prostrata Running postman Victoria
Stems prostrate, eventually forming mats to 90cm(3ft) wide. Leaflets rounded to oval, wavy-edged, 2cm($\frac{3}{4}$in) or more long. Flowers solitary or in pairs from the leaf axils, about 2.5cm(1in) long, scarlet with a yellow blotch at the base of the standard, borne in spring to summer.
K. rubicunda Dusky coral pea Victoria to Queensland
Climber to 3m(10ft), with dark green, ovate leaflets 3–10cm(1$\frac{1}{4}$–4in) in length. Flowers coral-red, distinct from those described for the above species, having long, curved keel petals reminiscent of *Clianthus*, about 4cm(1$\frac{1}{2}$in) long (from standard tip to keel tip), in small axillary umbels from spring to summer.

KENTIA See HOWEIA
KITCHINGIA See KALANCHOE
KLEINIA See SENECIO

KOCHIA

Chenopodiaceae

Origin: Central and southern Europe, Asia, Africa and Australia. A genus of 80 to 90 species of annuals, perennials and sub-shrubs, with slender undivided leaves and small, inconspicuous flowers. The species cultivated makes a good foliage pot plant. Propagate by seed in spring in gentle warmth and prick off singly into small pots as soon as the true leaves show. *Kochia* was named for Wilhelm Daniel Josef Koch (1771–1849), Professor of Botany at Erlangen in Germany.

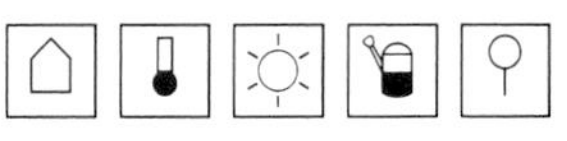

Species cultivated
K. scoparia Burning bush, Summer cypress S. Europe to Japan
Annual to 1.5m(4$\frac{3}{4}$ft). Leaves 3–6cm(1$\frac{1}{4}$–2$\frac{1}{2}$in) long, narrowly lanceolate to linear, pale green.
K.s. tricophylla (often mis-spelt *trichophila*)
The form normally grown, rarely exceeding 60cm(2ft) in height, egg-shaped with dense foliage which turns red in autumn. *K.s.t.* 'Childsii' is a more uniformly-sized selection.

Kohleria digitaliflora

KOHLERIA

Gesneriaceae

Origin: Mexico, Central America and northern South America. A genus of 50 species of shrubs and perennials growing from scaly rhizomes. Their leaves are lanceolate to ovate in whorls or pairs and the flowers are tubular with five rounded lobes carried singly or in clusters from the upper leaf axils. They make handsome pot plants. Keep them barely moist but not dry during the winter and re-pot annually in spring, separating the scaly rhizome and discarding the rest of the plant. Propagate in the same way or by seed if available. *Kohleria* was named for Michael Kohler, 19th-century Swiss teacher of natural history.

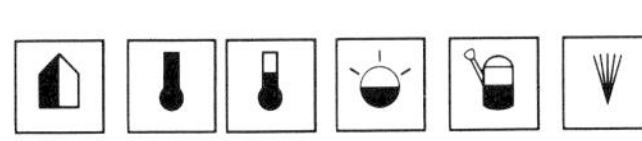

Species cultivated

K. amabilis (*Isoloma amabilis*) Colombia
Herbaceous perennial to 60cm(2ft) tall. Leaves to 10cm(4in) long, ovate, dark green with purple-brown veining and often silvery-marked above. Flowers 2.5–3cm(1–1¼in) long, deep rose-pink, opening to broad lobes which are marked with brick-red bars and spots; they nod from slender stalks to 6cm(2½in) long in summer. Originally mistakenly identified as *K. seemannii*, it has been hybridized with other species producing plants with deep red, yellow or white flowers; these are often sold under the name of *K. amabilis*.

K. bogotensis Colombia
Herbaceous; stems up to 45cm(1½ft) or more, erect. Leaves ovate, 8cm(3in) long, dark green, densely soft-hairy, sometimes with paler or darker markings above and red flushed beneath. Flowers 2.5cm(1in) long, red with a yellow base outside, red-dotted yellow within, solitary or borne in pairs from the upper leaf axils in late summer.

K. digitaliflora Colombia
Herbaceous; stems white-hairy, robust, erect to 60cm(2ft) or more. Leaves elliptic-lanceolate to ovate, up to 20cm(8in) in length, hairy, dark green, crenate. Flowers 3cm(1¼in) long, white and pink, densely white-hairy, petal lobes green with purple spots, borne in stalked clusters of up to six from the upper leaf axils in summer to autumn.

K. eriantha (*K. hirsuta*) Colombia
Shrubby, to 1.2m(4ft) tall. Leaves 7–13cm(2½–5in) long, ovate to elliptic, mid-green, woolly beneath and with densely red-hairy margins. Flowers to 5cm(2in) long, orange-red, the lower three lobes marked with yellow spots; they are borne singly or in clusters in summer. Hybrids of this species having the lobes spotted or lined with other colours are available.

K. hirsuta See *K. eriantha.*

K. lanata Mexico
Herbaceous, stems woolly, to 35cm(14in) or more tall. Leaves ovate to elliptic, to 13cm(5in) in length, toothed and brown-hairy beneath. Flowers about 4cm(1½in) long, densely hairy, orange-red, the purple spotted lobes arranged to create a 2-lipped appearance, solitary in the upper leaf axils, borne in late summer.

K. tubiflora Colombia
Herbaceous, stems sturdy, erect, densely short-hairy. Leaves ovate, to 9cm(3½in) long, toothed, dark green, hairy, the veins beneath red. Flowers about 3cm(1¼in) long, reddish-orange to scarlet, paling to yellow at the mouth, with tiny, dark red-spotted green lobes, borne in summer to autumn.

KOPSIA See OCHROSIA

L

LACHENALIA

Liliaceae

Origin: South Africa. A genus of 50 species of bulbous perennials, each bulb usually with two strap-shaped leaves, occasionally one or more than two. The cylindrical to bell-shaped flowers are carried in spikes above the foliage on leafless stalks. They are very good pot plants for the sunny window sill or the conservatory. Propagate by offsets removed at potting time, or by seed in spring. The bulbs need a cool dry rest once the leaves begin to yellow. *Lachenalia* was named for Werner de la Chenal (1736–1800), Professor of Botany at Basle in Switzerland.

 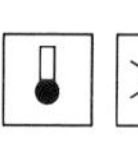 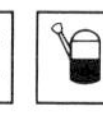 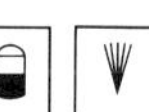

Species cultivated

L. aloides (*L. tricolor*) Cape cowslip Cape
Leaves 2–2.5cm($\frac{3}{4}$–1in) wide, lanceolate to strap-shaped, dull green often with reddish-brown spotting. Flowers 2.5cm(1in) long, yellow to greenish-yellow, orange-red at the base, the inner tepals edged with red-purple; they are pendulous and carried on 15–30cm(6–12in) long stems. *L.a.* 'Aurea' is a soft orange-yellow without markings; 'Nelsonii' has longer, looser racemes of bright yellow flowers which are tipped with green; 'Quadricolor' has red and yellow flowers, the outer tepals edged with green, the inner with dark red.

L. bulbifera (*L. pendula*)
Leaves to 5cm(2in) wide, narrowly lanceolate, dull green with no spotting. Flowers 3–4cm(1$\frac{1}{4}$–1$\frac{1}{2}$in) long, coral- to vermilion-red, sometimes tipped purplish-red and with a green marking at the apex, inner paler; they are borne on 15–25cm(6–10in) stems from late winter to spring. Forms with yellow colouring may be of hybrid origin.

BELOW *Lachenalia aloides*
BELOW RIGHT *Lachenalia reflexa*

L. contaminata
Leaves narrowly linear becoming cylindrical above, four to ten to each bulb. Flowers 6mm($\frac{1}{4}$in) long, widely bell-shaped with spreading corolla lobes; they are white, sometimes with a reddish flush and are carried in dense spikes on 15–20cm(6–8in) stems in spring.

L. glaucina Opal lachenalia
Leaves to 2.5cm(1in) wide, strap-shaped, glaucous green sometimes spotted with brown. Flowers 2cm($\frac{3}{4}$in) long, facing upwards, the inner tepals longer than the outer, white with an almost iridescent blue, red, pink or green suffusion; they are fragrant and borne on 15–30cm(6–12in) tall stems in early spring.

L. mutabilis Fairy lachenalia
Leaves lanceolate, to 4.5cm(1$\frac{3}{4}$in) wide, wavy. Flowers sterile at the top of the raceme, the lower fertile ones to 1.5cm($\frac{5}{8}$in) long, pale blue tipped with yellow-green becoming reddish-brown as they age; they open in late winter.

L. orchioides
Leaves usually two per bulb only, lanceolate, up to 35cm(14in) long when fully mature, darker spotted. Flowering stems erect, above the leaves, terminating in a dense raceme of horizontal or slightly upwards-pointing flowers; each bloom is about 1cm($\frac{3}{8}$in) long, variable in colour ranging from white to pink, yellow or blue, but usually with the outer tepals an iridescent blue-green and the inner ones cream-tinted mauve, in spring.

***L.* × 'Pearsonii'**
Vigorous hybrid, probably of *L. bulbifera*, with bright orange flowers edged with deep red and carried on stems to 45cm(1$\frac{1}{2}$ft) tall.

L. pendula See *L. bulbifera*.

L. reflexa
Leaves two per bulb, strap-shaped to lanceolate, up to 15cm(6in) in length, bright green, spreading to reflexed. Flowering stems shorter than leaves, terminating in a few-flowered raceme of erect flowers; each bloom about 2.5cm(1in) long, green-tinted yellow, borne in late winter to early spring.

L. tricolor See *L. aloides*.

LAELIA

Orchidaceae

Origin: Mexico to tropical South America. A genus of 30 species of epiphytic orchids closely allied to *Cattleya* and separated by the number of pollinia to each flower, *Cattleya* having four, *Laelia* eight. They are best in the conservatory but some species can be tried in the home. Propagate by division in spring or after blooming. *Laelia* was named after one of the Vestal Virgins.

  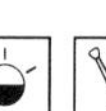 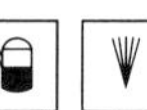

Species cultivated

L. anceps Mexico to Belize
Pseudobulbs 7–15cm(2$\frac{3}{4}$–6in) tall, ovoid-oblong,

Laelia anceps

flattened, either clustered or spaced apart on stout, creeping rhizomes. Leaves 15–23cm(6–9¼in) long, oblong-lanceolate, one from each pseudobulb. Flowers to 10cm(4in) wide, the tepals lilac-pink with a yellow-striped and purple-margined labellum; two to five blooms are carried together on 60–90cm(2–3ft) long racemes. Many forms and cultivars have been described.

L. autumnalis Mexico
Pseudobulbs to 10cm(4in) or more tall, ovoid to conical, clustered. Leaves 10–20cm(4–8in) long, lanceolate, two to three from each pseudobulb. Flowers three to nine, each to 7–10cm(2¾–4in) wide, fragrant, rose-purple with a yellow-lined labellum; they are carried on racemes from 60–90cm(2–3ft) long in autumn and winter.

L.a. gouldiana (*L. gouldiana*)
Broader, magenta labellum with purplish lines.

L. cinnabarina (*Cattleya cinnabarina*) Brazil
Pseudobulbs clustered, cylindrical, erect, 13–25cm(5–10in) tall. Leaves usually solitary, as long as the pseudobulbs, leathery-textured, narrowly oblong, often purple-flushed; each bloom 6cm(2½in) wide, with five equal-sized, spreading cinnabar-red tepals; they are carried on wiry, erect to arching racemes 30–60cm(1–2ft) long, bearing five to fifteen flowers from late winter to spring.

L. crispa Brazil
Pseudobulbs 18–30cm(7–12in) tall, club-shaped. Leaves 23–30cm(9¼–12in) long, oblong-lanceolate, one to each pseudobulb. Flowers four to nine, each 10–15cm(4–6in) wide, fragrant, with narrow, waved, white tepals which are varyingly flushed with purple; labellum yellow, streaked with purple inside; they are carried in short racemes in summer.

L. digbyana See *Brassavola digbyana.*

L. glauca See *Brassavola glauca.*

L. gouldiana See *L. autumnalis gouldiana.*

L. perrinii (*Cattleya perrinii*) Brazil
Pseudobulbs clustered, narrowly club-shaped, erect, 15–30cm(6–12in) tall. Leaves usually solitary, narrowly oblong, almost stiffly leathery, up to 30cm(12in) in length. Racemes to 13cm(5in) long, few-flowered; each bloom 13cm(5in) wide, with five equal-sized, spreading, pale rose-purple tepals; labellum small, purple-crimson and yellow; autumn to winter.

L. pumila Brazil
Pseudobulbs 5–10cm(2–4in) tall, cylindrical. Leaves 5–12cm(2–4½in) long, elliptic-oblong. Flowers in ones or twos, 7–10cm(2¾–4in) wide, fragrant, pinkish-purple, opening in autumn.

L. purpurata Brazil
Pseudobulbs 30–60cm(1–2ft) tall, club-shaped. Leaves singly from each pseudobulb, 30–37cm(12–15in) long, oblong. Flowers 15–20cm(6–8in) wide, fragrant, white or pale purple, sometimes flushed and streaked with darker purple; labellum yellow with purple lines inside, the lip a rich velvety purple; they are carried three to seven together on short racemes from early to late summer.

× LAELIOCATTLEYA

Orchidaceae

Origin: Natural and man-made hybrids between species of *Cattleya* and *Laelia*. Several thousands of hybrid cultivars of this origin are listed and many are available from orchid specialists around the world. They nicely combine the characters of their parent genera and are more popular than either, both in collections and as cut flowers. Culture as for *Cattleya*.

× *Laeliocattleya* 'Cherry Chips'

LAGENARIA

Cucurbitaceae

Origin: Tropics. A genus of six species of annual climbers related to gourds and cucumbers. They have alternate palmate leaves and climb by tendrils. The flowers are white and the fruits hard-shelled. When cleaned out, the latter become the calabash and bottle gourds still used in tropical countries. In the conservatory they make decorative plants which, when in fruit, provide an interesting talking point. Propagate by seed in spring. *Lagenaria* derives from the Greek *lagenos*, a flask.

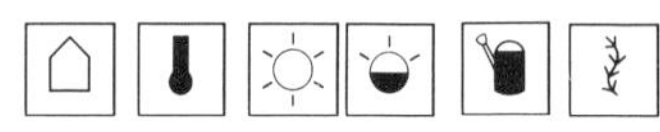

Species cultivated

L. siceraria (*L. vulgaris*) Bottle/Calabash gourd
Old World tropics
To 6m(20ft), or more if planted out in rich soil; less in containers. Leaves broadly ovate-cordate, with or without small pointed lobes, glandular pubescent, to 22cm(9in) or more long and wide. Flowers solitary in the upper leaf axils, to 10cm(4in) wide, opening at dusk during summer and early autumn. Fruits highly variable and many varieties or cultivars are known; the main sorts are either narrowly bottle-shaped or globular. The narrow ones can attain 60cm(2ft) or more in length; the globular sorts up to 30cm(1ft). When very young, fruits can be cooked as a vegetable.

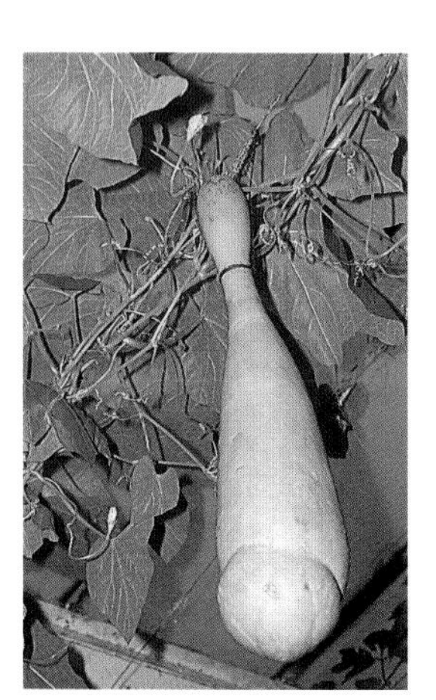

Lagenaria siceraria

LAGERSTROEMIA

Lythraceae

Origin: Old World tropics especially S.E. Asia and the Pacific Islands. A genus of 53 species of large shrubs or trees, some deciduous, some evergreen. They have pairs or whorls of entire leaves and showy clusters of white, pink, red or purple flowers. The species described below is a good large tub or pot plant for the conservatory and responds well to hard pruning in spring. The dwarf forms can be grown in a sunny window in the home. Propagate by cuttings in spring or summer, in heat. *Lagerstroemia* was named for Magnus von Lagerström (1691–1759), a Swedish merchant and friend of Linnaeus who named the genus.

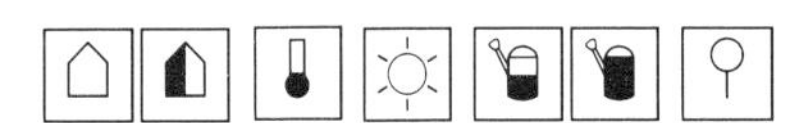

Species cultivated

L. indica Crepe myrtle China
Deciduous shrub, in the open ground making a tree to 8m(26ft), remaining shrubby in a tub. Leaves 3–7cm(1¼–2¾in) long, opposite to alternate on young branches, oblong-elliptic to obovate, stalkless or on very short stalks. Flowers 5–7cm(2–2¾in) wide, individual petals with a broad, heart-shaped, waved blade narrowing at the base into a slender stalk-like portion; they can be white, pink or purple, borne in terminal panicles to 20cm(8in) or more long in summer. A number of named cultivars and hybrids have been raised in the USA including dwarf sorts to 90cm(3ft) in height.

LAMIUM

Labiatae

Dead nettles

Origin: Old World, but some species widely naturalized. A genus of 40 to 50 species of tufted to clump-forming annuals and perennials spreading widely. They have 4-angled stems and opposite pairs of broad, toothed leaves. Their flowers are tubular with two lips, the upper one hooded. The species described below is a decorative pot plant, especially so in its silver foliage forms. Propagate by division or cuttings, the species also by seed. *Lamium* is the old Latin name for dead nettle, derived from the Greek *lamos*, a throat, alluding to the shape of the flower.

Species cultivated

L. maculatum Spotted dead nettle Europe
Leaves 4–8cm(1½–3in) long, ovate to triangular, cordate, coarsely toothed, dark green with a central silvery-white stripe. Flowers 2.5cm(1in) long, purplish-pink. Hardy. *L.m.* 'Album' has white flowers; 'Beacon Silver' has leaves with an all-over silvery patina made more obvious by a narrow, dark green margin, and pink-purple flowers; 'Roseum' has pink flowers and 'White Nancy' is like 'Beacon Silver', but with white flowers.

LAMPRANTHUS

Aizoaceae

Origin: South Africa. A genus of 160 species of succulent perennials and sub-shrubs closely allied to and once included in *Mesembryanthemum*. They may be prostrate or erect and have opposite pairs of fleshy, cylindrical leaves. Flowers appear daisy-like and are carried singly or in cymose clusters. They are easily grown in the conservatory or home where they must have a sunny window because the flowers will not open when away from sunlight. Propagate by cuttings from late spring to late summer or by seed in spring in warmth. *Lampranthus* derives from the Greek *lampros*, shining and *anthos*, flower, from the satiny sheen to the petals.

Species cultivated

L. aureus Cape (Saldanha Bay)
Bushy, erect, 30–40cm(12–16in) in height. Leaves to 5cm(2in) long, almost triangular in cross-section, slightly glaucous with tiny transparent dots. Flowers bright lustrous orange, about 6cm(2½in) wide, borne mainly in summer.

L. blandus Cape
Bushy, more or less erect, up to 45cm(1½ft) in height, the stems red. Leaves 3–5cm(1¼–2in) long, equally triangular in cross-section, pale grey-green studded with tiny transparent dots. Flowers 6cm(2½in) wide, pale to rich pink, borne mainly in summer.
L. haworthii Cape
Shrubby, more or less erect, to 30–60cm(1–2ft) in height. Leaves 2.5–4cm(1–1½in) long, semi-cylindrical, grey-green. Flowers 5–7cm(2–2¾in) wide, pale purple, opening in summer.
L. multiradiatus See *L. roseus*.
L. roseus (*L. multiradiatus*) Cape
Shrubby, erect to spreading, 30–60cm(1–2ft) tall. Leaves 2–3cm(¾–1¼in) long, triangular in cross-section, glaucous blue-green marked with translucent dots. Flowers to 4cm(1½in) wide, light pink to pale rosy-purple, opening in summer.
L. spectabilis Cape
Loosely mat-forming, the stems prostrate to semi-erect, up to 30cm(1ft) long. Leaves 5–8cm(2–3in) long, incurved, roughly triangular in cross-section, glaucous with a finely pointed red tip. Flowers 7cm(2¾in) across, purple, carried above the leaves in late spring to late summer.
L. stipulaceus Cape
Shrubby, to 40cm(16in) tall, producing many short shoots from the leaf axils. Leaves 4–5cm(1½–2cm) long, triangular to semi-cylindrical in cross-section, bright green with translucent dots. Flowers to 4cm(1½in) across, purple, opening in summer.

LANTANA

Verbenaceae

Origin: Tropical parts of the Americas and West Africa. A genus of 150 evergreen shrubs and perennials, those cultivated having opposite pairs of ovate to oblong, toothed leaves and small, tubular, 5-lobed flowers in dense, rounded, flattened heads. They make fine pot and tub plants for the conservatory and home. Propagate by cuttings in late summer, or in early spring in warmth. *Lantana* derives from an old Latin name for *Viburnum*, an unrelated genus with somewhat similar flowers.

   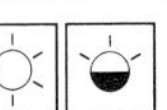  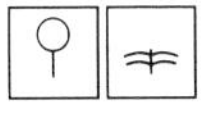

TOP LEFT *Lamium maculatum*
ABOVE *Lampranthus haworthii*
LEFT *Lantana camara*

Species cultivated

L. camara Yellow sage Tropical America
Shrub 1–2m(3–6½ft) tall, the stems sometimes prickly. Leaves 5–10cm(2–4in) long, ovate to oblong, dark green with a wrinkled appearance reminiscent of a sage leaf. Flowers in heads to 5cm(2in) wide, the central florets opening yellow and ageing red or white, opening from late spring to autumn. Many forms and cultivars are grown with colours ranging from yellow to red, pink and white, but changing colour as they age; those listed as *L.c. hybrida* are shorter.
L. montevidensis (*L. delicata*, *L. delicatissima*, *L. sellowiana*) South America
Trailing or pendulous species with stems to 90cm(3ft) long. Leaves to 3cm(1¼in) long, ovate, coarsely toothed. Flowers in long-stalked clusters 2–3cm(¾–1¼in) wide, rose-purple, each with a yellow eye. An attractive plant for a hanging basket.
L. sellowiana See *L. montevidensis*.

LAPAGERIA

Philesiaceae (Liliaceae)

Origin: Central Chile. A genus of one species found in woods and thickets; a semi-woody evergreen climber with handsome foliage and uniquely

Lapageria rosea 'Nashcourt'

beautiful flowers. It is the national flower of Chile where it is called *copihue*. A very rewarding plant for a conservatory where it is best trained on strings up the back wall and under the roof so that its flowers can hang down. After flowering, edible yellow-green, oblong fruits may develop, the seeds of which provide a ready means of increase. These should be soaked for two days before sowing. *Lapageria* was named for Napoleon's Empress, Josephine de la Pagerie, who was a keen gardener.

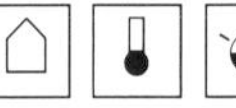 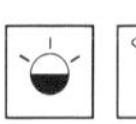  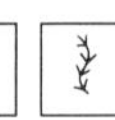

Species cultivated

L. rosea Chilean bell flower
Stems twining to 3m(10ft) or more. Leaves 8–14cm(3–$5\frac{1}{2}$in) long, alternate, ovate-lanceolate, leathery and deep glossy green. Flowers bell-shaped, to 7cm($2\frac{3}{4}$in) long with six waxy-textured petals, rose-pink, faintly spotted within, opening in summer and autumn. *L.r. albiflora* has white flowers; 'Nashcourt' has clear pink flowers. Other forms with larger, darker, or striped flowers are sometimes available.

LAPEIROUSIA (LAPEYROUSIA)

Iridaceae

Origin: Temperate areas of South Africa. A genus of about 50 species of cormous plants. The narrow, sword-shaped leaves are borne in two ranks and the branched flower spikes carry 6-petalled flowers usually in shades of pink or red. They flower from summer onwards and can be stored dry in their pots over winter, then re-potted and brought into growth in early spring. They are attractive flowering pot plants for the conservatory and can be brought into the house when ready to flower. Propagate by separating corms, or by seed in spring. *Lapeirousia* was named for Baron Philippe Picot de la Peyrouse (1744–1818), a Pyrenean botanist.

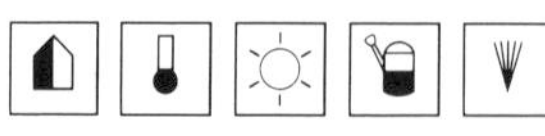

Species cultivated

L. corymbosa
Leaves usually only two per corm, sickle-shaped, to 10cm(4in) or more long. Flowers 1–1.5cm($\frac{3}{8}$–$\frac{5}{8}$in) wide, erect, deep purple-blue, borne in a compact cluster on a 15cm(6in) slender, erect stem.

L. laxa (*L. cruenta, Anomatheca cruenta*)
Leaves narrowly sword-shaped, to 20cm(8in) long, in fan-like tufts. Flowers to 2.5cm(1in) long, slender-tubed, carmine-crimson, the outer three of the six spreading lobes bearing darker red, basal blotches; they are carried in short racemes on slender stems to 25cm(10in) tall. *L.l.* 'Alba' has pure white flowers.

LATANIA

Palmae

Latan palms

Origin: Mauritius and adjacent island groups (Mascarene Islands). A genus of four species of elegant fan palms with erect, unbranched stems topped by a crown of deeply fingered leaves of rounded outline. The trees are dioecious, but the clusters of tiny flowers which are borne in the leaf axils and rather hidden, are never produced on small pot specimens. Like so many palms, they are decorative house or conservatory plants when

Lapeirousia laxa

young. Propagate by seed sown in spring, in warmth. *Latania* is a Latinized form of *latanier*, the local name used in Mauritius.

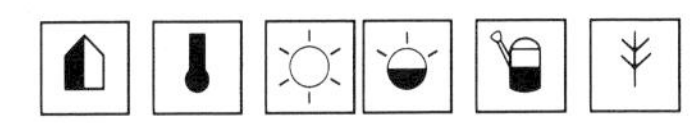

Species cultivated

L. aurea See *L. verschaffeltii.*
L. borbonica See *L. lontaroides.*
L. commersonii See *L. lontaroides.*
L. loddigesii Silver latan Mauritius
Leaves of adult plants to 1m(3ft) wide, much less on young pot plants, rigid and blue-grey, the veins and long stalks orange to red-brown. Slow growing as a pot plant, this handsome palm attains 10m(33ft) grown in the open.
L. lontaroides (*L. borbonica, L. commersonii*) Red latan Mauritius
Leaves of adult plants 2m(6½ft) or more wide, much less on potted specimens, the segments finely toothed, grey-green when mature, reddish or purplish-flushed when young, this colour retained on the veins and margins; stalks long, thorny, orange. This most attractive species reaches 15m(48ft) or more when grown in the open.
L. verschaffeltii (*L. aurea*) Yellow latan Rodriguez Islands
Leaves of mature plants 1m(3ft) or more wide, less on potted specimens, green above, white-downy beneath, the stalks smooth, long, and orange to yellow. This robust palm reaches 15m(48ft) outside, but makes a handsome container plant.

LAURENTIA See ISOTOMA

LAURUS

Lauraceae
Laurels

Origin: Mediterranean region and North Atlantic Islands (Macronesia). A genus of two species of evergreen trees of which one is commonly grown both as a decorative foliage plant and for its leaves which are used as a flavouring in cooking. It makes an excellent pot or tub specimen and can be pruned or cut back if it gets too large. For a sturdy plant, standing outside in summer is recommended. Propagate by cuttings in summer, or by seed. *Laurus* is the Latin name of this plant.

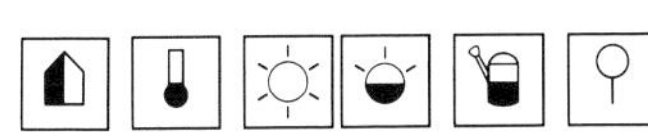

Species cultivated

L. nobilis Bay, Sweet bay Mediterranean region
To 10–20m(32–65ft) in the wild, but easily kept to 1.2m(4ft) in a pot. Leaves 5–8.5cm(2–3½in) long, narrowly oval to ovate, leathery, glossy deep green and strongly aromatic when crushed. Flowers unisexual, produced in axillary clusters in spring. Fruits a black ovoid berry-like drupe, 1–1.5cm(⅜–⅝in) long. Hardy. *L.n.* 'Aurea' has golden-tinted foliage.

LAVANDULA

Labiatae
Lavenders

Origin: Macronesia, Mediterranean region south and west to Somalia and India. A genus of about 20 species of evergreen shrubs which have strongly aromatic, linear to oblong-lanceolate leaves in opposite pairs and fragrant, 2-lipped, tubular flowers in dense spikes. The species described are attractive plants for the conservatory, the smaller ones for the home window sill. Propagate by cuttings in late summer or by seed in spring. *Lavandula* derives from the Latin *lavo*, to wash, referring to its use for perfuming scent, soap, shampoo and other toiletries.

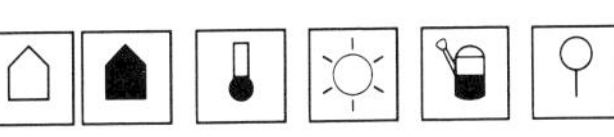

Species cultivated

L. angustifolia (*L. spica, L. vera*) English/Common lavender Mediterranean region
Shrub to 60cm(2ft) or more tall. Leaves 2–5cm(¾–2in) long, linear, white-downy when young. Flowers about 1cm(⅜in) long in 8cm(3in) spikes at the end of almost leafless stems to 30cm(12in) in height. Several cultivars are available, the smaller ones most suitable for pot culture: 'Hidcote' ('Nana Atropurpurea') is compact to 40cm(16in) in a container; 'Loddon Pink' is similar with pink flowers; 'Munstead' is compact with bright lavender-blue flowers 'Twickle Purple' is compact with very fragrant rich purple flowers. Hardy.
L. canariensis See *L. multifida canariensis.*
L. dentata S. Spain and the Balearic Isles
Shrub to 30cm(1ft) or more. Leaves linear to oblong, toothed or pinnately cut into many small lobes giving the appearance of small teeth. Flowers 8mm(⅓in) long, dark purple in spikes which are tipped by a cluster of purple bracts.
L. lanata Spain
Similar to *L. angustifolia*, but densely white-felted. *L.l.* 'Nana' is much smaller, a charming silvery-white shrublet.
L. multifida Western Mediterranean, Portugal
Fairly bushy shrub to 90cm(3ft) in height. Leaves about 4cm(1½in) long, ovate in outline but bipinnately divided into grey linear segments. Flowers blue-purple, about 1.2cm(½in) long, in solitary or 3-pronged spikes to 6cm(2½in) in length in summer.
L. pedunculata See *L. stoechas pedunculata.*
L. pinnata Madeira and Canary Islands
Bushy shrub eventually to 90cm(3ft) in height. Leaves to 6cm(2½in) long, broadly lanceolate to narrowly ovate in outline, pinnately divided into a dozen or more linear grey-white lobes. Flowers lavender-purple, 6mm(¼in) or more in length, in simple or branched spikes 5–8cm(2–3in) long.

Latania loddigesii

Laurus nobilis

ABOVE *Leptospermum scoparium* 'Red Damask'
ABOVE CENTRE *Leptospermum scoparium* 'Ballerina'
TOP *Lavandula stoechas*

L. spica See *L. angustifolia.*
L. stoechas French lavender Mediterranean region, Portugal
Shrub 30cm(1ft) or more tall. Leaves 1–4cm(⅜–1½in) long, linear to oblong-lanceolate, grey-green. Flowers 6–8mm(¼–⅓in) long, dark purple in 2–3cm(¾–1¼in) spikes topped by a tuft of purple bracts; flowering stems are leafy.
L. stoechas pedunculata (*L. pedunculata*) Central Spain, Portugal
A little larger in all its parts with relatively longer flowering stems – up to 25cm(10in) long.
L. vera See *L. angustifolia.*

LEEA

Leeaceae (*Vitidaceae*)

Origin: Tropical Africa, Asia and Australia. A genus of 70 species of shrubs and trees, one of which is sometimes available and an addition to containerized specimen plants. Propagate by cuttings in summer. *Leea* is named for James Lee (1715–1795), partner with Lewis Kennedy who owned The Vinyard, a Hammersmith (London) nursery.

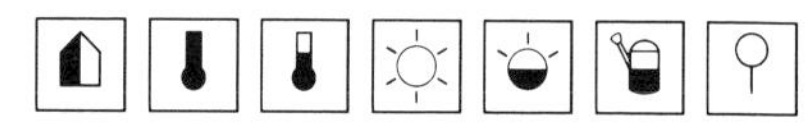

Species cultivated
L. coccinea Burma
Shrub to 2m(6½ft) in height, but fairly slow-growing and blooms when young. Leaves bipinnate or partially tripinnate, 30–60cm(1–2ft) long or more, the leaflets 5–10cm(2–4in) long, elliptic to obovate, strongly undulate, slender-pointed, rich glossy green when mature, bronzy when young. Flowers red in bud, opening pink, about 1cm(⅜in) wide, 5-petalled, in flattish terminal corymbs, 8–13cm(3–5in) across. It seems possible that this species is sometimes confused with the very similar *L. guineensis* from tropical Africa.

LEPISMIUM See RHIPSALIS
LEPTOSIPHON See LINANTHUS

LEPTOSPERMUM

Myrtaceae
Tea trees

Origin: Australia, New Zealand and Malaysia. A genus of 40 to 50 species of evergreen trees and shrubs. The leaves are small and linear to rounded and the flowers 5-petalled. The species described below are free-flowering and suitable for a sunny conservatory. The dwarf form of *L. scoparium* can be grown in the home. Propagate by heel cuttings in late summer or by seed, both in heat. *Leptospermum* derives from the Greek *leptos*, slender and *sperma*, seed.

 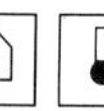   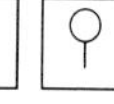

Species cultivated
L. nitidum Shiny tea tree Victoria, Tasmania
In the wild a large shrub, but easily kept to about 1.5m(5ft) in a tub. Stems slender, usually freely branching. Leaves oblong to oval, sharply pointed, leathery, glossy ciliate and more or less hairy beneath, 8–20mm(⅓–¾in) long. Flowers 2cm(¾in) or more wide, white with a red zone in the centre. 'Copper Sheen' has purple-flushed foliage and cream flowers; 'Jervis Bay' has extra large lavender blooms.
L. scoparium Manuka, New Zealand tea tree
New Zealand, Australia (Victoria, Tasmania)
Shrub 2–4m(6½–13ft) tall, occasionally a tree in the open ground, remaining shrubby in a pot. Leaves 8–15mm(⅓–⅝in) long, narrowly elliptic to oblanceolate, sharp-pointed, silky-hairy beneath. Flowers to 1.5cm(⅝in) across, solitary, white, opening in late spring. A number of cultivars are grown including: 'Chapmanii', bronze-flushed leaves and rosy-red flowers; 'Keatleyi', reddish-flushed young growth and large soft-pink flowers; 'Nanum', only 30cm(1ft) tall, the best for a window sill; 'Red Damask', double, cherry-red flowers; and 'Snow Flurry', double white flowers greenish centred, with reddish foliage.

LEUCADENDRON

Proteaceae

Origin: South Africa. A genus of 80 species of evergreen shrubs and trees allied to *Protea.* They have entire, usually leathery-textured leaves and terminal cone-like flower spikes surrounded by leaf- or petal-like, but not colourful, bracts. The one species described here is a spectacular foliage plant. Propagate by seed in spring. *Leucadendron* derives from the Greek *leukos*, white and *dendron*, a tree, referring to the silvery-white foliage of the species described below.

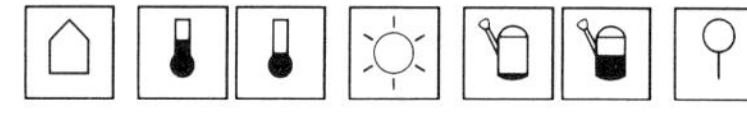

Species cultivated
L. argenteum Silver tree Cape (South west).
Medium-sized tree in the wild, but can be kept to about 2.5m(8ft) or so in a tub. Leaves 8–10cm(3–4in) long, lanceolate, firm-textured, covered with long, silky white hairs. Flower heads almost globular, surrounded by leaf-like bracts, but seldom on containerized specimens. Needs good ventilation and a sandy, peaty soil.

LEUCHTENBERGIA

Cactaceae

Origin: Mexico. A genus of one species of unique cactus having papery spines at the tips of curiously leaf-like tubercles. It can be grown in the home and conservatory where its un-cactus like appearance usually catches the eye of visitors. Propagate by seed in spring. *Leuchtenbergia* honours Prince Maximilian E.Y.N. von Beauharnais (1817–1852) of Leuchtenberg, Oberfpaz, Germany.

  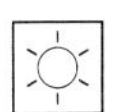  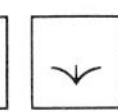

Species cultivated

L. principis Agave cactus Central and northern Mexico
In overall appearance rather like a small agave. Root parsnip-like; stem shortly ovoid, covered with slender tubercles; each tubercle sharply triangular in cross-section, 10–15cm(4–5in) long, bluish-green, erect and spreading. Spines irregular in length, six to 14 radiating from the tubercle apices, flattened, with soft tapering tips, the longest to 10cm(4in) long. Flowers satiny yellow, fragrant, funnel-shaped, 5–6cm(2–$2\frac{1}{2}$in) wide, borne on the young tubercle tips.

LEFT *Leuchtenbergia principis*
BELOW LEFT *Leucospermum cordifolium*

LEUCOCORYNE

Alliaceae (Amaryllidaceae)

Origin: Chile. A genus of five to six species of bulbous plants with basal channelled leaves and umbels of 6-tepalled flowers on erect stems, opening in spring. They make pretty flowering plants for the conservatory or sunny window sill. Propagate by offsets or by seed in spring. *Leucocoryne* derives from the Greek *leukos*, white and *koryne*, a club, referring to the all-white sterile stamens.

  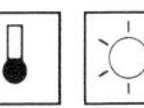 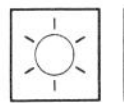  

Species cultivated

L. ixioides Glory of the sun
Leaves narrow, 20–30cm(8–12in) long. Flowers 3–4cm($1\frac{1}{4}$–$1\frac{1}{2}$in) across, starry, pale blue-white and very fragrant; they are borne in umbels of six to ten on slender stems 30–40cm(12–16in) tall.

L. purpurea
Very similar to *L. ixioides* and sometimes considered to be a variety of that species, the flowers lilac-purple with a reddish centre.

LEUCOSPERMUM

Proteaceae

Origin: South Africa. A genus of about 40 species of evergreen shrubs closely allied to *Protea*. They have densely borne, entire, leathery leaves and terminal rounded heads of crowded, tubular flowers with very long protruding styles. In some species the flower heads are spectacularly showy. They need a well ventilated conservatory and large pots or tubs of a peat and sand mixture to thrive. Fertilizers rich in phosphates must be avoided (see under culture of *Banksia*). *Leucospermum* derives from the Greek *leukos*, white and *sperma*, seed.

 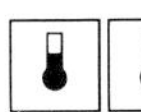 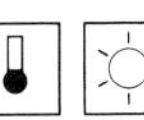 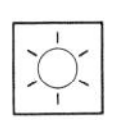  

Species cultivated

L. cordifolium (*L. nutans*) Nodding pincushion
Spreading shrub eventually to about 1.2m(4ft) in height and more in spread. Leaves elliptic to ovate-cordate, 3–8cm($1\frac{1}{4}$–3in) long, 3-toothed at their tips. Flower heads solitary, 8–10cm(3–4in) wide, the individual florets waxy, brick-red to orange, borne in summer.

L. reflexum
In its own country or planted out, an erect shrub to 3m(10ft) or more, but half this in a tub. Leaves lanceolate to oblong, to 4cm($1\frac{1}{2}$in) or more in length, downy, glaucous. Flower heads solitary or in pairs, about 4cm($1\frac{1}{2}$in) wide by twice as long, orange with red styles, borne in summer.

L. nutans See *L. cordifolium*.

LIBERTIA

Iridaceae

Origin: Australia, New Zealand, New Guinea and South America. A genus of 12 species of evergreen perennials with fan-like clusters of sword-shaped leaves. The 3-petalled flowers are borne in panicles on long wiry stems in summer. They are best in a conservatory, but can be brought into the home

Libertia formosa

when they flower. Propagate by division after flowering, or in spring, also by seed in spring. *Libertia* was named for Marie Libert (1782–1863), a Belgian botanist.

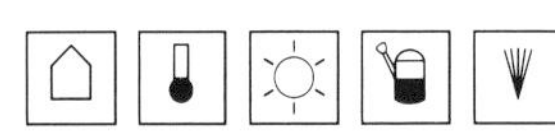

Species cultivated

L. formosa Chile
Leaves to 45cm(1½ft) long, narrow. Flowers 2cm(¾in) wide, with white petals; they are carried in narrow panicles on stems 60–120cm(2–4ft) high.

L. grandiflora New Zealand
Leaves to 50cm(20in) or more long. Flowers 2–3cm(¾–1¼in) wide, pure white, borne in loose panicles on stems to 90cm(3ft) high.

L. ixioides New Zealand
Leaves 30–50cm(12–20in) long, with a paler green midrib and with orange-brown along the margins. Flowers 1.5–2.5cm(⅝–1in) wide, white, yellow-flushed on the reverse and carried in leafy panicles on stems to 60cm(2ft) tall.

LIBONIA See JUSTICIA

Ligularia tussilaginea 'Aureo-maculata'

LICUALA

Palmae

Origin: S.E. Asia, Malaysia, Bismark and Solomon Islands. A genus of at least 100 species of mostly small palms. Solitary or cluster-stemmed, they have palmate or costapalmate (in between palmate and pinnate) leaves which may be almost entire or divided into lobes or longer segments. The panicles of small, tubular or urn-shaped, 3-lobed flowers are red or orange. Fruits are seldom produced on containerized specimens. The species described below makes an elegant specimen plant. Propagate by seed in spring. *Licuala* is a Latinized version of the Moluccan vernacular name *leko wala*.

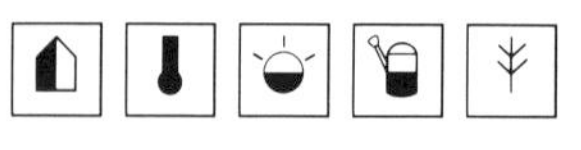

Species cultivated

L. elegans (*L. gracilis*) See *L. pumila*.

L. grandis (*Pritchardia grandis*) New Hebrides
Stems solitary, erect, fairly slender, eventually to 2m(6½ft) in height. Leaves palmate (fan-shaped), shallowly pleated, boldly toothed or shortly lobed, rich, bright, lustrous green, 60–90cm(2–3ft) wide on stalks of the same length.

L. pumila (*L. elegans*, *L. gracilis*) Java, Sumatra
Stems solitary or clustered, eventually to 1.5m(5ft), but often less. Leaves palmate, about 45cm(1½ft) wide, deeply divided into six to eight broad segments or many more narrower ones, rich green.

LIGULARIA

Compositae

Origin: Temperate Europe and Asia. A genus of 80 to 150 species of perennials most of which are deciduous, but the one described below is, however, evergreen. Unlike most of the other species it is not hardy, but makes a bold foliage plant for a large container, or can be planted out in the conservatory border. *Ligularia* derives from the Latin *ligula*, a strap, with reference to the shape of the ray florets of the daisy flowers.

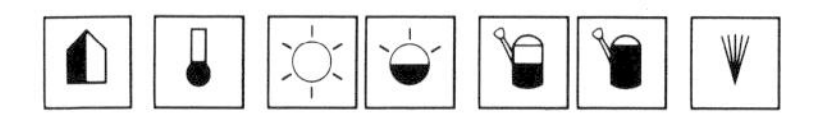

Species cultivated

L. tussilaginea (*Senecio kaempferi*, *Farfugium grande*, *F. japonicum*) China, Korea, Japan, Taiwan
Clump-forming, 90–120cm(3–4ft) across when planted out, but can be grown in 25cm(10in) pots if regularly divided. Leaves almost circular with faceted or very shallowly lobed margins, up to 30cm(1ft) wide, thick-textured, rich glossy green. Flower heads like 4–6cm(2–2½in) wide yellow daisies in a loose corymbose cluster above the leaves in autumn. In cultivation the normal green-leaved species is seldom seen, its place being taken by the

following cultivars: *L.t.* 'Argentea' ('Albovariegata', 'Variegata'), with leaves irregularly margined and segmented creamy-white and grey; 'Aureomaculata', popularly known as Leopard plant, with leaves bearing a scattering of large round yellow blotches; and 'Crispata', green with a parsley-like frill around the margins.

LILIUM

Liliaceae
Lilies

Origin: Northern hemisphere. A genus of about 80 species of perennial plants with bulbs made up of a number of separate overlapping scales. They have erect stems with narrow leaves and racemes of colourful flowers which can be trumpet-, funnel-, star-, bowl- or Turk's-cap-shaped. With the exception of the Easter lily (*L. longiflorum*), all the species and cultivars listed below are reasonably frost-hardy and usually thought of as plants for the open garden. However, almost all lilies respond to pot culture and, depending upon the available warmth, will bloom well ahead of their usual season in the unheated, cool, or temperate conservatory. When in bloom they can be brought into the home. Any standard compost is suitable and 20–30cm(8–12in) pots the best sizes, though single bulbs can be grown in 15cm(6in) containers. Place the stem-rooting species in the bottom third of the pot so that at least 10cm(4in) of compost is above the nose of the bulb. After blooming, plunge the containers outside in a sheltered site. If kept well fed and re-potted annually in autumn they will flower as regularly as any more orthodox pot plants. Propagate by separating bulblets when potting, by sowing stem bulbils in late summer, removing bulb scales in winter or, with the species only, sowing seed in spring. *Lilium* is a form of the Latin vernacular name for the Madonna lily.

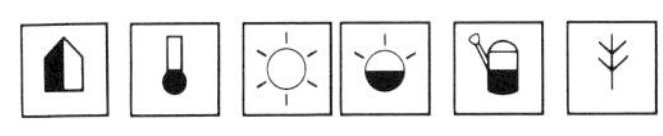

Species cultivated

L. auratum Gold-rayed lily Japan
Stems 1–1.4m(3–4½ft) tall. Leaves to 23cm(9¼in) long, ovate to lanceolate, scattered up the stem. Flowers 25–30cm(10–12in) across, bowl-shaped with widely spreading tepals, each white, with a golden-yellow band along the centre and red and yellow spots; they are very fragrant and up to ten or more are borne on each stem in late summer.

L.a. rubrum
Red bands on the tepals.

L. formosanum (*L. philippinense formosanum*, *L. longiflorum formosanum*) Taiwan
Stems 1–1.4m(3–4½ft) tall. Leaves to 20cm(8in) long, linear, dark green along the length of the stem, densest on the lower parts. Flowers 13–20cm(5–8in) long, funnel-shaped, white, marked outside with chocolate to purple, fragrant; they open in summer and autumn.

L. lancifolium
The most recent name for *L. tigrinum* (*q.v.*) and also an old synonym of *L. speciosum*.

L. longiflorum Easter/White trumpet lily
Japan
Stems to 90cm(3ft) tall. Leaves to 18cm(7in) long, lanceolate, pointed, glossy-green. Flowers 13–18cm(5–7in) long, pure white, fragrant, opening throughout the summer.

L. monadelphum Caucasian lily USSR (Caucasus)
Stems to 1.3m(4¼ft) tall. Leaves to 13cm(5in) long, lanceolate to oblanceolate, scattered along the stem. Flowers pendent and bell-shaped, to 9cm(3½in) long, tepals bright yellow, recurved, sometimes spotted inside with blackish-purple and with a reddish flush outside; they are fragrant, though not everyone finds this pleasant, and bloom in summer.

L. nepaulense Himalaya
Stems 60–90cm(2–3ft) in height, bearing fairly widely scattered lanceolate to oblong-lanceolate leaves 10–15cm(4–6in) long. Flowers solitary or two or three together, pendent, elegantly funnel-shaped, the tepal tips recurved, pale, glowing green-yellow, the basal half inside deep red-purple, borne in summer. Unlike most lilies the stem runs underground before emerging and so invariably when in containers the shoots come up at the edge. In the conservatory border shoots may emerge 60cm(2ft) from the bulb.

L. regale Regal/Royal lily Western China
Stems 75–120cm(2½–4ft) tall (more in the open ground). Leaves to 13cm(5in) long, linear, dark

Lilium formosanum

ABOVE *Limonium suworowii*
TOP LEFT *Lilium* 'Abendglut'
TOP RIGHT *Lilium* 'Manuela'

green, scattered along the stem. Flowers 12–15cm(4¾–6in), funnel-shaped, white flushed with yellow at the base inside, and with red-purple on the outside, fragrant, borne in umbel-like clusters in summer.

L. speciosum (*L. lancifolium*) Japanese lily
Japan, China, Taiwan
Stems to 1.4m(4½ft) tall. Leaves to 18cm(7in), broadly lanceolate along the length of the stems. Flowers 10–15cm(4–6in) wide, nodding, white spotted with rich crimson papillae, densest near the centre of the tepals which are slightly wavy and reflexed; fragrant, late summer and autumn.

L.s. album
Without the red papillae.

L.s. rubrum
Carmine tepals, edged with white.

L. tigrinum (*L. lancifolium*) Tiger lily China, Korea, Japan
Stems purplish with cobweb-like hairs, 90–180cm(3–6ft) tall. Leaves to 19cm(7½in) long, linear-lanceolate, scattered along the stem, with purple-black bulbils in their axils. Flowers to 10cm(4in) long, nodding, bright orange-red with black-purple spotting, tepals strongly reflexed and opening in summer and autumn.

Hybrid cultivars
Most of the modern breeding of lilies has concentrated on raising colourful hybrids using a wide range of species. These are classified into eight divisions according to their main parentage and consequent flower shapes. All are good for pot culture with the Asiatic hybrids such as the Mid-Century group perhaps the easiest.

LIMONIA See CITRUS

LIMONIUM (STATICE)

Plumbaginaceae
Statice, Sea lavenders

Origin: World-wide, especially in saline areas both coastal and semi-desert. A genus of some 300 species of annuals, perennials and sub-shrubs with chiefly basal, entire or pinnately lobed leaves. The flowers are small with a tubular calyx and 5-lobed corolla; they form spikelets grouped into stiff panicles which are borne on erect, wiry stems. Propagate by seed sown in autumn or spring. *Limonium* derives from the Greek *leimon*, a meadow, an allusion to its salt marsh habitat.

 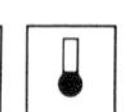 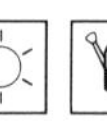 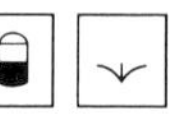

Species cultivated

L. sinuatum Notch-leaf/Winged statice
Northern coasts of the Mediterranean
Although a biennial or perennial, it is best grown as an annual. Leaves to 10cm(4in) long, deeply pinnately lobed and waved on a winged stalk. Flowers white, from a 1–1.2cm(⅜–½in) long, purple, funnel-shaped calyx, flowering in late summer and autumn though earlier from an autumn sowing. Cultivars and seed strains are available in shades of red, pink, orange, yellow, blue and lavender.

L. suworowii Russian/Rat's-tail statice
Caucasus, Iran to Central Asia
Leaves 15–25cm(6–10in) long, oblanceolate, lobed and waved. Flowers 4mm(⅙in) long, rose-pink, grouped into narrow, finger-like spikes, the whole making up panicles on stems to 45cm(1½ft) tall. If sown from autumn onwards a succession of flowers can be had from early summer to autumn.

LINANTHUS

Polemoniaceae

Origin: Western North America and Chile. A genus of 35 species of mainly small annuals, but also some perennials. They have usually deeply cut leaves and bell- to funnel-shaped flowers with five spreading petals. The species described below makes a pretty pot plant for summer flowering in a conservatory or on a sunny window sill. Propagate by seed in autumn or spring. *Linanthus* derives from the Greek *linon*, flax and *anthos*, a flower.

  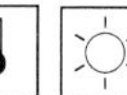  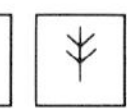

Species cultivated

L. androsaceus (*Gilia androsaceus*, *Leptosiphon androsaceus*, *Leptosiphon* French Hybrids [of seedsmen]) California
Erect, freely-branched plant 10–20cm(4–8in) tall. Leaves 1–3cm(⅜–1¼in) long, finely palmately-divided into five to nine linear to oblanceolate segments. Flowers to 1.5cm(⅝in) long, the slender tube opening to five rounded lobes at the mouth, pink, lilac, yellow and white, opening in summer and autumn.

LINARIA

Scrophulariaceae

Origin: Northern temperate zone mostly in the Mediterranean region. A genus of 100 species of annuals and perennials with narrow leaves and racemes of *Antirrhinum*-like flowers with a basal spur. Propagate by seed sown in spring for a summer-autumn flowering, in autumn for an earlier display. *Linaria* derives from the Greek *linon*, flax, because of the similarity of its leaves to those of flax.

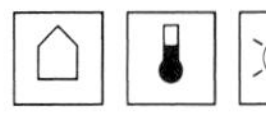 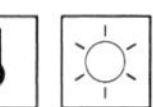  

Species cultivated

L. cymbalaria See *Cymbalaria muralis*.
L. maroccana Toadflax Morocco
Erect, branched annual to 30cm(1ft) or more tall. Leaves narrowly linear. Flowers 2–3cm($\frac{3}{4}$–$1\frac{1}{4}$in) long including the pointed spur, violet-purple, yellow in the throat. Cultivars and seed strains are available in a wide colour range including blues, reds, pinks and yellows, some of which are probably of hybrid origin. 'Fairy Bouquet' is a more dwarf selection.

Linaria maroccana

LIPPIA

Verbenaceae

Origin: North and South America, Africa. A genus of 200 species of shrubs and perennials with entire leaves and spikes or panicles of small, tubular flowers. The species described below is best in a conservatory, though it can be grown in a sunny window where it can be cut back hard in spring to keep it small and tidy. Propagate by cuttings in either spring or summer or by seed sown in spring. *Lippia* was named for Augustin Lippi (1678–1701), an Italian botanist killed in Ethiopia while plant collecting.

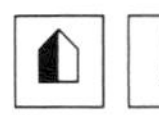 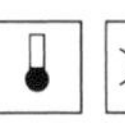   

Species cultivated

L. citriodora (*Aloysia citriodora, Verbena triphylla*) Lemon verbena Argentina, Chile
Deciduous shrub rarely more than 90cm(3ft) in a pot if cut back annually; to 3m(10ft) in the open garden. Leaves 6–10cm($2\frac{1}{2}$–4in) long, in opposite pairs, or whorls of three, lanceolate, strongly lemon-scented (source of verbena oil). Flowers very small, 2-lipped, white or very pale purple, borne in axillary spikes or terminal panicles in late summer and autumn.

LIRIOPE

Liliaceae

Lily-turfs

Origin: China, Japan and Vietnam. A genus of five or six species of evergreen perennials with narrowly linear, leathery leaves, 6-tepalled flowers in racemes and berry-like black fruits. They make attractive pot plants. Propagate by division in spring or by seed when ripe. *Liriope* was named after the wood nymph of Greek mythology who was the mother of Narcissus.

    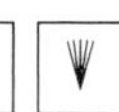

Species cultivated

L. exiliflora (*L. muscari exiliflora*) China, Japan
Tufted and rhizomatous, forming colonies if planted out in the conservatory border. Leaves strap-shaped, 30–40cm(12–16in) long, arching at the tips, dark green. Flowering stems above the leaves, purple-tinted, the raceme loose; each flower violet-purple, widely expanded to about 8mm($\frac{5}{16}$in) in late summer.
L. graminifolia China, Vietnam
Tufted and rhizomatous, forming turf-like colonies if planted out. Leaves to about 30cm(1ft) long, but usually less, very narrow and grass-like, strongly arching, usually deep green. Racemes to 20cm(8in) or more, slender and loose; flowers somewhat cupped, about 6mm($\frac{1}{4}$in) wide, light violet to white, borne in late summer. *L.g. minor* is somewhat smaller.
L.g. densiflora See *L. muscari*.
L. hyacinthiflora See *Reineckea carnea*.
L. muscari (*L. graminifolia densiflora, L. platyphylla*) China, Japan
Clump-forming, usually densely so. Leaves 40–60cm(16–24in) long, strap-like, slightly arching, deep green. Racemes level with and just above the leaves, long and dense; flowers somewhat bell-shaped to 1cm($\frac{3}{8}$in) wide, violet-purple, appearing in late summer to autumn. The most decorative species for pot culture and very shade tolerant. Among cultivars, 'Big Blue', 'Blue Spire' and 'Majestic' are all bigger and finer, the last two with the racemes more or less crested at the tips. 'Variegata' has the leaves with cream margins; 'Gold Banded' is similar, but with narrow yellow leaf margins; and 'Munroe White' has white flowers.

BELOW *Lippia citriodora*
BOTTOM *Liriope muscari*

L. platyphylla See *L. muscari.*
L. spicata China, Vietnam
Rhizomatous and vigorous, forming turf-like colonies when planted out in the conservatory border and providing good ground cover. Leaves 30–45cm(1–1½ft) long, narrowly strap-shaped, arching, rich green. Racemes about level with the leaves, moderately dense; flowers pale violet to almost white, about 8mm($\frac{5}{16}$in) wide, appearing in late summer.

LISIANTHUS See EUSTOMA

LITHOPS

Aizoaceae
Living stones

Origin: Southern Africa. A genus of about 50 species of solitary or clump-forming succulents, each stem made up of two very fleshy swollen leaves which are fused to form an obconical plant body, slit at the rounded or flattened top. Through the slit emerge in autumn the large, many petalled daisy-like flowers which often completely hide the plant body beneath. They make very good plants for a sunny window sill or conservatory, needing treatment as for cacti. Many species are available and the list below is only a selection of these. Propagate by separating individual plant bodies in summer and treating as for cuttings, or by seed in spring. *Lithops* derives from the Greek *lithos*, stone and *ops*, like.

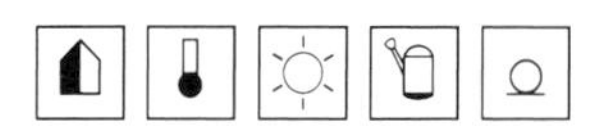

Species cultivated
L. bella Namibia
Plant bodies forming clumps 2.5–3cm(1–1¼in) tall, 2cm(¾in) wide, strongly convex at the top, brownish-yellow marked with darker branched lines. Flowers pure white.
L. erniana Namibia
Clump-forming, the plant bodies 2cm(¾in) high, obconical, oval and flattened at the top, the fissure dividing it into two unequal halves, reddish-grey with a pattern of maroon lines rather like heiroglyphics. Flowers white, 3cm(1¼in) wide.
L. lesliei Transvaal, Cape
Solitary or small clump-forming, the plant bodies 3–4.5cm(1¼–1¾in) tall, the top flattened, light brown or reddish-brown marked with dark greenish-brown spots and pits. Flowers golden-yellow.
L. pseudotruncatella Namibia
Clump-forming, the plant bodies to 3cm(1¼in) high, very broadly obconical with prominent fissures across the wide, flattish tops, light brownish-grey bearing a reddish-brown pattern of lines and dots. Flowers to 3.5cm(1½in) wide, rich yellow. A variable species in the patterning and colouring of the plant bodies.
L. salicola Orange Free State (Fauresmith – salt plain)
Plant bodies solitary or in small clumps, to 2.5cm(1in) high, obconical with a fairly wide, deep fissure, shallowly convex above, pale grey with transparent areas, almost obscured by a dark grey-green mosaic pattern. Flowers about 2.5cm(1in) wide, white.
L. schwantesii Namibia
Clump-forming, the plant bodies 3–4cm(1¼–1½in) high, the tops flattened to convex, greyish to yellowish-red, the reddish-brown dots and lines often bordered with rusty-yellow. Flowers yellow.
L. translucens See *L. herrei.*
L. turbiniformis Cape
Plant bodies solitary or in small groups, to 2.5cm(1in) high, obconical, almost flat on top with a narrow, shallow fissure and an elaborate network of sunken brown lines on a tan surface. Flowers up to 4cm(1½in) wide, bright yellow from whitish buds.

ABOVE *Lithops lesliei*
RIGHT *Lithops pseudotruncatella*

LITTONIA

Liliaceae

Origin: Africa and Arabia. A genus of six to eight species of tuberous-rooted herbaceous perennials related to *Gloriosa*. They have slender, climbing or sprawling stems and nodding, yellow to orange bell-shaped flowers followed by large capsules with large seeds coloured either brown or red. They are unusual pot plants and can be grown on canes or against the back wall of the conservatory. Propagate by separation of the tubers during the dormant season, or by seed in spring. *Littonia* was named for Dr Samuel Litton (1779–1847), Professor of Botany at Dublin.

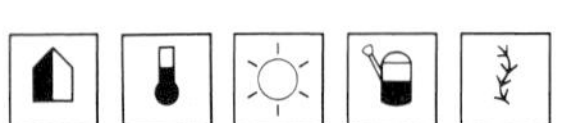

Species cultivated
L. modesta South Africa
Plant usually to 90cm(3ft) and occasionally to 1.5m(5ft). Leaves ovate to linear ending in slender tendrils, the lower ones whorled, the upper single or paired. Flowers 3–4cm(1¼–1½in) long, bright

orange, widely bell-shaped, borne singly in the upper leaf axils in summer.

LIVISTONA

Palmae
Fountain palms

Origin: Indo-Malaysia, Australia. A genus of about 30 species of mainly large fan palms. They have single, erect stems and a terminal cluster of costa-palmate (halfway between pinnate and palmate) leaves. Small, 3-lobed, cup-shaped flowers are borne on mature trees only, and are followed by berry-like drupes. Propagate by seed in spring. Young plants of the species described make effective containerized foliage plants. *Livistona* commemorates Patrick Murray, Baron of Livingstone, whose fine plant collections (prior to 1680) became the nucleus of the Royal Botanic Garden, Edinburgh.

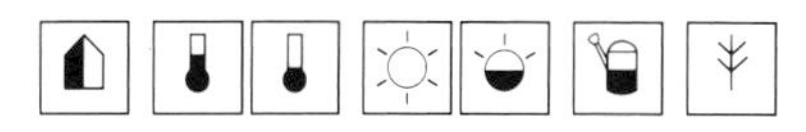

Species cultivated

L. australis Australian/Gippsland fountain palm Eastern Australia
In the wild a medium-sized tree, but fairly slow-growing in a pot. Leaves broadly oval to orbicular in outline, up to 1.2m(4ft) long on tubbed specimens (twice this in its native land), divided halfway to the midrib into numerous narrow lobes, rich, fairly glossy green; leaf stalks long, edged with spines.

L. chinensis Chinese fountain palm Southern Japan, perhaps also China.
In the wild a small tree, fairly slow-growing in containers. Leaves to 90cm(3ft) long in containers (at least twice this in the wild), rounded in outline, divided almost halfway to the midrib into many narrow, pendulous lobes, glossy rich green.

LOBELIA

Campanulaceae (Lobeliaceae)

Origin: World-wide, mostly in the tropics. A genus of about 375 species of annuals, perennials, shrubs and occasionally trees. The two species described below are short-lived perennials, but are best grown as annuals, flowering quickly from seed. They have alternate entire leaves and tubular flowers, basically 2-lipped, the upper divided into two narrow lobes, the lower into three. They make colourful pot plants for a window or conservatory. Propagate by seed in early spring, in warmth. *Lobelia* was named for the Belgian Matthias L'Obel (1538–1616), doctor to King James I and also a renowned botanist.

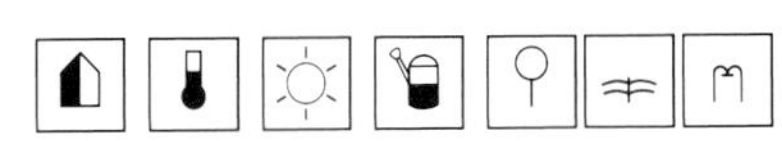

Species cultivated

L. erinus South Africa
Tufted, bushy plant to 15cm(6in) tall. Leaves 1–

Livistona australis

2.5cm($\frac{3}{8}$–1in) long, obovate to linear. Flowers 1.2–2cm($\frac{1}{2}$–$\frac{3}{4}$in) long, blue to purplish-blue with a white or yellowish eye, borne in summer. Cultivars are available in shades of blue, violet, red and white. The 'Cascade' cultivars have prostrate stems and are very attractive in hanging baskets.

L. longiflora See *Isotoma longiflora*.

L. tenuior Western Australia
Slender, erect to spreading plant to 30cm(1ft) tall. Lower leaves to 6cm($2\frac{1}{2}$in) long, linear, usually 3-lobed, upper leaves smaller and linear to lanceolate, without lobes. Flowers 2.5cm(1in) long, bright blue with a white or yellow eye, the centre lobe of the lower lip much wider than the outer.

Lobelia tenuior

LOBIVIA

Cactaceae

Origin: Andean South America. A genus of about 70 species of small to medium-sized rounded to shortly cylindrical cacti. They have funnel- to bell-shaped flowers which are relatively large and usually colourful, opening by day and closing at night. They make attractive pot or pan plants for the sunny window sill or conservatory. Propagate by

RIGHT *Lobivia pentlandii*
FAR RIGHT *Lophocereus schottii* and *L. s.* 'Monstrosus'

seed or offsets. *Lobivia* is an anagram of Bolivia where many species are found wild.

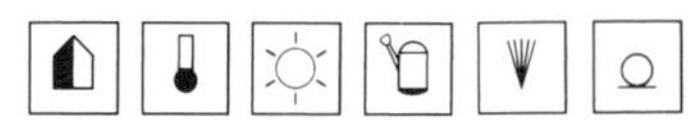

Species cultivated

L. aurea (*Echinopsis aurea*) Golden lily cactus W. Argentina
Stems globose to elongated, to 5–10cm(2–4in) tall. Ribs sharply prominent, 12 to 15, with brownish areoles. Radial spines brown, eight to ten, with a yellow-tipped central one to 3cm(1¼in) long. Flowers narrowly funnel-shaped, 9cm(3½in) long, bright yellow, freely produced.

L. hertrichiana S.E. Peru
Stems globular, to 10cm(4in) high, solitary or in small clusters, glossy bright green. Ribs about 11, notched above the round woolly areoles. Radial spines six to eight, yellow-brown, about 1.5cm(⅝in) long; one central spine, longer and paler. Flowers 6cm(2½in) long, bright scarlet.

L. jajoiana N. Argentina
Stems usually solitary, sometimes in small groups, globular at first, later ovoid, pale green. Ribs to 20, cut into prominent, sharply angled tubercles. Radial spines ten, reddish to whitish, 1cm(⅜in) long; one central spine, twice as long, deep brown to black. Flowers funnel-shaped, 6cm(2½in) long, dark purple-red, with a bluish sheen.

L. pentlandii (*Echinopsis pentlandii*) Bolivia, Peru
Stems globular to shortly cylindrical, 5–10cm(2–4in) tall, clump-forming, dark to greyish-green. Ribs ten to 15, prominent, divided into large, oblong tubercles. Radial spines seven to 12, sometimes as few as five, recurved, brown, to 3cm(1¼in) long; central spines one or missing, longer, straight or curved upwards. Flowers funnel-shaped, 5–6cm(2–2½in) long and almost as wide, carmine-pink to orange-red. Fruit globular, to 2cm(¾in) wide, green, edible and sweet. Several varieties are known, varying in stem shape and size and flower colour.

L. silvestri See *Chamaecereus silvestri*.

LOMARIA See BLECHNUM

LOPHOCEREUS

Cactaceae

Origin: Mexico and south-western USA. A genus of two species of tall cacti, one of which is commonly grown, making a good accent plant either in a container in the home or planted out in the conservatory. Propagate by seed in spring or by cuttings in summer. *Lophocereus* derives from the Greek *lophos*, a crest, and the allied genus *Cereus*; the stem tips bear crest-like tufts of hair.

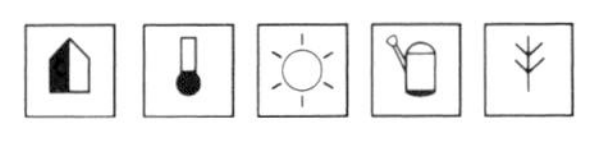

Species cultivated

L. schottii (*Pilocereus schottii*) W. Mexico, S. Arizona
Stems erect, columnar, in time branching from the base and reaching 3m(10ft) or more in height. Ribs about six or seven, sometimes less or more, prominent. Areoles large, the upper ones on flowering stems producing tufts of bristle-like, twisted grey spines; lower areoles and all those on young specimens produce four to ten thick, conical, black to grey spines. Flowers short-tubed, opening out flat to 4cm(1½in) across, greenish in bud, pink within, nocturnal, borne during spring and summer. *L.s.* 'Monstrosus' (Totem pole cactus) is almost spineless with curiously interrupted and irregularly swollen ribs.

LOPHOMYRTUS See MYRTUS

LOPHOPHORA

Cactaceae

Origin: USA (Southern Texas) and Mexico. A genus of two or three small species of spineless cacti, all very similar to each other and thought by some botanists expressions of one variable species. The one described below is a rewarding and easy plant for the home window sill or sunny conservatory. Propagate by seed sown in warmth in spring. *Lophophora* derives from the Greek *lophos*, a crest and *phoreo*, to bear, referring to the tufts of hair which are carried by each areole.

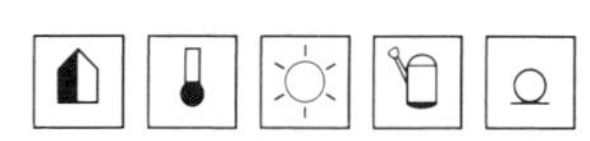

Species cultivated

L. williamsii Dumpling cactus, Peyote, Mescal button
Stem rounded, smooth, blue-grey, reaching 7cm(2¾in) wide by 5cm(2in) tall with age, but slow growing. It has a fleshy, turnip-shaped root. The five to 13 ribs are broad and flattened, bearing well-spaced areoles, with no spines, but an erect tuft of white wool (although seedlings do have very small spines at first). Flowers 2.5cm(1in) wide, pink, rarely white, arising from the centre of the plant and followed by red fruits about 1.5cm(⅝in) long. The plant contains poisonous, hallucinogenic alkaloids.

LOROPETALUM

Hammamelidaceae

Origin: Eastern Asia. A genus of one or two species of evergreen shrubs rather like a white-flowered *Hamamelis*, to which genus they are closely related. The species described is an attractive pot or tub plant for the cool conservatory or room. Propagate by cuttings in late summer, with bottom heat, and if possible, mist as they are not easy; alternatively by layering a branch into another pot, or by seed. *Loropetalum* derives from the Greek *loron*, a strap or thong and *petalon*, a petal.

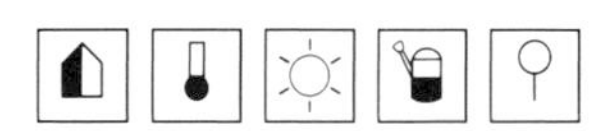

Species cultivated

L. chinense Assam, China, Japan
Rounded, freely branched shrub to 1.2m(4ft) or so tall. Leaves 2.5–6cm(1–2½in) long, alternate, pointed-ovate and asymmetrical. Flowers with 2–2.5cm(¾–1in) narrow petals, very effective *en masse*, opening in late winter and spring.

LOTUS

Leguminosae

Origin: Africa, Europe, Asia and North America, particularly in the Mediterranean region. A genus of about 100 species of perennials and sub-shrubs with trifoliate or pinnate leaves and typical pea flowers with strongly beaked keels, borne in axillary umbels. The species described are good pot plants for the conservatory or sunny window and are particularly decorative in a hanging basket. Propagate by seed in spring, or by cuttings in late summer. *Lotus* is from the Greek vernacular name *lotos* and is applied to various members of the pea family.

Loropetalum chinense

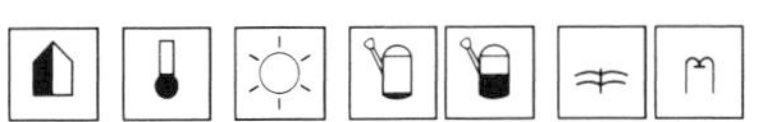

Species cultivated

L. berthelotii Coral gem, Parrot's beak, Winged pea Canary Isles (Tenerife), Cape Verde Is.
Silvery-hairy sub-shrub with arching or trailing stems to 60cm(2ft) long. Leaves pinnate, made up of three to seven linear leaflets 1–2cm(⅜–¾in) long. Flowers to 3cm(1¼in) long, scarlet to crimson, with a long-beaked keel and hood-shaped standard; they are borne singly or in umbels of two to three or more in summer.

L. mascaensis Canary Isles (Tenerife)
Silvery-hairy shrub to 30cm(1ft) tall, spreading more widely. Leaves pinnate, with three to seven linear leaflets 1.2–1.5cm(½–⅝in) long. Flowers 1.5–2cm(⅝–¾in) long, golden-yellow, the hooded standard flushed with red at the tip; they are borne in 3- to 7-flowered terminal umbellate clusters in late spring and early summer.

Lotus mascaensis

RIGHT *Luculia gratissima*
FAR RIGHT *Luffa cylindrica*

LUCULIA

Rubiaceae

Origin: Himalayas to China. A genus of five species of evergreen and deciduous shrubs with opposite pairs of undivided leaves and tubular flowers expanding to five spreading lobes which are borne in dense terminal panicles. The species described below are good plants for pots or tubs or can be grown in a ground-level bed in a conservatory. Prune back by two-thirds after flowering if a small plant is required. Propagate by heel cuttings in heat in late spring, or by seed. *Luculia* derives from the Nepalese vernacular name for *L. gratissima*.

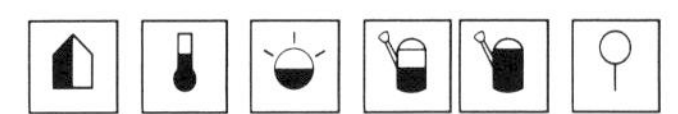

Species cultivated

L. grandifolia Bhutan
In the wild a small tree to 6m(20ft), but easily kept to a third of this height in a tub. Leaves elliptic to broadly ovate, the stalks and mid-veins red, the lateral veins prominent, up to 35cm(14in) in length. Flowers 6cm($2\frac{1}{2}$in) long, white, fragrant, in terminal clusters. Handsome in foliage and flower.

L. gratissima Himalaya
To 2m($6\frac{1}{2}$ft) or more in height, easily kept smaller in a pot. Leaves to 20cm(8in) long, ovate-oblong, with prominent veins. Flowers to 3cm($1\frac{1}{4}$in) wide, opening to five rounded lobes, pink, borne in large terminal panicles from late autumn to late winter.

L. pinceana Nepal, Assam
Shrub to 2m($6\frac{1}{2}$ft) and blooming at less than half this in a pot. Leaves up to 15cm(6in) long, white with pink tints, fragrant, borne in terminal clusters.

LUFFA

Cucurbitaceae

Origin: Tropical Asia. A genus of six species of annual climbers remarkable for their fruits which have a dense, fibrous skeleton which, in the species described below, provides the loofah. These are prepared by soaking the fruits in water to remove the skin, pulp and seeds. The plant is grown for curiosity and can be trained up the back wall of the conservatory, or on a framework of canes. Propagate by seed in warmth in spring. *Luffa* is the Latin form of the Arabic vernacular, loofah.

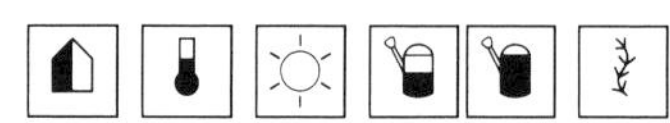

Species cultivated

L. cylindrica (*L. aegyptica*) Loofah, Sponge gourd Probably originally from India, but long widely grown in the tropics.
It is a tendril climber which can reach 4m(13ft), but can be kept smaller. Leaves 6–25cm($2\frac{1}{2}$–10in) long, broadly ovate to rounded, palmately 5- to 7-lobed, roughly hairy. Flowers 5–10cm(2–4in) across, monoecious, the five petals opening widely, the females borne singly, the males in 4- to 20-flowered axillary racemes. These are followed by the 30–60cm(1–2ft) long, cucumber-shaped fruits.

LYCASTE

Orchidaceae

Origin: Mountains of tropical America and Cuba. A genus of 40 species of epiphytic orchids with firm, conical to ovoid pseudobulbs with leaves from their tops and bases. These are oblong-lanceolate, deciduous and prominently veined or pleated. The flowers have three broad, spreading outer tepals (sepals) and three smaller inner ones (petals), often arching over the small 3-lobed labellum. They are borne singly, with or before the young foliage. Plants are successful in pots or on sections of tree bark. Propagate by division when re-potting. *Lycaste* is the name in Greek mythology for one of the daughters of King Priam of Troy.

  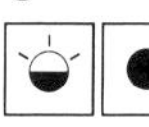  

Species cultivated

L. aromatica Mexico to Honduras
Pseudobulbs to 8cm(3in) tall. Leaves to 45cm(1½ft) long. Flowers to 7cm(2¾in) wide, yellow to orange-yellow, the sepals paler than the petals, the labellum having a wavy-edged middle lobe which is sometimes spotted with red; in texture they are waxy and have a strong lemony fragrance and are borne on 15cm(6in) stems mainly in spring.

L. cruenta Mexico, Guatemala, El Salvador
Pseudobulbs 8–10cm(3–4in) tall, long-ovoid and flattened. Leaves to 38cm(15in) long, pointed-elliptic. Flowers 7–10cm(2¾–4in) wide, sepals yellow-green, the smaller petals orange-yellow, the labellum spotted and marked at the base with dark red; they are very fragrant and borne on 45cm(1½ft) stems in spring.

L. deppei Mexico and Guatemala
Pseudobulbs 9cm(3½in) tall, flattened, ovate. Leaves to 45cm(1½ft) or more long, three or four together. Flowers to 11cm(4½in) wide, the sepals pale green with red or purple-spotting, the petals white to creamy white, sometimes similarly spotted near their bases, the labellum bright yellow with red markings; they are waxy in texture and fragrant, opening in spring and autumn on stems to 15cm(6in) tall.

L. skinneri Se *L. virginalis*.

L. virginalis (*L. skinneri*) Mexico to Honduras
Pseudobulbs 8–13cm(3–5in) tall, oblong-ovate, variably furrowed. Leaves to 67cm(27in) long and up to 15cm(6in) wide. Flowers 10–15cm(4–6in) wide, pure white, but sometimes with a pink to purplish flush, and a red-spotted lip; they are fragrant and open in autumn and winter.

LYCOPERSICON

Solanaceae

Origin: Western South America, Galapagos Islands. A genus of seven species of annuals and perennials, one of which, *L. lycopersicum*, is the tomato of commerce. Both species described below are good examples of a plant with decorative and edible fruit which can be grown in the conservatory and home. Propagate by seed in spring or by cuttings in summer. *Lycopersicon* derives from the Greek *lykos*, a wolf and *persicon*, a peach; originally the name of an Egyptian plant transferred to this genus by Philip Miller (1691–1771), gardener and botanist, Curator of the Chelsea Physic Garden.

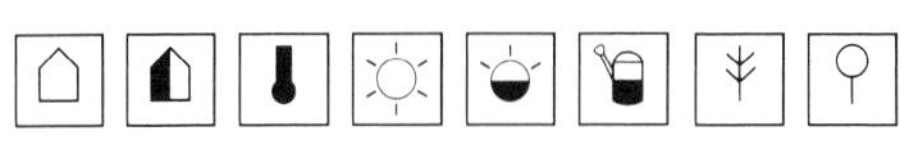

Species cultivated

L. lycopersicum (*L. esculentum*) Tomato
Peru, Ecuador
Sprawling, somewhat shrub-like, probably short-lived perennial in the wild, with stems 1–2m(3–6½ft) in length. Leaves 25–40cm(10–16in) long, irregularly pinnate, seven to nine major, ovate to oblong often lobed or partially pinnatifid leaflets and several pairs of very much smaller interjected leaflets in between. Flowers small, yellow, pendent, in lateral racemes, 5- to 6-petalled, with a central cone of stamens. Fruits variable in size and shape, ripening to shades of red, orange, yellow and bicoloured. Tomatoes were first grown in Britain as 'love apples', purely for their ornamental fruits which were considered unfit to eat despite their use as food in their homeland. The Spaniards introduced the tomato to Europe and then to the Philippines and other warm climate colonies. It later reached Africa and south-eastern Asia. As house and conservatory plants, tomatoes are best secured to canes. Alternately, the bush cultivars can be grown and these make effective hanging basket specimens.

L.l. cerasiforme (Cherry tomato)
Larger flower clusters and smaller fruits, the so-called salad tomatoes.

L.l. pyriforme
Ovoid to narrowly pear-shaped fruits.

L. pimpinellifolium Currant tomato Peru, Ecuador
Much like *L. lycopersicum*, but lacking its pungent smell and with more slender stems, smaller leaves and large trusses of 1.2cm(½in) wide, red or yellow fruits.

LYCOPODIUM

Lycopodiaceae
Club mosses

Origin: World-wide, but especially in the tropics and sub-tropics. A genus of about 450 species of epiphytic and terrestrial plants closely related to *Selaginella*, and like that genus, an ally of the ferns. They are erect to prostrate in habit with branched stems bearing small, usually crowded and overlapping leaves. The sporangia which contains the spores are carried terminally in leafy spikes. The species described provide something different in the way of foliage plants. They are essentially subjects for the conservatory, but the epiphytic species can be brought into the home for short periods if kept in humid conditions. Propagate by division or by cuttings in early summer. *Lycopodium* derives from the Greek *lykos*, a wolf, and *podion*, a foot, presumably from a fanciful resemblance of part of the plant to a wolf's foot.

Species cultivated

L. cernuum Tropics and sub-tropics (well scattered)
Terrestrial, clump- to colony-forming with erect stems about 30cm(1ft) tall (several times this in the wild where conditions are shady and warm). Each stem has many side branches and resembles a tiny, coniferous tree. Leaves small, awl-shaped, not overlapping.

L. phlegmaria Queensland tassel fern Tropics
Epiphytic, tufted to clump-forming. Stems erect

Lycopersicon lycopersicum

Lycopodium pinifolium

when young, then as they lengthen and branch becoming pendent. Leaves awl-shaped, glossy rich green, up to 2.5cm(1in) in length, not, or only slightly, overlapping.
L. squarrosum Tropical Asia
Epiphytic, having the same growth habit of *L. phlegmaria*, but with more crowded, narrower, bright green leaves. *L. pinifolium* has much the same appearance.
L. taxifolium Jamaica
Epiphytic, in overall appearance, much like *L. phlegmaria*, but with somewhat thickened stem-tips and brighter green leaves to 1.2cm($\frac{1}{2}$in) long.

LYCORIS

Amaryllidaceae

Origin: Eastern Himalaya to Japan. A genus of about ten species of deciduous bulbous plants. They have basal, strap-shaped leaves and terminal umbels of 6-tepalled, funnel-shaped flowers which open before or after the leaves. They grow well in pots, flowering in a conservatory or on a sunny window sill. Keep dry when dormant and only re-pot when essential as they resent root disturbance. Propagate by offsets, or by seed in warmth in spring. *Lycoris* commemorates the beautiful Roman actress of that name, mistress of Mark Antony.

 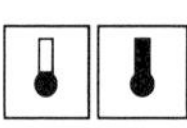 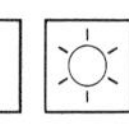 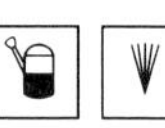

Species cultivated
L. alba See *L. radiata* 'Alba'.

Lygodium japonicum

L. africana (*L. aurea*) Golden spider lily
Japan, China, Taiwan and Vietnam
Stems to 30cm(1ft) or more tall. Leaves glaucous green to 45cm($1\frac{1}{2}$ft) long. Flowers to 8cm(3in) long, the tepals narrow, wavy-margined, golden-yellow, borne in late summer before the leaves develop. If kept at tropical temperatures and moist throughout the year it will continue growing.
L. aurea See *L. africana*.
L. radiata Red spider lily China, Japan
Stems to 45cm($1\frac{1}{2}$ft) tall. Leaves linear, blunt-tipped, glaucous green. Flowers 5–6cm(2–$2\frac{1}{2}$in) across, the tepals wavy, reflexed, bright red, opening in autumn. *L.r.* 'Alba' (*L. alba*) has white flowers.
L. squamigera Resurrection lily, Autumn amaryllis Japan
Stems 50–70cm(20–28in) tall, leafing in spring. Flowers 8cm(3in) across, pink to rosy-lilac, fragrant, very reminiscent of an amaryllis bloom; they open in autumn.

LYGODIUM

Schizaeaceae

Origin: Tropics to warm temperate regions. A genus of 40 species of climbing ferns. They form clumps of underground rhizomes from which arise slender fronds of unlimited growth, the stem-like midrib twining up any suitable support. The lateral leaflets (pinnae) are like individual fronds and may be pinnate or bipinnate. The sporangia are produced on greatly reduced pinnules which take on a spiky appearance. Either in pots or planted out in the conservatory border, these ferns provide elegant tresses of bright green which nicely complement richly hued flowering plants or those of bolder leafage. Propagate by spores or division in spring. *Lygodium* derives from the Greek *lygodes*, like a willow, alluding to the flexuous, twisting leaf stalks.

Species cultivated
L. japonicum Japan, eastern temperate Asia
Leaves to 2m($6\frac{1}{2}$ft) or more in length. Lateral leaflets pinnate, the narrowly ovate, five to eleven pinnules lobed and/or irregularly toothed. Formerly, and perhaps still, confused with *L. microphyllum* (*q.v.*)
L. microphyllum (*L. scandens*) Tropical Africa, Asia and Australia
Much like *L. japonicum*, but leaf pinnules broader and either un-lobed or with a few small basal lobes, usually with a bluish-green cast. Sometimes confused with *L. japonicum* (*q.v.*).
L. scandens See *L. microphyllum*.
L. volubile Cuba to Brazil
Leaves 2–3m($6\frac{1}{2}$–10ft) in length; lateral leaflets pinnate, composed of five to 13 lanceolate pinnules, each 5–15cm(2–6in) in length.

M

MACFADENYA See DOXANTHA

MACKAYA

Acanthaceae

Origin: South Africa. A genus of one ornamental flowering shrub well adapted to container culture. It is best in the conservatory but can be brought into the home at least while in bloom. Propagate by cuttings in spring or summer. *Mackaya* honours James Townsend Mackay (c. 1775–1862), a Scottish gardener and botanist who founded the Botanic Garden of Trinity College, Dublin.

  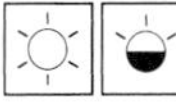  

Species cultivated

M. bella (*Asystasia bella*)
Up to 1.5m(5ft) or more in height, branching from the base and above. Leaves in opposite pairs, 8–13cm(3–5in) long, ovate, slender-pointed, usually a deep, fairly glossy green. Flowers lavender with darker veining, foxglove-shaped (tubular with five pointed lobes), about 5cm(2in) long, borne in terminal arching racemes in late spring to summer, sometimes later or even in winter.

MACLEANIA

Ericaceae

Origin: Central to western South America. A genus of 32 to 45 species (depending on the botanical authority consulted) of evergreen shrubs and scramblers. They have alternate, simple, often leathery-textured leaves and compact nodding clusters of colourful tubular flowers. The two species described here are scramblers with long, wand-like shoots which, in the wild, push through and are supported by other shrubs. In the conservatory they are best tied to a support and treated as climbers. Propagate by seed in spring, by cuttings in summer or by layering in autumn. *Macleania* commemorates John Maclean, a Scottish merchant in Lima, Peru during the 1830s who sent many native plants back to Britain.

Species cultivated

M. cordifolia Ecuador
Stems to about 3m(10ft) in length, branching from the base and sparingly above. Leaves 5–10cm(2–4in) long, ovate-cordate, leathery-textured, red-purple flushed when young. Flowers about 4cm($1\frac{1}{2}$in) long, scarlet, with whitish lobes, borne in pendent leafy racemes in spring and summer.

M. insignis Southern Mexico, Guatemala
In habit and size much like *M. cordifolia*. Leaves ovate, to 8cm(3in) in length, red-flushed when young. Flowers 3–4cm($1\frac{1}{4}$–$1\frac{1}{2}$in) long, deep scarlet, borne in terminal leafy racemes in summer.

ABOVE *Macleania cordifolia*
ABOVE LEFT *Mackaya bella*
FAR LEFT *Macropiper excelsum*

MACROPIPER

Piperaceae

Origin: Polynesia to New Guinea and New Zealand. A genus of possibly six species of evergreen shrubs formerly included in the climbing genus *Piper*. One species is cultivated and deserves to be given a wider trial as a foliage house plant. Propagate by seed in spring or by cuttings in summer. *Macropiper* derives from the Greek *macros*, large, and the allied genus *Piper*. In the wild they are robust shrubs with largish leaves in contrast to the slender climbing habit of *Piper*.

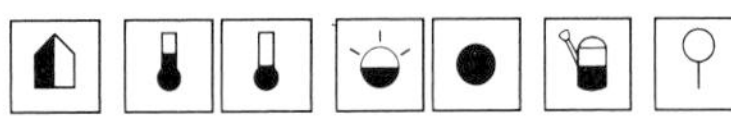

Species cultivated

M. excelsum New Zealand, adjacent islands
In the wild, ranging from a small to large shrub or even a small tree; as a pot plant, it can be grown to almost any size by pruning or regular propagation.

Leaves aromatic, broadly ovate to sub-orbicular, cordate at the base, pointed-tipped, usually deep lustrous green, 6–12cm(2½–4¾in) long. Flowers minute, in slender, unisexual spikes 2–8cm(¾–3in) in length.

M.e. majus (*M.e. psittacorum*) Islands to the north of New Zealand
Leaves to 20cm(8in) in length. Superior to the type species, but difficult to obtain.

MACROZAMIA

Zamiaceae

Origin: Australia. A genus of 12 to 14 species of dioecious palm-like plants and a member of the cycad group. Like all the cycads, it is a surviving representative of the primitive seed-bearing plants which flourished at the end of the Triassic and beginning of the Jurassic periods. All the species have hard-textured pinnate leaves in a dense rosette which may arise at ground level or from the top of a short trunk. Mature plants produce stalked, cone-like flowering spikes in the centre of the rosette. When young, the species described below make interesting pot plants. Propagate by seed in spring or as soon as obtained. *Macrozamia* derives from the Greek *macros*, large and the allied genus *Zamia* which contains smaller growing species.

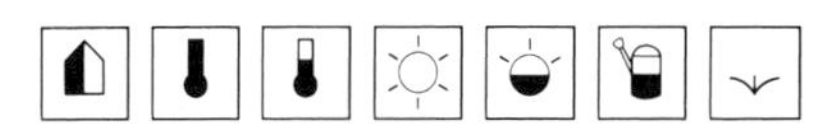

Species cultivated

M. moorei Queensland, New South Wales
Usually stemless until quite old; in its native country rare specimens with trunks 6m(20ft) tall are known, but are probably at least 300 years old. Leaves 1–2m(3–6½ft) in length, the leaflets linear, 20–30cm(8–12in) long, the lowest ones entirely reduced to spines, the upper ones spine-tipped, deep green.

M. peroffskyana (*Lepidozamia denisonii, L. peroffskyana*) Queensland, New South Wales
In the wild, one of the largest species with a trunk to 6m(20ft) or more in height; it is also somewhat faster growing than the other species described here, but still takes many years to develop a trunk of any size. Leaves 1.2–3m(4–10ft) long, the leaflets about 30cm(1ft) long, broadly linear, rich glossy green. The stalkless flowering cones with their hairy scale tips distinguish it from other macrozamias, and botanists nowadays call it *Lepidozamia*.

M. riedlei (*M. fraseri*) Western Australia
Usually stemless, but in the wild in the north and east of its range short trunks are formed. Leaves 1–2m(3–6½ft) or more in length, the linear leaflets to 25cm(10in) long, spine-tipped, glaucous green.

M. spiralis New South Wales
Stemless with leaves 60–90cm(2–3ft) long, the midribs more or less spiralled; leaflets 10–20cm(4–8in) in length. A small species, but lacking the architectural value of the other species.

ABOVE *Macrozamia moorei*
TOP *Malpighia coccifera*

MALACOCARPUS See WIGGINSIA

MALPIGHIA

Malpighiaceae

Origin: Tropical America, particularly central America and Mexico, West Indies. A genus of 30 to 35 species of evergreen shrubs and trees, a few of which provide ornamental plants for pots or the conservatory border. They have leaves in opposite pairs and flowers with five conspicuously clawed (stalked) petals. Propagate by cuttings in summer, or by seed in spring. *Malpighia* is named for Marcello Malpighi (1628–94), an Italian naturalist and Professor of Anatomy at Bologna.

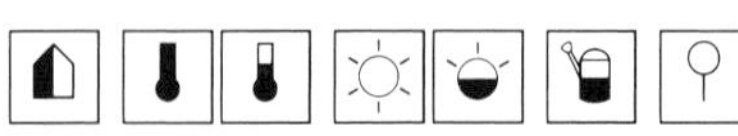

Species cultivated

M. coccigera Miniature holly West Indies
Bushy shrub 60–90cm(2–3ft) in height and width. Leaves elliptic to obovate, spiny-toothed, deep glossy green, to 2cm(¾in) long. Flowers 1.2cm(½in) wide, pink, from the upper leaf axils in summer, sometimes in profusion. Worth trying in the home, at least when in bloom. A prostrate form is recorded.

M. glabra Barbados cherry Texas, south to northern South America including West Indies
Well-branched shrub eventually to 3m(10ft) or more but less than half this height in containers. Leaves broadly elliptic-lanceolate, cuneate, dark glossy green, to 10cm(4in) long. Flowers about 1.5cm(⅝in) wide, pink to red-purple, borne in small axillary umbels in spring to autumn. Fruits red, about 1cm(⅜in) in diameter, globular, grooved, sweet and juicy. It seems likely that the shrub grown for its fruit in the West Indies and elsewhere under the name Barbados cherry is actually a hybrid between this species and the allied *M. punicifolia*.

MALVASTRUM See ANISODONTEA

MALVAVISCUS

Malvaceae

Sleeping hibiscus

Central and tropical South America. A genus of three species of evergreen shrubs grown for their showy red flowers. In general they resemble *Abutilon*, but the five petals create a cigar-shaped tube opening only partially at the top. The species described makes an excellent tub specimen. Propagate by cuttings in spring or summer.

Malvaviscus comes from *Malva* (mallow) and the Latin *viscus*, glue, referring to the sticky pulp around the seeds.

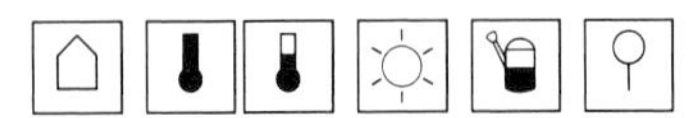

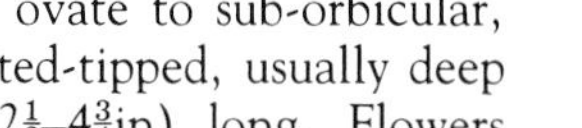

Species cultivated

M. arboreus (*M. mollis*) Mexico to Peru and Brazil
Large spreading shrub or eventually a small tree, but easily kept to under 2m($6\frac{1}{2}$ft) in containers. Leaves rounded to ovate, with or without lobes, softly hairy, 10cm(4in) or more long. Flowers pendulous, axillary, 2.5–5cm(1–2in) long. A variable species as to leaf lobing and flower length.
M.a. drummondii
Symmetrically lobed leaves with petals 3cm($1\frac{1}{4}$in) long.
M.a. mexicanus (*M. conzattii, M. grandiflorus, M. penduliflorus*)
Mainly hairless, unlobed, ovate to lanceolate leaves and petals 3–5cm($1\frac{1}{4}$–2in) long.

MAMMILLARIA

Cactaceae

Origin: S.W. North America to northern South America, but most in Mexico. A genus of 200 to 300 species of mostly small cacti with solitary or clustered, rounded to shortly cylindrical stems. They do not have ribs but are covered with prominent tubercles, sometimes spiralling around the plant. The areoles are woolly and the small funnel- or bell-shaped flowers often freely borne from late spring and through summer. They are followed by club-shaped, usually pink or red fruits. Many are excellent pot or pan plants for the conservatory or window sill and a great number are now commercially available. Those listed below are generally easy to grow and free-flowering. Propagate by offsets in summer or by seed in late spring. *Mammillaria* derives from the Latin *mammilla*, a teat, referring to the tubercles.

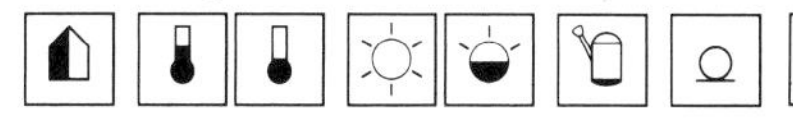

Species cultivated

M. bocasana Mexico
Stems clump-forming, globose to shortly cylindrical, to 4–5cm($1\frac{1}{2}$–2in) wide. Tubercles to 1cm($\frac{3}{8}$in) long, cylindrical. Radial spines white and bristle-like ending in silky hairs to 2cm($\frac{4}{5}$in) long, borne in abundance around the one to four needle-like central spines which are yellow-brown, hooked and 2cm($\frac{4}{5}$in) long. Flowers 1.5cm($\frac{5}{8}$in) long, pale yellow with red mid-veins.
M. bombycina Northern Mexico
Stems clustering or sometimes solitary, globose at first then becoming shortly cylindrical, to 20cm(8in) in height. Tubercles in spirals, cylindrical, their axils woolly. Radial spines 30 to 40, white, pectinate, to 1cm($\frac{2}{5}$in) long; centrals longer, dark tipped, the lower one hooked. Flowers 1cm($\frac{3}{8}$in) long, red.
M. camptotricha Birds nest cactus Central Mexico
Stems clustering, globose, 3–5cm($1\frac{1}{4}$–2in) high, dark green. Tubercles conical to cylindrical, the axils with a tuft of white bristles, spirally arranged. Radial spines only, four to eight, wavy, pale yellow, the longest one to 3cm($1\frac{1}{4}$in). Flowers 1.2cm($\frac{1}{2}$in) long, white and green.
M. carmenae Mexico (central Tamaulipas)
Stems solitary at first then clustering, globular to broadly obovoid, 5–10cm(2–4in) high. Tubercles conical, arranged in spirals, matt-green above, pinkish at the base. Radial spines more than 100, hair-like, yellow-white, like the pappus of a dandelion in miniature (in fact, each stem rather resembles a large dandelion clock); central spines absent. Flowers about 8mm($\frac{1}{3}$in) long, white or pink-tinted.
M. dealbata See *M. elegans dealbata*.
M. elegans Central Mexico
Stems solitary when young then clustering, cylindrical, pale to glaucous green. Tubercles ovoid to conical, angular at the base, the areoles white-woolly only when young. Radial spines 20 to 30, needle-like, white, about 5mm($\frac{1}{5}$in) long; central spines one to four, twice as long and brown-tipped. Flowers carmine to purple-red, to 1.5cm($\frac{5}{8}$in) long.
M.e. dealbata (*M. dealbata*)
Solitary yellow central spines, though some authorities list a black-spined plant under this name.
M. elongata Lace cactus Central Mexico
Stems branching and clump-forming, cylindrical,

ABOVE *Malvaviscus arboreus mexicanus*
TOP *Mammillaria bombycina*
ABOVE LEFT *Mammillaria bocasana*

to 10cm(4in) or more high, erect when young then reclining. Tubercles small, conical, in spirals, the axils slightly woolly. Radial spines 15 to 20, yellow, needle-shaped, recurved, to 1.2cm($\frac{1}{2}$in) long, central spines one or absent, a little longer. Flowers 1.5cm($\frac{5}{8}$in) long, white to yellowish, the outer tepals sometimes red-zoned. A variable species in most of its characteristics. M. *e.* 'Cristata' has stems monstrous and brain-like.

M. fragilis (M. *gracilis*) Central Mexico
Stems freely branching, forming dense mounds, cylindrical, to 10cm(4in) high, the numerous small side branches falling at a touch (hence its name *fragilis*) and affording an easy means of increase. Tubercles conical, arranged in spirals, their axils woolly. Radial spines 12 to 16, white, 5mm($\frac{1}{4}$in) or more long, arching, central spines three to five, brown, to 1.5cm($\frac{5}{8}$in) long. Flowers cream to yellow, about 1.2cm($\frac{1}{2}$in) long.

M.f. pulchella
More slender stems; fewer radial spines some of which are brown and no centrals. A very popular and easy species.

M. geminispina Central Mexico
Stems solitary at first, later forming dense clumps, cylindrical, to 15cm(6in) or more in height. Tubercles cylindrical to conical, in spirals, their axils woolly and bristly. Radial spines 15 to 20, white, needle-shaped, about 5mm($\frac{1}{5}$in) long; central spines two to four, brown-tipped, the longest to 2.5cm(1in) long. Flowers up to 2cm($\frac{3}{4}$in) long, red to carmine.

M. gracilis See M. *fragilis*.

M. hahniana Old woman cactus Mexico
Stem solitary at first, later branching and becoming clump-forming, globular and somewhat flattened with 6–9mm($\frac{1}{4}$–$\frac{1}{3}$in) long conical tubercles. The whole plant body is covered with white spines. Radial spines about 30; central spines one or two. Flowers 2cm($\frac{3}{4}$in) long, purplish-red.

M. longimamma See *Dolichothele longimamma*.

M. magnimamma See *Dolichothele longimamma*.

M. multiceps (M. *prolifera multiceps*) Southern Texas, northern Mexico
Hummock-forming, to 15cm(6in) or more across; stems shortly cylindrical to 6cm($2\frac{1}{2}$in) long, branching, covered with conical tubercles. Radial spines hair-like, white, numerous; central spines thicker, inclined, red-tipped when young. Flowers yellow to whitish-yellow, funnel-shaped to 1.5cm($\frac{5}{8}$in) long. Fruits 1.2cm($\frac{1}{2}$in) or more in length, bright rich red, edible.

M. pectinifera (*Solisia pectinifera*) Southern Mexico
Stems clustering, to 6cm($2\frac{1}{2}$in) high, shortly cylindrical. Tubercles small, conical, in spirals. Radial spines 20 to 40, appressed to tubercle, about 4mm($\frac{3}{16}$in) long, white. Flowers small, yellow.

M. plumosa Feather cactus Northern Mexico
Stems freely clustering, eventually forming highly decorative hummocks, globose, 8cm(3in) or more high. Tubercles cylindrical, to 1.2cm($\frac{1}{2}$in) long, their axils woolly. Radial spines only up to 40 or more, slender, branched and looking like minute feathers; the hairs interlock with their neighbours to form miniature umbrellas which in turn overlap at their edges to form a complete covering to the stem. Flowers 2cm($\frac{3}{4}$in) long, white, with red or green markings. A uniquely intriguing species easily grown in a limy compost.

M. prolifera (M. *pusilla*, M. *multiceps*) West Indies
Clump-forming with globose to shortly cylindrical stems 4–6cm($1\frac{1}{2}$–$2\frac{1}{2}$in) long, covered in 5–7mm($\frac{1}{8}$–$\frac{1}{4}$in) long, dark green tubercles. Radial spines white, bristle-like, up to 40; central spines five to nine, yellowish, to 8mm($\frac{1}{3}$in) long. Flowers 1.5cm($\frac{5}{8}$in) long, yellow flushed with green and followed by coral-red, strawberry-flavoured fruits.

M. pusilla See M. *prolifera*.

M. sphaerica See *Dolichothele sphaerica*.

M. tetracantha Ruby dumpling Central Mexico
Stems globose to cylindrical, eventually to 30cm(1ft) in height, clustering in the wild but usually solitary in cultivation. Tubercles more or less 4-sided, their axils with a few woolly hairs. Radial spines none, or a few deciduous bristles; central spines usually four in a cross-formation, straight or curved, 1–2.5cm($\frac{3}{8}$–1in) long, yellow-brown to reddish or greyish. Flowers 2cm($\frac{3}{4}$in) long, carmine-red.

M. zeilmanniana Rose pincushion Mexico
Stems solitary, later producing offsets, oval to shortly cylindrical, to 6cm($2\frac{1}{2}$in) in height, glossy green, with 6mm($\frac{1}{4}$in) long tubercles. Radial spines 15 to 18, white, soft and hair-like to 1cm($\frac{3}{8}$in) long; central spines four, red-brown, hook-tipped. Flowers 2cm($\frac{3}{4}$in) long, violet to purple-red.

RIGHT *Mammillaria geminispina*
FAR RIGHT *Mammillaria multiceps*

MANDEVILLA

Apocynaceae

Origin: Central and South America. A genus of 100 species of woody climbers often still grown under the name *Dipladenia*. They have entire leaves in whorls or opposite pairs, and racemes of funnel-shaped blooms opening to five spreading lobes, twisted in bud. They do well in pots or tubs and are best in a conservatory, not usually thriving or flowering well in small pots. Propagate by seed in spring, or by cuttings of short lateral shoots in summer with bottom heat. *Mandevilla* was named for Henry John Mandeville, British Minister in Argentina and the introducer of the first species into cultivation.

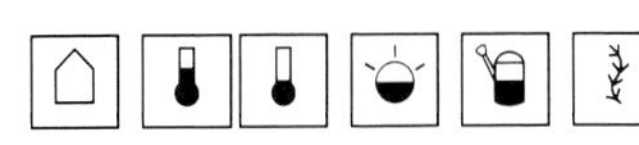

Species cultivated

M. × 'Alice du Pont' See under M. × *amabilis*.
M. × amabilis (M. × *amoena*, *Dipladenia* × *amabilis*) Garden origin
The parentage of this vigorous and floriferous hybrid is in some dispute, but M. *splendens* is almost certainly one and the hybrid favours it. A woody twiner to 3m(10ft) or so, it has ovate to elliptic-oblong leaves, 9–15cm(3½–6in) long. Flowers funnel-shaped, 9cm(3½in) long by 10–13cm(4–5in) wide, pink on opening, soon maturing rose-crimson. M. × 'Alice du Pont', which may have a similar parentage, produces larger trusses of smaller, glowing pink blooms.
M. × amoena See M. × *amabilis*.
M. boliviensis Bolivia
Climbing to 3m(10ft) or more, but can be kept smaller. Leaves 5–8cm(2–3in) long, oblong, slender-pointed, shining green. Flowers to 5cm(2in) wide, the petal lobes slightly twisted, white with a golden-yellow eye.
M. laxa (M. *suaveolens*) Chilean jasmine
Peru, Bolivia, Argentina
Vigorous twiner to 5m(16ft) or more. Leaves to 10cm(4in) long, oblong-ovate, cordate, with a slender point. Flowers to 5.5cm(2¼in) long, white and fragrant, in clusters of five to 15, opening in summer.
M. splendens Brazil
To 3m(10ft) in height. Leaves to 20cm(8in) long, broadly elliptic, thinner in texture, shining green. Flowers 10cm(4in) or more across, white, flushed with rose-pink.
M. suaveolens See M. *laxa*.

MANETTIA

Rubiaceae

Origin: Central and tropical South America, West Indies. A genus of 100 to 130 species of evergreen perennial and woody-stemmed climbers. They have opposite pairs of usually simple leaves and tubular flowers from the upper axils. The species

make decorative pot plants and bloom when young. Propagate by cuttings in spring and summer. *Manettia* commemorates Saverio Manetti (1723–1785), Prefect of the Florence Botanic Garden.

  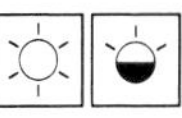 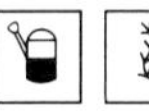 

Species cultivated

M. bicolor See M. *inflata*.
M. cordifolia South America, mainly Brazil
Stems twining, to 2m(6½ft) or more. Leaves ovate-cordate to oblong-lanceolate, 4–10cm(1½–4in) long, mid to deep green, lustrous above, downy beneath. Flowers narrowly funnel-shaped, about 3cm(1¼in) long, red, in small clusters.
M.c. glabra (M. *glabra*)
Identical but for the hairless leaf undersurfaces.
M. glabra See M. *cordifolia glabra*.
M. inflata (M. *bicolor*, M. *luteo-rubra*)
Firecracker vine Paraguay, Uruguay
Stem much branched, twining to 3m(10ft). Leaves 3–6cm(1¼–2½in) long, lanceolate to narrowly ovate, bright green, with or without short stalks. Flowers 2cm(¾in) long, tubular, but somewhat inflated at their bases, scarlet with yellow tips, thickly short-hairy. Some authorities keep M. *bicolor* as a separate species as it has almost stalkless leaves and erect or spreading sepals (M. *inflata* has recurved to reflexed ones).
M. luteo-rubra See M. *inflata*.

MANIHOT

Euphorbiaceae

Origin: North and South America. A genus of 160 species of perennials, trees and shrubs of which one

ABOVE *Mandevilla laxa*
ABOVE CENTRE *Mandevilla boliviensis*
TOP *Mandevilla splendens*

is widely cultivated in the tropics for food as cassava, manioc or tapioca. Its variegated form makes a highly decorative foliage plant. Propagate by cuttings of young tips, or of 20cm(4in) long leafless stem sections in summer with bottom heat. *Manihot* is the Latin version of the Brazilian vernacular name, manioc.

Species cultivated

M. esculenta (*M. utilissima*) Tapioca, Cassava, Manioc Mexico, Guatemala, Brazil
Tuberous-rooted shrub to 90cm(3ft) or so in a pot; to 3m(10ft) in the open ground. Leaves 4–20cm(1½–8in) long, long-stalked, palmately lobed almost to the base, the lobes 3–9cm(1¼–3½in) long, obovate-lanceolate, dark green above, glaucous underneath. Flowers petalless, with pale yellow or red-tinged sepals. *M.e.* 'Variegata' has dark green leaves with centres irregularly patterned yellow.

MARANTA

Marantaceae

Origin: Tropical America. A genus of about 20 species of rhizomatous, clump-forming perennials. They have undivided leaves with sheathing stalks. The flowers are small and 3-petalled with two larger petal-like staminodes, but are mostly not showy, the plants being cultivated for their decorative foliage. Arrowroot is obtained from the roots of *M. arundinacea*. They are excellent house and warm conservatory plants. Propagate by division when potting, or by cuttings with two or three leaves in summer in warmth. *Maranta* was named for Bartolommeo Maranti, an Italian physician and botanist living in Venice in 1559.

Species cultivated

M. amabilis See *Stromanthe amabilis*.
M. arundinacea Arrowroot Tropical America
Stems slender, 1–2m(3–6½ft) tall in a large pot. Leaves to 30cm(1ft) long, lanceolate, long-stalked. Flowers small, white, but not produced on small plants. *M.a.* 'Variegata' has patterning in three shades of green.

M. bicolor Brazil, Guyana
Tufted to clump-forming, the stems very short. Leaves ovate to oblong-elliptic, up to 15cm(6in) long, slightly wavy, bluish-green with an irregular pale zone down the centre of the leaf, bordered by two rows of dark to brownish-green blotches. Flowers small, white, with red-purple lines.
M. kerchoviana See *M. leuconeura*.
M. leuconeura Prayer plant Brazil
Stems short, usually branched and spreading, eventually to 30cm(1ft) high. Leaves to 15cm(6in) long, elliptic to oblong, usually blunt, green.
M.l. erythroneura (*M. tricolor*) Herringbone plant
Dark green veins with a silvery central zone and crimson veins curving to the leaf margins.
M.l. kerchoveana Rabbit tracks/foot
Light green leaves and a row of dark greenish-brown blotches between the veins.
M.l. leuconeura (*M.l. massangeana*)
Dark green veins with a silvery central zone and vein patterning of the same colour.
M. lubbersiana See *Ctenanthe lubbersiana*.
M. makoyana See *Calathea makoyana*.
M. porteana See *Stromanthe porteana*.
M. tricolor See *M. leuconeura erythroneura*.

MARSILEA

Marsileaceae

Pepperworts

Origin: Tropical to temperate regions; widespread. A genus of about 60 species of wet ground or aquatic fern allies grown for their ornamental foliage. They are more or less evergreen perennials (in the wild some species die down in the dry season) spreading by creeping rhizomes. The long-stalked quadrifoliate leaves (barren fronds) stand erect in wet soil or shallow water, and float in deep water. Sporangia are produced on small, greatly modified, almost stalkless, capsule-like leaves known as sporocarps. These are interesting plants for a conservatory pond or aquarium, or as pot plants stood permanently in saucers of water. Propagate by division in spring. *Marsilea* honours Count Luigi Ferdinando Marsigli (1658–1730), an Italian botanist.

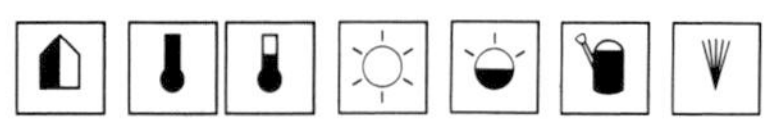

Species cultivated

M. drummondi Water clover Australia
Rhizomes branching freely, producing crowded tufts of foliage. Leaflets 2.5–4cm(1–1½in) long, broadly obovate-cluneate, silky hairy and wavy-edged. Temperate to tropical.
M. quadrifolia Water clover Europe and Asia; naturalized in North America
Rhizomes comparatively thick with few branches. Leaflets obovate-deltoid, 2cm(¾in) long, smooth-edged and glossy rich green. Cool to hardy, usually dying back in winter. Ideal for the unheated conservatory.

BELOW *Marsilea quadrifolia*
BELOW RIGHT *Maranta leuconeura erythroneura*

MASDEVALLIA

Orchidaceae

Origin: Central and tropical South America, West Indies. A genus of 275 species of chiefly epiphytic orchids which lack pseudobulbs. They have fleshy, linear to obovate leaves and flowers which, although solitary, are often carried in abundance. These have three large outer tepals often ending in tail-like tips and fused at the base to form a tube or cup, two inner tepals which are smaller, and a small labellum. These orchids are suitable for a conservatory and can be grown in the home as long as a humid atmosphere can be maintained. Propagate by division when potting or re-potting in spring or early autumn. *Masdevallia* was named for José Masdevall (*d.* 1801), a Spanish botanist.

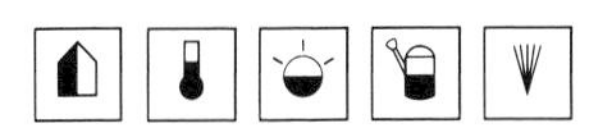

Species cultivated

M. amabilis Peru
Leaves narrowly oblanceolate, leathery-textured, glossy, 13–18cm(5–7in) long. Flowering stems one-flowered, erect, up to 30cm(1ft) in height; flowers about 2.5cm(1in) wide, upper sepal orange with red shading, lateral sepals red, all with tails, the larger upper one to 4cm(1½in) long, borne in winter to spring.

M. bella Colombia
Leaves 13–18cm(5–7in) long, oblong to lanceolate. Flowers to 23cm(9¼in) long including the long tails, pale yellow with red-brown spotting and a white labellum; they are carried on pendent scapes from winter to spring.

M. coccinea Colombia
Leaves 15–23cm(6–9¼in) long, obovate to lanceolate. Flowers 7–9cm(2¾–3½in) long, the upper sepal shorter than the two side ones, all crimson, scarlet, orange or yellow, occasionally white or magenta; they are borne on erect scapes to 30cm(1ft) tall in spring and summer.

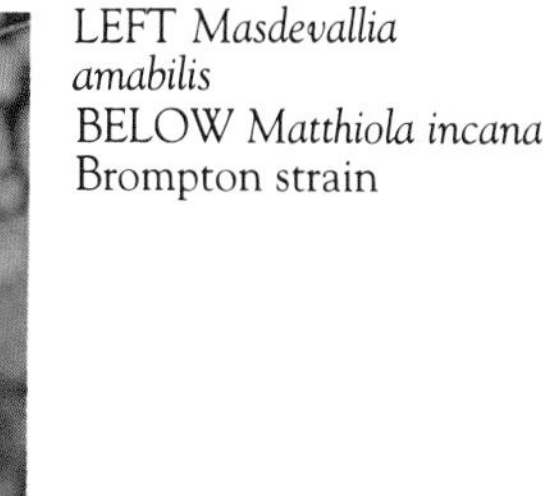

LEFT *Masdevallia amabilis*
BELOW *Matthiola incana* Brompton strain

MATTHIOLA

Cruciferae

Origin: Azores and Canary Islands westwards through the Mediterranean region to central Asia. A genus of 55 species of annuals, biennials, perennials and shrubs, with leaves borne alternately or in rosettes, usually undivided but sometimes pinnatifid. Flowers are 4-petalled and borne in erect racemes. The species described below makes an excellent pot plant which can be grown in the conservatory and brought into a cool room for flowering. Propagate by seed in spring for summer blooms; sow in late summer to flower the following winter and spring and provide temperate conditions. *Matthiola* was named for Pierandrea Mattioli (1500–1577), an Italian doctor and botanist.

  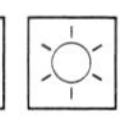  

Species cultivated

M. incana Stock S.W. Europe on coasts
Woody-based perennial 30–80cm(12–32in) tall, normally grown as an annual or biennial. Leaves 5–10cm(2–4in) long, linear to oblong-lanceolate, covered with grey-white felt. Flowers 2–3cm(¾–1¼in) across, purple, pink or white. In cultivation, red, purple-blue and yellow colours occur and the following strains are used: Ten-week stocks, and Brompton or East Lothian stocks. Ten-week stocks are treated as annuals and are up to 30cm(1ft) tall, branched, with single flowers. Also similar are Trysomic which have 85% double flowers, and All Double with 100% doubles. When these are in the seedling stage it is possible to identify and discard the single-flowered plants as they are much greener than the somewhat yellowish doubles. Column and Excelsior strains grow to 90cm(3ft) and are good for

cutting, but make less satisfactory pot plants. Brompton stocks are grown as biennials and reach 45cm($1\frac{1}{2}$ft) in height; they are well branched plants and the flowers can be single or double. East Lothian stocks are shorter, reaching about 37cm(15in) and can be grown as annuals or biennials.

MATUCANA See BORZICACTUS
MAURANDYA See ASARINA

MAXILLARIA

Orchidaceae

Origin: The Americas from Florida to Argentina. A genus of about 300 species of mainly epiphytic orchids. Most have conspicuous pseudobulbs and are clump-forming to rhizomatous. Their leaves are linear to elliptic and the flowers solitary, composed of five spreading tepals and a deeply 3-lobed labellum. These orchids are suitable for growing in pots or pans or attached to pieces of bark. Propagate by division when re-potting in spring. *Maxillaria* derives from the Latin *maxilla*, a jaw, the flowers being fancifully likened to the jaws of an insect.

   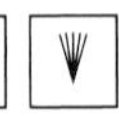

Species cultivated

M. picta Brazil
Pseudobulbs clustered or spaced along the rhizomes, 5–8cm(2–3in) tall, ovoid. Leaves one or two together, to 30cm(1ft) or more long, narrowly oblong and glossy green. Flowers to 6cm($2\frac{1}{2}$in) wide, creamy-white, heavily spotted and dotted with purple, the labellum white with purple markings, opening in winter and spring.

M. sanderana Ecuador
Clump-forming, with clustered, compressed ovoid pseudobulbs 5cm(2in) long. Leaves 18–30cm(7–12in) long, solitary, narrowly oblong. Flowers 10–15cm(4–6in) wide, ivory-white, blotched and spotted with deep red and crimson, the labellum similarly marked especially near its base; they open in summer and autumn. Considered by many to be the most rewarding species to grow.

M. tenuifolia Central America to Mexico
Pseudobulbs ovoid, to 4cm($1\frac{1}{2}$in) long on erect rhizomes. Leaves solitary, to 35cm(14in) long, linear. Flowers to 5cm(2in) wide, red to red-purple or red-brown, spotted yellow, labellum dark red at the base, shading to yellow with red spotting.

M. variabilis Mexico to Venezuela and Guyana
Rhizomatous, sometimes pendulous when mature. Pseudobulbs close together, elliptic-oblong to subcylindric or even narrower, 1–6cm($\frac{3}{8}$–$2\frac{1}{2}$in) long. Leaves solitary, more or less strap-shaped, 2-lobed at the apex, leathery-textured, 3–15cm($1\frac{1}{4}$–6in) in length. Flowers about 2cm($\frac{3}{4}$in) or more wide, usually purple-red, but very variable in the wild, ranging from white to yellow marked with red to entirely red or red-brown, borne intermittently throughout the year.

ABOVE *Maxillaria picta*
RIGHT *Medinilla magnifica*

MEDINILLA

Melastomalaceae

Origin: Tropical Africa, S.E. Asia and Pacific islands. A genus of 150 species of evergreen shrubs and climbers with pairs or whorls of undivided leaves and pink or white flowers in panicles or clusters, sometimes with showy bracts. The species described below is best in the warm conservatory, but can be grown as a short-term pot plant needing extra humidity. Propagate by cuttings in spring or summer with bottom heat. *Medinilla* was named for Jose de Medinilla y Pineda, Governor of the Marianas Islands in *c.* 1820.

 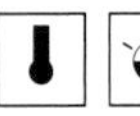 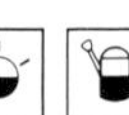 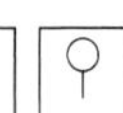

Species cultivated

M. magnifica Rose grape Philippine Islands
Robust shrub 1–2m(3–$6\frac{1}{2}$ft) high, with a 4-angled or winged stem. Leaves 20–30cm(8–12in) long, ovate to obovate, with strongly marked veins, shining rich green. Flowers about 2.5cm(1in) wide, pink to coral-red with yellow stamens and purple anthers; they are carried in pendent, pyramidal panicles, hanging beneath a number of bright pink, ovate bracts, to 10cm(4in) or more long, in spring and summer. *M.m.* 'Rubra' has darker coloured flowers; 'Superba' has darker and larger flowers.

MEDIOLOBIVIA See REBUTIA

MELALEUCA

Myrtaceae

Origin: Australia and Pacific islands. A genus of about 100 species of evergreen trees and shrubs the majority of which are native to Australia. Allied to *Callistemon*, they have simple, alternate leaves and small, many-stamened flowers in bottlebrush-like spikes. The species described below make useful and decorative specimen shrubs for containers or conservatory borders. Propagate by seed in spring or by cuttings in late summer. *Melaleuca* derives from the Greek *melos*, black and *leukos*, white; some species when mature have whitish papery-barked trunks and dark to blackish branches.

Species cultivated

M. fulgens Western Australia
Wiry-stemmed shrub about 90–180cm(3–6ft) tall, usually rather sparingly branched. Leaves to 2cm($\frac{3}{4}$in) or more long, lanceolate to linear. Flowers crimson, in dense lateral spikes, 2.5–6cm(1–2$\frac{1}{2}$in) long, in summer. One of the showiest species in bloom but of straggly habit when old.

M. hypericifolia New South Wales
Well-branched shrub to 2m(6$\frac{1}{2}$ft) tall, remarkably like a species of *Hypericum* when not in bloom. Leaves closely set, to 4cm(1$\frac{1}{2}$in) long, oblong-elliptic. Flowers red, in lateral spikes, 4–8cm(1$\frac{1}{2}$–3in) long, in summer.

M. leucodendron Cajeput, River tea tree
Northern Australia, New Guinea, Moluccas
In the wild, a tree of medium size with a sturdy trunk thickly covered with layers of pale, papery bark, but in containers this bark character is not so obvious. Leaves 5–15cm(2–6in) or more in length, lanceolate. Flowers usually white but also greenish-yellow, pink or purple, in spikes 8–20cm(3–8in) long. For temperate to tropical conditions. In cultivation plants under this name are likely to be the allied M. *quinquenervia*, *q.v.*

M. quinquenervia Eastern Australia, New Guinea, New Caledonia
Much like M. *leucodendron* and confused with it, but bark white, leaves not above 9cm(3$\frac{1}{2}$in) long and flowers always white.

MELIA

Meliaceae

Origin: Tropical and sub-tropical Asia and Australia. A genus of two to 20 species, depending on the classification followed. They are large trees and shrubs having pinnately divided leaves and panicles of small white or purple flowers followed by conspicuous fruits. One species is sometimes available, being suitable as a pot or tub plant for the conservatory. Flowering does not occur on a small plant but it has attractive foliage. Propagate by seed in warmth as soon as ripe, or by cuttings with bottom heat in summer. *Melia* is the Greek vernacular for ash, the name presumably taken for this genus because of its similar leaves.

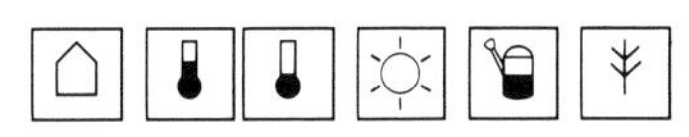

Species cultivated

M. azederach Bead tree, Persian lilac Asia
Deciduous tree to 12m(40ft) in the open, to 2m(6$\frac{1}{2}$ft) or so in a pot. Leaves 30–60cm(1–2ft) long, pinnate or bipinnate, the leaflets 4–5cm(1$\frac{1}{2}$–2in) long, ovate, toothed. Flowers 2cm($\frac{3}{4}$in) wide, 5-petalled, lilac-mauve, fragrant, borne in loose panicles 10–20cm(4–8in) long, from the leaf axils in summer. Fruits are yellow, berry-like drupes, about 1cm($\frac{3}{8}$in) long.

MELIANTHUS

Melianthaceae

Origin: South Africa and India. A genus of six species of shrubs or shrubby perennials with coarsely toothed, pinnate leaves. Two species are grown for their foliage in pots or tubs, suitable for use in a cool conservatory, the smaller species also for a cool room. Propagate annually from cuttings in summer with bottom heat, or by division, or suckers when possible in spring. *Melianthus* derives from the Greek *meli*, honey and *anthos*, a flower, from the nectar which is freely produced.

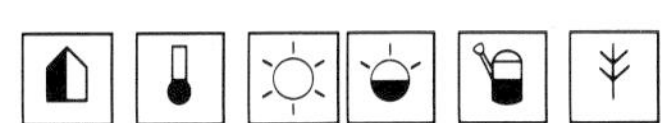

Species cultivated

M. major Honeybush South Africa
Sparingly branched shrub to 3m(10ft) tall. Leaves 25–45cm(10–18in) long, pinnately divided into seven to 13 ovate, glaucous to grey-green leaflets, each 5–13cm(2–5in) long. Flowers 2.5cm(1in) long, brownish-red, in terminal racemes to 30cm(1ft) long, in summer.

M. minor South Africa
Very similar to M. *major*, but smaller in all its parts, making a shrub 1.2–1.5m(4–5ft) high, but less in a container.

Melianthus minor

Melocactus

ABOVE *Michelia figo*
ABOVE CENTRE *Metrosideros excelsa* 'Kerm Radiant'

MELOCACTUS (CACTUS)

Cactaceae

Origin: Tropical America and West Indies. A genus of about 30 species of solitary, globose to shortly columnar cacti. They have ribbed and spiny stems like those of *Cereus*, but the flowers are small and borne on a terminal cephalium covered with wool and bristle-like spines. Propagate by seed in spring. *Melocactus* derives from the Latin *melopepo*, an apple-shaped melon, alluding to the shape of certain species.

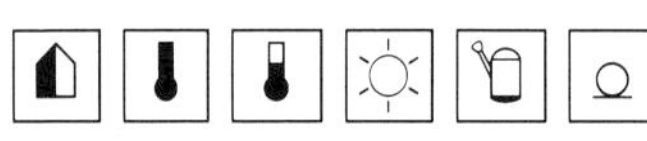

Species cultivated

M. bahaiensis Brazil
Stem wider than high, to 10cm(4in) high by 15cm(6in) across. Ribs ten to 12, prominent, each one bearing up to seven areoles. Radial spines ten, about 2.5cm(1in); central spines four, about half again as long, straight, brown. Flowers small, pink.

M. matanzanus Mexico
Stems slightly higher than wide, to 9cm(3½in) tall by 10cm(4in) across. Ribs eight to nine, thick, each one bearing about five areoles. Radial spines seven or eight; one central spine; all about 2cm(¾in) long, reddish to yellowish, awl-shaped, curved. Flowers pink, 2cm(¾in) long.

MESEMBRYANTHEMUM See CARPOBRUTUS, DOROTHEANTHUS & OSCULARIA

METROSIDEROS

Myrtaceae

Origin: South Africa, Malaysia to Polynesia, Australia and New Zealand. A genus of 60 species of evergreen trees, shrubs and climbers, with lanceolate to ovate leaves in opposite pairs and small, 5-petalled flowers in terminal racemes or cymes. In pots or as tub plants, the variegated cultivar is particularly rewarding. Propagate by cuttings with bottom heat in summer, or by seed (species only). *Metrosideros* derives from the Greek *metra*, core or hardwood and *sideros*, iron, referring to the hardness of the wood of some species.

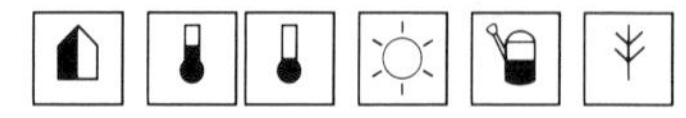

Species cultivated

M. collina Lehua Hawaii
Shrub or small tree 2–30m(6½–100ft) tall in the wild, but flowering well in a pot when quite small. Leaves 3–8cm(1¼–3in) long, broadly elliptic to almost rounded, deep glossy green, downy beneath. Flowers rich crimson with 2–3cm(¾–1¼in) long stamens, in freely borne cymes, in spring and summer. Temperate.

M. excelsa (*M. tomentosa*) New Zealand Christmas tree New Zealand
To 20m(65ft) in the wild, only 3m(10ft) in a pot. Leaves 4–10cm(1½–4in) long, broadly oblong to elliptic, glossy green above, grey-white downy beneath. Flowers small, with showy pendent stamens to 3–4cm(1¼–1½in) long, rich crimson, borne in cymes in late spring and summer. Cool. *M.e.* 'Kerm Radiant' has the leaves marked with a broad yellow central splash.

MICHELIA

Magnoliaceae

Origin: Tropical and temperate Asia. A genus of 50 species of evergreen trees and shrubs very closely related to *Magnolia*. The species described below makes an interesting pot plant and starts to flower when small. Propagate by cuttings in late summer with bottom heat. *Michelia* was named for Pietro Antonio Micheli (1679–1737), an Italian botanist.

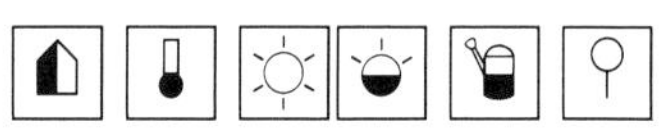

Species cultivated

M. figo (*M. fuscata*) China
Shrub to 3m(10ft) or so, but slow-growing. Leaves 4–10cm(1½–4in) long, elliptic to broadly ovate, deep glossy green. Flowers 2.5–4cm(1–1½in) across, the buds covered with brown hairs, the petals creamy-yellow, rimmed or sometimes flushed with dull purple, opening in spring to summer; they are strongly fragrant, the scent usually considered banana-like, yet some liken it to pear-drops.

MICROCOELIUM See SYAGRUS

MICROLEPIA

Dennstaedtiaceae (*Pteridaceae*)

Origin: Old World tropics, Japan and New Zealand. A genus of 45 species of ferns, several of which make pleasing and elegant foliage plants for the shady conservatory or room. They have creeping rhizomes, but in some species these are thick and slow-growing and the plants form clumps. The fronds are mostly tripinnate, the ultimate pinnules of the fertile ones bearing membranous crescent-shaped indusia. Propagate by spores or division in spring. *Microlepia* derives from the Greek *micros*, small and *lepis*, a scale, a reference to the small scale-like indusia.

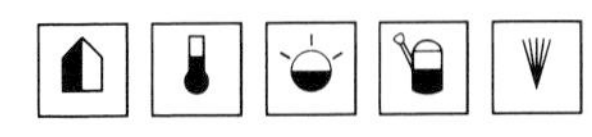

Species cultivated

M. hirta See *M. pyramidata* and *M. speluncae hirta*.

M. platyphylla India to Eastern Asia
Eventually widely clump-forming and 1.5–2.1m(5–7ft) in height if planted out; much smaller in containers with fronds to 90cm(3ft) or so tall. Each frond tripinnatifid, white-hairy when young, more or less hairless when mature, often slightly glaucous.

M. pyramidata (*M. hirta* of some authorities) Tropical Asia
Clump-forming, similar in size and general appearance to *M. platyphylla*. Fronds triangular in outline, often quadripinnate, usually hairy on the veins beneath and on the midrib (rachis).

M. speluncae Tropics and sub-tropics
Strong-growing, the fronds 1–2m(3–6½ft) in height on mature, well-grown specimens. Each frond pale green, soft textured, tri- or quadripinnatifid, white-hairy.

M.s. hirta (*M. hirta*)
Richer green fronds, no more than tri-pinnatifid, and purple-brown stalks. Said to be confused in cultivation with *M. strigosa*.

M. strigosa S.E. Asia, Polynesia
Widely clump-forming with age. Fronds 45–120cm(1½–4ft) in height, narrowly ovate in outline, bipinnatifid, light green, more or less hairless except at the base of the stalk. Reported to be confused in cultivation with *M. speluncae*.

MICROSORIUM

Polypodiaceae

Origin: Old World tropics. A genus of about 60 species of evergreen ferns related to, and formerly included in, *Polypodium*. They are mainly epiphytic, usually with far-creeping rhizomes and simple and pinnately lobed leaves. The species described below combine well with a collection of bromeliads or orchids, either on sections of bark or in hanging baskets. Propagate by division or by spores in spring. *Microsorium* derives from the Greek *micros*, small and *soros*, a heap, an oblique reference to the small clusters of sporangia.

Species cultivated

M. diversifolium (*Phymatodes diversifolium*) Australasia
Rhizome eventually to several metres (yards) in

Microsorium scolopendrium

length, branching, somewhat glaucous but covered with dark brown scales to 1cm(⅜in) long. Fronds with stalks 5–20cm(2–8in) long. Blades of three forms – simple, entire, lanceolate to elliptic-oblong; more or less lobed, usually rather irregularly; or regularly and closely pinnately lobed. These leaf blade forms also vary greatly, ranging from 10–40cm(4–16in), all being leathery-textured, dark glossy green. Withstands cool conditions and can be used as ground cover in the conservatory.

M. musifolium See *Polypodium musifolium*.

M. scolopendria (*Phymatodes scolopendria*)
Rhizomes to several metres (yards) in time, fleshy at first then almost woody. Fronds with stalks 4–20cm(1½–8in) long. Blades up to 60cm(2ft) in length, simple and oblong-lanceolate or, more usually, deeply pinnately lobed, leathery-textured, deep lustrous green.

MIKANIA

Compositae

Origin: Tropical America, West Indies, South Africa. A genus of about 250 species of mainly climbers and shrubs with a few erect perennials. One climbing species has become a popular house plant. Propagate by cuttings or layering in summer. *Mikania* honours Joseph Gottfried Mikan (1783–1814), Professor of Botany at Prague, or possibly his son Johann Christian who collected plants in Brazil and followed his father as a Professor at Prague.

Species cultivated

M. ternata (*M. apiifolia*) Brazil
Stems scrambling to 2m(6½ft) or much more if planted out, purple-woolly when young. Leaves in opposite pairs, hairy, digitate, composed of five stalked, rhomboidal to broadly obovate lobed and waved leaflets 1.5–4cm(⅝–1½in) long, deep purple beneath, purple-green above. This describes the foliage of the form commonly cultivated, but in the wild this is a variable species as regards leaflet numbers (three to seven) and shape and intensity of the purple colouring. Flowers small, groundsel-like, yellowish, in loose corymbs in summer.

Mikania ternata

Miltonia 'Limelight'

MILLA See IPHEION

MILTONIA

Orchidaceae
Pansy orchids

Origin: Central and tropical South America. A genus of 25 species of rhizomatous to clump-forming orchids with conspicuous pseudobulbs. Their leaves are usually in twos, occasionally in ones or threes and the flowers carried singly or in racemes. Upland species are suitable for cool conditions, whilst lowland species need temperate or warmer culture. All are evergreen and need to be kept moist throughout the year. They are suitable for the conservatory and are worth trying in the home. Propagate by division in autumn or early spring. *Miltonia* was named for Viscount Milton (1786–1851), later Lord Fitzwilliam, a patron of horticulture.

Species cultivated

M. candida Brazil (lowland)
Rhizomatous, with narrowly ovoid, somewhat flattened pseudobulbs 8–10cm(3–4in) high. Leaves in pairs, narrowly oblanceolate, glossy palish green, 22–39cm(9–15in) long. Flowering stems arching, to 60cm(2ft) in length, bearing three to seven flowers; each bloom about 8cm(3in) wide, tepals spotted and blotched red-brown, labellum white, fragrant, opening in late summer to autumn.

M. regnellii Brazil (lowland)
Clump-forming, with narrowly conical, flattened pseudobulbs 8–10cm(3–4in) tall. Leaves to 30cm(1ft) long, narrowly strap-shaped, glossy yellow-green. Flowering stems to 60cm(2ft) long bearing three to seven blooms; individual flowers 5–8cm(2–3in) long, tepals white with a pink flush especially near the base, labellum rounded to heart-shaped, lilac-pink, opening in summer to autumn.

Mimosa pudica

M. spectabilis Brazil (lowland)
Rhizomatous, with ovoid-oblong, flattened pseudobulbs 7–10cm($2\frac{3}{4}$–4in) long. Leaves 10–18cm(4–7in) long, narrowly strap-shaped. Flowers solitary to 8cm(3in) long, flat, tepals narrow, creamy-white, sometimes shaded with pink at the base, labellum rounded to obovate, red-purple marked with darker veins.

Hybrids

From a house plant point of view the modern man-made hybrids are the most amenable. They also embrace most of the large, broad-petalled, pansy-faced sorts which so much catch the eye at flower shows. All the many listed in specialists' catalogues, such as 'Limelight' are worth trying.

MIMOSA

Leguminosae

Origin: Tropical and sub-tropical America, Africa and Asia. A genus of 450 to 500 species of annuals, perennials, shrubs, trees and climbers. They have bipinnate leaves and tiny tubular flowers in globular heads made conspicuous by their protruding stamens. The species described is a popular house or conservatory plant, grown for interest or amusement. Propagate by seed in warmth in spring, pricking off as soon as the first true leaves appear and potting-on regularly.

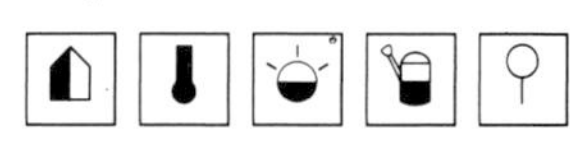

Species cultivated

M. pudica Sensitive plant, Humble plant
Tropical America
In the wild it is a sub-shrubby perennial, but grown as an annual to 90cm(3ft), usually less, with prickly stems. Leaves bipinnate, the four pinnae divided into 15 to 25 pairs of small, narrowly oblong leaflets. When touched, the leaflets close up, then the leaf stalks droop downwards. Recovery is fairly quick, but reaction is slower if it is touched again immediately. Flowers are tiny, pale pink, in heads from the leaf axils in early summer.

MIMULUS

Scrophulariaceae

Origin: Temperate regions of the world, but chiefly in the Americas (not present as a native in Europe). A genus of 100 or more species of mostly perennials but including some shrubs. They have ovate to linear-lanceolate leaves in opposite pairs and usually showy, tubular flowers opening to five lobes at the mouth. They make attractive plants for the conservatory and home, the perennial sorts being grown either as annuals from seed in spring, or as herbaceous perennials, in which case they must have a cool winter rest. Propagate by cuttings or by division in spring; shrubby species by cuttings in summer. *Mimulus* derives from the Latin *mimus*, mimic, the flowers of the first known species fancifully resembling the face of a monkey.

Species cultivated

M. aurantiacus (M. *glutinosus*, *Diplacus glutinosus*) California, Oregon
Evergreen shrub to 1m(3ft) or more. Leaves 5cm(2in) long, narrowly to oblong-lanceolate, stickily glandular. Flowers to 4cm(1½in) long, yellow or orange, funnel-shaped, borne in the upper leaf axils from spring to autumn.

M. cupreus S. Chile
Annual to 20cm(8in) tall, bushy. Leaves 2.5–3cm(1–1¼in) long, ovate to elliptic, coarsely toothed. Flowers 2.5–4cm(1–1½in) long, yellow, changing to coppery-orange as they age.

M. glutinosus See M. *aurantiacus*.

M. guttatus Monkey flower, Monkey musk Western North America and Mexico
Perennial 30–60cm(1–2ft) high. Leaves 2–15cm(¾–6in) long, oblong-lanceolate, coarsely irregularly toothed. Flowers to 5cm(2in) long, yellow, spotted red in the throat.

M. luteus Monkey musk, Monkey flower Chile
Perennial to 30cm(1ft), with sprawling to semi-erect stems. Leaves to 7cm(2¾in) long, broadly ovate, with few, regular teeth. Flowers to 4.5cm(1¾in) long, yellow, spotted with red in the throat and on the corolla lobes. Many popular cultivars are the result of hybridizing this species with M. *guttatus*.

M. moschatus Musk Western North America
Stickily-hairy perennial with 10–40cm(4–16in) creeping stems. Leaves 1–4cm(⅜–1½in) long, ovate or elliptic. Flowers to 2cm(¾in) long, funnel-shaped, pale yellow, sometimes dotted with light brown. Formerly grown for its musky fragrance, but the plants are now without scent and it is thought the original collection was from an aberrant, scented plant which for a while was propagated vegetatively. When seeds became available they replaced the more laborious method of propagation by cuttings, but being collected from plants without fragrance, the scented type was lost. It is said still to survive in the wild.

M. puniceus (M. *glutinosus puniceus*) California
Evergreen shrub to 1m(3ft) or more. Leaves to 5cm(2in), narrowly to oblong lanceolate. Flowers from 4.5cm(1¾in) long, red. This species has been hybridized with M. *aurantiacus*.

MINA See IPOMOEA

MITRARIA

Gesneriaceae

Origin: S. Chile. A genus of one species, an evergreen climbing shrub. It is suitable for the conservatory where it is best trained on canes or against the back wall, or for the home. It can also be grown in a hanging basket if there is room for its long trailing stems. Propagate by cuttings of non-flowering shoots in summer, or by division of large plants in spring.

Mitraria derives from the Greek *mitra*, a cap or mitre, alluding to the shape of the seed capsules.

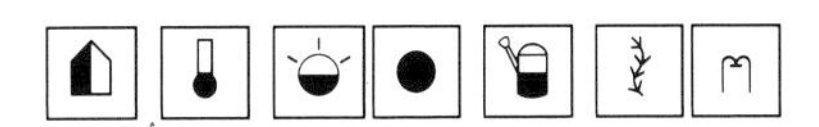

Species cultivated

M. coccinea
Semi-trailing, stems climbing and holding on by means of small roots; the stems can be to 2m(6½ft) long. Leaves 1–2cm(⅜–¾in) long, ovate, coarsely toothed and glossy green, borne in opposite pairs. Flowers to 3cm(1¼in) long, tubular, slightly inflated with five small spreading lobes at the mouth, bright scarlet; they are pendent and solitary, borne from the upper leaf axils from early summer to autumn.

ABOVE *Mitraria coccinea*
ABOVE CENTRE *Mimulus puniceus*
TOP *Mimulus puniceus* 'Currant Red'

MITRIOSTIGMA

Rubiaceae

Origin: Tropical and South Africa. A genus of three species of evergreen shrubs related to *Gardenia*. One species has long been grown in our greenhouses and makes a pleasing pot plant for the home, flowering when quite small. Propagate by cuttings in spring or summer.

Mitriostigma derives from the Greek *mitra*, a

ABOVE *Momordica balsamina*
ABOVE CENTRE *Monstera deliciosa* 'Variegata'
TOP *Mitriostigma axillare*

mitre and *stigma*, the pollen receptive organ attached to the ovary within a flower.

  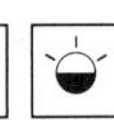  

Species cultivated

M. axillare (*Gardenia citriodora*) Wild coffee
South Africa, Natal
Well-branched spreading shrub, in the wild to 1.5m(5ft) tall, but easily kept to half this in containers. Leaves in opposite pairs, 5–10cm(2–4in) long, lanceolate to narrowly elliptic, leathery-textured, usually dark glossy green, somewhat wavy. Flowers 1.2cm(½in) wide with a tubular base and five rounded lobes, white, fragrant of orange blossom, borne in small clusters from the leaf axils in spring. In general appearance more like *Coffea* than *Gardenia*.

MOMORDICA

Cucurbitaceae

Origin: Tropical Africa and Asia. A genus of about 40 species of annual or perennial tendril climbers. They have alternate palmate or digitate leaves and unisexual flowers which can occur together on the same plant, or be on single-sexed plants. Momordicas are grown for their ornamental, often warty, rather gourd-like, rounded to oblong fruits, and can be used on the back wall of a conservatory or grown on a support of canes or wires. Propagate by seed in spring in warmth.

Momordica is derived from the Latin *mordeo*, to bite, referring to the irregularly notched seeds which look rather as if they have been gnawed.

  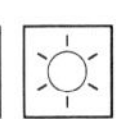  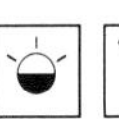 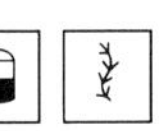

Species cultivated

M. balsamina Balsam apple Tropical Africa and Asia
To 3m(10ft) or more tall. Leaves to 10cm(4in) wide, rounded to heart-shaped in outline, the three to five lobes sharply toothed. Flowers 3cm(1¼in) wide, yellow, each on a slender stalk bearing a solitary, toothed bract; they are followed by ovoid to ellipsoid orange fruits to 8cm(3in) long. When ripe, fruits open into three segments showing the bright red arils of the seeds.
M. charantia Balsam pear Probably from tropical Africa, but now widespread
More vigorous than M. *balsamina*, to 4m(13ft) tall. Leaves rounded to heart-shaped in outline, but with deeper, blunt lobes.

Flowers similar, but the bract on the flower stalk not toothed and the fruits cylindrical to pear-shaped, 10–20cm(4–8in) long.

MONDO See OPHIOPOGON

MONSTERA

Araceae

Origin: Tropical America. A genus of 25 species of evergreen root climbers many of which are epiphytic in the wild. They have alternate, long-stalked leaves which can be ovate to oblong, and tiny petalless flowers carried on a spadix within a spathe typical of members of the arum family. They are familiar pot plants in conservatory or home, tolerant of poor light and a dry atmosphere, but looking much more healthy and attractive where humidity is provided. They grow particularly well on a moss stick. Propagate by stem tip or leaf bud cuttings in warmth in summer. *Monstera* may be derived from the Latin *monstrifer*, monster-bearing, referring to its large, oddly perforated leaves, but this is by no means certain.

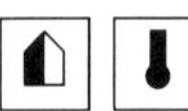    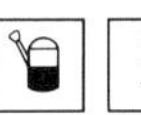

Species cultivated

M. deliciosa (M. *pertusa*, *Philodendron pertusum*) Ceriman, Swiss cheese plant, Mexican bread fruit Mexico to Central America
Very tall climber in warm countries, but rarely more than 2–3m(6–10ft) in a pot. Leaves 40–90cm(16–36in) long, broadly ovate, the margins deeply cleft, leaving curved, blunt-ended oblong lobes; on mature specimens they are also perforated with large elliptic to oblong holes. Spathes to 20cm(8in) or more long, creamy-white or green. Fruits ovoid, with a pineapple-like scent; they are edible but contain minute spicules which can prove an irritant to sensitive throats. M. *d.* 'Variegata' has variable yellowish-green markings, but is unstable and very liable to revert to dark green.
M. epipremnoides (M. *leichtlinii*) Costa Rica
Leaves to 90cm(3ft) long and 30cm(1ft) broad, but much smaller on young specimens; they are pinnately divided into 1.5–3cm(⅝–1¼in) wide lobes and perforated by one to three rows of holes. Flowering spathes are white and can reach 35cm(14in) in length.
M. latevaginata See *Rhaphidophora celatocaulis*.
M. leichtlinii See M. *epipremnoides*.
M. pertusa See M. *deliciosa*.

MORAEA

Iridaceae
Natal lily

Origin: Tropical and southern Africa. A genus of about 100 species of cormous perennials, at first glance with flowers like those of an iris but, in fact, without a perianth tube. They have linear leaves, one to several to each corm. Plants are suitable for a conservatory but can be brought into the house for flowering. Propagate by separating the corms when potting or by seed in spring. *Moraea* was named for

Robert More (1703–1780), a keen amateur botanist.

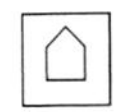   

Species cultivated

M. bicolor See *Dietes bicolor*.
M. catenulata See *Dietes vegeta*.
M. iridioides See *Dietes vegeta*.
M. neopavonia (*M. pavonia, Iris pavonia*) Peacock iris South Africa
Leaves linear, downy, to 60cm(2ft) long, one from each corm. Flowers to 6cm($2\frac{1}{2}$in) across, bright orange-red, the falls with a bluish to greenish black spot in the centre, its iridescent gloss like the 'eye' of the tail of the peacock; they are borne on branched stems to 60cm(2ft) tall.
M. pavonia See *M. neopavonia*.
M. ramosissima (*M. ramosa*) South Africa
Leaves to 75cm($2\frac{1}{2}$ft) long, linear, several to each corm. Flowers 5cm(2in) across, bright yellow, the falls with a brownish or greyish basal spot and carried on freely-branched stems 60–90cm(2–3ft) tall in early summer.
M. spathulata (*M. spathacea*) South Africa
Leaves to 60cm(2ft) long, sword-shaped, one to each corm. Flowers about 5cm(2in) across, bright yellow, fragrant, on stems 60–120cm(2–4ft) tall, opening in summer.

MUEHLENBECKIA

Polygonaceae

Origin: South America and Australasia, including New Guinea. A genus of 15 or 20 species, depending on the classification followed. They are climbing or prostrate shrubs with tough wiry stems and alternate leaves, in some species reduced to little more than scales. The flowers are tiny, greenish or whitish, carried in conspicuous clusters in the axils of the leaves. Those described below can be grown as foliage plants in the home or conservatory, *M. complexa* being best as a hanging basket subject and needing cooler conditions than *M. platyclada*. Propagate by seed when ripe or by cuttings in summer. *Muehlenbeckia* was named for Henry Gustave Muehlenbeck (1798–1845), a physician and botanist in Alsace, France.

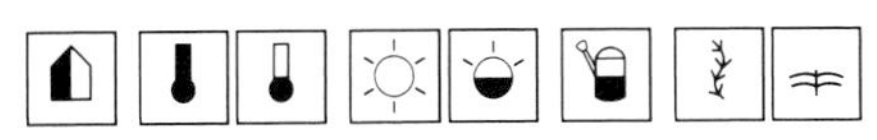

Species cultivated

M. complexa Wire-vine, Maidenhair vine New Zealand
Climbing or sprawling shrub with wiry, dark purple-brown stems, easily trained around supports. Leaves 5–10mm($\frac{1}{5}$–$\frac{3}{8}$in) long, rounded to heart-shaped, sometimes lobed, dull green often with purple-brown margins. Cool.
M. platyclada (*Homalocladium platyclados*) Solomon Is.
Sprawling shrub to 1.5–2m(4–6ft) high; leaves

Muehlenbeckia platyclada

reduced to scales, their place being taken by the stems which are flattened and ribbon-like. Tropical.

MURRAYA (MURRAEA)

Rutaceae

Origin: S.E. Asia, Indo-Malaysia, Pacific Islands. Depending on the botanical authority, a genus of four to 12 species of evergreen trees and shrubs related to *Citrus*. They have pinnate leaves with the leaflets arranged alternately or sub-oppositely, and 5-petalled flowers in terminal or axillary panicles. The globular fruits are colourful in some species but are not eaten. Propagate by cuttings in summer, or by seed in spring. The species described here make pleasing pot plants for the conservatory and are well worth trying in the home. The leaves of *M. koenigii* also have culinary uses. *Murraya* commemorates Johann Andreas Murray (1740–1791), a Swedish Professor of Botany and Medicine at Gottingen University and formerly a student of Linnaeus.

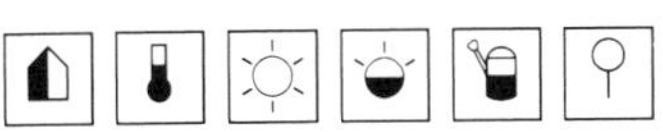

Species cultivated

M. exotica See *M. paniculata*.
M. koenigii Curry leaf India, Sri Lanka
In the wild or planted out, becoming eventually a large shrub or small tree, but can be kept at around 90cm(3ft) in a container. Leaves composed of 11 to 21 ovate to lanceolate leaflets, 2.5–5cm(1–2in) long, strongly and pungently aromatic when bruised. Flowers about 1cm($\frac{3}{8}$in) long or just under, white or cream, many together in terminal cymes. The leaves are a standard ingredient of curry and in India, the bark, leaves and roots are used as a tonic.
M. paniculata (*M. exotica*) Orange jasmine S.E. Asia, Malaysia
Large shrub or small tree, but easily kept to 1.2m(4ft) or so in a large pot or tub. Stems well branched, bearing leaves composed of three to nine obovate, 5–7cm(2–$2\frac{3}{4}$in) long, glossy rich green leaflets. Flowers 1.2–2cm($\frac{1}{2}$–$\frac{3}{4}$in) long, with pointed white petals, fragrant, borne in corymbs at intervals throughout the year if warm enough. Fruits ovoid, about 1.2cm($\frac{1}{2}$in) long, bright red.

MUSA

Musaceae

Origin: Tropical Asia and East Africa. A genus of 25 species of evergreen perennials, including the banana of commerce. They have trunk-like stems made up of sheathing leaf stalks closely wrapped around each other, and very large, oblong to elliptic leaves. The spike-like inflorescences are made up of many 6-tepalled flowers often with five fused and one free tepal, and carried within sometimes colourful bracts. The typical banana fruits are, correctly, cylindrical berries. They can be grown in the home and conservatory mainly as foliage plants when young, but M. *coccinea* and M. *velutina* will flower in a tub, and M. × *paradisiaca* 'Dwarf Cavendish' will fruit if given adequate space. Propagate by division, offsets or suckers when potting, or by seed of the species sown in warmth in spring. *Musa* is probably derived from the Arabic vernacular name *mouz* or *moz* for the banana, though it is sometimes said to be for Antonius Musa (63–14BC) Physician to the Roman Emperor Octavius Augustus.

  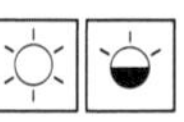  

Species cultivated

M. acuminata (M. *chinensis*) S.E. Asia
Up to 2m(6½ft) in large containers, with leaves 1m(3ft) or more in length and paddle-shaped, mottled and suffused with red when young. If planted out in a large conservatory it can get considerably larger and capable of reaching 6m(20ft) or more. When large it will produce arching to pendent flower spikes, with red to purple bracts, followed by 13cm(5in) long yellow bananas. See also M. × *paradisiaca*.
M. arnoldiana See M. *ensete*.
M. basjoo Japanese banana Ryuku Islands, Japan
To 2.5m(8ft) with leaves to 1.2m(4ft) long, green. When fully grown, produces an arching inflorescence, the flowers with yellow-green, brown-tinted bracts. Prefers cool temperature conditions.
M. cavendishii See under M. × *paradisiaca*.
M. chinensis See M. *acuminata*.
M. coccinea Flowering/Scarlet banana
Vietnam
To 1–1.2cm(3–4ft) tall, with leaves to 1m(3ft), dark green above, paler beneath. Erect flowering stems carry yellow blooms within showy scarlet, yellow-tipped bracts followed by 5–8cm(2–3in) long orange-yellow fruits which do not split open.
M. ensete (M. *arnoldiana*, *Ensete ventricosum*)
Ethiopia
Large plant to 2m(6½ft) or so in a container, with leaves almost as long. In the open it is capable of reaching 10m(33ft) or more and can have leaves to 5m(16ft) long. Flowers and fruits are not produced on potted specimens.
M. × paradisiaca (M. × *sapientum*) Common banana
A hybrid between M. *acuminata* and M. *balbisiana* which includes the cultivars of dessert and cooking bananas and is very like the parent species. M. × *p.* 'Dwarf Cavendish' (M. *cavendishii*, Canary Island banana) is the most rewarding banana plant to grow for fruit, being a dwarf mutant to about 2–2.5m(6½–8ft) tall, flowering and fruiting in a large tub if given room to reach mature size.
M. × sapientum See M. × *paradisiaca*.
M. velutina Assam
To 1–1.5m(3–5ft) tall, with dark green leaves to about 1m(3ft) long. Inflorescence erect, the flowers carried within pink bracts and followed by 5–8cm(2–3in) long, curved fruits which split open when ripe.

Musa × paradisiaca

MUSCARI

Liliaceae

Origin: Mediterranean region and S.W. Asia. A genus of 40 to 60 species of small, bulbous plants with linear, channelled leaves and racemes of bell- or urn-shaped flowers on erect, leafless stems. They are pretty spring-flowering pot plants for the cool conservatory and can be brought into the home for flowering. Hardy. Propagate by offsets or bulblets separated when dormant, or by seed as soon as ripe or in spring. *Muscari* derives from the Persian *mushk*, from the musk-like fragrance of the flowers of M. *moschatum*.

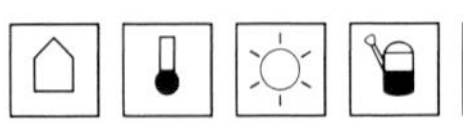

Species cultivated

M. armeniacum Grape hyacinth N.E. Turkey
Leaves six to eight, to 30cm(1ft) long. Flowers to 8mm(⅓in) long, bright purple-blue, fragrant, with reflexed petal tips; at the top of the spike are a few sterile flowers, paler blue and smaller than the lower ones; they are borne in racemes on stems 20–

30cm(8–12in) long. 'Blue Spike' has pale blue double flowers; 'Cantab' is single, blue; 'Heavenly Blue' is bright blue and very fragrant.

M. azureum See *Pseudomuscari azureum*.

M. botryoides Grape hyacinth Central southern Europe to the Caucasus
Leaves two to four, to 30cm(1ft) long. Flowers about 5mm(⅕in) long, almost spherical, bright blue, the few sterile ones paler, borne in short racemes on 15cm(6in) stalks. *M.b.* 'Album' has white flowers.

M. comosum Tassel hyacinth S. Europe, N. Africa
Leaves three to four, to 45cm(1½ft) long. Flowers of two distinct sorts in a spike, the lower ones 8mm(⅓in) long, dull, bluish-green on long stalks held at first pendent, then horizontally; the upper flowers sterile, bell-shaped and purple-blue with spreading to erect filaments. *M.c.* 'Monstrosum' ('Plumosum') has all the flowers sterile, giving a tassel-like effect to the whole flower.

M. macrocarpum Yellow musk hyacinth Turkey
Leaves five to six, to 30cm(1ft) long. Flowers all fertile, urn-shaped, yellow, on stems 20–25cm(8–10in) tall.

M. tubergenianum Oxford and Cambridge grape hyacinth N.W. Iran
Leaves two to three, to 25cm(10in) long. Flowers to 5mm(⅕in), rich deep blue, the upper sterile ones much paler; they are borne in racemes on stalks to 15cm(6in) or more tall.

MUSSAENDA

Rubiaceae

Origin: Tropical Africa and Asia, much cultivated elsewhere. A genus of 200 species of mainly shrubs and climbers. They have simple leaves in pairs or whorls of three, and tubular flowers with five petal lobes. The flowers are borne in clusters surrounded by what appear to be large white or coloured bracts like those of *Euphorbia* (spurges). However, each 'bract' is in fact a greatly enlarged sepal and strictly a part of the flower. The fruits are many-seeded berries. Mussaendas bring a real touch of the tropics into the warm conservatory and can be brought indoors when in bloom. Propagate by cuttings or air-layering in summer, or by seed in spring. *Mussaenda* is derived from the native vernacular name of a Sri Lankan species.

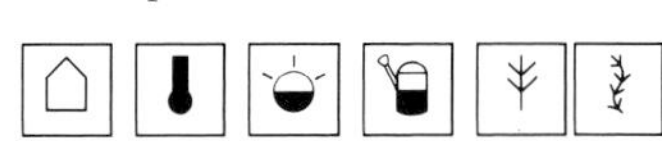

Species cultivated

M. erythrophylla Tropical West Africa
As a pot plant, a shrub to about 1m(3ft) or more; planted out or in a large container it becomes a climber – in the wild to 10m(3ft) in height. Leaves broadly ovate to elliptic, 8–15cm(3–6in) long, bright green. Flowers 4cm(1½in) long with red tubes and yellow lobes, in clusters, with several enlarged, ovate, red sepals 5–10cm(2–4in) in length. The fruits are egg-shaped, red and hairy. A showy and fairly easily grown species.

M. luteola Tropical Africa
Erect shrub to 1.5m(5ft) or more tall. Leaves narrowly ovate to lanceolate, to 5cm(2in) long. Flowers about 2.5cm(1in) long, yellow with an orange-red eye, the enlarged sepals white, broadly ovate, 2–6cm(¾–2½in) long, borne in late spring to early autumn.

ABOVE *Muscari macrocarpum*
ABOVE LEFT *Mutisia* sp.

MUTISIA

Compositae
Climbing gazanias

Origin: South America. A genus of 60 species of climbers and shrubs, most of which are evergreen. They have alternate leaves, in most species linear to oblong-ovate but occasionally pinnate. The flowers are in solitary daisy-like heads and open from distinctive, overlapping, cigar-like buds. Mutisias are best grown in a conservatory where they can be trained against the back wall, or through a shrub. Propagate by seed in spring or by cuttings in summer. *Mutisia* was named for José Celestino Mutis (1732–1809), a Spanish doctor and botanist who lived in Bogota, Colombia and collected plants, sending specimens to Linnaeus.

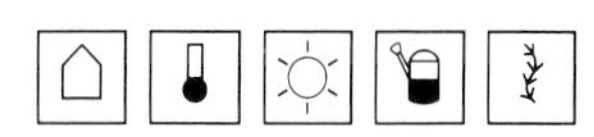

Species cultivated

M. clematis Colombia, Andes
Vigorous climber to 6m(20ft) or more, but easily kept to half this in a container. Leaves pinnate, composed of six to ten, 2–4cm(¾–1½in) long oblong-ovate entire leaflets. Flower heads to 6cm(2½in) wide, nodding to pendulous, bright orange-red, borne in late spring to autumn.

M. decurrens Chile, Argentina
Moderately vigorous climber to 2m(6½ft) or more, the slender stems sparingly branched or unbranched. Leaves narrowly oblong to oblong-lanceolate, 8–13cm(3–5in) long, the basal margins wing-like, extending down the stem (decurrent). Flower heads 10cm(4in) or more wide, bright orange.

M. ilicifolia Chile
Stems to 5m(16ft) with toothed wings. Leaves to 6cm($2\frac{1}{2}$in) long, stalkless, ovate and leathery, with holly-like teeth and a slender tendril from the leaf apex. Flower heads to 6.5cm($2\frac{5}{8}$in) across, the ray florets pink to pinkish-mauve, yellow at the centre, opening in summer and autumn.
M. oligodon Chile, Argentina
Semi-climbing to 1m(3ft) or more. Leaves 2.5–4cm(1–$1\frac{1}{2}$in) long, narrowly elliptic to oblong, hairy beneath, with small teeth especially near the apex and terminating in a branchless tendril. Flower heads 5–8cm(2–3in) across, with six to 12 rose to salmon-pink ray florets, opening in summer and autumn.

MYOSOTIS

Boraginaceae
Forget-me-nots

Origin: Temperate parts of the world, particularly in Europe, Asia and Australia. A genus of 50 or more species of annuals and perennials with alternate entire leaves and tubular flowers opening in five rounded lobes, borne in cymes. The species described makes a pretty flowering pot plant for the unheated room or conservatory. Propagate by seed in spring for late summer blooming, or in summer for plants in flower earlier in the next year. *Myosotis* derives from the Greek *mus*, mouse and *ous* or *otos*, ear, a name used for a number of plants with softly hairy, pointed leaves and chosen for this genus by Linnaeus.

  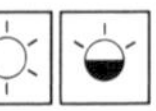  

Species cultivated
M. sylvatica Wood forget-me-not Europe, Asia
Biennial or short-lived perennial of tufted habit. Leaves to 8cm(3in) long, ovate to elliptic, softly and shortly hairy. Flowers 6–10mm($\frac{1}{4}$–$\frac{3}{8}$in) wide, bright blue, borne in branched cymes. Hardy. Of the cultivars available, often listed as M. *alpestris*, most belong to this species, while the more compact sorts are hybrids between the two species. 'Blue Ball' with blue flowers, and cultivars in shades of pink or white are of this origin.

Myosotis sylvatica

MYRICARIA See EUGENIA

MYRIOPHYLLUM

Haloragaceae
Milfoils

Origin: Cosmopolitan. A genus of 45 species of aquatic and bog plants with slender, flexible stems and narrow leaves, those underwater usually in whorls. Their spikes of tiny 4-petalled flowers are carried above the water. The species described below is suitable for use in an aquarium or greenhouse pool. Propagate by taking three or four shoots, fastening them together with a lead clip and pushing it into the sandy bottom ofthe pool or aquarium; they can also be put into pots of compost topped with gravel, though in the richer medium they will grow faster. *Myriophyllum* derives from the Greek *myrios*, many and *phyllon*, a leaf, referring to the dissected leaves of many of the species.

  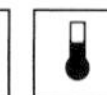  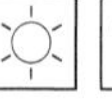 

Species cultivated
M. aquaticum (M. *brasiliense*, M. *proserpinacoides*) Water feather Brazil, Argentina and Chile
Stems capable of reaching 2m($6\frac{1}{2}$ft) in open water, far less in a container, mostly underwater but with its tips above. Leaves 2–3cm($\frac{3}{4}$–$1\frac{1}{4}$in) long, finely pinnatifid in whorls of four to six, blue-green below water, green above.

MYRTILLOCACTUS

Cactaceae
Myrtle cacti

Origin: Mexico, Guatemala. A genus of four species of columnar 'tree' cacti, one species of which is readily available and a handsome addition to a collection of succulents. Propagate by seed in spring, or by cuttings in summer. *Myrtillocactus* derives from the Latin *myrtillus*, a small myrtle, and the genus *Cactus*; a rather oblique reference to the fact that the berry fruits resemble those of a myrtle.

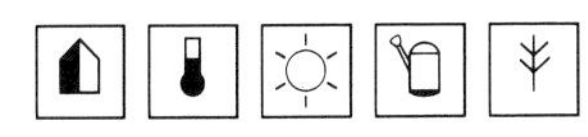

Species cultivated

M. geometrizans (*Cereus geometrizans*) Blue candle, Blue myrtle cactus Mexico
In the wild, a small tree with a well-branched crown; in containers to 2m(6½ft) if tip-pruned occasionally. Stems blue-green, jointed, cylindrical, with five to eight ribs bearing areoles about 2.5cm(1in) apart. Radial spines five to nine, 1.2cm(½in) or more long; central spines one, flattened, up to 8cm(3in) long. Flowers to 4cm(1½in) wide, white, in clusters of four to nine. Fruits purple, globular, 2cm(¾in) wide.

MYRTUS

Myrtaceae
Myrtles

Origin: Tropics and sub-tropics. A genus of 100 species of evergreen shrubs and trees with entire, lanceolate to obovate leaves carried in opposite pairs, and 4- to 5-petalled flowers followed by fleshy berries which in many species are edible. They make decorative pot plants. Propagate by seed in spring, or by heel cuttings in summer. *Myrtus* is the Greek vernacular name for this genus.

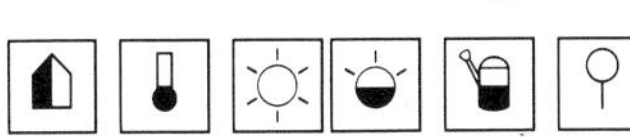

Species cultivated

M. bullata (*Lophomyrtus bullata*) New Zealand
Shrub to 1.5m(5ft) in a pot; shrub or tree to 5m(16ft) in the wild. Leaves 2–3cm(¾–1¼in) long, broadly ovate, strongly bullate, dark glossy green with a purplish tinge. Flowers 1.5cm(⅝in) wide, 4-petalled, white, solitary, borne in summer followed by dark red to black ovoid berries, 5–8mm(⅕–⅓in) across.

Myrtus communis

M. communis Common myrtle W. Asia
Shrub to 2m(6½ft) in a pot; to twice this in the open. Leaves 2.5–5cm(1–2in) long, lanceolate to ovate, glossy rich green. Flowers 2cm(¾in) wide, white, borne singly in summer and followed by 9–12mm(⅓–½in) long purple-black berries. *M.c.* 'Microphylla' is lower growing, with leaves to 2cm(¾in) long; 'Variegata' has the leaves margined with creamy-white.

M. obcordata (*Lophomyrtus obcordata*) New Zealand
Shrub to 2m(6½ft) in a pot, to 5m(16ft) in the open. Leaves 5–12mm(⅕–½in) long, heart-shaped with the stalk at the narrow end. Flowers about 6mm(¼in) wide, 4-petalled, white, carried singly and followed by dark red or violet berries.

N

NAEGELIA See ALOINOPSIS
NANANTHUS See ALOINOPSIS

NANDINA

Nandinaceae (Berberidaceae)

Origin: India and China. A genus of only one species; an evergreen shrub grown for its decorative foliage, graceful flowers and colourful fruits. It is best in a conservatory, but can be grown in the home. Propagate by seed when ripe or by heel cuttings with bottom heat in summer. *Nandina* is a latinized form of the plant's Japanese vernacular name *nanten*.

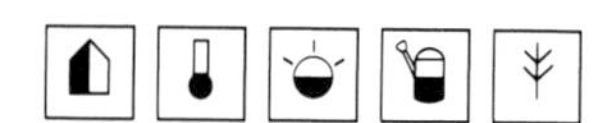

Species cultivated

N. domestica Heavenly bamboo, Sacred bamboo
Slender, erect shrub, sparingly branched, to 2m(6½ft) high. Leaves 30–45cm(1–1½ft) long, alternate, bi- to tripinnate with lanceolate leaflets 3–7cm(1¼–2¾in) long; they are a coppery-red in the spring, becoming green, then tinted red to purple in the autumn. Flowers to 6mm(¼in) long, made up of several whorls of tepals, the outer ones petal-like; they are borne in large terminal panicles 20–35cm(8–14in) long in summer and are followed by globular, 2-seeded red berries. *N.d.* 'Alba' has white fruits. Dwarf and purple-fruited cultivars have been named in Japan.

Nandina domestica

NARCISSUS

Amaryllidaceae
Narcissi, Daffodils

Origin: Europe, N. Africa and Asia, but chiefly S.W. Europe. A genus of 26 to 60 species of bulbous perennials with strap-shaped or filiform leaves and erect stems bearing one or a few flowers emerging from a usually papery spathe. The flowers have six wide spreading tepals joined at their base into a short funnel-shape, and a central corona which can range from large and trumpet-shaped to a shallow ring. They make excellent plants for the cool conservatory or room, but need to be kept in a cool place at night or they are very short-lived. Although they can be grown in bulb fibre they will be better in a standard potting mix and, if fed regularly, can be used again a second year or planted in the open garden for flowering. Hardy. Propagate species by seed, and cultivars by offsets when potting.

Narcissus was named after the Greek youth of mythology who was turned by the gods into a flower after becoming entranced by his own reflection in a pool.

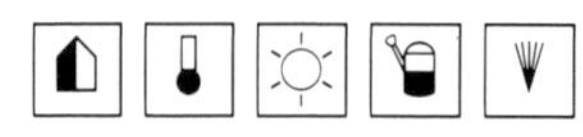

Species cultivated

N. bulbocodium Hoop-petticoat daffodil
France, Spain, Portugal, N.W. Africa
Leaves slender, almost rounded in cross-section, 10–30cm(4–12in) long, green. Flowers 2–3.5cm(¾–1⅓in) long, tepals very narrow and pointed, corona large and funnel-shaped, pale to deep yellow, borne on 10–15cm(4–6in) long stems. It is very variable in shape and colour, the following being distinct:

N.b. conspicuus
Flowers deep yellow, stronger growing.

N.b. obesus
Flowers with a large, deep yellow corona.

N. campernellii See *N.* × *odorus*.

N. cyclamineus Cyclamen-flowered daffodil/narcissus Spain, Portugal
Leaves flat, keeled, bright rich green, 10–20cm(4–8in) long. Flowers 4–5cm(1½–2in) long, corona 2–2.5cm(¾–1in) long, tubular, tepals swept backwards to about the same length, deep yellow.

N. jonquilla Jonquil S. Europe, Algeria
Leaves to 30cm(1ft) long, rush-like, dark green. Flowers 3–4cm(1¼–1½in) wide, rich yellow and very fragrant, borne in umbels of two to six.

N. juncifolius Rush-leaved jonquil N. Spain, Portugal, S.W. France
Similar to *N. jonquilla*, but much smaller, to 15cm(6in) high overall with one to five flowers, each to 2.5cm(1in) across.

N. obvallaris Tenby daffodil Europe
Leaves to 30cm(1ft) long, flat, bluish-green. Flowers 4–5cm(1½–2in) long, with the rather flat tepals and the corona about the same length, golden-yellow.

N.* × *odorus (*N. campernellii*) Campernelle jonquil
A hybrid between *N. jonquilla* and *N. pseudonarcissus*, rather similar to the first parent but with flattened leaves to 6mm($\frac{1}{4}$in) wide. Flowers with a corona to 1.5cm($\frac{5}{8}$in) and tepals to 4.5cm($1\frac{3}{4}$in) across, fragrant.
N. poeticus Poet's/Pheasant's-eye narcissus
Spain to Greece
Leaves narrowly strap-shaped, glaucous green. Flowers 4.5–7cm($1\frac{3}{4}$–$2\frac{3}{4}$in) wide, tepals pure white, often somewhat reflexed, corona a short, flat cup with an orange-red, crimped margin, very fragrant. One parent of the many small-cupped cultivars.
N. pseudonarcissus Wild daffodil, Lent lily
Europe
Leaves to 1.5cm($\frac{5}{8}$in) wide, grey-green. Flowers 5–7cm(2–$2\frac{3}{4}$in) long, somewhat nodding, tepals somewhat twisted, pale yellow, corona trumpet-shaped, varying from straight-sided to flared at the mouth, deep yellow.
N. rupicola Portugal, Spain
Leaves narrow, rounded, triangular in section to 20cm(8in) or more long, grey-green. Flowers about 2cm($\frac{3}{4}$in) wide, yellow, borne singly on stalks to 15cm(6in) tall.
N. tazetta Bunch-flowered/Polyanthus narcissus
Leaves 1–2cm($\frac{3}{8}$–$\frac{3}{4}$in) wide, 30cm(1ft) long, green. Flowers 2.5–4cm(1–$1\frac{1}{2}$in) wide, with white tepals and a short, cup-like yellow or white corona, four to eight borne together on stems 30–45cm(1–$1\frac{1}{2}$ft) high; very fragrant.
N. triandrus Angel's tears narcissus Portugal, W. Spain
Leaves narrowly linear, to 15cm(6in) or more tall, grey-green. Flowers to 3cm($1\frac{1}{4}$in) across, white, nodding, with a corona 1cm($\frac{3}{8}$in) long and broad, and swept-back tepals, borne one to three together on stems 15cm(6in) or more tall.

Hybrid groups

Daffodils or Trumpet narcissi (*Division 1*) Flowers with the corona as long as or longer than the tepals.
'Dutch Master', rich golden-yellow.
'Queen of the Bicolors', white tepals, yellow corona.
'Vigil', pure white.
Large-cupped narcissi (*Division 2*) Corona more than one-third as long as the tepals.
'Carlton', clear yellow.
'Ice Follies', white tepals, yellow corona fading to white.
'Salome', white tepals, salmon corona.
Small-cupped narcissi (*Division 3*) Corona less than one-third as long as the tepals.
'Angel', pure white with a green and yellow eye.
'Barret Browning', white tepals, orange-red corona.
Double-flowered narcissi (*Division 4*) Double flowers.
'Bridal Crown', tepals white, centre yellow.
'Dick Wilden', yellow.
'White Cheerfulness', three to four creamy-white

Narcissus 'Salome'

fragrant blooms to a stem.
'Yellow Cheerfulness', three to four yellow fragrant blooms to a stem.
Triandrus narcissi (*Division 5*) Several nodding flowers to each stem with reflexed or recurved tepals.
'Liberty Bells', creamy-yellow.
'Thalia', white.
Cyclamineus narcissi (*Division 6*) One nodding flower to each stem, tepals recurved.
'Dove Wings', white tepals, yellow corona.
'Peeping Tom', golden-yellow, very long corona.
'Tête à Tête', buttercup-yellow, flowers in twos.
Jonquil narcissi (*Division 7*) Several small flowers to a stem, very narrow leaves.
'Lintie', yellow, rounded tepals, small orange-edged corona.
Tazetta narcissi (*Division 8*) Several flowers to a stem, rather like *N. tazetta*.
'Grand Primo Citronière', white tepals, yellow corona, strong sweet scent.
'Grand Soleil d'Or', yellow, with darker corona, strong scent.
'Minnow', two to four small lemon-yellow flowers to each stem; only 15cm(6in) high.
'Paper White', pure white, sweet scent.

Narcissus 'Peeping Tom'

ABOVE *Nematanthus gregarius*
ABOVE CENTRE *Nautilocalyx lynchii*
TOP *Nelumbo nucifera*

NAUTILOCALYX

Gesneriaceae

Origin: South America. A genus of about 15 species of erect to decumbent, evergreen perennials. They have opposite pairs of usually handsome, bullate or quilted leaves and axillary clusters of tubular flowers with five rounded petal lobes, in some species emerging from coloured calyces. Propagate by seed in spring, or by stem and leaf cuttings in summer. *Nautilocalyx* derives from the Latin *nautilus*, the marine shell and *calyx*, the ring of sepals (free or fused) which protects the flower bud, but the similarity is hard to see.

Species cultivated

N. forgetii Peru
Stems erect, to 60cm(2ft) in height. Leaves elliptic, up to 15cm(6in) in length, quilted, with wavy and toothed margins, bright green with reddish-brown veining above, purple beneath. Flowers white with yellow throats, 4cm(1½in) long, from large red calyces.

N. lynchii Colombia
Stems erect, to 60cm(2ft) tall. Leaves broadly lanceolate, slender-pointed, lightly quilted, with short winged stalks, lustrous maroon above, purple beneath. Flowers pale yellow, 4cm(1½in) long, from red calyces.

NEANTHE See CHAMAEDOREA
NELUMBIUM See NELUMBO

NELUMBO

Nelumbonaceae (Nymphaeaceae)

Origin: Asia and North America. A genus of two species of rhizomatous, aquatic perennials with large peltate leaves which are held above the water on long stalks. The flowers are like those of a water lily and borne singly, followed by a large pepper-pot-like receptacle in which is embedded a cluster of one-seeded fruits. They are suitable for growing in a conservatory pool or in a tub of wet soil, ideally using a rich compost of equal parts loam and well-rotted manure. For a pool, plant in a wire basket at least 45cm(1½ft) in diameter. Propagate by division or by seed, putting these individually in pots of compost covered with gravel, and standing submerged in warm water at 18–24°C(65–75°F). *Nelumbo* is the vernacular name for *N. nucifera* in Sri Lanka.

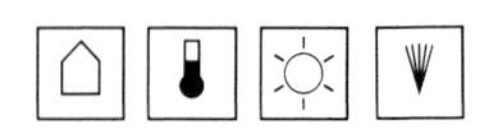

Species cultivated

N. lutea (*N. pentapetala, Nelumbium lutea*)
American lotus USA
Leaves circular, 30–60cm(1–2ft) wide, blue-green. Flowers to 25cm(10in) wide, cup-shaped, yellow, fragrant. *N.l.* 'Flavescens' has more abundant but smaller flowers.

N. nucifera (*N. speciosa*) Sacred lotus Asia
Of similar appearance to *N. lutea* but the leaves more glaucous-green and the flowers pink. Cultivars with carmine, white, striped and double flowers are known.

NEMATANTHUS

(inc. HYPOCYRTA)

Gesneriaceae

Origin: South America; those listed below all from Brazil. A genus of about 30 species of largely epiphytic shrubs and perennials. They are evergreen, with pairs of simple leaves closely set on semi-erect to prostrate stems. The axillary borne flowers are tubular, in many species the lower side of the upper half inflated to look like a pelican's pouch and constricted to a small mouth ringed with five tiny lobes. The species described here make splendid hanging basket subjects. *Nematanthus* derives from the Greek *nema*, a thread and *anthos*, a flower; several species have long flower stalks allowing the blooms to dangle beneath the leaves.

Species cultivated

N. gregarius (*N. radicans, Hypocyrta radicans*)
Stems trailing, 30–60cm(1–2ft) long. Leaves in pairs or whorls of three, 2–4cm(¾–1¼in) long, somewhat fleshy but firm, dark lustrous green. Flowers on short stalks, 2cm(¾in) long, orange with yellow lobes.

N. longipes (*Columnea splendens*)
Shrubby, with lax, branched stems to 60cm(2ft) long. Leaves obovate to elliptic, 5–10cm(2–4in) long, rich green above, paler or purple-flushed beneath. Flowers very long-stalked, 5cm(2in) or more in length, trumpet-shaped, wide-mouthed, red with an orange-mottled throat.

N. perianthomegus (*Hypocyrta perianthomega*)
Robust, with semi-erect to spreading rigid stems 60cm(2ft) or more in length. Leaves to 11cm(4½in) long, obovate to elliptic, rich lustrous green. Flowers in small clusters from leaf axils, about 4cm(1½in) long, yellow with contrasting longitudinal maroon stripes, from soft orange calyces.

N. radicans See *N. gregarius*.

N. strigillosus (*Hypocyrta strigillosa*)
Stems trailing, 30–45cm(1–1½ft) or more in length. Leaves elliptic, slightly cupped, 2–4cm(¾–1¾in) long, densely minutely hairy. Flowers 3cm(1¼in) long, orange to orange-red.

N. wettsteinii (*Hypocyrta wettsteinii*)
Stems trailing, 30–45cm(1–1½ft) in length. Leaves ovate, 2–2.5cm(¾–1in) long, dark green and brightly glossy. Flowers 2.5cm(1in) long, red to tangerine with yellow lobes.

Hybrids

In USA during the 1960s and 70s in particular, a number of hybrids were raised using several species, among them *N. gregarius*, *N. perianthomegus*, *N. strigillosus* and *N. wettsteinii*. Of the cultivars that have persisted, the following are well worth seeking:
'Bijou' (*N. wettsteinii* × *N. fritschii*), trailing, pink flowers.
'Butterscotch' (*N. strigillosus* × *N.* 'Tropicana'), trailing, bright yellow.
'Rio' (*N. gregarius* × *N. selloana*), erect habit, vermilion flowers.

NEMESIA

Scrophulariaceae

Origin: South Africa. A genus of 50 species of annuals, perennials and sub-shrubs, with 4-angled stems, pairs of lanceolate leaves and 2-lipped tubular flowers with a short spur or pouch, the upper lip divided into four and the lower entire or notched. One species is an attractive annual for conservatory or home. Propagate by seed, sowing in autumn or spring and providing support for the stems. *Nemesia* derives from the Greek vernacular name for a plant of similar appearance.

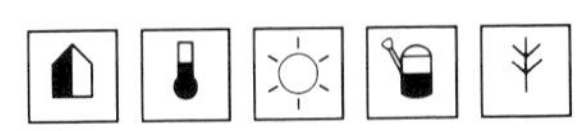

Species cultivated

N. strumosa
Erect, 30–60cm(1–2ft) tall. Leaves to 8cm(3in) long, pale green and coarsely toothed. Flowers 2cm($\frac{3}{4}$in) wide, pouched but not spurred, in a wide range of colours including yellows, purples and white, carried in terminal racemes in summer. *N. s.* 'Suttonii' is the most commonly grown form, usually lower growing, 20–30cm(8–12in) tall, in shades of yellow, orange, red, pink, blue and purple or white.

NEMOPHILA

Hydrophyllaceae

Origin: North America. A genus of 11 species of slender-stemmed, decumbent annuals. They have narrow leaves which can be toothed or pinnately lobed, and 5-petalled, cup-shaped or flat flowers in blue, purple or white, sometimes with conspicuous spotting, carried singly on long stalks from the upper leaf axils. Propagate by seed; an autumn sowing gives plants in flower in spring, and a spring sowing a display in the following summer. *Nemophila* derives from the Greek *nemos*, a wooded grove and *phileo*, to love, referring to the pine forest habitat of some species.

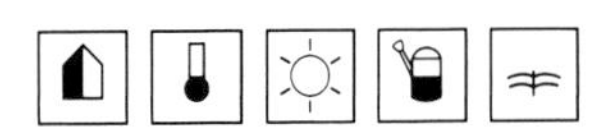

Species cultivated

N. maculata Five spot California
Plant to 15cm(6in) tall. Leaves 1–3cm($\frac{3}{8}$–$1\frac{1}{4}$in) long, deeply 5- to 9-lobed. Flowers 2.5cm(1in) wide, bow-shaped, with white petals, each having a deep purple spot in the centre of the petal margin.

N. menziesii (*N. insignis*) Baby blue-eyes California
Plant to 20cm(8in) tall. Leaves 2.5cm(1in) long, deeply 9- to 11-lobed, sometimes silvery-spotted. Flowers about 3cm($1\frac{1}{4}$in) wide, bowl-shaped, bright blue with a white centre, occasionally with black-purple spots and veins.

NEOCHILENIA See NEOPORTERIA
NEOGOMESIA See ARIOCARPUS

NEOPORTERIA

Cactaceae

Origin: Peru, Chile and Argentina. A genus of about 66 species of globose to shortly cylindrical cacti. Some botanists include the majority in *Neochilenia* leaving only 13 species here. They have

BELOW LEFT *Nemesia strumosa* 'Suttonii'
BELOW *Nemophila maculata*

Neoporteria subgibbosa

ribbed stems with woolly and spiny areoles and sometimes woolly crowns. Flowers are bell- to funnel-shaped and the fruits coloured, dry and hollow. Propagate by seed in spring in warmth or by offsets in summer. *Neoporteria* derives from the Greek *neo*, new and the genus *Porteria*, named for Carlos Porter, a Chilean entomologist.

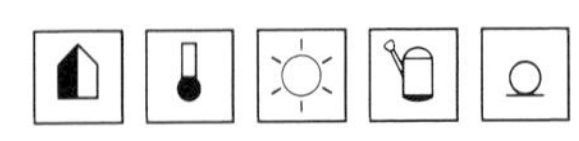

Species cultivated

N. chilensis (*Neochilenia chilensis*, *Nichelia chilensis*) Chile
Stems globular when young then cylindrical, up to 25cm(10in) in height. Ribs about 20, thick, green, slightly notched. Radial spines 20, yellow to whitish, about 1.2cm(½in) long; central spines six to eight, yellow to brownish, twice as long. Flowers funnel-shaped, about 5cm(2in) long and wide, bright pink to carmine-red.

N. napina (*Neochilenia napina*) Chile
Roots swollen and tuber-like. Stem 3–9cm(1¼–3½in) tall, globose to elongated, reddish to brownish-green with about 14 spiralling ribs deeply notched. Areoles with three to nine black or brown spines each to 3mm(⅛in) long. Flowers funnel-shaped, yellow, 3–3.5cm(1¼–1½in) long.

N. subgibbosa (*Chilenia acutissima*) Chile
Stems globose at first, becoming cylindrical, to 8–30cm(3–12in) or more tall, green to greyish-green. Ribs 14 to 16, deeply cleft with rounded woolly areoles and brownish spines to 3cm(1¼in) long. Flowers 4–5cm(1½–2in) long, broadly funnel-shaped, pink to light red.

Neoregelia carolinae 'Tricolor'

NEOREGELIA

Bromeliaceae

Origin: Tropical South America, chiefly from Brazil. A genus of 52 species of epiphytic perennials. They are evergreen and the broadly strap-shaped leaves grow in rosettes, their bases widening and overlapping to form a water-holding cup. After producing flowers, the rosette dies but is normally replaced by offsets. Propagate by removing offsets, or by seed. *Neoregelia* derives from the Greek *neo*, new and *Regelia*, an allied genus named for E. Albert von Regel (1815–1892), Director of the Imperial Botanical Garden in St. Petersburg.

Species cultivated

N. ampullacea Brazil
Rosettes small, tubular, formed of a few leaves, suckering freely. Leaves 13cm(5in) long, linear, red-tipped, red-brown banded beneath, finely toothed. Flowers just above the neck of the rosette tube, 1.2–2.5cm(½–1in) wide, blue.

N. carolinae (*Aregelia marechalii*, *Nidularium meyendorfii*) Brazil
Rosettes formed of 40cm(16in) long leaves which are strap-shaped, glossy-green and finely spine-toothed, the inner ones red to purplish, brightest at the centre. Flowers are blue-purple and arise from bright red bracts. *N.c.* 'Tricolor' (Blushing bromeliad) has longitudinal stripes of ivory-white, tinted with pink as they age.

N. marechalii See *N. carolinae*.

N. marmorata (*Aregelia marmorata*) Marble plant Brazil
Leaves wide-spreading, strap-shaped, to 40cm(16in) long, soft-textured, light green irregularly blotched red-brown, the tip spine red. Flowers to 2cm(¾in) wide, pale purple to lavender. Hybrids with *N. spectabilis* are often sold under this name; they have olive-green leaves with maroon blotches and a red tip.

N. meyendorfii See *N. carolinae*.

N. spectabilis Fingernail plant Brazil
Rosettes of leathery strap-shaped leaves to 40cm(16in) long, spine-toothed, dull green tipped with bright red. Flowers blue from green bracts marked with purple bands.

NEPENTHES

Nepenthaceae

Origin: S.E. Asia to Northern Australia. A genus of about 70 species of carnivorous plants, most of which are epiphytic climbers with narrow, evergreen leaves and very small, inconspicuous, greenish flowers borne in racemes. The insect-catching pitchers are formed at the ends of the terminal leaf tendrils; they contain a digestive fluid and have a lid. Insects are attracted to nectar glands which are positioned at the mouth of the pitcher. They land on the inner surface which is very slippery, fall into the pitcher, drown and their bodies are broken down by the fluid, the necessary minerals being taken in by the plant. The plants are suitable for conservatories or a glass case in the house, but are best in orchid baskets, growing in a standard compost mixed with an equal part of sphagnum moss. Keep them moist and syringe daily. Propagate by cuttings of stem tips in warmth with humidity, or by seed sown in spring in warmth. *Nepenthes* is the Greek name for another dissimilar plant used for this genus by Linnaeus because of its supposedly similar medicinal

  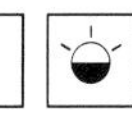 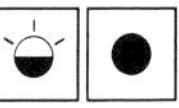  

Species cultivated

N. ampullaria Malaysia
Stems to 3m(10ft) or more. Leaves to 30cm(1ft) long, oblong, the pitchers ovoid, to 5cm(2in) long, green, sometimes with red blotching.

N. hookeriana Borneo
Similar in stems and foliage to *N. ampullaria*, but the pitchers are broadly oval to 13cm(5in) long and 8cm(3in) across, pale green, strongly spotted and blotched with purple.

N. khasiana Assam
Vigorous. Leaves lanceolate; pitchers rounded to tubular, up to 10cm(4in) long, green flushed with red. Not difficult to cultivate.

NEPHELIUM See LITCHI

NEPHROLEPIS

Oleandraceae (*Davalliaceae*)

Origin: Tropics and sub-tropics, extending to temperate Japan and New Zealand. A genus of about 30 species of rosette-forming ferns, often spreading by runners, with oblong-lanceolate, pinnate, evergreen fronds bearing their rounded sori on vein tips near the margins of the leaves. They can be grown in home or conservatory, those with arching ferns being particularly successful in hanging baskets. Propagate by division or by removing offsets; spores can be used for the species. *Nephrolepis* derives from the Greek *nephros*, a kidney and *lepis*, a scale with reference to the membraneous indusium which covers the spores.

  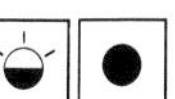  

Species cultivated

N. cordifolia Sword fern, Ladder fern
Tufted, often tuberous with short, scaly stems. Fronds 30–60cm(1–2ft) long, almost erect, the individual pinnae to 4cm(1½in) long, sharply toothed.

Nepenthes hookeriana

N.c. duffii
Much narrower fronds often forked at the tips, the pinnae rounded to 1.5cm(⅝in) wide.

N. exaltata Sword fern Tropics
Tufted, but without tubers. Fronds 60–120cm(2–4ft) long, erect at first then arching over, the many narrow pinnae to 8cm(3in) long. Many cultivars are more commonly grown than the true species:

'Bostoniensis' (Boston fern), the oldest and still the most frequently seen, its fronds drooping more strongly, to 15–20cm(6–8in) broad.

'Bostoniensis Compacta', similar but smaller.

'Elegantissima', fronds bi- to tripinnate giving a feathery effect.

'Rooseveltii', darker green fronds with wavy leaflets.

'Teddy Junior', much more compact sport of 'Rooseveltii'.

'Whitmannii' (Lace fern), bi- to tripinnate light green fronds to 45cm(1½ft) long, with small pinnules which give it a very lacy appearance.

NEPHTHYTIS

Araceae

Origin: Tropical Africa. A genus of four to five species of evergreen perennials similar to and much confused with *Syngonium*. The long-stalked arrow- or spear-shaped leaves are leathery and rise in tufts from creeping rhizomes. The flowers are arum-like, their spadix and spathe being green in colour, and are later followed by spikes of orange fruits. They make good foliage plants for the conservatory and home provided enough humidity can be given. Propagate by division. *Nephthytis* is derived from the goddess of the same name in Egyptian mythology, wife of Typhon and mother of Anubis.

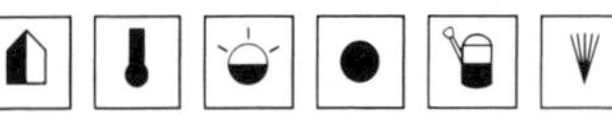

Species cultivated

N. afzelii Sierra Leone, Liberia
Plant to 1m(3ft) tall. Leaves to 35cm(14in) or more long, saggitate, carried on a stalk to 45cm(1½ft) long. Spathes 7cm(2¾in) long by 3cm(1¼in) wide, oblong, green, followed by obovoid berries about 1cm(⅜in) long.

N. hoffmannii See *Syngonium hoffmannii.*

N. liberica (of gardens) See *Syngonium podophyllum* 'Variegata'.

N. poissonii Cameroon, Gambia
Similar to *N. afzelii*, the leaves rich green and the 5cm(2in) long spathes green, finely spotted with brown. Berries to 3cm(1¼in) long, ellipsoid.

N. wendlandii See *Syngonium wendlandii.*

NERINE

Amaryllidaceae

Origin: South Africa. A genus of 20 to 30 species of bulbous plants with narrowly strap-shaped arching leaves and umbels of short-tubed, funnel-shaped flowers opening to six often narrow and waved tepals. They are best grown in the conservatory, needing good light, but can be brought into the home in late summer and autumn when they flower. Propagate by seed sown as soon as ripe – they are fleshy and quickly lose their viability when dry – or by offsets removed when dormant. *Nerine* derives from *Nereis*, a Greek water nymph; the name was given because the first species recognized was brought ashore from a ship wrecked or stranded in Guernsey.

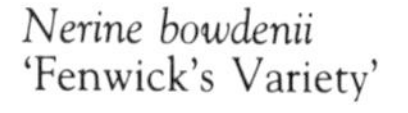

Nerine bowdenii 'Fenwick's Variety'

Nerine bowdenii 'Zeal Giant'

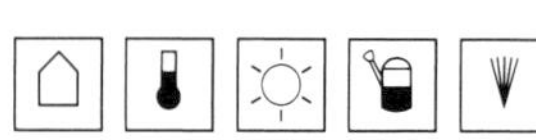

Species cultivated

N. bowdenii
Leaves to 30cm(1ft) high with the flowers. Blooms with waved tepals to 7cm(2¾in) long, pink, borne in umbels of six to 12 on stems to 45cm(1½ft) high. *N.b.* 'Fenwick's Variety' is taller and more vigorous, with deeper pink flowers. 'Zeal Giant' is a hybrid with large blooms of a deep carmine-pink.

N. filifolia
Leaves 15–20cm(6–8in) long, linear, almost grassy. Flowers with slender, recurved tepals to 2.5cm(1in) long, rose-red, and long conspicuous down-curved stamens; they are borne in umbels of six to ten blooms on a stem 20–30cm(8–12in) tall.

N. flexuosa
Leaves to 30cm(1ft) or more long. Flowers with waved tepals 5–8cm(2–3in) long, pink with a darker central line; they are carried in umbels of ten to 20 on tall stems 60–90cm(2–3ft) high. *N.f.* 'Alba' has white flowers.

N. sarniensis Guernsey lily
Leaves strap-shaped, to 30cm(1ft) long and 2cm(¾in) wide, green or glaucous green. Flowers with recurved, sometimes wavy tepals 3–4cm(1¼–1½in) long, usually crimson, scarlet or pink; they are carried in umbels of four to eight on stems 25–50cm(10–20in) tall. A number of named cultivars are available, some selections, some hybrids. Some of these variants have, in the past, been given specific rank, but are only forms of this species; they include *N. corusca*, stronger-growing with umbels of up to 30 salmon-pink flowers, and *N. curvifolia* with glaucous leaves and scarlet flowers.

N. undulata
Leaves 25–35cm(10–14in) long. Flowers have strongly waved pale pink petals 2–2.5cm(¾–1in) long in umbels of ten to 15 on 20–40cm(8–16in) stems. A white form is known. In cultivation it is sometimes grown erroneously as *N. crispa*.

NERIUM

Apocynaceae

Origin: Mediterranean region across Asia to Japan. A genus of two or three species of evergreen shrubs with pairs or whorls of three or four leaves and large showy flowers in terminal cymes. The species commonly grown world-wide, in the tropics and subtropics, makes a large, handsome pot or tub plant for the conservatory and, when small, in the home. Propagate by cuttings in summer with bottom heat or by seed in spring in warmth. *Nerium* is the ancient Greek name for oleander.

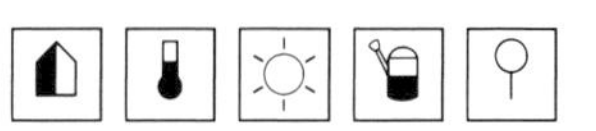

Species cultivated

N. oleander Oleander, Rose-bay
Erect bushy shrub to 2m(6½ft) in a container, to 5m(16ft) in the open. Leaves 6–20cm(2½–8in) long, linear-lanceolate, dark green with a paler central vein and on reverse side. Flowers 3–5cm(1¼–2in) across, funnel-shaped, opening to five flat lobes varying in colour from shades of pink, red and white to cream and yellow. Double-flowered forms are commonly grown. *N.o.* 'Variegatum' has pink flowers and cream-margined leaves.

NERTERA

Rubiaceae

Origin: S.E. Asia, Australia, New Zealand, islands of the Pacific and western Central and South America. A genus of 12 species of small, evergreen perennials with very slender, thread-like stems which form mats or hummocks. The leaves are small and borne in opposite pairs, while the minute, rather inconspicuous flowers are followed by orange or red 2-seeded fruits, correctly called drupes. They are delightful pan plants for the conservatory or home. Propagate by careful division or by seed sown in warmth in spring. *Nertera* derives from the Greek *nerteros*, lowly, referring to the mode of growth.

Species cultivated

N. granadensis Bead plant Mexico and Central America
Low, cushion-forming plant soon covering the top of a pan with foliage. Leaves 2–4mm($\frac{1}{12}$–$\frac{1}{6}$in) long, oblong to broadly ovate.
Fruit to 6mm($\frac{1}{4}$in) wide, bright orange, freely produced.

NICHELIA See NEOPORTERIA

NICODEMIA

Loganiaceae

Origin: Sri Lanka, Mascarene Islands. A genus of six species of shrubs formerly included in the genus *Buddleja*, but separated on the basis of fleshy, berry-like fruits (those of *Buddleja* are dry capsules). The species described here make showy tub or border plants for the conservatory. Propagate by cuttings in late summer, or by seed in spring or when ripe. *Nicodemia* honours Gaetano Nicodemo, (*d.* 1803), Italian botanist and curator of the Lyons Botanic Garden 1799–1803 when he committed suicide.

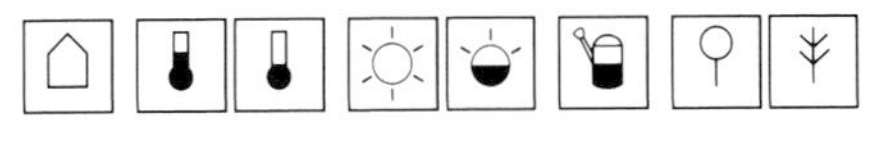

Species cultivated

N. madagascariensis (*Buddleja madagascariensis*) Malagasy
Erect habit, 2–3m($6\frac{1}{2}$–10ft) or more in height. Leaves lanceolate, to 13cm(5in) long, dark green above, white-felted beneath. Flowers small, bright orange, borne in terminal slenderly pyramidal panicles in winter to spring. Fruits purple.

NICOLAIA

Zingiberaceae

Origin: Indo-Malaysia. A genus of 16 to 25 species of large, rhizomatous perennials related to *Amomum* and *Alpinia*, one of which is widely grown in the tropics and deserves a place in the warm conservatory. Like some other members of the *Zingiberaceae* family, *Nicolaia* produces two sorts of stems, tall and leafy, and short and leafless. Only the latter bear flowering spikes. Propagate by division in spring, or just as new growth commences. *Nicolaia* honours the Russian Tzar Nicholas I (1796–1855).

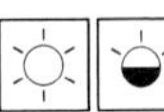
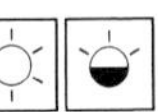

Species cultivated

N. elatior (*Alpinia magnifica*, *Phaeomeria magnifica*) Torch ginger, Philippine wax flower Celebes, Java
Leafy stems erect, unbranched, 3m(10ft) or more tall if planted out, less in a tub. Leaves to 60cm(2ft) long, linear to oblong-lanceolate. Flowering stems to 1.2m(4ft) or more in height, topped by a pyramidal spike of waxy red, white-edged, overlapping bracts which partially obscure the darker red flowers with their yellow-margined lips; late summer.

NIDULARIUM

Bromeliaceae

Origin: South America, chiefly Brazil. A genus of 22 species of rosette-forming epiphytic perennials with strap-shaped evergreen spine- or prickle-toothed leaves, the bases of which overlap to form a water-holding cup. The small flowers are clustered together and are usually borne close to the centre of the rosette, but are occasionally carried on longer stalks. They have erect, hooded tepals, a character which separates them from the closely allied *Neoregelia*. Propagate by separating offsets or by seed. *Nidularium* derives from the Latin *nidus*, a nest, an allusion to the shape of the plants.

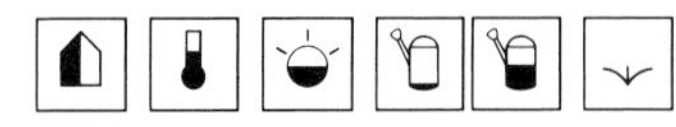

Species cultivated

N. billbergioides Brazil
Leaves more or less erect, sword-shaped, 20–40cm(8–16in) long, bright green, shortly spiny-toothed. Flowering stem erect, to 25cm(10in) in height, bearing scattered, sheathing bracts and topped by a cluster of triangular-ovate, yellow floral bracts with greenish flared tips; flowers 2cm($\frac{3}{4}$in) long, white, partially hidden by the bracts.
N. × chantrieri (*N. fulgens* × *N. innocentii*)
Much resembles a somewhat finer version of *N. fulgens* with white flowers.
N. fulgens (*N. pictum*, *Guzmania picta*) Blushing bromeliad Brazil
Leaves to 30cm(1ft) long by 4.5cm($1\frac{3}{4}$in) broad, the margins coarsely spine-toothed, shining pale green with darker mottling. Flowers blue, surrounded by bright scarlet bract-leaves in the centre of the rosette.

ABOVE *Nidularium billbergioides*
ABOVE CENTRE *Nicolaia elatior*
TOP *Nertera granadensis*

Nicodemia madagascariensis

Nolina recurvata

N. innocentii (*N.* "*amazonicum*") Bird's nest bromeliad Brazil
Leaves to 30cm(1ft) long by 5cm(2in) broad, the margins finely spine-toothed, shining deep purple above with a metallic lustre, red-purple beneath. Flowers white in a dense head surrounded by pink to brick-red bracts. *N. i.* 'Lineatum' has green leaves striped with many white lines, bracts red-tipped; 'Striatum' is similar but with fewer, broader white lines and red-purple bracts.
N. meyendorfii See *Neoregelia carolinae*.
N. pictum See *N. fulgens*.

NIEREMBERGIA

Solanaceae
Cup flowers

Origin: Mexico to South America. A genus of 30 to 35 species of perennials and sub-shrubs with linear to spathulate, entire leaves and 5-lobed, regular flowers carried singly or in terminal cymes. The species described below is most floriferous when grown as an annual. Propagate by cuttings taken in late winter, or by seed in early spring. *Nierembergia* was named for Juan Eusebio Nieremberg (1595–1658), Spanish Jesuit and a naturalist.

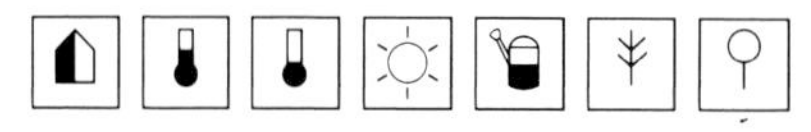

Species cultivated
N. hippomanica (*N. caerulea*) Argentina
Sub-shrub usually grown as an annual. Stems slender and freely branched. Leaves to 1.5cm($\frac{5}{8}$in) long, linear to spathulate. Flowers to 2cm($\frac{3}{4}$in) across, cup-shaped, white in the type. In general cultivation the species is replaced by *N. h. violacea*, which has mauve flowers with a yellow throat.

NOLINA

Agavaceae (*Liliaceae*)

Origin: Southern USA, Mexico. A genus of 30 species of evergreen perennials of which five are sometimes separated into the genus *Beaucarnea*. They have a short trunk-like stem and dense rosettes of narrow, linear leaves, sometimes held stiffly erect, sometimes softer and then arching to pendent. The panicles of white flowers are normally produced only on large specimens and rarely on those grown in containers. They are mostly grown for their very decorative shape. Propagate by offsets when re-potting, by cuttings of branch tips taken in summer, or by seed in late spring in warmth. *Nolina* was named for P.C. Nolin, a French agricultural writer in the mid 18th century.

Species cultivated
N. bigelowii (*Beaucarnea bigelowii*, *Dasylirion bigelowii*) Arizona, California (USA), Baja California (Mexico)
About 1.5m(5ft) in a large container (twice this planted out) but slow-growing, the erect, woody stem unbranched or rarely with a few branches when old. Leaves linear, straight, 60–90cm(2–3ft) long, in a dense radiating tuft or globular rosette.
N. recurvata (*N. tuberculata*, *Beaucarnea recurvata*) Pony tail, Elephant foot S.E. Mexico
Tree-like in the open ground, to 4m(13ft) or more tall; in a pot much smaller, but the trunk soon forming and becoming markedly swollen at the base. Leaves to 1m(3ft) long, recurved, mid-green.

NOPALXOCHIA See EPIPHYLLUM
NOTHOPANAX See PSEUDOPANAX

NOTOCACTUS

Cactaceae
Ball cacti

Origin: Southern South America. A genus of 15 to 25 species of small, solitary or clustered cacti formerly included in the allied genus *Echinocactus*. They have usually prominent ribs with tubercles divided by deep notches and often densely felted areoles. The flowers are shortly funnel- to bell-shaped and are carried at the top of the plant body. Propagate clustered species by removing offsets; all species by seed in spring. *Notocactus* comes from the Greek *noto*, southern and *Cactus*.

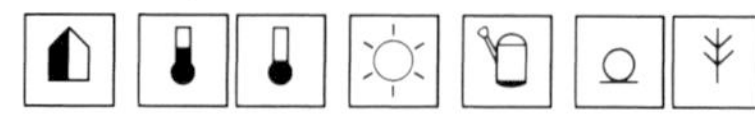

Species cultivated

N. leninghousii Golden ball cactus S. Brazil
Stem cylindrical, eventually to 1m(3ft) tall and 10–20cm(4–8in) thick, but taking many years to reach this size; solitary at first but branching from the base when mature. Ribs about 30, the areoles with up to 15 bristle-like, pale yellow radial spines and 4cm(1½in) long central spines reflexed at the tips. Flowers 4–5cm(1½–2in) long, yellow. *N.l.* 'Cristata' has fasciated stems forming a crest.
N. magnificus Brazil
Stems clustering, broadly ovoid to shortly cylindrical. Ribs about 12 or more, deep and sharply edged, bearing a continuous row of small areoles set with short wool and longer, yellowish hair-like spines. Flowers yellow.
N. scopa S. Brazil, Uruguay
Stem globose, later becoming cylindrical, remaining solitary. Ribs 30 to 35, notched, the areoles with up to 40 white, bristle-like spines to 5mm(⅕in) long; central spines four, longer and thicker, brown. Flowers to 4cm(1½in) long, crimson.

ABOVE *Notocactus scopa*
ABOVE CENTRE *Notocactus magnificus*
ABOVE LEFT *Nymphaea capensis* hybrid

NOTONIA See SENECIO

NYMPHAEA

Nymphaeaceae
Water lilies

Origin: Cosmopolitan. A genus of 35 to 50 species of aquatic perennials with thick rhizomes, sometimes tuberous, and rounded, floating leaves carried on stalks which can be from 15cm(6in) to 2m(6½ft) long according to the depth of water in which they grow. The cup-shaped flowers are made up of four sepals and many petals; in some species they float on the water, in others they are carried above the surface. The berry-like fruits mature beneath the water. Water lilies are suitable for the conservatory pool or a large tub, while the smaller *N. tetragona* and *N.* × *helvola* make unusual house plants for a large bowl or tank on a strong table or windowsill. Propagate by division or by seed when ripe (sow in pots which are just submerged in water). *Nymphaea* derives from *Nymphe*, one of the mythological Greek water nymphs.

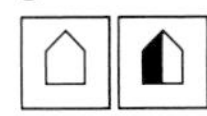 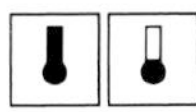 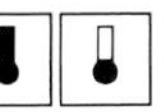 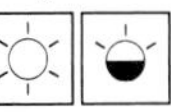 

Species cultivated

N. caerulea Egyptian blue lotus North and Central Africa
Vigorous, with leaves 30–45cm(1–1½ft) across, marked with purple beneath, the margins slightly wavy. Flowers 8–15cm(3–6in) across, the 14 to 20 pale blue petals white at the base, opening in the morning. Tropical.
N. capensis Blue water lily South and East Africa, Malagasy
Leaves 30–40cm(12–16in) across, toothed, green. Flowers with 20 to 30 blue petals, white at the base, opening all day. Tropical.
N. flava See *N. mexicana.*
N. × **helvola** (*N. pygmaea helvola*)
Hybrid between *N. tetragona* and *N. mexicana.* One of the smallest water lilies. Leaves 7–10cm(2¾–4in) across, green, blotched with purple-brown above. Flowers 4–6.5cm(1½–2½in) across, sulphur-yellow, opening in the afternoon. Will grow in a water depth as little as 15–30cm(6–12in). Cool.

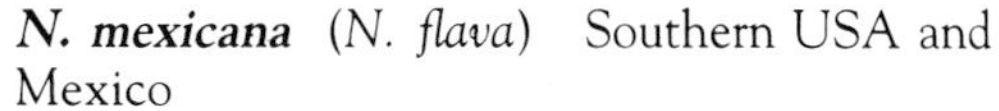

N. mexicana (*N. flava*) Southern USA and Mexico
Rhizome tuberous. Leaves 10–20cm(4–8in) across, slightly wavy at the edges, blotched with red-brown above and entirely dull crimson beneath. Flowers 10cm(4in) wide, bright yellow, opening from midday onwards. Tropical.
N. pygmaea See *N. tetragona.*
N. stellata Blue lotus South and East Asia
Leaves to 30cm(1ft) across, the margins irregularly waved and toothed, purple beneath, green above. Flowers 8–10cm(3–4in) wide, of 11 to 14 sky-blue petals, white at the base, opening from morning to early afternoon; fragrant. Tropical.
N. tetragona (*N. pygmaea*) Pygmy water lily Siberia to North America
Leaves 7–10cm(2¾–4in) across, elliptic, reddish beneath, blotched red-brown above when young, later all green. Flowers 4–6.5cm(1½–2½in) across, white, opening in the afternoon. Suitable for shallow water from 15–30cm(6–12in). Cool.

Hybrid cultivars

A large number of cultivars have been raised using the tropical species. Most are of American origin. Tropical.

O

ABOVE *Obregonia denegrii*
TOP × *Odontioda* 'Lettaford'

OBREGONIA

Cactaceae

Origin: Mexico. A genus of one species of cactus allied to *Leuchtenbergia*, but very different in appearance with its shorter tubercles. An interesting addition to a collection of succulents, its non-cactus appearance providing a point of conversation. Propagate by seed in spring. *Obregonia* honours Don Alvaro Obregon (1880–1928), a President of Mexico.

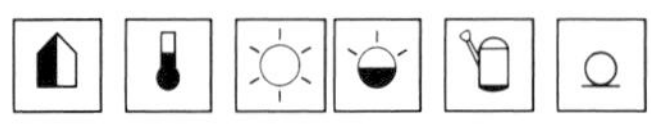

Species cultivated

O. denegrii (*Strombocactus denegrii*)
Stems usually solitary, occasionally clustering, covered with overlapping, flattened, triangular, leaf-like tubercles creating the impression of a houseleek (*Sempervivum*) rosette, 10–13cm(4–5in) across. Tubercle tips bearing an areole and several small, weak spines which soon fall. Flowers about 2cm(¾in) wide, white or pale pink.

OCHNA

Ochnaceae

Origin: Tropical and sub-tropical Africa and Asia. A genus of 85 to 90 species of trees and shrubs, with alternate, entire leaves and usually greenish or yellow flowers followed by often conspicuous fruits. One species is frequently cultivated being suitable for pots or tubs. Propagate by heel cuttings in summer with bottom heat, or by seed, ideally as soon as ripe, otherwise in spring with heat. *Ochna* is the Greek name for the wild pear.

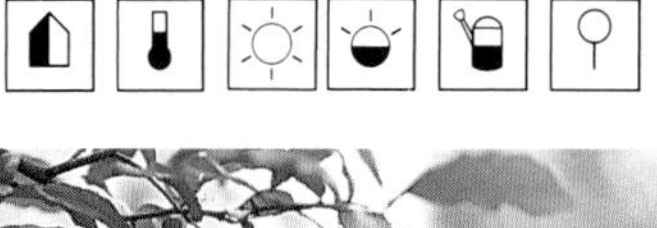

Ochna serrulata

Species cultivated

O. serrulata Mickey Mouse plant South Africa
Shrub 1–2m(3–6½ft) tall with finely warted twigs. Leaves to 6cm(2½in) or more long, narrowly elliptic and sharply toothed, evergreen and glossy. Flowers 1.5–2cm(⅝–¾in) wide, opening almost flat, bright yellow borne singly or in small clusters. After flowering the receptacle becomes thickened and red, the recurved sepals are also red and persist, making a colourful contrast to the glossy black berry-like drupelets which are carried by the receptacle.

This species is often confused with *O. atropurpurea* which has ovate leaves and deep purple sepals, but which is rarely seen.

× ODONTIODA

Orchidaceae

A group of hybrids raised by crossing species of *Miltonia* and *Odontoglossum*. They are closest to the latter species and have the same cultural requirements.

ODONTOGLOSSUM

Orchidaceae

Origin: Tropical America, many from cool upland forests of the Andes. A genus of about 250 species of epiphytic orchids with usually clustered, ovoid or rounded pseudobulbs on short rhizomes. The flowers are usually showy with wide-spreading similar petals and sepals (tepals) and a large flattened, 3-lobed and waved labellum. They are carried in erect or arching, rarely pendent racemes. These orchids are suitable for a conservatory and can be tried in the home.

Propagate by division when re-potting in spring or autumn. *Odontoglossum* derives from the Greek *odontos*, tooth and *glossa*, tongue, the labellum being toothed.

Species cultivated

O. bictoniense Mexico to Central America
Pseudobulbs 10–18cm(4–7in) long, ovoid and compressed. Leaves in twos or threes, each to 30cm(1ft) or more long, lanceolate to elliptic-oblong. Flowering stems erect, usually unbranched to 90cm(3ft) or more tall. Flowers to 4–5cm(1½–2in) across, the tepals yellowish-green, with strong chocolate-purple banding, the labellum large and heart-shaped, white, sometimes tinted with pale pink, crimped at the edges, opening in winter and spring.

O. crispum Lace orchid Colombia
Pseudobulbs to 10cm(4in) long, ovoid, flattened. Leaves in twos or threes, to 38cm(15in) long, linear. Racemes to 45cm(1½ft) long, arching, carry-

Odontoglossum 'Mimosa Canary'

BELOW *Odontoglossum rossii*
BOTTOM *Odontonema schomburgkianum*

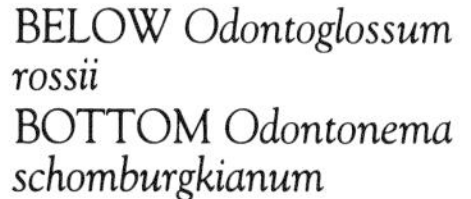

ing up to 20 flowers; each bloom 7–9cm($2\frac{3}{4}$–$3\frac{1}{2}$in) across and very variable in colour, the waved tepals white or pale pink, variably blotched and spotted with red to brown, the labellum pink with red spotting and a central yellow patch; they open in autumn and winter.

O. grande Mexico and Guatemala
Pseudobulbs clustered, 6–10cm($2\frac{1}{2}$–4in) tall, ovoid. Leaves in pairs, to 35cm(14in) long, ovate to lanceolate, glaucous green. Racemes 30–40cm(12–16in) long, usually erect with seven to eight blooms opening in autumn and spring; these are 10–15cm(4–6in) wide, the yellow outer tepals barred with reddish-brown and narrower than the two inner ones which are red-brown in their basal halves, yellow above; the white labellum is flecked and banded with red-brown.

O. lunatum See *Aspasia lunata.*

O. noetzlianum See *Cochlioda noetzliana.*

O. pendulum Mexico, Guatemala
Pseudobulbs 8–15cm(3–6in) tall, ovoid to rounded. Leaves in pairs, to 30cm(1ft) long, oblong strap-shaped. Racemes to 30cm(1ft), pendent, with up to 15 fragrant blooms; these are 5–7cm(2–$2\frac{3}{4}$in) across, white, sometimes flushed with pink, the basically bi-lobed lip pink, opening in autumn.

O. pulchellum Lily of the valley orchid
Mexico to Costa Rica
Pseudobulbs to 8cm(3in) long, oblong, flattened. Leaves usually in pairs, occasionally three, 30–40cm(12–16in) long, linear.

Racemes 30cm(1ft) or more long, pendent, with up to ten very fragrant flowers; each flower 2.5–3cm(1–$1\frac{1}{4}$in) wide, white, opening in autumn and spring.

O. roseum See *Cochlioda rosea.*

O. rossii Mexico to Nicaragua
Basically similar to *O. cervantesii*, but with racemes not exceeding 20cm(8in) and flowers to 8cm(3in) across, opening in winter and spring. Tepals white, the outer thickly barred with red-brown, the inner with brownish spotting at the centre, while the waved lip is white.

Hybrid cultivars

Many hybrids have been raised and orchid specialists offer an ever-changing list of names in an effort to produce the largest and most colourful blooms within the genus.

ODONTONEMA

Acanthaceae

Origin: Mexico to tropical South America, West Indies. A genus of about 40 species of evergreen shrubs and perennials allied to *Jacobinia*. One species is regularly grown though less commonly than formerly. It makes a charming and unusual specimen pot plant for the conservatory and can be brought into the home when in bloom. Propagate by cuttings in spring.

Odontonema derives from the Greek *odontos*, a tooth and *nema*, a thread; the stamens bearing toothed filaments.

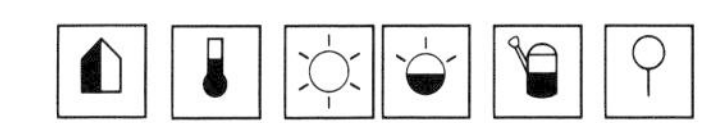

Species cultivated

O. schomburgkianum (*Thyrsacanthus rutilans*)
Colombia
Eventually a shrub to 2m($6\frac{1}{2}$ft) or more, but best if pruned hard annually after flowering and is then seldom more than 90cm(3ft) tall. Leaves in opposite pairs, oblong-lanceolate, to 20cm(8in) long, rich green, more or less corrugated. Flowers tubular, red or pink, 3cm($1\frac{1}{4}$in) long in slender cascading panicles 30–90cm(1–3ft) in length, borne in winter to early spring.

OENOTHERA See CLARKIA

OLEA

Oleaceae

Origin: Europe, Asia and North Africa. A genus of 20 evergreen trees and shrubs with opposite leaves and panicles or clusters of small white flowers. One species is grown as a pot or tub plant for the conservatory, mainly for its foliage interest, though large specimens may flower and fruit. Propagate by heel cuttings in late summer with bottom heat, or by seed also in warmth. *Olea* is the Latin vernacular name for this plant.

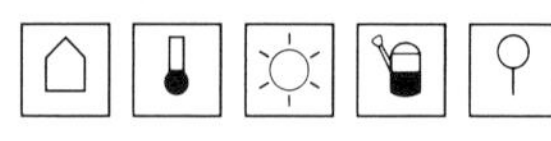

Species cultivated

O. europaea Olive
Shrub to 3m(10ft) in a tub, but in the open ground making a tree to 14m(45ft) in height. Leaves 2–8cm($\frac{3}{4}$–3in) long, lanceolate to narrowly obovate, dull dark green above, light grey-scaly beneath. Flowers shortly tubular opening to four small lobes, white and fragrant; they are borne in axillary and terminal panicles in late summer and are followed by plum-like, glossy black fruits 1–3.5cm($\frac{3}{8}$–1$\frac{1}{2}$in) long. This is the cultivated olive which is the form normally grown.

OLIVERANTHUS See ECHEVERIA

ONCIDIUM

Orchidaceae

Origin: Warm temperate regions of the Americas from Florida to Argentina. A genus of 350 to 400 species of epiphytic orchids closely allied to *Miltonia* and *Odontoglossum*. They have showy flowers somewhat like those of *Odontoglossum*, but generally with a larger and more conspicuous labellum. Plants are best displayed on slabs of bark or in hanging baskets. They are suitable for the conservatory and can also be grown in the home. Propagate by separating pseudobulbs when re-potting. *Oncidium* derives from the Greek *onkos*, a tumour, in this diminutive form referring to the swellings on the labellum.

ABOVE *Olea europaea*
RIGHT *Oncidium tigrinum*

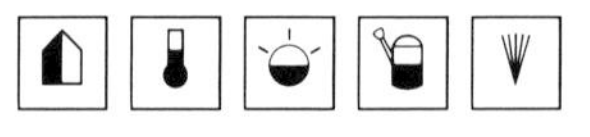

Species cultivated

O. flexuosum Dancing doll orchid Brazil, Paraguay, Uruguay
Pseudobulbs oval, flattened, 4–7cm(1$\frac{1}{2}$–2$\frac{3}{4}$in) long. Leaves in pairs, but sometimes solitary, 15–20cm(6–8in) long, linear-lanceolate. Flower stalks 60–90cm(2–3ft) tall, very slender, arching; flowers 2–3cm($\frac{3}{4}$–1$\frac{1}{4}$in) across, bright yellow, the tepals with cinnamon-brown streaking, especially near the base, the much larger labellum yellow with red spots, opening late summer and winter.

O. ornithorhynchum Dove orchid Mexico to Costa Rica
Pseudobulbs clustered, to 13cm(5in) long, ovoid and glaucous green. Leaves in pairs, to 30cm(1ft) long, narrowly lanceolate. Flowering stems to 60cm(2ft) long, arching or pendent; flowers 2cm($\frac{3}{4}$in) long, lilac-pink, fragrant, opening in autumn and winter.

O. papilio Butterfly orchid Venezuela to Brazil and Trinidad
Tightly clustered pseudobulbs to 5cm(2in) long, ovoid to rounded. Leaves usually solitary, 15–22cm(6–9in), erect, oblong-elliptic, dull green mottled with red-purple. Flowering stems 60–150cm(2–5ft) tall, jointed and winged in the upper section; flowers 10–13cm(4–5in), crimson-red, often yellow-barred, the outer tepals very narrow and pointing upwards, produced in succession throughout the year.

O. sphacelatum Mexico to Honduras
Pseudobulbs 10–18cm(4–7in) long, ovoid to oblong, clustered, or spaced along the stout rhizomes. Leaves in twos or threes, to 60cm(2ft) or more long, narrowly strap-shaped. Flowering stems 1–2m(3–6$\frac{1}{2}$ft) long, drooping; individual flowers about 3cm(1$\frac{1}{4}$in) long, yellow, barred and spotted with purplish-brown, spring and summer.

O. tigrinum Mexico
Pseudobulbs clustered, 7–10cm(2$\frac{3}{4}$–4in) long, rounded. Leaves in twos or threes, to 30cm(1ft) long, narrowly oblong, blunt-tipped. Flowering stems to 90cm(3ft) tall, usually erect, loosely branched; individual blooms about 7cm(2$\frac{3}{4}$in) long, bright yellow, the tepals blotched with brown, the lip yellow, opening in autumn and winter.

O. varicosum Golden butterfly orchid Brazil
Pseudobulbs clustered, 7–13cm(2$\frac{3}{4}$–5in) long, oblong to ovoid. Leaves in twos or threes, to

25cm(10in) long, lanceolate to strap-shaped. Flowering stems 90–150cm(3–5ft) long, nodding; flowers 4cm(1½in) across, the small inconspicuous tepals dull yellow barred with light reddish-brown, and a large kidney-shaped yellow labellum, opening in autumn and winter.

OPHIOPOGON

Liliaceae
Lily-turf

Origin: S.E. Asia from the Himalayas to Japan and the Philippines. A genus of ten or more species of tufted evergreen perennials spreading by means of stolons or rhizomes. Their leaves are grassy in shape, but leathery in texture. Racemes of small, nodding, bell-shaped flowers are followed by berry-like seeds. They make very good pot plants. Propagate by division or by seed sown when ripe. *Ophiopogon* derives from the Greek *ophis*, snake and *pogon*, beard, evidently from the Japanese vernacular *ja-no-hige* which means a snake's beard.

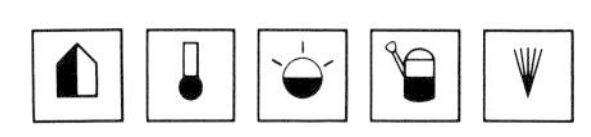

Species cultivated

O. arabicus See under *O. planiscapus*.

O. jaburan (*Mondo jaburan*) White lily-turf Japan
Tufted to clump-forming. Leaves 45cm(1½ft) or more long, only 6–12mm(¼–½in) wide, dark green, erect, then arching over. Flowers 7–8mm(¼–⅓in) long, white to lilac, carried on racemes on stems 30–50cm(12–20in) tall in summer, followed by violet-blue oblongoid fruits. Sometimes confused in cultivation with *Liriope* which, however, has flowers that always face outwards or up and are never nodding. *O.j.* 'Vittatus' ('Argenteo-variegatus', 'Aureo-variegatus') has leaves which are longitudinally striped with creamy-white.

O. planiscapus Japan
Tufted and mat-forming. Leaves 30–50cm(12–20in) long, deep green. Flowers 5–7mm($\frac{3}{16}$–¼in) long, white or pale purple, in racemes on stems 20–30cm(8–12in) high in summer. *O.p.* 'Nigrescens' ('Arabicus', *O. arabicus*) has the leaves and flowering stalks purple-black, showing up the white to pinkish-purple flowers and blue fruits.

OPLISMENUS

Gramineae

Origin: Tropics and sub-tropics. A genus of 15 to 20 species of grasses with slender trailing stems carrying narrowly ovate to lanceolate leaves and small flowers in spike-like racemes. The species described below is a useful house and conservatory plant particularly in its variegated form. It looks very attractive in a hanging basket. Propagate by cuttings in summer. *Oplismenus* derives from the Greek *hoplismos*, armed for war, with reference to the bristle-tipped awns on the spikelets.

LEFT *Ophiopogon jaburan* 'Vittatus'
BELOW LEFT *Oplismenus hirtellus* 'Variegatus'

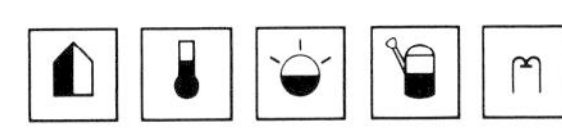

Species cultivated

O. hirtellus Basket grass Southern USA to Argentina
Wiry spreading stems up to 1m(3ft) in length. Leaves to 10cm(4in) long and 2cm(¾in) wide, narrowly ovate and pointed, green. Flowers insignificant (the panicles are best removed as soon as they appear). *O.h.* 'Variegatus' (*Panicum variegatum* of some gardens) has the leaves striped with white and sometimes also tinted with pink.

OPUNTIA

Cactaceae
Prickly pears

Origin: North and South America. A genus of 250 to 300 species of usually branched and shrubby cacti with distinctive jointed stems, the cylindrical or flattened sections known as pads. They have small, cylindrical or conical leaves which are usually fleeting, and abundant glochids and spines from the areoles. The short-tubed flowers have spreading petals which are often colourful and are followed by fleshy, frequently edible fruits. Prickly pears are suitable for conservatory or home when small and are easily propagated by seed in spring or by cuttings

ABOVE *Opuntia microdasya* 'Albispina'
ABOVE RIGHT *Opuntia decumbens*

of mature pads in summer. This makes it easy to replace a specimen which has outgrown its space. *Opuntia* was originally named for a Greek plant now unidentifiable, and used for this species for no obvious reason.

Species cultivated

O. basilaris Beaver tail cactus S.W. USA, northern Mexico
Bushy growth, to 90cm(3ft) in height. Pads broadly oval to obovate, 10–20cm(4–8in) long, bluish-green, sometimes red-suffused. Areoles numerous, red-brown, spines solitary or lacking. Flowers 6–8cm(2½–3in) wide, pale pink to carmine.

O. bergeriana Provenance unknown
Large, more or less erect bush to 3m(10ft) when planted out, but easily kept to a third of this height in containers. Pads narrowly oval to oblong, to 25cm(10in) long, pale to greyish-green. Areoles unequal, well spaced, with two or three irregularly-sized, awl-shaped spines, the longest to 4cm(1½in). Flowers 5cm(2in) wide, usually deep red, sometimes orange, freely produced.

O. compressa See *O. humifusa austrina.*

O. decumbens Mexico to Guatemala
Spreading shrub, eventually to about 30cm(1ft) or more high and three times as wide. Pads broadly oval to elliptic-oblong, 10–20cm(4–8in) long, dark green. Areoles woolly, often with a reddish or purplish halo; spines one or two but often absent, yellowish, to 4cm(1½in) long. Flowers 5cm(2in) wide, pale yellow, usually ageing reddish.

Opuntia phaeacantha

O. erinacea S.W. USA
Spreading to semi-prostrate, to 80cm(32in) high. Pads oval, 8–10cm(3–6in) long, bluish-green. Spines to 5cm(2in) long, flexible and almost hair-like, densely borne from the areoles. Flowers 6–8cm(2½–3in) across, yellow, pink or white.

O.e. ursina Grizzly bear cactus
Spines to 10cm(4in) long.

O. ficus-indica Indian fig Tropical America
Becoming shrubby with age and used as a hedge in warm countries where it has become widespread as a naturalized weed. Pads to 45cm(1½ft) long, the areoles with one to five spines each 2.5cm(1in) long. Spineless forms are grown. Flowers 4–5cm(1½–2in) across, bright yellow; fruits edible, yellow or red-streaked, 5–10cm(2–4in) long.

O. furiosa See *O. tunicata.*

O. humifusa (*O. compressa, O. opuntia*) USA
Loosely mat-forming, to 90cm(3ft) wide. Pads broadly obovate to orbicular, up to 15cm(6in) long, with well-spaced areoles. Spines absent or one to two from the marginal areoles only. Flowers yellow, 5–8cm(2–3in) or more wide.

O.h. austrina (*O. compressa* of gardens)
Erect stems to 90cm(3ft) in height with more spiny pads. A frost-hardy species.

O. microdasys (*O. pulvinata*) Bunny ears northern Mexico
Shrubby, to 90cm(3ft) tall. Pads oval, 8–15cm(3–6in), finely downy. Areoles with a dense boss of golden glochids, sometimes with a single spine, often without. Flowers 4–5cm(1½–2in) wide, yellow, followed by red, rounded fruit. *O.m.* 'Albispina' has smaller pads with silvery-white glochids.

O. opuntia See *O. humifusa.*

O. ovata Andes
Hummock-forming, 20–30cm(8–12in) tall. Pads 4–7cm(1½–2¾in) long, broadly cylindrical with five to nine awl-shaped spines to 4cm(1½in) long from each areole. Flowers yellow flushed with red, 4–5cm(1½–2in) wide, followed by small yellow fruits.

O. pentlandii (*Tephrocactus pentlandii*) Peru, Bolivia, Argentina (Andean region)
Stems densely clump-forming, eventually producing hummocks to 60cm(2ft) or more wide when planted out. Pads obovoid to cylindrical or very narrowly ovoid, to 5cm(2in) or longer, bright green, with broad, low tubercles. Spines two to ten from the upper areoles, but often absent, yellowish or brownish, up to 6cm(2½in) long but usually less. Flowers 6cm(2½in) wide, yellow or yellow ageing reddish.

O. phaeacantha Texas to California, northern Mexico
Variable in habit but usually prostrate and eventually forming mats. Pads broadly obovate to orbicular, to 15cm(6in) long or sometimes up to twice as large. Areoles well spaced bearing one to four spines, the longest one to 6cm(2½in), down-pointing. Flowers about 8cm(3in) wide, rich yellow, sometimes red in the centre.

O. pulvinata See *O. microdasys*.
O. robusta Central Mexico
Eventually a shrub with pads to 30cm(1ft) long, but smaller in a pot, rounded to oval, glaucous green, the areoles with up to twelve, 5cm(2in) long spines. Flowers yellow, 5–7cm(2–2¾in) wide, followed by red, rounded, edible fruits to 8cm(3in) long.
O. scheeri Mexico
Spreading, eventually 60–90cm(2–3ft) high. Pads 15–20cm(6–8in) long, rounded to oblong, grey to bluish-green, the areoles with eight to twelve 1cm(⅜in) long spines. Flowers to 10cm(4in) across, yellow ageing to red, followed by red fruit. Best in temperate conditions.
O. tunicata (*O. furiosa*) Texas to Mexico, south to northern Chile
Bushy, dwarf shrub up to 60cm(2ft) tall with stems in whorls. Pads oblongoid,to 15cm(6in) or more long with prominent tubercles, blue-green. Spines six to ten, about 4–5cm(1½–2in) long, barbed, pale yellow in loose papery sheaths. Flowers 3–5cm(1¼–2in) wide, yellow with a hint of green. A handsome species, but best placed where its very sharp barbed spines cannot damage the skin and tempers!
O. verschaffeltii Northern Bolivia
Low and spreading, rarely more than 30cm(1ft) tall. Pads 10–15cm(4–6in) or more long in cultivation, but seldom exceeding 4cm(1½in) in the wild, cylindrical with persistent leaves and one to three thread-like spines to 6cm(2½in) long, not always present on cultivated plants. Flowers 4cm(1½in) across, deep red.

OREOCEREUS See BORZICACTUS

ORNITHOGALUM

Liliaceae

Origin: Europe, Africa and Asia. A genus of 100 to 130 species of bulbous perennials with basal tufts of strap-shaped to linear leaves and 6-tepalled, starry flowers in racemes on leafless stems. They make good pot plants, the hardy species being treated as for *Crocus*, the rest needing some warmth when not dormant. Propagate by offsets, by division of congested clumps or by seed, ideally sown as soon as ripe, but failing this, in spring. *Ornithogalum* derives from the Greek *ornis*, a bird and *gala*, milk (apparently with reference to the colour of the excreta of doves); the flowers are mainly white.

Species cultivated
O. arabicum Star of Bethlehem
Mediterranean region
Leaves to 30–60cm(1–2ft) long by 2–2.5cm(¾–1in) wide, narrowly strap-shaped, glaucous green. Flowers 5cm(2in) across, facing upwards, pearly-white with a black-green pistil, and yellow anthers; they are carried in a roundish raceme of eight to 12 and, with the stem, are 40–60cm(16–24in) tall; very fragrant. Cool.
O. caudatum South Africa
Leaves 45–60cm(1½–2ft) long by 2–4cm(¾–1½in) wide, rather fleshy and recurving. Flowers starry, white with a green central stripe on the back of each petal, carried in cylindrical racemes of as many as 50 to 100 blooms on 30–60cm(1–2ft) stalks. Cool.
O. nutans Drooping star of Bethlehem Europe to western Turkey
Leaves 30–45cm(1–1½ft) long, linear. Flowers 2–3cm(¾–1¼in) long, nodding, silvery-white within, the backs of the petals greenish and conspicuous as the flower does not open widely; they are carried in racemes 20–40cm(8–16in) long which droop at the ends. Hardy.
O. saundersiae Giant chincherinchee South Africa
Similar to *O. arabicum*, but taller, having stems to 1–2m(3–6½ft). The white flowers have greenish-black ovaries and pistil, giving a conspicuous black centre to each bloom; anthers grey.
O. thyrsoides Chincherinchee South Africa
Leaves to 30cm(1ft) long, linear to lanceolate and somewhat fleshy. Flowers 3–4cm(1¼–1½in) wide, white to creamy-white, in pyramidal racemes, on stems to 45cm(1½ft) or more tall in summer; the blooms are very long-lasting.
O. umbellatum Star of Bethlehem Europe, North Africa
Leaves 15–25cm(6–10in) long, narrowly linear, with a conspicuous silvery-white midrib. Flowers 3–4cm(1¼–1½in) wide, white with a green stripe on the back of each sepal; they are borne in umbel-like racemes opening in spring and summer as the leaves fade, and, with their stems, are 10–30cm(4–12in) tall.

ORPHIUM

Gentianaceae

Origin: South Africa. A genus of one species of evergreen shrub grown for its showy flowers. It is a lovely pot plant for the conservatory which can be brought into the home when in bloom. Propagate by seed in spring, or by cuttings in spring or summer. *Orphium* is named for Orpheus, a character from Greek mythology.

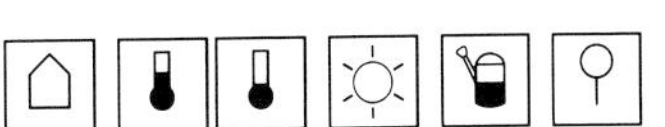

Species cultivated
O. frutescens South-western Cape
Erect shrub to 60cm(2ft) in height. Leaves sessile, linear to narrowly obovate, thick-textured, usually downy, 2–4cm(¾–1½in) long. Flowers in small terminal clusters, deep pink to red, 5-petalled, 3–4cm(1¼–1½in) wide. Grows by the sea in sandy ground in the wild, and must have well drained soil and plenty of sun to succeed.

BOTTOM *Orphium frutescens*
BELOW *Ornithogalum thyrsoides*

Oscularia caulescens

OSCULARIA

Aizoaceae

Origin: South Africa. A genus of five sub-shrubby flowering succulents two of which are widely grown and recommended for adding colour to a collection of succulents in spring. Propagate by seed in spring or by cuttings in summer. *Oscularia* derives from the Latin *osculum*, a little mouth, reputedly referring to the mouth-like appearance of the leaf pairs, but equally and more aptly applicable to the circular hole at the top of the tight cone-like mass of stamens.

  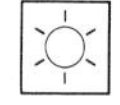  

Species cultivated

O. caulescens (*Mesembryanthemum caulescens*)
Cape, etc.
Low mound or thick mat-forming plant with reddish, branching stems to 25cm(10in) or more in length. Leaves in pairs, about 2cm($\frac{3}{4}$in) long, flattened on top, expanded and keeled in the upper half below, grey-green with reddish edges. Flowers 1.5cm($\frac{5}{8}$in) wide, mauve to lilac-pink, fragrant.

Osmanthus heterophyllus

O. deltoides (*Mesembryanthemum deltoides*)
Cape, etc.
Similar to *O. caulescens*, but the leaves more glaucous, smaller, chunkier and with a few teeth on the upper edges and a prominent keel. Flowers pink.
O.d. majus
Somewhat larger flowers.

OSMANTHUS

Oleaceae

Origin: East Asia, eastern USA, Mexico and Pacific Islands. A genus of 30 to 40 species of evergreen trees and shrubs with opposite, undivided, often leathery leaves, and 4-lobed, tubular flowers usually in axillary clusters which are mostly white or occasionally creamy-yellow and fragrant. The species described below are useful flowering and foliage pot plants for the conservatory and for short spells in the home. They can also be used on windowsills when young. Propagate by heel cuttings in late summer in bottom heat. *Osmanthus* derives from the Greek *osme*, fragrance and *anthos*, a flower.

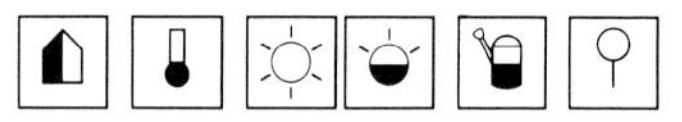

Species cultivated

O. fragrans China
To 2m($6\frac{1}{2}$ft) or more in a tub, capable of reaching 12m(40ft) in the open. Leaves 8–11cm(3–$4\frac{1}{2}$in) long, broadly lanceolate, usually with small, widely spaced toothing along the margins, dark green. Flowers small, bell-shaped, white or pale yellow in clusters, strongly fragrant.
O. heterophyllus (*O. ilicifolius*) Japan
To 2m($6\frac{1}{2}$ft) or more in a tub, to 8m(25ft) in the open. Leaves 4–6cm($1\frac{1}{2}$–$2\frac{1}{2}$in) long, ovate to elliptic, very holly-like with three to five spiny teeth along each margin. Flowers small, white, very

fragrant, borne in clusters in late summer. *O.h.* 'Aureo-marginatus' has yellow margins to the leaves; 'Purpureus' has the young leaves a purplish-black ageing to purplish-green.
O. ilicifolius See *O. heterophyllus*.

OSTEOSPERMUM

Compositae

Origin: South Africa to Arabia and St. Helena. A genus of about 70 species of perennials, sub-shrubs and shrubs with showy daisy flowers. Those described are on the borderline of frost hardiness and are grown outside in sheltered places. In cold areas they are best in the cool conservatory and will flower in winter if a minimum temperature if 10°C(50°F) is maintained. Propagate by cuttings in spring or summer and by seed in spring. They are best re-propagated annually as pot plants. *Osteospermum* derives from the Greek *osteon*, bone and *sperma*, a seed; the seed-like fruits are hard.

Species cultivated

O. barbariae See under *O. jucundum*.
O. ecklonis (*Dimorphotheca ecklonis*) South Africa
Sub-shrubby perennial to 60cm(2ft) high and wide. Leaves lanceolate to oblanceolate, toothed, deep green, 5–10cm(2–4in) long. Flower-heads 6.5–8cm(2¼–3in) wide, white above with a purple-blue centre, purple to blue-purple beneath. *O.e.* 'Prostrata' is prostrate and mat-forming with erect flowering stems to 22cm(9in) in height. *O.e.* 'Tauranga' ('Whirligig') has the habit of the type plant but has flowers with the ray florets constricted near their middles, creating a propeller-like shape.
O. jucundum (*Dimorphotheca jucunda*) South Africa (Drakensberg and adjacent mountains)
Bushy, sub-shrubby perennial to 45cm(1½ft) high and wide. Leaves alternate, oblong-lanceolate to spathulate, deep green, sometimes sparingly toothed, 4–8cm(1½–3in) long. Flower heads 5–6cm(2–2½in) wide, rose-purple above, dull purple beneath. *O.b.* 'Compactum' is more or less prostrate and mat-forming with the flowers on erect 15cm(6in) stems. For long, and still, grown as *O. barbariae* (*Dimorphotheca barbariae*), a more tender but almost identical species from the coastal areas of Eastern Cape. True *O. barbariae* has the tips of the ray florets bearded; hairless in *O. jucundum*.

Hybrids

Of recent years several hybrid cultivars, mostly from South Africa, have been introduced as summer bedding plants and all make fine pot subjects.

OTHONNA

Compositae

Origin: Tropical and South Africa. A genus of around 140 species of shrubs and perennials many of which have fleshy roots and/or leaves, resembling the succulent sorts of *Senecio*. One species is commonly grown and looks effective in a wide pan or in a hanging basket in the conservatory or sunny home window. Propagate by cuttings in spring or summer. *Othonna* is a Greek name once used for a now unspecified group of succulent plants and applied to this genus by Linnaeus.

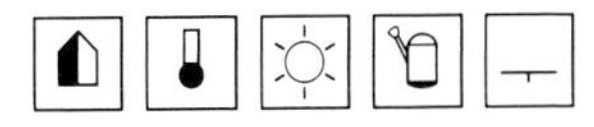

Species cultivated

O. capensis Cape
Mat-forming with slender stems to 60cm(2ft) or more long. Leaves cylindrical, 2.5–5cm(1–2in) long, greyish-green, sometimes purple-tipped. Flowers yellow daisies, 1.2–2.5cm(½–1in) wide, each one borne on a slender stem above the foliage in spring to autumn.

BELOW LEFT *Othonna capensis*
BELOW *Osteospermum ecklonis* 'Prostrata'

RIGHT *Oxalis deppei*
BELOW *Oxalis ortgiesii*
BOTTOM *Oxalis hirta*

OXALIS

Oxalidaceae
Wood sorrels

Origin: Cosmopolitan. A genus of about 850 species of annuals, perennials and shrubs some of which are bulbous or tuberous-rooted, some succulent. The leaves frequently have three leaflets, sometimes more, which fold up at night. Their flowers are 5-petalled. When the capsules open seeds are ejected from their arils with force. The plants are suitable for a conservatory or sunny windowsill in the home. Propagate by division when planting, by seed in spring or by cuttings of the succulent species in summer. *Oxalis* derives from the Greek *oxys*, acid or sour, referring to the taste of the leaves.

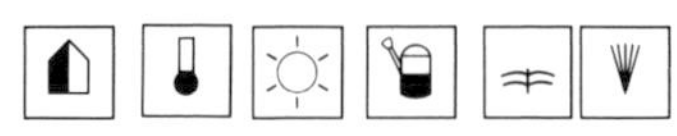

Species cultivated

O. bowei (*O. purpurata bowei*) South Africa
Bulbous perennial 20–30cm(8–12in) tall. Leaves with three rounded to broadly obovate leaflets to 5cm(2in) long. Flowers 2.5–4cm(1–$1\frac{1}{2}$in) long, pink to pinkish-purple, borne in umbels of three to 12 in summer.

O. brasiliensis Brazil
Bulbous perennial 15–25cm(6–10in) tall. Leaves with three obcordate leaflets to 2.5cm(1in) long. Flowers 3cm($1\frac{1}{4}$in) long, singly or in umbels of two or three, bright purple-red in colour, opening in summer.

O. cernua See *O. pes-caprae*.

O. deppei Lucky clover Mexico
Bulbous perennial with edible tubers which can reach 30cm(1ft) tall. Leaves with four broadly obovate leaflets to 4cm($1\frac{1}{2}$in) long. Flowers to 2cm($\frac{3}{4}$in) long, red to rich purple, carried in umbels of five to 12 in summer. There is a white-flowered form.

O. depressa (*O. inops*) South Africa
Bulbous, rhizomatous perennial 6–10cm($2\frac{1}{2}$–4in) tall. Leaves with three rounded to triangular-ovate leaflets 6–10m($\frac{1}{4}$–$\frac{3}{5}$in) long, grey-green. Flowers about 2cm($\frac{3}{4}$in) long, usually bright pink, but forms with purple or white flowers are known; borne singly in summer.

O. hirta South Africa
Trailing or decumbent bulbous perennial with stems to 30cm(1ft) long. Leaves with three linear to oblong leaflets 1–1.5cm($\frac{3}{8}$–$\frac{5}{8}$in) long, hairy. Flowers about 2.5cm(1in) long, pink to violet with a yellow eye, borne singly from the upper leaf axils in autumn.

O. inops See *O. depressa*.

O. lasiandra Mexico
Bulbous perennial 15–30cm(6–12in) tall, stemless. Leaves with five to ten narrowly oblong leaflets to 5cm(2in) long. Flowers 2cm($\frac{3}{4}$in) long, purple-crimson in colour, in umbels of up to 15 borne in summer.

O. ortgiesii Peru, Andes
Persistent-stemmed perennial to 45cm($1\frac{1}{2}$ft) in height, the erect stems usually sparingly branched or unbranched. Leaves long-stalked, trifoliate, the leaflets to 5cm(2in) long, obovate, divided at the tip into two large triangular lobes, olive-green above, red-purple beneath. Flowers about 1cm($\frac{3}{8}$in) wide, yellow with darker veins, in long-stalked cymes well above the leaves, intermittently all the year if warm enough.

O. pes-caprae (*O. cernua*) Bermuda buttercup
South Africa; widely naturalized in mild climates of the world
Tender bulbous perennial 20–30cm(8–12in) tall.

Leaves with three deeply obcordate leaflets to 2cm($\frac{3}{4}$in) long. Flowers 2–2.5cm($\frac{3}{4}$–1in) long, bright yellow, carried in umbels of three to eight in spring and summer. 'Flore Pleno' has double flowers.

O. purpurea South Africa (Cape)
Bulbous perennial to about 15cm(6in) in height. Leaves trifoliate, the leaflets orbicular to widely obovate, ciliate, to 2.5cm(1in) or more long on longish, white-downy stalks. Flowers to 3cm($1\frac{1}{4}$in) wide, red, purple or white with a yellow eye, solitary, at about the same height as the leaves. 'Grand Duchess' is a name given to several showy colour forms, with 5cm(2in) wide blooms in autumn to winter.

O. rubra Brazil
Clump-forming perennial with a semi-woody tuberous crown. Leaves long-stalked, trifoliate, the leaflets obcordate, notched-tipped, about 2cm($\frac{3}{4}$in) long. Flowers about 1.2cm($\frac{1}{2}$in) wide, red, pink or white, in umbel-like cymes above the leaves in spring to autumn. Cultivated forms may have coppery-tinted foliage. This species resembles *O. articulata* (*O. floribunda*) and is sometimes confused with it. It also occasionally parades under the name *O. rosea*, a seldom cultivated annual species.

O. siliquosa See *O. vulcanicola*.

O. succulenta Peru, Chile
Small succulent shrub usually under 15cm(6in) in height, often semi-prostrate. Leaf stalks fleshy, cylindrical to cigar-shaped, 2.5cm(1in) or more long. Leaves trifoliate, deciduous, broadly obovate, somewhat fleshy, to 1.2cm($\frac{1}{2}$in) long. Flowers yellow, about 1.5cm($\frac{5}{8}$in) wide, borne in small umbel-like cymes on long stems in spring.

O. vespertilionis Mexico
Bulbous perennial to 20cm(8in) or more in height, stemless. Leaves with three V-shaped leaflets each having two linear lobes to 3cm($1\frac{1}{4}$in) long. Flowers 2cm($\frac{3}{4}$in) long, purple, in umbels of three to ten in summer. Often confused with the rather similar *O. drummondii* which has lanceolate to ovate lobes to its leaflets, and violet flowers.

O. vulcanicola (*O. siliquosa*) Costa Rica
Bushy perennial having succulent persistent stems to 20cm(8in) or more tall with age, spreading and mat-forming. Leaves trifoliate, the leaflets obcordate, 1cm($\frac{3}{8}$in) or more long, red-flushed above, magenta beneath. Flowers about 1.2cm ($\frac{1}{2}$in) wide, yellow with purple-red lines within, in umbel-like cymes above the leaves in spring.

OXYPETALUM

Asclepiadaceae

Origin: South America. A genus of 125 to 150 species of perennials and sub-shrubs classified by some botanists as *Tweedya*. They have opposite leaves and clusters or umbels of shortly tubular flowers opening to five slender, spreading lobes. The species described below is most suitable for a conservatory or can be brought into the home when in flower. The stems are best pruned back to two buds each spring. Propagate by cuttings in late spring or summer with bottom heat, or by seed in spring also in warmth. *Oxypetalum* derives from the Greek *oxys*, sharp and *petalon*, a petal.

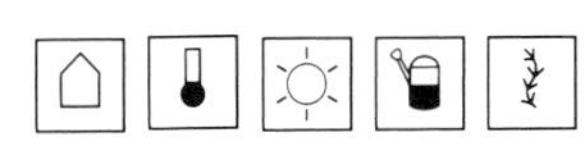

Species cultivated

O. caeruleum (*Amblyopetalum caeruleum*, *Tweedya caerulea*) Brazil, Uruguay
Weakly twining sub-shrub to 1m(3ft) or more. Leaves to 10cm(4in) long, oblong-lanceolate, cordate, hairy, in opposite pairs. Flowers 2.5cm(1in) wide, shortly tubular, opening widely, pinkish when in bud, opening to a clear sky-blue, ageing to purple; they are in loose, upward-facing axillary cymes, expanding in summer.

BELOW *Oxypetalum caeruleum*
BELOW LEFT *Oxalis succulenta*

P

PACHYPHYTUM

Crassulaceae
Moonstones

Origin: Mexico. A genus of 12 species of succulents allied to *Echeveria*. They have very thick leaves usually in attractive shades of grey to blue-green and somewhat bell-like 5-petalled flowers in cymes above the leaves. Propagate by stem and leaf cuttings or seed in late spring or summer. Those described here make welcome additions to a collection of succulents. *Pachyphytum* derives from the Greek *pachys*, thick and *phyton*, a plant; the leaves are distended with water storage tissue.

 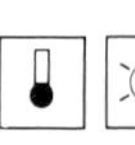 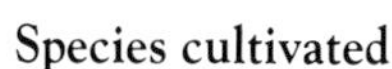  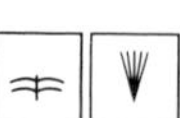

Species cultivated

P. amethystinum See *Graptopetalum amethystinum*.

P. brevifolium
Stems to 25cm(10in) or more, more or less erect, branching from the base. Leaves well spaced but crowded at the tips, obovate, blue-white, sometimes with a tinge of red. Flowers 1.2cm(½in) long, dark or carmine-red. It has been misidentified with the allied *P. glutinicaule* (having sticky glandular stems), at least in USA.

P. oviferum Moonstones, Sugar almond plant
Stems eventually to 10cm(4in) in length, sparingly branched, erect when young then prostrate or decumbent. Leaves crowded at the stem tips, obovate, 2–4cm(¾–1½in) long, oval in cross-section, usually with a blue-white patina, sometimes lavender-tinted or reddish. Flowers about 1.2cm(½in) long, rich red, in cymes 5–13cm(2–5in) tall in late winter and spring.

ABOVE × *Pachyveria scheideckeri*
TOP *Pachystachys lutea*
RIGHT *Pachyphytum oviferum*

PACHYSTACHYS

Acanthaceae

Origin: Tropical America. A genus of five or six species of shrubs and evergreen perennials closely related to *Jacobinia*. They have opposite pairs of entire leaves and tubular flowers with very unequal lobes which are borne in twos and threes within large bracts, many together making spike-like clusters or heads. They make very good pot plants. Propagate by cuttings in summer. *Pachystachys* derives from the Greek *pachys*, thick and *stachys*, a spike.

 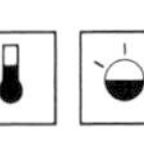 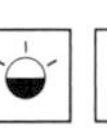 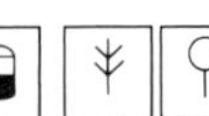

Species cultivated

P. coccinea (*P. cardinalis*, *Justicia coccinea*)
Cardinal's guard Northern South America, West Indies
Erect shrub, to 1.5m(5ft) if regularly pruned to keep bushy, but capable of making a loose plant to 2m(6½ft) in a large container. Leaves to 20cm(8in) long, ovate-elliptic, strongly veined. Flowers about 5cm(2in) long, bright scarlet, within four ranks of overlapping green bracts, making up 15cm(6in) spikes in winter.

P. lutea Lollipop-plant Peru
Shrub to 1m(3ft) or more. Leaves 10–18cm(4–7in) long, lanceolate to narrowly ovate, the upper surface a glossy dark green. Flowers 4–5cm(1½–2in) long, white, within golden-yellow bracts, making up flower spikes to 10cm(4in) long from late spring to autumn.

× PACHYVERIA

Crassulaceae

Origin: Hybrids between species of *Echeveria* and *Pachyphytum*. Several hybrids of this origin have been raised, using mainly *Pachyphytum bracteosum* with several, not always specified, *Echeveria* species. In general they are halfway between the parents in details of habit, foliage and flower and are more vigorous than either. They require the same cultural conditions as *Pachyphytum*, *q.v.* One of the best known sorts which is usually available commercially is × *P. scheideckeri* (*E. secunda* × *P. bracteosum*) with rosettes of grey-white leaves to 13cm(5in) wide, and orange, yellow-tipped flowers in spring.

PALISOTA

Commelinaceae

Origin: Tropical Africa. A genus of 25 species of evergreen perennials and shrubs. The perennials are rosette- and clump-forming, the shrubs sparingly branched with the leaves at the stem tips. In the perennials, the 3-petalled flowers are clustered within the rosettes, but in the shrubs they are carried in looser, more conspicuous inflorescences. The fruit is a berry, usually red in colour, but also purple to blue. Those described here provide foliage pot plants. Propagate by seed, cuttings or division in spring. *Palisota* honours Baron A. M. F. J. Palisot de Beauvois (1752–1820), a French botanist, traveller and author with a particular interest in grasses and mosses.

Species cultivated

P. barteri West Africa (Fernando Po.)
Rosette-forming; leaves broadly elliptic to obovate-oblong, sometimes narrowly so, lustrous deep green, 45–60cm(1½–2ft) long. Flowers white or purple-tinged, in clusters to about 10cm(4in) long. Fruits hairy, ovoid, beaked, bright orange-red, showy.

P. elizabethae See under *P. pynaertii.*

P. hirsuta West Africa (Ghana, Cameroons, Fernando Po., etc.)
Shrub 2–3m(6½–10ft) tall, but easily kept to half this in containers. Leaves oblanceolate to obovate, 15–30cm(6–12in) or more in length, more or less long-hairy at the base and on the margins. Inflorescence a loose raceme-like thyrse to 25cm(10in) long, the flowers white, followed by bluish fruits.

P. pynaertii West Africa
Rosette-forming; leaves 40–90cm(16–36in) long, narrowly ovate to lanceolate, wavy-margined, rich green with an edge of reddish hairs. Flowers white in an ovoid, dense inflorescence to 10cm(4in) tall. *P.p.* 'Elizabethae' (*P. elizabethae*) has a longitudinal, feathered, pale yellow band down the centre of each leaf; the most ornamental palisota in leaf and well worth seeking.

PAMIANTHE

Amaryllidaceae

Origin: South America (western Andean region). A genus of three species of bulbous plants very closely related to *Hymenocallis* and in flower virtually indistinguishable. The primary distinctive character is the seeds – in *Hymenocallis* they are globose to oblongoid and fleshy, whereas in *Pamianthe* they are flattened and winged. Propagate by seed or division in spring. The species described is a desirable addition to a collection of tender bulbs for the conservatory and can be brought into the home when in bloom. *Pamianthe* is named for Major Albert Pam (1875–1955) of Wormleybury, Hertfordshire, a financier and keen collector of bulbous plants.

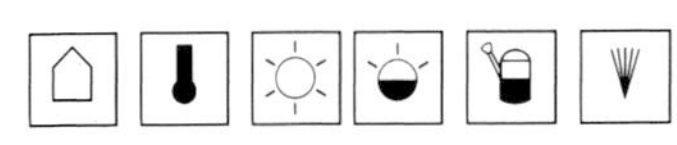

Species cultivated

P. peruviana Peru
Bulb fairly large, with a long, tapering neck or 'false stem'. Leaves strap-shaped, arching, 40cm(16in) or more long. Flowering stem above the leaves bearing an umbel of one to four blooms; each flower white, fragrant, composed of a tube to 13cm(5in) long, six tepals almost as long and a corona (staminal cup) to 8cm(3in) in length, usually borne in late winter to spring. Best if kept dry during the winter when some or all the leaves will die back.

PANCRATIUM

Amaryllidaceae

Origin: Mediterranean region to tropical Africa and Asia. A genus of about 15 species of bulbous perennials with basal strap-shaped leaves and umbels of somewhat daffodil-like flowers on long, leafless stems. The blooms are white and funnel-shaped, opening to six narrow, spreading lobes and within these a corona made up of the widened, joined stamen bases, the upper part of the stamens remaining free. They are suitable for the conservatory or for flowering in the home. Propagate by separating offsets, or by seed when ripe or in spring. *Panctratium* is from the Greek vernacular name for a bulbous plant, not necessarily of this genus.

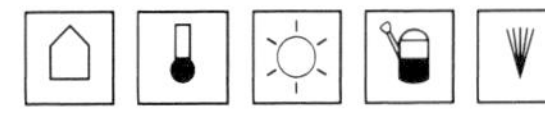

Species cultivated

P. caribaeum See *Hymenocallis caribaea.*

P. illyricum Spain, Corsica, Sardinia
Leaves deciduous, 45cm(1½ft) long, broadly strap-shaped, glaucous. Flowers 8cm(3in) long, having a short corona, six to 12 blooms in each umbel, on terete stems; very fragrant, opening in late spring.

P. maritimum Sea lily, Sea daffodil
Mediterranean shores
Leaves more or less evergreen, to 60cm(2ft) long, narrowly strap-shaped, glaucous. Flowers to 12cm(4¾in) long, with a deep cup-shaped corona, in umbels of 13 to 15 on a flattened stem; fragrant, opening in summer.

PANDANUS

Pandanaceae

Screw pines

Origin: Tropical Africa, Asia, Australasia mostly on sea coasts and islands. A genus of 600 to 650 species of evergreen trees, shrubs and climbers, often distinctive in the wild because of their prop roots which support the erect stem and branches. The leaves are usually crowded near the stem tips and arranged spirally giving them their vernacular name of screw pine. The leaves are strap-shaped and of thick texture, often with finely spiny margins. The dioecious flowers are petalless, the males carried in branched spikes, the females in rounded heads which develop into tight cone-shaped clusters of fruits. They are attractive pot or tub plants though care must be taken in placing

ABOVE *Pancratium illyricum*
TOP *Palisota pynaertii* 'Elizabethae'

those with prickle-edged leaves. *Pandanus* is a Latinized form of the Malaysian name *pandan*.

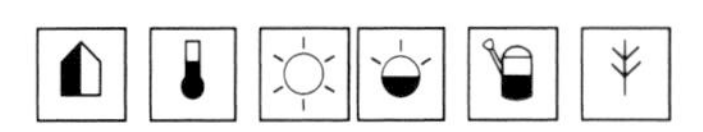

Species cultivated

P. baptisii Blue screw pine New Britain
Stem short, the arching leaves to 40cm(16in) or more long, deep blue-green with one or more longitudinal creamy-white or yellow stripes and spineless margins. Flowers and fruit have not been recorded.

P. odoratissimus Sri Lanka (perhaps Malaysia and Philippine Islands or naturalized there)
In the wild a small tree, but slow growing and making a good specimen for a container. Leaves 90–150cm(3–5ft) in length, linear, with sharp spiny teeth on the back of the midrib and the margins, and slender pendent tips.

P. sanderi Reputedly from Timor, Indonesia
Eventually to 2m($6\frac{1}{2}$ft) or more, but slow-growing. Leaves 45–75cm($1\frac{1}{2}$–$2\frac{1}{2}$ft) long, linear, minutely spiny toothed, rich green with a yellow central stripe.

P. utilis Common screw pine Malagasy
In the wild, a medium-sized tree, but when young easily maintained under 2m($6\frac{1}{2}$ft). Leaves 90–180cm(3–6ft) in length, stiff, slightly glaucous especially below, with small red spiny teeth.

P. veitchii Polynesia
Stem short, freely branching when well established, to 2m($6\frac{1}{2}$ft) or more tall. Leaves arching, 60–90cm(2–3ft) long with wide creamy-white margins which are spiny.

RIGHT *Pandanus sanderi*
BELOW RIGHT *Pandorea jasminoides*

PANDOREA

Bignoniaceae

Origin: Malaysia to Central Australia. A genus of eight species of evergreen, twining climbers with pinnate leaves in opposite pairs and axillary panicles of colourful funnel- to bell-shaped flowers opening to five spreading lobes. They are suitable for a conservatory making a good covering for the back wall and can be trained on strings or wires under the roof. Although amenable to pruning, they will not flower so well if cut back hard. Propagate by summer cuttings in bottom heat, or by seed in spring in warmth. *Pandorea* was named for Pandora, in Greek mythology the first woman sent to earth by Zeus.

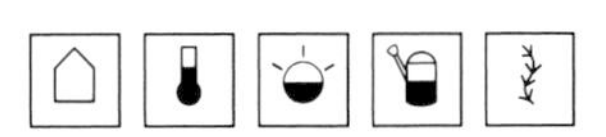

Species cultivated

P. jasminoides Bower plant Australia
To 5m(16ft) in a pot, much more in the open ground. Leaves with five to nine lanceolate to ovate leaflets, each 2.5–5cm(1–2in) long. Flowers 3–5cm($1\frac{1}{4}$–2in) long and 5cm(2in) across, white with pink in the throat, in few-flowered panicles, opening from winter to summer. *P. j.* 'Alba' has pure white flowers; 'Rosea' is pink with a darker throat.

P. pandorana Wonga-wonga vine Australia, New Guinea
Stems to 6m(20ft) or more. Leaves with three to nine ovate lanceolate leaflets, each 3–7cm($1\frac{1}{4}$–$2\frac{3}{4}$in) long; in the seedling and juvenile stages very deeply dissected and almost fern-like in appearance. Flowers to 2cm($\frac{3}{4}$in) long, creamy-white, spotted and streaked with purple to reddish-brown, carried in summer in many-flowered panicles. *P. p.* 'Rosea' has pink blooms.

P. ricasoliana See *Podranea ricasoliana*.

PANICUM See OPLISMENUS

PAPHIOPEDILUM

Orchidaceae
Slipper orchids

Origin: Southern Asia from the Himalaya east to New Guinea. A genus of 50 to 60 species including both terrestrial and epiphytic orchids, for a long time classified in *Cypripedium*, a genus now used only for a group of frost-hardy terrestrial orchids. They are without pseudobulbs having a tufted habit and thick evergreen basal leaves which are usually strap-shaped. Their flowers are borne singly or occasionally in racemes on long stalks. Each has an erect standard tepal, two fused ones appearing as one, pointing down, two smaller, narrower lateral or wing tepals which usually curve downwards, and a pouched lip fancifully like a helmet or the toe of a slipper, hence their vernacular name of slipper orchids. Propagate by division when re-potting.

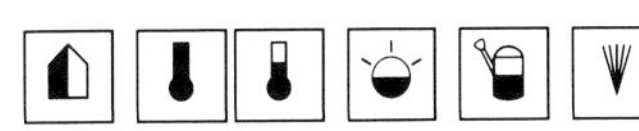

ABOVE *Paphiopedilum bellatulum*
LEFT *Paphiopedilum sukhakulii*

Species cultivated

P. bellatulum Burma, Thailand
Leaves narrowly elliptic to tongue-shaped, about 15cm(6in) long, dark green with paler spots or mottling above, red-purple beneath. Flowers about 6cm(2½in) wide, white with a variable amount of red-purple spotting, tepals very broad and overlapping to form a rounded flower, the labellum small, ovoid; flowering stem short, the flower sitting on or just among the leaves in spring. Tropical.

P. callosum Thailand, Cambodia
Leaves few, to 25cm(10in) or more long, mottled with darker green. Flowers to 10cm(4in) wide, the standard white, lined with purple-brown, the downswept wings purplish-green and the labellum a glossy brownish-purple; they are carried singly on stems 30cm(1ft) or more tall from winter to summer. Tropical.

P. fairieanum Himalaya
Leaves to 15cm(6in) long, pale green, unmarked. Flowers to 6cm(2½in) wide, tepals wavy, standard white, netted and lined with purple, the downswept curving wings white with purple striping, labellum purplish-green with darker veins; they are solitary on 25cm(10in) long stems in late summer and autumn. Temperate.

P. hirsutissimum Himalaya
Leaves to 30cm(1ft) long, green, unmarked. Flowers ciliate, to 10cm(4in) across, green, the standard densely spotted with purple black except for the unmarked upper third, wings held horizontally, waved, violet-purple at the tip and black hairy towards the base, labellum dull green shaded purple and with minute black warts; they are solitary, on purple, hairy stems to 30cm(1ft) high in spring. Temperate.

P. insigne Himalaya
Leaves to 30cm(1ft) long, green, unmarked. Flowers 10–13cm(4–5½in) wide, the standard green, waved, white at the tip, otherwise spotted and streaked with brown, wings yellow-green with brown veins, labellum green flushed with brown, glossy; they are usually solitary but sometimes in pairs, on stems to 30cm(1ft) tall, from autumn to spring. The amount of shading and spotting varies considerably and a number of forms are listed, some probably of hybrid origin. Temperate.

P. sukhakulii Thailand
Leaves to 20in(8in) long, green with darker mottling above. Flowers to 10cm(4in) or so wide, the standard white with regular green lines, wings flat, held horizontally, greenish purple with brown-purple spotting; they are usually solitary on stems 20–30cm(8–12in) high in winter. Tropical.

P. venustum Himalaya
Leaves 10–15cm(4–6in) long, green with darker mottling above, blotched with purple beneath. Flowers about 7cm(2¾in) wide, the standard white with regular green lines, wings narrow and waved, ciliate, greenish, the upper half with purplish lines, the lower with small blackish warts, labellum yellow-green with darker veining and a purple flush; they are solitary on purple stems 15–23cm (6–9¼in) high in autumn and spring. Temperate.

Hybrids

Garden origin. The greenhouse slipper orchids have long been popular among specialists and laymen alike and many hybrids have been bred. Hundreds of cultivars have been recorded, though a fairly limited number of these remain commercially available. On the whole, hybrids are a better bet as house plants, particularly those derived from P. × *insigne*. They tend to be more vigorous and to have

ABOVE *Parochetus communis*
ABOVE RIGHT *Paphiopedilum* 'Brownstone'

larger flowers than the species and generally bloom more regularly. Names to look out for are:
'Brownstone', large, cream to brown.
'Burfordiensis', strong grower, tawny-brown and green.
'Del-Rosi', rich pink.
'King Arthur Burgoyne', wine-red.
'Le Baudyanum', pink, brown and green.
'New Era' (*P. barbatum* × *P. sukhakulii*), shades of purple, green and white.
'Transvaal', dark red and green.
'Vanda M. Pearman', ivory-pink, darker spotted.

PAROCHETUS

Leguminosae

Origin: Mountains of Central Africa and Himalaya. A genus of one or perhaps two species of prostrate perennials, the species from Africa being suitable for a pot or hanging basket plant. Propagate by careful division, by cuttings in summer or by seed in spring. *Parochetus* derives from the Greek *para*, near or close to and *ochetos*, a brook, the plant growing in such situations in the wild.

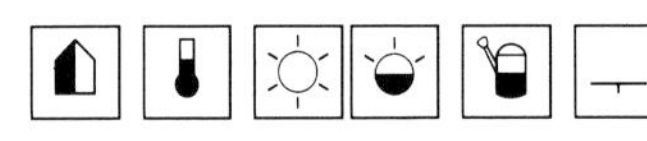

Species cultivated

P. communis Blue shamrock pea Mountains of East Africa
Prostrate, mat-forming. Leaves 2.5–4.5(1–1¾in) across, trifoliate with obcordate leaflets. Pea flowers 1.5–2cm(⅝–¾in) long, bright blue, usually singly from the leaf axils, appearing through most of the year but chiefly from autumn to spring when day length is twelve hours or less. (The Himalayan form blooms only in summer and is dormant in winter.)

PARODIA

Cactaceae

Origin: South America. A genus of 30 to 35 species of usually small globular or shortly cylindrical cacti which can be solitary or clump-forming. They have broadly funnel-shaped flowers and make excellent pot plants for the home windowsill or sunny conservatory. Propagate by seed in warmth, or by separating offsets in summer. *Parodia* derives from the Greek *para*, near to and *onyx*, a nail, a reference to its former reputation for curing whitlows.

Species cultivated

P. aureispina Golden Tom Thumb cactus
Northern Argentina
Stems usually solitary, globular, to 6cm(2½in) high. Ribs numerous, closely set, bluish-green, slightly spiralled, divided into tubercles. Radial spines bristle-like, about 40, white, 6mm(¼in) long; Central spines six or seven, bright yellow, about 1.2cm(½in) long, one of them hooked-tipped. Flowers yellow, about 3cm(1¼in) wide.

P. chrysacanthion Argentina
Stems to 6cm(2½in) tall, globular to slightly elongated, solitary or clump-forming. Ribs about 24, spiralled, bearing small yellow woolly areoles. Spines 30 to 40, radials 5mm(⅕in) long, bristle-like, very pale yellow; central spines (¾in) long, golden-yellow to brown. Flowers rich yellow to 2cm(¾in) long.

P. microsperma Argentina
Stems globular, clustered, 5–10cm(2–4in) tall with about 20 spiralling ribs. Radial spines ten to 20, glossy-white, to 5mm(⅕in) long, central spines three to four, red-brown to 2cm(¾in) long, the longest with a hooked tip. Flowers pale red outside, orange inside, 3.5–5cm(1½–2in) across.

P. mutabilis Northern Argentina (mountains)
Stems globular, to 8cm(3in) high, blue-green. Ribs indistinct, cut into tubercles with large, white-woolly areoles. Radial spines about 50, very slender, bristle-like, white, about 1cm(⅜in) long; centrals usually four, reddish, darker-tipped, somewhat longer, one with a hooked tip. Flowers bright yellow, sometimes with a red throat, about 5cm(2in) across.

PASSIFLORA

Passifloraceae

Passion flowers

Origin: Chiefly from the Americas, but a few species from Asia and Australia. A genus of 400 or more species of climbing plants, with alternate, often 3- to 5-lobed leaves, holding on to their supports by means of tendrils. The flowers are usually solitary from the upper leaf axils and are followed by rounded to ovoid berries which are sometimes edible. The flowers are large and often showy, tubular at the base with ten tepals which can be spread out flat, or overlapped to form a bowl-like shape. The ovary with its three styles and five stamens are borne together on a central stalk (androphore) which is surrounded by one or several rows of fleshy filaments forming a corona. Passion flowers can be grown against the back wall of a conservatory, or trained around and up a wire or stick. When young some are successful house plants, blooming when very small. Propagate by cuttings in summer with bottom heat, or by seed when ripe or in spring in warmth. *Passiflora* derives from the Latin *passio*, passion and *flos*, a flower, from the interpretation of the flower by Jesuit missionaries to South America.

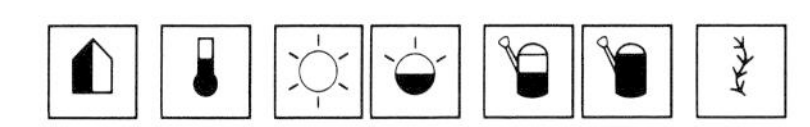

Species cultivated

P. alata North-eastern Peru (perhaps introduced), eastern Brazil
Superficially much like *P. quadrangularis*, but with leaf-stalks having two or four glands and narrow stipules. Flowers a little smaller, with the longest filaments to 4cm($1\frac{1}{2}$in), fragrant. Fruits yellow, up to 10cm(4in) long, edible.

P. × alatocaerulea (*P. alata × P. caerulea*)
Very much a blend of parental characters with a strong nod towards *P. alata*. Stems fairly robust, angled or narrowly winged. Leaves 3-lobed, 10–13cm(4–5in) long. Flowers 10cm(4in) wide, fleshy-textured, fragrant, the petals white outside, pinkish-violet tinted within; filaments 2.5cm(1in) long, white, zoned with pinkish and bluish-purple.

P. antioquiensis See under *P. mollissina.*

P. caerulea Common/Blue passion flower
Central America and western South America
Stems slender, grooved. Leaves 10–15cm(4–6in) wide, deeply 5- to 9-lobed, rather glaucous. Flowers 6–10cm ($2\frac{1}{2}$–4in) across with white or pinkish tepals and a tricoloured corona, the filaments purple at the base, white in the middle and tipped with blue; they open in summer and autumn and are followed by ovoid, pale orange fruits, 3–4cm($1\frac{1}{4}$–$1\frac{1}{2}$in) long. *P. c.* 'Constance Elliott' has white flowers.

P. × caeruleo-racemosa
Hybrid between *P. caerulea* and *P. racemosa*, very like the latter but it has larger 3-lobed leaves and purple flowers.

***P. × caponii* 'John Innes'** (*P. quadrangularis × P. racemosa*)
Stems robust, angled, eventually to 8m(26ft) or more in length. Leaves 3-lobed, up to 18cm(7in) long by 25cm(10in) wide, rich green. Flowers 10–13cm(4–5in) wide, bowl-shaped, nodding, white, flushed claret-purple with four zones of filaments, the longest sideways twisted and boldly banded purple and white from a crimson base.

P. cinnabarina Eastern Australia
Stem slender, to 3m(10ft) or more in length. Leaves 3-lobed, 5–10cm(2–4in) long, deep, almost glossy green. Flowers 5cm(2in) or a little more wide, scarlet with yellow filaments.

P. coccinea Red passion flower, Red granadilla
Venezuela to Bolivia and Brazil
Stems slender, slightly angled, red-downy, to 4m(13ft) or more. Leaves broadly oblong to orbicular, 6–13cm($2\frac{1}{2}$–5in) long, toothed. Flowers 8cm(3in) or more wide, scarlet, with crimson, pink and white outer filaments 2cm($\frac{3}{4}$in) long and shorter inner white ones. Fruits ovoid, about 5cm(2in) long, orange or yellow with six green stripes, edible.

P. edulis Passion fruit, Purple granadilla
Brazil (cultivated in warmer countries)
Stems to 3m(10ft) or more. Leaves deeply 3-lobed, toothed, to 10cm(4in) or more across. Flowers 5–7cm(2–$2\frac{3}{4}$in) wide, white, the corona of white filaments tipped with purple, followed by 5cm(2in) long ovoid fruits which are greenish-yellow to purplish.

P. × exoniensis (*Tacsonia × exoniensis*)
Hybrid between *P. antioquensis* and *P. mollissima*, two very similar species, which is more vigorous and free-flowering than either. Flowers 10–13cm (4–5in) wide, long-tubed, rose-pink, darker on the outside, with a small white corona.

P. ligularis Sweet granadilla Central Mexico to central Peru and western Bolivia
Stems rounded, to 5m(16ft) or more in length. Leaves broadly ovate, cordate, 8–15cm(3–6in) long, the stalks with four to six filiform glands. Flowers 6–10cm($2\frac{1}{2}$–4in) across, with white or pink-tinted petals and blue-tipped filaments banded with white and red-purple. Fruits ovoid, 6–8cm($2\frac{1}{2}$–3in) long, yellow, flushed or tinted with red or purple, sweet and juicy and by some considered to be the best of all the passion fruits.

P. manicata Western Venezuela, Colombia, northern Peru
Stems angled, 3m(10ft) or more in length. Leaves 3-lobed, 4–9cm($1\frac{1}{2}$–$3\frac{1}{2}$in) long, the stalks bearing four to ten short-stalked or sessile glands. Flowers 8cm(3in) or more wide, petals scarlet, the filaments blue and white. Fruits ovoid, to 5cm(2in) long, glossy dark green.

P. mollissima (*Tacsonia mollisima*) Banana passion fruit, Curuba Venezuela to Bolivia
Stems to 5m(16ft) or more. Leaves 5–10cm(2–4in) long, deeply 3-lobed, softly downy. Flowers 6–9cm($2\frac{1}{2}$–$3\frac{1}{2}$in) across, pink and white, the corona a small, purplish warted ring; they open in summer and are followed by ovoid, yellowish fruits to 6cm($2\frac{1}{2}$in) long. *P. antioquiensis* from Colombia is similar, but bears larger rose-red blooms.

P. quadrangularis Giant granadilla Tropical South America, but now widespread throughout the tropics
Strong-growing to 5m(16ft) and more, the stems 4-angled and sharply winged. Leaves 10–20cm(4–8in) long, ovate to ovate-cordate, entire. Flowers 8–11cm(3–$4\frac{1}{2}$in) across, bowl-shaped, white, pink or pale violet, the corona filaments longer than the tepals, strongly waved and banded with deep purple and white; they are fragrant and open in summer, being followed by ovoid or oblongoid, edible fruits 20–30cm(8–12in) long.

P. racemosa Red passion flower Brazil
Stems slender, to 5m(16ft) or more. Leaves 10cm(4in) wide, dark green and leathery, 3-lobed, the lobes all forward pointing. Flowers to 10cm(4in) wide, the tepals crimson-red and soon reflexing, the short corona banded white and

Passiflora edulis

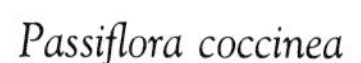

Passiflora coccinea

BELOW *Pavonia multiflora*
BOTTOM *Pelargonium betulinum*

purple; they are carried in pendent racemes to 10cm(4in) long in summer and autumn.
P. sarguinea See *P. vitifolia.*
P. vitifolia (*P. sanguinea*) Nicaragua to Peru
Stems rounded, brownish pubescent, to 5m(16ft) in length. Leaves 3-lobed, the largest 8–15cm(3–6in) long, more or less glossy deep green, with a pair of rounded, marginal glands at the base of the sinuses (between the lobes). Flowers fragrant, 13–19cm(5–7$\frac{1}{2}$in) wide, bright scarlet with short, red, yellow and white filaments. The finest of all the red passion flowers and widely cultivated in the tropics.

PAVONIA

Malvaceae

Origin: Tropics and sub-tropics. A genus of 150 to 200 species of shrubs and perennials one of which has long been grown in our greenhouses and conservatories. Propagate by cuttings in summer. *Pavonia* commemorates José Antonio Pavon (1754–1840), a Spanish botanist and plant collector, part author of the monumental but incomplete *Flora Peruviana.*

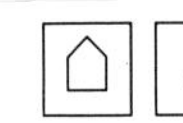 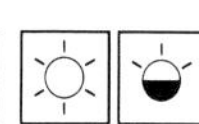  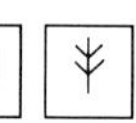

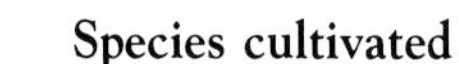

Species cultivated
P. multiflora (*Goethea multiflora, Triplochlamys multiflora*) Brazil
Eventually an erect shrub to 2m(6$\frac{1}{2}$ft), but if regularly pruned or grown as a short term plant seldom above 90cm(3ft) tall. Leaves narrowly ovate to oblong, 10–25cm(4–10in) in length, dark green. Flowers solitary in the upper leaf axils or in leafless terminal corymbs; each flower has a 2.5–4cm(1–1$\frac{1}{2}$in) long, 5-petalled, purple-red tubular corolla, a red tubular calyx and ten to 24 linear red bracts in one or two whorls radiating out from its base. A well-flowered specimen provides a striking sight and always excites comment.

PEDILANTHES

Euphorbiaceae

Origin: Florida and Mexico to tropical South America and West Indies. A genus of 14 species of succulent shrubs closely related to *Euphorbia.* One species is widely grown for its intriguing bird's-head or slipper-shaped flower heads. Propagate by seed in spring or by cuttings in summer. *Pedilanthes* comes from the Greek *pedilon,* a sandal and *anthos,* a flower.

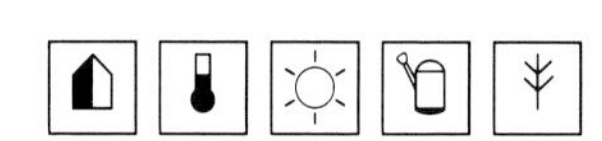

Species cultivated
P. tithymaloides Redbird cactus, Slipper cactus
Northern South America to Mexico and Florida, including West Indies
Stems erect with ascending, zigzag branches. Leaves 5–10cm(2–4in) long, ovate to broadly lanceolate, pointed, with a keeled midrib. Flower heads (cyathia, *q.v. Euphorbia*) in terminal clusters in summer and autumn, each one red, 2cm($\frac{3}{4}$in) long; viewed from the side resembling somewhat fancifully, a bird's head, or a slipper upside down. *P.t.* 'Nana Compacta' is dwarfer and more compact in habit; 'Variegatus' has the leaves white-margined often with a flush of pink.
P.t. smallii Jacob's ladder
Stems much more prominently zigzag.

PELARGONIUM

Geraniaceae
Geraniums

Origin: Widespread throughout warm temperate areas of the new world, concentrated in South Africa but also in north Africa and the adjacent Atlantic Islands and eastwards to Arabia and southern India, also in Australasia; those listed below from South Africa unless stated. Long ago included in *Geranium* and still often sold under that name. A genus of 250 to 280 species mainly of shrubs and sub-shrubs, but including annuals and perennials, most of which are adapted to areas of low rainfall. Some are succulent or almost so. Their main characteristics are the 5-petalled flowers which have a small nectary spur joined to the pedicel, and are carried in stalked, axillary umbels. The beaked fruit split open explosively when ripe into five sections each carrying with it a narrow segment of the style. The species described below are all good conservatory and house plants, flowering in summer. Propagate by seed in spring or by cuttings taken in spring or summer. *Pelargonium* is derived from the Greek *pelargos,* a stork, an illusion to the long-beaked fruit.

Species cultivated
P. abrotanifolium
Shrub to 1m(3ft) tall. Leaves to 2.5cm(1in) across, 3-lobed, each lobe finely cut, grey-downy and aromatic. Flowers 2cm($\frac{3}{4}$in) wide, the two upper petals broadest, white or pink, veined with red.
P. angulosum
Shrubby or sub-shrubby to 90cm(3ft) in height, usually well branched. Leaves rounded to 6cm(2$\frac{1}{2}$in) wide, 5-angled and toothed, firm-textured and somewhat scabrid. Flowers about 2.5cm(1in) wide or more, carmine-purple with darker veins, three to seven in stalked umbels.
P. betulinum
Shrubby, fairly erect and slender, eventually to 90cm(3ft) in height. Leaves broadly ovate, toothed, 1–2cm($\frac{3}{8}$–$\frac{3}{4}$in) long, leathery-textured and glaucous-tinted. Flowers 2–2.5cm($\frac{3}{4}$–1in) wide, white or pink with a carmine vein pattern at the base of the two upper petals, usually in two or three bloomed umbels.

P. capitatum Rose-scented geranium
Shrubby, with straggling, decumbent, white-hairy stems, 30–60cm(1–2ft) tall. Leaves to 5cm(2in) across, three to five lobed and toothed, softly hairy. Flowers 2cm(¾in) long, pink with red-purple veins, carried in umbels of eight to 24 on slender stalks. Often confused in cultivation with *P. karooense* which is more erect and has deeply lobed leaves.
P. carnosum See under *P. crithmifolium*.
P. × citrosum (*P. crispum* hybrid)
The origins of this plant are not recorded, but it greatly favours *P. crispum* in general appearance with less obviously 2-ranked, larger leaves. When bruised, the latter are strongly lemon-scented. Flowers similar to *P. crispum*, but lilac-purple with darker veining and spotting on the upper petals. Often listed as *P. citriodorum*, but that name is strictly a synonym of *P. acerifolium*.
P. crispum Lemon geranium, Prince Rupert geranium
Erect, branched shrub to 60–90cm(2–3ft) tall. Leaves 2–3cm (¾–1in) wide, shallowly 3-lobed, greyish, with crisped or curled edges, borne close together in two ranks up the stems, strongly lemon-scented. Flowers 2–2.5cm(¾–1in) long, white to pink with darker reddish veining, borne singly or in twos or threes. *P.c.* 'Variegatum' has yellow-blotched leaves.
P. crithmifolium
Shrubby, with thick, somewhat knobbly, fleshy stems to 20cm(8in) or more tall. Leaves bipinnatifid, glaucous, 5–10cm(2–4in) long. Flowers about 2.5cm(1in) wide, lilac-pink with crimson feathery blotches on the two upper petals, in umbels of three to seven. *P. carnosum*, with pink flowers, and *P. dasycaule*, with white flowers, are of similar appearance.
P. cucullatum
Freely-branched shrub to 2m(6½ft) tall, softly hairy. Leaves to 6cm(2½in) across, kidney-shaped, more or less cupped with finely scalloped or toothed margins. Flowers 4–5cm(1½–2in) across, purplish-red with darker veins. A pale purple sort is sometimes offered under this name and is a form or possibly a hybrid of this species.
P. dasycaule See under *P. crithmifolium*.
P. denticulatum
Shrubby, to 1m(3ft) tall. Leaves 4–8cm(1½–3in) long, bipinnate-partite, toothed, ferny in appearance with a balsam scent. Flowers 1.5–2cm(½–¾in) wide, lilac to rose-pink to pale purple with darker markings. 'Filicifolium' (*P. filicifolium*) has more finely dissected, narrower leaves.
P. × domesticum Regal/Fancy/Show geraniums (in US Martha/Lady Washington geraniums)
Hybrid group involving *P. angulosum*, *P. cucullatum*, *P. fulgidum*, *P. grandiflorum* and other species. They are mostly compact, shrubby, 40–60cm (16–24in) tall or more. Leaves 5–10cm(2–4in) wide, usually shallowly lobed, sometimes more deeply so, the margins toothed. Flowers 4–5cm(1½–2in) wide in shades of white, pink, red and purple, blotched with darker shades. Many cultivars have been raised and the numbers are being increased every year, some having strongly waved petals with or without picotee edges, or are bicoloured.

Pelargonium cucullatum

P. echinatum Cactus geranium Namaqualand
Tuberous-rooted shrublet with stems woody below and fleshy above, bearing persistent, prickle-like stipules. Leaves broadly ovate, more or less cordate, with five to seven shallow, toothed lobes. Flowers reddish-pink to purple, the upper petals bearing heart-shaped red blotches.
P. filicifolium See *P. denticulatum*.
P. × fragrans Nutmeg geranium
Hybrid between *P. exstipulatum* and *P. odoratissimum*, rather like the latter parent but its leaves are more or less 3-lobed and with a scent described by different authorities as being of nutmeg, lemon or pine. It has small white flowers, the upper petal with red veins.
P. fulgidum
Sub-shrubby, to 60cm(2ft) or more tall. Leaves 7cm(2¾in) long, heart-shaped, pinnately lobed, silvery-hairy. Flowers to 4cm(1½in) across, bright scarlet with darker veining.
P. graveolens Rose geranium
Shrubby, to 1m(3ft) tall. Leaves with five deeply lobed, toothed segments, three of the lobes cut to near the base, grey-green. Flowers 3–4cm(1¼–1½in) wide, rose-pink marked with purple. A number of forms of this species are probably of hybrid origin, *P. radens* being the other plant involved.
P. × hortorum Zonal geraniums
Complex hybrid group involving *P. inquinans*, *P. zonale* and other species. They are shrubby, to 1.5m(5ft) or more tall. Leaves 6–13cm(2½–5in) across, rounded to kidney-shaped, the margins scalloped and waved, usually mid-green with a darker brown or bronze horseshoe-shaped mark in the centre. Flowers 2–5cm(¾–2in) wide, in shades of red, pink, orange, white and purple, some bicoloured and some with a picotee edge; they can be single or double and may have quilled petals. Cultivars with variegated leaves are also grown, the markings varying from white and yellow to crimson or brown, and some have dark green or purplish foliage. The following cultivars are popular.
'Fleurette', 20cm(8in) or less in height, double salmon-red with dark leaves.

TOP *Pelargonium* × *hortorum* 'Mr Henry Cox'
ABOVE *P.* × *domesticum* 'Rembrandt'
ABOVE RIGHT *P.* × *hortorum* 'Startel Pink'

'Golden Lion', single, pale orange flowers.
'Hermoine', single, white flowers.
'Irene', double, crimson-red flowers.
'Irene Cal', double, pale salmon-pink flowers.
'Mr (Mrs) Henry Cox', colourful coppery-green leaves shading to a darker centre with a yellow border; flowers single pink.
'Red Black Vesuvius', to 20cm(8in) tall, single bright red flowers and black-green foliage.
'Snowbody', double white variegated leaves.
'Venus', single, pure white flowers.

RIGHT *Pelargonium peltatum*

P. karooense See under *P. capitatum.*

P. odoratissimum Apple geranium
Shrubby with sprawling stems to 45cm(1½ft) long. Leaves 2–3cm(¾–1¼in) across, ovate-cordate, wavy and velvety-hairy, with a sweet apple scent. Flowers to 2cm(¾in) across, white, spotted and veined with red.

P. papilionaceum
Shrubby with well-branched stems to 90cm(3ft) in height. Leaves to 10cm(4in) across, rounded, cordate, toothed. Flowers composed of two large, obovate, 2cm(¾in) long upper petals and three small, narrow, almost rudimentary lower ones; upper petals pink with a white base and a carmine blotched and veined pattern. The main flowering stem branches, carrying several umbels of five to ten blooms. It seems likely that at least some of the plants in cultivation under this name are of hybrid origin, perhaps with the allied *P. capitatum* and/or *P. vitifolium.*

P. peltatum Ivy-leaved geranium
Stems to 1m(3ft) or more long, slender and trailing. Leaves 5–7cm(2–2¾in) wide, somewhat fleshy, 5-angled or lobed, rather ivy-like but somewhat peltate, the leaf stalk joining the leaf a little way in from the margin. Flowers 3–4cm(1¼–1½in) wide, pink to carmine, with darker markings on the upper petals. Some of the named forms available are probably of hybrid origin, the following being long established and still popular:
'Abel Carrière, semi-double, soft rose-purple.
'L'Elégante', leaves with cream variegation and tinted with purplish-red; flowers single, white with purple veins.
'Mrs W. A. R. Clifton', leaves green, flowers fully double, scarlet.

P. quercifolium (*P. terebinthaceum*) Oak-leaved geranium
Shrubby, to 1.2m(4ft) tall and well branched. Leaves 5–10cm (2–4in) long, ovate, deeply pinnately lobed, waved and toothed, green, sometimes with brown markings. Flowers about 3cm(1¼in) across, pink to pinky-purple, with purple veins.

P. radens (*P. radula)*
Shrubby, to 1m(3ft) tall. Leaves 4–7cm(1½–2¾in) long, bipalmatipartite with slender, toothed segments which are rolled under at the margins and roughly hairy. Flowers about 3cm(1¼in) wide, pink to rose-pink with red-purple markings.

P. radula See *P. radens.*

P. scandens
Scrambling, shrubby species with sparingly branched stems to 1.5m(5ft) or more. Leaves more or less orbicular, crenate, greyish-green with a dark horseshoe-shaped zone. Flowers 2.5cm(1in) or more wide, pink to light red, in umbels of five to 12. An effective plant for a large hanging basket, or can be trained around a window embrasure or on the back wall of a lean-to conservatory.

P. splendidum See *P. violareum.*

P. terebinthaceum See *P. graveolens* and *P. quercifolium*; most of the plants in cultivation belong to the latter species.

LEFT *Pelecyphora asseliformis*
FAR LEFT *Pelargonium violareum*

P. tetragonum
Succulent shrubby species to 1m(3ft) tall, with jointed, 4-angled fleshy stems. Leaves 5-lobed, with rounded teeth. Flowers 2.5–3.5cm(1–1½in) across, with four pink petals, the upper two larger and deeper-coloured than the lower and feathered with purple veining.
P. tomentosum Peppermint-scented geranium
Shrubby with sprawling, partially erect stems to 1m(3ft) in length. Leaves to 13cm(5in) across, triangular, cordate with five shallow, rounded lobes, softly velvety-hairy and strongly peppermint-scented. Flowers about 1cm(⅜in) wide, white with red markings. *P.t.* 'Variegatum' has creamy-white margined leaves.
P. tricolor See *P. violareum*.
P. violareum (*P. tricolor*, *P. splendidum violareum*)
Shrubby, well branched, spreading to 45cm(1½ft) or so by 30cm (1ft) in height. Leaves usually ovate, sometimes lanceolate, irregularly toothed, occasionally with a few narrow lobes, grey pubescent, to about 4cm(1½in) long. Flowers about 2.5cm(1in) wide, composed of five rounded petals, the upper two red to red-purple, darker at their bases, the lower ones white or palest pink; the flowering stems branch, carrying two or more umbels each with two to six blooms, mainly in spring and summer.

PELECYPHORA

Cactaceae

Origin: Mexico. A genus of two species of small more or less globular cacti, with prominent rhombic tubercles flattened above and bearing elongated areoles with a comb-like arrangement of small spines. The short-tubed flowers are terminal and large for the size of the plant. These attractive little cacti are allied to *Ariocarpus*, but easier to grow. They are an endangered species in the wild and very deserving of a place in a collection of succulents in either the home or conservatory. Propagate by seed in spring. *Pelecyphora* derives from the Greek *pelekys*, a hatchet and *phoreo*, to bear, referring to the shape of the flattened tubercles.

   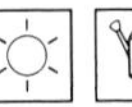 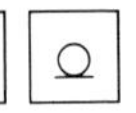

Species cultivated
P. asseliformis Central Mexico
Stems solitary or clustered, globose when young, eventually ovoid to 10cm(4in) high, greyish-green. Tubercles spirally arranged, hatchet-shaped, tipped by narrowly elliptic areoles. Spines blunt, eight to 60, to 5mm(⅕in) long. Flowers rose-purple, 3cm(1¼in) wide.
P. strobiliformis (*Encephalocarpus*/*Ariocarpus strobiliformis*) Northern Mexico
Stems globular, 4–6cm(1½–2½in) across, grey-green. Tubercles scale-like, triangular, incurving and overlapping, creating a cone-like structure. Flowers violet-pink to magenta, 3cm(1¼in) wide.

PELLAEA

Sinopteridaceae (Polypodiaceae)

Origin: Temperate and sub-tropical regions with the majority of species in the Americas. A genus of 80 species of evergreen, tufted or rhizomatous ferns. They have pinnate to tripinnate fronds with small linear to orbicular pinnae. The globular to oblongoid sori are usually close to the margins of the pinnae which are rolled over at the edge to cover them. They are tolerant of more dryness than many ferns. Propagate by spores, or by division when re-potting, both in spring. *Pellaea* derives from the Greek *pellaios*, dark, for the purple to black frond stalks of many species.

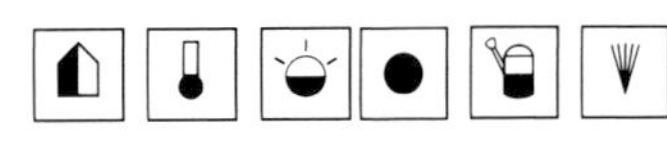

Species cultivated
P. adiantoides See *P. viridis*.
P. atropurpurea Purple-stemmed cliff brake
North America and Guatemala
Clump-forming with short rhizomes. Fronds to 30cm(1ft) long and 15cm(6in) wide, more or less erect, bipinnate below, pinnate at the ends, the pinnae 1–5cm(⅜–2in) long, lanceolate to linear, leathery, with shining deep purplish stalks.
P. cordata See *P. sagittata*.
P. hastata See *P. viridis*.
P. rotundifolia Button fern New Zealand, Australia and Norfolk Is.
Clump-forming with creeping rhizomes. Fronds to

ABOVE *Pellionia daveauana*
ABOVE RIGHT *Pellaea rotundifolia*

30cm(1ft) or more in length, arching to procumbent, the alternate pinnae 1–2cm(⅜–¾in) long, orbicular to broadly oblong, 20 to 60 to a frond; the dark stalks are covered in red-brown scales and bristly hairs.

P. sagittata (*P. cordata sagittata*) Texas to southern Mexico
Clump-forming. Fronds to 45cm(1½ft) in length, sometimes more, narrowly triangular in outline, bipinnate, firm-textured. Pinnules 1–2cm(⅜–¾in) long, sagittate to triangular-lanceolate, rich green.

P. viridis (*P. adiantoides, P. hastata*) Green brake fern South Africa (some authorities give West Indies)
Clump-forming. Fronds to 60cm(2ft) or more in length, ovate in outline, bipinnate (simply pinnate on young plants), poised on glossy purple-black stalks. Pinnules up to 2cm(¾in) long, sometimes more, ovate to broadly lanceolate, those near the midrib occasionally auricled.

P.v. macrophylla (*P. adiantoides*)
Fewer, broader pinnules.

PELLIONIA

Urticaceae

Origin: Tropical Asia and Polynesia. A genus of about 50 species of low-growing perennials with alternate leaves usually all lying on one plane, and small insignificant flowers. They are fine foliage plants for the conservatory and home and can be grown in pots, pans and hanging baskets. Propagate by division or by cuttings taken in late spring and summer. *Pellionia* was named for Adolphe Odet Pellion (1796–1868), a French Admiral who accompanied Louis Freycinet on a world voyage.

Species cultivated

P. daveauana (*P. repens*) Trailing water-melon begonia Burma, Vietnam, Malaysia
Stems rather succulent, to 60cm(2ft) long. Leaves 2.5–5in(1–2in) long, broadly elliptic, asymmetrical at the base and somewhat fleshy; their upper surfaces are bronze to olive-green with purple-suffused edges and a broad, paler green band along the centre of each leaf.

P.d. viridis
Dark green leaves with central silvery-grey area.

P. pulchra Vietnam
Stems somewhat succulent, to 60cm(2ft) long. Leaves to 5cm(2in) long, oblong-elliptic, asymmetrical at the base, purplish beneath, green above, with a network of dark brown veining.

P. repens Now the correct botanical name for *P. daveauana, q.v.*

PENTAPTERYGIUM See AGAPETES

PENTAS

Rubiaceae

Origin: Africa, Arabia and Malagasy. A genus of 35 to 50 species of shrubs or perennials having hairy, opposite, entire leaves and clusters of small, long-tubular flowers with five spreading lobes. The species described below makes an excellent pot plant. Cut back plants in late winter after flowering and either take the young shoots as cuttings in spring, or, if a large plant is required, pot-on the original plant. *Pentas* derives from the Greek *pentos*, five, the genus differing from most members of the Rubiaceae family in having its floral parts in fives.

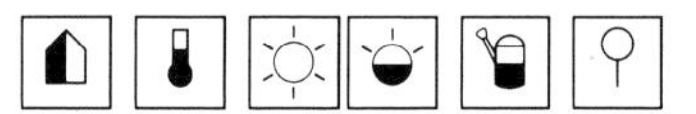

Species cultivated

P. lanceolata (*P. carnea*) (Egyptian) Star cluster Tropical East Africa to Southern Arabia
Sub-shrub, 90cm(3ft) or more tall though flowering at 30–60cm(1–2ft) in a pot. Leaves 4–9cm(1½–3½in) long, lanceolate to ovate, hairy, carried in opposite pairs. Flowers 2–4cm(¾–1½in) long, white, pink, red, lilac or magenta, in showy terminal corymbs in late summer and autumn.

Pentas lanceolata

PEPEROMIA

Piperaceae

Origin: Tropics and sub-tropics, world-wide. A genus of about 1,000 species of small annuals and perennials many of which are epiphytic in the wild.

ABOVE *Peperomia clusiifolia* 'Variegata'
LEFT *Peperomia fraseri*
FAR LEFT *Peperomia argyreia*

They have fleshy leaves which range from lanceolate to orbicular and are often decoratively marbled or veined, with spikes of small, usually insignificant flowers. They make good pot plants.

Propagate by division at planting time, or by seed or cuttings of leaves or stems in spring or summer in warmth. *Peperomia* derives from the Greek *peperi*, pepper and *homoios*, resembling, because of the likeness of some species to the related true peper, *Piper*.

Species cultivated

P. argyreia (*P. sandersii*) Water-melon begonia Tropical South America
Tufted, 15–20cm(6–8in) tall. Leaves 6–13cm (2½–5in) long, peltate and broadly ovate, dark green with silvery-grey areas between the veins.

P. arifolia Brazil to Argentina
Much like *P. argyreia*, but leaves lacking the silvery bands. Flower spikes simple, to 20cm(8in) in length (*P. argyreia* has shorter inflorescences with three to five branches).

P. berlandieri See under *P. hoffmannii*.

P. bicolor See under *P. velutina*.

P. caperata Emerald ripple Probably Brazil, but uncertain
Tufted plant, 10–15cm(4–6in) tall, densely covered with leaves; these are 2–5cm(¾–2in) long, broadly ovate with deeply impressed veins, glossy deep green. Flowers white, in tail-like spikes. *P.c.* 'Tricolor' has the leaves suffused and margined with pink and white.

P. clusiifolia Baby rubber plant West Indies, Venezuela
Plant to 20cm(8in) or more in height, the stems branching, red. Leaves obovate to oblanceolate, thickly fleshy, red-purple-margined and sometimes also suffused, up to 10cm(4in) long, sometimes more. *P.c.* 'Variegata' has leaves with cream borders and red edges.

P. crassifolia Uganda
Stems more or less decumbent and up to 30cm(1ft) long. Leaves orbicular to very broadly ovate, about 1.2cm(½in) long. Rarely seen in cultivation, the plant under this name with oval to obovate, 5–8cm(2–3in) long leaves is probably a form of *P. magnoliifolia*.

P. cubensis Cuba
Stems prostrate or semi-climbing, to 60cm(2ft) in length. Leaves ovate-cordate, slender-pointed to 8cm(3in) in length, long stalked. *P.c.* 'Variegata' has the leaves white margined.

P. dahlstedtii (*P. fosteri*) Southern Brazil
Stems prostrate or semi-climbing, stiff, angular, purplish, to 60cm(2ft) or so. Leaves in pairs, but sometimes in whorls of three or four, elliptic to obovate, 5cm(2in) or more long, dark green with three to five pale, sunken veins.

P. dolabriformis Peru
Stems stiff, erect, sparingly branched, to about 10cm(4in) in height. Leaves 4–6cm(1½–2½in) long, tightly folded and fused, broadly sickle-shaped, very fleshy, the folded edge like a transparent, greyish-green line. Flower spikes forming a long, narrow panicle.

P. fosteri See *P. dahlstedtii*.

P. fraseri (*P. resediflora*) Flowering mignonette Ecuador
Tufted plant. Leaves to 4.5cm(1¾in) long, broadly ovate to rounded cordate, the surface deep matt-green, shiny, bullate. Flowers white, in 5cm(2in) long fluffy spikes on 30cm(12in) or more long, branched stems. The only species that is grown

Peperomia glabella

chiefly for the attraction of its flowers.
P. galioides North-western South America
Stems erect, red, fairly succulent, branched, usually less than 30cm(1ft) in height, but sometimes much more. Leaves narrowly elliptic to oblanceolate, about 2–2.5cm(¾–1in) long, in whorls of three to six.
P. glabella Wax privet Tropical America
Stems reddish, erect to spreading, 15cm(6in) or more long. Leaves 3–5cm(1¼–2in) long, alternate, ovate to broadly elliptic, waxy in texture and bright green. *P.g.* 'Variegata' has broad creamy-white margins to the leaves.
P. griseoargentea (*P. hederifolia*) Ivy/Silver leaf peperomia Brazil
Tufted plant to 15cm(6in) tall. Leaves 4–6cm(1½–2½in) long closely covering the plant, broadly ovate-cordate, peltate, the upper surface bullate with a metallic silvery-green. *P.g.* 'Nigra' has black-green shading along the veins.
P. hederifolia See *P. griseoargentea.*
P. maculosa Radiator plant West Indies, Panama to northern South America
Plant to about 30cm(1ft) tall with several semi-erect stems. Leaves usually elliptic-ovate, but sometimes almost orbicular, to 15cm(6in) or more in length, very dark green with a glaucous sheen, an ivory to silvery mid-vein and a long, red-purple-spotted stalk. Flower spikes simple, to 30cm(1ft) tall. The largest-leaved species and possibly the most handsome.
P. magnoliifolia Desert privet West Indies, Panama and northern South America
Stems erect, freely branched, bending downwards when old, to 25cm(10in) or more long. Leaves alternate, to 10cm(4in) long, obovate to rounded, firm and fleshy in texture, dark green. *P.m.* 'Variegata' has cream young leaves becoming green as they age.
P. marmorata Silver heart Southern Brazil
Similar to *P. argyreia* but neater, the leaves with overlapping lobes and the stalk at the end, not peltate, the surface somewhat bullate.
P. metallica Peru
Bushy plant to about 10cm(4in), with erect, branching stems. Leaves crowded, lanceolate to narrowly ovate, about 2.5cm(1in) long, dark metallic green above with a grey midrib stripe, pale with red veins beneath.
P. nivalis Peru
Stems branching, semi-erect forming tufts or clumps to about 10cm(4in) or more. Leaves 1.2cm(½in) or a little more long, tightly folded and fused, boat-shaped, thick and fleshy, the folded edge transparent and forming a linear window. Allied to *P. dolabriformis*, but distinct and desirable.
P. nummulariifolia See *P. rotundifolia.*
P. pereskiifolia (*P. pereskiaefolia*) Venezuela, Colombia, Brazil
Stems reclining or semi-scandent, mainly branching from the base, to 45cm(1½ft) or more. Leaf internodes very long (up to 13cm[5in]) with leaves in whorls of three to six. Each leaf 4–8cm(1½–3¼in) long, narrowly obovate to elliptic, firm-textured, slightly up-folded with a back-curving point, mid-green, with reddish stalks. A distinctive species which makes an interesting hanging basket plant.
P. polybotrya Coin leaf Columbia and Peru
Stems erect or eventually reclining, to 30cm(1ft) in length. Leaves broadly ovate to almost orbicular, peltate, abruptly slender-pointed, glossy green, 5–10cm(2–4in) long, held almost vertically on longish red stalks.
P. prostrata See under *P. rotundifolia.*
P. pulchella See *P. verticillata.*
P. resediflora See *P. fraseri.*
P. rotundifolia (*P. nummulariifolia*) Tropical America (widespread)
Stems very slender, creeping, branching and forming mats to 30cm(1ft) across. Leaves usually orbicular but sometimes broadly elliptic to ovate, glabrous, about 1cm(⅜in) or a little more, wide.
P.r. pilosior (*P. prostrata*) Creeping peperomia
Leaves very broadly ovate, densely hairy when young, ciliate when mature, dark green to reddish-brown with a pale green vein pattern.
P. sandersii See *P. argyreia.*
P. scandens (of gardens) Cupid peperomia Possibly Peru
Stems trailing to 60cm(2ft) long. Leaves to 7cm(2¾in) long, broadly ovate, the base often cordate ending in a slender point, palmately 7-veined. *P.s.* 'Variegata' has irregularly creamy-white margined leaves.
P. serpens (a synonym of the true *P. scandens*, but not of the plant grown under this name [*q.v.*]) West Indies, Panama to Brazil
Stems trailing to 40cm(16in) or more in length. Leaves about 5cm(2in) long, orbicular to broadly rounded triangular, palmately 3- to 5-veined.
P. velutina Ecuador
Stems more or less erect, branched and somewhat zigzag, softly red-hairy. Leaves ovate to elliptic, usually broadly so, up to 8cm(3in) long, glabrous or velvety-pubescent, deep bronzy-green with five to

seven pale green parallel vein stripes above, red beneath. Mature plants may produce stolons. The plant listed as *P. bicolor* is very similar and may be no more than a variant of *P. velutina*; its leaves are a little smaller and more strongly coloured, with silvery-green stripes.
P. verticillata (*P. pulchella*) Whorled peperomia
Central America, West Indies, Jamaica
Stems erect, to 30cm(1ft) or more in height, robust below, slender above, red-hairy. Leaves in whorls of three to five, obovate to ovate or almost orbicular, 1.2cm(½in) or more long, rich green with pale veins when young, becoming fleshy and somewhat boat-shaped with age.

PERESKIA

Cactaceae

Origin: Mexico, West Indies, Central and South America. A genus of about 20 species of trees, shrubs and climbers often considered to be like the ancestral stock from which the true cacti of today evolved. Unlike most true cacti they have recognizable leaves, but bear areoles within the leaf axils. Most have showy, many-petalled flowers and fleshy fruits which are edible. The climbers need some support. Keep almost dry from the time the leaves yellow in autumn until spring. Propagate by cuttings in summer, or from seed in spring in heat. *Pereskia* was named for Nicholas Claude Fabre de Peiresc (1580–1637), a French naturalist.

Species cultivated
P. aculeata Barbados gooseberry Mexico to Argentina
Shrubby at first, then sending up long, scrambling shoots to 10m(33ft) in the open; they have hooked spines on the climbing stems arising from areoles. Leaves 5–10cm(2–4in) long, lanceolate to elliptic, with prominent midribs. Flowers 2.5–4cm(1–1½in) wide, white, yellow or pale pink with a fragrance not always thought pleasant; they are carried in panicles in autumn and are followed by pale yellow fruits, spiny at first, but smooth when ripe.

PERILLA

Labiatae

Origin: India to Japan. A genus of five or six species of annuals with square stems and undivided leaves carried in opposite pairs. The species cultivated is grown for its attractive ornamental foliage. Propagate by seed in early spring, or take cuttings from an over-wintered plant. *Perilla* may be derived from a diminutive form of the Greek *pera*, a bag, alluding to the fruiting calyces, but is also sometimes said to be from an Indian vernacular name.

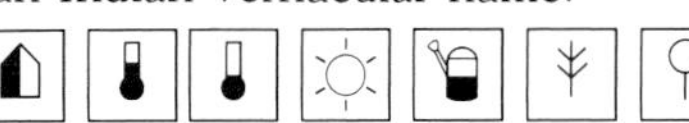

Species cultivated
P. frutescens (*P. ocimoides*) Himalaya to East Asia
Bushy, erect annual to 60–90cm(2–3ft) tall. Leaves to 11cm(4½in), broadly ovate, with marked veining. Flowers small, tubular, with five lobes, white and insignificant. *P.f.* 'Atropurpurea' has dark purple leaves; 'Crispa' (*P.f. nankinensis*, *P. laciniata*) has deeply cut bronze to purple foliage.

PERISTERIA

Orchidaceae

Origin: Tropical America. A genus of nine species of epiphytic orchids allied to *Acineta*. One statuesque species is readily available and well worth growing in the conservatory. Propagate by division after flowering or just as new growth starts. Plants need an almost dry rest period once the new pseudobulbs are mature. *Peristeria* derives from the Greek *peristera*, a dove, the central column of the flowers of *P. elata* being likened to that bird.

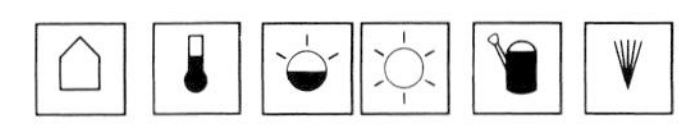

Species cultivated
P. elata Dove orchid, Holy Ghost plant
Costa Rica to Columbia and Venezuela
Pseudobulbs clustered, ovoid, to 13cm(5in) high. Leaves lanceolate, plicate, 60–90cm(2–3ft) long, arching. Flowers waxy-textured, white, to 6cm(2½in) wide, bowl-shaped, the labellum with small rose to lilac spots, fragant, in racemes of ten to 40 up to 1.2m(4ft) in height in late summer.

PERISTROPHE

Acanthaceae

Origin: Tropical Africa and Asia. A genus of about 15 species of perennials and sub-shrubs with opposite, entire leaves and tubular flowers carried in clusters. Propagate by cuttings in summer. *Peristrophe* derives from the Greek *peri*, around and *strophe*, to twist, from the twisted corolla lobes.

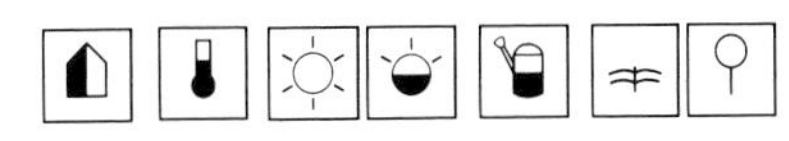

Species cultivated
P. angustifolia See *P. hyssopifolia.*
P. hyssopifolia (*P. angustifolia*) Java
Widely spreading, with stems 60–90cm(2–3ft) long. Leaves 5–8cm(2–3in) long, narrowly elliptic to lanceolate, dark green. Flowers small, red-purple, within two ovate bracts, one larger than the other, opening in summer. *P.h.* 'Aureo-Variegata' is an attractive foliage plant, with the leaves having a central yellow zone and veins, producing a feathery pattern.
P. speciosa India
Evergreen shrub to 60–90cm(2–3ft) tall. Leaves 10–13cm(4–5in) long, ovate to elliptic, slender-

BELOW *Pereskia aculeata*
BOTTOM *Peristrophe hyssopifolia* 'Aureo-Variegata'

pointed. Flowers about 4cm(1½in) long, showy, tubular, the lobes fused into two equal-sized lips, violet-purple, borne in clusters from the upper leaf axils in winter.

PERSEA

Lauraceae

Origin: Tropics and sub-tropics, chiefly South America and S.E. Asia. A genus of about 150 species of evergreen trees with alternate undivided leaves and small greenish, 6-tepalled flowers followed generally by fleshy berry-like fruits, the avocado being a familiar example. The species described is an interesting pot plant, grown chiefly for the fun of sowing the seed of the greengrocers' avocado and watching it develop. The resulting pot plant will not fruit until large. Propagate by seed, carefully removing the soft seed coat to assist germination. *Persea* was the Greek vernacular for an Egyptian tree, (*Cordia myxa*).

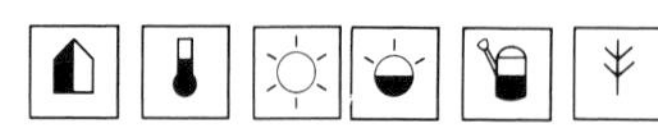

Species cultivated

P. americana (*P. gratissima*) Avocado
Central America (widely cultivated)
To 4m(13ft) in a tub, but to 20m(65ft) in the open. Leaves 5–20cm(2–8in) long, elliptic to oblong, dull green with lighter veins, glaucous beneath. Flowers in clusters, green, followed by large egg- or pear-shaped dark green or purple-tinted fruits.

PETREA

Verbenaceae

Origin: Tropical America and West Indies. A genus of about 30 species of shrubs and climbers only one of which is widely grown. This is a superbly showy plant for the larger conservatory. Propagate by cuttings in summer. *Petrea* honours Lord Robert James Petre (1713–1743), a patron of botany and horticulture who did much to introduce new plants to British gardens.

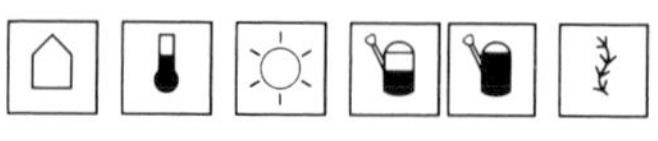

Species cultivated

P. volubilis Queen's/Purple wreath Mexico to Central America and West Indies
Stems vigorous, twining to at least 6m(20ft), but responds to regular light pruning after flowering. Leaves 5–20cm(2–8in) long, elliptic, rough-textured. Flowers 2.5–3.5cm(1–1⅜in) wide, composed of five deep violet petals and five longer, narrower, lilac-blue sepals in star formation; flowers carried in simple racemes up to 30cm(1ft) or more in length, but several arise together creating panicle-like inflorescences which make a spectacular sight from late winter to late summer depending on warmth.

Petunia × hybrida

PETUNIA

Solanaceae

Origin: South America. A genus of about 40 species of tufted annuals or perennials, often stickily glandular-hairy. They have entire leaves and solitary funnel-shaped flowers, in some species opening almost flat, and are most suitable for the conservatory, but can be brought into the house when blooming. Propagate by seed in spring in warmth, the double-flowered plants by summer cuttings. *Petunia* is a Latinized form of *petun*, the Brazilian vernacular name for tobacco.

Species cultivated

P. × hybrida
This is a group of cultivars raised by hybridization between *P. axillaris*, *P. inflata* and *P. violacea*. Although perennials, the best plants are obtained by raising them annually and most of the cultivars come true to type when raised from seed. Plants vary considerably in size, flower form and colour and are grouped as follows:
Multiflora Bushy plants, 15–30cm(6–12in) tall, flowers single or double about 5cm(2in) across.
Grandiflora Bushy plants, 15–30cm(6–12in) tall, with fewer, larger flowers to 10cm(4in) across, single or double.
Nana Compacta Low-growing, not exceeding 15cm(6in), flowers large, often with waved petals; recommended for pot culture.
Pendula Low-growing with long, trailing stems and very good for hanging baskets.

PHAEDRANASSA

Amaryllidaceae

Origin: Central and South America. A genus of six species of bulbous plants allied to *Hippeastrum*.

Only the one species described here is in general cultivation and not always easy to obtain, but it makes a handsome pot plant. Propagate by offsets when dormant. *Phaedranassa* derives from the Greek *phaidros*, gay and *anassa*, queen, referring to the gaily-coloured blooms.

    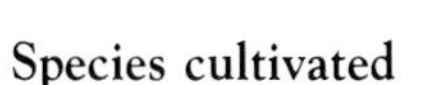

Species cultivated

P. carmiolii Usually stated to be a native of Costa Rica but almost certainly originating from Peru Leaves deciduous, in a basal tuft of one to three, oblanceolate, grey-green, 35–60cm(14–24in) in length. Flowers tubular, about 5cm(2in) long, pendent, crimson, the tepal lobes green, in umbels of six to ten on erect naked stems 40–60cm(16–24in) in height in late spring to early summer. Should be kept dry in winter when leaves will fade. Best in a loam-based compost.

PHAEOMERIA See NICOLAIA

PHAIUS

Orchidaceae

Origin: S.E. Asia to the Himalaya, Malagasy. A genus of 50 species of orchids most of which are ground dwelling. They have clustered pseudobulbs which can be short and thick or stem-like and leafy. Flower stems usually arise from the sides of the pseudobulbs and bear erect racemes of flowers with five similar tepals and a large, often trumpet-shaped labellum. They are best in the conservatory, but can be brought into the house for short spells if extra humidity can be given. Propagate by division when repotting or by severing back bulbs in spring. *Phaius* derives from the Greek *phaios*, dark, referring to the flower colour of the first species described.

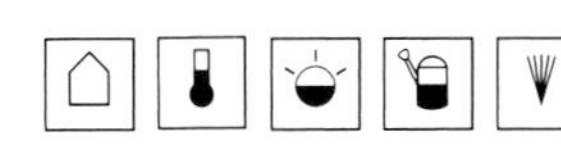

Species cultivated

P. flavus (*P. maculata*) S.E. Asia
Pseudobulbs to 10cm(4in) tall, narrowly conical. Leaves to 60cm(2ft) long, lanceolate, folded around each other to give a stem-like base, green sometimes spotted with yellow. Flowers 4–8cm(2–3in) across, pale yellow with a brownish lip, carried on stalks 60–90cm(2–3ft) tall.

P. tankervillae (*P. bicolor*, *P. blumei*, *P. grandiflorus*, *P. gravesii*, *P. wallichii*) Nun's hood
Himalaya, S.E. Asia to Northern Australia
A very variable orchid especially in its flower size and colour, hence the number of names that have been applied to different forms of it. Pseudobulbs 10cm(4in) tall, broadly ovoid. Leaves 60–90cm(2–3ft) long, narrowly elliptic to oblong, pointed. Flowers to 10–13cm (4–5in) across, tepals reddish brown with a yellowish margin inside and whitish

on the back, labellum trumpet-shaped, yellowish-brown inside with a red-purple throat patch; they are very fragrant and carried in racemes on stiffly erect stems 1m(3ft) or more in height.

ABOVE *Phaius tankervillae*
TOP *Phaedranassa carmiolii*

PHALAENOPSIS

Orchidaceae
Moth orchids

Origin: Tropical Asia and S.E. Asian islands to New Guinea and Northern Australia. A genus of 35 to 55 species of chiefly epiphytic orchids which do not produce pseudobulbs. Most are evergreen having short, thick stems and thick leathery leaves. Their flowers are showy with generally similar sepals and petals and a 3-lobed lip. They are best in a conservatory, needing warmth and humidity, but can be brought into the house for flowering, thriving in orchid baskets or on slabs of bark as well as in pots, and they need a well-drained compost. Propagate by division of established plants or by removing the plantlets which sometimes develop on the flowering stems.

Phalaenopsis derives from the Greek *phalaina*, a moth and *opsis*, like.

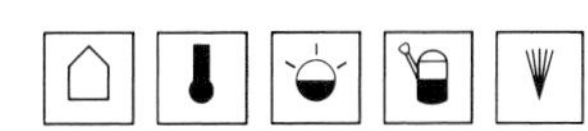 

Species cultivated

P. leuddemanniana Philippines
Leaves to 30cm(1ft) long, oblong-elliptic. Flowers 5cm(2in) wide, white, lined with brown and purple and having the labellum flushed with rose-purple; they are very fragrant and are carried on arching to pendent racemes to 30cm(1ft) long in spring. *P.l.* 'Ochracea' has yellow sepals with darker yellow-brown banding.

P. sanderiana Philippines
Leaves 35cm(14in) long, elliptic. Flowers 8cm(3in) wide, pink, the labellum spotted and streaked with red and purple, carried on arching, sometimes branched racemes in spring.

Phalaenopsis leuddemanniana

RIGHT *Phalaenopsis schilleriana*
FAR RIGHT *Phalaenopsis stuartiana*

P. schilleriana Philippines
Leaves 30cm(1ft) or more long, ovate-elliptic, green mottled with silver, often purple beneath. Flowers 8cm(3in) wide, rosey-lilac to white, the labellum spotted with red, carried on arching, branched racemes to 90cm(3ft) long in spring.
P. stuartiana Philippines
Leaves to 30cm(1ft) long, elliptic-oblong, green mottled above with silver-grey, red-purple beneath. Flowers to 5cm(2in) across, greenish-white, the outer spotted at the base with reddish-brown, the labellum flushed with yellow and spotted with red-purple; they are carried on erect to arching racemes, sometimes branched, to 90cm(3ft) long from autumn to spring.

PHANEROPHLEBIA See CYRTOMIUM
PHARBITIS See IPOMOEA

PHILODENDRON

Araceae

Origin: Tropical Central and South America. A genus of 200 or more species of evergreen shrubs and climbers, often epiphytic. Many have a distinct juvenile phase with leaves quite dissimilar to those of mature plants; plants grown in pots are often at this earlier stage of growth. They have arrow- to heart-shaped, often leathery leaves of a dark glossy green which are alternate on climbing species, crowded or in rosettes on free-standing ones. Their flowers are small and are borne typically of the arum family, on small spadices partly surrounded by white, green or red spathes. The plants are suitable for the home or conservatory, the climbing ones needing support, ideally of a moss stick. Propagate by cuttings of young tips, stem sections or leaf buds, all in summer. *Philodendron* derives from the Greek *philo*, love and *dendron*, a tree.

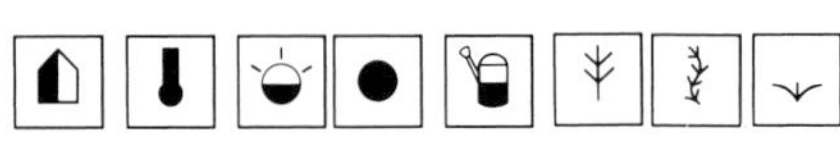

Philodendron erubescens

Species cultivated
P. andreanum See *P. melanochrysum*.
P. auriculatum Costa Rica
Much like a larger version of *P. martianum* (*q.v.*), with leaf blades up to 90cm(3ft) long, auricled or cordate at the base, the petiole less swollen.
P. auritum See *Syngonium auritum*.
P. bipennifolium (P. panduriforme) Fiddle leaf S.E. Brazil
Stems climbing. Leaves 30–80cm(12–32in) long, ovate-sagittate, cut into five to seven radiating lobes, the side ones appearing as if they have been cut away leaving a fiddle-shaped leaf.
P. bipinnatifidum Tree philodendron S. Brazil
Stem erect, self-supporting. Leaves to 90cm(3ft) long, ovate-sagittate, pinnatipartite, the narrow lobes themselves lobed.
P. 'Burgundy' A hybrid between *P. erubescens*, *P. hastatum*, *P. imbe* and *P. wendlandii*, with slowly climbing stems. Leaves to 30cm(1ft) long, intermediate in shape between those of the last two parents named above, flushed with red above, the undersides and the winged leaf-stalks a rich wine-red.
P. cannifolium See *P. martianum*.
P. cordatum Heart-leaf S. Brazil
Stems climbing. Leaves 30–45cm(1–1½ft) long, ovate-cordate, the basal lobes touching or overlapping, much confused in cultivation with *P. scandens oxycardium*.
P. crassinervium Brazil
Stems climbing, eventually several metres(yards) in length if planted out in the warm conservatory. Leaf blades a deep lustrous green, very narrowly elliptic to lanceolate, up to 60cm(2ft) in length; petioles shorter and inflated.
P. crassum See *P. martianum*.
P. domesticum (*P. hastatum* in cultivation) Spade-leaf, Elephant's ear Origin unknown
Stem climbing. Leaves to 60cm(2ft) long, narrowly triangular to sagittate.
P. eichleri Tree philodendron Brazil
Stems erect, to 2m(6½ft) or more in height. Leaves long-stalked, the blades 1m(3ft) long, narrowly triangular-hastate in outline, broadly lobed and waved, deep glossy green. One of the finest so-called tree philodendrons.
P. erubescens Blushing philodendron
Trinidad, Colombia and Venezuela
Stem climbing, red-purple when young. Leaves to 25cm(10in) long, ovate to triangular, dark green with a strong coppery flush beneath.

P. fenzlii Mexico
Like *P. tripartitum*, but with broader, narrowly obovate leaflets, the centre one appreciably longer than the laterals.

P. × 'Florida' (*P. pedatum* × *squamiferum*)
Vigorous, slender-stemmed climbing hybrid more or less halfway between the parents. Leaves basically like those of *P. squamiferum*, but the five lobes more deeply cut as in *P. pedatum*.

P. glaziovii Brazil
Much like *P. crassinervium*, but with slightly shorter though broader oblong leaves.

P. gloriosum Satin leaf Colombia
Stems prostrate to climbing, but slow-growing. Leaves long-stalked, heart-shaped, up to 40cm(16in) long, deep green with a satiny sheen, the main veins picked out in ivory-white, the margins reddish.

P. hastatum Brazil
The true species has leaves which are oblong to oblong-hastate and shallowly cordate. *P. hastatum* of cultivation is *P. domesticum*, *q.v.*

P. ilsemannii Brazil
Possibly a form of *P. sagittifolium*. Leaves narrowly triangular-oblong, sagittate, creamy-white marbled with grey and dark green.

P. imbe Brazil
Stems climbing. Leaves to 33cm(13in) long, ovate-oblong, sagittate, thin but firm in texture, green above, flushed with red beneath.

P. lacerum Jamaica, Cuba, Hispaniola
Stems climbing, to several metres (yards) in height. Leaves on young (juvenile) specimens broadly ovate-cordate, entire, to 60cm(2ft) or more long; on older plants deeply pinnately lobed.

P. laciniatum See *P. pedatum*.

P. mamei Ecuador
Stems prostrate to climbing, slow-growing. Leaves ovate, cordate, to 40cm(16in) long, somewhat quilted, waxy-textured, bright green to greyish, marbled with grey-silver. Allied to *P. gloriosum*, but with narrower leaves lacking the satiny lustre.

P.* × *mandaianum (*P. domesticum* × *erubescens*) Red-leaf philodendron
Very much like *P. erubescens*, but with stems and leaves a more intense, metallic purple-red. One of the first philodendron hybrids, U.S.A. 1936.

P. martianum (*P. cannifolium*, *P. crassum*)
S.E. Brazil
Stem short. Leaves crowded, appearing rosette-like, at least for the first several years; leaf blades lanceolate, to 45cm(1½ft) or more long, petioles swollen, cigar-shaped. The plant formerly separated as *P. cannifolium* has somewhat longer, less swollen petioles and is like a smaller version of *P. auriculatum*.

P. melanochrysum (*P. andreanum*) Black gold/Velour philodendron Colombia
Stems climbing. Leaves 45–75cm(1½–2½in) long, ovate-cordate, sagittate, dark olive-green with an iridescent coppery gloss and paler veins.

P. micans See under *P. scandens*.

P. oxycardium See under *P. scandens*.

P. panduriforme See *P. bipennifolium*.

P. pedatum (*P. laciniatum*) Guyana to Venzuela and Brazil
Stems climbing. Leaves 30–80cm(12–32in) long, ovate to oblong-sagittate, cut deeply into five to seven lobes which can be shallowly lobed.

P. pertusum See *Monstera deliciosa*.

P. pinnatifidum Venezuela, Trinidad
Stem erect, thick, covered with fibrous remains of scale leaves and petiole bases. Leaf blades triangular-ovate, cordate, to 60cm(3ft) in length, deeply pinnately lobed, glossy green; petioles as long as or longer than blade.

P. quercifolium
A juvenile phase, almost certainly of *P. pedatum*, with leaf lobes less deeply cut.

P. sagittifolium (*P. sagittatum*) Mexico
Stems climbing, but fairly slow-growing with closely set leaves and plenty of aerial roots. Leaves narrowly triangular-oblong, sagittate, firm-textured, lustrous bright green, up to 60cm(2ft) long. In its homeland the aerial roots are gathered and woven into baskets.

P. scandens Heart leaf, Sweetheart plant
Tropical America
Stems climbing. Leaves 10–15cm(4–6in) when juvenile, to 30cm(12in) when mature, ovate-cordate with a slender point, dark green with a silky sheen, sometimes flushed purple-red beneath.

P. s. micans (*P. micans*)
Leaves of juvenile phase purple-red beneath.

P. s. oxycardium
The commonest philodendron, grown as a house

ABOVE *Philodendron martianum*
LEFT *Philodendron scandens*

plant, with leaves glossy green above and below.
P. selloum Lacy tree philodendron Brazil
Probably only a form of *P. bipinnatifidum*, differing from that species in having the leaf lobes only waved, toothed or shallowly lobed.
P. sellowianum
Botanically this is a synonym of *P. imbe*, but most plants under this name in the trade are referable to *P. selloum* or *P. bipinnatifidum*.
P. squamiferum French Guiana, Surinam and adjacent Brazil
Stems climbing, eventually to several metres (yards) in height. Leaves to 30cm(1ft) long, basically ovate-cordate in outline, but cut into five dissimilar lobes; terminal lobe largest, ovate; the two middle ones next largest, broadly sickle-shaped, pointing forwards; two basal lobes smallest, ovate, pointing backwards. Leaf stalks bear red to green bristles. A distinctive species allied to *P. pedatum* and *P. panduriforme* and having the rich green lustrous foliage of those species.
P. trisectum See under *P. tripartitum*.
P. verrucosum Costa Rica to Ecuador
Stems climbing, eventually to several metres (yards) high. Leaf blades up to 60cm(2ft) in length, ovate-cordate, satiny dark to bronze-green with paler zones along the main veins and a bright green margin, purple-red beneath; the leaf stalks, which are about as long as the blades, are reddish and thickly covered with green bristles.
P. warscewiczii Mexico to Panama
Stems semi-climbing, eventually sinuous but thick and trunk-like. Leaves ovate-cordate in outline, to 60cm(2ft) or more long, bipinnatifid, all the lobes of irregular size but finely pointed, soft-textured, bright green; stalks as long as the blades. In the wild this species is deciduous or semi-deciduous, but in cultivation it retains foliage for at least a year.
***P.* × 'Wend-imbe'**
A hybrid between *P. wendlandii* and *P. imbe*, favouring *P. wendlandii*, but forming an elongating rosette. Leaf blades oblong to ovate, shortly cordate, to 50cm(20in) long, firm-textured, glossy rich green; stalks shorter, flattened, entirely lacking the spongy thickening characteristic of the *P. wendlandii* parent.
P. wendlandii Nicaragua to Panama
Short-stemmed, erect, forming a rosette. Leaves to 60cm(2ft) or more long, narrowly oblanceolate, on short, thick spongy stalks.

Phoenix roebelenii

PHLEBODIUM See POLYPODIUM

PHOENIX

Palmae
Date palms

Origin: Tropical and sub-tropical Africa and Asia. A genus of 17 species of palms having terminal rosettes of narrowly pinnate leaves with the lowest leaflets often spine-like. The flowers are small and dioecious, borne in dense panicles arising among the foliage. They are followed by fleshy fruits each containing a single, grooved seed. When young, date palms make useful and decorative pot plants. Propagate by seed sown in spring at not less than 21°C (70°F), or by suckers if available treated as cuttings and grown on in warmth. *Phoenix* is the Greek vernacular name for the date palm.

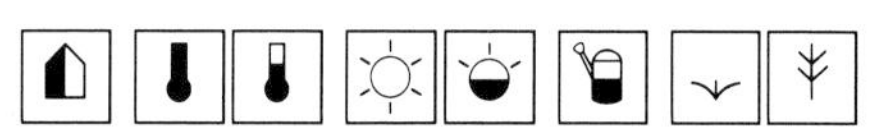

Species cultivated
P. canariensis Canary Islands date palm
Canary Isles; widely grown in warm temperate countries
Tree to 16m(50ft) or more in the open, not suckering. Leaves flat, finely divided, feather-like, to 5m(16ft) long, but less than half this in a container, in time forming a dense crown.
P. dactylifera Date palm Probably originally from North Africa or S.W. Asia, but cultivated for thousands of years
Tree to 30m(95ft) or so high, suckering at the base. Leaves to 5m(16ft) long, less than half this in a container, the stiff, linear leaflets grey-green. Not so graceful as most other species of *Phoenix*, but easily raised from date stones.
P. roebelenii Pygmy date palm Assam to Vietnam
Eventually 2–4m(6–13ft), sometimes producing suckers. Leaves arching, 1–1.2m(3–4ft) long, flat and feathery with dark, glossy-green leaflets. Flowers in panicles to 45cm($1\frac{1}{2}$ft) long, followed by ovoid black fruits 1.2cm($\frac{1}{2}$in) long. Best at tropical temperatures with extra humidity.

PHORMIUM

Agavaceae (Liliaceae)
New Zealand flax

Origin: New Zealand and adjacent islands. A genus of two species of clump-forming, evergreen perennials with leathery, sword-shaped leaves and tubular dark red or bronze flowers in stiff panicles. The fruits are cylindrical capsules containing flattened, almost papery, shining black seeds. Plants are suitable for the conservatory, especially the smaller-growing cultivars. Propagate by division or by seed in spring. *Phormium* derives from the Greek *phormion*, a mat; the leaf fibres of the plant are used in New Zealand to make mats, cloth, baskets, etc.

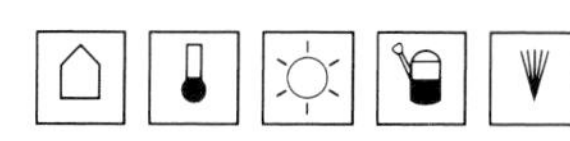

Species cultivated

P. cookianum (*P. colensoi*) Mountain flax
Leaves usually 1–1.5m(3–4½ft) long, erect at first then arching over and drooping when mature, glossy green. Flowers 2.5–4cm(1–1½in) long, greenish to orange on stems 1.5–2m(4½–6½ft) high; capsules to 10cm(4in) long, twisted and markedly pendent. Several cultivars are available, some of hybrid origin; those of small growth include: 'Bronze Baby', with bronze foliage, rather glaucous beneath, reaching only 60cm(2ft) in a pot; 'Cream Delight', compact, to 90cm(3ft), leaves green with a central cream band.

P. tenax New Zealand flax
Leaves to 1.5–2m(4½–6ft), often more in the open ground, erect, slightly arching, glaucous beneath, dull green or brownish-green, often with an orange-red border. Flowers 2.5–5cm(1–2in) long, dull red, on erect stems to 2.5m(8ft) or much more – to 5m(16ft) in the open. Fruits are erect capsules 5–10cm(2–4in) long, not twisted. Cultivars include: 'Maori Surprise', 70–90cm(28–36in) tall, leaves pinkish striped with bronze; 'Thumbelina', only 30cm(1ft) tall with bronze-purple leaves; 'Yellow Wave', 60–75cm(2–2½ft) tall, leaves green with a broad yellow band.

PHRAGMIPEDIUM
(PHRAGMIPEDILUM)

Orchidaceae

Origin: South and Central America. A genus of about twelve species of epiphytic and terrestrial orchids, very closely related to *Paphiopedilum*. They differ from this genus in having more leafy fans of foliage and, in the best known species, petals with tail-like tips. Those described here are worthy additions to a group or collection of orchids in the conservatory. Propagate by division in spring. *Phragmipedium* derives from the Greek *phragma*, a fence, reputedly referring to the ovary structure.

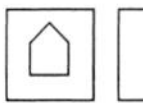  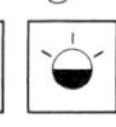  

ABOVE & RIGHT
Phormium tenax hybrid cultivars

Species cultivated

P. caudatum Mexico to Peru
Leaf fans clustered and forming clumps, each leaf leathery-textured, arching, to 60cm(2ft) long with a bilobed tip. Flowering stems erect, 30–60cm (1–2ft) in height, bearing one to six blooms in spring to summer; individual flowers to 25cm(10in) wide, cream and green with a yellow and reddish-brown slipper-shaped labellum; remarkable for the slender tails to the petals which can exceed 60cm(2ft) in length.

P. schlimii Colombia
Leaf fans clustered, the leaves to 30cm(1ft) long, strap-shaped, arching, usually bright green above, purplish beneath. Flowering stems 45cm(1½ft) or more tall bearing five to eight blooms; each flower about 5cm(2in) wide, velvety-downy, green-tinted white, flushed rose-pink with a white and rose-carmine labellum; petals pointed, but lacking tails.

P. × sedenii (*P. longifolium × schlimii*)
Tends to favour the latter species, but is larger in all its parts, with spirally twisted, pendent, tailed petals to 8cm(3in) or more in length.

PHYLLANTHUS See BREYNIA
PHYLLITIS See ASPLENIUM

Phragmipedium × sedenii

PHYLLOSTACHYS

Gramineae

Origin: Temperate eastern Asia and the Himalaya. A genus of 30 to 40 species of rhizomatous, clump-forming bamboos. A characteristic is that the internodes of the rather zig-zag stems are alternately flattened and grooved. Two lateral branches arise at each node with lanceolate leaves. The 2- to 4-flowered spikelets of greenish, grass-like flowers are not produced regularly. Propagate by division in spring or by cuttings of young rhizomes in a cool place in late winter. *Phyllostachys* derives from the Greek *phyllon*, a leaf and *stachys*, a spike.

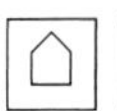  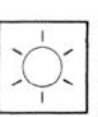 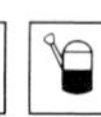 

ABOVE *Pilea peperomioides*
TOP LEFT *Phyllostachys aurea*
TOP RIGHT *Pilea involucrata* 'Norfolk'

Species cultivated

P. aurea Golden bamboo, Fishpole bamboo
China; long cultivated and naturalized in Japan
To 2m(6½ft) in a large container, to 6m(20ft) or more in the open ground. Canes have a swollen line beneath each node making them appear as if jointed, bright green at first, becoming golden-yellow as they mature. Leaves 5–11cm(2–4½in) long, linear, mid-green.

PHYMATODES See MICROSORIUM & POLYPODIUM

PILEA

Urticaceae

Origin: Tropics and sub-tropics world-wide. A genus of between 200 and 400 species of annuals and perennials with opposite pairs of obovate to ovate leaves sometimes with decorative patterning and tiny, rather inconspicuous panicles or cymes of petalless flowers. They make excellent container plants, thriving best with extra humidity in summer. Propagate by cuttings from late spring to late summer with bottom heat. *Pilea* derives from the Latin *pileus*, a cap.

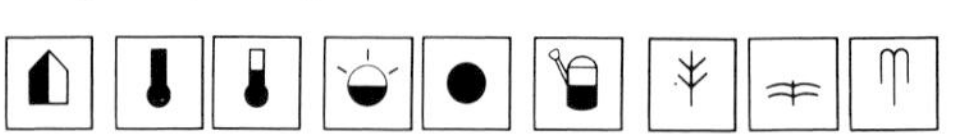

Species cultivated

P. cadierei Aluminium plant Vietnam
Branching perennial to 30cm(1ft) high. Leaves to 7cm(2¾in) long and 5cm(2in) broad, ovate, with an abrupt, acuminate tip, dark green, the areas between the veins raised with a silvery-white metallic patina. *P.c.* 'Minima' is similar, but smaller in all its parts.

P. crassifolia Jamaica
Stems erect, up to 60cm(2ft) or more in height. Leaves usually narrowly ovate, 5–10cm(2–4in) long, slightly corrugated, lustrous green, sometimes bronze-flushed.

P. depressa Puerto Rico
Stems prostrate, rooting, forming mats to 30cm(1ft) wide. Leaves succulent, broadly obovate to rounded, bright green, about 6mm(¼in) wide. A useful plant for small hanging baskets.

P. involucrata Panamiga, Friendship plant
Panama to Northern South America
Stems decumbent to ascending, 10–15cm(4–6in) tall by up to twice as wide. Leaves 4–5cm(1½–2in) long, broadly ovate to obovate, rounded-tipped, hairy crenate, corrugated between the veins, bronze-green above, reddish below. *P.i.* 'Norfolk' has silvery-grey zones between the main leaf veins. For *P.i.* 'Silver Tree' see *P.* 'Silver Tree'.

P. microphylla (*P. muscosa*) Artillery plant
Tropical America northwards to Florida
Bushy, freely branched to 30cm(1ft) in height with somewhat fleshy stems. Leaves tiny, 3–9mm(⅛–⅓in) long, bright green, densely covering the stems to give a ferny effect. The flowers shed their pollen explosively like mini-cannon fire, hence the vernacular name.

P. muscosa See *P. microphylla.*

P. nummulariifolia Creeping Charlie Panama and West Indies, Northern South America
Much like *P. depressa*, but the leaves always orbicular and usually a little larger.

P. peperomioides China (Yunnan [Hunnan])
Clump- to mound-forming, 20–30cm(8–12in) or more high and wide, formed of numerous suckers direct from the roots. Each stem unbranched, becoming woody with age, with a tuft of foliage at the top. Leaves broadly ovate to almost rounded, peltate, 6–11cm(2½–4½in) long, fleshy, bright sub-lustrous green. Flowers greenish-yellowish, tiny, in fluffy panicles, but insignificant and often sparingly produced (it seems to need a cool winter period to flower well). This description is of the material generally cultivated.

P. repens Black leaf panamiga West Indies
Stems decumbent, branched, forming low mounds to about 10cm(4in) in height. Leaves to 3cm(1¼in) long, ovate to obovate, broadly blunt-pointed, with large, almost lobe-like crenations, pilose, coppery brown-green and lustrous above, purplish beneath.

P. 'Silver Tree' Caribbean area
Stems erect, to 15cm(6in) or so, branching. Leaves 3–8cm(1¼–3in) long, ovate, blunt-pointed, crenate-toothed, lightly quilted, rich green above with a broad, silvery, longitudinal central zone, reddish beneath. 'Silver and Bronze' and 'New Silver and Bronze' appear to be alternative names. Perhaps a form or hybrid of *P. involucrata.*

P. spruceana Peru, Bolivia
Much like *P. involucrata* and perhaps conspecific, but having oblong to oblong-ovate leaves with broadly pointed tips. Rarely seen in cultivation, plants under this name being *P. involucrata*. For *P.s.* 'Norfolk' see *P. involucrata 'Norfolk'*.

PILOCEREUS See BORZICACTUS, CEPHALOCEREUS, CLEISTOCACTUS & LOPHOCEREUS

PINGUICULA

Lentibulariaceae
Butterworts

Origin: World-wide, chiefly in temperate regions. A genus of 45 species of carnivorous perennials with rosettes of linear to ovate leaves. These have glands on their upper surface which produce a sticky, muscilaginous fluid to which small insects are attracted. When one becomes stuck the leaf rolls inwards and an enzyme is secreted which breaks down the soft parts of the creature allowing minerals to be absorbed into the plant. The violet-blue spurred flowers are solitary. Grow in pans of an all-peat compost mixed with sphagnum moss either standing in a saucer of water or growing inside a glass case. For the home, the latter method is far the best. Propagate by division, by seed in spring or by leaf cuttings in summer. *Pinguicula* is derived from the Latin *pinguis*, fat, from the thick greasy texture of the leaves.

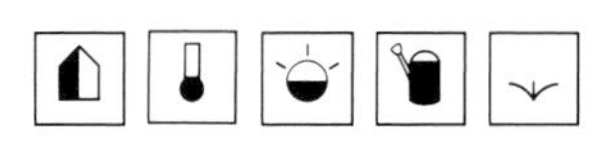

Species cultivated

P. bakerana See *P. caudata*.

P. caudata (*P. bakerana*) Mexico
In the summer it is loosely rosetted with leaves to 10cm(4in) long, obovate to rounded; in winter the leaves are narrow and fleshy in a dense rosette, to 5cm(2in) or so across. When in this stage the plant should be kept drier. Flowers to 5cm(2in) long, including the very long spur, violet-purple to rich carmine, on stems to 18cm(7in) tall.

P. grandiflora Greater butterwort Western Europe
Summer rosettes loose, of ovate-oblong leaves, 2–8cm($\frac{3}{4}$–3in) long, in winter dying back to a dormant bud. Flowers 2cm($\frac{3}{4}$in) across with a spur of the same length or somewhat longer, violet-purple, on stems 8–20cm(3–8in) tall.

P. vulgaris Butterwort Arctic and temperate regions of the northern hemisphere
Summer rosettes loose, of ovate to elliptic leaves, 2–6cm($\frac{3}{4}$–$2\frac{1}{2}$in) long, in winter a dormant bud. Flowers 1–1.5cm($\frac{3}{8}$–$\frac{5}{8}$in) wide, purple, on stems 5–15cm(2–6in) tall.

PIPER

Piperaceae
Peppers

Origin: Pantropical. Depending upon the botanical authority consulted, a genus of 1,000 to 2,000 species of climbers, shrubs and small trees. They have alternate, simple, often ovate or heart-shaped leaves, in some cases beautifully marbled with white, pink, purple or red. The tiny, petalless, insignificant flowers are borne in cylindrical spikes and are followed by more or less fleshy, berry-like drupes. Propagate by cuttings in summer or by seed in spring. *Piper* derives from the Greek *peperi*, itself derived from an Indian name for pepper (*P. nigrum*).

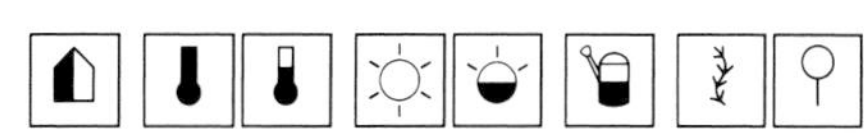

Species cultivated

P. crocatum (*P. ornatum crocatum*)
Ornamental pepper Peru
Climbing, with wiry stems 2–3m($6\frac{1}{2}$–10ft) in length if planted out. Leaves heart-shaped, peltate, 8–13cm(3–5in) long, lustrous dark olive-green with bands of silver-pink dots along the veins, deep purple beneath.

P. nigrum Black pepper India (Western Ghats), but widely grown throughout India and Sri Lanka since ancient times and naturalized in Burma, Assam and elsewhere
Stems climbing to 4m(13ft) or more in height, clinging to their supports by aerial roots. Leaves heart-shaped, pointed-tipped, thick-textured, dark green and boldly veined, 10cm(4in) or more long. Fruits 4–6mm($\frac{1}{6}$–$\frac{1}{4}$in) wide, ripening red then black. For white pepper the fleshy coat is removed by retting in running water, and the seeds then dried and ground. For black pepper the whole fruits are dried and ground. Not a highly decorative plant, but an interesting and amenable house plant. It is also one of the most important and oldest of the spices; there was a Guild of Pepperers in London as early as 1180.

P. ornatum Celebes
Much like *P. crocatum*, but leaves relatively broader and more evenly marbled with silvery-pink which ages white.

Piper ornatum

PISONIA

Nyctaginaceae

Origin: Tropics and sub-tropics, widespread. A genus of about 50 species of shrubs and trees one of which makes a handsome foliage pot or tub plant for the conservatory and is worth trying in the home. Propagate by seed, cuttings and layering in spring or summer. *Pisonia* commemorates William Pison (*d.* 1648), a Dutch doctor and naturalist.

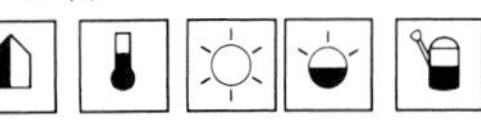

ABOVE LEFT *Pitcairnea maidifolia*
ABOVE RIGHT *Pistia stratiotes*
TOP *Pisonia umbellifera* 'Variegata'

Species cultivated

P. umbellifera (*Heimerliodendron brunonianum*) Parapara, Bird-catching tree Australia
Small tree in the wild, but amenable to pruning and maintainable at around 2m(6½ft) in height, or less in containers. Leaves usually in opposite pairs, sometimes in whorls of three or alternate, ovate-oblong, up to 25cm(10in) or more long, glossy green, firm-textured when mature. Flowers tiny, greenish, yellowish or pinkish, funnel-shaped, 4- or 5-lobed, in terminal panicles. Fruits 2.5cm(1in) long, narrowly oblongoid, 5-ribbed, very sticky. *P.u.* 'Variegata' has leaves marbled green, with a broad, irregular cream margin which is often shot with pink – a most handsome foliage plant.

PISTIA

Araceae

Origin: World-wide tropics and sub-tropics. A genus of one species, a free-floating evergreen perennial rooting when growing in very shallow water over mud. It is an attractive plant for a pool in the conservatory, or a shallow glass tank in the home. Water temperature needs to be kept at about 21°C(70°F). Propagate by division in summer or by seed in warmth in a pot of sand stood in water. *Pistia* derives from the Greek *pistos*, watery.

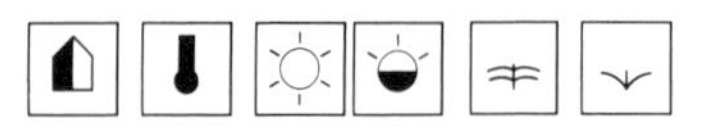

Species cultivated

P. stratiotes Water lettuce
Stoloniferous species producing many neat, floating rosettes with roots which hang free in the water. Leaves 5–13cm(2–5in) long, broadly obovate, truncate, pale green, velvety in texture. Flowers minute, petalless, one female and several males carried together in a spathe to 2cm(¾in) long.

PITCAIRNEA

Bromeliaceae

Origin: Tropical America with one species in West Africa (*P. feliciana*). A genus of about 250 species of evergreen, normally ground dwelling bromeliads. They have rosettes of strap-shaped to elliptic or linear leaves often with spiny margins, and tubular, asymmetrical, 3-petalled flowers in racemes or panicles.

Propagate by division or seed in spring. *Pitcairnea* was named for William Pitcairn (1711–1791), a London physician who kept a private botanic garden at Islington.

Species cultivated

P. integrifolia Venezuela, Trinidad
Leaves linear, 60–90cm(2–3ft) long, tapering to a thread-like point, matt-green above, grey-white-scaly beneath. Inflorescence a pyramidal panical to 45cm(1½ft) or more long; flowers 3–4cm(1¼–1½in) long, bright red.

P. maidifolia Costa Rica to Colombia, Guyana and Surinam
Leaves lanceolate, tapering to a stalk, 45–90cm(1½–3ft) in length, green. Inflorescence a simple spike-like raceme 30–45cm(1–1½ft) long; flowers green and yellow or white, 5cm(2in) or more in length, each one in the axil of a broadly oval, green or yellow bract about 3cm(1¼in) long.

PITTOSPORUM

Pittosporaceae

Origin: Eastern Asia, Australasia, the Pacific Islands and Africa. A genus of 150 species of evergreen trees and shrubs with alternate, entire leaves, usually ovate or obovate to oblong. Flowers are 5-petalled, often shortly tubular and are carried singly or in clusters from spring onwards. The seeds are surrounded by a sticky muscilage contained in rounded to ovoid capsules. They are excellent tub plants for the conservatory and, when small, for the home windowsill and can be stood outside in the summer. Propagate by seed in spring or by cuttings in late summer, all in warmth. *Pittosporum* derives from the Greek *pitta*, pitch and *spermum* a seed, referring to the sticky coating of the seeds.

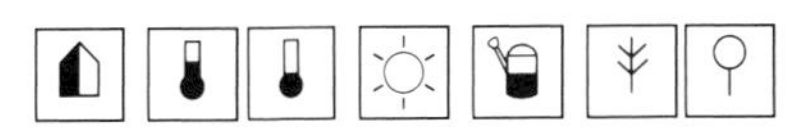

Species cultivated

P. crassifolium Karo New Zealand
Erect shrub to 3m(10ft) or more in a pot, but reaching 9m(30ft) and becoming tree-like in the wild. Young twigs and leaves white-downy. Leaves 5–7cm(2–2¾in) long, obovate to oblanceolate with thick margins, dark glossy green above, remaining downy beneath. Flowers about 1cm(⅜in) wide, dull red, in terminal umbels; fruits small, dark green to black.

P. eugenioides New Zealand
Erect to spreading shrub, to 3m(10ft) or more in a pot, in the wild eventually to 12m(40ft). Leaves 5–10cm(2–4in) or more long, oblong-elliptic with wavy margins, dark glossy green and aromatic. Flowers 5–7mm(⅕–¼in) across, yellow, making up umbel-like clusters and having a honey scent. *P.e.* 'Variegatum' has creamy-white leaf margins.

P. tenuifolium Kohuhu New Zealand
Erect shrub or small tree to 3m(10ft) or more in a pot, to 10m(33ft) in the wild, with dark purplish-black twigs. Leaves 2.5–6cm(1–2½in) long, oblong to ovate, rich glossy green, undulate. Flowers 1cm(⅜in) across, dark purple to almost black, fragrant, carried in small clusters in the leaf axils. *P.t.* 'Atropurpureum' has dark purple leaves; 'Garnettii' has leaves with white variegation and an overall pink flush; 'Golden King' has a golden suffusion to the leaves, particularly noticeable in winter; and 'Silver Queen' has a marked silver suffusion.

P. tobira Japanese pittosporum, Tobira
China, Japan
Rounded shrub eventually reaching 3m(10ft) in a pot, but slow growing to 6m(20ft) in the open. Leaves 5–10cm(2–4in) long, obovate, thick-textured, shining deep green. Flowers 1cm(⅜in) or so across, creamy-white, fragrant, in showy umbels. *P.t.* 'Variegatum' has leaves marked silvery-grey with an irregular cream margin.

P. undulatum Victorian box S.E. Australia
Shrub to 4m(13ft) in a pot, making a tree to 12m(40ft) in the wild. Leaves 6–13cm(2½–5in) long, lanceolate to oblong, undulate. Flowers 1cm(⅜in) across, white, in terminal umbel-like clusters, sweetly fragrant. Fruits orange-brown.

P. viridiflorum South Africa
Shrub to 3m(10ft) in a pot, to 10m(30ft) in the wild. Leaves 2.5–10cm(1–4in) long, obovate, thick-textured, wavy-margined, aromatic. Flowers 6mm(¼in) across, yellow-green, in crowded terminal clusters.

PLATYCERIUM

Polypodiaceae

Origin: Africa, Australia, Malaysia, South America. A genus of 17 species of large, evergreen epiphytic ferns of uniquely imposing appearance. They have leathery fronds of two types, sterile and fertile. The sterile ones are usually simple, overlapping each other and pressed firmly against their support (a tree branch in the wild). Dead leaves and other organic debris collects behind them, decaying down to provide a nutritious humus. In the home or conservatory it is worthwhile inserting a little peaty compost to take the place of this natural humus supply. The fertile fronds are in most, but not all, species branched like those of a deer or elk horn, either arching, hanging down, or erect. Brown sporangia completely cover discrete areas of the undersides of the fertile frond lobes. The common stag's horn fern (*P. bifurcatum*) is a proven house plant and the other species described here are worth trying in the home. All are easily grown in the warm conservatory. Propagate by spores or division in spring. A bromeliad type compost gives the best results and these ferns can be grown in hanging baskets or, more effectively, pinned to slabs of tree bark or tree fern stem. *Platycerium*

ABOVE *Pittosporum tobira* 'Variegatum'
LEFT *Pittosporum eugenioides* 'Variegatum'

derives from the Greek *platys*, broad and *keros*, a horn, alluding to the shape of the fronds.

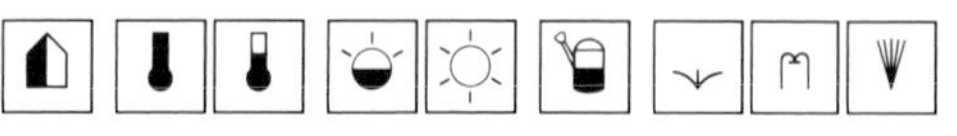

Species cultivated

P. alcicorne See *P. bifurcatum*.

P. angolense Elephant's ear fern Tropical Africa

Sterile fronds rounded to broadly oblong, 30–60cm(1–2ft) long, wavy-edged. Fertile fronds ovate, cuneate, to 45cm(1½ft) in length.

P. biforme See *P. coronarium*.

P. bifurcatum (*P. alcicorne*) Common stag's horn fern Eastern Australia to Polynesia

Sterile fronds rounded to kidney-shaped, undulate or slightly lobed, downy when young, 15–30cm(6–12in) wide. Fertile fronds spreading to drooping, long-cuneate at the base, branching dichotomously into several narrow lobes above, bright green. A very variable species with several distinctive forms described as cultivars: 'Majus' is somewhat larger with erect, brighter green fertile fronds; 'Netherlands' (*P. alcicorne* 'Regina Wilhelmina') has grey-green fronds which tend to be shorter and broader and radiate in all directions; and 'Ziesenherne' is smaller and slower growing (plants still grown as *P. alcicorne* are usually this cultivar).

P. coronarium (*P. biforme*) Crown stag's horn S.E. Asia

Sterile fronds erect, deeply lobed, up to 60cm(2ft) long. Fertile fronds pendulous to 2m(6½ft) long, divided into forked, thong-like lobes, bright green.

P. grande Regal elkhorn fern Australia, Java, Malaysia

Sterile fronds up to 60cm(2ft) or more wide, broadly cuneate-ovate, pointing upwards with a boldly lobed top, glossy bright green. Fertile fronds fan-shaped, divided into long, wide lobes, arching to pendulous, 45–150cm(1½ft–5ft) long.

P. hillii Elk's horn fern Australia

In overall effect like *P. bifurcatum*, but with erect, blunt-tipped, broader leaf lobes.

ABOVE *Platycerium grande*
RIGHT *Platycerium bifurcatum*

PLATYCLINIS See DENDROCHILUM

PLECTRANTHUS

Labiatae

Origin: Africa to eastern Asia, south to Australasia and the Pacific Islands. A genus of 250 species of evergreen perennials and sub-shrubs, with 4-angled stems, opposite pairs of entire leaves and whorls of 2-lipped tubular flowers carried in racemes or panicles. They make good container plants for the home and conservatory. Propagate by cuttings from late spring to late summer in warmth, or by division if possible when re-potting. *Plectranthus* derives from the Greek *plectron*, spur and *anthos*, flowers.

 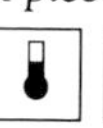 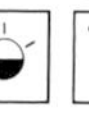  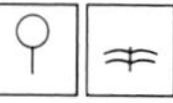

Species cultivated

P. coleoides Candle plant S.W. India
Bushy plant to 1m(3ft) when planted out, but rarely over 30cm(1ft) in a pot. Leaves 5–7cm(2–2¾in) long, ovate, dark green. Flowers white and purple in erect racemes up to 30cm(1ft) long. 'Marginatus' is the form usually seen in cultivation, having creamy-white, scalloped margins and a grey flush.

P. nummularius South Africa
Stems decumbent to prostrate, 30cm(1ft) or more in length. Leaves broadly ovate to orbicular, 4–6cm(1½–2½in) long, coarsely toothed, somewhat fleshy, metallic green above, grey-green, purple-veined beneath. Flowers white to pale lavender, 1cm(⅜in) long, in racemes to 30cm(1ft) in length.

P. oertendahlii Swedish ivy, Brazilian coleus South Africa
Prostrate to decumbent, the 30–60cm(1–2ft) stems trailing and rooting. Leaves to 6cm(2½in) long, rounded, with a shallowly scalloped margin, dark green above with a pattern of white veins, purplish beneath. Flowers 2cm(¾in) long, pale lavender, in erect spikes. An effective plant for a hanging basket.

PLEIOBLASTUS See ARUNDINARIA

PLEIONE

Orchidaceae

Origin: S.E. Asia, from India to China. A genus of ten species of orchids closely allied to, and once included in, *Coelogyne*. They are epiphytic, though in the wild more frequently found on mossy rocks and logs than on tall trees. They have small pseudobulbs which produce one or two elliptic, somewhat pleated leaves and live for one season only. The solitary flowers are often large for the size of the plant and have five similar tepals and a usually 3-lobed, trumpet-shaped labellum. They are suitable for the conservatory or home but need to be kept cool in winter, growing best in shallow pans. Keep dry from late autumn when the foliage yellows until the new leaves unfold in spring. Propagate by dividing the clumps or by potting separately any pseudobulbs that are formed on top of the clump. *Pleione* derives from the Greek goddess of that name, mother of the seven Pleiades.

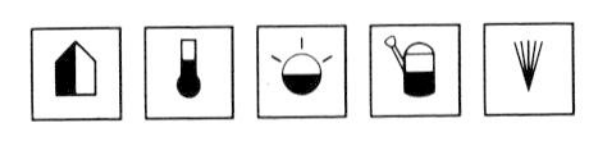

Species cultivated

P. bulbocodioides (including *P. delavayi*, *P. formosana*, *P. pogonoides*, *P. limprichtii* and *P. pricei*) Tibet, China, Taiwan
Pseudobulbs almost rounded, dull green, sometimes partly blackish-purple. Leaves appear with the flowers, becoming 3–6cm(1¼–2½in) wide and 20cm(8in) long when mature. Flowers 7–10cm(2¾–4in) wide, pale or deeper mauve with a paler or white, fringed labellum spotted with red, purple or yellow and opening in early summer. Many cultivars and forms are grown, some of hybrid origin:
'Blush of Dawn' (*P. formosana*) has pale lilac-pink tepals and a white labellum tinged with pale mauve; 'Limprichtii' has magenta-purple tepals with a paler labellum thickly red-spotted and streaked; 'Oriental Splendour' (sometimes grown as *P. pricei*) has pseudobulbs more flask-shaped, black-purple, tepals pale violet with a white labellum finely lined with orange; 'Polar Sun' has pure white, unmarked flowers.

Pleione bulbocodioides 'Limprichtii

P. forrestii S.W. China, Burma
Pseudobulbs narrowing gradually to the top, blooms opening after the leaves have died back. Flowers yellow or orange-yellow with reddish-brown markings on the labellum, borne in summer.

PLEIOSPILOS

Aizoaceae

Origin: South Africa. A genus of about 35 species of virtually stemless succulents, some of which mimic stones in the same way as *Lithops*. They may be clump-forming or solitary, each very short stem bearing two to four, sometimes more pairs of very thickly fleshy leaves. The daisy-like flowers are unusually large for the size of the plant and showy. The species described here are desirable additions to a collection of succulents. *Pleiospilos* derives from the Greek *pleios*, many and *spilos*, a speck or spot; most, if not all, the species have conspicuously dotted leaves.

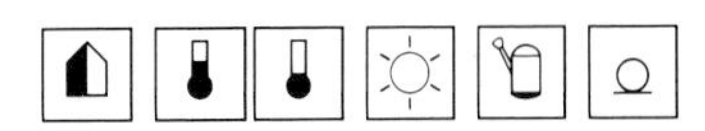

Species cultivated

P. bolusii Living rock Cape, Karroo, etc.
Solitary or, with age, forming small clusters. Leaves usually one pair, sometimes with two; each leaf 5cm(2in) long, roughly half ovoid but thickened towards the apex like a broad upturned chin, grey-green with numerous dark dots, often reddish-flushed in bright light. Flowers 6–8cm(2½–3in) wide, rich bright yellow, opening in the afternoon in summer.

Pleiospilos nellii

P. nellii Cape
Similar to *P. bolusii*, but having usually two pairs of somewhat smaller, more nearly hemispherical leaves and flowers of a pale salmon-orange with a white centre.

PLEOMELE See DRACAENA

PLUMBAGO

Plumbaginaceae
Leadworts

Origin: Widespread in the warmer regions of the world. A genus of 12 to 20 species of shrubs, sub-shrubs, perennials and annuals, with alternate, entire leaves and slenderly tubular flowers opening to five broad lobes and carried in terminal spikes or panicles. They are good container plants, flowering when small. *P. auriculata* needs support for its partially climbing stems and responds to hard pruning in late winter, having its side branches cut back to within a few centimetres of the main stem. *P. indica* should also be hard pruned annually in spring, cutting back all stems to 10–15cm(4–6in) above soil level. Propagate the former by cuttings of non-flowering shoots in warmth in summer, the latter by basal cuttings in late spring also in warmth; both also by seed in spring. *Plumbago* derives from the Latin *plumbum*, lead, one species having the reputation to cure lead poisoning.

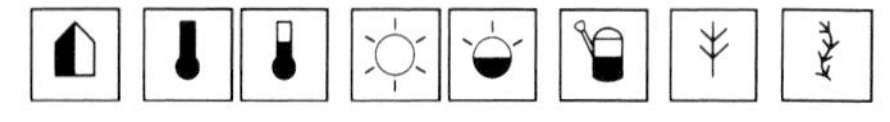

Species cultivated

P. auriculata (*P. capensis*) Blue Cape plumbago/leadwort South Africa
Straggling, scandent shrub, easily kept to 3m(10ft) in a pot, to 6m(20ft) in the open, its slender stems needing support and best treated as for a climber. Leaves 5cm(2in) long, obovate to oblong, spathulate, evergreen. Flowers to 4cm($1\frac{1}{2}$in) long, sky-blue, opening from summer to autumn. *P.a.* 'Alba' has white flowers. Best grown in temperate conditions, but will survive cool conditions.
P. capensis See *P. auriculata.*

Plumbago auriculata

P. indica (*P. rosea*) Scarlet leadwort S.E. Asia
Shrub with slender stems to 60cm(2ft) if pruned hard each year, to 2m(6ft) or so in the open. Leaves 5–10cm(2–4in) long, pink to red, opening in summer if pruned, but if unpruned and kept warm it will flower in late winter. Tropical.
P. rosea See *P. indica.*

PLUMERIA

Apocynaceae

Origin: Tropical America. A genus of seven species of shrubs and trees with thick, succulent branches and large entire leaves. The flowers are tubular with five overlapping corolla lobes carried in terminal panicle-like clusters of 20 to 60 blooms which are very fragrant. Grow in large pots or tubs in a conservatory or a sunny window, keeping almost dry through the winter after leaf fall. Propagate by stem cuttings in late spring before the leaves appear, keeping in warmth. *Plumeria* was named for Charles Plumier (1646–1704), a French Franciscan monk who travelled widely and made accurate drawings of the plants he saw.

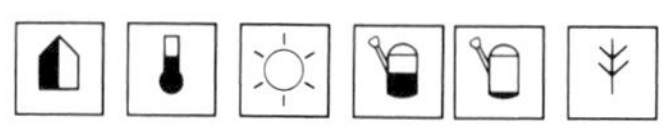

Species cultivated

P. rubra (*P. acuminata*) Frangipani Central America
Sparingly branched deciduous shrub to 3m(10ft) in a pot; shrub or tree to 8m(25ft) in the open. Leaves alternate, 15–30cm(6–12in) long, elliptic to oblong, oblanceolate, pointed at both ends, the veins running at almost right angles to the mid rib and joined to a vein which lies just inside the leaf margin. Flowers 5cm(2in) or more wide, the perianth tube to 2.5cm(1in) long, rose-pink to red with a yellow eye, borne in summer and autumn, followed by seed pods 15–25cm(6–10in) long containing plumed seeds. Named cultivars are rarely seen except in warm climates.
P.r. acutifolia (*P. acutifolia*) Pagoda tree
Large white flowers with more obvious yellow eyes.
P.r. lutea
Yellow flowers, sometimes tinged with pink.

PODRANEA

Bignoniaceae

Origin: Tropical and South Africa. A genus of two species of woody-stemmed twining climbers, one of which is in cultivation though not very readily available. Well-grown, it can garnish a conservatory roof to perfection. Propagate by cuttings in summer or by seed in spring. *Podranea* is an anagram of *Pandorea*, the genus to which it is most closely related and in which it was once included.

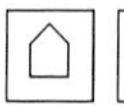  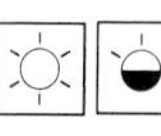  

Species cultivated

P. ricasoliana (*Pandorea ricasoliana*) South Africa (Cape)
Evergreen climber to 4m(13ft) or more in length. Leaves in opposite pairs, pinnate, composed of seven to nine (sometimes less or more) lanceolate to ovate leaflets each one about 5cm(2in) in length, wavy-edged and dark green. Flowers 5cm(2in) long, tubular bell-shaped with five large rounded petal lobes, rose-pink, darker veined, fragrant, in loose terminal panicles. A beautiful and free-flowering species deserving to be seen more often. Will stand slight frost and is then partially or wholly deciduous.

POINCIANA See CAESALPINIA

POLIANTHES

Agavaceae (Amaryllidaceae)

Origin: Mexico. A genus of 12 species of rhizomatous perennials with more or less fleshy leaves, mostly basal and erect flower stalks and spike-like racemes of long-tubed, funnel-shaped blooms with six similar tepals. The species described below makes a decorative flowering plant. Because they rarely flower well after the first year it is usual to acquire fresh rhizomes annually. They can be propagated by division or offsets, but it is usually difficult to get them to flower satisfactorily. *Polianthes* derives from the Greek *polios*, white or grey and *anthos*, flower.

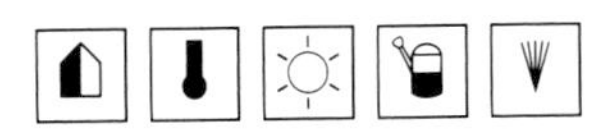

Species cultivated

P. tuberosa Tuberose Unknown in the wild
Rhizomes tuberous, almost bulb-like. Leaves mostly basal, to 45cm(1½ft) long, becoming smaller up the stem. Flowers to 7cm(2¾in) long, white and very fragrant, on erect, unbranched stems to 1m(3ft) tall, opening in summer and autumn. 'The Pearl' (*P.t. plena*) is double-flowered and is the form most commonly available.

POLYGALA

Polygalaceae

Milkworts

Origin: World-wide except for New Zealand and Polynesia. A genus of 500 to 600 species of trees, shrubs, perennials and annuals with entire leaves and flowers superficially like those of the pea family. These have five sepals and five petals, two of the sepals being large and petal-like, forming the wings and the petals of which only three are normally fully developed and united at the base; the central one is folded to form a keel and is crested at the tip, the side ones sometimes cleft into two. The species described makes an attractive pot plant, needing pruning in spring to keep it from getting leggy. Propagate by cuttings in late spring or summer in warmth. *Polygala* derives from the Greek *poly*, much and *gala*, milk; certain species were thought to increase milk flow.

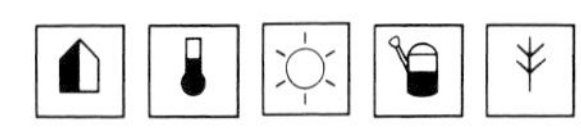

Species cultivated

P. myrtifolia South Africa
Erect, evergreen shrub reaching 2m(6½ft), though slow growing and easily kept smaller by pruning. Leaves alternate, 2–3cm(¾–1¼in) long, elliptic to ovate-oblong, glaucous green. Flowers to 2cm(¾in) long, greenish-white, the wing petals margined and veined with pinkish-purple and the keel petal purple with a white crest; they are carried in short, leafy racemes from spring to autumn.

P.m. grandiflora
Larger, richer purple flowers to 2.5cm(1in) long. The form most commonly met with in cultivation.

ABOVE *Polygala myrtifolia*
TOP *Podranea ricasoliana*

POLYPODIUM

Polypodiaceae

Origin: Cosmopolitan. A genus once including over 1,000 species of fern, but now reduced by modern botanists to 75. They are terrestrial and epiphytic in growth and mostly evergreen, with branched rhizomes and usually pinnate or pinnately lobed fronds, though a few species have almost entire fronds. The spores on the undersides of the fronds have no coverings (indusia). Propagate by cuttings of rhizomes, division or spores in spring. *Polypodium* derives from the Greek *polys*, many and *podium*, a foot, referring to the branched rhizomes.

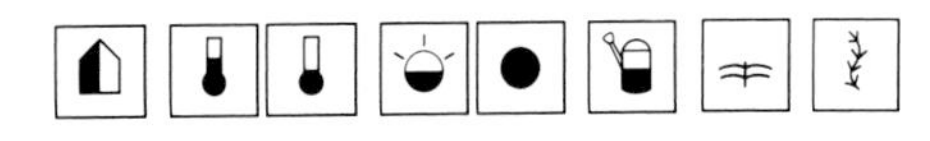

Species cultivated

P. angustifolium Narrow-leaved strap/ribbon fern Tropical America
Epiphytic species forming dense clumps of rhizomes. Fronds linear, entire, 30–60cm(1–2ft) long, arching, dark green, leathery-textured, densely borne. A very un-fern-like fern, ideal for a hanging basket.

P. aureum (*Phlebodium aureum*) Hare's foot fern Florida to Argentina and Australia
Epiphytic or ground dwelling with thick, creeping rhizomes densely covered by hair-like red-brown scales. Fronds long-stalked, pinnatifid, 30–90cm(1–3ft) or more in length, the lobes strap-shaped, waved and pointed, pale to yellow-green.

P.a. areolatum
Smaller, more leathery-textured glaucous fronds. This is the commonest variety in cultivation, often sold under such botanically invalid names as *P. glaucum*, *P. glaucophyllum* and *P. glaucopruinatum*. *P.a. areolatum* 'Mayi' has more arching fronds with strongly undulate lobes.

P. glaucum (*P. glaucophyllum*, *P. glaucopruinatum*) See *P. aureum areolatum*.

P. integrifolium See *P. punctatum*.

P. musifolium (*Microsorium musifolium*, *Phymatodes musaefolia*) Malaysia, New Guinea
Epiphytic, with robust, semi-woody rhizomes. Fronds stalkless or almost so, 45–90cm(1½–3ft) in length, narrowly oblanceolate, widely undulate, pale green with an attractive pattern of dark veins. Suitable for pots or baskets.

P. phyllitidis (*Campyloneurum phyllitidis*) Strap/Ribbon fern Tropical America
Epiphytic, with a short, thickish creeping rhizome. Fronds almost stalkless, entire, narrowly lanceolate to strap-shaped, lustrous, brightish green, 30–90cm(1–3ft) long. Suitable for pots, baskets or growing on a section of tree bark.

P. piloselloides Tropical America
Epiphytic, with fairly slender, far-creeping rhizomes. Fronds variable, entire, 2.5–8cm(1–3in) long, the sterile ones ovate to obovate, leathery-textured and glossy, the fertile spore-bearing ones somewhat narrower and longer. Needs a hanging basket, section of tree branch or tree fern stem to contain its wandering rhizomes.

P. polycarpon See *P. punctatum*.

P. punctatum (*P. integrifolium*, *P. polycarpon*, *Phymatodes irioides*) Tropical Africa, Asia, Australia and Polynesia
Epiphytic, with thick, dark brown, scaly rhizomes. Fronds entire, narrowly lanceolate to strap-shaped, leathery-textured, light- to yellow-green, more or less glossy, 30–90cm(1–3ft) in length.

P. subauriculatum Lacy pine fern Tropical Asia to Australia
Usually epiphytic, with creeping rhizomes. Fronds to 90cm(3ft) or more in length, spreading to pendulous, lanceolate in outline, pinnate, each of the pinnae entire or toothed, wavy, bright glossy green. *P.s.* 'Knightiae' is a little smaller and slower growing with the pinnae cut into several narrow, pointed lobes; in overall character it resembles a nephrolepis fern. Both *P. subauriculatum* and *P.s.* 'Knightiae' make excellent subjects for a hanging basket.

P. vulgare Common polypody Northern Hemisphere, South Africa, Kerguelen Is.
Evergreen, with densely brown-scaly rhizomes from which the fronds arise singly. Fronds ovate, linear-lanceolate or oblong in outline, varying considerably throughout their entire geographical range, pinnatisect or pinnatipartite, 10–30cm(4–12in) or more long; the rounded yellow-brown sori are carried in single rows which occur parallel to the leaf margins on the fertile fronds. Many hundreds of mutants have been recorded, some of which are very decorative with more finely dissected, crested or plumed fronds.

BELOW LEFT
Polypodium piloselloides
BELOW RIGHT
Polypodium aureum

POLYSCIAS

Araliaceae

Origin: Tropical Asia and Polynesia. A genus of 80 or so species which have also been classified in *Aralia*. They are distinctive foliage plants having alternate leaves which are usually palmate or pinnate, and umbels of small, greenish flowers each made up of four to eight perianth segments, but these are, however, seldom produced on potted specimens. Propagate by stem tip cuttings or by leafless sections of stems in summer.

Polyscias derives from the Greek *polys*, many and *skias*, a sunshade or canopy, referring to the flowering umbels.

Species cultivated

P. balfouriana Dinner plate aralia
New Caledonia
Bushy shrub rarely exceeding 1.5m(5ft) in a pot, though to 8m(25ft) in the wild. Leaves trifoliate and sometimes unifoliate, the stalked leaflets 5–10cm(2–4in) long, broadly ovate to rounded, with toothed margins.. *P.b.* 'Pennockii' has each leaf composed of one broadly ovate leaflet up to 13cm(5in) long, basically creamy-white variably mottled pale green between the veins and

irregularly splashed deep glossy green.
P. b. marginata
Grey-green leaves irregularly margined with white.
P. filicifolia Fern-leaf aralia Polynesia
Shrub to 1m(3ft) or so, to 2.5m(8ft) in the open. Leaves to 30cm(1ft) long, pinnately divided, the leaf-lobes themselves cleft and spiny-toothed, bright green with purplish mid-ribs when young. Mature specimens have broader, less toothed leaflets.
P. guilfoylei Wild coffee Polynesia
Shrub to 1m(3ft) or so in a pot, to 5m(16ft) in the wild. Leaves 25–40cm(10–16in) long, pinnate with ovate to orbicular leaflets, sometimes obscurely lobed and toothed. A very variable species usually represented in cultivation by one of its forms or cultivars. 'Laciniata' has deeply lobed leaflets with white margins; 'Monstrosa' has irregularly toothed leaflets which are blotched with grey and margined white; 'Quinquefolia' has coarsely lobed leaflets which are a dark, coppery-green; and 'Victoriae' (Lace aralia) is slower growing and compact with bipinnate leaves having small white-bordered leaflets.
P. paniculata Mauritius
Shrub 1–2m(3–6ft) high in a pot, to 4m(13ft) in the wild. Leaves pinnately divided; leaflets broadly elliptic to oblong, toothed, the terminal one the largest. *P.p.* 'Variegata' is blotched with pale green and cream.

LEFT *Polyscias filicifolia*
BELOW LEFT *Polystichum setiferum* 'Acutilobum'

POLYSTICHUM

Polypodiaceae
Shield/Holly ferns

Origin: Cosmopolitan. A genus of 120 to 135 species of rhizomatous ferns with clusters or rosettes of pinnate or bipinnate fronds. The fertile pinnae have two rows of rounded sporangia. Propagate by spores or division in spring.

Polystichum derives from the Greek *polys*, many and *stichos*, a row, referring to the rows of sori on the fronds.

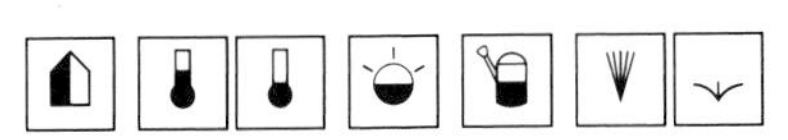

Species cultivated
P. acrostichoides Christmas fern North-eastern America
Fronds evergreen to 60cm(2ft) long, pinnate, the pinnae lanceolate with a distinct ear or lobe at the base; fertile pinnae much smaller and narrower than the sterile ones, forming the upper part of the fertile fronds. Spores closely set in two rows on each side of the mid vein of the pinnae. A number of crested or forked mutants are known. Hardy.
P. aculeatum Hard/Prickly shield fern Europe, Asia and South America
Fronds evergreen, 1–3m(3–10ft) long, lanceolate, bipinnate, firm to almost leathery, the toothed pinnules with prickle-tips. Frequently confused with *P. setiferum*. Hardy.
P. aristatum (*Arachnoides aristata*) East Indian holly fern Southern Asia to Australia
Rhizomes creeping and branching, the fronds forming colonies if planted out in the conservatory border; each frond up to 60cm(2ft) long, tri- to quadri-pinnatifid, broadly triangular-ovate in outline, usually a rich green. *P.a.* 'Variegatum' has the pinnae banded with palest green.
P. falcatum See *Cyrtomium falcatum.*
P. munitum Giant holly fern, Western sword fern Alaska to Montana and California
Much like *P. acrostichoides*, but the fronds to 105cm(3½ft) in length and more upstanding. A splendid plant for a cool or unheated conservatory, being reasonably frost hardy.
P. setiferum Soft shield fern Temperate regions of northern and southern hemisphere, absent in North America
Fronds 30–60cm(1–2ft) long, arching, midribs and stalks with golden-brown chaffy scales, bipinnate, similar to *P. aculeatum* but less rigid and the pinnules with fewer, larger, but not bristle-tipped teeth. Frost hardy. *P.s.* 'Acutilobum' ('Proliferum', Plume/Mother fern) develops tiny plantlets on the upper surface of the fronds, especially along the midrib, and has narrower and rather denser pinnules. Other cultivars with variable frond divisions are available. Hardy.
P. tsus-simense Tsusina holly fern Asia, mainly China and Japan
Neatly clump-forming, the arching fronds to 30cm(1ft) or sometimes more in length; each frond blade triangular-lanceolate in outline, bi- or tri-pinnate, at least in the lower half, deep green. A hardy species useful for unheated conservatories.

Portulaca grandiflora

PORTULACA

Portulaceae

Origin: Tropics and warm temperate regions, particularly the Americas. A genus of 100 to 200 species, depending upon the classification followed, of annuals and perennials. Their leaves are alternate, or almost opposite, ovate to linear and often fleshy. The flowers open their five spreading petals only in sun. The species described is a colourful annual for a pot in a sunny conservatory or home window sill. Propagate by seed in spring. *Portulaca* is from the Latin vernacular name for purslane.

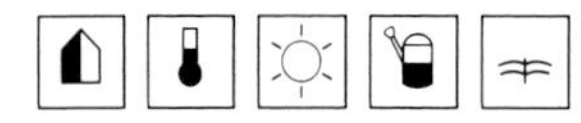

Species cultivated

P. grandiflora Sun plant Brazil, Uruguay, Argentina
Tufted annul to 15cm(6in) or more in height with spreading to ascending stems. Leaves 2.5cm(1in) long, cylindrical and fleshy. Flowers 2–3cm($\frac{3}{4}$–$1\frac{1}{4}$in) across, opening widely in sun, in a wide range of colours including reds, pinks, yellows, purples and white, both single and double forms being grown; they have a central boss of bright yellow stamens and each flower sits above a leafy collar.

POTHOS

Araceae

Origin: Indo-Malasia, Malagasy. A genus of 70 to 75 species of evergreen climbers. They have aerial roots and an ivy-like mode of growth with widely diverging alternate leaves having winged stalks like those of *Citrus*. The small *Anthurium*-like greenish floral spathes are seldom seen on pot specimens. Propagate by cutting in summer. *Pothos* is a Latinized version of the Sinhalese name *potha*.

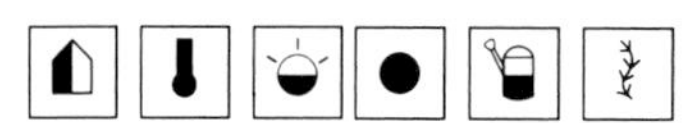

Species cultivated

P. argyraeus See *Scindapsus pictus* 'Argyraeus'.
P. aureus See *Epipremnum aureum*.
P. jambea A name of no botanical standing sometimes used for *P. seemanii*, *q.v.*
P. scandens Malaysia
Stems fairly slender, at least when young, branching, eventually to several metres (yards) in length, but easily kept in check. Leaves to 10cm(4in) or more long, deep more or less glossy green, composed of an oblong-obovate to lanceolate winged stalk and an ovate blade of almost equal length.
P. seemannii Southern China, Formosa
Habit and size similar to *P. scandens*. Leaves com-

posed of a winged stalk to about 2.5cm(1in) long and an ovate to obovate-oblong to lanceolate blade about 8cm(3in) in length.

PRIMULA

Primulaceae

Origin: World-wide, but mainly in the northern hemisphere and particularly Asia. A genus of about 400 species of tufted to clump-forming evergreen and deciduous perennials. They have oblanceolate to orbicular leaves in basal rosettes and umbels of tubular flowers opening to five broad lobes carried on leafless stems (scapes). In some species the scape is so reduced that the individual stalks of the flowers which make up the umbel arise near the base of the plant and appear separate; this is the case with the primrose (*P. vulgaris*). The petals are usually obovate and notched at the tip. Individual plants can have the stigma longer than the stamens (pin-eyed), or vice versa (thrum-eyed). Primulas suitable for conservatory or home culture fall into two groups. The hardy *P. vulgaris* and *P* × *tommasinii* (Polyanthus) are potted from late autumn onwards and will flower in about a month from being brought into warmth. They must be returned to the open garden after flowering. The tender species are full-time house and conservatory plants and, although short-lived perennials, are normally treated as annuals. They are sown from spring to late summer and will flower in winter and spring, 9–15 months later. Pot on regularly, the best specimens being obtained from plants in 13cm or 15cm pots. For winter flowering, plants require temperate conditions. Propagate by seed or by division of hardy species.

Primula derives from the Latin *primus*, first, referring to the early spring primrose.

   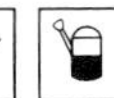 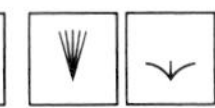

Species cultivated

P.* × *kewensis
Hybrid between *P. floribunda* and *P. verticillata* of which the cultivated form comes true from seed. Leaves to 20cm(8in) long, obovate, waved and toothed with a fine dusting of white meal (farina). Flowers 2cm(¾in) across, buttercup-yellow, in whorls, two to five on each of the scapes, which are up to 30cm(1ft) tall.

P. malacoides Fairy primrose Western China
Leaves to 7cm(2¾in) long, ovate to oblong-elliptic, cordate, toothed, hairy beneath, on long stalks. Flowers 1–2cm(⅜–¾in) across, fragrant, light mauve in the original species, with three to six whorls on the 30–45cm(1–1½ft) scapes. Now represented in cultivation by forms with single or double flowers in shades of red, pink, purple and white. Dwarf forms 15–20cm(6–8in) tall are also available. Most are sold as colour mixes, but single colours are sometimes offered. Recommended forms are: 'Carmine Pearl', dwarf, carmine-red; 'Fire Chief', brick-red; 'Lilac Queen', soft lilac, double; 'Mars', deep lavender; and 'White Pearl', dwarf, white.

P. obconica Northern China
Leaves 6–10cm(2½–4in) long, ovate-cordate, glandular-hairy, with long stalks. Flowers 2.5cm(1in) wide, lilac-purple to pink, on scapes 15–18cm (6–7in) tall, opening in winter. Represented in cultivation by a range of often larger, more robust plants sometimes listed as Grandiflora. These have flowers to 4cm(1½in) across in shades of red, pink, mauve, blue, apricot and white. Usually available as mixes, but the following recommended types are sometimes offered: 'Apricot Brandy' ('Appleblossom'), cream in bud opening to apricot; 'Coerulea', blue; 'Snowstorm', white; and 'Wyaston Wonder', bright crimson with extra large flowers.

P. polyantha See *P.* × *tommasinii.*

P. sinensis Chinese primrose China
Leaves 7–13cm(2¾–5in) long, broadly ovate to rounded with several toothed lobes, softly hairy, on long stalks. Flowers 2.5–4cm(1–1½in) wide with notched petals on scapes 15–25cm(6–10in) long, in shades of pink, red, purple and white. 'Stellata' has starry flowers in 2-tiered umbels.

Do not over-pot as the roots are susceptible to rot.

P.* × *tommasinii Polyanthus
Hybrid between *P. veris* and coloured forms of *P. vulgaris*. Leaves to 25cm(10in) long, obovate, wrinkled and often downy beneath. Flowers to 2.5cm(1in) or more across, in umbels on scapes 15cm(6in) or more long; they are available in

ABOVE *Primula malacoides*
ABOVE LEFT *Primula obconica*
LEFT *Primula* × *tommasinii*

shades of red, pink, purple, blue, yellow and white. 'Gold Laced' has smaller, dark red flowers neatly edged with gold patterning; 'Pacific' has very large flowers to 5cm(2in) across.

P. vulgaris Primrose Europe, Asia
Leaves to 25cm(10in) long, oblanceolate to obovate, wrinkled, usually downy beneath, Flowers 2–3cm($\frac{3}{4}$–$1\frac{1}{4}$in) across on very short scapes, yellow in west Europe but in shades of pink, red and purple farther east, usually with a yellow eye. Double forms are available. From this species and *P. juliae* have been raised a number of short, compact primroses often with a ring of red around the yellow eye. These are excellent short-term pot plants and are available under such names as: 'Colour Magic', 15cm(6in) high, mixed colours; 'Julian Bicolor', 10cm(4in) high, mixed; and 'Panda', 10–12cm(4–5in) high, mixed.

PRITCHARDIA See LICUALA & WASHINGTONIA

PROSTANTHERA

Labiatae
Mint bushes

Origin: Australia. A genus of 50 species of aromatic, evergreen shrubs with opposite, entire leaves and tubular 2-lipped flowers, the upper lip made up of two lobes, the lower of three. They are carried in leafy racemes or panicles in spring and summer. The species described make attractive pot or tub plants and can be brought into the home for flowering. Propagate by seed in spring in warmth, or by cuttings in summer with bottom heat. *Prostanthera* is derived from the Greek *prosthema*, an appendage and the Latin *anthera*, an anther, because of the spur-like outgrowths on the anthers.

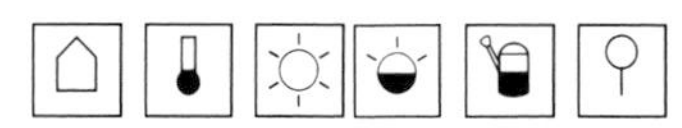

Species cultivated

P. melissifolia Balm mint bush Victoria
Slender shrub 1–3m(3–10ft) tall, flowering when small. Leaves 2–5cm($\frac{3}{4}$–2in) long, ovate, coarsely toothed and strongly aromatic. Flowes 1.5cm($\frac{5}{8}$in) long, lavender to purple, wreathing the stems in spring and early summer.

P. nivea Snowy mint bush S.E. Australia
Shrub to 2m($6\frac{1}{2}$ft) in a pot, more than this if trained to a wall; (stems slender, 4-angled. Leaves 2.5–4cm(1–$1\frac{1}{2}$in) long, linear-lanceolate. Flowers about 1.5cm($\frac{5}{8}$in) long, white or flushed with lavender, carried in abundance in leafy racemes from late spring to summer.

P. rotundifolia S.E. Australia, Tasmania
Bushy shrub to 2m($6\frac{1}{2}$ft) in a pot, to twice this in the wild. Leaves 5–10mm($\frac{3}{16}$–$\frac{3}{8}$in) long, rounded, sometimes crenate-margined. Flowers 1cm($\frac{3}{8}$in) long, purple-blue to lavender, in dense terminal racemes in late spring.

PROTEA

Proteaceae

Origin: Tropical and southern Africa. A genus of 100 to 130 species of evergreen shrubs and trees with alternate, entire, leathery leaves and dense terminal heads made up of many relatively small 4-lobed flowers and surrounded by petal-like bracts giving the whole inflorescence the appearance of being one large flower. All species bloom in summer. Grow in neutral to acid soil. Fertilizers rich in phostates and nitrates must not be used; give occasional light feeds of magnesium sulphate and urea only. Propagate by seed in spring sowing singly in 5cm or 7.5cm pots, so that pricking off and subsequent root disturbance is not necessary, or by cuttings with bottom heat in summer. *Protea* derives from Proteus, the Greek Sea God.

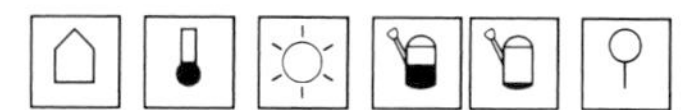

Species cultivated

P. compacta
In the wild, a large shrub to 3m(10ft) tall, but rarely half this in containers. Leaves to 10cm(4in) or more long, ovate, densely borne and overlapping. Flower heads 10–14cm(4–$5\frac{1}{2}$in) long, narrowly cup-shaped, pink to carmine.

RIGHT *Prostanthera melissifolia*
FAR RIGHT *Protea compacta*

P. mellifera See *P. repens.*

P. neriifolia

In the wild, a shrub to 3m(10ft) in height, but not usually much more than half this in containers. Stems downy; leaves 10cm(4in) or more long, narrowly oblong-lanceolate, fairly crowded. Flower heads to 13cm(5in) or more long, cup-shaped, red, pink or white; bract tips black-bearded.

P. repens (*P. mellifera*) Sugarbush, Honey flower

To 2m(6ft) tall. Leaves 10cm(4in) or more long, linear to narrowly spathulate. Flower heads to 15cm(6in) wide, not opening widely, pink-red to white.

PSEUDERANTHEMUM

Acanthaceae

Origin: Tropics – widespread. Depending upon the botanical authority, a genus of 60 to 120 species of shrubs and perennials very closely allied to *Eranthemum*. Several species have long been popular conservatory plants and can be cultivated successfully in the home. Some are best as foliage plants, others for their flowers. Erect to spreading in habit, they have opposite pairs of mainly simple leaves, and flowers with slender tubes and five prominent petal lobes. Propagate by cuttings in spring or summer. *Pseuderanthemum* derives from the Greek *pseudo*, false and the allied *Eranthemum*.

Species cultivated

P. atropurpureum (*Eranthemum atropurpureum*) Probably Polynesia, but naturalized in tropical America

Shrubby erect habit, to 1.2m(4ft) in height if not regularly pruned. Leaves ovate to elliptic, 10–15cm(4–6in) long, usually strongly purple-flushed, but sometimes green or patterned with yellow, pink, white and green. Flowers white or purple-flushed with a red-purple eye, about 2cm($\frac{3}{4}$in) wide, in short terminal and axillary spikes in summer. *P.a.* 'Tricolor' has deep purple leaves irregularly splashed with white, pink and green.

P. reticulatum (*Eranthemum reticulatum*) Probably Polynesia, but widely grown and naturalized throughout the tropics

Erect, shrubby, to 90cm(3ft) or more in height. Leaves narrowly ovate to lanceolate, 10–25cm(4–10in) long, rich green with an elaborate network of bright yellow veins. Flowers 2.5cm(1in) or more wide, white with a red-purple throat and spotting on the lower petal, in small terminal panicles in summer. *P.r.* 'Eldorado' has ovate leaves and makes a more substantial foliage pot plant.

P. sinuatum (*Eranthemum cooperi*) Probably New Caledonia

Small spreading shrub to about 30cm(1ft) tall and rather more across. Leaves linear or narrowly lanceolate, irregularly shallowly lobed, olive and grey-green mottled above, purplish beneath. Flowers almost 4cm($1\frac{1}{2}$in) across, white, with the lowest petal purple-spotted, in short terminal racemes; flowers borne when plants are small.

ABOVE *Pseuderanthemum atropurpureum* 'Tricolor'
ABOVE LEFT *Pseuderanthemum reticulatum*

PSEUDOPANAX

Araliaceae

Origin: China, New Caledonia, Tasmania, New Zealand and Chile. A genus of about 20 species of evergreen trees and shrubs, with alternate leaves which are totally different in juvenile and adult stages. Seedlings produce solitary stems with long, linear, usually toothed or lobed leaves which are spirally arranged and hang downwards. As they age, the stems branch at the top and develop shorter, broader leaves which are often digitate with three to five leaflets. The flowers are small, 5-petalled and carried in irregular compound umbels followed by purple-black berry-like fruits. Propagate by seed when ripe. *Pseudopanax* derives from the Greek *pseudo*, false and the genus *Panax*.

Species cultivated

P. arboreus (*Nothopanax arboreus*) New Zealand

Large shrub or small tree, 3–5m(10–16ft) tall in the open, half this in a container, with only a brief juvenile phase in the seedling stage. Leaves long-stalked, digitate with three to seven stalked leaflets 8–20cm(3–8in) long, oblong-ovate to oblong-elliptic, coarsely toothed, glossy green with a leathery texture. Flowers greenish, in large terminal clusters. Handsome when in fruit.

P. colensoi New Zealand

More compact than *P. arboreus*, but otherwise very similar, with the leaves divided into three to five sessile leaflets.

P. discolor Bushy shrub to 2m($6\frac{1}{2}$ft) in height when in a container, but much more if planted out. Leaves composed of three to five narrowly obovate, cuneate, sharply toothed, yellow to bronzy-green leaflets, 4–8cm($1\frac{1}{2}$–3in) in length. The neat, well branched and fairly slow-growing habit of this species makes it one of the best for containers.

BELOW *Pseudopanax colensoi*
BOTTOM *Pseudopanax discolor*

P. ferox New Zealand
Similar to *P. crassifolius*, but with shorter juvenile leaves to only 50cm(20in) in length, with small, hooked marginal spines. Adult leaves simple, 5–15cm(2–6in) long, oblong-linear to oblanceolate. Flowers yellowish, the males in terminal racemes, the females in dense, simple umbels.

PSEUDORHIPSALIS See DISOCACTUS
PSEUDOSASA See ARUNDINARIA
PSEUDOWINTERA See DRIMYS

PTERIS

Pteridaceae
Brake ferns

Origin: Cosmopolitan. A genus of about 250 species of tufted to clump-forming ferns with long-stalked pinnate to bipinnate fronds with marginal spores covered by their rolled-under edges. They make good container plants for home or conservatory. Propagate by division at potting time or by spores in spring at 18°C(66°F).

Species cultivated
P. argyraea See *P. quadriaurita* 'Argyraea'.
P. cretica Ribbon fern, Table fern, Cretan brake Tropics and sub-tropics
Fronds 30–45cm(1–1½ft) long, pinnate, with five to 13 narrowly lanceolate to strap-shaped pinnae, 7–15cm(2¾–6in) long, the basal ones often divided into two or three lobes. Many cultivars are available including; 'Albo-Lineata', pinnae slightly broader with a wide central whitish band; 'Childsii', pinnae lobed, waved and frilled, light green; 'Major', robust, with the lower pinnae often deeply cut; 'Major Cristata' ('Wimsettii'), fronds irregularly lobed and toothed ends of the pinnae often crested; and 'Mayi', like a dwarf 'Albo-Lineata', with crested tips to the pinnae.
P. ensiformis Sword brake E. Asia to Australia
Fronds 30–50cm(12–20in) long, bipinnate with lanceolate to linear pinnae; sterile fronds smaller, with somewhat broader lobes. 'Victoriae' has an irregular white band along the centre of the pinnae.
P. multifida (*P. serrulata*) Spider fern E. Asia
Fronds 45–60cm(1½–2ft) in height by 20–30cm(8–12in) wide, pinnate, the upper pinnae linear and entire, the lower ones twice or thrice forked; midrib of frond winged between the pairs of pinnae. A fern of slender elegance once much grown and now making a modest come-back, especially in the more robust, crested cultivar 'Rochfordii'.
P. serrulata See *P. multifida*.
P. quadriaurita (*P. biaurita*) Tropics
Robust species with fronds to 60cm(2ft) long, more in the wild; they are bi- to tri-pinnate, the ovate to lanceolate pinnae being finely toothed. 'Argyraea' (*P. argyraea*) is the form usually grown, the pinnae having a central white band.
P. vittata Ladder fern Tropics and sub-tropics of Asia, Africa, Australia; naturalized elsewhere.
Fronds to 90cm(3ft) or more in height, simply pinnate, each pinna oblong to strap-shaped, slender-pointed, 10–15cm(4–6in) long, toothed and dark green. Fast growing and of easy culture.

ABOVE *Pteris cretica* 'Albo-Lineata'
RIGHT *Pteris cretica* 'Childsii'

PTYCHOSPERMA

Palmae

Origin: Northern Australia, New Guinea, Solomon Islands. A genus of 30 to 38 species of palms, one of which makes an elegant pot or tub plant. Propagate by seed in spring. *Ptychosperma* derives from the Greek *ptyche*, a fold and *sperma*, a seed, alluding to the grooved seeds of some species.

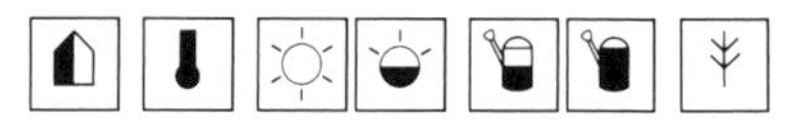

Species cultivated
P. alexandrae See *Archontophoenix alexandrae*.
P. cunninghamiana See *Archontophoenix cunninghamiana*.
P. elegans (*Seaforthia elegans*) Alexander/Solitare palm Queensland
Slender-stemmed, graceful small palm in the wild, fairly slow-growing, taking some time to reach 2m(6½ft) in a container. Leaves 60–120cm(2–4ft) in length (almost twice this on mature trees), pinnate, composed of many linear leaflets. The small green and white fragrant flowers and red, berry-like fruits are not produced on young plants.

PUNICA

Punicaceae

Origin: S.E. Europe to Himalaya, and Socotra. A genus of two species of deciduous small trees and shrubs with often spine-tipped branches and opposite or almost opposite entire leaves, which are coppery when young, becoming shining green. The flowers have five to eight rather crinkled petals emerging from leathery, tubular calyces with the same number of lobes as petals. The species described below is a good pot plant for the conservatory or home window sill, especially the dwarf form. Propagate by seed in spring or summer, or by softwood cuttings in summer. *Punica* derives from the Latin *malum punicum*, Apple of Carthage.

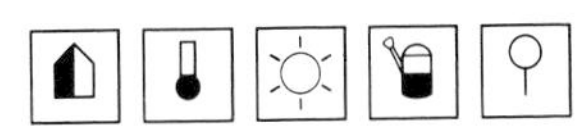

Species cultivated

P. granatum Pomegranate S.E. Europe to the Himalayas
Deciduous shrub to 2m(6½ft) or so in a pot, sometimes over 5m(16ft) in the wild. Leaves 2–8cm(¾–3in) long, oblong to lanceolate. Flowers to 4cm(1½in) across, cup-shaped, orange-red, opening in summer and followed by rounded fruits 6–10cm(2½–4in) wide with a leathery, brownish-yellow skin flushed with red. *P.g.* 'Nana' is the form best suited to window sill culture, flowering when less than 30cm(1ft) tall; 'Flore Pleno' has double red flowers; 'Legrellei' ('Mme Legrelle', 'Variegata') has orange-red flowers with whitish streaks.

PYCNOSTACHYS

Labiatae

Origin: South and central Africa, Malagasy. A genus of about 35 species of annuals and perennials related to *Coleus*. They have mostly erect, 4-angled stems, opposite pairs of leaves and terminal dense spikes of tubular, 2-lipped flowers. The species described have long been grown in greenhouses and conservatories for their winter blooming and are worth trying in the home. Propagate by cuttings. *Pycnostachys* derives from the Greek *pyknos*, dense and *stachys*, a (flower) spike.

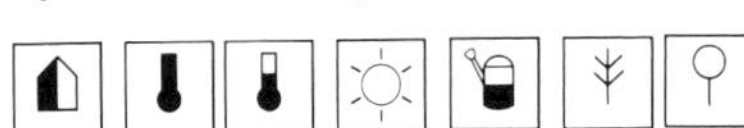

Species cultivated

P. dawei Central Africa
Stems sturdy, 1.2–1.8cm(4–6ft) tall, branching. Leaves lanceolate to narrowly so, 13–30cm(5–12in) long, pointed, toothed, red-brown, glandular beneath. Flowers 2cm(¾in) long, rich bright blue, in ovoid spikes to 8cm(3in) or more in length.
P. urticifolia Central Africa
Stems 1.2–1.8cm(4–6ft) in height, strong, but slimmer than those of *P. dawei*. Leaves to 10cm(4in) long, ovate, prominently toothed, densely pubescent. Flowers 2.5cm(1in) long or a little more, bright blue, in conical spikes to 10cm(4in) or more long.

PYROSTEGIA

Bignoniaceae

Origin: Tropical South America. A genus of four of five species of tendril climbers formerly included in *Bignonia*. The one cultivated species makes a splendid adornment to the roof space of the larger conservatory. Propagate by cuttings in summer. *Pyrostegia* derives from the Greek *pyr*, fire and *stege*, a roof, reputably from the colour and shape of the upper lips of the flowers, though the overall dense display of fiery blossom would seem more apt.

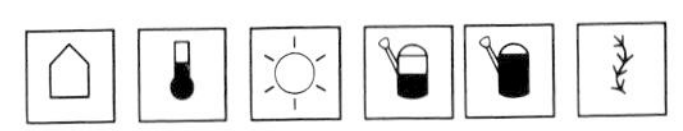

Species cultivated

P. venusta (*P. ignea, Bignonia venusta*) Flame flower/vine, Flaming trumpets Brazil and Paraguay
Evergreen climber to 10m(33ft) or more in height. Leaves composed of two or three ovate leaflets, to 8cm(3in) long, and a trifid tendril. Flowers tubular with five rolled back petal lobes fringed with white hairs, bright reddish-orange, in profuse clusters from autumn to spring, sometimes also in summer depending on temperature.

PYRRHEIMIA See SIDERASIS

PYRROSIA

Polypodiaceae

Origin: Tropical and temperate regions of Asia and Africa. A genus of about 100 species of rhizomatous ferns closely related to *Polypodium* and *Microsorium*. They are mostly epiphytes roaming along mossy trunks and branches, with scaly rhizomes and alternate, largely simple fronds. The fertile fronds bear interveinal rows of sori which may be large and close together. The species described make a nice contrast among dissected, fronded ferns and can be effectively grown in hanging baskets. Propagate by cuttings from the rhizomes, or spores in spring. *Pyrrosia* derives from the Greek *pyrros*, flame-coloured, referring to the reddish, hair-like scales on the rhizomes.

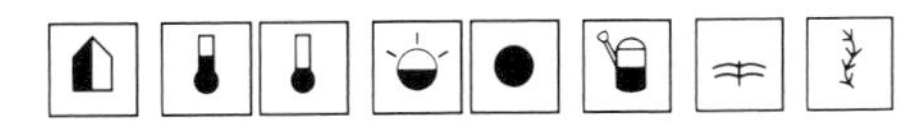

Species cultivated

P. lingua China, Japan, Taiwan
Rhizomes slender, eventually to 1.2m(4ft) or more, branching, and easily pruned to size, covered with rusty scales. Fronds 10–25cm(4–10in) long, lanceolate, wavy, brown-felted when young then rich green above. 'Cristata' ('Obake') has broad leaf tips which are several times deeply forked. 'Montrifera ('Lacerata', 'Hagoromo') has deeply and irregularly fringed leaf margins. 'Nankin-shishi' is similar to 'Cristata' but with more finely and densely crested leaf tips. 'Variegata' and 'Nakogriri-ba' produces fronds with rounded teeth and a pattern of yellowish streaks and bands.
P. longifolia Australia, Malaysia, Polynesia
Rhizomes usually shorter and thicker than those of *P. lingua*. Fronds strap-shaped, 20–60cm(8–24in) in length, thick-textured, grey-hairy beneath, subglossy above, arching to pendent.

Pyrrosia longifolia

QR

ABOVE *Ravenala madagascariensis*
TOP *Quisqualis indica*

QUISQUALIS

Combretaceae

Origin: Tropical and South Africa to Indo-Malaysia. A genus, depending on the botanical authority, of four or 17 species of evergreen climbers one of which is widely grown in the tropics and in greenhouses and conservatories in cooler climates. It is a most decorative plant, but requires plenty of room to grow and flower well. Propagate by cuttings in spring or summer. *Quisqualis* is derived from the Latin *quis*, who? and *qualis*, what?, a translation of the native Malay name expressing surprise at the change from shrub to climber.

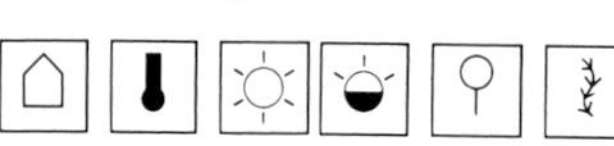

Species cultivated

Q. indica Rangoon creeper Burma, Malaya, New Guinea, Philippines
When planted out stems of 7.5m(25ft) can be expected, not much more than half this in a tub. Leaves elliptic to oblong, slender-pointed, 8–15cm(3–6in) long, red-brown pubescent when young; at least some of the stalks remain as thorn-like stubs when old leaves fall. Flowers fragrant, slenderly tubular, 4–7cm(1½–2¾in) long, formed of a combined calyx and corolla topped by five small sepals and five larger petal lobes. The petals expand in the evening and are then white; the following day they change to pink which gradually deepens in tone; on the succeeding day they become deep carmine to blood-red (a dramatic change). Seed pods are about 3cm(1¼in) long, ellipsoidal with five narrow wings along their entire lengths. Plants from seed start off as non-climbing bushes.

RADERMACHERA

Bignoniaceae

Origin: India to China, Philippines, Java. A genus of 40 species of evergreen trees and shrubs with opposite pairs of pinnate to tripinnate leaves and panicles of funnel-shaped blooms with five rounded lobes. One species has recently become popular as a young specimen for its foliage alone (in the same way as *Jacaranda*) and makes a handsome specimen for the home and conservatory. Propagate by imported seed. *Radermachera* commemorates J.C.M. Radermacher (*d.* 1783), a Dutch amateur botanist resident in Java.

Species cultivated

R. sinica (*Stereospermum sinicum*) China
In the wild, a small tree, but easily kept to 1.2m(4ft) or so by pruning. Leaves tripinnate, partially bipinnate when young, the leaflets 3–6cm(1¼–2½in) long, ovate, abruptly slender-pointed, prominently veined, lustrous rich green, especially when young. The yellow flowers are not produced on young containerized specimens. Sometimes wrongly sold as *S. suaveolens*, a distinct, simply pinnate-leaved species.

RAPHIDOPHORA See RHAPHIDOPHORA
RAPHIOLEPIS See RHAPHIOLEPIS

RAVENALA

Strelitziaceae (*Musaceae*)

Origin: Malagasy. A genus of one species of palm-like plant allied to *Musa* and *Strelitzia*. When young it makes an imposing foliage plant for the conservatory or large room. Propagate by seed in spring.

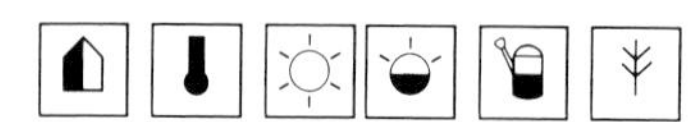

Species cultivated

R. madagascariensis Travellers' tree
In the wild, eventually a tree to 10m(30ft) tall with an erect palm-like trunk and fan-shaped head of long-stalked leaves to 3m(10ft) long. In containers the trunk takes several years to develop and remains short. Leaves to 1.5m(5ft) or so, the blade oblong, lustrous rich green. Somewhat strelitzia-like white flowers arise in short clusters in the leaf axils, but not when grown in a container. The bases of the closely overlapping leaves are hollow and hold rainwater which can be tapped and drunk in an emergency, hence the name, travellers' tree.

REBUTIA

Cactaceae

Origin: Bolivia and Argentina. A genus of 27 species of small cacti with globular or somewhat elongated, solitary or clustered stems. They have prominent areoles and freely borne colourful flowers opening around the base of the stem in late spring, making good pot plants for the home window sill or conservatory. Propagate by seed in spring or by offsets in summer. *Rebutia* was named for Mons. P. Rebut, a French cactus dealer at the end of the nineteenth century.

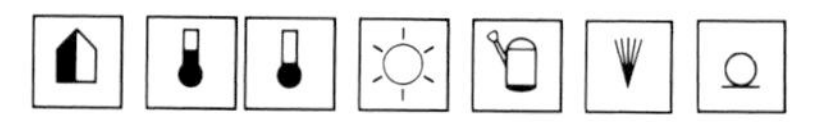

Species cultivated

R. aureiflora (*Mediolobivia aureiflora*)
N. Argentina
Stems clustering, globular to shortly cylindrical,

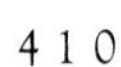

about 4cm(1½in) tall, bearing spiralling rows of 6mm(¼in) tubercles. Radial and central spines similar, bristle-like, 6–9mm(¼–⅓in) long, white or brownish. Flowers 4cm(1½in) long and wide, yellow with a white throat.

R. chrysantha Argentina
Stems 5–6cm(2–2½in) tall, globular, becoming shortly cylindrical, clustered; tubercles spiralling around the stem. Radial spines 25 to 30, bristle-like, 1–2cm(⅜–¾in) long, whitish yellow. Flowers 2cm(¾in) long, brick-red from an orange tube.

R. deminuta Argentina
Stems globular, clustering, 5–6cm(2–2½in) tall; tubercles in 11 to 13 spiralling rows. Radial spines ten to 12, white with brown tips, 6–10mm(¼–⅜in) long. Flowers to 3cm(1¼in) long, deep orange-red.

R.d. grandiflora
Similar to the species, but with larger flowers.

R.d. pseudominuscula
One to four short central spines and 2.5cm(1in) long deep purple flowers.

R. fiebrigii Bolivia
Stems globose, to 6cm(2½in) tall, clustering; tubercles in 18 rows. Radial spines 25 to 40, bristle-like, whitish, to 2cm(¾in) long. Flowers 3.5–4cm (1⅓–1½in) long, bright yellow-red.

R. krainziana Bolivia, exact origin uncertain
Stems globular, to 5cm(2in) tall, clustering; tubercles small. Radial spines eight to 12, small and bristle-like, white. Flowers about 3.5cm(1⅓in) long, orange-red.

R. kupperiana Argentina
Stems cylindrical, to 4cm(1½in) thick, branching; tubercles small. Radial spines 14 to 18, needle-like, brown, to 5mm(⅕in) long. Flowers to 4cm(1½in) long, brick-red.

R. marsoneri Argentina
Stems broadly globular, 5cm(2in) tall and up to 10cm(4in) broad, solitary, dark green. Radial spines 30 to 35, small, bristly, white. Flowers 4–5cm(1½–2in) long, golden-yellow.

R. minuscula Red crown cactus, Mexican sunball N.W. Argentina
Stems solitary or clustering, globular, somewhat flattened and sunken at the top, 4–5cm(1½–2in) thick; tubercles in about 20 spiralling rows, pale-green. Radial spines about 30, white, 3–6mm (⅛–¼in) long. Flowers to 4cm(1½in) long, crimson.

R. senilis Fire crown cactus Argentina
Stems globular, to 7cm(2¾in) tall and wide, clump-forming, pale green; tubercles spiralling. Radial spines 25 to 40, glassy-white tipped with brown, to 1.3cm(½in) long. Flowers to 5cm(2in) long, carmine-red.

R. spinosissima Argentina
Stems globular, to 4cm(1½in) tall, clustering; tubercles spiralling. Radial spines numerous, whitish, to 1cm(⅜in) long. Flowers 4cm(1½in) long, deep pink to brick-red.

R. violaciflora Argentina
Stems globular, 4–5cm(1½–2in) tall, clustering; tubercles spiralling, in 20 to 25 rows. Radial spines 15 to 25, white to yellowish, 5–10mm($\frac{3}{16}$–⅜in) long. Flowers to 4cm(1½in) long, magenta-red.

R. xanthocarpa Argentina
Stems globular, 4–5cm(1½–2in) tall, clustering. Radial spines numerous, to 5mm(⅕in) long, bristle-like and glassy-white. Flowers 2cm(¾in) long, red.

LEFT *Rebutia aureiflora*
BELOW LEFT *Rebutia minuscula*

RECHSTEINERIA See SINNINGIA

REHMANNIA

Scrophulariaceae (*Gesneriaceae*)

Origin: China. A genus of eight species of tufted, evergreen perennials with alternate, pinnately lobed leaves and tubular to bell-shaped flowers in spike-like racemes. The species described is a good plant for the conservatory and for flowering in the

Rebutia violaciflora

home. Propagate by seed sown in late spring, potting-on regularly for blooming in the following summer. They are best raised afresh in this way treating them as biennials, though the plants can be cut for flowering in the next season, but results are rarely as good. *Rehmannia* was named for Joseph Rehmann (1753–1831), a German botanist.

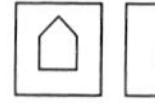 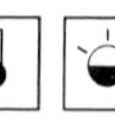  

Species cultivated

R. elata (*R. angulata* of gardens) China
Tufted perennial to 30cm(1ft) or more tall. Lowest leaves the longest, to 20cm(8in), elliptic to ovate, with several pairs of toothed, acute lobes. Flowers 6cm(2½in) long, tubular, the upper lip divided into two lobes, the lower into three, the tube yellow, red-dotted within, the lobes rose-purple.

REINECKEA

Liliaceae

Origin: Japan and China. A genus of one species of grassy perennial which is suitable for the conservatory and home window sill and also makes a good ground cover for a floor-level bed. Propagate by division in spring or by seed when ripe. *Reineckea* was named for Johann Heinrich Julius Reinecke (1799–1871), a German gardener.

 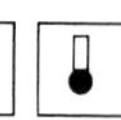 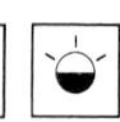 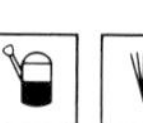

Reineckea carnea

Species cultivated

R. carnea (*Liriope hyacinthiflora*) Japan, China
Rhizomatous perennial forming dense clumps. Leaves 10–40cm(4–16in) long, all from ground level, linear to lanceolate, very dark green. Flowers 8–12mm(⅓–½in) long, bell-shaped, with six reflexed lobes, pale pink, in 8cm(3in) long spikes in late summer, followed by red, berry-like fruits. *R.c.* 'Variegata' has leaves with clear, longitudinal creamy-white stripes.

REINWARDTIA

Linaceae

Origin: N. India and China. A genus of two species of evergreen sub-shrubs with alternate, undivided leaves and a long succession of short-lived, tubular, symmetrical flowers with five spreading petals. A good conservatory and house plant. Propagate by cuttings in spring with bottom heat, keeping cool when potted until flowering time. They can be cut back when flowering finishes and re-potted when they start into growth, but the best plants are those raised annually from cuttings. *Reinwardtia* was named for Caspar Carl Reinwardt (1773–1854), Professor of Botany at Leiden, Holland, and the founder of the Bogor Botanic Garden, Java.

 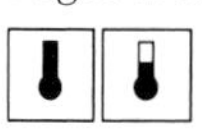  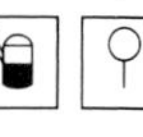

Species cultivated

R. indica (*R. trigyna*) Yellow flax
Softly shrubby, 60–90cm(2–3ft) tall. Leaves 5–10cm(2–4in), alternate, elliptic to oblong-obovate ending in a fine point. Flowers 3–5cm(1¼–2in) across, tubular, opening wide, bright yellow, borne singly or in small terminal clusters from late autumn onwards to spring.

RHAPHIDOPHORA

(RAPHIDOPHORA)

Araceae

Origin: Indo-Malaysia, New Caledonia. Depending upon the botanical authority, a genus of 60–100 species of creeping and climbing plants allied to *Epipremnum*, *Scindapsus* and *Monstera*. The species described here climb by aerial roots and have alternate, ovate leaves which in some cases are pinnately cut to the midrib. Some species, not described here, have their leaves perforated in the same way as *Monstera*. The seldom-produced petalless flowers are borne in a spadix, subtended by an open boat-shaped spathe. In the conservatory and home this genus has the same uses and similar effect as *Philodendron*. Propagate by cuttings of stem tips or one-leaved stem sections. *Rhaphidophora* derives from the Greek *rhaphis*, a needle and *phoros*, to bear, referring to the sharp crystals of calcium oxalate in the plant cells, particularly in those of the fruits.

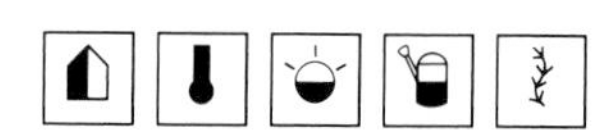

Species cultivated

R. aurea See *Epipremnum aureum*.

R. celatocaulis (*Monstera latevaginata*) Shingle plant Borneo
High climbing in the wild, but to 2m(6½ft) or so in containers. Leaves of juvenile plants elliptic-ovate, cordate, glaucous, on short winged stalks to 30cm(1ft) or so in length, closely set and partially overlapping (fancifully like the shingles of a roof); adult leaves larger, pinnatifid, green.

R. decursiva Shingle plant India to Vietnam
Stiffly climbing, eventually to several metres (yards) in height. Juvenile leaves similar to those of *R. celatocaulis*, the adult ones to 40cm(16in) long, pinnatifid, dark lustrous green.

RHAPHIOLEPIS (RAPHIOLEPIS)

Rosaceae

Origin: Sub-tropical areas of East Asia. A genus of about 14 species of evergreen shrubs with alternate, undivided leathery leaves which are a dark, glossy green, and 5-petalled flowers in terminal racemes or panicles. They grow well in pots and tubs and are suitable for a conservatory, flowering when small; they can also be brought into the home when in bloom and stood outside on a patio in summer.

Propagate by seed when ripe or by cuttings in late summer with bottom heat. *Rhaphiolepis* derives from the Greek *rhaphis*, a needle, referring to the pointed ends to the bracts below the inflorescences.

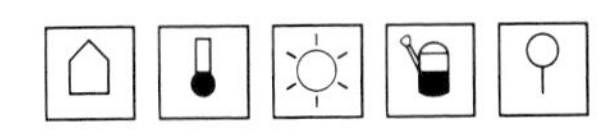

Species cultivated

R.* × *delacourii
Hybrid between *R. indica* and *R. umbellata* making a rounded shrub to 2m(6½ft) or more in height. Leaves 4–9cm(1½–3½in) long, obovate, toothed above the middle. Flowers 1.3–2cm(½–¾in) across, rose-pink, in pyramidal panicles to 10cm(4in) long, from late winter and spring to late summer.

R. indica S.W. China
Rounded shrub 1–1.5m(3–5ft) tall. Leaves 5–8cm(2–3in) long, oblong to obovate-lanceolate, toothed along the margins. Flowers about 1.5cm(⅝in) wide, white with a pinkish flush and deeper pink stamens, carried in terminal racemes to 8cm(3in) long from late winter through spring and summer to autumn.

R. umbellata Korea, Japan
Rounded shrub to 3m(10ft) or more. Leaves 4–9cm(1½–3½in) long, obovate to broadly elliptic, thick in texture, with in-rolled margins. Flowers 2cm(¾in) across, pure white, fragrant, in stiff racemes or panicles 7–10cm(2¾–4in) long in summer, followed by blue-black, pear-shaped fruits.

RHAPIS (RHAPHIS)

Palmae

Origin: Southern China to Java. A genus of nine to 15 species of very small clump-forming palms with bamboo-like stems. Those described below thrive in containers, making elegant and neat foliage plants. They have erect, unbranched stems which sucker freely at the base, and palmate/digitate leaves, the stalks of which are fibrous and sheath the main stem. The flowers are dioecious, yellowish, and borne in short clusters among the leaves, but they are not conspicuous. Propagate by suckers, division and seed in spring. *Rhapis* derives from the Greek *rhapis*, a needle, presumably referring to the narrow leaflets or their pointed tips.

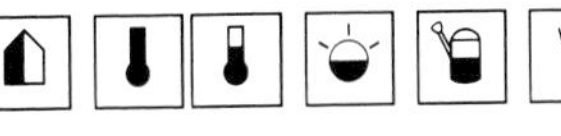

Species cultivated

R. excelsa Ground rattan, Bamboo/Miniature fan palm, Little lady palm Southern China
In containers, to about 1.5m(5ft) in height (to twice this in the wild). Stems covered with coarse leaf sheath fibre. Leaves composed of three to ten lanceolate, deep lustrous-green, somewhat puckered leaflets 20–30cm(8–12in) long. A very variable species in height and leaflet number. Long cultivated in China and Japan; in the latter country there is a society devoted to collecting and cultivating variants of this and *R. humilis*, at least a hundred of which have been named. Some of these are now in cultivation in the West. *R.e.* 'Variegata' has the leaflets striped ivory-white.

R. humilis Southern China
Much like a smaller, slimmer version of *R. excelsa* with reed-like stems and leaves composed of nine to 20 narrower, shorter, smooth-surfaced leaflets. Dwarf and variegated forms are cultivated in Japan and some have been tried in the West.

ABOVE *Rhaphiolepis indica*
ABOVE LEFT *Rhaphiolepis* × *delacourii*

Rhapis excelsa

RHINOPETALUM See FRITILLARIA

RHIPSALIDOPSIS

Cactaceae

Origin: Brazil. A genus of two species of woody-based, somewhat shrubby epiphytic cacti with leafless stems which are branched and jointed. Between the joints they are flattened and appear leaf-like. The flowers are tubular at the base, the many narrow tepals opening to a wide, funnel-

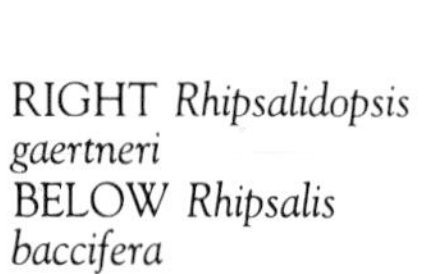

RIGHT *Rhipsalidopsis gaertneri*
BELOW *Rhipsalis baccifera*

shaped bloom, the stigmas having narrow, spreading lobes. They make particularly rewarding hanging basket plants. Propagate by cuttings, taking pieces with one to three joints, preferably putting three cuttings around a 7.5cm(3in) pot and growing them together to make more of a display than a single shoot can. *Rhipsalidopsis* derives from the genus name *Rhipsalis* and the Greek *opsis*, like, referring to the resemblances between the genera.

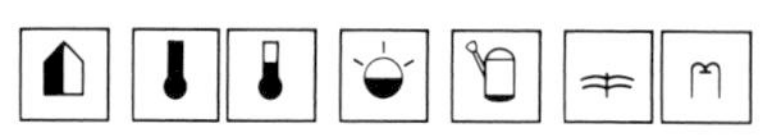

Species cultivated

R. gaertneri (*Schlumbergera gaertneri, Epiphyllopsis gaertneri*) Easter cactus
Stems spreading, eventually pendulous, forming a plant 20–30cm(8–12in) high. Leaf-life internodes 4–8cm(1½–3in) long, with a crenate margin, purplish-green, carrying minute areoles with a few white bristles. Flowers 6–9cm(2½–3½in) wide, bright scarlet, the long narrow tepals ending in a point, opening in spring.

R × graeseri
Hybrid between *R. gaertneri* and *R. rosea* and generally intermediate between the two species. Flowers usually 3–4cm(1¼–1½in) wide, red, opening in spring and summer.

R. rosea
To 15cm(6in) high and wide, pendent. Leaf-like internodes 2.5–4cm(1–1½in) long, crenate, green but sometimes flushed with red margins. Flowers 3–4cm(1¼–1½in) wide, rose-pink to magenta-pink, very freely borne in late spring and early summer. A rewarding house plant.

RHIPSALIS

Cactaceae

Origin: The Americas from southern USA to northern South America, with a few species in Africa and Sri Lanka though these may be introductions. A genus of 60 species of somewhat shrubby, freely branched epiphytic cacti. They have often pendent, leafless stems which are cylindrical or flattened and blade-like. The small, funnel-shaped flowers have few tepals and are followed by rounded, slightly sticky berries which are often white and semi-transparent. The plants are best in hanging baskets, but can be gown in pots or pans. Propagate by cuttings in summer or by seed in spring.

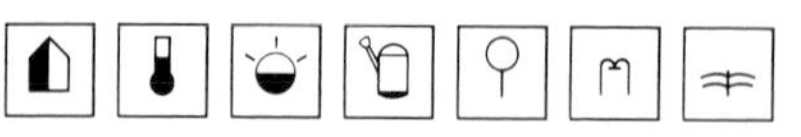

Species cultivated

R. anceps (*Lepismium anceps*) E. Brazil
Stems freely branching, about 45cm(1½ft) in length, usually arching to pendent, flattened or 3-angled, shallowly notched, often purplish-margined. Flowers small, violet and white.

R. baccifera (*R. cassutha, R. cassytha*) Mistletoe cactus Florida to Brazil and Peru, Africa, Sri Lanka
Stems pendent, capable of reaching 2m(6½ft) or more in length, cylindrical, light green. Flowers 6mm(¼in) across, greenish, followed by translucent whitish berries 4–5mm(⅙–⅕in) wide.

R. capilliformis E. Brazil
Stems hanging vertically, freely branching, to 90cm(3ft) or more in length; main stem joints 10–15cm(4–6in) long by about 3mm(⅛in) thick, lateral joints much shorter and somewhat thinner, often borne in worls of three to seven, pale green. Flowers about 6mm(¼in) long, lustrous greenish-white. Fruits white, globular, 6mm(¼in) wide.

R. cassutha (*R. cassytha*) See *R. baccifera*.

R. crispata Brazil
Stems more or less erect, to 40cm(16in) or more; stem joints elliptic to oblong, crenate, 8–13cm(3–5in) long, pale to yellowish-green. Flowers in clusters of one to four, pale yellow, 1.2cm(½in) wide. Fruits white, globular, 6mm(¼in) wide.

R. cruciformis (*Lepismium cruciforme*) Brazil
Stems trailing to pendent, to 60cm(2ft) or more long; stem joints 15–30cm(6–12in) long, 3- to 5-ribbed or winged, tapering towards the tips, shallowly notched with some white bristles, deep, almost glossy green. Flowers in clusters of one to five, 1.2cm(½in) wide, white. Fruits almost the same width, globular, rich purplish-red.

R. houlletiana Snowdrop cactus Brazil
Stems pendent, eventually to 1m(3ft) or more long, flattened, 2.5–5cm(1–2in) wide, toothed or with sharp tooth-like lobes. Flowers 2cm(¾in) long, bell-shaped, white, often with a red centre, followed by 6mm(¼in) wide, red fruits.

R. micrantha Ecuador, Peru
Stems pendent or scrambling, to 1m(3ft) or more long, flattened, sometimes 3- to 4-angled. Flowers 6mm(¼in) across, white, followed by 7mm(⅓in) wide, white or reddish fruits.

R. pachyptera E. Brazil
Stems erect to spreading, eventually to 60cm(2ft) or more in length; stem joints flattened, thick, broadly elliptic to oblong, with reddish, notched margins, 8–20cm(3–8in) long. Flowers 1.5cm(⅝in) long, bright yellow with red tepal tips. Fruit globular, red.

R. pilocarpa (*Erythrorhipsalis pilocarpa*) E. Brazil
Stems erect when young, then pendent to 90cm(3ft) or more, cylindrical; stem joints often

whorled, 5–12cm(2–4¾in) in length, grey-white bristly-hairy. Flowers about 2cm(¾in) wide, white or pale pink. Fruits broadly ovoid, purple-red with tufts of white bristles, 1.2cm(½in) long. A pretty species both in flower and fruit.

R. rosea See *Rhipsalidopsis rosea.*

R. salicornioides (*Hatiora salicornioides*, *Hariota salicornioides*) Brazil

Stems erect at first, spreading and arching with age, 30–40cm(12–16in) high or long, very freely branching; stem joints in whorls of three to five, club-shaped (clavate), 1–3cm(⅜–1¼in) long by 4–8mm(⅙–⅓in) thick. Flowers 1.2cm(½in) long, greenish-yellow flushed red in bud, opening bright yellow. Fruits obovoid, translucent white. A plant of unique appearance seemingly formed of tiny green bottles joined end to end. In the USA at least, known as Dancing bones, Drunkard's dream and Spice cactus.

R. shaferi Paraguay, N. Argentina

Stems cylindrical, arching or pendent, to 1m(3ft) or more in length. Flowers 9mm(⅜in) across, white, followed by small 3mm(⅛in) wide, white or pink fruits.

RHODOCHITON

Scrophulariaceae

Origin: Mexico. A genus of one species of climbing plant grown for its fascinating and colourful

FAR LEFT *Rhipsalis cruciformis*
LEFT *Rhipsalis salicornioides*

flowers. It is best kept in the conservatory, but can be tried in the home. Propagate by seed in early spring.

Rhodochiton derives from the Greek *rhodo*, red and *chiton*, a cloak, referring to the large, flared calyx which, however, is rather more like a wide bell than a cloak.

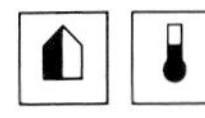 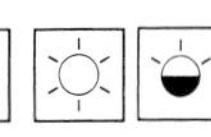  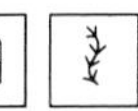

Species cultivated

R. atrosanguineum (*R. volubile*) Purple bell-vine

Capable of attaining 3m(10ft) in height, but often no more than half this in containers; stems slender, becoming woody with age. Leaves alternate, ovate-cordate, 5–8cm(2–3in) long, with small widely spaced teeth, borne on slender stalks which act as tendrils like those of *Tropaeolum*. Each flower composed of a 3–4cm(1¼–1½in) wide, red calyx and a 4–5cm(1½–2in) long, deep purple-red tubular corolla with five spreading, rounded lobes of equal size, intermittently through the year if warm enough. Apt to be short-lived as a perennial, but will flower in the first year from seed and can be treated as an annual.

RHODODENDRON

Ericaceae

Origin: Temperate regions of the northern hemisphere, particularly the Himalayas to China, extending to the tropics in the mountains of S.E. Asia, from Malaysia to New Guinea, with one species in N.E. Australia. A genus of about 800 species of evergreen and deciduous shrubs with linear to ovate, entire, often leathery leaves. Their flowers are tubular, funnel- or bell-shaped with five to eight lobes, and can be solitary but are usually in umbel-like clusters terminating the stems. Half-hardy species make splendid tub specimens for the cool or unheated conservatory, while the tender species are equally effective under cool to temperate conditions. Most of the tender species are epiphytes in the wild and native to rain forests in the islands and adjacent mainland of S.E. Asia. They are often known as Vireya rhododendrons,

RIGHT *Rhodendron javanicum*
BELOW *Rhododendron macgregoriae.*
BELOW RIGHT *Rhododendron veitchianum*

but in fact very few are strictly members of this botanical section of the genus *Rhododendron* and they are best known as Malaysian rhododendrons. They thrive best in a bromeliad compost placed in orchid baskets, but will grow in pots of a standard all-peat, so-called 'ericaceous' compost ideally mixed with equal parts of coarse perlite, chopped bark or moss. Ordinary ground-dwelling, half-hardy rhododendrons thrive in almost any well drained, acid compost so long as the pH is around 5–6. A proprietary ericaceous mix is ideal. Although strictly conservatory plants and needing a humid atmosphere to thrive, individual specimens can be brought indoors when in bloom. *Rhododendron* (from *rhodon*, a rose and *dendron*, a tree) was the Greek vernacular name for the pink-flowered oleander, later chosen by Linnaeus as the scientific name for this genus.

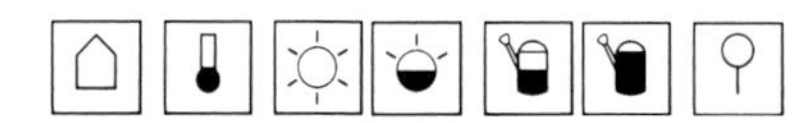

Species cultivated

R. bullatum See *R. edgeworthii.*

R. ciliatum Eastern Himalaya
Usually 60–90cm(2–3ft) tall and of wider spread, the stems covered with bristly hairs. Leaves oval to obovate, 5–10cm(2–4in) long, bristly-hairy above. Flowers in umbels of three to five, widely bell-shaped, to 6cm(2½in) across, cerise in bud opening pink and ageing white, early spring.

R. × cilipinense (*R. ciliatum × moupinense*)
Forms a compact rounded bush, but is otherwise similar to its first parent.

R. cubittii See under *R. veitchianum.*

R. edgeworthii (*R. bullatum*) Himalaya (lower slopes)
About 1.2m(4ft) in containers, but twice this in the wild, tending to get thin and untidy with age. Leaves 5–11cm(2–4½in) long, ovate to broadly elliptic, the upper surface finely puckered, boldly veined and a deep glossy green, lower surface thickly buff-felted. Flowers pink-tinted in bud, opening white, waxy-textured, funnel-shaped, fragrant. Handsome in foliage and flower.

R. javanicum Java
To about 1.2m(4ft) in height, or possibly more in time. Leaves 5–18cm(2–7in) long, lustrous green. Flowers in umbels up to 15cm(6in) across, each with a narrow tube and five orange-yellow petal lobes which open out to 4–6cm(1½–2½in) wide.

R. lochae Queensland, Australia (mountains)
In cultivation, usually under 60cm(2ft), but with a potential of several times this height (it is sometimes a large shrub in the wild); of rather scraggy habit with age unless judiciously tip-pruned when young. Leaves in whorls, ovate, 5–8cm(2–3in) long, more or less glossy rich green. Flowers bell-shaped, about 4cm(1½in) long, waxy-red in few-flowered umbels. The only rhododendron native (endemic) to Australia, and flowering regularly when small.

R. macgregoriae New Guinea
To 1.2m(4ft) or more in containers, with age somewhat scraggy, unless judiciously tip-pruned when young. Leaves usually in whorls of three or more, elliptic to ovate, up to 8cm(3in) in length, usually a rich green and sub-glossy. Flowers in umbels of seven to 17, each with a shortish tube and five obovate, apricot to clear yellow lobes opening out to 5cm(2in) wide. In cultivation, more robust, sometimes hybrid, forms with orange or reddish-orange flowers occur. One of the easiest and most showy of the Malaysian rhododendrons, having flushes of bloom throughout the year.

R. simsii (*R. indicum* in part, *R. indicum simsii, Azalea indica*) Azalea Southern China
Semi-deciduous to fully evergreen, bushy spreading habit to 1.5m(5ft) in the wild, but rarely more than 60–90cm(2–3ft) in containers. Leaves of two sorts: spring ones to 5cm(2in) long, elliptic-oblong; summer ones to 4cm(1½in) long, obovate to oblanceolate. Flowers grow up to 5cm(2in) wide, broadly funnel-shaped, in shades of red usually with darker markings. The primary parent of all the so-called Indian azaleas, which are popular as winter pot plants and with a wider colour range including shades of pink, purple and white.

R. veitchianum Burma, Thailand, Laos (mountains)
In the wild, a shrub to 3m(10ft) or more, but about half this in containers. Leaves elliptic to obovate, sometimes narrowly so, rich green above, scaly beneath, 5–10cm(2–4in) long. Flowers in umbels

of two to five, widely funnel-shaped, to 8cm(3in) wide, white, green-tinted in bud, the five lobes waved or crimped. *R. cubittii* (northern Burma) is closely allied, but has some bristles on the stem, oblong to lanceolate leaves with bristly-hairy margins and pink flowers with darker throats borne in spring.

RHODOHYPOXIS

Hypoxidaceae

Origin: South Africa. A genus of two species of small, rhizomatous perennials which make colourful pan plants for the conservatory or home window sill when flowering. Propagate by separating the rhizomes (the only method for named cultivars) or by seed sown in spring. Plants raised from seed often show a range of shades of pink, light red and white. *Rhodohypoxis* derives from the Greek *rhodo*, red and the genus *Hypoxis*.

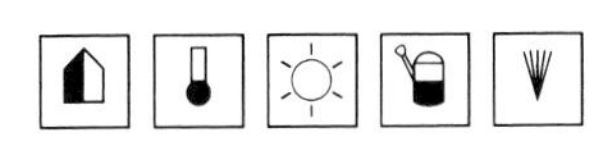

Species cultivated

R. baurii South Africa
Clump-forming perennial with ovoid, corm-like rhizomes. Leaves 5cm(2in) or more in length, all rising from the base of the plant, linear, pointed and white-hairy. Flowers 2cm(¾in) wide, pink to red, solitary, on stems equalling or shorter than the leaves, opening from summer to autumn. 'Allbrighton Red' is deep red; 'Fred Broome' is a creamy pink; 'Margaret Rose' is a bright pink; *platypetala* has white or pink-tinted flowers.

RHOEO

Commelinaceae

Origin: Mexico, Guatemala and the West Indies. A genus of one species of evergreen perennial which, together with its variegated forms, makes an undemanding conservatory and house plant. Propagate by seed in spring in warmth, or by cuttings of basal shoots in summer, the latter method only for the variegated forms. *Rhoeo* is of unknown derivation.

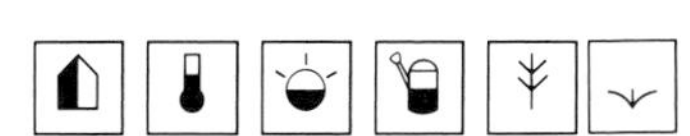

Species cultivated

R. spathacea (*R. discolor*) Boat lily, Moses-in-a-boat
Stems to 20cm(8in) long, solitary or branching at the base, erect, often becoming decumbent. Leaves to 30cm(1ft) long, overlapping at the base, linear to strap-shaped, rather succulent, dark glossy green above, purplish beneath. Flowers about 1.5cm(⅝in) across, 3-petalled, white, opening from deep, boat-shaped bracts periodically throughout the year. *R. s.* 'Variegata' ('Vittata') has the leaves longitudinally striped with pale yellow.

RHOICISSUS

Vitidaceae

Origin: Africa. A genus of ten to 12 species of evergreen climbers with alternate leaves and stem tips modified to form tendrils. The flowers are tiny, 5-petalled and greenish and are followed by small grape-like edible berries. Propagate by cuttings in summer with bottom heat or by seed when ripe, in warmth. *Rhoicissus* is of dubious derivation, perhaps from the Latin *rhoicus*, similar to *Rhus* (sumach) together with the Greek *kissos*, ivy. Alternatively, the first element may be *rhoia*, a pomegranate, though for unknown reasons.

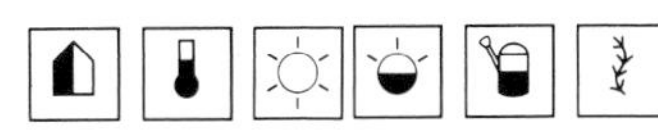

Species cultivated

R. capensis (*Cissus capensis*, *Vitis capensis*) Cape grape South Africa
Climber to 2m(6½ft) in a pot, but capable of reaching 5m(16ft), easily kept smaller by pruning. Leaves 20cm(8in) across, simple, cordate, deeply crenate, shining green. Fruits seldom appear on potted specimens.

R. rhomboidea South Africa
Climber to 6m(20ft), with trifoliate leaves, the leaflets rhomboid with the lateral ones asymmetrical. Tendrils unbranched. This species is probably not in cultivation, the plant sold under the name being *Cissus rhombifolia* which is easily separated by having forked tendrils.

RHYNCHOLAELIA See BRASSAVOLA
RICHARDIA See ZANTEDESCHIA

RICINUS

Euphorbiaceae

Origin: Africa, but now grown and naturalized throughout warm regions of the world. A genus of one species, a tender evergreen shrub whose seeds contain castor oil which is used not only medicinally, but in the manufacture of plastics, paints, varnishes, soaps, dyes and as a lubricant. The seed coats are very poisonous and in a home with small

BELOW *Rhoeo spathacea*
BELOW LEFT *Rhodohypoxis baurii*

children it is best to pick off any fruits which form (they do so only on large specimens). The plants can be grown in the home or conservatory for their handsome foliage and are often raised annually as short term foliage plants. Propagate by seed in spring in warmth. *Ricinus* is the Latin for a tick, an allusion to the size and shape of the seeds.

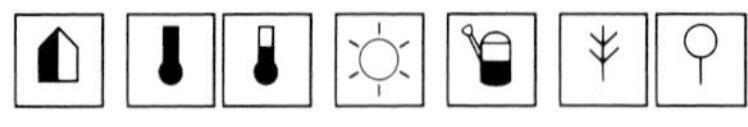

Species cultivated

R. communis Castor oil plant
In pots, to 1.5m(4ft) in height, but in the open to 5m(16ft) or more, making a shrub or small tree. Leaves alternate, 15–50cm(6–20in) in length, deeply palmately lobed, with five to 11 toothed lobes, often bronze or red-tinted when young, glossy. Flowers unisexual, very small, in clusters making up terminal panicles and followed by softly spiny, woody fruits which open explosively to release three mottled, beetle-like seeds. Several cultivars are available: 'Cambodgensis' has dark red-purple foliage; 'Gibsonii' is lower growing with shining dark red leaves; and 'Zanzibarensis' has bright green leaves with white veining.

RIVINA

Phytolaccaceae
Bloodberry

Origin: Tropical America. A genus of one to three bushy perennials one of which has long been cultivated as a pot plant for its chains of red or yellow berries. It is best in the conservatory but will grow in the home. Propagate by seed or cuttings in spring. *Rivina* is named for August Quirinus Rivinus (1652–1722), a Professor of Botany at Leipzig.

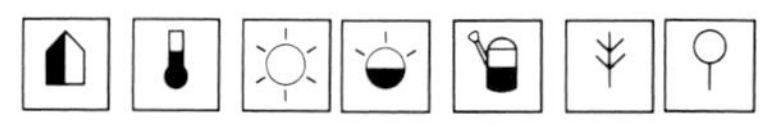

Species cultivated

R. humilis (*R. laevis*) Rouge plant Southern USA, Central America, West Indies
When well grown, erect and bushy, reaching 90cm(3ft) or more in height, but is usually less than this in pots. Leaves alternate, ovate to lanceolate, 5–10cm(2–4in) long. Flowers small with four greenish or pinkish sepals and no petals, in axillary racemes longer than the leaves. Fruits pea-sized, red, superficially like currants and usually borne in profusion. Best grown as an annual from seed, resulting in plants full of berries by autumn, these persisting through the winter.

R.l. aurantiaca
Yellow to orange berries.

BELOW *Rochea coccinea*
BELOW RIGHT *Romulea bulbocodium*

ROCHEA

Crassulaceae

Origin: South Africa. A genus of three to four succulent, evergreen shrubs closely allied to *Crassula* (stonecrops). They have pairs of entire leaves and 5-petalled flowers. The species described below is an attractive flowering pot plant. If kept more than one season the stems need cutting back to about 5cm(2in) in late winter. *Rochea* honors Daniel de la Roche (1743–1813), a Swiss doctor and botanist.

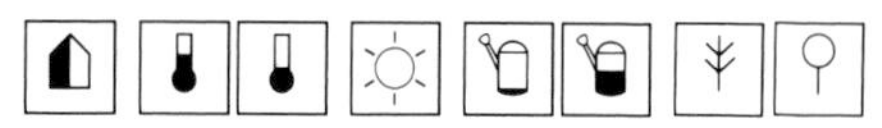

Species cultivated

R. coccinea (*Crassula coccinea*)
Stems erect, branching from the base, to 30–60cm(1–2ft) tall. Leaves 2.5–4cm(1–1½in) long, oblong to obovate, in four close ranks around the stem. Flowers 4cm(1½in), tubular, opening to five spreading lobes, carmine-red, borne in dense terminal clusters in summer.

ROMULEA

Iridaceae

Origin: Europe, the Mediterranean region including adjacent Atlantic islands, and South Africa. A genus of 75 to 90 species of rather crocus-like cormous perennials. They differ from crocus in having short-tubed flowers carried well above the ground on a stem, and their narrow leaves are entirely green. The species described is suitable for the conservatory and can be brought into the home for flowering. Propagate by offsets when planting or by seed in spring. *Romulea* was named for Romulus, supposedly the founder of Rome, and its first king.

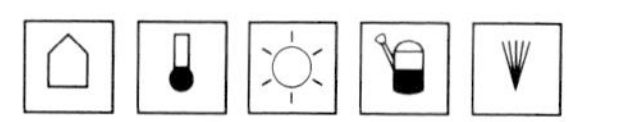

Species cultivated

R. bulbocodium Mediterranean, Western France
Leaves usually in fours, almost cylindrical, 5–8cm(2–3in) long. Flowers to 3cm(1¼in) wide, purplish to bluish-lilac with a yellow or white throat on 5–7cm(2–2¾in) stems in spring.

RONDELETIA

Rubiaceae

Origin: Tropical and sub-tropical America, West Indies. A genus of 100 to 120 species of evergreen

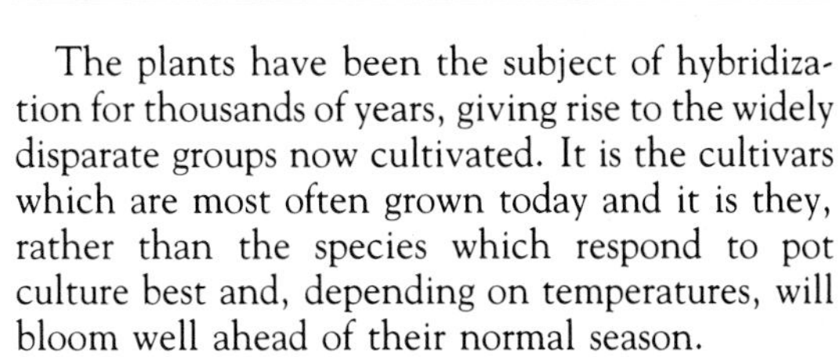

LEFT *Miniature rose*
FAR LEFT *Rondeletia amoena*

trees and shrubs, a few of which make decorative flowering specimens for the conservatory. They have opposite pairs of mainly simple leaves with stipules at their bases and small tubular flowers with four to five petal lobes arranged in dense terminal or lateral panicles. The fruits are dry and capsular. Propagate by cuttings in summer or by seed in spring. *Rondeletia* commemorates Guillaume Rondelet (1507–1566), Chanceller and Professor of Medicine and Natural History at the University of Montpellier.

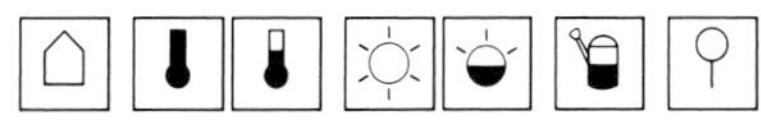

Species cultivated

R. amoena Southern Mexico to Panama
In the wild, a large shrub or small tree, but easily kept to 90cm(3ft) or so by pruning or regular propagation. Leaves ovate, 8–15cm(3–6in) long, densely brown-pubescent beneath. Flowers about 1cm($\frac{3}{8}$in) long, pink with a yellow-bearded throat, in panicles 5–15cm(2–6in) long in summer.

R. odorata (*R. speciosa*) Cuba, Panama
Usually a well branched shrub 90–180cm(3–6ft) in height. Leaves ovate to oblong-elliptic, about 5cm(2in) long, slightly convex and bullate, rich green. Flowers 1cm($\frac{3}{8}$in) long and wide, orange to brick-red with yellow throats, in corymb-like, terminal panicles 5–10cm(2–4in) wide in autumn. Reputedly fragrant, but not to all noses, or perhaps confused with its shorter-flowered variety *R. o. breviflora* which is scentless.

ROSA

Rosaceae
Roses

Origin: Northern temperate zone and mountains farther south. A genus of 100 to 250 species, depending on the classification followed. They are shrubs and woody climbers with alternate, pinnate leaves formed of toothed leaflets which are mostly deciduous though some are more or less evergreen. The flowers are 5-petalled, opening widely and are followed by rounded to ovoid hips (or heps) which are hollow receptacles filled with small, nutlet-like achenes (seeds).

The plants have been the subject of hybridization for thousands of years, giving rise to the widely disparate groups now cultivated. It is the cultivars which are most often grown today and it is they, rather than the species which respond to pot culture best and, depending on temperatures, will bloom well ahead of their normal season.

Potting should be carried out in autumn, choosing containers large enough to accommodate the root system. The best compost is a loam-based one such as John Innes Potting No. 2 or 3 (from a reputable source). The all-peat formulations can also be used, but feeding must be attended to weekly during the growing season. Once potted, the plants should be plunged in a sheltered site outside or placed in a cold greenhouse. Early in the new year the plants can be brought into a cool or temperate conservatory and hard pruned, removing at least two-thirds of the top growth. This treatment refers to Hybrid Tea (Large-flowered Bush), Floribunda (Cluster-flowered Bush), Dwarf Polyantha, Miniature and similar hybrid groups.

Species roses require light pruning only, removing very thin stems and dead or spindly tips. After blooming (first flush in the case of Hybrid Teas and other remontant-flowered cultivars) the plants should be dead-headed and put outside again. Repot annually or at least each other year with top-dressing in between. Prune annually in autumn or winter just prior to bringing into the conservatory. Miniature roses can be kept indoors when in bloom or until autumn when they must have a cool rest.

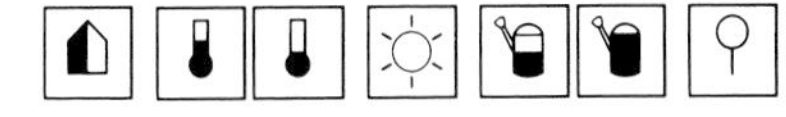

Species cultivated

See text above for cultivar groups. Consult a specialist catalogue for the vast range of named roses.

ROSEOCACTUS See ARIOCARPUS

ROSMARINUS

Labiatae

Origin: Mediterranean region. A genus of three species of evergreen, aromatic shrubs having

ABOVE *Roystonea regia*
ABOVE RIGHT
Rosmarinus officinalis

opposite pairs of leaves with rolled margins. The 2-lipped flowers are in shades of purple. They make attractive pot plants for the conservatory, adding a bonus of leaves for culinary purposes. Propagate by cuttings in late summer in cool conditions.

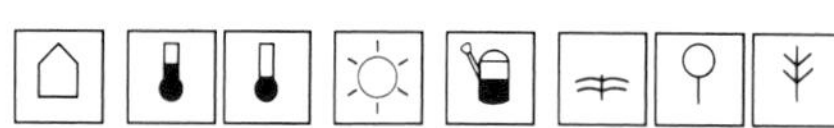

Species cultivated

R. lavandulaceum (*R. officinalis prostratus*)
Prostrate, mat-forming species. Leaves 1–1.5cm ($\frac{3}{8}$–$\frac{5}{8}$in) long, dark green above, grey-white-hairy beneath. Flowers 2–3cm($\frac{3}{4}$–1$\frac{1}{4}$in) long, violet-blue, in dense clusters from spring to summer.

R. officinalis Rosemary
Erect or semi-erect shrub, rarely over 1m(3ft) in a pot, more than this in the open garden. Leaves 2–5cm($\frac{3}{4}$–2in) long, dark green above, closely white-hairy beneath. Flowers to 2cm($\frac{3}{4}$in) long, blue to lavender-blue, in clusters of two or three from the leaf axils in spring and summer, sometimes in the autumn also. 'Albiflorus' has white flowers; 'Benenden Blue' is shorter with brighter blue flowers; 'Fastigiatus' ('Miss Jessop's Upright') has erect stems and strong growth; 'Roseus' has lilac-pink flowers; and 'Severn Sea' is dwarf with arching branches and bright blue flowers. Hardy.

ROYSTONEA

Palmae

Royal palms

Origin: Central and Tropical South America, West Indies and Florida. A genus of 14 to 18 species of large, fast-growing, statuesque palms, two of which are sometimes grown in containers when young. Adult specimens have smooth grey trunks, sometimes swollen at the base or the middle. The leaf blades are tightly rolled around each other forming a green stem-like structure known as a crown shaft. The pinnate leaves, which can reach 5–6m(15–20ft) in length, arch outwards, the lowest almost pendent. Large panicles of tiny flowers are borne immediately below the crown shafts. Propagate by seed in spring. *Roystonea* commemorates General Roy Stone (1836–1905), an American army engineer who served in Puerto Rico during the Spanish/American war.

Species cultivated

R. oleracea Trinidad, Barbados, Venezuela, Colombia
Leaves to 2m(6$\frac{1}{2}$ft) in length or more on tub-grown specimens; leaflets linear, rich green.

R. regia Cuban royal palm Cuba and perhaps adjacent islands
Leaves to 2m(6$\frac{1}{2}$ft) or more on containerized specimens; leaflets linear, arranged in four ranks.

RUELLIA

Acanthaceae

Origin: Tropics and sub-tropics, and North America. A genus of about 250 species of perennials and shrubs, split up into several genera by some botanists, but here considered as one. They have narrowly oblong to ovate leaves in opposite pairs and tubular to funnel-shaped flowers opening to five broad, spreading lobes at the mouth, borne in spikes or singly in the axils of the upper leaves. Propagate by division if possible, by cuttings with bottom heat in spring, or by seed also with heat in spring. *Ruellia* was named for Jean Ruel (1475–1537), a French physician and herbalist serving King Francois I.

Species cultivated

R. affinis Brazil
Small shrub to about 90cm(3ft) in height if allowed to grow naturally, no more than half this if pruned annually. Leaves elliptic, 8–13cm(3–5in) long, short-stalked. Flowers pale to deep red, 6–9cm (2$\frac{1}{2}$–3$\frac{1}{2}$in) long, solitary in the upper leaf axils.

R. amoena See *R. graecizans*.

R. devosiana Brazil
Sub-shrubby, producing spreading stems 20–45cm(8–18in) in height. Leaves up to 8cm(3in) long, narrowly elliptic, velvety-hairy, the upper surface a deep green bearing a white vein pattern, the lower surface purple. Flowers 4cm(1$\frac{1}{2}$in) in length, white, the throat lilac and sometimes with streakings of the same colour elsewhere.

R. graecizans (*R. amoena*) Brazil
Sub-shrubby with spreading stems 30–60cm(1–2ft) tall. Leaves oblong-lanceolate to ovate, obscurely toothed, 5–13cm(2–5in) long. Flowers 2.5cm(1in) long, bright red, several together on slender peduncles from the leaf axils.

R. mackoyana Monkey plant Brazil
In size, habit and foliage, much like *R. devosiana*, but with bright carmine to purple-red flowers.

R. macrantha Brazil
Shrub 1–2m(3–$6\frac{1}{2}$ft) high if allowed to grow naturally, but easily kept to less than half this if pruned annually. Leaves 7–15cm($2\frac{3}{4}$–6in) long, ovate-lanceolate. Flowers 5cm(2in) long, rose-purple with darker veins, borne singly from the upper leaf axils intermittently throughout the year, but chiefly in summer.

R. portellae Brazil
Spreading, short-lived perennial, sometimes only an annual, to 30cm(1ft) tall. Leaves 4–7cm($1\frac{1}{2}$–$2\frac{3}{4}$in) long, elliptic to obovate, dark green above with a white vein pattern, red-purple beneath. Flowers to 4cm($1\frac{1}{2}$in) long, rose-pink, solitary from the upper leaf axils and opening in winter. Needs tropical conditions.

RUMOHRA

Davalliaceae (*Polypodiaceae*)

Origin: Southern hemisphere in tropical to warm temperate regions. A genus of one species of rhizomatous evergreen fern allied to, and formerly included in, *Polystichum*, but resembling a *Davallia*. Provides a durable and attractive foliage plant, being tolerant of low humidity and fluctuating temperatures. Propagate by spores or division in spring. *Rumohra* is named for Dr. Carolus de Rumohr Holstein.

Species cultivated

R. adiantiformis Leather fern
Rhizomes far creeping, densely clad in pale brown ovate scales. Fronds 60–120cm(2–4ft) or more long including the long stalks; blade of frond ovate to triangular in outline, bi- or tri-pinnate, leathery-textured, light green; ultimate pinnules oblong, bearing beneath large sori, each one covered with an umbrella-like crenate-margined indusium. It can be grown terrestrially or epiphytically in a hanging basket.

RUSCUS

Ruscaceae (*Liliaceae*)

Origin: Atlantic Islands, through Mediterranean Europe to the Caucasus and Iran. A genus of three to seven species of evergreen, clump-forming or tufted, rhizomatous shrubs. They have green stems which broaden to form leaf-like cladodes. The plants are dioecious and the small 6-tepalled flowers are carried in the axils of the leaves, reduced to tiny scales sitting in the centre of the cladodes. To the uninitiated the flowers, and later the red and yellow berries, appear to grow from the centre of a leaf. The species described is an interesting foliage plant for the conservatory. Propagate by division in spring or by seed when ripe. *Ruscus* is the Latin name of butcher's broom (*R. aculeatus*).

    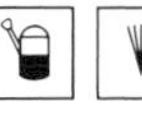

LEFT *Ruellia mackoyana*
BELOW LEFT *Rumohra adiantiformis*

Species cultivated

R. hypoglossum *Butcher's broom* Southern Europe
Clump-forming shrub 8–40cm(3–16in) tall with flexible, arching stems. Cladodes 5–11cm(2–4½in) long, elliptic to oblanceolate, shining mid-green. Flowers 7–10mm($\frac{1}{3}$–$\frac{3}{8}$in) wide, greenish-yellow, in clusters of three to five and followed by 1–2cm ($\frac{3}{8}$–$\frac{3}{4}$in) long scarlet fruits.

BELOW *Russelia sarmentosa*
RIGHT *Russelia equisetiformis*

RUSSELIA

Scrophulariaceae

Origin: Mexico to tropical South Ameica. Depending upon the botanical authority, a genus of 20 to 50 species of shrubs and sub-shrubs, a few of which provide the conservatory with some highly decorative pot plants. Their stems may be erect, semi-climbing or pendent, in some species green and rush-like with small scale-like leaves that soon fall. The flowers are tubular to funnel-shaped and slightly 2-lipped, usually freely produced. Propagate by cuttings or layering in spring or summer, or by division in spring. *Russelia* honours Dr Alexander Russell (c. 1715–1768), resident doctor to the English Factory at Aleppo and a keen natural historian.

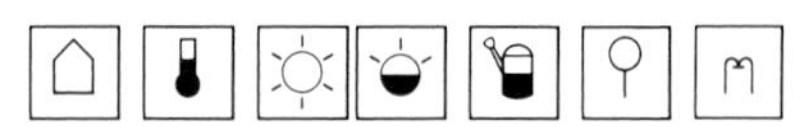

Species cultivated

R. equisetiformis (*R. juncea*) Coral plant, Fountain bush Mexico, naturalized widely elsewhere in the tropics and sub-tropics
Freely branched shrub to 90cm(3ft) or more in height; stems 4-angled, rush-like, bright green, pendent on mature plants. Flowers tubular, 2.5cm(1in) long, bright red, borne in summer.

R. sarmentosa Mexico and Cuba to Colombia
Well branched shrub to 1.2m(4ft) or more tall; stems 4- to 6-angled, bearing opposite pairs of toothed, ovate leaves 4/8cm(1½–3in) in length. Flowers red, 2.5cm(1in) long, in summer.

S

SACCOLABIUM See ASCOCENTRUM

SAINTPAULIA

Gesneriaceae
African violets

Origin: East Africa, most from one mountain range in Tanzania. A genus of 21 species of evergreen herbaceous perennials which are rosette-forming or tufted; some are stemless, others have long stems. Their leaves are obovate to almost orbicular, often cordate and scalloped at the margins and usually with short and long hairs on the leaf surface. The flowers are shortly tubular, opening to five lobes, and can be blue, violet, pink, red or white. Until the early 1970s most cultivars were forms of *S. ionantha* and were typically rosetted stemless plants with single or double flowers. During the last ten or so years other species have been hybridized with it, bringing in the long-stemmed character which has resulted in the pendulous and trailing cultivars, while smaller species such as *S. pusilla* have been used to develop a race of mini African violets. All are excellent house and warm conservatory plants. Propagate by leaf cuttings in summer, species only also by seed in warmth in spring. *Saintpaulia* was named for Baron Walter von Saint Paul Illaire (1860–1910), who discovered the first species.

Species cultivated

S. amaniensis See *S. magungensis*.
S. confusa Tanzania (eastern Usambara Mts.)
Rosetted and clump-forming, with age developing a short thick stem. Leaves to 4cm(1½in) long, elliptic to ovate, rounded-tipped, hairy, long-stalked. Flowers 2–3cm(¾–1¼in) wide, blue-violet. In the past confused with the similar *S. diplotricha* and *S. kewensis* (a synonym of *S. ionantha*).
S. diplotricha Tanzania (Usambara Mts.)
Much like *S. confusa*, but with the leaf blades purple-flushed above and bearing erect hairs (those of *S. confusa* are appressed).
S. grotei Tanzania (eastern Usambara Mts.)
Tufted, the prostrate stems eventually to 10cm(4in) or more in length. Leaves to 9cm(3½in) long, orbicular, coarsely rounded-toothed, prominently veined, very long stalked. Flowers to 3cm(1¼in) wide, pale mauve with a darker eye.
S. intermedia Tanzania (eastern Usambara Mts.)
Tufted, with prostrate, rooting and branching stems to 15cm(6in) or more in length. Leaves broadly ovate, to 5cm(2in) long, crenate, dark green and white-hairy above, purple- or red-flushed beneath, long-stalked. Flowers 2.5cm(1in) wide, deep violet, in small clusters from the leaf axils.
S. ionantha Coastal Tanzania
Rosette-forming, often in clumps. Leaves 4–6cm (1½–2½in) long, long-stalked, broadly ovate to rounded, somewhat fleshy-textured with silky hairs. Flowers 2–2.5cm(¾–1in) across, held above the leaves on long scapes 5–10cm(2–4in) tall, blue, violet, pink, red and white, intermittently throughout the year. Very many cultivars have been raised.
S. magungensis (*S. amaniensis*) Tanzania (Usambara Mts.)
Tufted, the prostrate stems eventually to 15cm(6in) or more long. Leaves to 7cm(2¾in) long, broadly ovate to orbicular, obscurely crenate, shortly stalked. Flowers 2cm(¾in) wide, purple, in clusters of two to four from the leaf axils.
S.m. minima
Plant half size of species above or a little more; flowers two-thirds size.
S. orbicularis Tanzania (Usambara Mts.)
Rosetted and clump-forming, with age sometimes developing a short stem. Leaves orbicular to broadly ovate, 3–6cm(1¼–2½in) long, coarsely toothed, pale green, long-stalked. Flowers 2–2.5cm(¾–1in) wide, light blue, or violet to white with a blue eye, in clusters of eight to ten from the leaf axils.
S. pendula Tanzania (Usambara Mts.)
Similar to *S. intermedia*, but with faster growing stems and solitary blue-purple flowers with darker throats.
S. pusilla Tanzania
Rosetted and clump-forming in neat small hummocks. Leaves ovate, 2–3.5cm(¾–1⅜in) long, pale green above, often purple-flushed beneath. Flowers about 1cm(⅜in) wide with blue upper petals and white lower ones, in small clusters in the leaf axils. The smallest known species, used in recent years by plant breeders to produce a race of mini African violets.
S. tongwensis Tanzania (Usambara Mts.)
Rosetted and clump-forming, with age sometimes developing a short, thick stem. Leaves ovate to

BELOW *Saintpaulia magungensis*
BELOW LEFT *Saintpaulia pendula*

elliptic, pointed-tipped, to 8.5cm(3¼in) long, coarsely crenate, dark green with a pale mid-vein on stalks of equal length. Flowers lavender, about 3cm(1¼in) wide, in clusters of four to six from the leaf axils.

SALPIGLOSSIS

Solanaceae

Origin: South America. A genus of five to 18 species (depending on the classification followed) of annuals, biennials and perennials. Plants are erect in habit and often somewhat stickily hairy, with alternate leaves and 5-lobed, colourful flowers. They are suitable for the conservatory and can be brought into the home for flowering. Propagate by seed in spring for summer to autumn flowering, and in early autumn for winter and spring flowering. *Salpiglossis* derives from the Greek *salpinx*, trumpet and *glossa*, tongue, referring to the shape of the style.

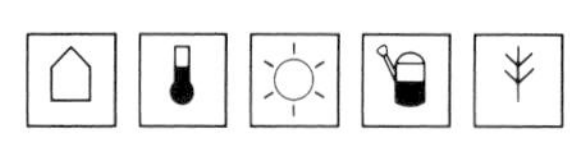

Species cultivated

S. sinuata Painted tongue Chile
Branched annual 60–90cm(2–3ft) tall. Leaves largest at the base, to 10cm(4in) long, decreasing up the stems, elliptic, deeply or shallowly toothed, with wavy margins. Flowers to 6cm(2½in) long and 5cm(2in) across, funnel-shaped, in shades of red, purple, blue, yellow and cream, often veined in contrasting colours. Named strains and cultivars usually have a wide range of colours.

BELOW *Salvia involucrata*
BOTTOM *Salpiglossis sinuata*

SALVIA

Labiatae

Sages

Origin: World-wide. A genus of 700 to 750 species of annuals, biennials, perennials, shrubs and sub-shrubs. Most are erect in habit with 4-angled stems, opposite leaves and spike-like inflorescences made up of whorls of tubular, 2-lipped flowers opening in summer and autumn. They make attractive pot and tub plants for the conservatory and include the culinary sage which will provide leaves for use throughout the year. Propagate by cuttings in late spring to late summer or by seed in spring. *Salvia* is derived from the Latin *salvus*, safe or well, from the medicinal use of several species.

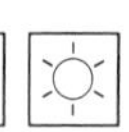

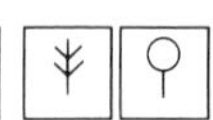

Species cultivated

S. elegans (*S. rutilans*) Pineapple sage Mexico
Sub-shrub to 1m(3ft). Leaves 5–10cm(2–4in) long, ovate-cordate, pointed, toothed and softly hairy, smelling of pineapples when crushed. Flowers to 4cm(1¼in) long, with protruding stamens, bright scarlet.

S. greggii Mexico, USA (Texas)
Sub-shrub 60–90cm(2–3ft) high. Leaves 2–4cm(¾–1½in) long, linear-oblong to spathulate, entire, dull green. Flowes 2.5–4cm(1–1½in) long, the corolla tubes swollen, bright red-purple in short racemes.

S. involucrata Mexico, Central America
Sub-shrub to 90cm(3ft) in pot, to 120cm(4ft) or more in the open. Leaves 5–10cm(2–4in) long, ovate with crenate margins, mid-green. Flowers 4–5cm(1½–2in) long, pink to magenta-red, opening from conspicuous reddish bracts which completely enclose the buds. 'Bethelii' is more robust with larger flowers and bracts.

S. leucantha Mexico
Sub-shrub 60–90cm(2–3ft) tall; stems and calyces densely covered with violet-purple hairs. Leaves 6–13cm(2½–5in) long, oblong-lanceolate, pointed, finely toothed with a wrinkled surface, white-hairy beneath. Flowers 2cm(¾in) long, white in marked contrast to the dark of the calyces.

S. officinalis Common sage S. Europe
Shrub to 60cm(2ft). Leaves 2.5–6cm(1–2½in) long, oblong, with a grey-green wrinkled surface. Flowers 1.5–2cm(⅝–¾in) long, light purplish-blue to almost white. The forms with coloured leaves are equally good for culinary purposes: 'Icterina' has golden variegated leaves; 'Purpurascens' has dark purple leaves; 'Tricolor' has leaves which are irregularly splashed with creamy-white.

S. rutilans See *S. elegans*.

S. splendens Scarlet sage Brazil
Sub-shrub to 2m(6½ft) tall, but usually grown as an annual, then 15–40cm(6–16in) tall. Leaves to 9cm(3½in) long, ovate, acuminate, mid-green. Flowers to 4cm(1½in) long, with showy bracts, both bright scarlet. Cultivars with the flowers and bracts white, pink or purple are grown.

SALVINIA

Salviniaceae

Origin: Tropics and sub-tropics world-wide. A genus of about ten species of aquatic plants closely related to the ferns. They are free-floating without roots, their slender stems bearing whorls of three leaves, two of which are flat and floating, the third deeply pinnately dissected into slender segments which remain below the water and take the place of the non-existent roots. They are good shelter plants for fish in aquaria and for a conservatory pool. Propagate by division in summer and plant by dropping the plants into the water. In cultivation, the three species below are much confused, their appearance and size varying with water conditions and temperature. *Salvinia* was named for Antonio Maria Salvini (1633–1729), Professor of Greek at Florence University.

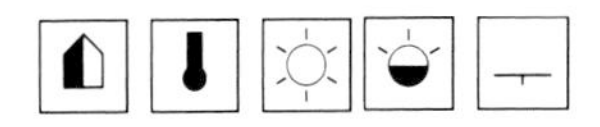

Species cultivated

S. auriculata Floating moss Mexico, Cuba to Argentina
To 25cm(10in) across. Surface leaves to 2.5cm(1in) long by 4.5cm($1\frac{3}{4}$in) wide, boat-shaped, the upper surface clothed with erect, stiff hairs in fours from minute papillae.

S. natans Central and S.E. Europe, N. Africa, Asia
To 15cm(6in) across. Surface leaves 1–1.5cm ($\frac{3}{8}$–$\frac{5}{8}$in) long by 6–9mm($\frac{1}{4}$–$\frac{1}{3}$in) wide, the papillae with three to four hairs.

S. rotundifolia Tropical America
To 7cm($2\frac{3}{4}$in) wide. Surface leaves almost circular, 1.5cm($\frac{5}{8}$in) long by 2cm($\frac{3}{4}$in) wide; the upper surface has many papillae, each with four hairs.

SANCHEZIA

Acanthaceae

Origin: Tropical South America. A genus of 60 species of perennials, shrubs and climbers only one of which is in general cultivation. It provides a striking foliage and flowering plant for the conservatory and can be used in the home, at least in the short term. Propagate by cuttings in spring or summer. *Sanchezia* honours José Sanchez, a nineteenth-century Professor of Botany at Cadiz, Spain.

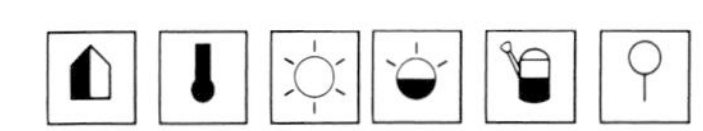

Species cultivated

S. nobilis See under *S. speciosa*.

S. speciosa (*S. glaucophylla*, *S. nobilis glaucophylla*) Probably Ecuador or Peru
Shrubby, eventually to 1.5m(5ft), but easily kept to half this height in a container. Leaves in opposite pairs, oblong-ovate to elliptic, pointed, 15–30cm(6–12in) long, glossy rich green with all the main veins boldly yellow-banded. Flowers tubular, to 4–5cm($1\frac{1}{2}$–2in) long, with five short reflexed lobes at the mouth, yellow, subtended by red bracts, borne in short, dense, terminal panicles. Long known as *S. nobilis*, a rarely if ever cultivated species from Ecuador, and perhaps only a form of it.

SANDERSONIA

Liliaceae

Origin: South Africa. A genus of one species of tuberous-rooted perennial which makes a distinctive pot plant. Propagate by seed in spring or by separating the tubers when re-potting; it will flower in 2–3 years from seed. *Sandersonia* was named for John Sanderson (1820–1881), secretary of the Horticultural Society of Natal and the discoverer of this plant.

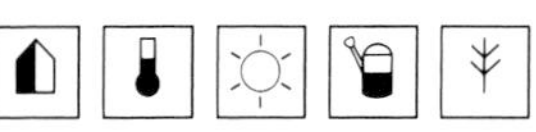

Species cultivated

S. aurantiaca South Africa
Erect, reclining or partially climbing stems, 30–60cm(1–2ft). Leaves to 10cm(4in) long, alternate, linear to lanceolate, the ends sometimes terminating in hooks. Flowers 2.5cm(1in) long, the six tepals joined, almost globose in form, dividing to six very shallow lobes at the constricted mouth, the whole very reminiscent of a nodding Chinese lantern; they are orange in colour and carried on slender-stalks.

SANSEVIERIA

Agavaceae (*Liliaceae*)
Snake plant

Origin: Africa and southern Asia. A genus of about 60 species of succulent evergreen perennials some of which are ground-dwelling in the wild, others epiphytic. They have rosettes or clumps of rather stiff sword-shaped or cylindrical leaves and somewhat tubular flowers in axillary racemes or panicles followed by usually red berries. They are best left in the same pot until congested before re-potting. Propagate by division or by cuttings of leaf sections in summer. *Sansevieria* was named for Prince Raimond de Sanagrio de Sanseviero (1710–1771).

Species cultivated

S. aethiopica South Africa
Leaves to 40cm(16in) long, linear-lanceolate with a white, awl-like tip, more or less glaucous, usually dark green cross-banded. Flowers white, in a raceme to 30cm(1ft) tall.

S. cornui See *S. senegambica*.

S. cylindrica Tropical Africa
Leaves to 45cm($1\frac{1}{2}$ft) in pots, to twice this in the

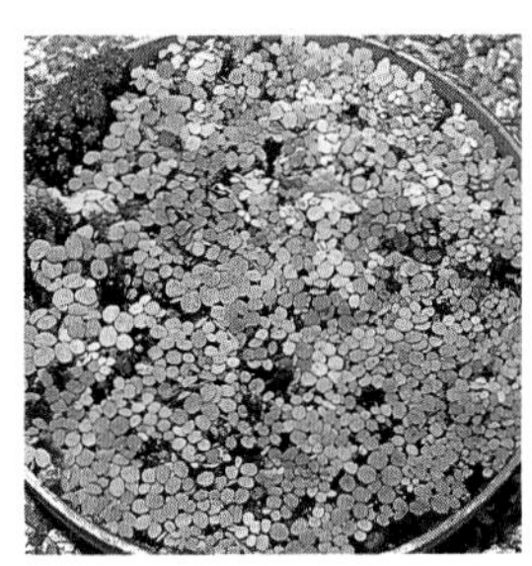

ABOVE *Salvinia natans*
CENTRE *Salvinia auriculata*
TOP *Sandersonia aurantiaca*

open, cylindrical, tapering to a point, grooved along the length of the leaf and cross-banded with grey-green when young. Flowers to 4cm(1½in) long, white or pink-tinted and carried in racemes 30–90cm(1–3ft) long, but not often produced.

S. grandicuspis Star sansevieria Tropical Africa
Leaves 40–50cm(16–20in) long, narrowly lanceolate with a soft awl-shaped point, mid to deep green cross-banded with grey and having a brown horny margin, arching. Often grown as S. *kirkii*, *q.v.*

S. grandis East Africa
Epiphytic, producing offsets on long stout runners. Leaves ovate to lanceolate, to 20cm(8in) long, dark green, silvery cross-banded and red-edged. Flowers white, in narrow panicles well above the leaves.

S. guineensis See S. *hyacinthoides*.

S. hahnii See under S. *trifasciata*.

S. hyacinthoides (S. *guineensis*, S. *spicata*, S. *thyrsiflora*) South Africa
Leaves lanceolate, to 45cm(1½ft) in length, tapering to a channelled petiole, dark green with a yellow margin and pale cross-bands. Flowers fragrant, greenish-white, in racemes to 75cm(2½ft) tall.

S. kirkii East Africa
Leaves on well grown plants to 1.8m(6ft) tall, greyish-green with mottled bands of pale green. Seldom grown, plants under this name usually being S. *grandicuspis*, *q.v.*

S. senegambica (S. *cornui*) West Africa

Sansevieria trifasciata

Leaves oblanceolate, sometimes broadly so, 45–60cm(1½–2ft) long, slender-pointed, matt-green sometimes with darker markings. Flowers whitish, in racemes equal to or above the leaves.

S. spicata See S. *hyacinthoides*.

S. thyrsiflora See S. *hyacinthoides*.

S. trifasciata Mother-in-law's tongue, Snake plant South Africa
Clump-forming. Leaves to 45cm(1½ft) long in a container, twice this in the open ground, linear-lanceolate, stiffly erect, deep green cross-banded with paler greens. Flowers to 1.5cm(⅝in) long, greenish-white and night-scented. *S.t.* 'Hahnii' is a dwarf mutant with leaves to 15–30cm(6–12in) long and somewhat broader than those of the type. 'Golden Hahnii' is similar in form, but has broad yellow margins to the leaves.

'Laurentii' is a yellow-margined form of the type species; it can only be increased by division because it reverts to green if propagated from stem sections.

S. zeylanica Devil's tongue Sri Lanka
Leaves lanceolate, about 30cm(1ft) in length, with a soft spine-shaped tip, grey-green, cross-banded with dark green, somewhat channelled above and keeled below. In cultivation this species is sometimes confused with S. *hyacinthoides* and forms of S. *trifasciata*, *q.v.*

SARMIENTA

Gesneriaceae

Origin: Chile. A genus of one species of small shrubby perennial well suited to accompany ferns in a cool, shady conservatory. Propagate by seed in spring, or by cuttings in summer or autumn. *Sarmienta* was named for the Spanish botanist M. Sarmiento.

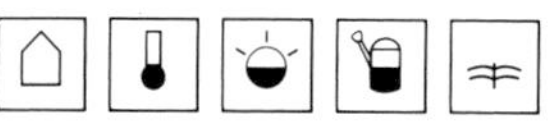

Species cultivated

S. scandens (S. *repens*)
Stems wiry, creeping and rooting, to 30cm(1ft) or more in length. Leaves obovate to elliptic, to 2.5cm(1in) long, shallowly to deeply 3- to 5-toothed at the tips. Flowers 2–2.5cm(¾–1in) long, tubular, inflated like lop-sided amphorae, scarlet, pendent on very slender stalks from the leaf axils in summer.

SARRACENIA

Sarraceniaceae

Pitcher plants

Origin: Eastern North America. A genus of eight to ten species of rosette-forming carnivorous plants. They have erect or reclining leaves modified into pitcher-like traps with the tops acting as partly open lids. The pitchers contain a digestive liquid into which insects can fall and drown, essential minerals coming from the bodies being absorbed by the plant. The flowers have five petals and five sepals and a curious umbrella-shaped growth over the stigmas. They are nodding and borne singly at the tops of long stalks. In the wild, sarracenias are native to boggy habitats, so in cultivation they should be grown in a compost of peat and moss stood in a shallow tray of water. Keep drier and very cool in winter. Propagate by seed sown in a similar compost (never allow to dry out). *Sarracenia* was named for Michel Sarrasin de l'Etang (1659–1734), a French-Canadian doctor and botanist who discovered the genus.

Species cultivated

S. drummondii See *S. leucophylla.*

S. flava Yellow pitcher plant, Huntsman's horn Florida to Virginia
Leaves erect, 30–60cm(1–2ft) tall in a container, to twice this in the open, slender, trumpet-shaped, yellowish green marked with red in the throat and sometimes flushed red on the outside. Flowers to 10cm(4in) wide, yellow, borne on stems usually shorter or equalling the leaves in spring.

S. leucophylla (*S. drummondii*) Georgia, Florida, Alabama
Leaves 45–90cm(1½–3ft) or more in height, very slenderly trumpet-shaped, rather like those of *S. flava* but the top of the pitcher white with an elaborate network of dark red veins. Flowers dark red, to 10cm(4in) wide, borne at about the same height as the leaves in spring.

S. psittacina Georgia, Florida to Louisiana
Leaves decumbent, like those of *S. purpurea* but rather more tubular, green with white blotches and purple veins. Flowers 5cm(2in) wide, red and purple, on erect to 20cm(8in) high in spring.

S. purpurea Common pitcher plant Nebraska and Labrador, south to Florida
Leaves to 15cm(6in) or more long, erect to decumbent, inflated, green with a touch of red-purple and veining of the same colour. Flowers 6cm(2½in) across, deep purple, borne on stems 30cm(1ft) high in spring.

SAUROMATUM

Araceae

Origin: Tropical Africa and Asia to western Malaysia. A genus of four to six species of tuberous-rooted perennials, each with one solitary one-stalked leaf and an arum-like spadix within a spathe. The species described below is the only one generally available and can be flowered without soil, but after this treatment they are usually discarded. If, however, they are potted in the normal way in a good compost, the tuber will produce its attractive leaf after the flowers have died down. Once the leaf yellows dry off the tuber until early spring when it will start into new growth. The plant is very worthwhile growing for its foliage alone. When blooming it gives off a strong smell of bad meat and is best put out of the living area for a few days. Propagate by separating offsets. *Sauromatum* derives from the Greek *sauros*, a lizard, referring to the mottled appearance of the inside of the spathe.

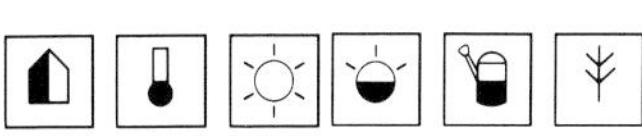

Species cultivated

S. guttatum Voodoo lily, Monarch of the East India
The rounded, flattened tuber can reach 6–13cm(2½–5in) across; it produces a solitary leaf 60–90cm(2–3ft) long of which two-thirds or so is stalk. Leaf blade divided into three lobes, the side ones again divided into lanceolate to oblong lobes. Flowers small and petalless, borne at the base of a stiff, tail-like, deep purple spadix longer than the 30cm(1ft) spathe which is greenish on the back and yellowish inside with deep purple spotting and blotching; it remains erect for only a matter of hours, then reflexes.

ABOVE *Sauromatum guttatum*
TOP *Sarracenia leucophylla*

ABOVE *Schefflera arboricola* 'Variegata'
ABOVE RIGHT *Saxifraga stolonifera* 'Tricolor'

SAXIFRAGA

Saxifragaceae
Saxifrages

Origin: Northern temperate zone and South America chiefly in the mountains. A genus of about 350 species most of which are perennials, but including a few annuals. They are mostly tufted, with rosettes of linear to orbicular leaves. The flowers are 5-petalled. Those species described below are suitable for an unheated room or porch, or a cool conservatory. Propagate by seed when ripe or in spring, by division when re-potting or by plantlets in late summer. *Saxifraga* derives from the Latin *saxum*, a rock and *frago*, I break, from the habit of some species which grow in rock crevices and appear to have split the rock.

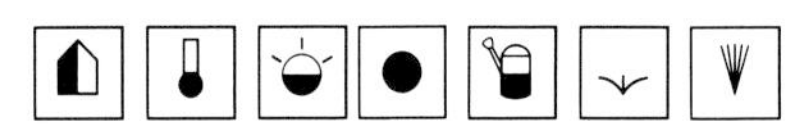

Species cultivated

S. cuscutiformis Probably China, but exact provenance not known
Much like a miniature *S. stolonifera* with a profusion of thread-like branching red runners like dodder (*Cuscuta*) stems. Leaves elliptic to broadly so, 2–5cm(¾–2in) long, medium green with whitish veins. Flowers like those of *S. stolonifera*, but smaller, not freely produced. An intriguing plant for a small hanging basket.

S. sarmentosa See *S. stolonifera.*

S. stolonifera (*S. sarmentosa*) Mother of thousands, Strawberry geranium Eastern Asia
Mat-forming, spreading by slender, branching red stolons which produce plantlets. Leaves to 10cm(4in) across, rounded, coarsely toothed, the upper surface silvery-hairy, the lower reddish. Flowers about 2cm(¾in) wide, white, with two petals, sometimes only one, much longer than the rest. *S.s.* 'Tricolor' has smaller leaves with white markings and an overall pink flush.

S. umbrosa Pyrenees
Hummock to mat-forming. Leaves to 6cm(2½in) long, obovate to oblong, broadly crenate, leathery in texture, stalks ciliate. Flowers 8mm(⅓in) across, white with red dots, borne in panicles 30–45cm(1–1½ft) high in summer. Hardy. Much confused with the following hybrid.

S × urbium London pride
Hybrid between *S. spathularis* and *S. umbrosa*; more vigorous than *S. umbrosa* and capable of thriving in total shade. Provides contrast to ferns in a shady, cool or unheated conservatory. Hardy.

SCHEFFLERA

Araliaceae

Origin: Tropical and warm temperate areas of Asia, Australasia and the Pacific Islands. A genus of about 150 to 200 species of evergreen shrubs and trees with alternate, long-stalked, digitate leaves. The small 5-petalled flowers are grouped into umbels which together make up racemes or panicles, but are not produced on small pot-grown specimens. They are good house and conservatory plants. Propagate by seed when ripe or in spring in warmth, by cuttings with bottom heat in summer or by air layering in spring. *Schefflera* was named for J.C. Scheffler, a nineteenth-century German botanist.

Species cultivated

S. actinophylla See *Brassaia actinophylla.*

S. arboricola (*Heptapleurum arboricola*)
S.E. Asia
To 1m(3ft) or more in a pot, becoming a large shrub in the wild. Leaves with stalks 10–15cm(4–6in) long, digitate, with seven to 16 arching, stalked leaflets, 8–15cm(3–6in) long, the margins slightly folded upwards with down-curving, pointed tips. A handsome pot plant. Tropical. *S.a.* 'Variegata' has leaves with a yellow variegation.

S. digitata New Zealand
To 2m(6½ft) in a pot, a shrub or small tree to 8m(26ft) in the open. Leaves with seven to ten obovate, unstalked leaflets to 18cm(7in) long, sharply toothed, shining dark green. Better in a cooler room than *S. arboricola.* Cool to temperate conditions.

SCHINUS

Anacardiaceae

Origin: Mexico to Argentina and Chile. A genus of about 30 species of evergreen, dioecious shrubs and trees, a few of which are widely grown in warm temperate and tropical regions and as foliage plants under glass. Those described here make good tub specimens for the larger conservatory. They have simple or pinnately compound, alternate leaves, panicles of tiny 5-petalled flowers and berry-like fruits. Propagate by seed in spring or by cuttings in summer. *Schinus* derives from the Greek name *schinos* used for the mastic tree (*Pistacia lentiscus*); some species yield a similar gum or resin.

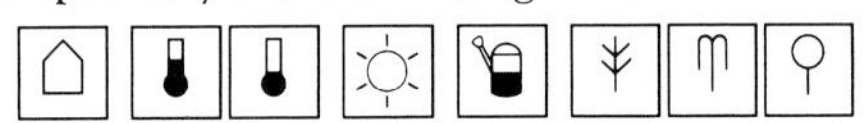

Species cultivated

S. molle Peruvian pepper/mastic tree Peru (Andes region)
Small to medium-sized tree with pendent branchlets, twigs and leaves, easily kept to about 2m($6\frac{1}{2}$ft) in a container. Leaves 10–22cm(4–9in) long, pinnate, composed of 15 to 41 narrowly lanceolate, rich green leaflets. Flowers whitish. Fruits like tiny rose-red peas, but not unless a plant of each sex is present.

S. terebinthifolius Brazil; widely naturalized elsewhere
Large shrub or small tree responding well to container culture and pruning. Leaves pinnate, 8–18cm(3–7in) long, composed of three to 13 broadly lanceolate to obovate leaflets. Flowers white. Fruits 5mm($\frac{1}{5}$in) wide, globular, red, but only if a plant of each sex is present.

LEFT *Schinus molle*
BELOW *Schlumbergera* × *buckleyi*

SCHIZANTHUS

Solanaceae

Origin: Chile. A genus of ten species of annuals, biennials and short-lived perennials with erect, branched stems, deeply divided leaves and 2-lipped, colourful, somewhat orchid-shaped flowers in terminal racemes in spring. The hybrid described below is an attractive flowering pot plant for the conservatory or sunny window sill. Propagate by seed, sowing in late summer for winter to early spring flowering (provide temperate growing conditions), or in spring for summer to autumn flowers (cool growing conditions). *Schizanthus* derives from the Greek *schizo*, to divide and *anthos*, a flower, from the deeply cleft corolla.

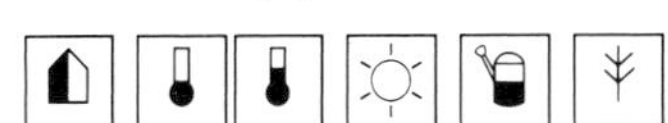

Species cultivated

S. pinnatus Poor man's orchid
To 60–90cm(2–3ft) tall. Leaves pinnate to bipinnate, ferny. Flowers 2.5–4cm(1–$1\frac{1}{2}$in) wide, tubular, the two lips broadly lobed and spreading, the upper with two lobes, the lower with three; they are available in shades of pink, red, mauve and white, the upper lip usually with a contrasting colour at the centre, most often yellow, spotted and streaked with purple. Various strains are available, the dwarf sorts most useful for pots. Some are sold as *S.* × *wisetonensis* but appear identical to the above species.

SCHIZOCENTRON See HETEROCENTRON

SCHLUMBERGERA

Cactaceae

Origin: Brazil. A genus to two species of epiphytic cacti allied to *Rhipsalidopsis* and sometimes still known as *Zygocactus*. They have thin green stems flattened into conspicuously jointed leaf-like segments which have toothed margins and small, bristly areoles. The tubular flowers are solitary or in pairs from the ends of the stems, having somewhat reflexed petals and protruding stamens. All are excellent house and conservatory plants, particularly effective in hanging baskets. Propagate by cuttings of one or two stem sections in summer. *Schlumbergera* was named for Frederick Schlumberger, a Belgian horticulturalist of the late nineteenth century.

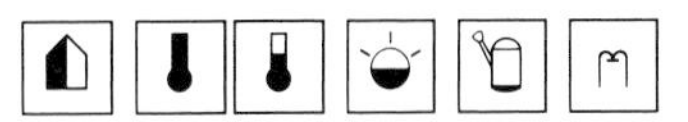

Species cultivated

S.* × *bridgesii See *S.* × *buckleyi*.

S.* × *buckleyi (*S.* × *bridgesii*) Christmas cactus
Hybrid between *S. russelliana* and *S. truncata*, to 30cm(1ft) tall with arching branches, the stem segments to 5cm(2in) long by 2.5cm(1in) wide, rounded at the ends and crenate. Flowers 5.5–6.5cm($2\frac{1}{5}$–$2\frac{2}{3}$in) long, the petals reflexing at the mouth, magenta to rosy-purple, opening in winter. Several named cultivars are now commercially available, the following being recommended: 'Bicolor' (shades of rose-red), 'Joanne' (red with a purple throat), 'Noris' (red and purple), 'Weinachtesfreude' (pale and dark red with a purple throat) and 'Westland' (shades of rose-red).

S. gaertneri See *Rhipsalidopsis gaertneri.*
S. russelliana
To 15cm(6in) or more tall. Stem segments to 4cm($1\frac{1}{2}$in) long by 1–1.3cm($\frac{3}{8}$–$\frac{1}{2}$in) wide with one or two marginal notches. Flowers 5–6.5cm(2–$2\frac{1}{2}$in) long, the petals scarcely reflexed, magenta, opening in late winter and spring.
S. truncata
To 30cm(1ft) or more tall with arching branches. Stem segments to 6.5cm($1\frac{2}{3}$in) long, by 3cm($1\frac{1}{4}$in) wide, with two to four sharp teeth along the sides, and one at each of the upper corners. Flowers to 8cm(3in) long, petals very strongly reflexed, bright rose-pink, paler inside, opening in late autumn and winter.

ABOVE *Scindapsus pictus*
ABOVE CENTRE *Scilla socialis* 'Violacea'
TOP *Schlumbergera truncata*

SCILLA

Liliaceae
Squills

Origin: Temperate Europe and Asia, also tropical and southern Africa. A genus of about 80 species of bulbous plants with basal tufts of linear to strap-shaped leaves and bell- to star-shaped flowers on leafless scapes. The small South African species are now sometimes classified as *Ledebouria*. They make attractive pot plants, the hardier species needing treatment as for crocus species. Propagate by seed when ripe or as soon after as possible, or by offsets separated when dormant. *Scilla* is the Greek vernacular for sea squill (*Urginea maritima*), included in this genus by Linnaeus.

Species cultivated
S. adlamii South Africa
Bulbs producing stolons. Leaves to 20cm(8in) or more long, linear-lanceolate, fleshy in texture, greenish-brown with darker striping. Flowers 3mm($\frac{1}{8}$in) long, bell-shaped, purplish-mauve on scapes about 15cm(6in) long in spring. Cool to temperate.
S. mischtschenkoana (*S. tubergeniana*) Iran, Caucasus
Rather similar to *S. sibirica*, but begins to flower at ground level before the leaves develop, the scape elongating as the blooms develop. Flowers starry and opening wider, pale blue with dark central vein to each tepal. Hardy.
S. peruvianus Portugal, Spain, Italy
This European species was given its name by Clusius in the mistaken belief that it had been sent to him from Peru. Leaves to 30cm(1ft) or more long by 3cm($1\frac{1}{4}$in) wide, somewhat fleshy. Flowers 2–2.5cm($\frac{3}{4}$–1in) across, starry, mauve to purple, opening in early summer. Cool. *S.p.* 'Alba' has white flowers.
S. sibirica Siberian squill Turkey, Iran, Caucasus
Leaves two to five together, to 15cm(6in) long, strap-shaped, glossy, not reaching full length after the flowers fade. Flowers 2cm($\frac{3}{4}$in) across, bell-shaped, nodding, brilliant deep blue, opening in early spring. Hardy. *S.s.* 'Alba' has white flowers; 'Spring Beauty' is more vigorous and earlier to flower.
S. socialis (*Ledebouria socialis*) South Africa
Bulbs forming clumps or hummocks, 15cm(6in) or more wide, above ground level. Leaves lanceolate to narrowly ovate, 5–8cm(2–3in) long, whitish-green with irregular pale olive-green spots above, somewhat fleshy-textured. Flowers broadly bell-shaped, green and white, about 5mm($\frac{1}{5}$in) wide, in racemes 8cm(3in) in height in summer. Cool to temperate. *S.s.* 'Violacea' (*S. violacea*) has leaves purple beneath and silvery above with dark olive-green blotches; flowers flushed reddish at the base, bearing prominent purple stamen filaments.
S. tubergeniana See *S. mischtschenkoana.*
S. violacea See *S. socialis.* Cultivar Violacea.

SCINDAPSUS

Araceae

Origin: S.E. Asia from China to Malaysia. A genus of about 20 species of evergreen climbers allied to *Philodendron*. They produce aerial roots and have entire leaves, the bases of the stalks sheathing the stem. The flowers are arum-like. The species described below is a good conservatory and house plant. Propagate by cuttings, or by layering in summer. *Scindapsus* is the ancient Greek name for a climbing plant akin to ivy.

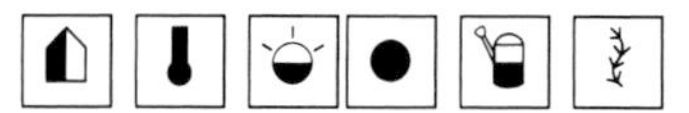

Species cultivated
S. aureus See *Epipremnum aureus.*
S. pictus Malaysia, Indonesia
To 2m($6\frac{1}{2}$ft) in pots, climbing to 12m(40ft) or more in the wild. Leaves 10–15cm(4–6in) long, ovate-cordate, dark green patterned with lighter green. *S.p.* 'Argyraeus' (*Pothos argyraeus*, Silver vine) is probably a juvenile form, having slender stems and

smaller leaves with silvery spots; this is the form most commonly available as a house plant.

SCIRPUS

Cyperaceae

Origin: Cosmopolitan in wet places. A genus of 250 to 300 species of tufted, often colony-forming rushes with leaves either grassy or reduced to basal sheaths, their functions being taken over by rounded or triangular green stems. The tiny petalless flowers are clustered into compact spikelets. They need permanently moist soil and can be grown either in a conservatory pool or in a pot stood in a saucer of water. Propagate by division. *Scirpus* is the Latin name for a rush.

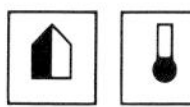  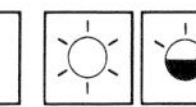  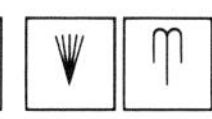

Species cultivated

S. cernuus (sometimes grown as *Isolepis gracilis*)
Densely tufted evergreen rush to 15cm(6in) tall with filiform leaves and stems erect at first, then drooping. Flowers 5mm($\frac{1}{5}$in) long, greenish or brownish, in solitary spikelets in summer.

SCUTELLARIA

Labiatae

Origin: Almost cosmopolitan—alpine to tropical. A genus of about 300 species of perennials and shrubs, one showy tropical member having long been popular in our conservatories. Propagate by cuttings in spring and summer, or by seed in spring. *Scutellaria* derives from the Latin *scutella*, a small dish, alluding to the shape of the fruiting calyces.

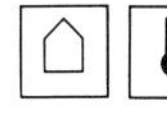 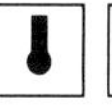 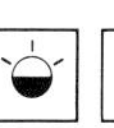  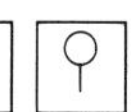

Species cultivated

S. costaricana Costa Rica
Shrubby, usually about 60cm(2ft) in height, but more than double this is possible when planted out. Stems erect, deep purple, 4-angled. Leaves in opposite pairs, elliptic to obovate, slender-pointed, somewhat puckered, deep green, 8–15cm(3–6in) long. Flowers tubular, 2-lipped, the upper one hooded, to 6cm($2\frac{1}{2}$in) long, orange-scarlet, paler to yellow within, carried in dense racemes in summer.

SEAFORTHIA See ARCHONTOPHOENIX & PTYCHOSPERMA

SEDUM

Crassulaceae
Stonecrops

Origin: Northern hemisphere. A genus of 500 to 600 species, most of which are succulent perennials and sub-shrubs but including a few annuals and biennials. They are prostrate to erect in habit with entire linear, cylindrical, ovoid or rounded, succulent leaves and 5-petalled flowers in terminal cymes. Propagate by seed in spring, or by leaf or stem cuttings in summer. *Sedum* derives from the Latin *sedo*, I sit, referring to the way some species appear to sit upon rocks and walls.

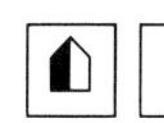    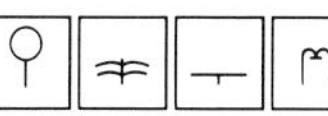

Species cultivated

S. adolphii See under *S. nussbaumerianum*.

S. allantoides Mexico (Oaxaca)
Sub-shrubby, branching mainly near ground level,

Scutellaria costaricana

the erect robust stems 20–30cm(8–12in) in height. Leaves 2.5cm(1in) or more long, clavate, incurving, with a waxy-white patina. Flowers 5-petalled, to 1.6cm($\frac{2}{3}$in) across, white with a greenish tinge, borne in small terminal panicles in summer.

S. burrito Eastern Mexico
Closely related to *S. morganianum* and with the same pendent growth habit, though the stems are stiffer. Leaves more crowded and with a somewhat bluish cast, elliptoid, blunt-tipped, to 1.5cm($\frac{5}{8}$in) long. Flowers more or less bell-shaped, to 8mm($\frac{1}{3}$in) long, the ovate pink petals with darker lines, in terminal clusters of ten to 30. A new species first collected in 1972 and described in 1977, it has the same appeal as *S. morganianum*, but unlike that species it can be handled without the fragile-based leaves falling off.

S. dendroideum Mexico (Sierra Madre del Sur, Guatemala mountains)
Sub-shrubby, usually with a trunk-like main stem from which the branches arise, erect, 20–30cm(8–12in) in height. Leaves mainly at the stem tips, about 3cm($1\frac{1}{4}$in) long, obovate to spathulate, flat but thickly fleshy, glossy bright green, the margins bearing minute reddish glands. Flowers 5-petalled, 1.2cm($\frac{1}{2}$in) wide, bright yellow, many together in loose, terminal panicle-like cymes, in spring but

ABOVE *Sedum morganianum*
TOP *Sedum hintonii*

not readily produced. Confused with the larger *S. prealtum q.v.*

S. hintonii Mexico (Michoacan, Pinzan)
Tufted, small hummock-forming perennial with short prostrate to decumbent stems to 10cm(4in) or more long. Leaves to 2.5cm(1in) long, crowded, rosetted, oblongoid to obovoid, flattish above, rounded below, slightly incurving, very fleshy, pale grey-green densely covered with short white, bristly hairs. Flowers 5-petalled, about 1cm($\frac{3}{8}$in) across, white, several in a dense, narrow, terminal, leafy cyme to 10cm(4in) or more in height, in spring. Perhaps the most desirable of all the stonecrops and certainly the most intriguing, having the same sort of appeal as *Echeveria setosa*. Reputedly difficult to grow but this is not so if a very gritty compost is used and the plant is allowed to thoroughly dry out between waterings.

S. morganianum Burro's/Donkey's tail
Mexico, though exact location somewhat uncertain
Stems prostrate or pendent to 60cm(2ft) or more in length, covered with overlapping leaves. These are to 2.5cm(1in) long, almost cylindrical, tapering to a point which curves inwards, pale glaucous-green. Flowers about 1cm($\frac{3}{8}$in) long, red, in small clusters in spring. An excellent plant for a hanging basket.

S. nussbaumerianum Mexico (Vera Cruz, Zacuapan)
Sub-shrubby, branching from the base and above, eventually forming hummocks to 15cm(6in) in height. Stems yellowish-green with red streaks. Leaves mainly at the stem tips, elliptic to oblong-elliptic, to 3cm($1\frac{1}{4}$in) long, flattish above, rounded beneath, very fleshy, yellow-green. Flowers 5-petalled, 1.6cm($\frac{2}{3}$in) wide, yellow, in dense lateral cymes, in winter to spring. Much confused with *S. adolphii* and frequently parading under that name. The true *S. adolphii* from Mexico has larger, 4–5cm($1\frac{1}{2}$–2in) long, less yellow-tinted leaves with distinctly keeled margins; the inflorescence is a loose paniculate cyme and the flowers are white. One wonders how two such distinct species became so mixed up.

S. oaxacanum Mexico (Oaxaca, high mountains)
Mat-forming perennial to 30cm(1ft) across, composed of interlacing, decumbent stems to about 10cm(4in) in height. Leaves fairly crowded, obovate, 5–8mm($\frac{1}{5}$–$\frac{1}{3}$in) long, obscurely keeled beneath and waxy-white-farinose above when young. Flowers to 1.5cm($\frac{5}{8}$in) wide, bright yellow, in small terminal clusters in late spring and summer.

S. pachyphyllum Mexico (mountains of Oaxaca, San Luis and Sierra Madre del Sur)
Sub-shrubby, the main stems arising from the base, sometimes decumbent at first, then erect to somewhat spreading. Leaves clavate, cylindrical in cross-section, to 2.5cm(1in) long, abruptly mucronate, grey-green, usually with reddish tips. Flowers 5-petalled, to 2cm($\frac{3}{4}$in) wide, bright yellow, in dense lateral cymes in spring.

S. palmeri Mexico (Sierra Madre, etc.)
Sub-shrubby, but only woody near the base, with erect to spreading stems up to 15cm(6in) in height. Leaves spathulate, about 2.5cm(1in) long, flat but thickly fleshy, glaucous, mainly at the stem tips on mature plants. Flowers 5-petalled, 1cm($\frac{3}{8}$in) wide, rich yellow (almost orange) in lateral, branched cymes in early spring, then intermittently to autumn.

S. prealtum Mexico, Guatemala (mountains)
When young, much like *S. dendroideum*, but the leaves to 5cm(2in) or more in length and lacking the red glandular margins. Older plants can attain 60cm(2ft) or more in height, with lax branches which arch down to the soil and root. Flowering takes place regularly in late spring to early summer. Stands light frost. Frequently met with in collections as *S. dendroideum*, *q.v.*

S. sieboldii Japan
Tufted, the rootstock with small carrot-like tubers. Stems 15–25cm(6–10in) long, semi-prostrate to slightly arching. Leaves sessile, 1–2cm($\frac{3}{8}$–$\frac{3}{4}$in) long, in whorls of three, almost round, glaucous blue-green. Flowers about 1cm($\frac{3}{8}$in) across, pink, in compact, leafy cymes in autumn. Hardy. *S.s.* 'Medio-Variegatum' has leaves with a central creamy yellow splash.

S. stahlii Mexico (mountains)
Loosely mat-forming perennial with slender, prostrate and decumbent branches to 20cm(8in) in length. Leaves fairly crowded along the stem, about 1cm($\frac{3}{8}$in) long, ovoid, circular in cross-section, minutely pubescent and suffused brownish-red. Flowers 5-petalled, 1.2cm($\frac{1}{2}$in) across, bright yellow, in terminal 2-forked cymes in late summer. Will stand light frost.

S. treleazii Mexico (Sierra Madre del Sur, etc.)
Sub-shrubby with one to a few erect main stems arising from the base, these in turn branching fairly low down, to 20cm(8in) in height. Leaves crowded along the stem, oblong, very fleshy, glaucous, 2–3cm($\frac{3}{4}$–$1\frac{1}{4}$in) long, incurving, Flowers 1.2cm($\frac{1}{2}$in) wide, yellow, in erect, lateral, compact, cymose

panicles in spring.
S. weinbergii See *Graptopetalum paraguayense.*

SELAGINELLA

Selaginellaceae

Origin: Mainly circum-tropical, with a few species extending into temperate and arctic zones. A genus of about 700 species of spore-bearing plants closely allied to ferns. They have prostrate, erect or climbing stems which branch freely, in some cases in one plane to produce frond-like growths. The leaves are tiny, scale-like, usually crowded and arranged distichously; the sporangia appear in compact, terminal, cone-like spikes but are more or less insignificant. Propagate by division in spring or by cuttings from spring to autumn. The species described below make charming foliage plants for the conservatory and indoor bottle gardens. *Selaginella* is the diminutive of the Latin *selago*, once used for *Lycopodium selago* (the common or fir club moss of the northern temperate zone) and chosen by Linnaeus for this genus.

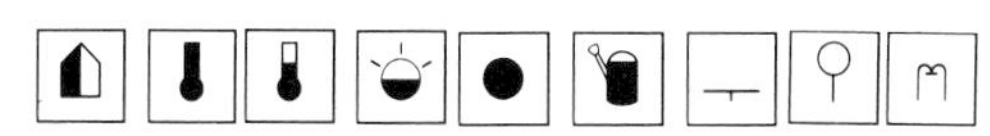

Species cultivated

S. cuspidata See *S. pallescens.*

S. denticulata Macronesia, Mediterranean region to Syria
Prostrate, mat-forming, to 30cm(1ft) or more wide, pale green, reddening in bright light especially when mature. Largest leaves are broadly ovate, about 2mm($\frac{1}{16}$in) long. Sometimes confused in cultivation with *S. kraussiana*, but that species has the main leaves narrow and twice as long. Cool.

S. emmeliana (*S. emiliana*) See *S. pallescens.*

S. kraussiana South Africa
Prostrate, mat-forming, up to 60cm(2ft) across, bright green, main leaves oblong-lanceolate, 3–4mm($\frac{1}{8}$–$\frac{1}{6}$in) long. Cool. *S.k.* 'Aurea' has golden green foliage; 'Variegata' is cream variegated; 'Brownii' forms congested mossy hummocks to 15cm(6in) across.

S. lepidophylla Resurrection plant, Rose of Jericho Texas, Arizona to El Salvador
Branches stiff and sparing, to 10cm(4in) long, radiating from a short central stem to form a rosette; the branches roll inwards when dry forming a ball, expanding again when moistened. A 'fun' plant, but also attractive when the bright green rosette opens out flat. Cool.

S. martensii Mexico
Main stems erect, tufted to clump-forming, freely branching above, 15–30cm(6–12in) tall in time, sprawling on and rooting into the soil. Upper branches frond-like and ferny. Temperate. A very variable species. *S.m.* 'Variegata' ('Albovariegata') has some of the branchlets white splashed.

S. pallescens (*S. cuspidata, S. emmeliana, S. emiliana*) Mexico to Colombia and Venzuela
Tufted to clump forming, the main stems erect, with frond-like branches 20–30cm(8–12in) high, pale green. Temperate. *S.p.* 'Aurea' has bright yellow-green foliage.

ABOVE *Selaginella kraussiana* 'Aurea'
ABOVE LEFT *Selaginella kraussiana* 'Brownii'
TOP *Selaginella martensii* 'Variegata'

S. uncinata Rainbow fern, Peacock moss China
Stems densely tufted, radiating outwards, at first ascending then more or less horizontal, 30–60cm(1–2ft) long, sending out aerial roots at intervals. Leaves an astonishing shade of metallic blue, especially in fairly bright light. An excellent hanging basket plant. Cool.

S. vogellii West Africa
Stems more or less erect, 30–60cm(1–2ft) tall, branching, frond-like above, bright green, taking on bronzy tints in a good light. Tropical.

S. willdenovii Tropical Africa and Asia
Stems tufted, climbing, to 3m(10ft) or more. Branches frond-like, similar to those of *S. uncinata*, the leaves having the same metallic blue sheen.

SELENICEREUS

Cactaceae
Night blooming sereus

Origin: Southern USA to South America. A genus of about 20 species of epiphytic cacti which flower at night. They have long ribbed stems which climb

ABOVE *Senecio confusus*
ABOVE CENTRE *Senecio bicolor cineraria*
TOP *Selenicerus hamatus*

or trail and often have aerial roots. Most species have spine-bearing areoles. The flowers are usually very large and sweetly scented, mostly white in colour and followed by round, fleshy red fruits. These plants are easy to grow, need space for their long stems which can be trained on wires or allowed to sprawl through shrubs. Propagate by cuttings of stem sections 10–15cm(4–6in) long in summer. *Selenicereus* derives from the Greek *selino*, the moon and *Cereus*, the cactus genus, referring to its appearance and night flowering.

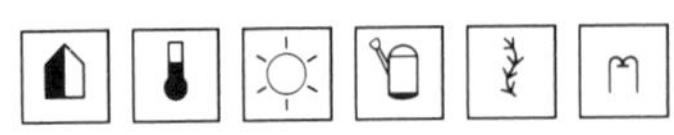

Species cultivated

S. grandiflorus Queen of the night Jamaica, Cuba
Stems fast growing, branching and rooting to 5m(16ft) in a large tub or planted out, less in a pot; each stem has five to eight low ribs and white-hairy areoles with yellow spines to 1.5cm($\frac{5}{8}$in) long. Flowers 18–25cm(7–10in) long, white inside, buff-yellow on the reverse, fragrant, opening in summer.

S. hamatus Southern and eastern Mexico
A fast growing species with slender, mainly 4-ribbed stems to 5m(15ft) in length if planted out in the conservatory border. The ribs, which usually number four, bear short, down-pointing spurs at about 5cm(2in) intervals; each spur has a brown areole and four to six bristle-like whitish spines. Flowers 20–25cm(8–10in) long and almost as wide at the mouth, white to yellow, aromatically fragrant, borne in summer and autumn.

S. pteranthus King/Princess of the night Mexico and widely cultivated in tropical America and elsewhere
Stems 4- to 6-ribbed, comparately thick and stiff, eventually 2–3m(6–10ft) in length, somewhat glaucous-green, often purple-flushed, producing aerial roots freely. Flowers to 30cm(1ft) long, woolly in bud, opening white to cream, fragrant, borne in summer.

SEMELE

Ruscaceae (Liliaceae)

Origin: Canary Islands, Madeira. Depending upon the botanical authority, a genus of one or five species of evergreen climbers related to *Asparagus* and *Ruscus*. One species is widely cultivated and is a perfect candidate for clothing the back wall of a conservatory with a curtain of lustrous laurel green. Propagate by seed or division in spring.

Semele is named for Semele, the mother of Dionysos.

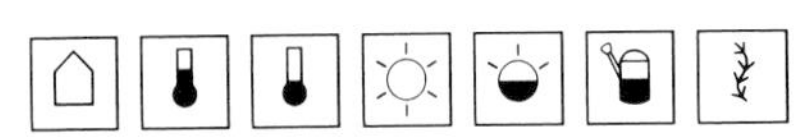

Species cultivated

S. androgyna Canary Islands
Stems twining, high-climbing in the wild, but easily maintained at about 3m(10ft). What appear to be leaves are flattened stems or cladodes (cladophylls), each one arising in the axil of a scale leaf. Cladodes ovate to narrowly so, slender-pointed, 5–10cm(2–4in) long, leathery-textured, glossy bright green when young then darker. Flowers small, the six spreading tepals greenish-white, borne in clusters of two to six on the margins of the cladodes. Berry fruits 1cm($\frac{3}{8}$in) wide, globular, greenish then black.

SENECIO

Compositae

Origin: Cosmopolitan. A genus of between 2,000 and 3,000 species of annuals, perennials, climbers, trees and shrubs, including some succulents – the largest genus of flowering plants. All species have alternate leaves and daisy-like flowers, sometimes without ray florets, most frequently borne in terminal clusters. The shrubs thrive in tubs or large pots, the climbers can be grown on the back wall of a conservatory or trained on a framework in a pot. *Senecio* × *hybridus* (cineraria) should be sown in spring for winter flowering and in summer for the following late spring to early summer. They require temperatures to 10°C(50°F) for over-wintering, but to 15°C(60°F) for winter blooms. Succulents require only light watering, but all others require medium watering. Propagate all species by seed; the succulents by cuttings in summer.

Senecio derives from the Latin *senex*, an old man, referring to the white-hairy pappus of some species.

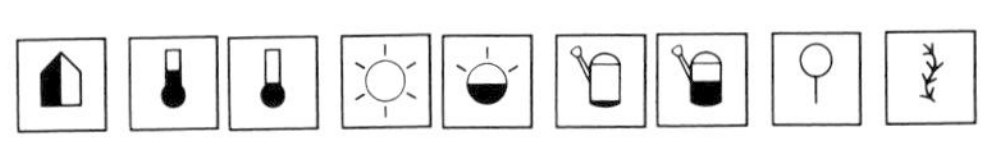

Species cultivated

S. articulatus (*Kleinia articulatus*) Candle plant South Africa
Succulent shrub 30–60cm(1–2ft) tall. Stems 1.5–2cm($\frac{5}{8}$–$\frac{3}{4}$in) across, cylindrical, divided with marked joints, grey-green. Leaves to 5cm(2in) long, deeply 3- to 5-lobed, short-lived, appearing briefly in the winter on young stems only. Flower heads about 1.5cm($\frac{5}{8}$in) long, yellowish-white, rather like those of groundsel. Propagate by stem sections which can be removed, or they sometimes fall off, at the joints.

S. bicolor cineraria (*S. cineraria*, *S. maritimus*, *Cineraria maritima*) Dusty miller West and central Mediterranean
Shrub to 50cm(20in) or more tall in a pot. Leaves 4–10cm(1$\frac{1}{2}$–4in) long, deeply toothed to pinnatisect, densely silvery-white-woolly. Flowers in heads to 1.5cm($\frac{5}{8}$in) across, yellow. Often sold under the cultivar names 'Silverdust' or 'White Diamond'.

S. cineraria See *S. bicolor cineraria.*

S. citriformis South Africa (Cape)
Succulent perennial forming radiating tufts of fairly thick stems to 10cm(4in) or more long. Leaves

ABOVE RIGHT *Senecio haworthii*
ABOVE LEFT *Senecio elegans*
LEFT *Senecio grandifolius*

usually crowded, erect, lemon-shaped, 1.5–2cm($\frac{5}{8}$–$\frac{3}{4}$in) long, glaucous with longitudinal translucent lines. Flowering stems up to 15cm(6in) in height bearing cream to palest yellow groundsel-like blooms.

S. confusus Mexican flame vine Mexico to Honduras
Twining climber to 3m(10ft) or more, bearing rather thick-textured, evergreen, narrowly ovate to lanceolate toothed leaves 4–7cm($1\frac{1}{2}$–$2\frac{3}{4}$in) in length. The daisy-like 5cm(2in) wide flower heads have bright orange-yellow ray florets which age to orange-red; they are borne in terminal and axillary clusters mainly in summer.

S. cruentus (of gardens) See *S.* × *hybridus*.

S. elegans Purple ragwort South Africa
Annual, with erect, branching stems 30–60cm(1–2ft) in height. Leaves groundsel-like, stem-clasping, deeply pinnately lobed and toothed, bright green, to 8cm(3in) long. Flower heads daisy-like, about 3cm($1\frac{1}{4}$in) wide, ray florets bright rose-purple to rose-lilac, in terminal corymbs. A showy and useful pot plant for late spring and summer blooming if sown in winter.

S. fulgens (*Kleinia fulgens*) South Africa, Natal
Sub-shrubby succulent from a tuberous base, to 60cm(2ft) or more in height. Leaves obovate to spathulate, 5–10cm(2–4in) long, pale glaucous green. Flower heads groundsel-like, orange to red, about 2.5cm(1in) long, solitary or in pairs terminating 20–30cm(4–8in) tall stalks.

S. grandifolius Mexico
Shrub, to 2m($6\frac{1}{2}$ft) or more in height with robust purple stems. Leaves 20–45cm(8–18in) long, ovate, with widely spaced, shallow teeth. Flower heads small, with five rays, borne in dense terminal corymbs to 30cm(1ft) across in winter. Best in temperate conditions. This species can be grown annually from spring cuttings and then does not exceed 90cm(3ft) in height.

S. haworthii (*Kleinia haworthii*, *K. tomentosa*) South Africa
Erect, sparingly branched succulent shrub to 30cm(1ft) in height, the fairly robust stems densely white-felted. Leaves more or less erect, 2–4cm($\frac{3}{4}$–

$1\frac{1}{2}$in) or more long, tapering to each end, also densely white-felted. Flower heads groundsel-like, orange, usually solitary on long stalks.

S. herreianus (*Kleinia gomphophylla*) Namibia (Buchu mountains)
Prostrate succulent perennial similar to the better known *S. rowleyanus*. Stems more or less freely branching and rooting, bearing fairly well spaced leaves; each leaf erect, broadly ovoid, 1–2cm($\frac{3}{8}$–$\frac{3}{4}$in) long, green with several translucent, longitudinal stripes. Flower heads small, solitary. An intriguing plant for a hanging basket.

S.* × *hybridus (*S. cruentus* of gardens)
Cineraria
A group of hybrid cultivars derived from several Canary Islands species including the true *S. cruentus* and *S. heritieri*, and popularly grown and sold as cinerarias. They are short-lived perennials grown as biennials, 30–90cm(1–3ft) tall. Leaves 10–20cm(4–8in) long, rounded-cordate with shallow lobes and teeth, downy beneath. Flower heads 3–8cm($1\frac{1}{4}$–3in) across in a wide variety of colours in the pink, red, purple and blue range, together with white, often bicoloured. Three strains are readily available: Multiflora, growing 40–50cm(16–20in) tall, with dense rounded heads of flowers; Multiflora Nana (Grandiflora Nana), growing about 30cm(1ft) tall, compact, with dense rounded heads; and Stellata, reaching 50–75cm(20–30in) tall with more starry flowers in looser clusters.

Senecio macroglossus 'Variegatus'

S. kaempferi See *Ligularia tussilaginea.*
S. kleinia (*Kleinia neriifolia*) Canary Islands
Succulent shrub, 2–3m($6\frac{1}{2}$–10ft) in the wild, but easily kept to half this in containers; stems thick, erect, segmented, sparingly branched. Leaves lanceolate to narrowly so, fleshy, 6–15cm($2\frac{1}{2}$–6in) long, grey-green, appearing in autumn and falling at the onset of the following summer. Flower heads groundsel-like, cream, about 2cm($\frac{3}{4}$in) long in terminal corymbs of 30 or more in winter or spring.
S. macroglossus Cape/Natal ivy, Wax vine South Africa
Evergreen, semi-succulent twining climber to 3m(10ft) or more in height; stems succulent at first, eventually becoming woody. Leaves triangular-hastate, usually with three to five pointed lobes, to 6cm($2\frac{1}{2}$in) long, fleshy-textured, dark glossy green. Flower heads daisy-like, 5–6cm(2–$2\frac{1}{2}$in) wide, ray florets white, disk yellow, solitary and terminal, borne mainly in winter. *S.m.* 'Variegatus' has the leaf margins irregularly white-bordered.
S. maritimus See *S. bicolor cineraria.*
S. mikanioides German ivy South Africa
Perennial evergreen climber with twining stems to 3m(10ft) or more in length if planted out, less in containers. Leaves rounded, with a widely cordate base and five to seven broad but pointed-tipped lobes, somewhat softly fleshy, bright green. Flowers groundsel-like, small, yellow, in axillary and terminal corymbs, together sometimes forming quite large panicles and then almost showy (but not on small plants), borne in autumn to early winter. Best grown cool and will stand light frost.
S. petasites Mexico
Evergreen shrub to 2m($6\frac{1}{2}$ft) or more, erect when young then spreading. Leaves to 20cm(8in) long, ovate to almost orbicular with eight to 14 shallow lobes, velvety-hairy on both surfaces, greyish beneath. Flowers daisy-like, small, both disk and ray florets yellow, borne in large and showy terminal panicles in winter to spring. Needs 10°C(50°F) to flower well.
S. radicans (*Kleinia radicans*) S. Africa (Cape)
Perennial succulent with much branched, very slender stems forming mats 30–60cm(1–2ft) wide. Leaves more or less erect, ovoid, pointed-tipped, cylindrical in cross-section, 2–3cm($\frac{3}{4}$–$1\frac{1}{4}$in) in length, grey-green bearing darker, translucent longitudinal lines. Flower heads groundsel-like, white, terminal, solitary or in pairs, but rarely produced in cultivation.
S. repens See *S. serpens.*
S. rowleyanus String of beads S.W. Africa
Succulent perennial with slender, prostrate or pendent stems 60–90cm(2–3ft) long. Leaves 5–12mm($\frac{1}{5}$–$\frac{3}{8}$in) across, almost spherical, bright green with a translucent band around one side and a small apiculate point. Flower heads 1.5–2cm($\frac{5}{8}$–$\frac{3}{4}$in) long made up of long white disk florets with purple stamen tubes, giving the effect of tiny shaving brushes. Very good in a hanging basket. Grows well if kept warm, but flowers best after a cool spell in early winter.
S. serpens (*S. repens, S. succulenta, Kleinia repens*) South Africa (Cape)
Shrubby succulent, the erect stems to 30cm(1ft) in height, branching mainly from the base. Leaves to 4cm($1\frac{1}{2}$in) or more in length, almost cylindrical, tapering to each end, grey-green with a bluish waxy patina. Flowers white to cream, in small cymes. Best grown in hanging baskets.
S. stapeliiformis (*S. stapeliaeformis, Kleinia gregorii, K. stapeliiformis*) South Africa (Cape)
Succulent shrub, to 25cm(10in) or so in height; stems cylindrical, branching mainly at the base, 5- to 7-angled or ribbed, greyish-green. Each rib bears a row of dark green leaf cushions or bases which, when young terminate in tiny awl-shaped leaves which soon wither. Flower heads red, brush-like, lacking ray florets, 2.5–4cm(1–$1\frac{1}{2}$in) across, solitary on long slender stalks.

A remarkable *Stapelia* mimic, this is easily grown providing it is kept dry during its summer dormancy period.
S. succulenta See *S. serpens.*
S. tomentosus Misapplied to *S. haworthii q.v.*

RIGHT *Serissa foetida* 'Variegata'

SERISSA

Rubiaceae

Origin: South-eastern Asia. A genus of one species of small evergreen shrub which makes a useful and modestly decorative pot plant. Propagate by cuttings in summer. *Serissa* is a Latinized version of the vernacular East Indian name.

   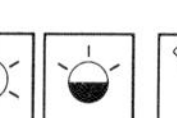 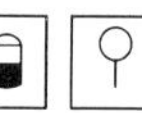

Species cultivated

S. foetida (*S. japonica*)
Wiry-stemmed, bushy shrub 30–60cm(1–2ft) tall and of greater width. Leaves in opposite pairs, ovate, 6–12mm($\frac{1}{4}$–$\frac{1}{2}$in) long, deep green. Flowers about 12mm($\frac{1}{2}$in) wide, funnel-shaped with four to six spreading white petals. A double-flowered form, 'Plena', is in cultivation; it has flowers like tiny white roses. *S.f.* 'Variegata' has the leaves prettily cream-margined.

SETCREASEA

Commelinaceae

Origin: Mexico to USA (Texas). A genus of six species of perennials allied to *Tradescantia*. They are tufted to clump-forming, somewhat fleshy with alternate ovate to oblong leaves and 3-petalled flowers borne within boat-shaped bracts. The species described below is a familiar conservatory and house plant. Repot annually in late spring and slightly shade from full summer sun. Propagate by division or by cuttings from spring to autumn. *Setcreasea* is a name of unknown derivation.

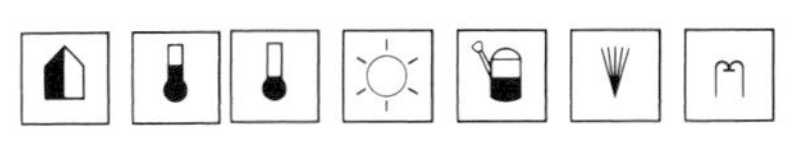

Species cultivated

S. pallida Mexico
Erect to trailing plant with slender stems to 40cm(16in) long. Leaves 8–15cm(3–6in) long, oblong, fleshy. Flowers about 2cm(¾in) across, mauve to lavender-pink, clustered at the ends of the stems. 'Purple Heart' (*S. purpurea*, Purple heart) is a form with dark, violet-purple leaves and red-purple flowers.
S. striata *See Callisia elegans.*

SIDERASIS

Commelinaceae

Origin: Brazil. A genus of one species of evergreen perennial formerly included in *Tradescantia*, with ornamental foliage and flowers. Propagate by seed or division in spring. *Siderasis* derives from the Greek *sideros*, iron, referring to the rust-coloured hairs which cover all but the petals.

Species cultivated

S. fuscata (*Pyrrheimia fuscata*, *Tradescantia fuscata*)
Tufted to small clump-forming. Leaves elliptic, 10–20cm(4–8in) long, dark green above with a longitudinal band of silvery white, red-purple-flushed beneath. Flowers 3-petalled, 2.5cm(1in) across, rose-purple, borne in summer.

SINARUNDINARIA See ARUNDINARIA

SINNINGIA

Gesneriaceae

Origin: Mexico to Argentina. A genus of about 75 species, including plants formerly classified as *Corytholoma*, *Gesneria*, *Gloxinia* and *Rechsteineria*. Most grow from somewhat woody tubers and have erect to ascending stems. Their leaves are ovate and the tubular, 5-lobed flowers can be solitary or borne in cymes from the upper leaf axils. Those described

ABOVE *Setcreasea pallida* 'Purple Heart'
ABOVE LEFT *Siderasis fuscata*

below need warmth to thrive. Pot tubers in spring and dry off when the leaves yellow in autumn, storing in their pots at not less than 12°C(52°F) for the winter. Propagate by cuttings of young shoots having at least two pairs of leaves, preferably with a sliver of tuber attached; these need bottom heat of 21°–24°C(70–75°F) and are best in a propagating case. Tubers can be cut into sections, each with a young shoot. Dust the cut area with a fungicide and pot at once. Seed can be sown in warmth in spring. *Sinningia* was named for Wilhelm Sinning (1794–1874), a German gardener.

Species cultivated

S. cardinalis (*Gesneria cardinalis*) Cardinal flower Brazil
Stems erect, 15–25cm(6–10in) tall. Leaves 10–15cm(4–6in) long, ovate-cordate, densely covered with short hairs, the veins deeper green. Flowers 5cm(2in) long, tubular, markedly 2-lipped, bright scarlet, opening in summer.
S. concinna Brazil
Small rosetted species springing from a tiny rounded tuber. Leaves ovate, 2cm(¾in) long, on slender stalks. Flowers on slender erect stems from the leaf axils, 2cm(¾in) long, in shape rather like those of a *Streptocarpus* with an oblique tube and five rounded lobes, lilac, purple and white, borne in summer.
S. × fyfiana See under *S. speciosa.*
S. leucotricha (*Rechsteineria leucotricha*) Brazil
Stems erect or reclining, to 25cm(10in) tall.

Sinningia leucotricha

Leaves to 15cm(6in) long, broadly ovate to obovate, densely covered with silvery hairs. Flowers 3–4cm(1¼–1½in) long, tubular, in clusters of three to five, salmon-red, opening in summer.
S.* × *maxima See under *S. speciosa.*
S. pusilla Brazil
Akin to *S. concinna*, but somewhat smaller and even daintier, the petal lobes lilac with purple lines. This species and *S. concinna* have been crossed in USA with *S. eumorpha* and other larger species to create a race of small cultivars of great charm. Names to look out for are: 'Bright Eyes' and 'Wood Nymph' (*pusilla* × *concinna*), 'Cindy' (*concinna* × *eumorpha*), 'Dollbaby' (*pusilla* × *eumorpha*) and 'Freckles' (*pusilla* × *hirsuta*).
S. regina Brazil
Erect, to 10–15cm(4–6in) tall. Leaves 10–15cm(4–6in) long, ovate to elliptic, velvety hairy in texture, bronze-green with a pattern of white veins, deep red beneath. Flowers nodding, solitary, 5cm(2in) long, violet-purple, open in summer.
S. speciosa (*Gloxinia speciosa*) Gloxinia Brazil
The true species is 15–25cm(6–10in) tall. Leaves 15–20cm(6–8in) long, ovate to oblong and velvety-hairy, dark green above, flushed with red beneath. Flowers 4–5cm(1½–2in) long, solitary, nodding, rather like a foxglove but with a fleshy texture, red to violet, opening in summer. Most frequently cultivated under this name are two hybrid groups of involved parentage (sown at intervals with sufficient warmth they can be had in bloom throughout the year): *Maxima group* (*Gloxinia* × *maxima*) Plants have larger, more fleshy-textured nodding flowers which range from pinks and reds to purple and white or can be bicoloured. *Fyfiana group* (*Gloxinia* × *fyfiana*, *G.* × *hybrida*) Similarly larger, fleshier flowers than the species which are more erect and bell-shaped and sometimes have waved or crimped petals; they can be single or double and come in a similar range of colours as the Maxima group.
S. tuberosa (*Rechsteineria tuberosa*) Brazil
Leaves and flowers direct from a tuber. Leaves one or two only, ovate to oblong, 25–45cm(10–18in) in length, crenate and slightly hairy. Flowers about 3cm(1¼in) long, cylindrical, tipped by five short lobes, the upper two pointing forward, red with a yellow throat, in clusters of six or so in summer.
S. tubiflora (*Achimenes tubiflora*) Argentina, Paraguay, Uruguay
Stems erect, up to 60cm(2ft) in height, rigid and usually with branches from the lower half. Leaves 8–13cm(3–5in) long, oblong-elliptic, crenate, finely hairy and prominently veined. Flowers in terminal racemes, white and slender tubed, to 9cm(3½in) long, bearing five broad lobes creating a face to 4cm(1½in) wide, fragrant of tuberose.

BELOW *Smilax aspera*
BOTTOM *Sinningia speciosa*

SMILAX

Smilacaceae (Liliaceae)

Origin: Tropics and sub-tropics; widespread. Depending upon the botanical authority, a genus of 200 to 300 species of perennial scramblers and climbers. They have alternate, simple leaves, in some species with two tendrils at the base which may represent modified stipules. In many species the stems have hooked prickles which also aid in climbing. The small, 6-petalled flowers are carried in axillary umbels or racemes and are followed by black, purple, blue or red berries. Propagate by seed or division in spring, or by cuttings in summer. The species described here can nicely embellish the back wall or roof space of a conservatory.

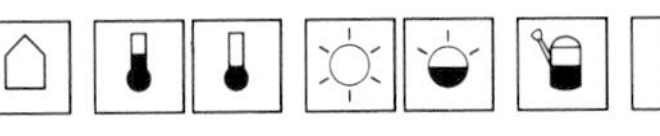

Species cultivated
S. asparagoides See *Asparagus asparagoides.*
S. aspera Southern Europe, Western Asia
Tendril climber, the stems usually with some hooked thorns, 2m(6½ft) or more in length. Leaves evergreen, lanceolate to narrowly triangular with a heart-shaped base, 4–10cm(1½–4in) long, leathery-textured, glossy rich green with a variable amount of pale flecking. Flowers yellow-green, fragrant. Berries red-purple to red, globular, 6mm(¼in) wide.
S. ornata Mexico
Evergreen tendril climber to 6m(20ft), sometimes twice this in the wild, with 4-angled stems bearing scattered hooked thorns. Leaves ovate-cordate to broadly elliptic, 10–30cm(4–12in) long, rich glossy green, the smaller ones on the lateral stems grey-flecked or blotched. Temperate.

SMITHIANTHA

Gesneriaceae

Origin: Mexico. A genus of four species of perennials formerly named *Naegelia*. They have fleshy, tuber-like rhizomes and erect stems which carry pairs of broadly ovate, velvety leaves and nodding, tubular, 5-lobed flowers rather like foxgloves, on long stalks in summer and autumn. They make handsome pot plants. Propagate by separating the rhizomes when potting in late winter and early spring. Potting some every three to four weeks will extend the flowering season. *Smithiantha* was named for Matilda Smith (1854–1926), a botanical artist working at Kew.

 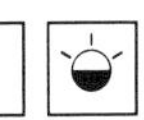   

Species cultivated

S. cinnabarina Temple bells Mexico
From 40–60cm(16–24in) tall with red-hairy leaf stalks. Leaves 8–15cm(3–6in) long, deep green and purple with velvety red-brown hairs giving a glossy effect. Flowers 4cm(1½in) long, brick-red to scarlet often lined with creamy-yellow.
S. multiflora Mexico
To 75cm(2½ft) tall. Leaves to 15cm(6in) long, dark green. Flowers 4cm(1½in) long, creamy-yellow to white with no spotting or striping.
S. zebrina Mexico
To 75cm(2½ft) tall. Leaves to 18cm(7in) long, dark green, patterned with purplish-brown. Flowers 4cm(1½in) long, the tube scarlet above, golden beneath, upper lobes orange-yellow, lower yellow.

Hybrid cultivars

These three species and the orange-red *S. fulgida* have been hybridized together to produce a race of strong growing cultivars with freely borne flowers forming pyramidal panicles. They are available in a wide colour range including scarlet, carmine, pink, orange, yellow and white, with varying amounts of bronzing on the leaves.

SOLANDRA

Solanaceae

Origin: Mexico and West Indies to tropical South America. A genus of about ten species of scrambling climbers much prized in tropical horticulture for their large trumpet-shaped flowers. They are robust plants with long, thrusting shoots and alternate simple leaves. Those species described below make handsome pot or tub plants, but are seen to best effect when planted out in the conservatory. *Solandra* honours the Swedish Daniel Carl Solander (1736–1782), a pupil of Linnaeus, botanist on Captain Cook's first voyage to the Pacific (1768–1771) and later Keeper of the Natural History Department of the British Museum.

  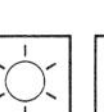  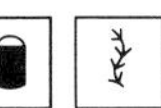

Species cultivated

S. grandiflora West Indies
Vigorous, with stems 5–10m(16–33ft) in length. Leaves elliptic to obovate, 13cm(5in) or more long, glabrous, usually rich, sub-lustrous green. Flowers to 15cm(6in) long, opening white and ageing tan-yellow with ten purple lines inside, each one from a calyx to 8cm(3in) long.

LEFT *Smithiantha* 'Karen'
BELOW LEFT *Solandra maxima*

S. guttata Cup of gold Mexico
Similar but less vigorous than *S. grandiflora*, with pubescent, elliptic-oblong leaves 8–15cm(3–6in) long. Flowers 15–22cm(6–9in) long, deep yellow spotted or feathered with purple. Much confused with *S. maxima*.
S. hartwegii See *S. maxima*.
S. longiflora Cuba, Hispaniola
Much like *S. grandiflora*, but with a calyx about half as long.
S. maxima (*S. hartwegii*, *S. nitida*) Cup of gold Mexico
Vigorous, with stems 5–10m(16–33ft) long. Leaves glabrous, elliptic, to 15cm(6in) long, usually deep green and lustrous. Flowers to 22cm(9in) long, yellow with five purple lines.
S. nitida See *S. maxima*.

SOLANUM

Solanaceae

Origin: Cosmopolitan. A genus of about 1,700 species of annuals, perennials, shrubs and climbers with alternate leaves. The flowers, which can be solitary or in clusters, look superficially to be on lateral shoots, but on close examination it can be

Solanum elaeagnifolium

seen that they do not arise in the leaf axils. They are, in fact, terminal, growth being continued by an axillary stem which grows beyond the previous main stem. Flowers have the corolla fused at the base, opening to five spreading lobes, with a prominent central cone of yellow stamens. They are followed by rounded berries, poisonous in some, edible in others. The climbers described are best in a conservatory and, being vigorous, it is advisable to prune them every spring, the rest are suitable for the house. Propagate shrubs and climbers by cuttings in summer or by seed in spring. *S. capsicastrum* and *S. pseudocapsicum* are usually propagated annually, sowing seed in spring in warmth, pricking off and potting-on as soon as they are large enough, finally using 13cm or 15cm (5in or 6in) pots. *Solanum* is an old Latin name probably for *S. nigrum*, the black nightshade.

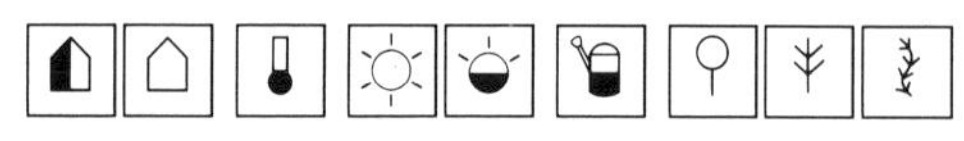

Species cultivated

S. aviculare Kangaroo apple New Zealand, S.E. Australia
Shrub to 2m($6\frac{1}{2}$ft) in a tub, more in the open. Leaves 15–30cm(6–12in) long, lanceolate to narrowly lanceolate; both entire and pinnatifid leaves are produced. Flowers 2–3.5cm($\frac{3}{4}$–$1\frac{1}{3}$in) wide, mauve to pale violet, singly or in cymes in summer and autumn, followed by yellowish, ovoid berries 2–2.5cm($\frac{3}{4}$–1in) long.

S. capsicastrum Winter cherry Brazil
Evergreen shrub to 60cm(2ft) or more, but usually grown as an annual and then rarely exceeding 30cm(1ft). Leaves 4–7cm($1\frac{1}{2}$–$2\frac{3}{4}$in) long, obovate to oblong-lanceolate, with short, branching hairs, each large leaf with a small one at its base. Flowers about 1cm($\frac{3}{8}$in) wide, white, carried singly or in pairs, followed by ovoid, pointed, scarlet berries 1.5–2cm($\frac{5}{8}$–$\frac{3}{4}$in) long. Often confused with *S. pseudocapsicum*. For *S.c.* 'Nanum' see *S. pseudocapsicum* 'Nanum'.

S. crispum Chilean potato tree Chile
Scrambling, evergreen shrub to 4m(13ft). Leaves 6–10cm($2\frac{1}{2}$–4in) long, ovate-lanceolate, often waved (crisped). Flowers 2cm($\frac{3}{4}$in) across, blue-purple, borne in dense clusters in summer and autumn, followed by small, rounded yellowish-white berries. A white form is known. *S.c.* 'Glasnevin' is free-flowering.

Solanum crispum

S. elaeagnifolium Central USA to Mexico
Branched, short-lived perennial to 1m(3ft) or more tall. Stems more or less erect, slightly prickly or unarmed. Leaves narrowly to oblong-lanceolate, 4–9cm($1\frac{1}{2}$–$3\frac{1}{2}$in) long, densely and finely silvery-white hairy. Flowers 2–3cm($\frac{3}{4}$–$1\frac{1}{4}$in) wide, purple to blue, in cymes. Berries globose, 1–1.5cm($\frac{3}{8}$–$\frac{5}{8}$in) wide, brownish to yellow.

S. jasminoides Potato vine, Jasmine nightshade Brazil
Semi-twining, slender-stemmed climber to 5m(16ft). Leaves 3–7cm ($1\frac{1}{4}$–$2\frac{3}{4}$in) long, entire or with irregular-pinnate lobes. Flowers to 2.5cm(1in) across with broad lobes, pale blue to light mauve, in branched cymes in summer and autumn. *S.j.* 'Album' has pure white flowers.

S. macranthum Brazilian potato tree Brazil
In its homeland, a small tree, but in cultivation, a shrub to 2m($6\frac{1}{2}$ft) or less. Branches erect, at least at first, downy and spiny. Leaves broadly ovate, to 35cm(14in) long, wavy-lobed or toothed. Flowers 5cm(2in) wide, deep blue-purple ageing to almost white, in lateral cymes to 13cm(5in) long.

S. melongena Aubergine, Egg-plant Tropical Asia, much cultivated elsewhere
Short-lived, woody-based perennial usually grown as an annual. Stems erect, forking at intervals, to 90cm(3ft) in height. Leaves ovate, 10–20cm(4–8in) long, slightly wavy and sometimes angled or lobed. Flowers purple, solitary, lateral, to 4cm($1\frac{1}{2}$in) across. Fruits pendent, ovoid, 10–15cm(4–6in) long, dark purple, edible. Represented in cultivation by *S.m. esculentum* with fruits 15cm(6in) or more in length, varying in shape and colour, often black-purple, ivory or striped and narrowly obovoid to oblong, usually slightly curved. Grown for its edible fruit, but also makes a decorative pot plant.

S. pseudocapsicum Jerusalem cherry Eastern South America, but widely naturalized in the tropics and sub-tropics
Erect shrub to 1m(3ft) or more tall. Leaves 5–10cm(2–4in) long, oblong to lanceolate, glossy green, hairless, in pairs often of unequal size. Flowers 1.5cm($\frac{5}{8}$in) across, white, singly or in twos or threes, followed by globose berries 1–1.5cm($\frac{3}{8}$–$\frac{5}{8}$in) across, red, poisonous. Distinguished from *S. capsicastrum* (with which it is much confused) by its smaller, round berries and smooth leaves. 'Nanum' is dwarf, compact and bushy; 'Patersonii' is dwarf but of a spreading habit.

S. rantonnetii Blue potato bush Argentina to Paraguay
Shrub to 2m($6\frac{1}{2}$ft) in height, usually freely branching. Leaves ovate, pointed, more or less wavy, 6–10cm($2\frac{1}{4}$–4in) long. Flowers in small cymes, about 1.2cm($\frac{1}{2}$in) wide, deep purple-blue. Fruits heart-shaped, to 2.5cm(1in) long, red. *S.r.* 'Grandiflorum' is the usual form in cultivation with flowers 2–2.5cm($\frac{3}{4}$–1in) wide.

S. wendlandii Giant potato vine Costa Rica, widely grown in the tropics
Showy scrambler to 5m(16ft) or more, with fairly robust stems, climbing by hooked prickles. Leaves pinnate, 10–30cm(4–12in) long, composed of seven to 13 leaflets, usually prickly on the stalks and midribs. Flowers lilac-blue, to 6cm(2½in) across, in often large panicle-like terminal clusters in late summer. Fruits ovoid to globose, about 1cm(⅜in) long, orange, but seldom produced in cultivation under glass.

SOLEIROLIA

Urticaceae

Origin: Corsica and Sardinia. A genus of one evergreen, creeping, mat-forming perennial grown for the effect of its dense foliage. It can be used as ground cover in a conservatory border or under a large plant in a tub, and is not without charm in pans in the conservatory and home. Propagate by division at almost any time. *Soleirolia* was named for Joseph Francois Soleirol (*d.*1863) who collected plants in Corsica.

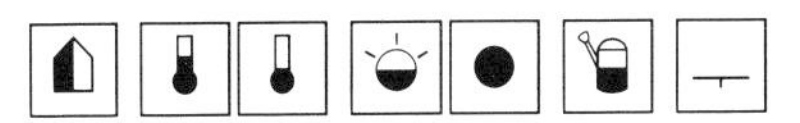

Species cultivated
S. soleirolii (*Helxine soleirolii*) Mind your own business, Baby's tears
Slender filiform stems form a neat interlacing mat, scarcely 2.5cm(1in) high. Leaves 2–6mm($\frac{1}{16}$–¼in) long, more or less rounded, sparsely hairy with short stalks, alternately borne to form a dense covering to the stems. Flowers minute, unisexual, singly in the leaf axils. *S.s.* 'Argentea' has silvery-green leaves; 'Aurea' has yellow-green leaves.

SOLLYA

Pittosporaceae

Origin: Western Australia. A genus of two species of perennial twining climbers allied to *Billardiera*. They have wiry stems, alternate, simple leaves, 5-petalled, rather bell-like flowers in lateral nodding cymes, and purple berries. Perfect small climbers for the cool conservatory, they can look charming winding through a shrub, e.g. *Callistemon*. Propagate by seed in spring or by cuttings in summer. *Sollya* honours Richard Horsman Solly (1778–1858), an English botanist and anatomist.

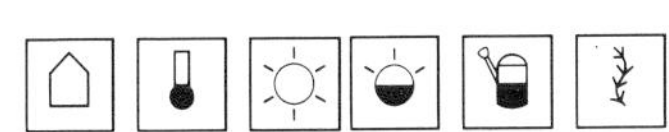

Species cultivated
S. fusiformis See *S. heterophylla*.
S. heterophylla (*S. fusiformis*) Australian bluebell creeper/bluebell
Stems 1.2–1.8m(4–6ft) in height. Leaves 2–5cm(¾–2in) long, ovate to lanceolate, pale green. Flowers about 1.2cm(½in) wide, sky-blue, borne in

ABOVE *Soleirolia soleirolii*
TOP *Solanum wendlandii*
LEFT *Sollya heterophylla*

summer to autumn. Fruits oblongoid, purple.
S. parviflora
Much like *S. heterophylla*, but a little smaller in stature and with consistently lanceolate leaves and darker blue flowers.

SONERILA

Melastomataceae

Origin: Tropical Asia. Depending upon the botanical authority consulted, a genus of 100 to 175 species of evergreen perennials and small shrubs. Only one species is widely cultivated for its beautifully patterned leaves and attractive flowers. It is essentially a plant for the warm humid conservatory, but responds well to plant case culture in the home. Propagate by cuttings in spring or summer. *Sonerila* is the Latinized version of the native Malabar name *soneri-ila*.

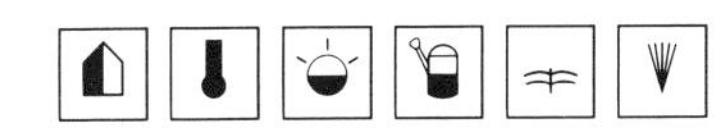

Species cultivated
S. margaritacea Java to Burma
Stems tufted, erect to decumbent, branched, red, to 20cm(8in) or more in height. Leaves in opposite

Sonerila margaritacea

pairs, ovate to lanceolate, 5–8cm(2–3in) long, rich coppery-green, neatly overlaid above with a raised, pearly-silver pattern, purple-red beneath. Flowers 1.2cm(½in) across, rosy-mauve, in racemes. This plant is usually met with in cultivation as one of its more heavily silver-patterned leaf forms, e.g. *S.m. argentea*, 'Hendersonii', 'Marmorata' and 'Mme Baextele', the last mentioned being smaller and more compact.

BELOW *Sparmannia africana*
BOTTOM *Sophronitis coccinea*

SOPHRONITIS

Orchidaceae

Origin: Brazil. A genus of six species of small epiphytic orchids related to *Cattleya* but much smaller. They have small pseudobulbs clustered together, each with one or rarely two leaves. The flowers have a small protruding lip and spreading tepals; they are usually solitary on stalks from the tops of the pseudobulbs. Plants are suitable for the home or conservatory, but require extra humidity. Propagate by division in spring or early autumn. *Sophronitis* derives from the Greek *sophron*, modest, an allusion to the small size of the flowers.

Species cultivated
S. coccinea (*S. grandiflora*)
Pseudobulbs 2.5–4cm(1–1½in) long, narrowly ovoid, tightly clustered. Leaves solitary, 6–8cm(2½–3in) long, oblong-lanceolate, dark glossy green with a somewhat leathery texture. Flowers 4–7cm(1½–2¾in) wide, bright scarlet, outer tepals elliptic to oblong, inner two rounded and wing-like, labellum narrow, orange to yellow, borne in autumn and winter.

SPARAXIS

Iridaceae

Origin: South Africa. A genus of about five species of cormous perennials. As the old corm withers at the end of each season a new one is formed above, and from it arises a narrow fan of strap- to sword-shaped leaves and wiry, branched stems carrying 6-tepalled, often strikingly coloured flowers. They are suitable for a conservatory and can be brought into the home for flowering. Propagate by separating offsets when dormant or by seed in spring. *Sparaxis* derives from the Greek *sparasso*, to tear, an allusion to the torn, papery bracts beneath the flowers.

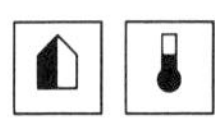 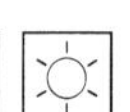   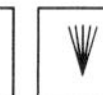

Species cultivated
S. tricolor Harlequin flower, Wand flower
To 30–45cm(1–1½ft) tall. Leaves to 30cm(1ft) long and 2cm(¾in) wide, sword-shaped to linear. Flowers 5cm(2in) across, tube yellow, often black-bordered, tepals spreading, in shades of purple, red, yellow and white, opening from late spring into summer. Some of the forms available are of hybrid origin.

SPARMANNIA

Tiliaceae

Origin: Africa, including Malagasy. A genus of three to seven species of trees and shrubs with large, usually palmately lobed leaves and umbels of white flowers. They make handsome container plants for the conservatory and can be flowered in the home, grown as shrubs, pruning after flowering, or as annuals, cuttings back in late winter and taking the resulting shoots as cuttings in spring, rooting them with bottom heat. Pinch out the tips of young plants to encourage bushiness. *Sparmannia* was named for Dr Andreas Sparrman (1748–1820), a Swedish botanist who collected plants in the Pacific region when on Capt. Cook's second voyage and in South Africa with Thunberg.

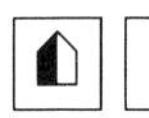 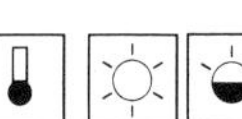  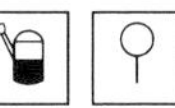

Species cultivated
S. africana African hemp South Africa
Shrub, easily kept at 1–1.5m(3–5ft) in a large pot

or tub, but to 3–6m(10–20ft) in the wild; grown as an annual, to about 60cm(2ft) tall. Leaves alternate, 12–15cm(4¾–6in) long, ovate-cordate, toothed, slender-tipped and long-stalked. Flowers 3–4cm(1¼–1½in) wide, white with yellow and red-purple stamen filaments, borne in stalked umbels from the upper leaf axils from late spring to summer. *S. a.* 'Flore-Plena' has double flowers.

SPATHIPHYLLUM

Araceae

Origin: Tropical areas of Central and South America, and the islands of S.E. Asia. A genus of about 35 species of evergreen perennials with long-stalked ovate to lanceolate leaves forming dense clumps. They are good tub or pot plants for the warm conservatory or home needing a humid atmosphere. Propagate by division when repotting. *Spathiphyllum* derives from the Greek *spathe*, bract and *phyllon*, leaf, the spathes being leaf-like in shape.

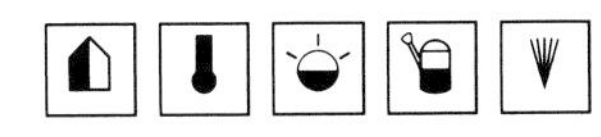

Species cultivated

S. cannifolium Northern South America, Trinidad
Leaf blades oblanceolate or elliptic, corrugated, dark green, 20–30cm(8–12in) long, on longer stalks. Spathes lanceolate, up to 25cm(10in) in length, green and white, on stems 60cm(2ft) or more in height.

S. × 'Clevelandii' (*S. kochii* of some gardens)
Hybrid of uncertain parentage closely allied to *S. wallisii*, but larger than that species. To 60cm(2ft) in height, it has lanceolate, somewhat wavy-margined leaves to 30cm(1ft) or more long. Spadices white; spathe to 15cm(6in) long, pure white at first with a green central line on the back.

S. cochlearispathum S.E. Mexico
Leaf blades broadly oblong, 30cm(1ft) or more in length, wavy, corrugated, light green, heavy-textured on stalks to 60cm(2ft) in length. Spathes about 30cm(1ft) long, oblanceolate, somewhat hooded, creamy-green on stems 1.2–1.5m(4–5ft) in height.

S. floribundum Colombia
To 30cm(1ft) or more tall. Leaves to 15cm(6in) long, elliptic-lanceolate, dark shining green. Spathes to 8cm(3in) long, oblong-lanceolate, pure white; spadices green and white.

S. × 'Mauna Loa'
Hybrid between *S. floribundum* and probably *S.* 'McCoy', itself the product of crossing *S. cochlearispathum* with *S.* × 'Clevelandii', so this cultivar has a very involved parentage. It is vigorous and compact, 45–60cm(1½–2ft) tall, the leaves oblong-lanceolate to 30cm(1ft) long, dark green. Spathes 13–20cm(5–8in) long, white.

S. wallisii Peace lily, White sails Colombia and Venezuela

Spathiphyllum wallisii

30–45cm(1–1½ft) tall. Leaves 15–25cm(6–10in) long, oblong-lanceolate, glossy green. Spathes to 13cm(5in) or more long, ovate, white.

SPIRONEMA See CALLISIA & HADRODEMAS

SPREKELIA

Amaryllidaceae

Origin: Mexico. A genus of only one species—a handsome bulbous plant which is suitable for growing in the conservatory or home where it needs positioning on a sunny window sill. Pot in late winter or early spring leaving the neck of the bulb above soil level. Dry off as the leaves yellow in autumn and keep dry at about 7–10°C(45–50°F) until late winter.

Propagate by offsets or seed. *Sprekelia* was named for Johann Heinrich von Sprekelsen (1691–1764), a German lawyer who was also a keen amateur botanist and gardener.

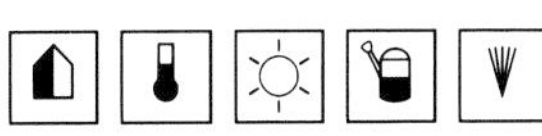

Species cultivated

S. formosissima (*Amaryllis formosissima*)
Jacobean/Aztec lily
Leaves to 30cm(1ft) tall, strap-shaped. Flowers to

Sprekelia formosissima

10cm(4in) long with six tepals; the upper one is broad and stands erect or curving back at the tip, those at the sides are narrower, curved and spreading, the bottom three overlapping and forming a shape like the labellum of an orchid; they are bright crimson and borne singly on 30cm(1ft) scapes in spring or summer as the new leaves develop.

STANHOPEA

Orchidaceae

Origin: Mexico to Brazil. A genus of about 25 to 45 species of epiphytic orchids with mainly ovoid, clustered pseudobulbs 5–7cm(2–2¾in) tall, each carrying one pleated leaf 30–60cm(1–2ft) long and two to six fragrant flowers usually in pendent racemes. The blooms have reflexed tepals, the three outer ones (sepals) wider than the inner, and a rounded to shoe-shaped labellum terminating in a tongue-like tip, and sometimes with two horn-like protruberances near the base. The long column arches over almost touching the tip of the labellum. The species described below are best in hanging baskets or on slabs of bark to accommodate the pendent flowers. Propagate by careful division in spring. *Stanhopea* was named for Philip Henry Stanhope (1781–1855), 4th Earl of Stanhope and President of the Medico-Botanical Society of London.

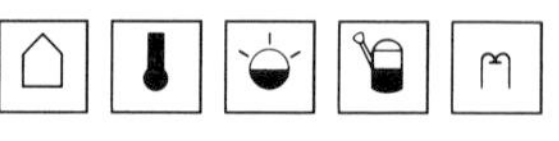

Species cultivated

S. costaricensis See *S. graveolens.*

S. devoniensis Mexico
Flowers to 10cm(4in) across, usually pale yellow with red-brown to deep red-purple spotting, labellum ivory to cream, flushed and blotched with purple and with two horns, opening in late summer.

S. eburnea See *S. grandiflora.*

S. ecornuta Central America
Flowers 10–12cm(4–5in) wide, creamy-white, tepals usually with purple spotting at the base, labellum yellow, shading to orange at the base as it ages, without horns, borne in summer.

S. grandiflora (*S. eburnea*)
Flowers 13–18cm(5–7in) wide, ivory-white, labellum also ivory but with purplish markings, borne in summer.

S. graveolens (*S. costaricensis*) Mexico to Brazil
Flowers 10–13cm(4–5in) long, yellow to white, spotted red-brown or purple, sometimes with ring markings, labellum yellow-white with two lateral purple 'eye' blotches, borne in spring to early summer.

S. hernandezii See *S. tigrina.*

S. tigrina (*S. hernandezii*) Mexico
Flowers to 20cm(8in) across, orange-yellow with purple-brown blotchings, labellum horned, borne in summer.

S. wardii Mexico to Panama
Flowers to 13cm(5in) wide, pale to deep yellow, heavily purple spotted, labellum horned, opening in summer to autumn.

STAPELIA

Asclepiadaceae
Carrion flowers

Origin: Tropical and southern Africa. A genus of 75 to 90 species of perennial succulent plants with fleshy, cylindrical, 4-angled stems which can be grooved or winged, in appearance rather like cacti but sometimes with spine-tipped scale leaves. Their flowers are bell-shaped or rotate with five, often hairy lobes; most have a carrion smell to attract the flies which pollinate them. The centre of each flower is made up of a two whorled corona, the outer five segments flat and spreading, the inner five erect and horn-like. Plants are suitable for the home or conservatory, though when in flower the smell may be unpleasant indoors. Propagate by division when re-potting or by cuttings in early summer; allow the cuttings to dry for several days before inserting them. Seeds sown in spring can germinate within 24 hours. *Stapelia* was named for

Johannes Bodaeus van Stapel (*d. circa* 1636), a Dutch physician.

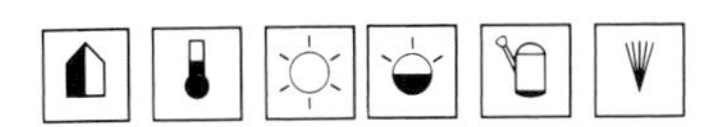

Species cultivated

S. gigantea Natal, Transvaal, Zimbabwe
Stems 15–20cm(6–8in) long, ascending, the angles wing-like and toothed, pale green. Flowers 15–35cm(10–14in) and sometimes more across, divided into large ovate lobes with tail-like tips, pale ochre-yellow, bearing numerous fine, transverse crimson wrinkles, purple-hairy on the upper surface, white-hairy on the margins.

S. hirsuta Hairy toad plant South Africa
Stems to 20cm(8in) tall, slender. Flowers one to three, 10–20cm(4–8in) across with slender-pointed lobes, yellowish at the base, becoming yellowish-red and transversely red-lined, purple-hairy.

S. nobilis South Africa, Mozambique
Much like *S. gigantea*, but the flowers 20–25cm(8–10in) across, usually with a darker ochre-yellow ground colour and more profusely purple-hairy.

S. plantii South Africa (Cape)
Stems 15–20cm(6–8in) long, erect, the angles compressed and set with small well-spaced teeth, minutely pubescent. Flowers 10–13cm(4–5in) wide, lobed to about halfway, each lobe ovate-lanceolate and taper-tipped, somewhat reflexed, brownish-purple, bearing narrow, transverse, yellow wrinkles and edged with long purple hairs.

S. variegata South Africa
Stems ascending, 5–10cm(2–4in) long, greyish-green, sometimes with a purplish mottling. Flowers one to five together, 5–8cm(2–3in) wide, lobes broadly triangular-ovate, transversely wrinkled, buff to yellow with purple-brown spotting and blotching, very variable in the amount of marking.

STENANDRIUM

Acarthaceae

Origin: Tropical and sub-tropical America. Depending on the botanical authority consulted, a genus of 30 to more than 60 species of mainly evergreen perennials, only one of which is cultivated. This requires warm humid conditions, but can be grown in plant cases or terraria in the home. Propagate by division or seed in spring, or by cuttings in summer.

Stenandrium derives from the Greek *stenos*, narrow and *aner*, a man, referring to the very slender stamens.

Species cultivated

S. igneum See *Chamaeranthemum igneum*.

S. lindenii Peru
Tufted, eventually forming leafy hummocks to 20cm(8in) or more wide. Leaves in opposite pairs, elliptic to broadly so, 2.5–8cm(1–3in) long, coppery-green with a bold, yellow, pinnate vein pattern, the leaf tissue corrugated between.

Flowers tubular, 5-lobed and 2-lipped at the mouth, yellow, 2cm($\frac{3}{4}$in) long, in spikes 5cm(in) or more long.

Stapelia variegata

STENOLOBIUM See TECOMA

STENOMESSON

Amaryllidaceae

Origin: Tropical America. A genus of about 20 species of bulbous plants allied to *Hippeastrum* and *Phaedranassa* and superficially resembling the latter. One species is sometimes cultivated and well worth while looking out for. It is best in the conservatory, but can be grown in the home. Propagate by seed in spring or by separating offsets when re-potting in late winter. *Stenomesson* derives from the Greek *stenos*, narrow and *messon*, middle; the flowers have narrow tubes, in some species slightly waisted.

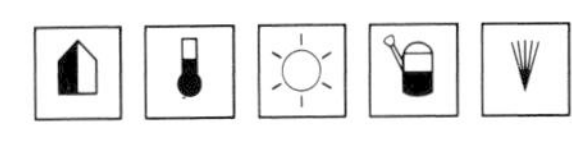

Species cultivated

S. incarnatum Ecuador, Peru
Leaves several per bulb, linear, more or less erect to 45cm(1$\frac{1}{2}$ft) long by 2.5cm(1in) wide, slightly glaucous. Flowers narrowly funnel-shaped with six flaring tepal lobes, about 10cm(4in) long, red, with each of the lobes bearing a central dark green stripe, nodding, in umbels of three to six on stems a little longer than the leaves, borne in late summer.

STENOTAPHRUM

Gramineae

Origin: Tropics and sub-tropics; widespread. A genus of seven species of annual and perennial grasses mainly from coastal regions. One species, mainly in its variegated form, makes a durable and attractive foliage plant for pots and hanging baskets. Propagate by division in spring, or by cuttings in spring and summer.

Stenotaphrum derives from the Greek *stenos*, narrow and *taphros*, a trench, referring to the grooves on the flowering stems in which the spikelets are partially imbedded.

Species cultivated

S. secundatum St. Augustine grass Southern USA and tropical America

Stems creeping, branching, to 60cm(2ft) or so. Leaves linear, up to 15cm(6in) long, a pleasing mid-green. Flowering spikelets 5mm($\frac{1}{5}$in) long, green and white, in one-sided racemes to 10cm(4in) long. *S.s* 'Variegatum' has the leaves longitudinally striped creamy-white. The green-leaved species is grown as a lawn grass in southern USA, especially near the sea; it is also used to stabilize sand dunes and sandy shores.

RIGHT *Stenotaphrum secundatum* 'Variegatum'
BELOW *Stephanotis floribunda*

STEPHANOTIS

Asclepiadaceae

Origin: Malagasy, southern Asia east to Malaysia and Peru. A genus of about 15 species of woody climbers with opposite, undivided, usually leathery-textured leaves and jasmine-like tubular flowers opening to five spreading lobes. The species described below is a handsome container plant, needing canes, wires or a trellis to support the long, twining shoots. Prune in winter before new growth is made. Propagate by seed or semi-hardwood cuttings in summer, both in heat, or by layering. *Stephanotis* derives from the Greek *stephanos*, crown and *otis*, ear, referring to the crown of stamens which have outgrowths fancifully like ears.

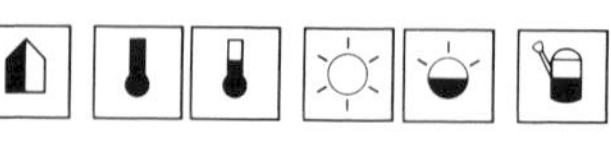

Species cultivated

S. floribunda Madagascar jasmine Malagasy

Evergreen climber which can be kept to 60cm(2ft) by twining around a low support, but is capable of reaching 5m(16ft) or more. Leaves in opposite pairs, 5–10cm(2–4in) long, broadly elliptic with a mucronate tip, glossy dark green. Flowers about 4cm($1\frac{1}{2}$in) long, white, waxy-textured and fragrant, borne in axillary cymes from spring to autumn.

STEREOSPERMUM See RADERMACHERA

STIGMAPHYLLON

Malpighiaceae

Orchid vine

Origin: West Indies, tropical America. A genus of 60 to 70 species of mainly twining, woody-stemmed climbers. They have opposite pairs of simple leaves with two glands at the tops of the stalks, and flowers with five long-clawed petals in umbel-like corymbs. The winged fruits are rather like miniature maple samaras. A few species are worthy of inclusion in the conservatory where they will garnish the roof space and provide shade below. Propagate by cuttings in spring or late summer. *Stigmaphyllon* derives from the Greek *stigma*, the pollen receptive part of the pistil and *phyllon*, a leaf; the stigmas have leaf-like appendages.

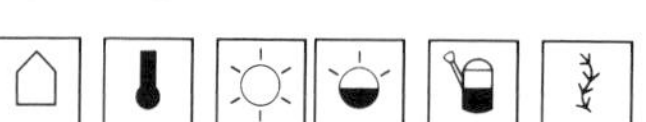

Species cultivated

S. ciliatum Golden creeper/vine, Butterfly vine Tropical America

Stems slender, to 5m(16ft) or more. Leaves to 8cm(3in) long, heart-shaped, light green, ciliate. Flowers rich bright yellow; borne in autumn; petals to 2.5cm(1in) long, the blades circular, fimbriate and crinkle-edged; in some or most blooms one or two petals are markedly smaller than the rest. Fruits winged, up to 2.5cm(1in) long.

S. ellipticum Mexico to Colombia
Slender-stemmed, 3–6m(10–20ft) in length. Leaves to 10cm(4in) long, elliptic, mid-green. Flowers similar in size and shape to those of S. *ciliatum*, but more crinkle-edged and lacking cilia, primrose-yellow aging pale lettuce-green.

STRELITZIA

Strelitziaceae (*Musaceae*)
Bird of Paradise flower

Origin: Sub-tropical to southern Africa. A genus of four to five species of evergreen perennials and small trees of palm-like form having long, undivided, leathery leaves usually in two ranks. The flowers are asymmetrical with three narrow sepals and three petals, two of which are joined to form a tongue-shaped structure. Several flowers are grouped together within a horizontally-borne boat-shaped bract. Propagate by division or by seed sown in warmth, both in spring. *Strelitzia* was named for Charlotte of Mecklenberg-Strelitz (1744–1818), who became the wife of George III of England.

Species cultivated

S. alba (*S. augusta*) South Africa
Much like *S. nicolai*, but seldom above 5m(16ft) in height, with smaller leaves and white flowers arranged in solitary boat-shaped bracts.

S. augusta See *S. alba*.

S. nicolai South Africa
Palm-like, the erect, clustered, woody stems usually branching only at ground level, to 8m(25ft) tall when mature. Leaves oblong-ovate 90–120cm(3–4ft) or more in length on stalks almost as long, glossy rich green. Flowers pale blue to light mauve or white. Floral bracts boat-shaped, 30cm(1ft) or more long, reddish-brown, several arranged on a stout peduncle.

S. reginae Bird of paradise flower South Africa
Clump-forming, to 1m(3ft) tall. Leaves long-stalked, the blades 25–45cm(10–18in) long, oblong-lanceolate, grey to blue-green, glaucous beneath, leathery in texture. Flowers with three orange or yellow sepals, one small orange petal and the joined pair blue, borne within a 15–20cm(6–8in) long, narrowly red-edged, boat-shaped bract in spring and summer. *S.r. juncea* has robust, rush-like leaf stalks usually without (but sometimes with very small) terminal blades; *S.r. intermedia* is half-way between variety *juncea* and the type species, having much reduced leaf blades; and *S.r. humilis* is much dwarfer than the type.

STREPTANTHERA

Iridaceae

Origin: South Africa. A genus of two species of cormous plants allied and similar to *Ixia* and *Sparaxis*. Each corm produces a fan of sword-shaped leaves and a simple or branched spike of strikingly bicoloured, wide open, 6-tepalled flowers with shortly tubular bases. Both species make excellent pot plants. Propagate by removing offsets when re-potting in autumn, or by seed in spring. *Streptanthera* derives from the Greek *streptos*, twisted and *anthera*, an anther.

Strelitzia reginae juncea

Species cultivated

S. cuprea Orange kaleidoscope flower
Leaves five to nine, narrow, pointed, up to 15cm(6in) long. Flowers 5cm(2in) wide, coppery-pink with a purple eye and an adjacent purple-black zone, up to four per spike in early summer. *S.c. coccinea* has bright orange petals.

S. elegans White kaleidoscope flower
Similar to *S. cuprea*, but the leaves blunt-tipped and with white blooms bearing yellow and purple eyes.

STREPTOCARPUS

Gesneriaceae
Cape primroses

Origin: Tropical to southern Africa and Malagasy. A genus of 132 species of evergreen perennials, monocarps and annuals which can be grouped into three distinct growth forms. These are tufted or clump-forming with basal leaves; shrub-like with pairs of small leaves on branched stems; or unifoliate, a curious development with each plant producing one very much enlarged cotyledon and occasionally a few small basal leaves, the plant usually dies after flowering (monocarpic). The flowers are normally tubular to funnel-shaped, opening to five rounded, spreading lobes and followed by slender, pod-like capsules which are spirally twisted. The species described below are suitable for the conservatory or home, needing extra humidity in summer. Propagate the tufted sorts by leaf or leaf section cuttings, the shrubby species by stem cuttings, both in summer and all species by seed in spring. *Streptocarpus* derives from the Greek *streptos*, twisted and *karpos*, a fruit; the slender seed pods are longitudinally spirally twisted.

Species cultivated

S. caulescens East Africa
Shrubby, 40–75cm(16–30in) tall. Leaves to 6cm (2½in) long, elliptic to ovate, velvety-hairy and dark green. Flowers 2cm(¾in) long, the tube slender, violet or white with violet striping, carried in axillary cymes.

S. dunnii Red nodding bells Southern Africa
Unifoliate, the single leaf to 30cm(1ft) or more in length with a crenate, wavy margin sometimes rolled inwards, grey-green and velvety-hairy. Flowers to 5cm(2in) long with a narrowly funnel-shaped tube, brick-red, borne in stalked clusters arising from the base of the plants.

S. holstii East Africa
Shrub-like, allied and similar to *S. caulescens* in height and general appearance, but stems more slender though swollen at the nodes, the leaves less hairy and the flowers longer, to 2.5cm(1in) long or sometimes a little more.

S. primulifolius South Africa
Tufted to clump-forming, with tongue- to strap-shaped, arching leaves, 30–45cm(1–1½ft) in length. Flowers 6–11cm(2½–4½in) long, narrowly funnel-shaped with blue-purple lobes, one to four on 25cm(10in) high peduncles.

Streptocarpus primulifolius

S. rexii Cape primrose South Africa
Tufted to clump-forming habit. Leaves 10–25cm(4–10in) long, narrowly oblong, wrinkled and hairy, the margins crenate. Flowers 4–7cm(1½–2¾in) long, funnel-shaped, mauve to blue-purple, carried in clusters of one to six on 10–15cm(4–6in) tall stems. Confused in cultivation with its hybrids.

S. saxorum False African violet Tanzania
Softly shrubby, forming low mounds of slender, hairy stems 15–30cm(6–12in) in length. Leaves in pairs or whorls of three, about 3cm(1¼in) long, elliptic to ovate, matt-green and softly hairy. Flowers 3–4cm(1¼–1½in) wide, the three lower lobes larger than the upper two, lilac with a white mouth and tube, singly or in pairs on slender stalks from the upper leaf axils.

Hybrids

S. × hybridus Garden origin
Under this name are grouped the varied results of crossing a dozen or so species of streptocarpus. In appearance, the plants resemble *S. rexii*, but the tufted leaves are larger, the funnel-shaped flowers also larger and variably coloured. The first hybrid strains or races appeared at least 100 years ago and since then plant breeders have greatly increased the flower size and colour range. These are mainly sold as seed mixtures under such names as Mixed Hybrids, Prize Strain, Triumph Hybrids, 'Concord' and 'Melody', the last two being F_1 hybrids. The colour range of these mixtures covers shades of purple, blue, red, pink and white, usually with throat stripes of a contrasting colour. Individual flowers are trumpet-shaped with five broadly rounded lobes and can reach 8cm(3in) across.

STREPTOSOLEN

Solanaceae

Origin: Andean Colombia and Ecuador. A genus of one species of evergreen, semi-scrambling shrub with abundant colourful flowers. It is a good container plant for the conservatory where it can be grown on the back wall, in free-standing pots or in hanging baskets. Cut back old stems by one-third in late winter to keep the plants bushy. Propagate by cuttings of young stems in late spring. *Streptosolen* derives from the Greek *streptos*, twisted and *solen*, a tube, from the twisted corolla tubes.

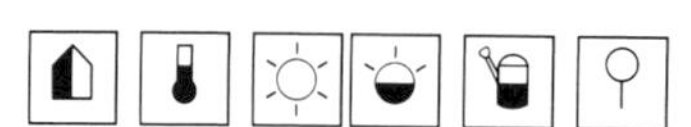

Species cultivated

S. jamesonii Marmalade bush Colombia, Ecuador
Capable of reaching 1.8m(6ft), but easily kept smaller. Leaves alternate, to 3cm(1¼in) long, ovate, deep green, with a finely wrinkled surface. Flowers 3–4cm(1¼–1½in) long, the twisted tube opening to a widely flared, 5-lobed mouth; they are bright orange with a paler tube and carried in terminal panicles 10–20cm(4–8in) long from late spring to late summer.

STROBILANTHES

Acanthaceae

Origin: Tropical Asia and Malagasy. A genus of about 250 species of perennials, some of which are more or less shrubby, and true shrubs. One species is grown for its highly ornamental foliage. It is best in the conservatory, but can be used in the home at least for short periods. Propagate by cuttings in spring or summer.

Strobilanthes derives from the Greek *strobilos*, a cone and *anthos*, a flower, referring to the bracted flower spikes of some species.

Species cultivated

S. dyerianus Persian shield Burma
Shrubby perennial 60cm(2ft) or more in height, the stems erect and sparingly branched. Leaves in opposite pairs, narrowly ovate to lanceolate, slender-pointed and running into a winged stalk, to 20cm(8in) long; the upper surface of each leaf is a blend of green, purple and silver with a shimmering iridescence, purple beneath. Flowers 2.5–4cm(1–1½in) long, narrowly funnel-shaped with an oblique mouth and five rounded lobes, pale blue or lavender in short, dense, terminal spikes. The slender-stemmed, willow-leaved cultivar 'Exotica' is probably a distinct, un-named species.

STROMANTHE

Marantaceae

Origin: Tropical America. A genus of about a dozen species of perennials related to *Maranta* and *Calathea*. They are rhizomatous evergreen plants with ornamental simple leaves and small, 3-petalled asymmetrical flowers which are largely insignificant, but are carried in the axils of red bracts. Propagate by division in spring and summer. *Stromanthe* derives from the Greek *stroma*, a bed and *anthos*, a flower, the shape of the inflorescence.

ABOVE *Streptosolen jamesonii*
ABOVE LEFT *Streptocarpus* × *hybridus* 'Albatross'
LEFT *Strobilanthes dyerianus*

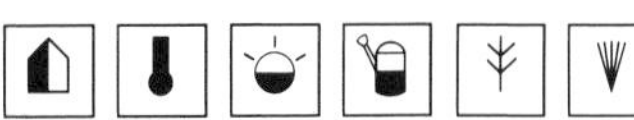

Species cultivated

S. amabilis (*Calathea amabilis, Maranta amabilis*) Brazil
Clump-forming, with stems to 90cm(3ft) or more in height. Leaves elliptic-oblong, to 30cm(1ft) or more long, the upper surfaces with grey zones between the veins, the lower surface red.

S. porteana (*Maranta porteana*) Brazil
More or less clump-forming, with stems to 2m(6½ft)

Stromanthe sanguinea

in a large container or if planted out. Leaves oblong, to 45cm($1\frac{1}{2}$ft) in length, blood-red beneath, slightly puckered above, dark glossy green with a silvery-white vein pattern.

S. sanguinea Brazil
Clump-forming, with stems to 1.5m(5ft) tall. Leaves elliptic-oblong to lanceolate, 25–40cm(10–16in) long, lustrous rich olive-green above, purple-red or striped green beneath.

BELOW *Strongylodon macrobotrys*
BOTTOM *Strophanthus divaricatus*

STROMBOCACTUS See OBREGONIA

STRONGYLODON

Leguminosae

Origin: Malagasy, Mascarene Islands, New Guinea, Philippines and adjacent islands. A genus of about 20 species, including shrubs and tall jungle lianas. The two species described here make spectacular climbers for the large, warm conservatory. They have twining, woody stems, alternate trifoliate leaves and long, pendent racemes of large pea flowers each with a prominent, pointed, strongly up-curving keel. The large woody seed pods which follow are not without a fascination all of their own. Propagate by seed in spring. *Strongylodon* derives from the Greek *strongylos*, round and *odous*, tooth, referring to the rounded calyx teeth.

Species cultivated

S. macrobotrys Philippines
In the wild, a liane to 20m(65ft) in length, but under glass can be kept at (and will flower on) plants half this size. Leaflets ovate-oblong, to 13cm(5in) or so in length, glossy green. Flowers a unique shade of luminous blue-green, 8cm(3in) long, in racemes 45–90cm($1\frac{1}{2}$–3ft) or sometimes more in length. Pods cylindrical, 13cm(5in) or more long, containing six to twelve seeds.

S. lucidus Hawaii
Similar to *S. macrobotrys*, but with red flowers 2.5cm(1in) long and flattened pods containing one or two seeds only.

STROPHANTHUS

Apocynaceae

Origin: Africa, Malagasy, Indo-Malaysia. A genus of about 60 species of shrubs, trees and climbers. They have mainly simple leaves in opposite pairs and tubular to bell-shaped flowers, the petal lobes of which are long and slender, and sometimes tail-like. The shrubby climbing species described here make specimen tub plants for the conservatory which provide a talking point with their uniquely appealing flowers. Propagate by seed in spring, or by cuttings in summer. *Strophanthus* derives from the Greek *strophos*, twisted and *anthos*, a flower, the tailed corolla lobes being spiralled.

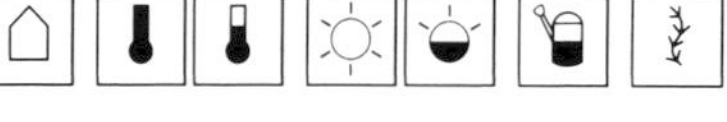

Species cultivated

S. caudatus Java to Burma
Shrub or semi-climber, 3m(10ft) or more in height. Leaves elliptic-oblong to obovate, to 15cm(6in) long, leathery-textured. Flowers white or yellow, flushed purple, the tubular bases to 2.5cm(1in) long, bearing tails 10–18cm(4–7in) in length.

S. divaricatus S.E. China
Semi-climbing shrub, 2–3m($6\frac{1}{2}$–10ft) tall. Leaves oblong-elliptic up to 8cm(3in) in length. Flowers greenish-white with crimson spots and streaks in the throat of the 1.2cm($\frac{1}{2}$in) long tubular base; tails whitish to pale yellow, 4–8cm($1\frac{1}{2}$–3in) long. A vigorous specimen produces twining stems, but by pinching these out it can be grown as a shrub.

S. speciosus South Africa
Somewhat rambling shrub, to 2m($6\frac{1}{2}$ft) or more in height. Leaves oblong-lanceolate, 4–9cm($1\frac{1}{2}$–$3\frac{1}{2}$in) long, olive-green, sub-lustrous, in whorls of three or four. Flowers cream to yellow or orange, spotted with red, the tubular bases about 1.2cm($\frac{1}{2}$in) long, bearing tails to 4cm($1\frac{1}{2}$in) in length.

STROPHOLIRION See BRODIAEA

STYLIDIUM

Stylidiaceae
Trigger plants

Origin: Mainly Australia, also S.E. Asia. A genus of about 135 species of annuals, perennials, sub-shrubs and climbers. Propagate by seed or division in spring. *Stylidium* derives from the Greek *stylos*, a column, referring to the central organ of the flower which combines stamens and style (as in the *Orchidaceae* family.

The column arches over and, when touched at its base by an insect, springs up rapidly, showering the visitor with pollen which is then transported to another folwer and dusted off on the stigma before it is triggered off.

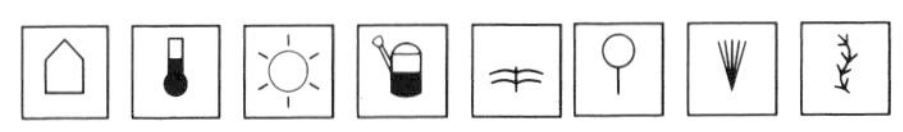

Species cultivated

S. adnatum Australia
Tufted perennial, usually branching freely, mainly from the base; stem 10–15cm(4–6in) long, erect then decumbent. Leaves crowded along the stems, linear 1.5–4cm(⅝–1½in) long. Flowers tiny, pink, in fairly dense spike-like racemes in summer.

S. crassifolium Western Australia
Perennial with tufts of fleshy, linear, basal leaves 10–20cm(4–8in) in length. Flowers 2cm(¾in) across, pink, in narrow, spike-like panicles up to 60cm(2ft) in height.

S. graminifolium Common/Grass trigger plant Eastern Australia
Densely tufted perennial with all leaves arising from ground level. Leaves linear to lanceolate, 10–15cm(4–6in) long, usually deep green. Flowers pink to magenta, 1–1.5cm(⅜–⅝in) long, in spikes above the leaves in summer. The hardiest species, tolerating light frost.

S. laricifolium Western Australia
Sub-shrub, to 30cm(1ft) tall. Leaves fairly crowded along the stems, linear, 1.2–2.5cm(½–1in) long. Flowers small, pink, in loose racemes or panicles.

S. scandens Western Australia
Stems climbing, to 60cm(2ft) or more in height. Leaves linear, 2.5–5cm(1–2in) long, in distant, whorl-like tufts. Flowers small, pink, in terminal racemes.

SUBMATUCANA See BORZICACTUS

SUTHERLANDIA

Leguminosae

Origin: South Africa. A genus, depending upon the botanical authority, of six, or only one variable species. It is an evergreen shrub reminiscent of bladder senna (*Colutea*) but much more showy in bloom and making a very decorative tub specimen. Propagate by seed in spring.

Sutherlandia commemorates James Sutherland (c.1639–1719), Superintendent of the Botanic Garden, Edinburgh.

Sutherlandia frutescens

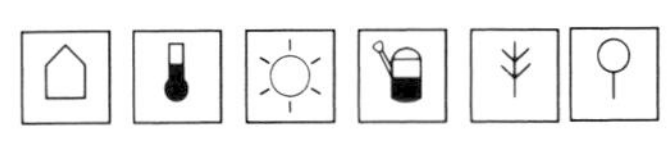

Species cultivated

S. frutescens
About 1.5m(5ft) in height in a container, but much more if planted out. Leaves pinnate, composed of 13 to 21 oblong to lanceolate, hairy leaflets each 1–2cm(⅜–¾in) long. Flowers 3cm(1¼in) long, pea-shaped, bright red, in racemes in late spring. Seed pods 5cm(2in) or more long, inflated, yellowish or reddish-flushed.

S.f. microphylla
Smaller leaves, hairless above.

S.f. tomentosa
Leaves densely grey-white-hairy.

SWAINSONA

Leguminosae

Origin: Mainly Australia; one in New Zealand. A genus of about 60 species of annuals, perennials sub-shrubs and shrubs, a few of which make very ornamental pot plants. They have alternate, pinnate leaves on usually erect stems, and upright racemes of pea flowers with very broad standard petals. Propagate by seed in spring or by cuttings in summer.

Swainsona commemorated Isaac Swainson (1746–1812), a London doctor who maintained a private botanic garden at Twickenham, London.

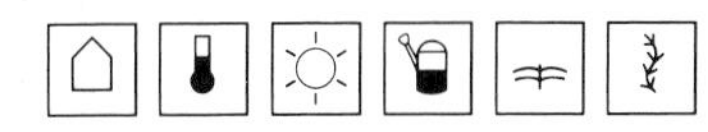

Species cultivated

S. galegifolia Eastern Australia
Sub-shrubby, with lax or semi-climbing stems 60–120cm(2–4ft) in height. Leaves composed of 11 to

21 oblong leaflets each about 2cm(¾in) long. Flowers about 2.5cm(1in) long, ten or more forming a loose raceme, very variable in colour, usually orange-red, but also pink, purple, blue, yellowish and brownish-red to maroon, borne in summer.
S. greyana Darling River pea Eastern and South Australia
Sub-shrubby, with branched erect stems, 60–90cm(2–3ft) tall. Leaves composed of 11 to 23, oblong, 2–4cm(¾–1½in) long leaflets. Flowers pink, with densely white-hairy calyces, 2cm(¾in) in length, in long racemes in summer.

SYAGRUS

Palmae

Origin: South America. A genus of 30 to 50 species of pinnate-leaved palms closely related to the coconut (*Cocos*). They are low-growing, some species never building up the false trunks so typical of most palms. They are good container plants remaining dwarf for many years. Syringe the foliage regularly in summer. Propagate by seed in spring at not less than 21°C(70°F).

Syagrus derives from the Greek *syagros*, the name of a type of date palm.

Species cultivated

S. weddelliana (*Cocos weddelliana, Microcoelium weddellianum*) Brazil (Organ Mts.)
Remaining small for many years though eventually capable of reaching 2–3m(6½–10ft). Leaves 30–60cm(1–2ft) long in pots, arching, borne in a terminal rosette; in time building a slender stem. Flowers and fruits rarely occur on potted plants. In cultivation, sometimes confused with the tougher *Chamaedorea elegans*.

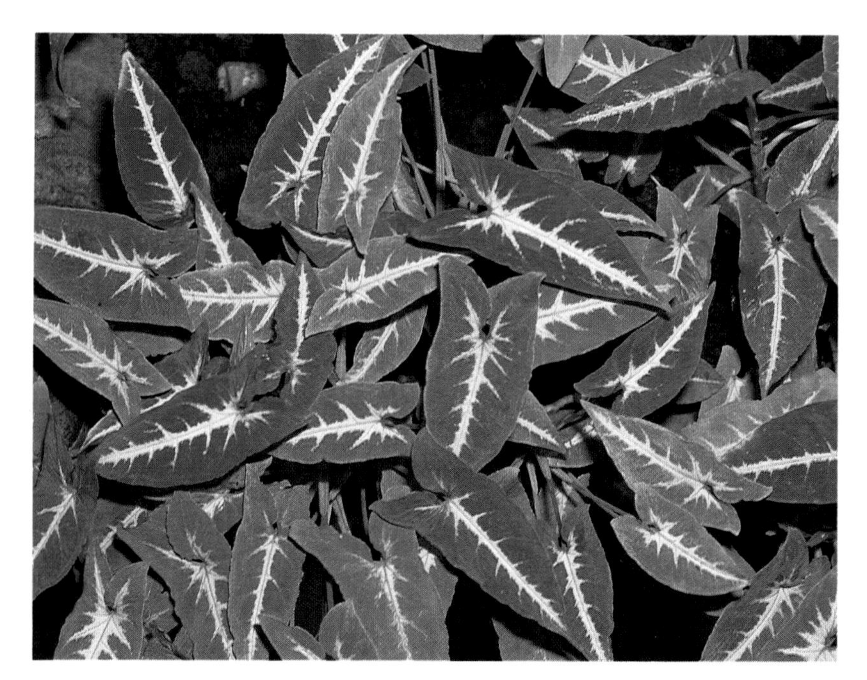

Syngonium wendlandii

SYNGONIUM

Araceae

Origin: Tropical areas of the Americas and the West Indies. A genus of 20 species of evergreen climbers, normally epiphytic in the wild. Their leaves vary in size and shape from juvenile to adult form, usually starting entire or arrow-shaped, later becoming trifoliate or pedate. Flowers, which are like those of the arum, are produced only on adult plants, the spathes being white, green or purple. They make very good house or warm conservatory plants, looking their best when in a humid atmosphere but tolerating drier air. As they mature they need support for the climbing stems, a moss stick being very suitable. Propagate by cuttings of stem sections in summer. *Syngonium* is derived from the Greek *syn*, together and *gone*, womb, an allusion to the united ovaries.

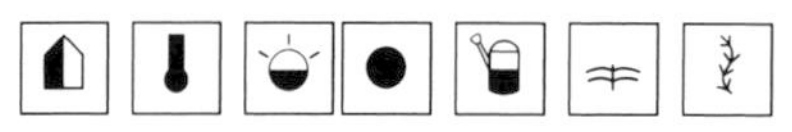

Species cultivated

S. auritum (*Philodendron auritum*) Five fingers Jamaica
Leaves to 20cm(8in) long, trifoliate or pedately divided into five leaflets, the central lobe always the longest; the side lobes have small ear-like lobes at their bases. Spathes to 27.5cm(11in) long, greenish-yellow. 'Fantasy' has leaves variegated with cream and white.
S. hoffmannii (*Nephthytis hoffmannii*) Costa Rica
Stems eventually 2–3m(6½–10ft) or more. Leaves sagittate in the juvenile phase, trifoliate when adult with the central lobe largest, to 15cm(6in) or more long, grey-green with silvery-white veining and central area. Spathes to 12cm(4¾in), white with a purplish throat.
S. podophyllum Goosefoot plant, Arrowhead vine Mexico to Panama
Small and compact while juvenile, later producing climbing stems to 2m(6½ft) or more. Juvenile leaves sagittate, often with silvery-white veining or suffusion; mature leaves trifoliate then pedate with five to nine leaflets, the central one the longest. Spathe to 30cm(1ft) long, whitish.

S.p. 'Albovirens' has creamy-silver bands along the veins; 'Atrovirens' is darker green with silvery veins and a silvery suffusion; 'Trileaf Wonder' produces trifoliate leaves when juvenile which are green with a silvery-grey vein patterning.

S.p. 'Variegatum' (*Nephthytis liberica* of gardens) has sagittate leaves irregularly splashed with creamy white.
S. wendlandii (*Nephthytis wendlandii*) Costa Rica
Juvenile leaves sagittate to hastate, sometimes entire, velvety deep green with white veins; adult leaves all trifoliate, the central leaflet to 18cm(7in) long, dark green without the white vein patterning. Spathes 13cm(5in) long, green outside, purplish within.

ABOVE *Syzygium jambos*
ABOVE LEFT *Syzygium paniculatum*

SYZYGIUM

Myrtaceae

Origin: Tropical Africa and Asia. A genus of 400 to 500 species of evergreen trees and shrubs very closely related to *Eugenia* and separated from that species only on small floral characters. A few species provide pleasing tub plants. They have opposite pairs of simple leaves, many-stamened flowers with four or five petals, in some species united into a cap and falling on opening as in *Eucalyptus*. The fruit is an edible berry, usually red, purple or white in colour. Propagate by seed in spring or cuttings in summer. *Syzygium* derives from the Greek *suzugos*, joined, referring, according to various authorities, either to the united petals or branchlets, the former seeming the most likely.

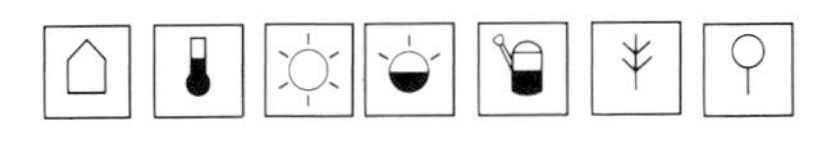

Species cultivated

S. jambos (*Eugenia jambos*) Rose apple, Malabar plum S.E. Asia; naturalized elsewhere
In its native country, a tree to 12m(40ft), but with annual pruning easily kept to 2–3m(6½–10ft). Leaves lanceolate, 13–25cm(5–10in) long, deep lustrous green when mature, red-purple as they unfold. Flowers white, 5–8cm(2–3in) wide, with a pompon-like mass of stamens. Fruits pale yellow, fragrant, ovoid, hollow, to 4cm(1½in) long; the fruit reminds one of the scent of roses and they are eaten fresh, candied, and used for flavouring jellies.

S. paniculatum (*Eugenia australis, E. paniculata, E. myrtifolia*) Brush cherry Queensland
Small tree or large shrub; this species responds well to pruning and container culture. Leaves broadly elliptic, 8–15cm(3–6in) long, glossy mid-deep green. Flowers about 1cm(⅜in) long in quite large terminal panicles. Fruits broadly obovoid, to 2cm(¾in) long, red to red-purple. *S.p. australis*, 'Antone Dwarf' and 'Compacta' are smaller-growing and to be preferred as container plants.

T

TABERNAEMONTANA

Apocynaceae

Origin: Tropical Africa and Asia. A genus, depending upon the botanical authority, of 100 to 150 species of evergreen trees and shrubs. One species is widely planted in the tropics and makes a handsome tub plant for the warm conservatory. Propagate by cuttings in summer or by layering in spring.

Tabernaemontana honours Jakob Theodor von Bergzabern (*d.*1590), who Latinized his name to Tabernaemontanus under which pseudonym he wrote the classic herbal *Neuw Kreuterbuch*.

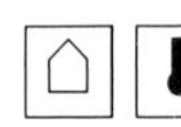 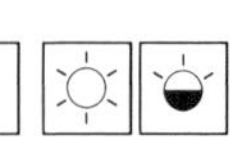  

Species cultivated

T. divaricata (*T. coronaria, Ervatamia coronaria*) Cape jasmine, East Indian rosebay India
Well branched shrub, to 2m(6½ft) or more in height. Leaves in opposite pairs, 8–15cm(3–6in) long, elliptic to lanceolate or narrowly obovate, lustrous. Flowers in small clusters, with a tubular base and five large, white, wavy petals to 4cm(1½in) wide, fragrant at night, borne mainly in summer, but intermittently throughout the year if warm enough. Forked, leathery pods containing fleshy-coated red seeds may follow the flowers.

BELOW *Tabernaemontana divaricata*
BELOW RIGHT *Tacca chantrieri*
BOTTOM *Tacitus bellus*

TACCA

Taccaceae

Origin: Tropical America, Africa, Asia, Australia, Pacific Islands. A genus, depending upon the botanical authority, of ten or 30 species of rhizomatous or tuberous-rooted perennials. Those described are clump-forming with attractive basal leaves and bizarre, green to maroon umbels of 6-lobed, bell-shaped flowers mixed with long thread-like organs which may be bracts or modified flower stalks. Each umbel is backed by two or more broad leaf-like bracts of a similar colour. A well grown plant is intriguing and becomes a talking-point in the conservatory or home. Propagate by division or seed in spring. *Tacca* is derived from the Indonesian vernacular name *taka* in Latin form.

 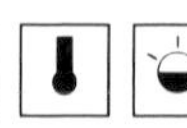  

Species cultivated

T. chantrieri Bat flower, Cat's whiskers
S.E. Asia
Rhizomatous; leaves stalked, 30–45cm(1–1½ft) tall, arching, the blade broadly elliptic, corrugated, lustrous green. Umbels composed of four ovate-cordate bracts, several nodding flowers 2–3cm(¾–1¼in) long and a large number of pendent whiskers, all purple-brown to maroon, sometimes greenish.

T. leontopetaloides (*T. pinnatifida*) East Indian/South Sea arrowroot Probably S.E. Asia, but formerly widely cultivated in the tropics of Africa, Asia and the Pacific and naturalized there Tuberous; leaves like those of *Carica*—stalked, 60–90cm(2–3ft) tall, deeply trilobed, each lobe again pinnately lobed. Umbels composed of four to 12, ovate to lanceolate green bracts, 20 to 40 1.2cm(½in) wide, yellow-green to black-purple flowers and a number of purple to brown whiskers. The plant dies down in autumn and should be stored dry and warm until spring. The globular tubers can attain 15–20cm(6–8in) in diameter and are rich in starch, but bitter and inedible. The starch (arrowroot) is extracted by grating the tubers, washing several times to extract the starch granules, then straining through a coarse cloth to remove the gratings. The starch then settles and is washed again to clear the last of the bitter principle before drying.

T. pinnatifida See *T. leontopetaloides*.

TACITUS

Crassulaceae

Origin: Mexico. A genus of one species of rosette-forming succulent perennial combining features of the allied *Echeveria* and *Sedum*. It is a highly desirable plant being showy when in bloom and with attractive foliage the year round. Propagate by division or seed in spring. *Tacitus* is Latin for silent, used by the authors of the name and first description (Reid Moran & Jorge Meyran) because the corolla tube is 'very close-mouthed'.

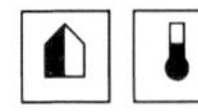 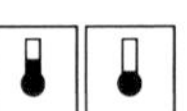 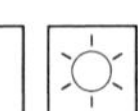  

Species cultivated

T. bellus
Rosettes compact, flattened, 4–8cm(1½–3in) wide, clustering and eventually forming hummocks.

Leaves to 3cm(1¼in) long, obovate-cuspidate, thickly fleshy, darkish grey-green, slightly reddened at the margins. Flowers starry, 5-petalled, 3–4cm(1¼–1½in) wide, usually bright carmine-pink but a paler form is in cultivation, three to ten in wide-spreading panicles just above the leaves in summer.

TACSONIA See PASSIFLORA

TAGETES

Compositae
Marigolds

Origin: Southern USA to Argentina. A genus of 30 to 50 species, depending upon the classification followed. Those in cultivation are familiar bedding annuals, though the genus does include a few perennials. The species described here are erect, freely branched plants with pinnately lobed leaves and terminal, solitary or clustered flower heads made up of disk and ray florets. They can be grown in the conservatory and home as short term pot plants. Propagate by seed in early spring, pricking off singly into pots. *Tagetes* was probably named for the Etruscan god Tages, but this is not certain.

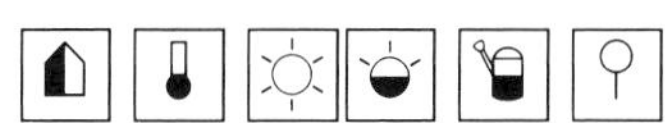

Species cultivated

T. erecta African/American marigold Mexico to Central America
Strong-growing annual to 90cm(3ft) with pinnate leaves. Flower heads 5–13cm(2–5in) across, yellow to orange; most cultivated forms are double or semi-double often with crimped, quilled or 2-lobed florets. Many cultivars are available.

T. patula French marigold Mexico, Guatemala
Bushy annual, 15–45cm(6–18in) tall with pinnate leaves. Flower heads to 5cm(2in) wide, yellow to orange, often marked or patterned with darker orange to brown. Many cultivars are available. Hybrids between this species and *T. erecta* have blurred the differences and many commercial cultivars are now of this origin.

T. signata See *T. tenuifolia*.

T. tenuifolia (*T. signata*) Tagetes, Signet marigold Mexico, Central America
Slender, branched annual to 60cm(2ft) tall with pinnate leaves. Flower heads to 2.5cm(1in) across, bright yellow, very freely borne. *T.t. pumila* is 15–25cm(6–10in) tall.

TALINUM

Portulacaceae
Fame flowers

Origin: Mainly Mexico and North America, but also South America, Africa and Asia. A genus of about 50 species of more or less succulent annuals, perennials and sub-shrubs. They have alternate, simple, often cylindrical leaves, in some species aggregated into terminal clusters or rosettes. Their flowers are 5-petalled, borne in cymes or panicles, and short-lived but opening in succesesion. The species described add variety to a collection of succulents. Propagate by seed in spring or by cuttings in summer. *Talinum* is presumably derived from an African vernacular name, but the exact origin and meaning appears to be unrecorded.

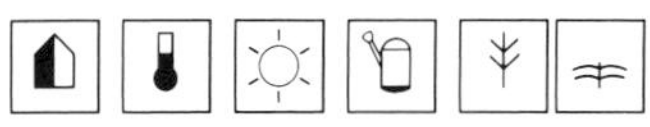

Species cultivated

T. guadalupense Guadalupe Island (Mexico)
Root fleshy, globose to cylindrical, bearing thickened, knotted stems to 60cm(2ft) in height. Leaves spathulate, fleshy, blue-green with a red edge, about 5cm(2in) long, forming terminal clusters or rosettes. Flowers 2–2.5cm(¾–1in) wide, cyclamen-pink, in panicles well above the leaves, each bloom lasting three to four days. A shrubby succulent of unusual appearance.

T. paniculatum (*T. patens*) Southern USA to Central America
Root tuberous, bearing one to several fleshy green stems 30–60cm(1–2ft) in height. Leaves obovate to lanceolate or elliptical, 5–10cm(2–4in) long, fleshy, usually rosetted at the stem tips. Flowers 1.2cm(½in) wide, rose-red, sometimes yellow, in panicles to 25cm(10in) tall. *T.p.* 'Variegatum' has the leaves boldly edged white and tinted pink.

ABOVE *Tagetes patula* 'Yellow Jacket'
TOP *Talinum paniculatum* 'Variegatum'

TAPEINOCHEILOS

(TAPEINOCHILUS)

Costaceae (*Zingiberaceae*)

Origin: Moluccas, New Guinea to Queensland. A genus of about 20 species of mainly large evergreen perennials, one of which makes a handsome addition to the warm conservatory. Propagate by division in spring. *Tapeinocheilos* derives from the

ABOVE *Tapeinocheilos ananassae*
ABOVE RIGHT *Tecomaria capensis*

Greek *tapeinos*, low and *cheilos*, lip, referring to the short labellum-like petal.

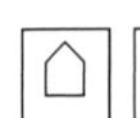    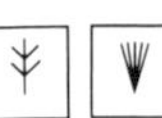

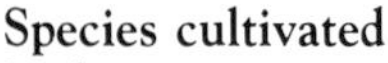

Species cultivated
T. ananassae Moluccas
Clump-forming with two sorts of stems—leafy and non-flowering, and leafless and flowering. Non-flowering stems 1.5–2.5m(5–8ft) tall, erect, unbranched, the tips ending in a half spiral; leaves scattered below, denser above, narrowly obovate, to 15cm(6in) in length, rich green. Flowering stems 45–120cm($1\frac{1}{2}$–4ft) in height, direct from ground level, terminating in a cone-shaped spike 15–20cm(6–8in) long, composed of recurved, firm-textured, crimson to orange-red bracts. Flowers small, yellow, hidden at the base of the bracts in summer.

TASMANNIA See DRIMYS

TECOMA

Bignoniaceae

Origin: Florida and Mexico south to Argentina. A genus of 16 species of shrubs and small trees. They have pinnate or simple leaves in opposite pairs, and racemes or panicles of sizeable, more or less funnel-shaped flowers with five rounded lobes. Propagate by seed in spring or by cuttings in summer. *Tecoma* is derived from its Mexican vernacular name *tecomaxochitl*.

 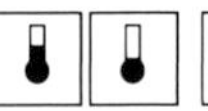  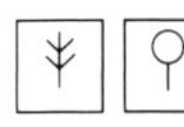

Species cultivated
T. capensis See *Tecomaria capensis.*
T. garrocha (*Stenolobium garrocha*) Argentina
Erect shrub about 2m($6\frac{1}{2}$ft) tall. Leaves pinnate, composed of seven to 11 oblong-ovate, toothed leaflets each to 5cm(2in) long. Flowers 5cm(2in) long, the funnel scarlet and the lobes yellow or salmon-pink.
T. stans (*Bignonia stans, Stenolobium stans*)
Yellow elder, Yellow bells Florida, Mexico, West Indies, South America
In its homeland, a large shrub or small tree, but with annual pruning easily kept to 2m($6\frac{1}{2}$ft) in height. Leaves pinnate, composed of five to 13 lanceolate to oblong-ovate, toothed, bright green leaflets 5–10cm(2–4in) long. Flowers 5cm(2in) long, bright yellow, in terminal panicles.

TECOMARIA

Bignoniaceae

Origin: Eastern and southern Africa. A genus of two or three species of shrubs and scramblers closely related to *Tecoma*. One species is widely grown in warm countries. Propagate by seed in spring or by cuttings in summer. *Tecomaria* derives from *tecoma*, indicating the close botanical relationship.

   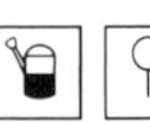 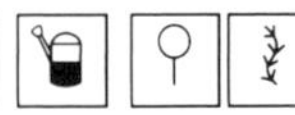

Species cultivated
T. capensis (*Bignonia capensis, Tecoma capensis*)
Cape honeysuckle South Africa
In an open site, a bushy shrub to 1.2m(4ft) or more, but among other shrubs or in partial shade it produces lax stems to twice this length and becomes a semi-climber. Easily maintained as a bush by pinching or cutting out all extra long shoots. Leaves pinnate composed of five to nine orbicular to ovate or rhomboid, dark green, toothed leaflets 2.5–5cm(1–2in) long. Flowers 5cm(2in) long, narrowly funnel-shaped, curved, with five rounded, reflexed lobes, orange-red, borne in spring to autumn. *T.c.* 'Aurea' ('Lutea') has yellow blooms.

TECOPHILAEA

Amaryllidaceae

Origin: Chile. A genus of one species of cormous-rooted perennial from semi-arid areas of the Andes. It is a most attractive plant for a pan in a cool conservatory and can be brought into the home for flowering. Propagate by seed in spring or offsets at potting time. *Tecophilaea* was named for Tecophila Billiotti, a botanical artist and daughter of Luigi Colla (1766–1848), Professor of Botany at Turin.

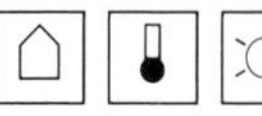 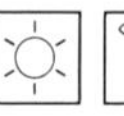 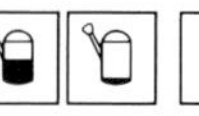 

Species cultivated
T. cyanocrocus
Plant 10–15cm(4–6in) tall. Leaves 7–12cm($2\frac{3}{4}$–$4\frac{3}{4}$in) long, narrowly linear, usually two or three arising from each corm. Flowers 3–4cm($1\frac{1}{4}$–$1\frac{1}{2}$in) long, broadly bell-shaped, somewhat crocus-like, deep blue with a white eye, opening in spring.
T.c. leichtlinii
Paler blue with a larger white eye.
T.c. violacea
Deeper blue-purple, no white eye.

Tecophilaea cyanocrocus

TELINE See CYTISUS

TELOPEA

Proteaceae

Origin: Eastern Australia and Tasmania. A genus of four species of evergreen shrubs with alternate, entire, leathery leaves and terminal globular racemes of flowers surrounded by colourful bracts. The individual flowers are tubular, splitting to four petal-like segments which cohere and reflex back like one petal. The woody fruits contain winged seeds. Suitable only for the conservatory. Propagate by seed in spring or by layering. *Telopea* derives from the Greek *telepos*, seen from afar, with reference to the brightness of the flowers.

Species cultivated

T. speciosissima Waratah New South Wales
Rarely over 1.5m(5ft) in a tub, but to three times this in the wild. Leaves 13–23cm(5–9¼in) long, obovate, coarsely toothed, narrowing sharply to their stalk. Flowers 2.5cm(1in) long, in rounded to ovoid heads 8–10cm(3–4in) across and surrounded by petal-like bracts to 8cm(3in) long, bright red, borne in summer.

TEPHROCACTUS See OPUNTIA

TETRANEMA

Scrophulariaceae

Origin: Guatemala and Mexico. A genus of two to three species of low, woody-based perennials with opposite pairs of dark green leaves and tubular, 2-lipped flowers opening to five lobes. The species described here is a house or conservatory plant. Propagate by seed in spring in warmth, or by division. *Tetranema* derives from the Greek *tetra*, four and *nema*, a thread, for the four stamens.

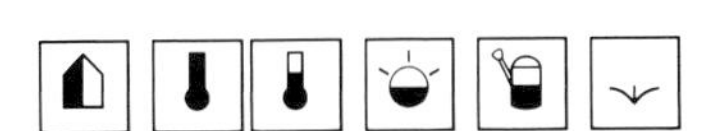

Species cultivated

T. roseum (*T. mexicanum*, *Allophyton mexicanum*) Mexican violet/foxglove Mexico
Leaves 7–15cm(2¾–6in) long, obovate, slightly toothed, arising from near the base of the stems and almost appearing to be in a rosette. Flowers violet-purple with a paler throat and tube, the lower lip whitish; they are carried in umbel-like clusters almost throughout the year if a temperature above 13°C(55°F) is maintained. 'Album' is white.

TETRAPANAX

Araliaceae

Origin: Southern China, Taiwan. A genus of one species of shrub closely related to, and formerly included in, *Fatsia*. It is a magnificent foliage plant

ABOVE *Tetrapanax papyriferus*
LEFT *Tetranema roseum* 'Album'

for the larger conservatory guaranteed to provide a tropical look in less than tropical temperatures. Propagate by suckers in winter or spring. *Tetrapanax* derives from the Greek *tetra*, four and the allied genus *Panax* differing in having flowers with floral parts in fours instead of fives.

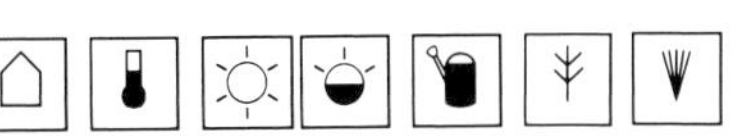

Species cultivated

T. papyriferus (*Aralia papyrifera*, *Fatsia papyrifera*) Rice-paper plant
Suckering shrub with robust, sparingly branched, erect stems 2–3m(6½–10ft) in height. Leaves alternate, long-stalked, the blades 30–60cm(1–2ft) wide, orbicular, with five to 14 broad pointed and toothed lobes each with a prominent mid-vein. Flowers small, whitish to yellowish in spherical umbels. Fruit a small berry. The white stem pith is the source of Chinese rice-paper. 'Album' is white.

TETRASTIGMA

Vitidaceae

Origin: S.E. Asia, Indo-Malaysia, Australia. A genus of about 90 species of woody-stemmed climbers allied to *Cissus* and *Vitis* (grape vine). One species is grown for its handsome foliage. Propagate by cuttings in summer or by layering in spring. *Tetrastigma* derives from the Greek *tetra*, four and *stigma*, the pollen-receptive part of the ovary; the stigma is 4-parted or lobed.

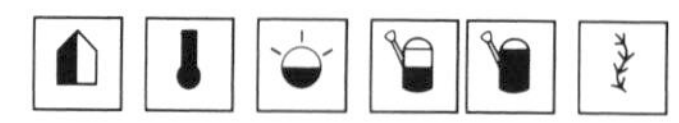

Species cultivated

T. voinierianum (*Cissus voinierianum*, *Vitis voinierianum*) Chestnut vine Laos
High-climbing vine in the wild, but by pruning and/or frequent propagation it can be kept under 1.2–2m(4–6½ft) in height. Leaves composed of three to five leaflets, the individual leaflets rhombic to obovate, 10–20cm(4–8in) long, usually with fairly prominent teeth and wavy margined, lustrous above, tawny to rusty-hairy beneath. Shoot tips and young leaves entirely covered with similar hairs, adding a further touch of distinction. Flowers tiny, greenish, with four petals and sepals, borne in axillary clusters. Fruit an acid grape-like berry.

THALIA

Marantaceae

Origin: Tropical America and Africa. A genus of seven to 11 species of aquatic or wet ground perennials one of which makes an elegant focal point for the edge of a conservatory pool. Almost hardy. Propagate by seed or division in spring. *Thalia* commemorates Johannes Thal (1542–1583), a German doctor who wrote a flora of the Hartz Mountains (*Sylva Hercynia*), published posthumously in 1588.

Thalia dealbata

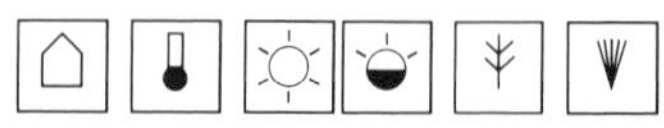

Species cultivated

T. dealbata S.E. USA
Clump-forming; leaves oblong-ovate to narrowly elliptic, 20–40cm(8–16in) long, covered with a white waxy powder and carried on long stalks. Flowering stems wand-like, 2m(6½ft) or more in height, with a few basal leaves and branching to form a dense panicle at the top; flowers composed of three small petals and a large petal-like staminode about 1.2–1.5cm(½–⅝in) long, all purple, borne summer.

THELOCACTUS

Cactaceae

Origin: Southern USA, Mexico. A genus of 40 or more species of mainly smallish cacti resembling *Coryphantha* and now including members of *Ancistrocactus*, *Hamatocactus* and allied genera. They are attractive cacti which usually flower regularly and are easily grown in the conservatory and home. Propagate by seed in spring, or by offsets where possible in summer. *Thelocactus* derives from the Greek *thele*, a nipple and *cactus*, referring to the shape of the prominent tubercles.

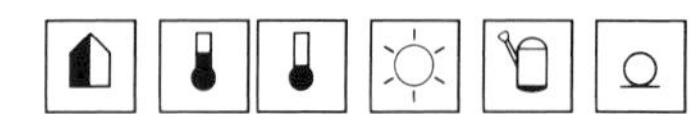

Species cultivated

T. bicolor Glory of Texas S. Texas, N. Mexico
Stem globose to conical, 15–25cm(6–10in) high. Ribs about eight, slightly spiralling, broken up into square-based tubercles. Radial spines red and yellow, to 3cm(1¼in) long, needle-like; central spines about four, to 5cm(2in) long, spreading to erect. Flowers bell-shaped, fuchsia-purple, to 6cm(2½in) long.

T. rinconensis (*Echinocactus rinconensis*) Mexico
Stem oblate, to 8cm(3in) high by 13cm(5in) thick. Ribs about 13, broken into very prominent, laterally flattened tubercles with more or less diamond-shaped bases. Spines one to four, not differentiated into radials and centrals, awl-shaped, the longest exceeding 13mm(½in) long. Flowers about 4cm(1½in) wide, white, yellow in the centre.

T. setispinus (*Echinocactus setispinus*, *Hamatocactus setispinus*, *Ferocactus setispinus*) Strawberry cactus S. Texas, N. Mexico
Stems globose to ovoid, eventually 20–30cm(8–12in) in height, producing offsets only when old. Ribs about 13, somewhat spiralling, broken into large areoles. Radial spines ten to 15, needle-like, to 4cm(1½in) long; central spines one to three,

thicker and somewhat longer, at least one with a hooked tip. Flowers funnel-shaped, to 8cm(3in) long, yellow in colour with a crimson centre.

THEVETIA

Apocynaceae

Origin: Tropical America, West Indies. A genus of eight or nine species of trees and shrubs, one of which makes a colourful specimen shrub. Propagate by seed in spring or by cuttings in summer. *Thevetia* honours Andre Thevet (1502–1592), a French monk who journeyed in Brazil and Guyana.

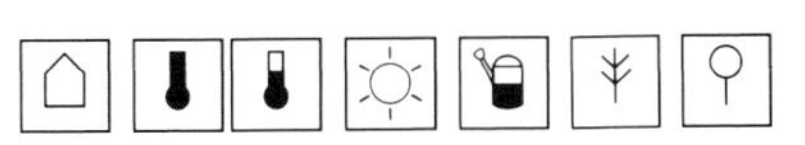

Species cultivated

T. peruviana Yellow oleander Tropical America
Eventually a small tree in its native land, but usually seen as a shrub 2–3m(6½–10ft) in height, easily kept to size with pruning. Leaves alternate, narrowly lanceolate to linear, 8–15cm(3–6in) long, usually rich green and semi-glossy. Flowers 5-petalled, funnel-shaped, about 8cm(3in) long, bright yellow, sometimes orange-yellow. Semi-fleshy, triangular-ovoid seed pods contain one or two hard nut-like seeds which are worn or carried as lucky charms.

THRINAX

Palmae
Thatch palms

Origin: Mexico to Belize, Florida, West Indies. A genus, depending on the botanical authority, of four to 12 species of palms. They are mostly of small to medium size with smooth trunks and rounded heads of palmate to digitate leaves. When young, the species described below make attractive foliage pot plants. Propagate by seed in spring. *Thrinax* is the Greek name for a trident and reputedly alludes to the forked tips of the leaflets or lobes.

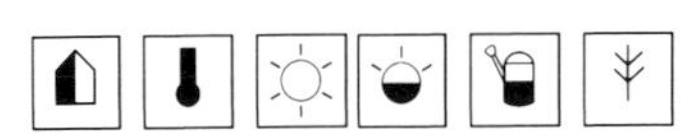

Species cultivated

T. microcarpa (*T. morrisii*) Key palm S. Florida, West Indies
Up to 10m(33ft) in the wild, but decorative when young and fairly slow growing in containers. Leaves orbicular in outline, 60–100cm(2–3ft) wide, divided into 20 to 30 lanceolate leaflets or lobes which are silvery-white beneath.
T. morrisii See *T. microcarpa.*
T. parviflora Palmetto thatch palm Jamaica
Similar to *T. microcarpa* when young, but the leaves lack the silvery-white undersurfaces. Sometimes confused with the similar but larger *T. radiata.*

THRIXANTHOCEREUS See ESPOSTOA

THUNBERGIA

Thunbergiaceae (*Acanthaceae*)
Clock vines

Origin: Central and southern Africa, Malagasy and southern Asia. A genus of between 100 and 200 species of annuals and perennials many of which are of twining habit. They have opposite pairs of entire leaves and irregularly 5-lobed, tubular flowers which can be solitary from the leaf axils or in racemes. They are really best for the conservatory in large pots or tubs, the climbers trained on strings or other supports, but the smaller species can be grown in the home. Propagate by seed in spring or by cuttings with bottom heat in early summer. *Thunbergia* was named for Carl Peter Thunberg (1743–1828), a Dutch botanist and doctor who worked in South Africa and Japan. He was a student of Linnaeus and later became Professor of Botany at Uppsala.

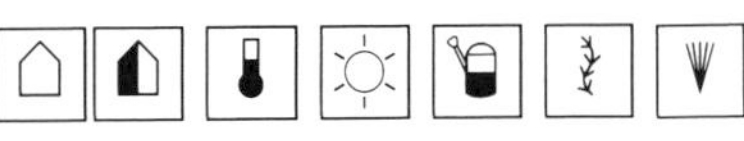

Species cultivated

T. alata Black-eyed Susan Tropical Africa
Perennial climber usually grown as an annual; stems to 3m(10ft) long. Leaves to 8cm(3in) long, triangular-ovate with winged leaf stalks. Flowers 4cm(1½in) long, white to deep creamy-yellow or pale orange with a chocolate centre, borne in the leaf axils in summer.
T. gibsonii See *T. gregorii.*
T. grandiflora Blue trumpet vine N. India
Stems woody, to 6m(20ft) in length. Leaves 10–20cm(4–8in) long, ovate with angular teeth and a rough-textured surface. Flowers to 7cm(2¾in) long and wide, blue, carried in pendent racemes in summer.
T. gregorii (*T. gibsonii*) Tropical Africa
Very like *T. alata*, but with ovate leaves and flowers to 4.5cm(1¾in) long, orange, without the central marking.
T. mysorensis India (Nilghiri Hills)
Vigorous climber to 6m(20ft) or more in length. Leaves oblong-lanceolate to elliptic, 10–15cm(4–6in) long, usually but not invariably toothed, rich green. Flowers to 5cm(2in) long, very asymmetrical, like elongated gaping yellow mouths, with five reflexed, red-purple lobes, emerging from a pair of purple-red calyx-like bracts; the pendent racemes, which may attain 45cm(1½ft) in length, appear from spring to autumn.
T. natalensis South Africa, Natal
Sub-shrubby, but best grown as an herbaceous perennial by cutting back to ground level annually each winter. Stems 4-angled, erect, 60–90cm(2–3ft) in height. Leaves ovate, toothed, 8cm(3in) long, pointed, sessile. Flowers to 5cm(2in) long, pale blue or blue-purple with a yellow throat, borne singly from the upper leaf axils in summer.

THYRSACANTHUS See ODONTONEMA

Thevetia peruviana

BELOW *Thrinax microcarpa*
BOTTOM *Thunbergia natalensis*

Tibouchina urvilleana

TIBOUCHINA

Melastomataceae

Origin: Central and South America. A genus of about 350 species of shrubs, sub-shrubs, climbers and perennials, with opposite, undivided leaves and usually showy flowers with five (occasionally four) petals and sepals. The species described makes a fine and attractive pot or tub plant. To keep plants small, cut back their stems annually to about 15cm(6in) in late winter or early spring. If allowed to grow freely they can be trained to cover the back wall of a conservatory, then needing the lateral stems cut back to two pairs of buds in spring to prevent the plant from becoming too leggy. Propagate by cuttings in spring or summer with bottom heat.

Tibouchina is a Latin form of the vernacular Guyanan name.

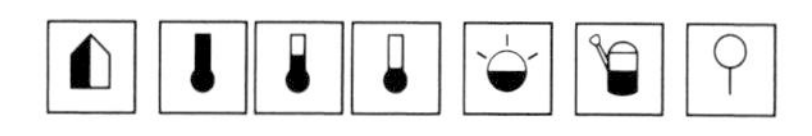

Species cultivated

T. urvilleana (*T. semidecandra*) Glory bush
Brazil
Shrub, 3–5m(10–16ft) tall, easily kept below half this; evergreen in tropical temperatures when it will flower intermittently throughout the year; deciduous in cool conditions, then flowering in summer and autumn only. Stems 4-angled, finely covered with red hairs. Leaves 7–15cm($2\frac{3}{4}$–6in) long, ovate to oblong with three to seven sunken longitudinal veins, rich green, velvety hairy. Flowers 7–10cm($2\frac{3}{4}$–4in) wide, satiny royal-purple, the stamens darker and curiously hooked, opening from rose-red buds.

TILLANDSIA

Bromeliaceae

Origin: The Americas from southern USA to South America; one species in West Africa. A genus of about 400 species of evergreen perennials most of which are epiphytic. Most form rosettes, but a few species have slender, pendent stems with small, well spaced leaves. Almost all species are covered in silvery scale hairs able to absorb and retain moisture. The flowers are small, 3-petalled, often within colourful bracts; they can be solitary, but are more usually in spikes or panicles. Tillandsias can be grown in pots or on a tree branch. Propagate the rosette species by removing offsets in summer, the pendent sorts by division. *Tillandsia* was named for Elias Til-Landz (*d.*1693), a Swedish botanist and Professor of Medicine.

Species cultivated

T. plumosa Mexico
Rosettes eventually forming large clumps. Leaves dense, linear, channelled, covered with feathery white scales, the expanded bases swollen with water storage tissue and collectively creating a bulk-like base to the rosette. Flowers 2cm($\frac{3}{4}$in) long, green, subtended by pink-tipped bracts and arranged in simple or clustered head-like spikes above the leaves.

T. pulchella See *T. tenuifolia*.

T. purpurea Peru
A variable species, either rosette forming or more usually with a branching stem to 30cm(1ft) or more in length. Leaves crowded, spirally arranged,

slenderly triangular, arching, thickly grey-scaly, 20–30cm(8–12in) long. Flowers fragrant, 2cm(¾in) long, white and violet-blue, subtended by lavender to purple bracts and arranged in branched spikes. A terrestrial species forming extensive colonies in the coastal desert zone where it acts as a sand binder and dune former.

T. seleriana Mexico, Guatemala
Superficially like *T. bulbosa*, but the 'bulbous' base to the rosette is larger and the leaf blades shorter, erect and more or less straight. In the wild, the inflated leaf bases are inhabited by a certain species of ant.

T. tenuifolia (*T. pulchella*) Cuba to Northern Argentina
Rosettes usually short-stemmed. Leaves 10–15cm(4–6in) long, narrow, awl-shaped. Flowers 2cm(¾in) long, white, pink or blue within red bracts and carried in an erect spike.

T. usneioides Spanish moss S.E. USA to Argentina and Chile
Stems pendent, slender, branched, to 1m(3ft) or so in a conservatory, but up to six times this length in warm, humid conditions in the wild. Leaves 2–5cm(¾–2in) long, narrowly linear, densely silvery-white-scaly (as are the stems). Flowers 1cm(⅜in) long, yellow-green to blue, carried singly and relatively insignificant.

TITANOPSIS

Aizoaceae

Origin: South Africa. A genus of six to eight species of succulents which mimic the limestone rock of their homeland. They are hummock-forming when mature, made up of rosette-like shoots of opposite leaf pairs. Each leaf is spathulate with an obliquely triangular tip thickly covered with white to grey-white low tubercles of irregular size. The autumn-borne flowers are large and showy. The species described here are easily grown and deserve to be added to a collection of succulents. *Titanopsis* derives from *Titan*, the Greek sun god and *opsis*, like; the mainly yellow flowers are likened to the sun.

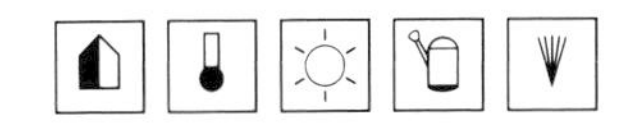

Species cultivated

T. calcarea Cape
Leaves about 2.5cm(1in) long, the tips broadly and bluntly triangular, the grey-white tubercles sometimes with a faintly bluish tinge. Flowers 2cm(¾in) wide, bright golden-yellow, sometimes almost orange.

T. crassipes See *Aloinopsis spathulata.*

T. fulleri Cape
Similar to *T. calcarea*, but the leaves about 2cm(¾in) long, with brownish-grey tubercles against a purplish ground colour. Flowers 1.5cm (⅝in) wide, dark yellow.

T. setifera See *Aloinopsis setifera.*

Titanopsis calcarea

T. spathulata See *Aloinopsis spathulata.*

TOLMIEA

Saxifragaceae

Origin: Western North America. A genus of one species of low evergreen perennial which makes an excellent pot or hanging basket plant. Hardy. Propagate by division or by removing the plantlets from the leaves and rooting them as for cuttings. *Tolmiea* was named for Dr. William Fraser Tolmie (1812–1886), surgeon at the Hudson Bay Company depot at Fort Vancouver.

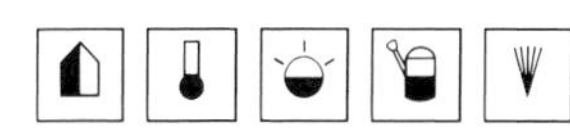

Species cultivated

T. menziesii Pick-a-back/Piggyback plant
Clump forming, to 40–60cm(16–24in) tall when in bloom. Leaves 4–10cm(1½–4in) long, long-stalked, broadly ovate-cordate with palmate lobing, toothed and hairy. A feature of this plant which immediately separates it from *Tiarella* is the presence of small plantlets which arise where leaf stalk and blade join, these will root when they touch the soil. Flowers 1.2–2cm(½–¾in) long, tubular, split open along one side and with five greenish-purple, petal-like calyx lobes and four thread-like deep brown petals which are borne in slender racemes in summer. A variegated form is grown.

Tolmiea menziesii 'Variegata'

TORENIA

Scrophulariaceae

Origin: Tropical and sub-tropical Asia and Africa. A genus of 40 to 50 species of annuals and perennials with opposite, undivided leaves and tubular flowers in short racemes or solitary from the upper leaf axils. The species described below makes an attractive pot plant for the conservatory or home window sill. Propagate by seed in spring. *Torenia* was named for the Reverend Olaf Toren (1718–1753), Chaplain to the Swedish East India Company in India and China.

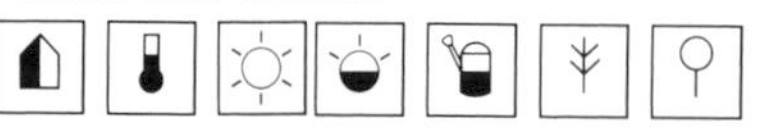

Trachycarpus fortunei

Species cultivated

T. fournieri Wishbone flower Vietnam
Erect bushy annual to 30cm(1ft) or more tall. Leaves 3–5cm(1¼–2in) long, narrowly to broadly ovate, toothed. Flowers 3–4cm(1¼–1½in) long, tubular, opening to four lobes, the upper one large, pale lilac-blue, the other three deep velvety blue-purple with a central yellow spot on the lower one. *T.f.* 'Grandiflora' has larger flowers than the type species.

TRACHELIUM

Campanulaceae

Origin: Mediterranean region. A genus of about seven species of perennials some of which are woody-based. They have erect stems, alternate entire leaves and corymbs or panicles of small, tubular, 5-lobed flowers. The species described below makes a good pot plant for the conservatory or sunny window sill in the home, and for this purpose is best raised annually. Propagate by seed in warmth in spring, pinching them once to make shorter, bushier plants. *Trachelium* is derived from the Greek *trachelos*, neck; *T. caeruleum* once being used for treating infections of the throat.

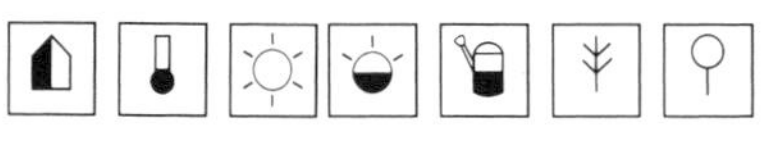

Species cultivated

T. caeruleum Common throatwort West and central Mediterranean
To 60cm(2ft) as an annual. Leaves 3–7cm(1¼–2¾in) long, ovate, double-toothed. Flowers 6mm(¼in) long, starry, lavender-blue and slightly fragrant; they are carried in large corymbose panicles terminating the branching stems.

TRACHYCARPUS

Palmae

Origin: Himalaya to S.E. Asia. A genus of six to eight species of small- to medium-sized palms. They have fan-shaped leaves divided to more than halfway into narrow, linear segments. The plants are surprisingly frost-hardy and are suitable for a cold or cool conservatory or room. Propagate by seed in spring in warmth. *Trachycarpus* derives from the Greek *trachys*, rough and *karpos*, fruit.

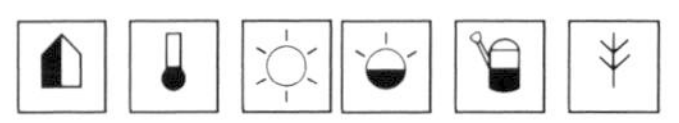

Species cultivated

T. fortunei Windmill/Chusan palm Borneo to East China
To 2m(6½ft) or more in a container, to 10m(33ft) in the open. Stem solitary, covered with brown fibrous leaf bases. Leaves to 60cm(2ft) long in a pot, but to twice this in the open, fan-shaped to almost rounded, divided deeply into many narrow, pleated segments. Flowers not produced when young.

TRADESCANTIA

Commelinaceae

Spiderworts

Origin: North and South America. A genus of 20 to 60 species (depending on the classification followed) of perennials, both deciduous and evergreen. They have somewhat fleshy, alternate, entire leaves and 3-petalled flowers within leaf- or spathe-like bracts; the flowers have bearded stamens. The species described make undemanding house or conservatory plants growing well in pots or hanging baskets. Propagate by cuttings in spring and summer. *Tradescantia* was named for John Tradescant (1608–1662), Gardener to King Charles I.

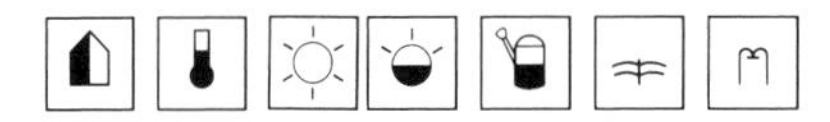

Species cultivated

T. albiflora Wandering Jew South America
Stems trailing, to 60cm(2ft) or more in length. Leaves 3–5cm(1¼–2in) long, oblong-elliptic, glossy light green. Flowers 8mm(⅓in) across, white. *T.a.* 'Albovittata' with white striped leaves is the form most frequently seen.

T. blossfeldiana Argentina
Stems to 20cm(8in) long, purple, erect or sprawling. Leaves 5–10cm(2–4in) long, elliptic, deep glossy green above, purple and hairy beneath. Flowers 1cm(⅜in) across, pinkish-purple with a

white eye, in compact umbels.
T. dracaenoides See *Callisia fragrans.*
T. fluminensis Wandering Jew South America
Stems trailing, to 60cm(2ft) or more in length. Leaves ovate, glossy, almost bluish-green above, purple-flushed beneath. *T.f.* 'Variegata' has irregularly white and cream striped leaves. It is sometimes confused with *Zebrina pendula.*
T. fuscata See *Siderasis fuscata.*
T. geniculata (*Gibasis geniculata*) Tropical America
Prostrate to decumbent species with stems 30–45cm(1–1½ft) or more in length. Leaves lanceolate to narrowly ovate, to 10cm(4in) long, more or less softly hairy. Flowers small, white, carried in loose terminal panicles on erect leafy stems to 22cm(9in) high in summer.
T. navicularis Mexico, questionably Peru
Tufted to mat-forming with prostrate stems to 15cm(6in) or more in length. Leaves crowded, distichous, boat-shaped, thickly fleshy, ciliate, grey-green. Flowers 1.2–2cm(½–¾in) wide in small cymes, mainly in summer but occasionally at other times.
T. pexata See *T. sillamontana.*
T. reginae See *Dichorisandra reginae.*
T. sillamontana (*T. pexata, T. velutina, Cyanotis sillamontana, C. veldthoutiana* of gardens)
Mexico
Stems semi-erect, branched, to 30cm(1ft) tall, the whole plant covered with dense white wool. Leaves 5–7cm(2–2¾in) long, elliptic to ovate, crowded on the stems. Flowers 2cm(¾in) across, bright magenta-rose, borne in summer and autumn. Best dried off in late autumn and kept in cool conditions until starting into growth in spring.
T. velutina See *T. sillamontana.*
T. warszewicziana See *Hadrodemas warszewiczianum.*

ABOVE *Tradescantia albiflora* 'Albovittata'
LEFT *Tradescantia fluminensis* 'Variegata'
BELOW LEFT *Trevisia palmata*

TREVESIA

Araliaceae

Origin: Indo-Malaysia and Pacific. Depending upon the botanical authority, a genus of four to ten species of shrubs and small trees some of them with wonderfully shaped leaves as though modelled on a snowflake crystal. One species is outstanding in this respect and provides a very specially decorative-foliaged pot or tub plant for the conservatory and is also worth trying in the home. Propagate by cuttings in summer or by seed in spring. *Trevesia* honours the family of Trèves de Bonfigli of Padua, Patrons of botany in the eighteenth century.

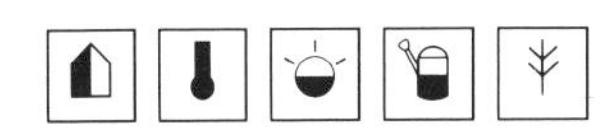

Species cultivated

T. palmata Snowflake aralia N. India to S.W. China
Small tree in its homeland, but maintainable at 2m(6½ft) or less in a container; stems erect, sparingly branched, sometimes prickly. Leaves alternate, long-stalked, circular in outline, digitate, divided into seven to 11 leaflets which are attached to the edge of a central duck's-foot-like web of leaf tissue; when young the leaflets are ovate to oblong-ovate with a few teeth, but when mature they become deeply cut into bluntly triangular lobes, all a deep glossy green and lightly corrugated; depending upon the age of the plant, the leaves vary from

25–60cm(10–24in) across. Flowers yellowish, small, in rounded umbels arranged in erect panicles, but only on mature plants.

TRICANTHA See COLUMNEA & ECHINOPSIS

TRICHOCEREUS

Cactaceae

Origin: South America. A genus of 25 species of mainly large columnar cacti. Planted out in the larger conservatory, the taller species will in time form splendid accent points. Propagate by seed in spring or by cuttings in summer. *Trichocereus* derives from the Greek *trix*, hair and the allied genus *Cereus*. It resembles *Cereus*, but differs in the ovary and tubular bases of the flowers being densely hairy and lacking spines or bristles.

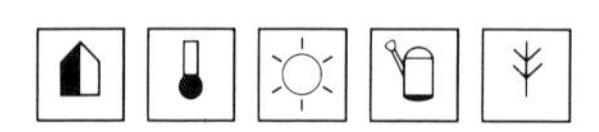

Species cultivated

T. candicans (*Cereus candicans*) Torch cactus
N.W. Argentina
Stems to 1m(3ft) in height, branching from the base. Ribs nine to 11, broad and rounded, bearing large white-woolly areoles. Radial spines ten to 14, yellow or white, to 4cm(1½in) long; central spines one to four, darker, to 10cm(4in) in length. Flowers funnel-shaped, to 25cm(10in) long, white and fragrant, but only borne on large specimens.

T. chiloense (*Cereus chiloensis*) Chile
In the wild, stems 3–8m(10–25ft) in height, branching above; old specimens tree-like. Ribs ten to 15, rounded, notched, slightly grey-green, bearing large, whitish areoles. Radial spines eight to 12, thick, awl-shaped, brownish-yellow to grey, 2–4cm(¾–1½in) long; central spines one or two, sometimes up to four, to 6.5cm(2½in) in length. Flowers white, to 15cm(6in) long. Fairly slow-growing as a pot plant. Despite its name, this species does not occur wild on the island of Chiloe.

Trichocereus chiloense

T. spachianus (*Cereus spachianus*)
W. Argentina
Stems to 1.5m(5ft) in height, branching from the base. Ribs ten to 15, rounded, bright green bearing yellowish to grey areoles. Radial spines eight to ten, bristle or needle-like, to 1cm(⅜in) long; central spines one or two, longer and thicker, yellow to brown. Flowers white, to 20cm(8in) long, opening at night, but only on sizeable plants.

T. werdermanniana See *Echinopsis werdermanniana*.

TRICHODIADEMA

Aizoaceae

Origin: South Africa (Cape). A genus of about 30 species of succulent plants formerly classified in *Mesembryanthemum*. They are tufted, eventually forming hummocks or mats. The narrow fleshy leaves are in opposite pairs, the tip of each one crowned by a tuft of slender bristles which gives all members of this genus such a distinctive air. The autumn- to spring-borne daisy-like flowers are generally showy but only open in sunshine. The species described are worthy additions to a collection of succulents. Propagate by seed in spring or by cuttings in summer, the latter not as easy as for most members of the *Aizoaceae* family with a similar habit. *Trichodiadema* derives from the Greek *trichos*, a hair and *diadema*, a crown or diadem, referring to the tufts of bristles on the leaf tips.

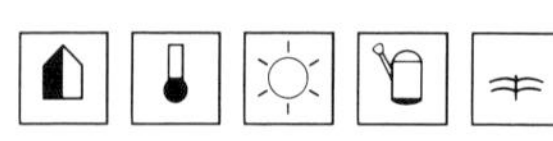

Species cultivated

T. densum
Roots thickened and fleshy. Stems short, forming rounded tufts, bearing crowded pairs of leaves, each one 1.5–2cm(⅝–¾in) long, greyish-green, crowned by spreading white bristles. Flowers 3–5cm(1¼–2in) across, carmine to purple-red. One of the choicest and best-known species.

T. intonsum
Loosely tufted, distinctive in having the leaf pairs well separated by visible lengths of stem; each leaf about 1.2cm(½in) long, semi-cylindrical, slightly curved outwards, densely covered with sharp papillae and tipped by short, erect, dark brown bristles. Flowers about 2cm(¾in) wide, pink or white.

T. mirabile
Bushy hummock-forming species to 8cm(3in) or more in height with stems bearing white bristles. Leaf pairs united at their bases, 1.2–2.5cm(½–1in) long, somewhat flattened above, papillose, crowned by erect, dark brown bristles. Flowers 4cm(1½in) wide, pure white.

T. stellatum
Roots fleshy. Stems forming hummocks 5–10cm(2–4in) tall. Leaves about 1cm(⅜in) long, papillose, grey-green, tipped by spreading white bristles. Flowers purple-red, about 3cm(1¼in) across.

TRICHOSANTHES

Cucurbitaceae

Origin: Indo-Malaysia, Polynesia, Australia. Depending upon the botanical authority consulted, a genus of 15 to 40 species of tendril-bearing climbing annuals and perennials which, superficially at least, have affinities with the cucumber and balsam apple (*Momordica*). The species described have curiosity value and are not without beauty of flower. When in fruit they become talking points of some interest. Propagate by seed in spring. *Trichosanthes* derives from the Greek *trichos*, a hair and *anthos*, a flower; the petals are margined with thread-like outgrowths.

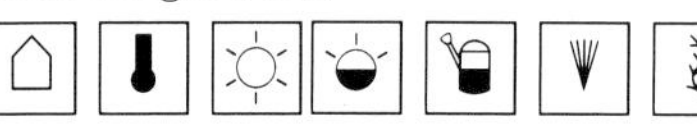

Species cultivated

T. anguina Snake gourd India, widely cultivated in the tropics
Annual, with stems 3–5m(10–16ft) or more in length, supported by branching tendrils. Leaves alternate, broadly ovate to triangular-ovate, sometimes shallowly 3- to 7-lobed or angled, toothed, about 22cm(9in) long. Male flowers in axillary racemes, females solitary, 5-petalled, white, to 5cm(2in) across, each petal bearing a long and elaborate fringe of filaments. Fruits to 60cm(2ft) or rarely to 2m(6½ft) long, green and white striped when young, ripening bright orange, often curiously twisted or even coiled, hence the vernacular name. In tropical countries the fruits are harvested before they are half grown and used as a vegetable, usually with curry.

T. cucumeroides Japan
Perennial, from a tuberous root producing stems to 3m(10ft) in length. Leaves ovate-cordate to kidney-shaped, to 10cm(4in) long, usually 3- to 5-lobed or angled. Flowers dioecious, the males in racemes, the females solitary, 5-petalled, white; male flowers tubular, about 6cm(2½in) long. Fruits to 8cm(3in) long, ovoid to oblong, ripening cinnabar-red, formerly dried and used as a soap substitute.

TRICHOSPORUM See AESCHYNANTHUS
TRIPLOCHLAMYS See PAVONIA
TRIPOGANDRA See HADRODEMAS
TRISTAGMA See IPHEION
TRITELEIA See BRODIAEA & IPHEION

TRITONIA

Iridaceae

Origin: Tropical and southern Africa. A genus of about 50 species of cormous perennials. They have sword-shaped or falcate leaves carried in fan-shaped, flattened tufts. The flowers have six tepals which are joined at the base to form a funnel-shaped tube. Plants are suitable for a conservatory and can be flowered in the home. Propagate by division when re-potting or by seed in spring. *Tritonia* derives from the Greek *triton*, in this case meaning a weathercock; an allusion to the positions of the stamens.

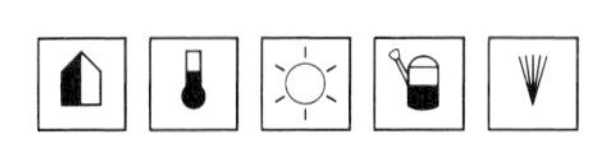

Species cultivated

T. crocata South Africa
Leaves to 20cm(8in) long, curved. Flowers 5–6in(2–2½in) wide. Bright tawny-yellow, stems 30–45cm(1–1½ft) tall in summer.

T.c. miniata
Smaller; bright red flowers.

T. hyalina
Allied and similar to *T. crocata*, but with broader leaves and orange flowers composed of spathulate

Trichosanthes cucumeroides

tepals bearing almost transparent (hyaline) margins. (The tepals of *T. crocata* are broadly cuneate to obovate.)

T. rosea See *T. rubrolucens*.

T. rubrolucens (*T. rosea*)
Leaves to 30cm(1ft) long, linear. Flowers about 5cm(2in) across, rose-red, the inner tepals with darker basal spots, carried in branched spikes on stems 40–60cm(16–24in) tall in summer.

TROPAEOLUM

Tropaeolaceae

Origin: Mexico, Central and South America. A genus of 50 to 90 species, depending upon the classification followed. They are annual and perennial climbers, often with tuberous roots and alternate leaves which can be rounded to shield-shaped to digitate, the stalks capable of twisting and acting as tendrils. The flowers have five sepals, the uppermost with a long nectary spur, and five broad petals. They are pot plants for the conservatory or window sill. Propagate the perennial kinds by seed when ripe, by offsets or division, or by cuttings of basal shoots removed close to the tuber in spring and rooted with bottom heat. Propagate annuals by seed sown in spring. *Tropaeolum* derives from the Latin *tropaeum*, a trophy, a plant of *T. majus* in flower on a support fancifully resembling a trophy pillar hung with shields (leaves) and bloody helmets (flowers) of the defeated enemy!

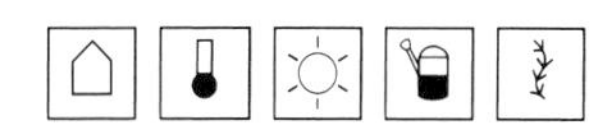

Species cultivated

T. canariensis See *T. peregrinum*.

T. ciliatum Chile
Rhizomatous, with stems to 2m(6½ft) or more in length. Leaves 2–3cm(¾–1¼in) wide, digitate, composed of five obovate leaflets. Flowers somewhat mimulus-like, about 2.5cm(1in) long with five rounded, stalked petals, the lower three yellow, the upper two beige-yellow, with a pattern of purple-brown veins, borne in summer.

T. hookerianum Chile
Small tuberous perennial to 60cm(2ft). Leaves to 2cm(¾in) across, digitate, with five to seven obovate leaflets. Flowers to 2.5cm(1in) or so

RIGHT *Tropaeolum majus* 'Hermine Grasshof'
BELOW RIGHT *Tropaeolum tricolorum*

across, yellow, on long erect stalks in early spring.
T. lobbianum See *T. peltophorum*.
T. majus Nasturtium Peru
Strong-growing annual climber capable of reaching 3m(10ft) in height if grown in a large pot. Leaves 5–15cm(2–6in) wide, almost circular, peltate, glabrous and bright green. Flowers 5–6cm(2–$2\frac{1}{2}$in) wide with long spurs, in shades of red, orange and yellow, opening in summer and autumn. Many cultivars are available, some the result of hybridization with *T. minus*. Most useful for pot culture are the dwarf strains such as Tom Thumb which rarely exceeds 25cm(10in) and can be single- or double-flowered, the Gleam strain of semi-trailing habit which are splendid in hanging baskets, and a group of fully double cultivars which have to be propagated from cuttings and have a trailing habit. 'Burpeei' and 'Hermine Grasshof' come into this last category.
T. peltophorum (*T. lobbianum*) Colombia, Ecuador
Annual similar to *T. majus*, but the undersurfaces of its leaves are downy. Flowers 2–2.5cm($\frac{3}{4}$–1in) wide, the lower petals toothed, orange-red.
T. pentaphyllum South America
Tubers large, irregular, corky-skinned. Stems 2–3m($6\frac{1}{2}$–10ft) in length. Leaves 2–3cm($\frac{3}{4}$–$1\frac{1}{4}$in) wide, digitate, composed of five leaflets, the central one largest. Flowers 3cm($1\frac{1}{4}$in) long, two-thirds being a thick red spur; sepals pale green, spotted and striped purple, larger than the tiny scarlet petals; borne summer. Despite the description, a remarkably pretty-flowered species.
T. peregrinum (*T. canariensis*) Canary creeper Peru, Ecuador
Annual or short-lived perennial with climbing stems to 4m(13ft) long. Leaves rounded in outline but deeply five- to seven-lobed. Flowers 2–2.5cm($\frac{3}{4}$–1in) across, the upper two petals much bigger than the rest, erect, divided almost to halfway into many narrow segments giving a fringed appearance, canary-yellow with a small red spot at the base of the petals, opening in summer.
T. tricolorum (*T. tricolor*) Chile, Bolivia
Tuberous-rooted perennial, the very slender stems climbing, 60–90cm(2–3ft) long. Leaves 1.2–2cm($\frac{1}{2}$–$\frac{3}{4}$in) wide, digitate, with five to seven obovate leaflets. Flowers 2–3cm($\frac{3}{4}$–$1\frac{1}{4}$in) long, the sepals and spur crimson, surrounding short, yellow, usually, but not always, maroon-edged petals, opening in spring.
T. tuberosum Bolivia, Peru
Tuberous-rooted perennial, 2–3m($6\frac{1}{2}$–10ft) long. Leaves rounded, palmate, having three to five broad lobes, greyish-green, 4–6cm($1\frac{1}{2}$–$2\frac{1}{2}$in) wide. Flowers up to 4cm($1\frac{1}{2}$in) long, elegantly poised on long red stalks singly from the upper leaf axils in autumn; usually the spur and sepals are orange-red, the petals deep to orange-yellow, brown-veined within. The potato-like tubers are pear-shaped, pale yellow, with crimson marking; in its native country they are an item of food known as anu. *T. t.* 'Ken Aslett' is a large-tubered cultivar which begins to flower in June; 'Sidney' is a rhizome-like tubered cultivar with somewhat smaller, entirely orange flowers.

TULBAGHIA

Alliaceae

Origin: Tropical and southern Africa. A genus of about 25 species of perennials having tuberous, rhizomatous or cormous rootstocks, allied to *Agapanthus* but smelling of garlic when bruised. They have strap-shaped basal leaves and naked flowering stems topped by umbels of 6-tepalled flowers with tubular bases. Those described make ornamental pot plants for the conservatory and can be brought into the home when in bloom – if the garlic odour is not objected to! Propagate by division or seed in spring.

Tulbaghia commemorates Rijk Tulbagh (1699–1771), a popular Dutch governor of the Cape of Good Hope.

Species cultivated
T. fragrans Sweet garlic South Africa, Transvaal
Clump-forming; leaves to 30cm(1ft) in length. Flowering stems 45cm($1\frac{1}{2}$ft) or more tall carrying umbels of 20 to 40 flowers; each bloom 1.2cm($\frac{1}{2}$in) long, lilac to mauve, fragrant, mainly summer

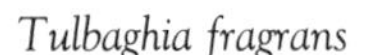

Tulbaghia fragrans

flowering, but also at other times even in winter in temperate warmth.

T. violacea South Africa
Clump-forming; leaves 20–30cm(8–12in) in length. Flowering stems 60cm(2ft) or more in height, bearing umbels of seven to 20 flowers; each bloom 1.2–2cm(½–¾in) long, lilac-pink, borne in late spring to autumn. *T. v.* 'Silver Lace' ('Variegata') has grey-green leaves edged and striped white, sometimes with a pink flush.

TULIPA

Liliaceae
Tulips

Origin: Europe and Asia, the largest number of species from the Steppe country of central Asia. A genus of about 100 species of bulbous plants with linear to ovate, stalkless leaves and 6-tepalled usually cup-shaped flowers opening widely as they mature. Their bulbs are replaced annually, the new one forming at the base of the stem. Tulips make colourful short term pot plants for the conservatory or home, but are best kept in a cool place at night to prolong flowering. The plants can be grown in bulb fibre, but if grown in a standard potting mix and fed regularly they will flower a second year or can be planted in the garden for flowering. Propagate species by seed or by offsets. *Tulipa* is a Latin form of the Arabic for a turban.

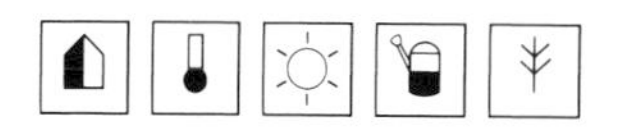

Species cultivated

T. batalinii Turkestan
To 15cm(6in) tall. Leaves to 13cm(5in) long, mostly basal, linear, glaucous and slightly wavy. Flowers to 5cm(2in) long, tepals pointed, creamy-yellow. Very similar to the red *T. linifolia* and considered by some to be just a coloured variant of that species. Often represented in gardens by hybrids between the two with creamy-yellow flowers flushed with bronze; 'Bronze Charm' is most frequently seen.

T. daystemon (of gardens) See *T. tarda.*

T. fosteriana Uzbekistan to Tadjikistan, USSR
25–45cm(10–18in) tall. Leaves to 20cm(8in) long, broadly ovate, glaucous. Flowers to 10cm(4in) long, brilliant scarlet with a cream-edged black blotch at the base of the inside of each tepal. Named cultivars are available, some hybrids.

T. greigii Turkestan
To 25cm(10in) tall. Leaves to 20cm(8in) long, broadly ovate, glaucous green, strongly streaked and mottled with dark purple. Flowers to 8.5cm(3½in) long, bright orange-scarlet with a yellow-edged black blotch at the base of each tepal. Many cultivars are available.

T. hageri Greece, W. Iran
To 30cm(1ft) tall. Leaves to 20cm(8in) long, lanceolate. Flowers about 4.5cm(1¾in) long, one to four on a stem, soft red, with the inside base of each tepal greenish-black with a yellow margin and with green markings on the outside.

T. humilis Asia Minor
10–13cm(4–5in) tall. Leaves to 15cm(6in) long, linear. Flowers 3cm(1¼in) long, one to three on a stem, rose-purple inside, with cream to yellow centres. *T. pulchella* and *T. violacea* are similar and probably only forms of *T. humilis.*

T. kaufmanniana Water-lily tulip Turkestan (USSR)
To 20cm(8in) tall. Leaves to 25cm(10in) long, broadly oblong, glaucous green. Flowers to 8 (3in) long, opening flat, creamy-yellow, the outer tepals flushed and streaked with red. Many cultivars are available, several of hybrid origin, making up Division 12 of the standard tulip classification.

T. linifolia Turkestan (USSR)
To 15cm(6in) tall. Leaves to 13cm(5in) long, linear, somewhat glaucous, wavy. Flowers to 5cm(2in) long with pointed tepals, dark red on the outside, scarlet inside.

T. praestans Central Asia
To 30cm(1ft) tall. Leaves to 25cm(10in) long, narrowly lanceolate, folded upwards from the midrib. Flowers 5.5cm(2¼in) long, one to four on a stem, bright orange to scarlet.

T. pulchella See under *T. humilis.*

T. tarda (*T. dasystemon* of gardens) E. Turkestan (USSR)
10–15cm(4–6in) tall. Leaves 15–18cm(6–7in) long, linear to lanceolate, mainly basal. Flowers to 4cm(1½in) long, three to seven to a stem, tepals bright yellow with white tips inside, suffused with red outside.

T. violacea See under *T. humilis.*

Hybrid cultivars

Most tulip cultivars can be grown in pots, but for home the single and double earlies (Divisions 1 and 2), listed in bulb catalogues, are the most rewarding.

BELOW *Tulipa hageri*
BELOW CENTRE *Tulipa fosteriana* 'Red Emperor'
BOTTOM *Tulipa* 'Apricot Beauty' (Div. 1)

BELOW *Turnera ulmifolia*
BOTTOM *Turraea floribunda*

TURNERA

Turneraceae

Origin: Tropics and sub-tropics. A genus of about 60 species of shrubs and perennials one of which resembles a yellow mallow and makes a showy plant for the conservatory. Propagate by seed in spring or by cuttings in summer. *Turnera* honours William Turner (*c.* 1508–1568), doctor, herbalist and clergyman, now affectionately known as the 'Father of English Botany'.

  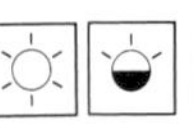 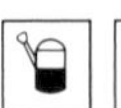 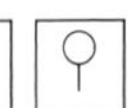

Species cultivated

T. ulmifolia (*T. trioniflora*) Sage rose
Mexico, West Indies, Tropical America
Shrubby, 60–120cm(2–4ft) tall, of somewhat spreading habit and freely branching. Leaves oblong-ovate, sharply toothed, deep green above, white-hairy beneath, 5–10cm(2–4in) long. Flowers 5-petalled, about 5cm(2in) wide, pale to

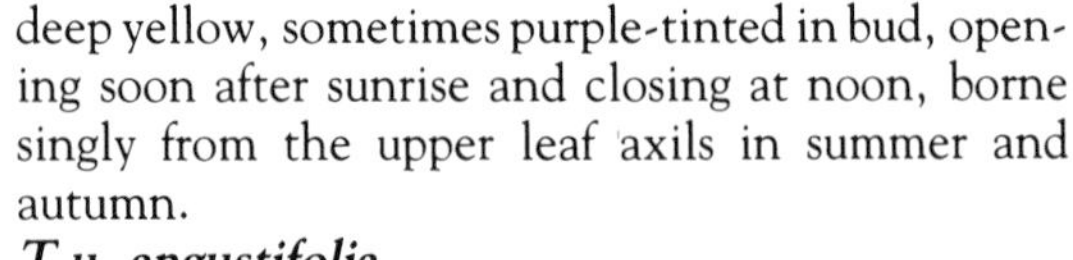

deep yellow, sometimes purple-tinted in bud, opening soon after sunrise and closing at noon, borne singly from the upper leaf axils in summer and autumn.

T.u. angustifolia
Lanceolate leaves and narrower petals.

TURRAEA

Meliaceae

Origin: Tropical to southern Africa, Malagasy, tropical Asia, Australia. A genus of about 90 species of trees and shrubs, a few of which make effective containerized plants and specimen flowering shrubs for the conservatory. They have alternate, simple leaves and 5-petalled starry flowers from their axils. Propagate by seed in spring or by cuttings in summer. *Turraea* commemorates Giorgia della Turre (1607–1688), a Professor of Botany at Padua and author of several books.

    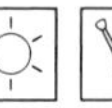

Species cultivated

T. floribunda South Africa
Shrub to 2m(6½ft). Leaves ovate, prominently veined, glossy rich green, 8–13cm(3–5in) long. Flowers fragrant, palest yellow-green, with linear petals 3–4cm(1¼–1½in) long, borne in twos and threes from the leaf axils in spring to early summer.

T. obtusifolia South Africa
Shrub, 90–180cm(3–6ft) in height. Leaves obovate to oblanceolate, 2.5–5cm(1–2in) long, glossy rich green. Flowers 4–5cm(1½–2in) across, the well spaced obovate petals arching, white, borne in spring to summer and sometimes into autumn.

TWEEDYA See OXYPETALUM
URBINIA See ECHEVERIA

V

VALLOTA

Amaryllidaceae

Origin: South Africa. A genus of one species of bulbous perennial closely allied to *Cyrtanthus* and looking like a slender *Hippeastrum*. It is a good pot plant for the conservatory or sunny room. Only re-pot when essential as it does not grow so well after its roots are disturbed. Propagate by seed or offsets. *Vallota* was named for Pierre Vallot (1594–1671), a French doctor and botanist.

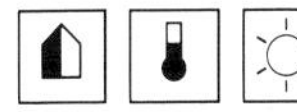 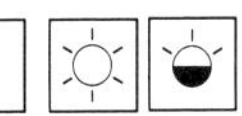 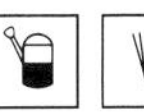 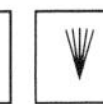

Species cultivated

V. speciosa (*V. purpurea*) Guernsey/ Scarborough lily
Leaves evergreen, 30–50cm(12–20in) long, broadly linear, basal. Flowers 7–10cm(2¾–4in) long, funnel-shaped with six bright red tepals, in terminal umbels of three to ten on 30–60cm(1–2ft) stems in late summer and autumn. *V.s.* 'Alba' is white flowered; 'Delicata' has pale salmon-pink blooms.

VANDA

Orchidaceae

Origin: Tropical Asia. A genus of 60 species of epiphytic orchids without pseudobulbs. They have erect, often rather woody based stems, usually unbranched, carrying linear leaves in two parallel ranks which can be cylindrical (terete) or flattened. The showy flowers have five similar tepals and a 3-lobed, spurred labellum, and are borne in erect to arching racemes. They make intriguing conservatory and house plants as long as they can be given good light. The terete-leaved species thrive in full sun and extra humidity. All grow well in pots or hanging baskets. Propagate by removing the top of a leggy plant and taking the shoots from the cut-back plant as cuttings when they have produced aerial roots. *Vanda* is the Indian (Hindi) name for this orchid.

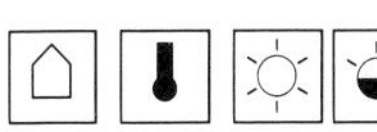   

Species cultivated

V. amesiana Vietnam, Laos, Thailand and Cambodia
Stems short and thick. Leaves 20–30cm(8–12in) long, fleshy, semi-cylindrical, deeply grooved on their upper sides. Flowers to 5cm(2in) across, white flushed with rose-pink, the labellum amethyst to deep rose; 12 to 35 blooms are borne on more or less erect racemes to 75cm(2½ft) long, mainly in summer, but intermittently throughout the year.

V. coerulea Himalaya, Burma and Thailand
Stems to 1m(3ft) or more tall. Leaves to 25cm(10in) long, linear, channelled above, the tip irregularly toothed and leathery, yellow-green in colour. Flowers to 10cm(4in) wide, usually blue to lavender-blue, occasionally white or pink with a darker network of fine veining; the erect to arching racemes are up to 60cm(2ft) long and carry five to 15 blooms which open in autumn and winter.

V. cristata Himalaya
Stems to 20cm(8in) tall. Leaves to 15cm(6in) long, linear, channelled and recurved with a 3-toothed apex. Flowers fragrant, 5cm(2in) wide, fleshy in texture with narrow yellow-green to greenish-brown tepals and a buff to tawny-yellow labellum marked with deep blood-red; the erect racemes carry two to seven flowers and are shorter than the leaves, opening in spring and summer.

V. parishii (*Vandopsis parishii*) Burma, Thailand
Stems robust, 15–20cm(6–8in) tall. Leaves 15–25cm(6–10in) long, oblong-elliptic, the tips bilobed. Flowers fragrant, to 6cm(2½in) wide, fleshy, the tepals yellow to yellow-green spotted with reddish-brown, the labellum small, magenta, opening in summer and autumn.

V. × rothschildiana
Hybrid between *V. coerulea* and *V. sanderiana*, similar to *V. coerulea* but with flowers to 13cm(5in) wide with a 2.5cm(1in) long lip, blue to purplish-blue with a network of darker veins. Racemes of ten to 20 blooms, opening from autumn into winter.

Vanda × rothschildiana

V. sanderiana Philippines
Stems erect and leafy, to 60cm(2ft) long. Leaves 25–40cm(10–16in) long, strap-shaped, the tip often with two or three small teeth. Flowers 8–13cm(3–5in) across, the tepals overlapping producing a flat, circular bloom; the upper sepal and petals are mauve to pink, the other sepals buff to yellow with red-brown veins, the small labellum dull red. They are carried in 30cm(1ft) racemes made up of six to 15 fragrant blooms opening in autumn.

V. suavis See *V. tricolor suavis.*

V. teres Thailand to Burma, E. Himalaya
Stems long and branched, in the wild growing through and over bushes, reaching 4m(13ft) in length, but in a pot rarely more than 1.4m(4½ft) and needing support. Leaves 10–20cm(4–8in) long, cylindrical. Flowers 8–10cm(3–4in) wide, white flushed with rose-purple, the labellum tawny-yellow spotted and shaded with magenta; three to six carried together on racemes to 30cm(1ft) long mostly in summer.

V. tricolor Java, Bali
Branched stems to 2m(6½ft). Leaves 30–45cm(1–1½ft) long, strap-shaped, leathery and channelled, the tip bilobed. Flowers 6–8cm(2½–3in) wide, pale yellow with reddish-brown spotting and a white

Vanda tricolor

labellum with the central lobe magenta, fragrant; seven to 12 carried on an arching scape to 30cm(1ft) long, opening in summer and winter.
V.t. suavis
Larger and more freely borne flowers which are white with less heavy magenta to purplish spotting.

VANDOPSIS See VANDA

VANILLA

Orchidaceae

Origin: Circum-tropical. A genus of 90 to 120 species of climbing orchids, best known as the source of the vanilla flavouring. The species described here climb by aerial roots and have alternate, thick-textured leaves. The relatively large flowers have similar sepals and petals and a tubular labellum. Although not highly decorative as orchids go, they are well worth trying in the conservatory. Propagate by cuttings of stem sections with three to five leaves in summer. *Vanilla* derives from the Spanish *vainilla*, a small pod; unlike most orchids, the seed capsules resemble bean pods.

Species cultivated
V. planifolia Mexico, Southern Florida, West Indies, Central and Northern South America
Stems to 10m(33ft) or more in the wild, sparingly branched but easily kept to 2–3m(6½–10ft) with careful pruning. Leaves oblong-elliptic to lanceolate, slender-pointed, 8–20cm(3–8in) long, sub-lustrous, dark green. Flowers about 5cm(2in) long, greenish-yellow, in short axillary racemes, intermittently all the year if warm enough. Pods cylindrical, obscurely 3-angled, 10–25cm(4–10in) in length. *V.p.* 'Variegata' has the leaves striped and boldly margined white. The pods of this species provide the main source of natural vanilla.
V. pompona Mexico, Central America to Brazil
Similar to *V. planifolia*, but the stems more robust and the leaves longer. Flowers 8cm(3in) or more long, the labellum white or orange. Pods are shorter but considerably thicker.

Vanilla pomponia

VEITCHIA

Palmae

Origin: Philippines, New Caledonia, New Hebrides, Fiji. A genus of about 18 species of palms. Propagate by seed in spring. *Veitchia* honours James Veitch (1815–1869) and his son John Gould Veitch (1839–1870), famous nurseryman of Chelsea and Exeter.

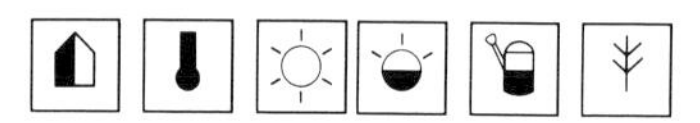

Species cultivated
V. merrillii Christmas palm Palawan Islands
In the wild, a small tree 5m(16ft) or more, but fairly slow-growing as a containerized specimen. Leaf bases tightly overlapping to form a cylindrical crown shaft; leaves 90–180cm(3–6ft) long, pinnate, ascending and strongly arching; leaflets linear, numerous, which are also ascending and arching, giving the leaf a distinctive stance. Young plants have flatter leaves. Large panicles of small flowers are carried just below the crown shaft and are followed by bright red, egg-shaped fruits 2.5–4cm(1–1½in) long, but only on sizeable plants.

VELLOZIA

Velloziaceae

Origin: Tropical America, Africa, Malagasy. A genus of about 100 species of woody based evergreen perennials, shrubs and small trees, one of which makes a modestly attractive pot plant. Propagate by seed, division or cuttings in spring. *Vellozia* is named for Jose Marianno Conceicao Velloso (1742–1811), a monk living in Brazil.

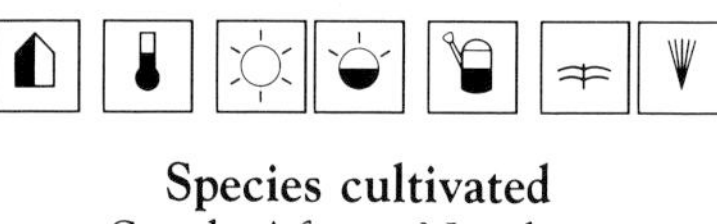

Species cultivated
V. elegans South Africa, Natal
Clump to mat-forming perennial, 15–30cm(6–12in) high, eventually by several times as wide. Leaves lanceolate to linear, 10–20cm(4–8in) long, sharply keeled, leathery-textured, dark, fairly glossy green. Flowers white, 6-tepalled, starry, about 3cm(1¼in) wide, solitary on slender wiry

stems above the leaves, appearing intermittently for most of the year if warm enough.

VELTHEIMIA

Liliaceae

Origin: South Africa. A genus of two species of bulbous plants with basal rosettes of fleshy, bright green, oblong-lanceolate leaves and dense racemes of tubular flowers on stout, erect scapes very reminiscent of those of a red-hot poker. They make handsome perennial pot plants. Pot in summer, the deciduous *V. capensis* needing a dry resting period in summer once the leaves begin to turn yellow in late spring. The evergreen *V. bracteata* should be kept moist throughout the year. Propagate by separating offsets when re-potting or by leaf cuttings taken as the leaves reach their full size. *Veltheimia* was named for August Ferdinand von Veltheim (1741–1801), a German Patron of botany.

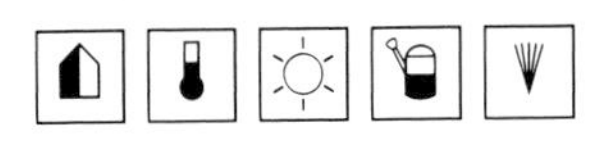

Species cultivated

V. bracteata (*V. undulata*, *V. viridiflora*, sometimes erroneously grown in gardens as *V. capensis*)
Bulbs broadly ovoid to rounded. Leaves evergreen, to 38cm(15in) long by 10cm(4in) wide, broadly strap-shaped with wavy margins, glossy, up to twelve arising from each plant. Flowers 3–4cm(1¼–1½in) long, pinkish-purple, faintly speckled with yellow, borne in dense clusters on 30cm(1ft) tall, glaucous green, purplish-mottled stems in spring. 'Rosalba' has white unspeckled flowers flushed rosy-pink at the base.

V. capensis (*V. glauca*)
Bulbs narrowly ovoid. Leaves deciduous, to 30cm(1ft) long and 2.5cm(1in) wide, lanceolate, glaucous-green, with wavy margins. Flowers 2.5cm(1in) long, pale pink, tipped with green. Yellow and red-flowered forms are recorded in the wild but do not seem to be in cultivation. They are carried on 30cm(12in) tall glaucous green, purple-spotted stems in spring. There is much confusion between the species, *V. bracteata* being the most common in cultivation though often under the wrong name.

V. glauca See *V. capensis*.

V. undulata (*V. viridiflora*) See *V. bracteata*.

× VENIDIO-ARCTOTIS

Compositae

Origin: A group of hybrid cultivars between *Venidium fastuosum* and *Arctotis breviscapa* or *A. grandis*, including a number of named forms. They are freely branched erect plant to 50cm(20in) tall. Leaves oblong-lanceolate, lobed, grey-green above and white-felted beneath. Flowers 6–8cm(2½–3in) across, daisy-like, in shades of ivory-white, cream,

BELOW *Veltheimia bracteata*
LEFT × *Venidio-Arctotis* 'Mahogany' cultivar

yellow, orange, bronze, red and purple, often with a darker patch near the base of each floret. Propagate by cuttings in late summer or spring. The name *Venidio-Arctotis* is derived from those of the parent species.

  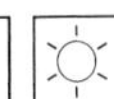 

VERBENA

Verbenaceae

Origin: Mainly North and South America, a few in Asia and Europe. A genus of about 250 species of annuals, perennials, shrubs and sub-shrubs, with opposite pairs or whorls of toothed or lobed leaves and clusters of small, tubular 5-lobed flowers. They can be grown in a conservatory or on a sunny window sill. Propagate by seed in spring or by cuttings in late spring or late summer (including the perennial sorts which are best raised afresh annually). *Verbena* is the old Latin name for *Verbena officinalis*, the medicinal herb vervain.

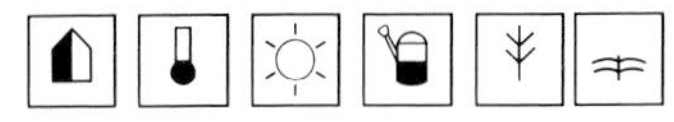

Species cultivated

V. × hybrida
Hybrid involving *V. peruviana* and perhaps three other species. It is a perennial, but is normally grown as an annual. Stems prostrate or ascending, reaching 20–30cm(8–12in) in height. Leaves 5–10cm(2–4in) long, oblong-lanceolate to ovate, lobed. Flowers 1–2cm(⅜–¾in) across, grouped in dense, head-like spikes to 6cm(2½in) across, in

shades of red, purple, mauve, blue, pink, yellow and white, fragrant.

Single colour cultivars and mixed seed strains are usually available.

V. peruviana (*V. chamaedrifolia* and *V. chamaedrioides* in some gardens) S. Brazil to Argentina
Prostrate perennial, the stem tips turning up to a height of 10cm(4in). Leaves to 5cm(2in) long, oblong-lanceolate to ovate, toothed. Flowers 1cm($\frac{3}{8}$in) wide in small head-like spikes, bright scarlet, opening in summer and autumn.

V. rigida (*V. venosa*) S. Brazil to Argentina
Erect perennial, 30–60cm(1–2ft) tall. Leaves 5–8cm(2–3in) long, oblong, stiff in texture and roughly hairy. Flowers 5mm($\frac{1}{5}$in) wide, purple to red-purple, in dense, head-like spikes in summer.

V. triphylla See *Lippia triphylla*.

V. venosa See *V. rigida*.

VERONICA See HEBE

RIGHT *Viburnum tinus*
ABOVE RIGHT *Verbena* × *hybrida* 'Pink Beauty'

VESTIA

Solanaceae

Origin: Chile. A genus of one species of evergreen shrub allied to *Cestrum* and equally vigorous and floriferous. It makes a fine tub plant for the unheated or just frost-proof conservatory. Propagate by seed in spring or by cuttings in late summer. *Vestia* is named for Professor Lorenz Chrysanth von Vest (1776–1840) of Graz, Austria.

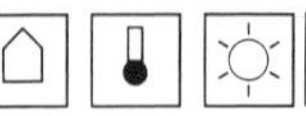 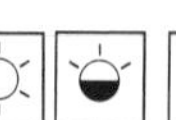 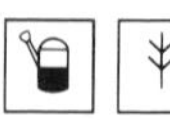 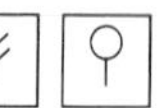

Species cultivated

V foetida (*V. lycoides*)
Erect habit, to 2m(6$\frac{1}{2}$ft) or more. Stems downy, usually branching quite freely. Leaves to 5cm(2in) long, alternate, oblong-elliptic to narrowly obovate, slightly fleshy, dark sub-lustrous green. Flowers tubular, 3cm(1$\frac{1}{4}$in) long, with five short, spreading to recurved petal lobes, pale yellow, from bell-shaped calyces, in late spring to summer. From a distance, a plant in bloom has something of the appearance of a yellow-flowered fuchsia.

VIBURNUM

Caprifoliaceae

Origin: Chiefly northern hemisphere, but also occurring in South America and Java. A genus of about 200 species of evergreen and deciduous shrubs, a few making small trees. Their leaves are entire and in opposite pairs and the small tubular 5-lobed flowers are often in flattish, umbel-like cymes or panicles. The three hardy species described make good evergreen shrubs for the conservatory or a large room. Propagate by seed when ripe or in spring (seeds which have become dry may take eighteen months to germinate), also by layering in late winter or by cuttings with bottom heat in late summer.

Viburnum is the Latin name for the wayfaring tree (*V. lantana*).

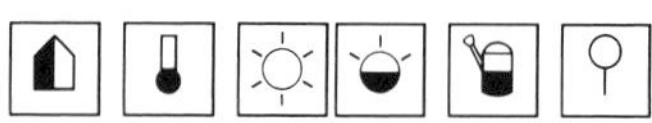

Species cultivated

V. cinnamomifolium China
Evergreen shrub to 3m(10ft) or more. Leaves 13cm(5in) long, elliptic to oblong, with three prominent veins, leathery and glossy dark green. Flowers 5mm($\frac{1}{5}$in) wide, off-white, in flattish clusters 10–15cm(4–6in) wide with a slight honey scent, opening in summer. Fruits 5mm($\frac{1}{5}$in) long, ovoid, glossy blue-black.

V. odoratissimum Japan, India
Evergreen shrub to 3m(10ft) or more, partly deciduous in cold conditions. Leaves 8–20cm(3–8in) long, elliptic to oblong, dark glossy green. Flowers small, white, slightly fragrant, carried in pyramidal clusters to 10cm(4in) long in summer. Fruits 5mm($\frac{1}{5}$in) long, red at first, becoming black.

ABOVE *Viola hederacea*
ABOVE LEFT *Viola* × *wittrockiana* 'Irish Molly'

V. tinus Laurustinus S.E. Europe
Evergreen shrub to 3m(10ft) tall, much more than this in open ground, sometimes making a small tree. Leaves 4–9cm(1½–3½in) long, ovate-oblong, shiny dark green above, paler beneath. Flowers 5–6mm(⅕–¼in) wide, pink in bud, opening to white, carried in flattened cymes 5–10cm(2–4in) wide from late autumn to spring. Frost hardy and a good subject for the unheated conservatory.

'Eve Price' is smaller and neater growing, with deeper pink buds; 'Variegatum' has creamy-yellow marked leaves.

VINCA See CATHARANTHUS

VIOLA

Violaceae

Origin: Cosmopolitan, the chief centres of distribution being the north temperate zone and the Andes. A genus of 500 species of annuals, perennials and sub-shrubs of which the one species and hybrid listed below are suitable for the conservatory or home. *V. hederacea* can be grown as ground cover in conservatory beds, in pans or hanging baskets; propagate by division in spring. *V* × *wittrockiana* is a useful short term flowering plant for the cool conservatory; propagate by seed, sowing in late summer for winter and spring flowering, in spring for summer to autumn flowers, or by basal cuttings at the same times. *Viola* is the Latin name for a violet, perhaps originally from the Greek.

Species cultivated

V. hederacea Ivy-leaved violet S.E. Australia
Stoloniferous, mat-forming perennial 30–60cm(1–2ft) wide. Leaves 2.5–4cm(1–1½in) wide, kidney-shaped, light green. Flowers to 2cm(¾in) wide, a typical violet but with broader petals, the upper two the largest, white with the inner third dark purple, opening chiefly in spring, but producing some flowers throughout the year.

V.* × *wittrockiana (*V.* × *hortensis*)
Garden pansy
A group of cultivars and seed strains derived by hybridization from *V. lutea*, *V. tricolor* and *V. altaica* and, like those species, a short-lived perennial. It is usually grown as an annual, to 20cm(8in) in height, with erect to decumbent, rather soft stems. Leaves to 8cm(3in) long, ovate, crenate. Flowers 4–12cm(1½–4¾in) across in a wide range of colours usually with a black blotch at the centre fancifully making a 'funny face'. Colours range from blue, purple, red, orange and yellow to buffs and white. Seed strains are divided into the Hiemalis or Winter-flowering group, Spring- to Early summer-flowering and those opening from summer onwards; they are also readily available as single colours or mixtures.

VITIS See CYPHOSTEMMA, RHOICISSUS & TETRASTIGMA

VITTARIA

Vittariaceae (*Polypodiaceae*)

Origin: Tropics and sub-tropics. A genus of 50 species of highly distinctive but not over decorative epiphytic ferns with linear, almost grass-like but firm-textured fronds. One is good for a hanging basket. Propagate by division or by spores in spring. *Vittaria* derives from the Latin *vitta*, a band or ribbon, descriptive of the fronds.

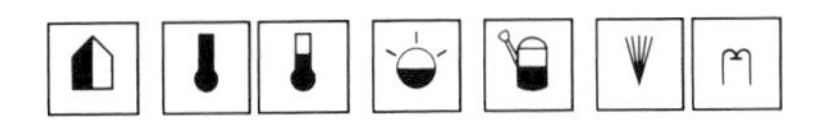

Vittaria lineata

Species cultivated

V. lineata Florida ribbon fern Tropical America including Florida
Rhizomes short and clustered, eventually forming dense clumps. Leaves crowded, erect at first then drooping, 30–90cm(1–3ft) in length by 6mm(¼in) wide, glossy green above; sori form a continuous line near the margins of each fertile frond.

VRIESEA

Bromeliaceae

Origin: Tropical America. A genus of 190 to 245 species, depending on the classification followed, of epiphytic perennials. They have strap-shaped, leathery leaves in rosettes, which are overlapping at the base, arching stiffly outwards, and often striped, netted or marbled with darker shades. The flowers are 3-petalled and rather small, borne within coloured, boat-shaped bracts making up simple or branched spikes. They can be grown either in pots or wired to a tree branch. Propagate by detaching well rooted offsets in summer. *Vriesea* was named for the Dutch botanist Willem Hendrick de Vriese (1806–1862) Professor at Leiden and Amsterdam.

Species cultivated

V. botafogensis See *V. saundersii.*

V. carinata Lobster claws Brazil
Leaves to 20cm(8in) long, pale green, unmarked. Flowers 5cm(2in) long, yellow within bracts which are scarlet below and yellow above and borne in a broad, flattened spike on a stem to 30cm(1ft) tall.

ABOVE *Vriesea carinata*
RIGHT *Vriesea hieroglyphica*

V. fenestralis Brazil
Rosette solitary, at least until after flowering. Leaves broadly strap-shaped to 40cm(16in) or more in length, spreading and recurved at the tips, yellow-green with a conspicuous network of dark green veins and cross-lines above, purplish-marked beneath. Flowers 4.5cm(1¾in) long, yellow-green or sulphur-yellow, opening at night and very fragrant, arranged in a green bracted, flattened spike to 30cm(1ft) in length.

V. fosteriana Brazil
Leaves numerous, broadly strap- or tongue-shaped, arching, 45–60cm(1½–2ft) in length, dark to yellow-green with irregular maroon cross-banding. Flowers 4cm(1½in) long, greenish-yellow tipped brownish-red, surrounded by broad, red-brown spotted green sepals, in a distichous spike to 25cm(10in) long on an erect stem to 90cm(3ft) tall. The flowers open at night and have a smell of rotten fruit; in the wild they are pollinated by bats.

V. gigantea (*V. tesselata*) Brazil
Leaves numerous forming a wide funnel-shaped rosette to 90cm(3ft) across; they are about 60cm(2ft) long, strap-shaped, dark green with a yellowish checkered pattern. Flowers 4cm(1½in) long, greenish-yellow with green bracts, in a well branched inflorescence 90–150cm(3–5ft) in height. An attractive foliage plant with the appeal of *V. fenestralis.*

V. hieroglyphica King of the bromeliads Brazil
Leaves to 45cm(1½ft) long and 10cm(4in) wide, yellow-green above marked with brownish-green, horizontal stripes with feathered edges fancifully like hieroglyphics; under surface purplish. Flowers to 3cm(1¼in) long, dull yellow on stems more than 1m(3ft) tall. A striking foliage plant.

V. incurvata Brazil
Leaves many, forming a rosette 40–60cm(16–24in) across; each leaf about 40cm(16in) long, strap-shaped, arching, pointed, bright lustrous green. Flowers 5cm(2in) long, yellow with green tips, protruding from overlapping, keeled, yellow-margined red bracts which form a blade-like spike up to 25cm(10in) long on a somewhat shorter stem. A showy species akin to the familiar flaming sword (*V. splendens*). Sometimes confused with *V.* × *poelmannii* (see under Hybrids below).

V. psittacina Painted feather Brazil, Paraguay
Rosettes loose, 50–60cm(20–24in) wide when well grown; leaves sword-shaped, to 40cm(16in) long, yellow with green tips, subtended by red and green bracts arranged in a distichous spike 20–30cm(8–12in) long on a stem to twice as tall.

V. rodigasiana Brazil
Leaves to 30cm(1ft) long, green above with brown suffusion near the base and spotted with brown beneath. Flowers to 3.5cm(1½in) long, pale green within waxy yellow bracts to 2.5cm(1in) long, borne in a branched spike on stems to 45cm(1½ft) tall.

V. saundersii (*V. botafogensis*) Brazil
Leaves numerous, arching, strap-shaped, to

45cm($1\frac{1}{2}$ft) long, grey-green with irregular red-brown spots, particularly beneath. Flowers 5cm(2in) long, yellow, subtended by greenish-yellow bracts, well spaced in a stiff panicle 20–30cm(8–12in) long, carried well above the leaves.
V. splendens Flaming sword Guyana, Trinidad, Venezuela
Leaves to 40cm(16in) long, dull green, with broad horizontal brown bands. Flowers to 6cm($2\frac{1}{2}$in) long, yellow, within flattened scarlet bracts which make up a 45cm($1\frac{1}{2}$ft) long, sword-like spike carried on a slightly shorter stem. 'Major' is brighter coloured and more robust.
V. tesselata See *V. gigantea.*

Hybrids

Vriesea is perhaps the most hybridized of all the genera in the *Bromeliaceae*. The first hybrids were produced about 100 years ago and there has been a resurgence of interest during the past 40 years. On the whole, the hybrid cultivars are just as decorative as the species, in some cases more so, and they tend to be easier to grow as house plants. The following are worth looking out for:
V. × erecta (*V. × poelmannii × rex*) Red feather, glossy light green foliage, deep red waxy floral bracts and yellow flowers.
V. × 'Favourite' (*V. ensiformis* hybrid), lustrous deep green leaves, maroon floral bracts and yellow flowers.
V. × mariae (*V. carinata × barilletii*) Painted feather, light green, pink-tinted foliage, salmon-red and yellow bracts and yellow flowers.
V. × poelmannii (*V. gloriosa × vangeertii*), light green leaves, flattened spikes of crimson bracts and yellow flowers.
V. × retroflexa (*V. psittacina × simplex*), small rosette of waxy pale green leaves, flowers and bracts yellow in a pendent spike.

× VUYLSTEKEARA

Orchidaceae

Origin: A group of hybrid cultivars derived from crossing together species of *Cochlioda*, *Miltonia* and *Odontoglossum*. They have clustered pseudobulbs and lanceolate to linear leaves. The blooms, which generally favour *Odontoglossum*, have similar rounded-ovate tepals and a larger, flattened, waved labellum.

Vuylstekeara was named for Charles Vuylsteke (1844–1927), a Belgian nurseryman who specialized in growing orchids.

× *Vuylstekeara* Edna 'Stamperland'

Cultivars

× *V.* Cambria *'Plush'* has flowers 7–10cm($2\frac{3}{4}$-4in) across, the tepals dark red, the labellum white with red markings; they are carried in racemes of three to nine in winter and spring.
× *V.* Edna *'Stamperland'* is similar with carmine-red sepals and a paler, strongly waved lip.

W

WASHINGTONIA

Palmae

Origin: Southern California and Arizona, northern Baja California. A genus of two species of large palms which, when young, make handsome foliage container plants. They have costapalmate leaves (halfway between pinnate and palmate) with spiny petioles. Trunks are seldom formed on containerized specimens, but in the wild they attain 26–32m(80–100ft) in height clothed with great columnar petticoats of dead leaves. The whitish flowers are in large axillary panicles which arch out among the leaves, exceeding them in length. Later when bearing the black berry-like fruits, they hang down. Propagate by seed in spring. *Washingtonia* commemorates George Washington (1732–1799), the first President of the USA.

  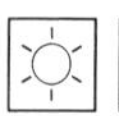 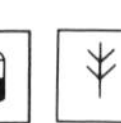

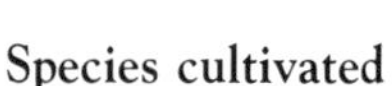

Species cultivated

W. filifera (*Pritchardia filifera*) Desert fan palm, Petticoat palm California, Arizona
Leaf stalks green, the blade grey-green, 1–2m(3–6½ft) long overall in containers. Leaf lobes linear, more or less flaccid with filamentous margins. When mature it is more robust than the next species.

W. gracilis See *W. robusta.*

W. robusta (*Pritchardia robusta, W. gracilis*)
Baja California
Leaf stalks and blades bright green, 1–2m(3–6ft) long in containers. Leaf lobes linear, more or less firm-textured, with few or no filaments on their margins. Despite its name, when mature this palm is more slender than *W. filifera.*

Watsonia beatricis

Washingtonia filifera

WATSONIA

Iridaceae

Origin: Southern Africa. A genus of 60 to 70 species of clump-forming cormous perennials. They have sword-shaped leaves growing in a fan-like formation and 6-lobed, funnel-shaped flowers with curved tubes, borne in spikes largely above the leaves on long slender stems. In appearance, midway between gladiolus and montbretia. They are best grown in a conservatory, but can be brought into the home for flowering. Propagate by division in spring, when re-potting, or by seed in spring. *Watsonia* was named for Sir William Watson (1715–1787), a British physician and scientist.

 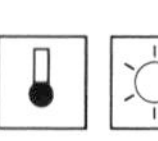   

Species cultivated

W. ardernei
Leaves 60cm(2ft) long, deciduous, erect, narrowly strap-shaped. Flowers about 4cm(1½in) long, white, on branched spikes to 1m(3ft) in height or more, in early summer.

W. beatricis
Leaves 60–75cm(2–2½ft) long, evergreen, sword-shaped, somewhat lax. Flowers 5cm(2in) long, orange-red, on stems 1–1.2m(3–4ft) tall from late summer to autumn.

W. marginata
Leaves to 75cm(2½ft) long, deciduous, narrowly strap-shaped. Flowers 3–4.5cm(1¼–1¾in) long, rose-pink, fragrant, on stems 1–1.5m(3–5ft) tall in early summer.

W. pyramidata (*W. rosea*)
Leaves to 75cm(2½ft) long, deciduous, narrowly sword-shaped. Flowers 8–10cm(3–4in) long, deep pink to rose-red, on stems 1.2–1.5m(4–5ft) tall in summer.

W. rosea See *W. pyramidata.*

W. versfeldii
Leaves to 90cm(3ft) long, deciduous. Flowers to 5cm(2in) or more long, rose-pink, on stems to 1.5m(5ft) high in early summer.

WATTAKAKA See DREGEA

WESTRINGIA

Labiatae

Origin: Australia. A genus of about 20 species of evergreen shrubs, some of which superficially resemble rosemary. One species is widely grown and makes a pleasing pot or tub plant for the conservatory and can be used in the home for short periods. Propagate by seed in spring or by cuttings in summer. *Westringia* is named for Johan Peter Westring (1753–1833), Doctor to the King of Sweden.

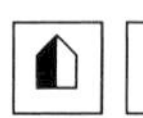 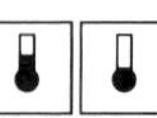 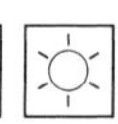  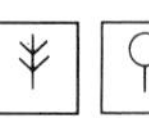

Species cultivated

W. rosmariniformis (*W. fruticosa*) Australian rosemary New South Wales (mainly coastal)
Erect, bushy species 90–150cm(3–5ft) in height. Leaves crowded, usually in whorls of four, linear to lanceolate, 1.5–2.5cm($\frac{5}{8}$–1in) long, leathery-textured, dark green above, white-felted beneath. Flowers tubular, with five prominent oblong petal lobes, the upper two longest, white to palest blue, borne from the upper leaf axils in late spring to early autumn, sometimes later.

WIGGINSIA

Cactaceae

Origin: Tropical South America. A genus of 13 species of globose to ovoid stemmed cacti closely related to *Notocactus* and having the same appeal. The species described here are easily grown indoors. Propagate by seed in spring. *Wigginsia* is named for Dr. Ira L. Wiggins (*b.*1899), an American taxonomic botanist.

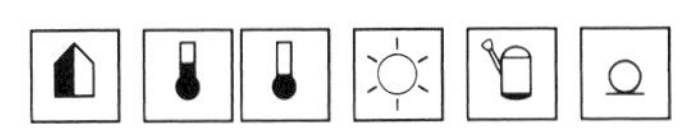

Species cultivated

W. corynodes (*Malacocarpus corynodes*)
S. Brazil, Uruguay, N. Argentina
Stems globular at first, then shortly cylindrical, to 20cm(8in) tall, woolly at the top. Ribs 13 to 16, narrow, notched, bearing round white-woolly areoles. Radial spines seven to 12, awl-shaped, yellow with darker zones, up to 2cm($\frac{3}{4}$in) long; central spines one or missing, slightly longer. Flowers funnel-shaped, 5cm(2in) wide, canary-yellow.

W. erinaceus S. Brazil, Uruguay, N. Argentina
Stems globular to shortly cylindrical, to 15cm(6in) tall, woolly at the top. Ribs 15 to 20, spiralling, blunt, notched, with white-woolly areoles. Radial spines six to eight, brownish, 1–2cm($\frac{3}{8}$–$\frac{3}{4}$in) long; one central spine, about 2.5cm(1in) long. Flowers funnel-shaped, to 7cm(2$\frac{3}{4}$in) wide, bright yellow.

WILCOXIA

Cactaceae

Origin: South-west USA, Mexico. A genus of six to eight species of cacti having tuberous roots and more or less slender trailing stems with small bristle-like spines and comparatively large, funnel-shaped flowers. They provide something different to the cactus scene and are worthwhile additions to a collection of succulent plants. Propagate by seed in spring or by cuttings in summer. *Wilcoxia* honours Brigadier General Timothy E. Wilcox (*d.*1932), an American enthusiast of succulent plants.

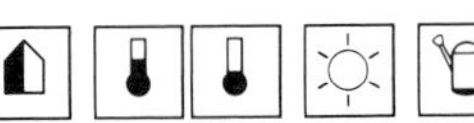

ABOVE *Westringia rosmariniformis*
LEFT *Wigginsia corynodes*
BELOW LEFT *Wilcoxia albiflora*

Species cultivated

W. albiflora Mexico
Stems up to 20cm(8in) long by about 6mm($\frac{1}{4}$in) thick, with several shallow ribs, grey-green. Spines ten to 12, tiny, whitish. Flowers 4–6cm(1$\frac{1}{2}$–2$\frac{1}{2}$in) wide, white.

W. poselgeri (*Echinocereus poselgeri, E. tuberosus*)
S. Texas, Mexico
Stems 30–90cm(1–3ft) in length by 1–1.5cm($\frac{3}{8}$–$\frac{5}{8}$in) thick, with about eight low ribs, dark green. Spines ten to 13, white to greyish, 3–6mm($\frac{1}{8}$–$\frac{1}{4}$in) long. Flowers to 5cm(2in) wide, deep pink, lasting for several days. Sometimes called dahlia cactus owing to the likeness of the tuberous roots.

WINTERA See DRIMYS
WINTERANIA See DRIMYS

WOODWARDIA

Blechnaceae (*Polypodiaceae*)

Origin: Northern hemisphere. A genus of about 12 species of ferns with erect or creeping rhizomes from which the fronds arise. They are pinnately lobed or twice pinnate and carry their sporangia in rows or chains parallel to their central veins, hence their vernacular name. Handsome ferns for an unheated conservatory, tolerating cold and shade. Propagate by division in spring. *Woodwardia* was named for Thomas Jenkinson Woodward (1745–1820), a British botanist.

Woodwardia radicans

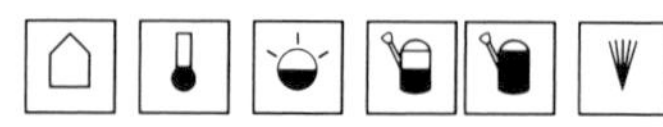

Species cultivated

W. orientalis Temperate eastern Asia
Very much like *W. radicans*, but with narrower pinnae and many small bulbils (buds or gemmae) scattered over the fronds. Cool.

W. radicans Southern Europe to Canary Islands
Rhizomes robust, eventually forming wide clumps. Fronds 1–2m(3–6½ft) long, bipinnatifid, the blade triangular to ovate-lanceolate in outline. Main pinnae to 30cm(1ft) in length, broadly lanceolate to narrowly ovate, pinnately lobed almost to the base, each lobe finely toothed and curving gently forwards. At the frond tips large bulbils (buds or gemmae) are borne and if pegged down to the soil soon root and grow into young plants. Cool.

W. virginica Chain fern North America
Fronds to 1m(3ft) long, blades 60cm(2ft), pinnate, the pinnae deeply pinnately lobed. Spores in two chain-like rows, one on either side of the mid-vein of each lobe. Hardy.

WORSLEYA See HIPPEASTRUM

XYZ

XANTHERANTHEMUM See CHAMAERANTHEMUM

XANTHORRHOEA

Xanthorrhoeaceae (*Liliaceae*)
Grass trees/Blackboy

Origin: Australia. A genus of 12 to 15 species of woody based perennials or small trees. From a distance the tree types resemble large tufts of coarse grass atop a short, dark to blackish palm-like trunk. When in flower, which they seldom do under glass, they produce tall candle-like spikes of small 6-tepalled, white flowers followed by explosive seed capsules. The species described here provide foliage pot plants with a difference. They are best in the conservatory, but can be tried in the home at least in the short term. Propagate from seed in spring. *Xanthorrhoea* derives from the Greek *xanthos*, yellow and *rheo*, to flow, from the yellow resin exuded.

Species cultivated

X. arborea Botany Bay gum Queensland, New South Wales
Trunk on mature specimens to 3m(10ft) or more in height, but takes many years to form and then extends in height very slowly. Leaves dense, linear, flattened, 60–120cm(2–4ft) long by 4–6mm($\frac{1}{6}$–$\frac{1}{4}$in) wide, erect then arching and hanging down. Flowers spikes to 120cm(4ft) in height.

X. australis Blackboy, Australian grass tree New South Wales, Victoria, Tasmania
Similar to *X. arborea* in overall appearance, but leaves narrower (to 3mm[$\frac{1}{8}$in] wide) and the trunk shorter. Flower spikes wider and more conspicuous in bloom.

X. preissii Common blackboy Western Australia
Similar to *X. arborea* with leaves like those of *X. australis*. Vigorous, old specimens sometimes have trunks with one to several branches.

Xanthorrhoea australis

XANTHOSMA

Araceae

Origin: Mexico, tropical America including the West Indies. A genus of 40 to 45 species of tuberous and rhizomatous perennials resembling *Alocasia* and *Colocasia*. Several are grown in the tropics for their edible tubers and leaves. Those described below make handsome container plants, having usually arrow- to spearhead-shaped leaves on long stalks from ground level and typically arum-like spathes bearing separate zones of tiny petalless male and female flowers. Propagate by division in spring or early summer or when re-potting. *Xanthosma* derives from the Greek *xanthos*, yellow and *soma*, a body, referring to the inner tissues of the tubers and rhizomes of some species.

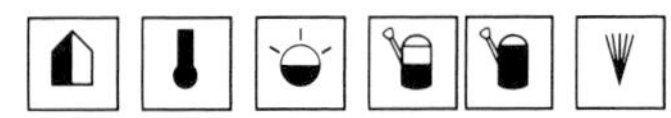

Species cultivated

X. lindenii Indian kale Colombia
Tuberous-rooted. Leaf blades arrow-shaped, to 30cm(1ft) in length, deep green with a bold pattern of broad white veins. Spathes about 13cm(5in) long, green and white.

X. violaceum (*Alocasia violacea*) Blue taro, Yautia West Indies
Leaf blades to 45cm(1$\frac{1}{2}$ft) long on stems of twice this length, triangular-ovate to arrow-shaped, deep green with purple veins and stalks. Spathes to 30cm(1ft) long, glaucous-purple. In its homeland, it is grown for its high quality starchy tubers.

Xanthosma lindenii

ABOVE *Yucca whipplei*
RIGHT *Yucca* 'Tricolor'

YUCCA

Agavaceae (Liliaceae)

Origin: Southern USA, Mexico and the West Indies. A genus of about 40 species of evergreen trees and shrubs, some of which are stemless while others develop with age erect, sturdy stems sometimes with a few branches. Their leaves are linear and pointed, forming rosettes. The white to cream bell-shaped flowers are carried in large panicles on sturdy stems usually overtopping the leaves. In the wild, yuccas are pollinated by the small moth *Pronuba* which lays its eggs in the ovary of a flower and at the same time purposefully pollinates it. Although the caterpillars feed on the developing seeds, enough survive to fall and germinate. Plants are suitable for a large conservatory. Propagate from seed when available or by removing suckers in spring.

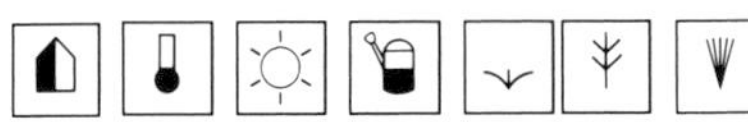

Species cultivated

Y. aloifolia Southern USA, Mexico, West Indies
In the wild, a small tree, but fairly slow growing and easily maintained at 1–2m(3–6½ft) in height. Stems robust, erect, sparingly branched or unbranched. Leaves sword-shaped, 60–75cm(2–2½ft) long,

sharp-pointed and rigid, the margins finely toothed. Flowers 8–10cm(3–4in) wide, white or purple-flushed, in panicles to 60cm(2ft) long, but rarely in containers. 'Marginata' has the leaves yellow-margined; 'Tricolor' ('Quadricolor') bears a central cream or white leaf stripe which is pink-suffused when young; 'Variegata' is probably a misnaming of 'Marginata'.

Y.a. draconis
More freely branching stem and curved, pliable leaves.

Y. brevifolia Joshua tree Western USA (California to Utah)
In the wild, a small- to medium-sized desert tree of singular appearance, rather like a caricature of a tree with enormously crooked branches terminating in bunches of green 'daggers'. Leaves narrow, hard, spine-tipped and minutely toothed, dark green, to 35cm(14in) long. Flowers green, greenish-yellow or cream, 7cm(2¾in) long, in panicles to 50cm(20in) long, but seldom if ever on containerized specimens. Best with light watering.

Y.b. brevifolia
Grows tall and starts to branch not less than 2–3m(6½–10ft) above the ground. An interesting slow-growing foliage plant to accompany cacti.

Y.b. herbertii
Rarely exceeds 5m(15ft) when mature and branches from ground level.

Y. elata Soap tree, Palmella Texas, Arizona, Mexico
When mature, a tree to 6m(20ft) tall, but easily kept at 2m(6½ft) or so in a container. Stem robust, erect, usually unbranched. Leaves to 90cm(3ft) long, linear, pliable, pale green, usually with whitish, thread-bearing margins. Flowers 5cm(2in) long, white to green, often pink-tinted, in large panicles or peduncles, 2–3m(6½–10ft) in height, but not on containerized specimens.

Y. elephantipes (*Y. gigantea*, *Y. guatemalensis*) Spineless yucca Mexico, much grown in Guatemala
In the wild, a tree to 10m(3ft), but when young makes a good foliage pot plant to 2m(6½ft) or more. Stems robust, erect, usually unbranched when young. Leaves 90–120cm(3–4ft) in length, linear, rough-edged but spineless. The large panicles of white or cream flowers are not produced on plants in containers. *Y.e.* 'Variegata' has creamy-white margined leaves.

Y. filamentosa South-east USA
Clump-forming and almost stemless. Leaves 35–50cm(14–20in) long in a pot, to 75cm(30in) in the open ground, somewhat glaucous, the margins with many curly threads. Flowers 5cm(2in) long, white to cream, in panicles 2–5m(6½–16ft) tall. *Y.f.* 'Variegata' has leaves with marginal yellow bands and stripes.

Y. flaccida South-east USA
Very similar to *Y. filamentosa*, but with more flexible, tapered leaves, the margins with straight threads.

Y. gigantea See *Y. elephantipes.*

Y. glauca USA (S. Dakota to New Mexico)
Clump-forming. Leaves 60cm(2ft) long by 1–1.5cm(⅜–⅝in) wide, linear with a few threads along the margins. Flowers to 6cm(2½in) long, greenish-cream often with a reddish-brown tinge and opening in summer on stems to 1m(3ft) tall.

Y. gloriosa Spanish dagger South-eastern USA
Trunk-forming, to 2m(6½ft) or more in height. Leaves to 75cm(2½ft) long, stiff, dark green. Flowers 6–10cm(2½–4in) long, white, often with a red tinge, in close panicles on stems about 2m(6ft) tall, opening in late summer.

Y. guatemalensis See *Y. elephantipes.*

Y. recurvifolia South-eastern USA
Similar to *Y. gloriosa*, but with less rigid leaves which recurve as they age. Flowers in loose, more branched panicles. *Y.r.* 'Variegata' has a central yellow stripe to its leaves.

Y. whipplei (*Hesperoyucca whipplei*) Our Lord's candle California, Mexico (Lower California)
Stem short, often less than 30cm(1ft), usually topped by a solitary rosette (but see varieties below). Leaves many, linear, rigid, spine-tipped, 45–60cm(1½–2ft) in length, glaucous, the margins finely toothed and brownish or yellowish. Flowers 6cm(2½in) long, pale cream, sometimes purple-flushed, fragrant, in a tapered, cylindrical panicle to 2m(6½ft) long on a stem of equal or greater height. A fine spectacle which can be achieved in a conservatory of sufficient height. After flowering, the plant dies.

Y.w. caespitosa
Many offsets produced from an early age; the plant does not die after blooming.

Y.w. intermedia
Between the two extremes above, each plant producing a few offsets after flowering. Plants take several years to build up a flowering sized rosette, but the actual inflorescence grows rapidly. There was, however, no truth in a newspaper article of 1952 which reported that the plant grew 30cm(1ft) per second! This was the result of a practical joke which Dr. Allbrecht of Taft College, California played on his colleagues using time lapse photography. It fooled a lot of people at the time.

ZAMIA

Zamiaceae (Cycadaceae)
Sago Palm

Origin: Tropical America, West Indies. A genus of about 40 species of woody-based evergreen perennials. They are members of the primitive cycad group, one of the earliest families of Gymnosperms (seed-bearing plants). Unlike most other genera of cycads, zamias either lack trunks or have only very short ones. The leathery, even hard-textured leaves are pinnate, and the separate-sexed cone-like flowering and seeding heads are carried on short stalks at the bases of the leaves and can be produced by quite small plants in pots. Propagate by seed in spring or summer. *Zamia* derives from

Yucca flaccida

zamiae, a mis-rendering of *azaniae*, pine cones, in the original texts of the Roman soldier, observer and writer Pliny the Elder (AD 23–79).

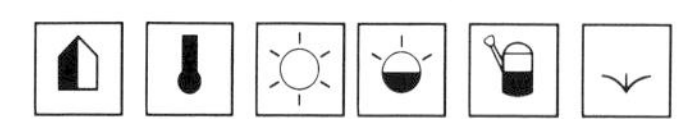

Species cultivated

Z. floridana Coontie, Seminole bread Florida
Trunk tuber-like, often partially or wholly underground. Leaves to 60cm(2ft) long including the stalk, leaflets dense, 28 to 44, linear, 9–15cm($3\frac{1}{2}$–6in) long, with sparingly toothed revolute margins, glossy deep green. Male flowering cones 4–6cm($1\frac{1}{2}$–$2\frac{1}{2}$in) long; females densely downy, up to 15cm(6in) in length. The macerated roots were used by the local indians as soap.

Z.f. portoricensis
Taller with widely-spaced leaflets. An intermediate habited plant is known; all are good house plants.

Z. furfuraceae See *Z. pumila.*

Z. pumila (*Z. furfuracea*) Florida arrowroot Florida, West Indies, Mexico
Trunk underground or partly exposed, to 15cm(6in) in height. Leaves 60–120cm(2–4ft) long including the prickly stalks, leaflets fairly close, four to 26, oblong-ovate, the margins toothed and revolute, glossy green above, brown-scurfy beneath. Male flowering cones to 10cm(4in) long, often in small clusters; females broader and somewhat longer. A good house plant.

Zantedeschia aethiopica 'Green Goddess'

ZANTEDESCHIA

Araceae

Callas, Arum lilies

Origin: Tropical to southern Africa. A genus of about six species of fleshy rhizomatous perennials with long-stalked, usually arrow-shaped or lanceolate leaves. The typical arum flowers are small and petalless, clustered on a spadix and surrounded by a large, conspicuous spathe. They are good pot plants and, with the exception of *Z. aethiopica*, are best given a dry rest in summer when the leaves begin to yellow after flowering. Propagate by seed in spring in warmth or by offsets separated at potting time. *Zantedeschia* was named for Francesco Zantedeschi (1797–1873), an Italian botanist.

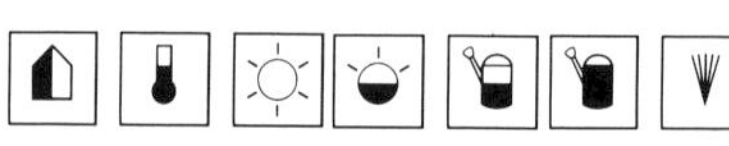

Species cultivated

Z. aethiopica (*Z. africana, Richardia africana*) Arum lily South Africa, widely naturalized in frost-free areas of the world
To 1m(3ft) tall. Leaves 25–45cm(10–18in) long, sagittate, shining deep green. Spathes 13–25cm(5–10in) long, white, surrounding a conspicuous yellow spadix. For cool conditions. *Z.a.* 'Childsiana' and 'Minor' are smaller forms; *Z. a.* 'Green Goddess' has green flushed and lined spathes.

Z. africana See *Z. aethiopica.*

Z. albomaculata (*Z. melanoleuca, Richardia albomaculata*) Spotted calla South Africa to Zambia
To 60cm(2ft) tall. Leaves 20–45cm(8–18in) long, narrowly triangular, shining dark green with translucent white spots. Spathes to 10cm(4in) long, creamy-white to greenish-yellow inside, with a dark red-purple blotch at the base. Forms with pinkish spathes are known.

Z. elliottiana (*Richardia elliottiana*) Yellow arum South Africa
60–90cm(2–3ft) tall. Leaves 15–30cm(6–12in) long, ovate-cordate, dark green with translucent white spotting. Spathes to 15cm(6in) long, bright golden-yellow inside, suffused with green on the outside. Temperate to Tropical.

Z. melanoleuca See *Z. albomaculata.*

Z. rehmannii (*Richardia rehmannii*) Pink/Red arum South Africa
40–60cm(16–24in) tall. Leaves 15–20cm(6–12in) long, lanceolate, tapering to a slender point. Spathes 7–13cm($2\frac{3}{4}$–5in) long, red-purple to deep red, sometimes light pinky-red. Hybrids between this and other species with spathes in a variety of colours are sometimes available.

ZEBRINA

Commelinaceae

Origin: Mexico, Guatemala. A genus of two to five species of decumbent or trailing evergreen perennials with alternate, ovate to oblong leaves and 3-

Zebrina pendula

petalled flowers fused at the base to form a tube and sitting between a pair of leaf-like bracts. It is a very rewarding pot or hanging basket plant and can be used as ground cover under larger plants in tubs or a conservatory border. Propagate by cuttings which root readily in warmth. *Zebrina* derives from the name of the animal zebra because of the plant's striped leaves.

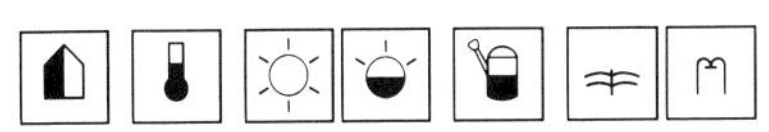

Species cultivated

Z. pendula Silvery inch plant, Wandering Jew Mexico
Stems prostrate or trailing to 60cm(2ft) or more in length. Leaves 4–8cm($1\frac{1}{2}$–3in) long, ovate-oblong, green, longitudinally striped with two broad silvery bands, bright red-purple on the reverse. Flowers about 1cm($\frac{3}{8}$in) wide, rose-purple from two unequal-sized bracts. *Z.p.* 'Discolor' has coppery-green leaves with a purplish suffusion overlaid by the two silvery bands. *Z.p.* 'Purpusii' (*Z. purpusii*, Bronze inch plant) is stronger growing, the leaves olive-green with a purplish flush and without the silvery bands (for strong purplish colour they must be kept in a good light); flowers lavender pink. *Z.p.* 'Quadricolor' has irregular pink-, red- and white-banded leaves.

ZEPHYRANTHES

Amaryllidaceae
Zephyr lilies

Origin: Warmer parts of the Americas. A genus of about 40 species of bulbous perennials with all-basal evergreen leaves and solitary tubular flowers, the six tepals opening widely. They make good pot plants for the conservatory and can be flowered in the home on a sunny window sill. They should not be re-potted until essential, as the congested clumps flower more freely. Keep dry in winter. Propagate by removing offsets or by seed in spring. *Zephyranthes* derives from the Greek *zephryos*, the west wind and *anthos*, a flower, a reference to their origins in the western hemisphere.

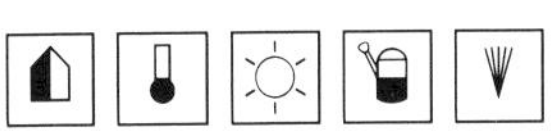

Species cultivated

Z. candida Flower of the western wind
Argentina, Uruguay
Leaves 20–30cm(8–12in) long, rush-like and

Zephyranthes candida

slightly fleshy. Flowers 3–5cm($1\frac{1}{4}$–2in) long, white, occasionally pinkish, shaded to green near the base; they open widely in summer and autumn on stems 10–20cm(4–8in) tall. Will withstand some frost.

Z. citrina South America
Leaves to 30cm(1ft) long, narrowly linear, channelled and blunt-tipped. Flowers 4–4.5cm($1\frac{1}{2}$–$1\frac{3}{4}$in) long, bright yellow-gold, green at the base; they are carried on stems 15–25cm(6–10in) tall in summer and autumn.

Z. grandiflora Mexico, Guatemala and West Indies
Leaves 25–40cm(10–16in) long, narrowly linear. Flowers 6–10cm($2\frac{1}{2}$–4in) long, pink to rosy-red, the petals to 2cm($\frac{3}{4}$in) across, like a small lily; they open in late summer and autumn on stems to 20cm(8in) tall. Much confused with *Z. rosea* which has similar coloured flowers, but with a maximum length of 4cm($1\frac{1}{2}$in).

Z. robusta See *Habranthus tubispathus*.

Z. tubispatha See *Habranthus tubispathus*.

ZINGIBER

Zingiberaceae
Gingers

Origin: Eastern Asia, Indo-Malaysia, Northern Australia. A genus of about 85 species of large perennials with tuber-like aromatic rhizomes, in some species the source of ginger. They form wide clumps when planted out, with erect unbranched stems, the non-flowering ones tall and leafy, the flowering ones short and leafless. The tubular flowers have three corolla lobes, the upper one usually broadest, and a large petal-like staminode which acts as a labellum. They are borne in the axils of large, firm-textured bracts which in turn form compact cone-like spikes. The species described here make handsome container plants for the conservatory where they can also be planted out to form statuesque clumps of tropical foliage. Propagate by division in late spring or just as they start into new growth.

Zingiber derives from the Greek *zingiberi*, a version of the Sanskrit *singabera*, which in turn perhaps comes from the Malayan vernacular *inchiver* (from *inchi*, a root).

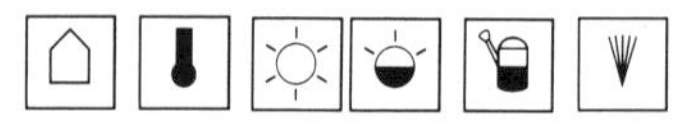

Species cultivated

Z. officinale Common ginger Probably originally from India, but cultivated in tropical Asia since ancient times and no longer known truly wild
Deciduous, with leafy stems to 90cm(3ft) or more tall; leaves linear-lanceolate, 15–30cm(6–12in) long. Flowers yellow-green with a purple, yellow-striped labellum, carried in green-bracted spikes to 8cm(3in) long, but not freely produced. The much branched tuber-like rhizomes are the source of the ginger root of commerce. Plants are best grown in rich, limy soil and are harvested as the leaves turn yellow for the production of ground or dried ginger. For preserved ginger, harvesting is carried out earlier while the foliage is still green.

Z. spectabile Malaysia
Leafy stems evergreen, robust, 2m($6\frac{1}{2}$ft) or more in height. Leaves oblong-lanceolate to narrowly ovate, deep green, about 30cm(1ft) long. Flowers cream with a pale and deep yellow, black-tipped labellum, carried in yellow- to red-bracted spikes 15–30cm(6–12in) in length.

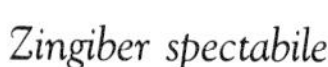

Zingiber spectabile

ZYGOCACTUS See SCHLUMBERGERA

ZYGOPETALUM

Orchidaceae

Origin: Central and tropical South America. A genus of about 20 species of mainly terrestrial orchids which not only have beauty of flower, but are easy to grow. They form clumps of pseudobulbs, narrow leaves and racemes of comparatively large, often colourful blooms. Each flower has five similar radiating tepals and a large fan-shaped labellum of a contrasting colour or pattern. Although best in the conservatory, those described are also worth trying indoors. Propagate by division in spring or after flowering. *Zygopetalum* derives from the Greek *zygos*, a yoke and *petalon*, petal; a swelling at the base of the labellum seems to join or yoke the petals together.

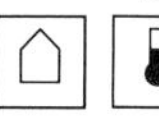  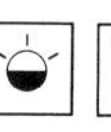 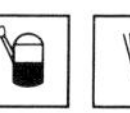

Species cultivated

Z. intermedium Brazil
Pseudobulbs ovoid-conical, bright green with a basal sheath, to 9cm(3½in) tall. Leaves in clusters or fans of three to five, lanceolate-elliptic to almost strap-shaped, bright green, arching, 30–45cm(1–1½ft) in length. Flowers about 8cm(3in) wide, waxy-textured, long-lasting, fragrant, with tepals usually yellow-green blotched red-brown and of more or less equal length; labellum white with a radiating pattern of purple-red branching veins, the margin waved or crimped; borne in autumn and winter. Frequently grown under the name of *Z. mackayi*.

Z. mackayi Brazil
Much like *Z. intermedium* and confused with it, but having flowers with the two inner tepals (true petals) noticeably shorter than the others and the veins of the labellum near to true red.

Hybrids

In recent years several fine hybrid cultivars have been raised using *Z. intermedium/mackayi*, the similar *Z. crinitum* and other less common species. Names to look for are *Z.* × *clayi*, Helen Ku, John Banks 'Wyld Court' and *Z. perrenoudii* 'Marshwood'.

Zygopetalum perrenoudii 'Marshwood'

Glossary

Acuminate Tapering to a slender point.
Adpressed Pressed flat, e.g. of hairs against a leaf or stem.
Adventitious A plant or organ that arises from an unexpected position, e.g. roots from an aerial stem and shoots direct from a true root.
Alternate One leaf at each node in a staggered formation up the stem.
Amplexicaule Stem clasping; the base of the leaf stalk or blade partially circling the stem.
Annual A plant which grows from seed and blooms within twelve months.
Anther The male part of a flower usually consisting of two lobes or 'sacs' which contain the pollen grains. *See* stamen.
Apiculate Of leaf tips which terminate abruptly in a short, often firm and sometimes sharp point.
Appressed The same as adpressed, *q.v.*
Areole A tiny hump-like organ found in all true members of the cactus family (Cactaceae) which bears bristles, spines, hairs or wool. It arises in what is technically a leaf axil and is considered to be a highly modified shoot.
Aril An extra external coating around a seed, often fleshy and brightly coloured.
Aristate Bearing bristles or awns, often appearing as if bearded.
Asymmetrical Lop-sided, usually of leaves which have one half larger than the other.
Auriculate Ear-shaped; used of lobes at the base of a bract, leaf or stipule.
Awl-shaped Quite literally like an awl, a short, broad needle-like tool usually flattened on one face.
Axillary Growing from the point where a leaf or bract joins the stem.

Beard or **bearded** Referring to a tuft or zone of hairs on or within a flower.
Biennial A plant which requires two growing seasons to complete its life cycle, producing leaves only in the first year, flowers and seed the next.
Bigeneric A plant derived from two genera, e.g. × *Fatshedera*, a cross between *Fatsia* and *Hedera*.
Bipinnate Of leaves, bracts and stipules which are twice or doubly pinnate, i.e. with leaflets which are again pinnate.
Bipinnatisect Of leaves, bracts or stipules which are twice or doubly pinnatisect, i.e. with leaf lobes which are again lobed.
Blade The main expanded part of a leaf as distinct from its stalk.
Bract A modified leaf usually associated with an inflorescence, in the axils of which flowers arise. Some bracts are scale-like and insignificant, others are large and coloured as in *Euphorbia pulcherrima*.
Bracteole A little bract.
Bud A shoot in embryonic form containing within it a miniature stem, leaves and sometimes flowers. Buds are typical of frost-hardy deciduous trees, shrubs and perennials. They respond rapidly to warmth and enable the plant to get away to a rapid start as soon as spring arrives. *Bud* also describes a flower before it opens.
Bulb A storage organ with a bud-like structure, usually underground. It is formed of a short flattened stem called a base plate which bears roots beneath and fleshy leaves or leaf bases above. The latter are of two forms, scaly and tunicated. Scaly bulbs have separate scales as in *Lilium*, while the tunicated sort has very broad scales which wrap tightly around each other, e.g. *Nerine*, *Hippeastrum*, *Narcissus*.
Burr Fruits or seeds with hooked or barbed hairs or bristles which cling to animal fur or feathers and are thus dispersed well away from the parent plant.

Calyx The whorl of sepals which protects the flower while in the bud stage.
Campanulate Bell-shaped.
Capitulum A head of tiny florets which looks like a single bloom, e.g. all members of the daisy family (Compositae).
Capsule A dry, often box-like fruit containing many seeds and opening by pores or slits or explosively as in *Impatiens* and *Viola*.
Carnivorous Used for a group of plants which catch insects and other small animals, e.g. *Darlingtonia*, *Nepenthes*, *Pinguicula*, *Dionaea* and *Sarracenia*. Sometimes also called insectivorous plants, they are capable of digesting the animal tissue and utilizing the nitrogen and other substances.
Carpel A tiny folded, specially modified leaf in the centre of a flower which contains the ovules. It may be separate or fused together with others to form a capsule.
Catkin A dense spike of usually petalless and unisexual tiny flowers, often flexuous and pendent.
Caudex A low trunk or part of the root above ground, usually thickened or swollen, with leaves or herbaceous stems only at the summit.
Channelled Used of a narrow leaf or leaflet with up-turned margins forming a channel or gutter-shape. Botanically known as *canaliculate*.
Chimaera A form of mutation (*q.v.*) in which the changed tissue forms either a glove-like surface layer or discrete sections into deeper tissue.
Ciliate Of leaves, bracts, sepals or petals fringed with hairs.
Cladode or **cladophyll** A flattened, leaf-like stem or joint as in *Ruscus*.
Clavate Club-shaped.
Clone A group of identical individuals all raised by vegetative propagation from one original plant, e.g. *Streptocarpus* × 'Constant Nymph'.
Column The central part of an orchid flower which combines ovary, stigma and pollinia.
Compound Used for parts of a plant that are in multiples, i.e. pinnate leaves and floral umbels.
Cone A spike-like structure or strobilus which bears seeds (conifers) or spores (clubmosses).
Conifer A colloquial name for all members of the Coniferales, a primitive plant group which covers firs, spruces, pines, cypresses, etc.
Connate Joined together, e.g. leaf bases and sometimes flowers.
Cordate Heart-shaped; usually referring to a more or less ovate leaf with two basal rounded lobes.
Coriaceus Having a stiff or leathery texture, especially of leaves.
Corm An underground storage organ derived from a stem base, e.g. *Gladiolus*.
Corolla The petals of a flower which may be separate or fused to form a funnel, trumpet or bell.
Corona An outgrowth of petal or perianth tissue in the centre of narcissus and daffodil flowers. (The false corona found in *Hymenocallis* and *Pancratium* flowers are formed by the fusion of stamen filaments.)
Corymb A racemose flower cluster with the stalks of the lower flowers longer than the upper ones creating a flattened or domed head.
Costa-palmate Used of palms with leaves which are halfway between palmate and pinnate.
Cotyledon Seedling leaves which are present in embryonic form within the seed. In some plants the cotyledons remain below ground serving as food stores, e.g. *Sophora*. This situation is termed hypogeal germination. In epigeal germination the cotyledons expand above ground and become the first leaves of the plant. Cotyledons are frequently of a different shape from those of the mature plant, e.g. *Ipomoea*.
Crenate Having shallow, rounded teeth.
Crest A ridge or flange of tissue, sometimes toothed or bearing hairs or bristles usually found on petals or styles, e.g. *Iris*.
Crisped Waved or wavy; of hairs, leaf and petal margins.
Crown The basal part of a non-woody perennial situated at or just below ground level and from which roots and stems grow.
Crownshaft Used of those palms which have broadly expanded leaf bases closely wrapped around in leek-like fashion, e.g. *Roystonea*.
Cultivar Short for cultivated variety and referring to a particular variant of a species or hybrid maintained in cultivation by vegetative propagation or carefully controlled seed production. Such a plant may be purposefully bred by man, or arise spontaneously as a mutation either in the garden or the wild. *See also* Clone.
Cuneate Wedge-shaped, usually of leaves and petals which taper from a broad tip to a very narrow base.
Cyme A compound infloresence made up of repeated lateral branching. In the monochasial cyme each branch ends in a flower bud and one lateral branch. In the dichasial cyme, each branch ends in a flower bud and two opposite branches.

Deciduous Becoming leafless during the dormant or resting season.
Decumbent Having prostrate stems with the tips erect.
Deliquescent Appearing to melt or become fluid as in the petals of *Commelina*.
Dentate Toothed, usually with outwards pointing teeth.
Denticulate Having many very small teeth.
Dichotomous Of stems which repeatedly fork into two diverging branches.
Digitate A compound palmate leaf with the leaflets radiating from the top of the stalk.
Dioecious Having single-sexed flowers on individual plants.
Disk/Disc The central flattened portion of a daisy flower-head.
Dissected Cut into several deep narrow lobes, usually of leaves and leaflets but also of sepals, petals and bracts.
Distichous Having leaves or flowers in two opposite facing ranks.
Double Of flowers with many extra petals usually derived from stamens.
Drupe A fleshy fruit usually but not invariably with one central seed.

Elliptic Usually of leaves which are broadest in the middle and taper evenly to the base and tip.
Emarginate Of petals and leaves with a notched tip.
Endemic A plant originating in one circumscribed area or country and found nowhere else in a natural state.
Entire Smooth edged.
Epicalyx A ring of bracts just below the calyx or sepals and resembling them.
Epiphyte A plant which perches upon another, as orchids and bromeliads grow on trees. They are not parasitic, gaining moisture from rain and air, and food from humus filled bark crevices.
Evergreen Plants that stay leafy the year round having leaves that last at least one full year before falling.
Exserted Projecting, as stamens protrude from the mouth of a flower.
Eye The centre of a flower, particularly when of a contrasting colour.

Falls The outer arching or pendent petals of an iris flower.
Farina Latin for flour, referring to the waxy powder found on some stems, leaves and flowers.
Fastigiate Of erect habit with all or most stems steeply ascending, creating an overall shape taller than wide.
Fertilization The moment when the two generative nuclei from the pollen grain enter the embryo sac within the ovule, *q.v.* *See also* Pollination.
Filament The stalk which attaches the anthers to the base of the flower.
Filiform Thread-like.
Fimbriate Bearing a fringe, usually of petals, sepals and bracts.
Flore-pleno Double-flowered, usually with stamens and sometimes with carpels transformed to petals or petaloids, *q.v.*
Floret Tiny flowers, usually when aggregated to form larger ones as in a daisy.
Flower The specialized organ of seed-bearing plants concerned with reproduction. They are usually formed of sepals (calyx), petals (corolla), stamens and pistils or carpels, but in some examples one or more of these may be missing.
Form A variant of a wild species, but not distinctive enough to be called a subspecies or variety. Sometimes loosely applied to horticultural variants or any plant which varies from the norm.
Frond An alternative name for the leaf of a fern or palm.
Fruit The pistils, carpels or ovaries when fertilized and grown to maturity. Fruits may be dry capsules or fleshy berries.

Genus A category or classification of all living things which group together all species with characters in common. Of the basic two scientific names which most plants have, the first is the generic, the second the species (*q.v.*).
Glabrous Smooth or hairless.
Glandular Bearing glands, tiny secretory organs often like tiny pores which exude water or volatile oil, etc.
Glaucous Blue-green or with a grey-blue patina over the basic green of leaf or stem.
Glochid A tiny barbed bristle found on the areole of certain cacti, notably *Opuntia*.
Grex A group of individual plants which differ slightly or much from each other but with a common origin. Usually applied to a batch of hybrid plants of the same parentage, particularly orchids.
Gymnosperm Members of the Gymnospermae, one of the two divisions of the seed-bearing plants which do not have true flowers but bear naked ovules and pollen sacs on spikes of flattened scales (primitive carpels) which later become cones.

Habit The general or overall appearance of a plant, e.g. erect, bushy, mat-forming, etc.
Hastate Shaped like a spear-head, i.e. ovate with two basal, spreading, barb-like lobes.
Herbaceous Used for plants with non-woody stems which last for one growing season only, dying down once flowering and seeding has taken place. Annual and biennial plants die totally, but perennials spring up from the root at the start of each growing season.
Hermaphrodite Of flowers which have functional stamens and pistils in the same flower.
Heterophyllus Of plants with differently shaped leaves on the same individual, e.g. ivy.
Hirsute Hairy.
Hoary Applied to leaves and stems covered with very short, usually branched or star-shaped white hairs.
Humus Technically organic matter in the form of a jelly-like colloid which coats soil particles, and absorbs water and dissolved minerals important to the plant growth. From a purely gardening point of view, humus is the brown crumbly substance resulting from the partial breakdown of plant and animal remains by bacteria, as typified by well-made garden compost or rotted farmyard manure.

Imbricate Overlapping like tiles on a roof, usually referring to closely set leaves on a stem.
Incised Sharply or deeply cut or toothed, mainly of leaf margins.
Indusium The membranous covering over the sporangia of ferns which gives each sorus its particular shape.
Inflorescence That part of a plant which bears the flowers consisting of one or more blooms and their stalks and pedicels. *See also* Corymb, Cyme, Panicle, Raceme, Spike and Umbel.
Insectivorous *See* Carnivorous.
Internode The length of stem between two nodes.
Involucre A whorl of bracts surrounding a cluster of flowers or florets as in a daisy, or a single flower as in *Anemone*.

Joint A colloquial name for node, *q.v.*

Keel The two lower petals of a pea flower (members of the Leguminosae) which are pressed together around the pistil and stamens.

Labellum The lowest of the three petals in the orchid flower and known colloquially as the lip. It is modified in a wide variety of ways to aid pollination by insects.
Labiate Members of the family Labiatae (or Lamiaceae) which have tubular flowers divided at the mouth into two lip-like structures to aid pollination by insects.
Laciniate Cut or slashed into narrow lobes.
Lanate Covered with thick woolly hairs.
Lanceolate Leaves, stipules, bracts or petals shaped like the head of a lance. For standardized botanical use, it denotes a shape broadest below the middle and between three and six times as long as wide.
Lateral Shoots which arise on the sides of main or leading stems.
Ligule A collar-like outgrowth at the junction of the sheath-like stalk and blade of the leaves of most grasses.
Linear Of very narrow leaves, petals, etc., with more or less parallel sides and at least twelve times as long as wide.
Lip *See* Labellum.
Lobe A section of a leaf, bract, etc., which is partially separated from the main part of the organ like a cape or isthmus. It is used also for the petal-like divisions at the mouth of a tubular flower.
Lyrate Lyre-shaped; mainly used of leaves with a broad, rounded apex and several small lateral lobes towards the base.

Meristem The extreme tip of a stem or root where the cells are actively dividing and growing.
Midrib The thickest central vein of a leaf.
Monocarpic Of plants that flower and seed once then die. Technically, annuals and biennials are monocarpic, but horticulturally the term is used for plants which live more than two years before flowering, e.g. some species of *Agave*.
Monoecious Having separate male and female flowers on the same plant, e.g. *Ricinus*.
Monotypic A genus containing one species only.
Mutant or **Mutation** Spontaneous changes that take place in whole plants or in small localized parts. Each mutation arises from a single meristem cell which becomes altered by cosmic radiation, radio activity or drugs such as colchicine. Often these mutations are small, but sometimes they can be large and startling as when a flower changes colour, a leaf becomes variegated or a tall plant stays small. *See also* Chimaera.

Nectary A small gland which secretes a sugary liquid (nectar). Nectaries are found mainly in flowers but sometimes elsewhere as those on the leaf stalks of *Passiflora*.
Node That part of a stem where the leaf is joined and a lateral shoot grows out.
Nut In the strict sense, a fruit with a tough or woody coat, i.e. the original carpel walls become hardened. In a loose sense, all fruits and seeds with woody or leathery coats can be termed nuts.
Nutlet A tiny nut as found in members of the Labiatae (Lamiaceae).

Obconical Reversed conical, i.e. with the stalk at the narrow end.
Obcordate Reversed cordate.
Oblanceolate Reversed lanceolate.
Oblong Applied in a botanical sense to leaves with more or less parallel sides and no more than twice as long; narrowly oblong equals no more than three times as long as wide.
Obovate Reversed ovate.
Obovoid Reversed ovoid.
Obtuse With a blunt or rounded tip, mainly of leaves and petals.
Opposite Of leaves or other organs borne in pairs on opposite sides of a stem.
Orbicular Disk-shaped or almost so.
Ovary A carpel or fused group of carpels which contain one to many ovules.
Ovate Egg-shaped in outline, often more or less pointed, with the broadest point below the middle and one and a half times as long as wide; narrowly ovate, twice as long as wide.
Ovoid Egg-shaped.
Ovule A minute ovoid mass of cells called the nucellus surrounding the embryo sac which, after fertilization, becomes a seed. One or more are contained within an ovary, *q.v.*

Palmate Used for leaves shaped like outspread hands and formed of five or more leaflets. Also known as digitate, *q.v.*
Palmatifid Rounded leaves lobed in a palmate fashion to half their length.
Palmatisect Rounded leaves palmately lobed almost to the base.
Panicle An inflorescence made up of several racemes or cymes.
Papillae or **Papillose** Small, usually soft, nipple-like protuberances on a leaf, stem or fruit.
Pappus In general, the tuft of hairs on a seed or fruit to assist distribution by wind. Strictly botanically, the bristles, hairs or scales should be derived from the calyx as in members of the Compositae.
Pedate A palmatisect leaf with at least the two basal lobes again lobed.
Pedicel The stalk of an individual flower.
Peduncle The main stem of an inflorescence usually bearing branches and several flowers, but sometimes as a solitary bloom as in most tulips.
Peltate A circular leaf with the stalk seeming to join at a point beneath, not at the edge as is usual, e.g. *Tropaeolum majus*.
Perennial Any plant that lives from several to many years, but in the strict sense applied to non-woody species which produce new stems annually from at or near ground level.
Perfoliate A leaf or bract which appears to have the stem growing through it because the basal lobes extend around the stem and fuse together.
Perianth The two outer whorls (calyx and corolla or sepals and petals) which first protect and then display the generative parts. In a general way perianth is used when the petals and sepals look alike as in a tulip.
Petal Modified, usually coloured leaves which form part of the flower. *See also* Corolla.
Petaloid Extra petals derived from stamens, or more rarely, carpels also.
Petiole The stalk of a leaf; a petiolule is the stalk of a leaflet in a compound leaf.
Phylloclade Another term for Cladode, *q.v.*
Phyllode A flattened leafstalk which assumes the form and function of a leaf blade as in *Acacia longifolia* and its allies.
Picotee Of petals with a narrow margin of a contrasting colour.
Pilose Covered with long ascending hairs.
Pinna/Pinnae The primary leaflets of a fern or palm frond.
Pinnate Of a leaf composed of two ranks of leaflets on either side of the midrib. Leaves which terminate in a solitary leaflet are termed *imparipinnate*, those with an equal number of leaf pairs only, as *paripinnate*.
Pinnatifid A leaf divided pinnately up to halfway to the midrib.
Pinnatisect A leaf divided pinnately to the midrib or almost so, but not rounded off into leaflets.
Pinnule The secondary or more complex division of a fern or palm frond.
Pistil The ovary of a flower plus a style if present, and a stigma. *q.v.*
Plena/Plenus Full; used of double and semi-double flowers.
Plicate Pleated or folded lengthwise; mainly of leaves.
Plumose Of a feathery appearance; usually but not invariably of an inflorescence.
Pollen The male sex cells of flowering plants borne in the anther lobes of the stamens. It is usually yellow or cream, but in some species can be orange, red, purple or blue.
Pollination The landing of pollen grains on the stigma. When this has happened the grains germinate and a root-like tube grows down into the ovary carrying the generative nuclei; *see* Fertilization.
Procumbent Lying flat on the ground; in the strict sense, of stems which do not root as they grow.
Prothallus A tiny leaf-like body which develops from the spore of a fern. Upon it are borne the sex cells or gametes which, upon fusion, give rise to a new fern plant.
Pseudobulb A swollen aerial stem typical of epiphytic orchids.
Pubescent Covered with short, soft hairs; in *puberulent*, minutely so.

Raceme An inflorescence composed of a central or main stem bearing stalked blooms at intervals.
Rachis The main stem or stalk of an inflorescence or compound leaf.
Ray floret The strap- or tongue-shaped outer florets of a daisy flower; also known as *ligulate florets*.
Receptacle The usually enlarged stem tip which bears the floral whorls (petals, sepals, etc.); also the greatly flattened stem tip which bears the florets of a daisy or scabious bloom.
Reflexed Bent back abruptly; of petals and other organs.
Repent/repens Growing flat on the ground; usually of stems that root at intervals.
Reticulate Bearing a conspicuous network of veins.
Revolute Rolled back at the edges; usually of leaves and petals.
Rhizome A more or less underground stem which produces roots and aerial stems. In some cases they are slender and fast growing, in others fleshy with storage tissue and then elongating slowly.
Rhomboid Roughly diamond-shaped.
Rosette A crowded whorl of spreading leaves as in palms, cycads, ferns, bromeliads, *Aeonium* and others.
Rosulate Bearing a rosette.
Rugose Having a wrinkled, puckered or corrugated appearance; mainly of leaves.
Runner A slender, prostrate stem bearing plantlets at the well-spaced nodes, e.g. *Saxifraga stolonifera*.

Sagittate Shaped like an arrowhead, e.g. sharply triangular in outline.
Scabrid Rough to the touch, usually with short, hard hairs.
Scale leaf *See* Bract.
Scandent Having climbing or scrambling stems.
Sepal The outer whorl of the perianth of a flower, usually green but sometimes coloured and then petal-like, e.g. *Abutilon megapotamicum, Fuchsia, Gloriosa, Vallota*, etc.
Sericeus Silky-hairy.
Serrate Edged with sharp teeth like that of a saw.
Serrulate Finely saw-toothed.
Sessile Without a stem or stalk, or apparently so.
Simple Undivided; mainly of leaves which are not divided into leaflets, but also of inflorescences composed of one flower.
Solitary Used of flowers when they are borne one to a stem or leaf axil.
Sorus A group of sporangia usually of rounded or oval shape on the back of a fern frond. Each sorus may or may not be covered with an indusium, *q.v.*
Spadix A thick fleshy flower spike with the small flowers imbedded in pits or sitting flush with the surface; typical of arum lilies and other members of the Araceae.
Spathe A green or coloured petal-like bract which surrounds or encloses the spadix, *q.v.*
Spathulate (spatulate) Like a spatula; usually of narrow leaves or petals which broaden out at the tip.
Species A group of individual plants which breed together and have the same constant and distinctive characters, though small differences of flower colour,

hairiness, etc., may occur.
Spike An unbranched inflorescence formed of a stem and several to many sessile flowers.
Spikelet A very small, usually compact spike; generally used for the basic unit of a grass inflorescence.
Spine A hard, sharp pointed structure derived from a modified twig.
Sporangium An asexually formed spore produced by ferns and certain fungi.
Spore Minute reproductive bodies formed of one or a few cells together, which give rise to new individuals, either directly as in fungi, or indirectly as in ferns. *See also* Prothallus.
Stamen The male unit of a flower comprising two anther lobes joined together at the top of a filament (stalk).
Staminode A rudimentary stamen, sometimes functioning as a petal or nectary, but usually producing no viable pollen.
Standard The upper or top petal of a pea flower, or the three inner usually upstanding petals of an iris bloom. Also a gardening term for a tree-like plant with an unbranched main stem and a head of branches.
Stellate Star-like, with radiating petals or branches. Some branched hairs have this form.
Stigma The receptive part of a pistil, often sticky with a sugary fluid where the pollen is held and germinates.
Stipe An alternative name for the leaf stalks of ferns; also used for the stems of mushrooms and toadstools.
Stipule Outgrowths from the base of a leaf stalk (petiole). In some cases they are like leaflets, in others tiny scales or hard and spine-like.
Stolon Aerial stems which root when they touch the soil, a common occurrence in *Rubus*.
Stomata Small openings or pores in the leaf which allow air and humidity in and out of the leaf tissue. Each stoma is guarded by two sausage-like cells (guard cells). When distended with water they are open, but when water is in short supply they become flaccid and the openings close.
Strain An improved selection of an existing species, variety, hybrid race or cultivar maintained by seed.
Strobilus *See* Cone.
Style The stalk which joins the pistil to the stigma.
Subshrub A small shrub which is woody at the base only, the upper part, particularly flowering stems, dying back each winter. From a gardening point of view the term is also used loosely for any low-growing softish-stemmed shrub.
Subspecies A distinct, true breeding form of a species, often isolated geographically from the species itself and differing more significantly than a variety.
Succulent A plant with either swollen leaves, stems or roots containing water storage tissue. They are mainly native to arid regions or where water is in short supply for a part of the year.
Sucker A shoot arising below ground level either from a buried stem, base or root.

Tendril Stems, leaves, stipules, leaflets and leaves modified to thread-like organs which twist around any suitable support, holding the plant in position.
Tepal Used of petals and sepals combined when they look exactly alike, e.g. tulip, crocus, narcissus.
Terete A stem which is smooth and rounded in cross-section.
Ternate In whorls or clusters of three; usually of leaves.
Terrestrial Growing on or in the ground; used of orchids which are not epiphytic, *q.v.*
Tesselated Chequered or with a pattern of roughly squared spots or marks.
Throat The lower inside area of a tubular, funnel- or bell-shaped flower.
Thyrse A panicle which is broadest in the middle and tapers to base and apex.
Tomentose Having a dense covering of short, firm, matted hairs.
Trifoliate Mainly of leaves divided into three leaflets, but sometimes used for whorls or groups of three leaves. Strictly speaking, a trifoliate leaf has the centre leaflet on a distinct petiolule (*q.v.*), whereas the term *trifoliolate* should be used when all three leaflets arise equally from the same point.
Truncate Ending abruptly or bluntly.
Truss A vernacular gardening name for a usually bold or prominent cluster of flowers (inflorescence) or fruits (infructescence).
Tuber Usually underground storage organs derived from stems and roots. Root tubers as typified by dahlia do not produce buds, new growth arising from the base of the existing or old stems. Stem tubers, e.g. *Tropaeolum tuberosum* and potato, bear buds (the eyes of a potato) at intervals some of which grow out to form the next season's stems.
Tubercle A small wart or knob-like projection on a stem, leaf, or fruit, etc.
Tunic The papery or netted covering of corms, e.g. *Crocus, Gladiolus*, derived from widely expanded leaf bases.
Type Original or wild type specimen. When a plant new to science is found and described, that first specimen is preserved, usually dried, stored in an approved herbarium and becomes the type specimen for reference.

Umbel An inflorescence of stalked flowers all of which arise and radiate from the tip of the peduncle.
Uncinate Hooked, as in the spines of certain cacti, e.g. *Mammillaria bocasana, Ferocactus wislizenii.*
Undulate Wavy, e.g. the margins of leaves, bracts, petals, etc.
Unisexual Having single-sexed flowers, though both may be on the same plant as in *Ricinus. See also* Dioecious.

Variegated The white to cream or yellow markings on leaves due to lack of chlorophyll. Sometimes there are also tints of red, pink or purple. There are three primary causes: mutation, virus infection and a deficiency of an essential mineral which upsets the formation of chlorophyll.
Variety A naturally occurring true breeding variety of a wild species. *See also* Cultivar.
Vegetative A gardener's propagation term for all methods of increasing plants except by seed. It covers division, layering, cuttings, offsets, runners and grafting.
Verrucose Covered with small wart-like outgrowths, usually on stems.
Verticil An alternative name for whorl, *q.v.*
Villous Covered somewhat shaggily with long soft hairs.
Viviparous Producing bulbils or young plants in place of, or mixed with, flowers and seeds or spores, e.g. *Asplenium bulbiferum* and *A. viviparum, Chlorophytum comosum.*

Weeping A gardening term for plants which have strongly down-curving or hanging stems or branches. Plants with normally prostrate stems grown in hanging baskets can also be described as weeping, e.g. *Tradescantia, Senecio rowleyanus, Aporocactus flagelliformis.*
Whorl Of leaves, bracts or flowers arranged in a ring of three or more.
Woolly The same as *lanate* – woolly-hairy.

Xerophyte Plants adapted to living in arid climates. Some survive by forming water storage organs as in succulents, others by reducing the surface area through which water is lost, e.g. the rolled leaves of some grasses. Other ways of cutting down water loss include leaves with a thick waxy cuticle, a dense covering of hairs and sunken stomata. A bulbous or tuberous rootstock or seeds (of annuals) which only grow when the rains come also assures survival in dry regions.

Zygomorphic Applied to flowers with asymmetrically arranged petals, e.g. members of the Labiatae, Leguminosae and Acanthaceae families.

INDEX OF COMMON NAMES

ACKNOWLEDGMENTS

All photographs by Gillian and Kenneth Beckett except the following:
6 Michael Boys
8*t* Mary Evans Picture Library
8*c* Mary Evans Picture Library
8*b* Mary Evans Picture Library
9*tl* The Image Bank/Miguel
9*tc* Michael Boys
9*tr* Tania Midgley
14 Michael Boys
15 Michael Boys
16*c* Swallow Editions/Nick Meers
19*tr* Harry Smith Horticultural Photographic Collection
33*br* Michael Warren
41 RHS Wisley/K.M. Harris, P. Becker, J. Maynard
42 RHS Wisley/J. Maynard, K.M. Harris
43*t* Harry Smith Horticultural Photographic Collection

The late Dr Geoffrey Herklots: 10 (6), 12-13 (14), (20), (22), 79*br*, 379 (2), 387*tr*, 441, 450*bl*

All illustrations by Will Giles except p. 36 by Sandra Pond

t = top, *c* = centre, *b* = bottom, *l* = left, *r* = right